14th Asia-Pacific Bioinformatics Conference (APBC 2016)

San Francisco, California, USA
11 - 13 January 2016

ISBN: 978-1-5108-2790-5

Printed from e-media with permission by:

Curran Associates, Inc.
57 Morehouse Lane
Red Hook, NY 12571

Some format issues inherent in the e-media version may also appear in this print version.

Printed by Curran Associates, Inc. (2016)

For further information, please contact BioMed Central Ltd
at the address below. The collection is available online at
https://bmcbioinformatics.biomedcentral.com/articles/supplements/volume-17-supplement-1

BioMed Central Ltd
236 Gray's Inn Road
London WC1X 8HB
UK

Phone: +44 (0)20 3192 2000
Fax: +44 (0)20 3192 2010

Email: info@biomedcentral.com
Web: www.biomedcentral.com

Additional copies of this publication are available from:

Curran Associates, Inc.
57 Morehouse Lane
Red Hook, NY 12571 USA
Phone: 845-758-0400
Fax: 845-758-2633
Email: curran@proceedings.com
Web: www.proceedings.com

TABLE OF CONTENTS

Jain et al. BMC Bioinformatics 2015, 17(Suppl 1):1
DOI 10.1186/s12859-015-0844-1

BMC Bioinformatics

PROCEEDINGS Open Access

Weakly supervised learning of biomedical information extraction from curated data

Suvir Jain[1†], Kashyap R.[1†], Tsung-Ting Kuo[2], Shitij Bhargava[1], Gordon Lin[1] and Chun-Nan Hsu[2*]

From The Fourteenth Asia Pacific Bioinformatics Conference (APBC 2016)
San Francisco, CA, USA. 11 - 13 January 2016

Abstract

Background: Numerous publicly available biomedical databases derive data by curating from literatures. The curated data can be useful as training examples for information extraction, but curated data usually lack the exact mentions and their locations in the text required for supervised machine learning. This paper describes a general approach to information extraction using curated data as training examples. The idea is to formulate the problem as cost-sensitive learning from noisy labels, where the cost is estimated by a committee of weak classifiers that consider both curated data and the text.

Results: We test the idea on two information extraction tasks of Genome-Wide Association Studies (GWAS). The first task is to extract target phenotypes (diseases or traits) of a study and the second is to extract ethnicity backgrounds of study subjects for different stages (initial or replication). Experimental results show that our approach can achieve 87 % of Precision-at-2 (P@2) for disease/trait extraction, and 0.83 of F1-Score for stage-ethnicity extraction, both outperforming their cost-insensitive baseline counterparts.

Conclusions: The results show that curated biomedical databases can potentially be reused as training examples to train information extractors without expert annotation or refinement, opening an unprecedented opportunity of using "big data" in biomedical text mining.

Keywords: Biomedical text mining, Natural language processing, Information extraction, Database curation, Machine learning

Background

Text mining from the scientific literature has been considered promising for creating and updating structured databases of biomedical knowledge [1] but it often falls short and, currently, manual curation by experts is still the standard practice for this task [2–5]. Some argue that text mining or natural language processing (NLP) becomes unnecessary when researchers report results following a standardized template [6]. Others argue that crowdsourcing may yield better performance than state-of-the-art NLP solutions [7–9]. However, given that scientific publications are still written in free-text and grow geometrically, automatic or semi-automatic approaches may still be necessary for scalable and sustainable data curation [10–13].

Annotation vs. curation

Machine learning has shown its potential in biomedical NLP and text mining [14–17]. However, supervised statistical learning algorithms require a large number of *annotated* training examples. Annotation and curation are fundamentally different processes. An annotation is a label applied to a span of text. Hence, an annotation appears verbatim in the source text. Annotated databases are rather labor-intensive albeit ideal for text mining applications. On the other hand, curated databases are prepared by domain experts who use a common terminology to describe entities in text. Curated databases require in-depth domain knowledge to produce but the

*Correspondence: chunnan@ucsd.edu
†Equal contributors
[2]Department of Biomedical Informatics, School of Medicine, University of California, San Diego, 9500 Gilman Drive, 92093 La Jolla, USA
Full list of author information is available at the end of the article

result often requires post-processing before being useful as training examples.

Numerous biomedical databases are available in the public domain. Many of them contain data derived directly from published literature either by curation by teams of experts or submitted by authors or other scientists. A survey estimated that in 2013, a total of 290 papers on biomedical databases that were published that also provided open URL links to access the data. Among these 290 databases, 77.59 % of them collected data from the literature and contained citations as supportive information [18]. However, these data cannot be readily used as training examples because the curated data rarely provide any information of where and how the data were derived from the text.

Catalog of GWAS

Consider the Catalog of Genome Wide Association Studies (GWAS) [19, 20], an online database developed by the National Human Genome Research Institute (NHGRI) by a curation team of experts. On a weekly basis, epidemiologists from NHGRI and more recently from the European Bioinformatics Institute (EBI) manually curate study-level fields of information from published GWAS and add them to the catalog. As of May 21, 2015, the Catalog of GWAS has been inserted with approximately 29,000 entries extracted from nearly 2200 distinct articles.

Figure 1 (upper panel) shows an example entry in the Catalog of GWAS. Each entry represents an observed association reported in an article, specifying that an association between a genetic variant, given in the data field `Strongest SNP`, and a phenotype, given in `Disease/Trait`, was observed from this study from an initial stage sample, given in `Initial Sample Size`. The entry also specifies that the observation was validated with a replication sample, given in `Replication Sample Size`. Other data fields include information of where the genetic variant resides in the genome and statistical strength of the observation.

Figure 1 (lower panel) shows the matching result of the entry in the Catalog of GWAS with the actual passages in the text of the article for three data fields. This example illustrates why curated data can be both useful and not useful as training examples. They are useful because matching the data to the text will create training examples. They are not useful because the matching is not trivial. As shown in Fig. 1, matching between the data and text requires background knowledge. In fact, curated data rarely provide verbatim copies of what is mentioned in the source article. For the purposes of easy searching, categorization, summarization, and data integration, curators usually adopt to a standardized terminology different from the text. Also, human curated data inevitably contain typos and inconsistencies in following standards.

Field	Value	Field	Value
PubMed ID	21764829	First Author	Png E
Date	7/15/2011	Journal	Hum Mol Genet
Study (title)	A genome-wide association study of hepatitis B vaccine response in an Indonesian population reveals multiple independent risk variants in the HLA region	Disease/Trait	Response to HBV vaccine
Initial Sample Size	1683 Indonesian individuals	Reported Gene(s)	HLA-DR
Replication Sample Size	1931 Indonesian individuals	Mapped Gene(s)	BTNL2 - HLA-DRA
Region	Chr 6:32389648	Strongest SNP	rs3135363
Context	Intergenic	Risk Allele	NR
p-Value	6.53E-22	Risk Allele Frequency	NR
OR or beta-coefficient	1.53	95% CI (text)	[1.35-1.74]
Platform [SNPs passing QC]	Illumina [455,508]	CNV	N

Entity Type	Catalog of GWAS	In the text
Disease / Trait	Response to HBV vaccine	"**hepatitis B vaccine response**"
Initial Sample Size	1683 Indonesian Individuals	1. "We performed a two-stage genome-wide association study (GWAS) of antibody titer in 3614 hepatitis B vaccine recipients from **Indonesia's** Riau Archipelago"
Replication Sample Size	1931 Indonesian Individuals	2. "In the **first stage**, following extensive quality-control (QC) filtering, we analyzed **1683** vaccine recipients..." 3. "After extensive QC filtering, we analyzed genotypes for 1706 SNPs in a second set of **1931** vaccine recipients from the same study. In **this second stage**, we replicated..."

Fig. 1 An examplar entry in the Catalog of GWAS and its currated data. Example of an entry in the Catalog of GWAS (*upper panel*) and after matching to the curated data in the text of the source paper [54] (*lower panel*)

Even when an exact match with curated data is found, the passage might be about a review of previous results but not the correct location from which the data should be extracted. To sum up, curated data are useful but imperfect as training examples.

Learning from curated data

This paper describes a general approach to using curated data from existing biomedical databases as training examples for information extraction from full-text research articles. Our approach is based on results of research in agnostic learning from data with noisy labels [21–27]. Among the results that fit our need here is *importance reweighting* [21], where training examples are weighted according to the reliability of their labels, which, however, is unknown in practice and must be estimated. Our solution is to employ a committee of weak classifiers that match candidate passages in the input text with curated data and then apply the EM algorithm to estimate the reliability of the labels. Then we can use a *cost-sensitive* learning algorithm that learns from training examples weighted by their misclassification costs derived according to the estimated reliability to develop accurate information extractors.

The applicability of this approach is not limited to the Catalog of GWAS. A large number of biomedical databases are available in the public domain, and many contain data derived directly from published literature either through manual curation by teams of experts or structured information submitted by authors or researchers. We intend for our approach to be generalizable across these databases as well.

The problem of learning from imperfect training examples is closely related to *entity normalization*, where the training examples available are normalized entity names rather than annotations in the text. For example, the gene normalization task in the BioCreative II and III initiative [28–32] aiming at extracting standardized gene IDs mentioned in an input article provided only the standardized gene IDs for each article as the training examples. The problem is also related to learning from data with noisy labels and learning from crowds [23, 33–35], where crowd inputs are considered noisy.

The test tasks

We implement the approach and apply it to two problems of information extraction from the full-text research reports of GWAS. Task 1 is to identify target phenotypes (disease/trait) examined in a GWAS study. This task is different from well-studied disease mention tagging and normalization [36–38] in that not all mentions but only the study targets need to be identified and that GWAS targets include not only diseases but a wide range of traits like eye color, response to ximelagatran treatment, sleeping habits, reading and spelling ability, education attainment, political ideology, etc. Usually, a GWAS targets a single phenotype, but a study may examine more than one phenotype. This is often the case when the study target is a complex disease, such as obesity, for which scientists may seek genetic associations with related traits such as body mass index, waist-height ratio, waist circumference, etc., in addition to direct association with obesity.

Task 2 is to extract stage (`"initial"` or `"replication"`) and ethnic group of the study samples. A GWAS involves study samples drawn from one or more ethnicity groups. Task 2 is concerned with the problem of extracting these ethnicity groups that the experiment pertains to. There are conventionally two *stages* in the study: initial and replication, and each of these can be associated with several distinct sample populations. However, many articles may not specified all information clearly in the text. Figure 2 illustrates some examples for this task. The first row gives an ideal case for information extraction, where both stages and their corresponding ethnicity information are specified in a single sentence, though it is still nontrivial to normalize terms in the sentence to a common terminology (i.e., mapping "screening" to "initial" and "France", "Italy", and "Sweden" to "European", the top-level ethnicity group, according to the curation guideline [39]). Figure 2 also contains examples of sentences where an ethnicity mention is not explicitly linked to either stage, or is specified for one stage and must be inferred for the other. This can be exacerbated when these mentions are far apart in a full text.

Task 1 is an entity recognition problem while Task 2 requires jointly extracting different attribute types and identifying linkage between them. For both tasks we develop information extractors by applying the same general approach to learning from the curated data in the Catalog of GWAS as the training examples. We are able to achieve 87 % of extraction in Precision-at-2 (P@2) for the Task 1 and 0.83 in F1-Score for the Task 2. Both outperform their baseline, cost-insensitive counterparts. The remainder of this paper describes the method and the results in details.

Methods

We start by describing our general approach and then follow by presenting the implementation of the general approach for the two test tasks.

The approach

Figure 3 shows the five components and the workflow of the whole learning approach. The input is a large corpus of research articles for training. For each article, **Step (A)** identifies the passages that may contain the information

Mentions	Curated Data	Comments
"A total number of 1040 unrelated patients from France, Italy and Sweden were included in the 'screening' dataset and 873 unrelated and patients from France and Sweden in the independent 'replication' dataset."	Initial stage: European Replication stage: European	In this case, a single sentence contains all information to be extracted.
"The first stage was conducted in 1,999 healthy Chinese men, and the second stage was performed in an additional 1,496 subjects."	Initial stage: East Asian Replication stage: East Asian	Ethnicity mention is specified for the initial stage but is implied for the replication stage.
"Population substructure was assessed to ensure that all included samples were of European ancestry." "Multivariate Cox proportional hazard analyses were performed on data from 1,304 Caucasian subjects..."	Initial stage: European Replication stage: European	Sentences with required mentions are in different, possibly long-range, sections of article. Further, ethnicity mention is specified for overall experiment and must be linked to both stages.
"We recruited 397 patients aged 18 to 75 years who were admitted to the hospital of the Max Planck Institute of Psychiatry (MPI), Munich, Germany..."	Initial stage: European Replication stage: European	Initial stage inferred from terms like "recruited" and "GWAS" and not explicitly mentioned.
"After quality control procedures (described below), the final sample included 1287 individuals of Northern European ancestry evaluated between 11/8/1985 and 9/27/2010."	Initial stage: European	Ambiguity introduced due to terms like "quality control" (most often associated with the replication stage) used in description of initial stage, which is not explicitly mentioned and must be inferred.

Fig. 2 Example sentences describing sample ethnicity groups and stages. Example of sentences in free text from which the system extracts study targets

to be extracted in the text. The identification of passages should be inclusive in the sense that any candidate passages will be extracted and no passage is missed.

Step (B) pairs each passage with a piece of matched curated data and creates a feature vector for the pair as the input to the committee classifiers. For example, we pair passage 2 in Fig. 1 (lower panel) to data item `"1683 Indonesian Individuals"` from the Catalog of GWAS, because passage 2 is likely from where the data item was derived. Again, the matching should be inclusive to contain all potential pairs.

Step (C) then sends the feature vectors to a committee of classifiers (diamonds on top of Fig. 3). Each classifier classifies each pair into positive, if the passage is deemed to contain the information given in the curated data, or negative otherwise. The classifiers can be as "weak" as simple decision rules, like "whether the passage contains a substring that exactly matches the curated data". Therefore, each committee member classifier provides noisy positive-negative labels of the passages extracted from the text. Combining the classification results of all committee members for all extracted passages creates a a large matrix

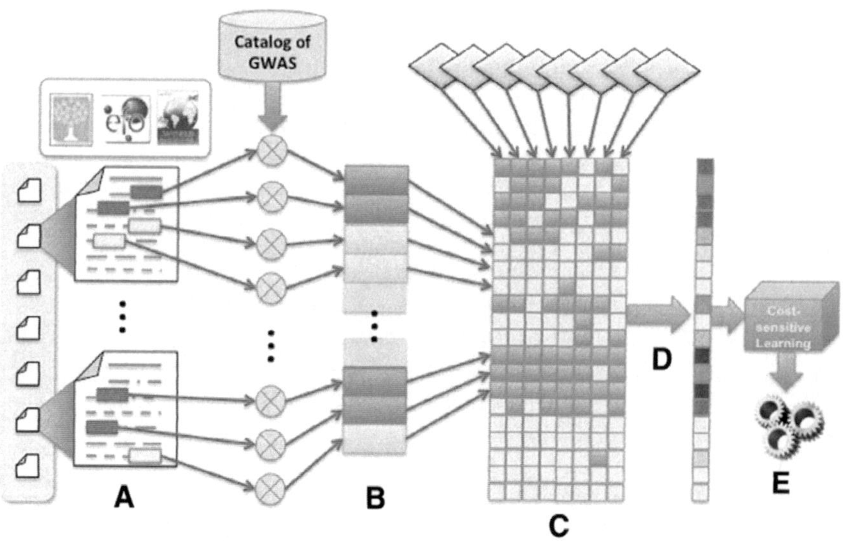

Fig. 3 System architecture. Our proposed system architecture summarizing the steps (**a**-**e**) in the machine learning training process

of yes-or-no votes, where each element (i, j) contains the vote from classifier i for candidate passage j.

Step (D) estimates from the matrix the probability that candidate passage j is truly positive by a label estimator that applies an Expectation-Maximization (EM) algorithm to compute maximum likelihood estimation of the probabilities, which can then be treated as the weight, or the reliability of a candidate training example. A similar approach was used in the BioCreative III gene normalization task [29] to create a silver standard. The EM algorithm works as follows:

1. **Input**: matrix $\boldsymbol{M}$ of committee (column)-passages (row), where each element in the matrix is either positive $(= 1)$ or negative $(= 0)$;
2. Let p_i be the probability that the i-th passage should be positive, e_j be the error rate of the j-th committee classifier; Let $t = 0$;
3. Initialize $e_j(0) = 0$ for all j;
4. Update for all i, $p_i(t) = \frac{\sum(1-e_j(t-1))\boldsymbol{M}_{ij}+k}{J+K}$, where J is the number of weak classifiers in the committee, k/K is the Laplace prior;
5. Update for all j, $e_j(t) = \frac{\sum p_i(t)\boldsymbol{M}_{ij}+k'}{I+K'}$, where I is the number of the passages, k'/K' is the Laplace prior;
6. $t = t + 1$ and repeat update steps until convergent;
7. **Output**: $\hat{p}_i = p_i(t)$ for all i and $\hat{e}_j = e_j(t)$ for all j as the final values.

With the estimated probability of each candidate passage, we can assign it a *cost*, and train a cost-sensitive learner [40–42] using the candidate passages as the cost-weighted training examples to learn to select correct passages that contain the desired information as **Step (E)**. The cost that we use here is derived according to Lemma 1 in [21], where the problem of classification with noisy labels is solved by importance reweighting. They show that an error bound can be achieved if the misclassification cost of a training example (x, y) is set to $p(y|x)/p_\rho(y|x)$, where ρ denotes sampling from a noise-perturbed distribution. Though neither $p(y|x)$ nor $p_\rho(y|x)$ are known, we can approximate $p(y = \text{"+"}|x)$ by $\hat{p}$ from the EM algorithm above and $p_\rho(y = \text{"+"}|x)$ by $p(\hat{p}(y = \text{"+"}|x) > 0.5)$ for a training example estimated as positive and analogously for a negative one. That is, let $y_i = \text{round}(\hat{p}_i)$. If $y_i = 1$ then the cost of misclassifying the i-th passage as negative is $c_i = \frac{\hat{p}_i}{\sum_i y_i/I}$, else $c_i = \frac{\hat{p}_i}{1-\sum_i y_i/I}$.

We note that this cost-sensitive classifier may use a completely different set of features to characterize a passage.

After the cost-sensitive learning completes, to extract desired data from a new article, we apply the same **Step (A)** to extract candidate passages and send them to the trained cost-sensitive classifier and extract data from the candidate passages classified as positive.

Task 1: Identifying target phenotype (disease/trait)

Data

The curated GWAS target phenotype data are in the form of a spreadsheet dated May 2014 available for download at [43]. Each row in the spreadsheet contains a column `DISEASETRAIT`, reporting a phenotype term chosen by curators as the study target of the paper with its ID given in column `PUBMEDID`. As we discuss earlier, these phenotype terms do not always exactly match what is mentioned in the text. Column `EFOTRAIT` contains the phenotype term mapped to a concept defined in Experimental Factor Ontology (EFO) [44], the standard terminology of diseases and traits for the Catalog of GWAS. EFO classifies concepts into 17 high-level categories as given in column `PARENT`. For example, `DISEASETRAIT` for the paper of PubMed ID 18849991 is "Male-pattern baldness". The corresponding `EFOTRAIT` and `PARENT` are "androgenetic alopecia" and "Other disease", respectively. These columns constitute the curated data that we intend to use to train our information extractor.

The spreadsheet contains data for 1742 unique PubMed IDs of GWAS articles. Among them, 307 papers have full text available in the NXML format [45] from PubMed Central. NXML provides useful metadata for NLP and text mining such as section headings to distinguish main text from references and other elements but currently only about one third of all PubMed papers have their NXML version available from PubMed Central. For other papers, we collected 965 full texts in PDF. These PDF files were all transcribed into XML using bioPDFX [46], a tool that we built on top of PDFX [47], to prepare PDF papers for biomedical text mining.

For the purpose of evaluation, we manually augmented the data with the corresponding terms that actually appear in the text of the 307 NXML articles to serve as our hold-out gold standard. We kept the data of the 965 PDF papers intact and used the data for training.

The spreadsheet contains a total of 2645 rows for 1742 unique papers. That is, each paper has on average 1.51 target phenotypes. Three hundred thirty three papers (19.11 %) have more than one target phenotypes in the curated data. We therefore measure the performance based on the precision-at-2 (P@2) metric, that is, if either of the top 2 extracted phenotypes match the gold standard, the extraction is considered correct.

Implementation

Step (A): Passage extractor. In this step, we identify all mentions of any disease or trait using an exact string matching approach, which is based on a dictionary of all diseases and traits from the search menu of the web query interface of the Catalog of GWAS at [48].

After string matching, we extracted 117,384 mentions in the training and 72,914 in the test data. Note that these numbers are the total mentions of any disease or trait in all articles in the data set, because a paper usually contains multiple mentions.

Step (B): Feature creator. The following features are generated for each training sample:

- *Token-based features*: Character-level n-grams of the mention.
- *Context-based features*: Word-level and character-level n-grams for up to 10 words before and after the mention.
- *Position-based features*: The location of each mention can be indicated using positional tags (e.g., ⟨article-title⟩ and ⟨*abstract*⟩) in the converted XML papers; therefore, whether a mention is located within positional tags are extracted as binary features. These tags, however, are not always available from the PDF-transcribed XML versions.

For each mention, token-based and context-based features are represented as normalized TF-IDF vectors. Together with position-based features, each mention has approximately 120,000 features.

Step (C): Committee of classifiers. To build the committee matrix, we design five rule-based binary classifiers as follows.

- *Title or Abstract*: Whether disease/trait mention occurs in the title or abstract of the paper; this is a simple yet strong indicator of a disease mentioned being the actual target of the paper.
- *Exact match*: Whether disease/trait mention exactly matches the target given by human curator.
- *Sub-string match*: Whether disease/trait mention partially matches the target given by the human curator (e.g., a mention of "Diabetes" would be classified as positive, if the human curator determines the disease as "Type-2 Diabetes").
- *Synonym*: Whether disease/trait mention is an exact or partial match of a synonym of the target determined by the human curator. The synonyms are collected from UMLS [49]; for a given disease or trait mention, all UMLS concepts that shared the same `CONCEPT-ID` are considered to be synonymous. To reduce noise, we only keep the synonyms which are in English and are preferred terms (i.e., the `IS-PREF` flag set to `Y` in UMLS).
- *Compound token*: Whether the mention has multiple tokens separated by a space or hyphen (e.g., "Parkinson's disease" would be classified as positive because it consists of compound tokens separated by a space).

It should be noted that although some rule-based classifiers are extremely weak (e.g., compound token), our idea is to show that multiple weak classifiers can actually contribute to a strong committee and make accurate predictions.

Step (D): Label estimator. We apply the EM method described previously to label each pair of passage and curated disease or trait with an estimated confidence (i.e., the conditional probability given the pair is positive).

Step (E): Cost-sensitive learner. In this step, we utilize the estimated confidence generated by the Label Estimator to assign the cost to train a cost-sensitive variant of Support Vector Machine (SVM).

Post-processing. Each paper may contains multiple mentions of various disease and traits. For example, a paper may contain 10 mentions of "Diabetes", and 30 mentions of "Hypertension". However, our cost-sensitive classifier may predict only part of them to be positive. This is reasonable because even though two sentences mention the same phenotype, it is not always the case that both are stating that the phenotype is the study target. We consider the following two scores for the post-processing, inspired by TF and IDF, respectively:

1. $P_{TF} = \frac{V_{pi}}{V_{pi}+V_{ni}}$, where V_{ni} is the number of negative votes assigned to the i-th candidate.
2. $P_{IDF} = \frac{V_{pi}}{\sum V_{pi}}$, where V_{pi} is the number of positive votes assigned to the i-th candidate.

To combine P_{TF} and P_{IDF}, we apply two mean computations, namely arithmetic and harmonic, to calculate the final scores and determine our predicted disease/traits. The harmonic mean better represents the mean value of these two metrics. That is because the P_{TF} and P_{IDF} values are often quite small and include outlying values. Harmonic mean is a more sensitive measure in such cases.

Task 2: Identifying stage and ethnicity of study samples

We represent this problem as that of extracting tuples of the form ⟨`stage, ethnicity`⟩ from the free text of a GWAS article, with the entities in the tuple corresponding to stage and ethnicity of the study sample.

Data

Again, our articles are selected from the Catalog of GWAS. The data in the form of a spreadsheet is available for download at [48].

We selected articles that satisfy the following criteria:

- *Curated data available*: 2185 PubMed articles were curated with the data available.
- *NXMLs or PDFs available*: We used NXML versions of the articles if they are available through PubMed Central. These versions have high-quality text. Otherwise, we transcribed PDF versions of the remaining articles to text. This leaves 1861 articles.

- *No missing values or "NR"*: The characteristics of the samples are available for whichever stage is mentioned in the article, and the curated data contain no blank entries. Also excluded are those for which curators were unable to find a conclusive ethnicity group for the sample and the entries state "NR" ("not reported"). This leaves 1674 articles.
- *Ethnicity mentions in text*: Terms that correspond to ethnicity groups must be available in text (but not inferred from affiliations of authors, for example).
- *Do not contain errors*: The curated data was found to contain errors in the entries for some articles. Those were excluded.

The final dataset consists of 1311 articles, comprising 2357 ⟨`stage, ethnicity`⟩ tuples.

The curated data is normalized to remove spelling errors and inconsistent wording primarily to ensure that there is only one top-level term for a given ethnicity entity. For example, ethnicity group entries in the curated data stating "North African/Middle East" or "Middle East/North African" are both considered to correspond to "Middle East/North African", with this choice of the eventual top-level entry being made arbitrarily.

Implementation

We applied the same pipeline given in Fig. 3 but we employed two committees to extract tuples with two elements:

Step (A) *Passage extraction*: Mentions in the text corresponding to ethnicity entities are tagged and their surrounding passages extracted. These instances are (weakly) labeled according to curated data as positive or negative.

Step (B) *Feature creator*: The ethnicity instances are featurized and made suitable for classification.

Step (C) *Committee of positive/negative classifiers*: A committee of weak learners are exploited to generate noisy labels, for cost-sensitive learner to classify ethnicity instances as positive or negative instances.

Committee of initial/replication classifiers: Ethnicity instances classified as positive are further classified into the initial and replication experimental stages of the GWAS.

For both committees, we perform **Step (D)** *Label estimator* using EM algorithm, followed by **Step (E)** *Cost-sensitive learner* to predict the ethnicity and stage of the mentions.

Post-processing: Instances of *ethnicity* classified into a particular *stage* are grouped as ⟨`stage, ethnicity`⟩ and duplicates removed. The performance of this method is evaluated upon this final set of results.

These steps are described in detail below.

Step (A): Passage extractor. Mentions in text are generally not exact string matches (or even exact synonyms) of ethnicity groups, necessitating a dictionary mapping of mentions in text (e.g., "German") to the top-level ethnicity entity (e.g., "European"). Mentions in the text that correspond to a top-level ethnicity entity are mapped to their corresponding entities and tagged with the help of a constructed dictionary as described below, followed by passage extraction. Mentions corresponding to a stage are not tagged.

We construct the dictionary of ethnicity mappings as follows. A multitude of terms can refer to the ethnicity of an individual, including the country of origin (e.g., "Germany"), the specific ethnicity group (e.g., "European"), an adjectival for the country (e.g., "German"), a demonym for the country (e.g., "Germans"), and similar sets of terms for cities and other regions. We handle these terms through the conventions:

- *Country/region name*: Not every mention of a region, say, a country, maps to a specific ethnicity term. The NHGRI curation guideline [39] stating that a given set of individuals belong to an ethnicity group only if it is directly stated in the study, or if at least 90 % of the population of the region is known to belong to a single ethnicity group, with this knowledge being based on the CIA World Factbook [50].
- *Adjectivals and demonyms*: An extensive list of the adjectivals and demonyms for countries are obtained from Wikipedia and a dictionary is constructed to map the terms to their corresponding countries. These countries are then mapped to the corresponding ethnicity group (or discarded if no mapping exists).

The final dictionary comprises 449 terms that map to 14 top-level ethnicity groups. These terms cover a majority of the mentions in text, and we omit publications that do not contain language that can be matched to this dictionary. A more comprehensive dictionary may include lists of tribes and indigenous peoples of the world.

This dictionary is used to match mentions in text to ethnicity entities through string matching. The tagged instances are extracted along with their corresponding passages, which consist of the 10 words on either side of the entity in the sentence.

For training and testing, these instances are weakly labeled from curated data by checking if, for a given article, the ethnicity group is present in either experimental stage, initial or replication. If so, this is considered a positive instance, and negative otherwise.

Step (B): Feature creator. The following types of features are generated for each instance:

- *Token-based features*: A set of binary features each of which turn on for a specific ethnicity entity (e.g., a feature will be 1 for "East Asian").
- *Context-based features*: These include normalized term frequency-inverse document frequency (TF-IDF) representations of unigrams and bigrams of 10 words in either direction of the ethnicity mention, as long as the words are within the same sentence. The words are stemmed using the Porter stemmer [51].
- *Position-based features*: These include features like section title (also in TF-IDF form), the distance (normalized) of the ethnicity mention from the start of the article, or from the start of the section.
- *Additional features*: These include features that do not fit into the above categories, such as the number of times the ethnicity entity was observed (tagged) in the same article.

This results in sparse feature vectors of approximately 80,000 dimensions. The features are normalized by removing the mean and scaling to unit variance across the values of each feature, or dimension of feature vector. As the feature set is mostly composed of various TF-IDF vectors, truncated Singular Value Decomposition (SVD), or Latent Semantic Analysis, was explored as a feature selection technique to reduce the dimensionality of the feature vectors. However, this did not affect the performance significantly (and in fact degraded performance slightly) and hence the complete feature vectors were retained. The complete list of the features are given in (Additional file 1).

Step (C): Committee of positive/negative classifiers. We use the cost-sensitive learning approach described previously to classify instances as positive or negative. The committee members of weak labelers include:

- *Binary classifier*: the results of a Logistic Regression binary classifier trained on the weak labels from curated data.
- *Rule-based classifier*: this classifier predicts a positive example if the features meet any criteria, such as the presence of words that are commonly found in descriptions of a sample (e.g., "stage", "cohort"). 65 such terms are used in total.
- *Weak labels from curated data*: the labels obtained by exact-matching the ethnicity to the curated data.

The committee matrix obtained from concatenating the outputs of all the members is used to estimate the cost to be assigned to each training instance, as in previous sections. These costs are used to train a cost-sensitive, L_2-regularized, linear support vector machine (SVM) classifier to classify instances as positive or negative instances.

Committee of initial/replication classifiers. Training a cost-sensitive classifier to classify positive instances of ethnicity entities into the corresponding stage of a study is performed in a similar fashion to that for the ethnicity instance classifier.

In this case, the positive training instances are now relabeled "initial" or "replication". The committee members are:

- *Binary classifier*: as above, but trained to distinguish "initial" from "replication" instances.
- *Rule-based classifier:* the rule-based classifier is modified to use the presence of stage-specific words to make its prediction (e.g., "discovery" for the initial stage, or "follow-up" or "second stage" for replication). 8 such terms are used.
- *Weak labels from curated data:* as above, but containing classes "initial" and "replication" instead.

The outputs of the members are used to construct the committee matrix and estimate the cost assigned to each training instance, which is then used to train a cost-sensitive, L_2-regularized, linear SVM to classify the test data into the initial or replication stages with the same set of features.

The output of this step is a classification of each positive *ethnicity* instance into a specific *stage* (initial or replication).

Post-processing. Either stage in a GWAS may have multiple ethnicity groups. Hence, the extraction can possibly result in multiple tuples of the form ⟨`stage`, `ethnicity`⟩ for each study. We compile a list of such tuples for each article, with duplicates being discarded.

Results and discussion

Results of task 1: Identifying target phenotypes (disease/trait)

We use precision at 2 (P@2) to measure the performance because a GWAS may examine one or more phenotypes. If either of the top 2 extracted phenotypes match the gold standard, the extraction is considered correct.

We compare our cost-sensitive approach with a cost-insensitive baseline. Also, we attempted the following additional alternatives to improve the cost-sensitive learner:

- *BIOADI*: Since in many articles, the target phenotype only appears once in its full form and all following mentions are in its abbreviation, we identify and normalize the abbreviations in the input text using the BIOADI system [52] to pre-process the text in an attempt to improve the performance.
- *Conditional Random Field (CRF)*: In order to deal with new diseases and traits that do not appear in our training dictionary, we also tried to apply CRF in the Passage Extractor step. The design of the features for the CRF is based on the method described in [53]; we

use a mixture of general linguistic, orthographic, contextual, syntactic dependency, and dictionary lookup features. By using this CRF model, we discover 59,648 mentions in test data, which is, however, less than the number of mentions using dictionary matching.

We train all systems with the data of 965 articles and test the trained systems with the hold-out data of 307 articles. Table 1 shows the performance results of these systems. The cost-sensitive learner outperforms the cost insensitive learner and the harmonic averaging outperforms arithmetic averaging. However, the alternatives to improve passage extraction (BIOADI and CRF) fail to improve the result of the cost-sensitive learner.

We experiment with more than 2 extractions as well. If we consider top five extractions, the accuracy of at least one of them being correct is higher than 93 %. If we consider a single extraction, the accuracy drops by a few percentage points but stays above 82 %.

Discussion of the results of task 1

The experimental results show that learning from curated data is feasible to accomplish a P@2 up to 87 % for the task of extracting study target and that the cost-sensitive learning approach outperforms the cost-insensitive baselines.

We analyze the errors and summarize that many errors are from pharmacogenomic studies, which examine phenotypes such as "Response to antipsychotics" as given in the curated data. Antipsychotics are a class of psychiatric medication instead of a specific medication, while the disease targets of the medication, such as "schizophrenia", "major depressive disorder", etc., may present stronger signal as the study target than the response of a medication for an information extractor. Studies of complex diseases also pose a main challenge because a complex disease may have many associated measurable traits.

Results of task 2: Identifying stage and ethnicity of study samples

We evaluate the performance by comparing them with the ⟨stage, ethnicity⟩ tuples known to correspond to each GWAS article. Further, we also compare the results with the following alternative approaches. The evaluation methodology and metrics are described below.

1. *Baseline*: All ethnicity instances tagged by the dictionary in an article are assigned to both experimental stages, and the results measured.
2. *Cost-insensitive classification*: the framework described above is used in a cost-insensitive fashion by excluding the committees and directly training the classifiers on the curated labels. This provides a candidate for comparison to the cost-sensitive approach for evaluating the performance of cost-based learning.
3. *Cost-sensitive classification*: the framework described above, including committee classification, is used.

In each of the methods (excluding the baseline), five-fold article-based cross validation (5-fold CV) is performed. The articles in the dataset are randomly shuffled, and each fold of the 5-fold CV utilizes all ⟨stage, ethnicity⟩ tuples belonging to 80 % of the articles in the dataset as training data, and the tuples in the remaining 20 % of articles as test data.

The results from each fold are then collected to obtain ⟨stage, ethnicity⟩ tuples for all the articles in the dataset. These results are compared against the curated data and the F1 score calculated in the standard way:

- If a ⟨stage, ethnicity⟩ tuple in the result for a specific article is present in curated data for that article, it is considered a true positive (TP); otherwise, it is considered a false positive (FP).
- If a ⟨stage, ethnicity⟩ tuple in the curated data for a specific article does not have a counterpart in the extracted results, it is considered a false negative (FN).

Using this, we calculate the precision, recall and the Macro F1 score for each method on 1,311 articles comprising 2,357 ⟨stage, ethnicity⟩ tuples from approximately 35,000 mentions of ethnicity-related terms. The resultant values are tabulated in Table 2.

Table 3 presents the macro precision, recall and the F1 score for the methods. The precision and recall are calculated as above, but for each article individually, and then averaged to obtain the macro precision and recall. The

Table 1 Precision-at-2 (P@2) of identifying target disease/trait mention of a GWAS study

Method	Arithmetic	Harmonic
Cost-insensitive	68.65 %	79.62 %
Cost-sensitive	78.05 %	87.46 %
BIOADI+Cost-sensitive	75.57 %	87.29 %
CRF+Cost-sensitive	65.79 %	75.24 %

Table 2 Performance of stage-ethnicity extraction (micro average)

Method	Precision	Recall	F1 Score
Baseline	0.4898	1.0000	0.6576
Cost-insensitive	0.6965	0.7077	0.7020
Cost-sensitive	0.7471	0.7711	0.7589

Table 3 Performance of stage-ethnicity extraction (macro average)

Method	Precision	Recall	F1 Score
Baseline	0.5972	1.0000	0.7478
Cost-insensitive	0.7408	0.7943	0.7666
Cost-sensitive	0.7893	0.8757	0.8302

harmonic means of these two values for each method are the respective Macro F1 scores.

The results in Tables 2 and 3 indicate that the cost-sensitive approach is able to significantly outperform the similar but cost-insensitive approach, which performs only close to a brute-force baseline. Not only is the cost-sensitive approach able to achieve a much higher degree of recall, but the improvement is accompanied by an increase in overall precision as well.

As the recall gets closer to the limit, the results also indicate that further improvements to the method will be gained by focusing not on extracting relevant ethnicity groups, but on eliminating the ones that are irrelevant to the article.

Discussion of the results of task 2

The results show that a cost-sensitive committee learning approach reliably outperforms a similar, cost-insensitive approach. This holds true even when the additional committee members are weak classifiers that encode real-world domain knowledge and patterns as rules, which can compensate to some extent for the lack of data, as it is not presumable that all patterns are present in the data in significant quantity as to be learned by a model.

Some of the challenges faced in the task of extracting ethnicity groups of sample populations from the Catalog of GWAS are described below.

- *Entity normalization*: There are various ways of representing the same entity, and it is necessary to normalize these representations to a single representative entity. However, there exist degrees of difficulty with respect to normalization; for example, it is relatively easy to equate "African American" to "African-American", but much harder to equate the two represntations with "American citizen of African origin".
- *Studies with several target entities to extract*: Many GWAS in the U.S. use a highly ethnically diversified study sample with, for example, "52 % Caucasian, 24 % Latino, 11 % African, 9 % Eastern Asian and 4 % Indigenous Americans", and studies may also divide into more than two stages. How to flexibly identify and deal with these situations is challenging.
- *Varying concept granularity*: Mentions of ethnicity terms might not be correctly tagged in an article as the authors may report ethnicities in specific terms such as names of tribes and indigenous people, etc., which may not map perfectly to a top-level ethnicity group. This introduces ambiguity which can be an issue for ethnicity background identification.
- *Inadequate reporting of ethnicity data*: Often, the importance of a study to a specific population or application only becomes apparent after the article is published. Hence, the text in the article may never refer to the specific ethnicity group that their experimental sample was drawn from, but simply describe it in terms of the city, state or region, or even the hospital that the population was recruited at. This is doubly challenging as it requires an indefinite expansion of the dictionary of ethnicity-related terms, and also a standardized mapping from each such term to a top-level ethnicity group.

Conclusions

The large number of curated biomedical databases available in the public domain provides an unprecedented opportunity to train NLP systems to comprehend biomedical publications. In this paper, we present an approach to take advantage of this opportunity. The approach applied methods from learning from noisy-label and committee classifiers to assign costs to train cost-sensitive classifiers. We tested our approach for two challenging biomedical information extraction tasks. The results show that our approach is effective and outperforms alternative approaches. We will continue to investigate if it is possible to define standard passage extractors and weak learners applicable to extract biomedical entities, attributes and relations of common interest to enable rapid development and portability between domains for biomedical literature mining.

Declarations

Research reported in this manuscript and the publication costs for this manuscript were supported by the National Human Genome Research Institute (NHGRI) of the National Institutes of Health under award number U01HG006894. The content is solely the responsibility of the authors and does not necessarily represent the official views of the National Institutes of Health.

Additional file

Additional file 1: Complete list of the features used for Task 2.
(CSV 1.16 kb)

Competing interests
The authors declare that they have no competing interests.

Authors' contributions

CNH conceived, designed, and supervised the study and secured funding. SJ developed the software system and designed and performed the experimental evaluation for Task 1. KT developed the software system and designed and performed the experimental evaluation for Task 2. SB developed the software system to transcribe part of test articles in PDF into XML. SJ, KT, SB, TTK, and GL were involved in the debugging and optimization of the software systems. CNH, SJ, KT, and TTK drafted the manuscript and all authors contributed substantially to its editing. CNH takes responsibility for the manuscript as a whole. All authors read and approved the final manuscript.

Acknowledgements

We would like to thank Dr. Lucia Hindroff of NHGRI for her generous support of this project by providing us PDF copies of test articles and the curation guidelines of the Catalog of GWAS to make this research possible. We also would like to thank Dr. Helen Parkinson, Dr. Jackie MacArthur and their team members at EBI for their assistance.

Declarations

This article has been published as part of BMC Bioinformatics Volume 17 Supplement 1, 2016: Selected articles from the Fourteenth Asia Pacific Bioinformatics Conference (APBC 2016). The full contents of the supplement are available online at http://www.biomedcentral.com/bmcbioinformatics/supplements/17/S1.

Author details

[1]Department of Computer Science and Engineering, Jacobs School of Engineering, University of California, San Diego, 9500 Gilman Drive, 92093 La Jolla, USA. [2]Department of Biomedical Informatics, School of Medicine, University of California, San Diego, 9500 Gilman Drive, 92093 La Jolla, USA.

Published: 11 January 2016

References

1. Manning CD, Schütze H. Foundations of Statistical Natural Language Processing, 1st edn. Cambridge, MA: The MIT Press; 1999.
2. Wiegers TC, Davis APP, Cohen KB, Hirschman L, Mattingly CJ. Text mining and manual curation of chemical-gene-disease networks for the comparative toxicogenomics database (CTD). BMC Bioinformatics. 2009;10(1):326. doi:10.1186/1471-2105-10-326.
3. Davis AP, Wiegers TC, Roberts PM, King BL, Lay JM, Lennon-Hopkins K, et al. A CTD-Pfizer collaboration: manual curation of 88,000 scientific articles text mined for drug-disease and drug-phenotype interactions. Database. 2013;2013:080. doi:10.1093/database/bat080.
4. Altman RB, Bergman CM, Blake J, Blaschke C, Cohen A, Gannon F, et al. Text mining for biology–the way forward: opinions from leading scientists. Genome Biol. 2008;9(Suppl 2):7. doi:10.1186/gb-2008-9-s2-s7.
5. Hettne K, Williams A, van Mulligen E, Kleinjans J, Tkachenko V, Kors J. Automatic vs. manual curation of a multi-source chemical dictionary: the impact on text mining. J Cheminformatics. 2010;2(1):3. doi:10.1186/1758-2946-2-3.
6. Mons B. Which gene did you mean? BMC Bioinformatics. 2005;6(1):. doi:10.1186/1471-2105-6-142.
7. Burger JD, Doughty E, Khare R, Wei C-HH, Mishra R, Aberdeen J, et al. Hybrid curation of gene-mutation relations combining automated extraction and crowdsourcing. Database: J Biol Databases Curation. 2014;2014:.
8. Good BM, Su AI. Crowdsourcing for bioinformatics. Bioinformatics. 2013;29(16):1925–33. doi:10.1093/bioinformatics/btt333 1302.6667.
9. Snow R, O'Connor B, Jurafsky D, Ng A. Cheap and fast – but is it good? Evaluating non-expert annotations for natural language tasks. In: Proceedings of the 2008 conference on empirical methods in natural language processing. Honolulu, Hawaii: Association for Computational Linguistics; 2008. p. 254–63.
10. Baumgartner WA, Cohen KB, Fox LM, Acquaah-Mensah G, Hunter L. Manual curation is not sufficient for annotation of genomic databases. Bioinformatics (Oxford, England). 2007;23(13):41–8. doi:10.1093/bioinformatics/btm229.
11. Kim S, Lu Z, Wilbur W. Identifying named entities from PubMedR; for enriching semantic categories. BMC Bioinformatics. 201557. doi:10.1186/s12859-015-0487-2.
12. Czarnecki J, Nobeli I, Smith A, Shepherd A. A text-mining system for extracting metabolic reactions from full-text articles. BMC Bioinformatics. 2012;13(1):172. doi:10.1186/1471-2105-13-172.
13. Xu R, Wang Q. Comparing a knowledge-driven approach to a supervised machine learning approach in large-scale extraction of drug-side effect relationships from free-text biomedical literature. BMC Bioinformatics. 2015;16(Suppl 5):6.
14. Krallinger M, Morgan A, Smith L, Leitner F, Tanabe L, Wilbur J, et al. Evaluation of text-mining systems for biology: overview of the Second BioCreative community challenge. Genome Biol. 2008;9(Suppl 2):1–9. doi:10.1186/gb-2008-9-s2-s1.
15. Zweigenbaum P, Demner-Fushman D, Yu H, Cohen KB. Frontiers of biomedical text mining: current progress. Brief Bioinform. 2007;8(5): 358–75. doi:10.1093/bib/bbm045.
16. Simpson M, Demner-Fushman D. Biomedical text mining: a survey of recent progress In: Aggarwal CC, Zhai C, editors. Mining text data. Springer; 2012. p. 465–517. doi:10.1007/978-1-4614-3223-4_14, http://dx.doi.org/10.1007/978-1-4614-3223-4_14.
17. Zhu F, Patumcharoenpol P, Zhang C, Yang Y, Chan J, Meechai A, et al. Biomedical text mining and its applications in cancer research. J Biomed Eng. 2013;46(2):200–11. doi:10.1016/j.jbi.2012.10.007.
18. Koh YZ, Ling MH. Catalog of biological and biomedical databases published in 2013. Comput Math Biol. 2014;3(3):. https://www.iconceptpress.com/journals/computational-and-mathematical-biology/13000001/1402001128.pdf.
19. Welter D, MacArthur J, Morales J, Burdett T, Hall P, Junkins H, et al. The NHGRI GWAS Catalog, a curated resource of SNP-trait associations. Nucleic Acids Res. 2014;42(Database issue):1001–6. doi:10.1093/nar/gkt1229.
20. Hindorff LA, Sethupathy P, Junkins HA, Ramos EM, Mehta JP, Collins FS, et al. Potential etiologic and functional implications of genome-wide association loci for human diseases and traits. Proc Natl Acad Sci. 2009;106(23):9362–7. doi:10.1073/pnas.0903103106.
21. Liu T, Tao D. Classification with noisy labels by importance reweighting. IEEE Trans Pattern Anal Mach Intell. 2005;1:. doi:10.1109/tpami.2015.2456899, 1411.7718.
22. Natarajan N, Dhillon IS, Ravikumar PK, Tewari A. Learning with noisy labels In: Burges CJC, Bottou L, Welling M, Ghahramani Z, Weinberger KQ, editors. Advances in neural information processing systems 26. Red Hook, NY: Curran Associates, Inc.; 2013. p. 1196–204.
23. Sheng VS, Provost F, Ipeirotis PG. Get another label? improving data quality and data mining using multiple, noisy labelers. In: Proceedings of the 14th ACM SIGKDD International conference on knowledge discovery and data mining, KDD '08. New York, NY, USA: ACM; 2008. p. 614–22. doi:10.1145/1401890.1401965, http://dx.doi.org/10.1145/1401890.1401965.
24. Frénay B, Verleysen M. Classification in the presence of label noise: a survey. IEEE Trans Neural Netw Learn Syst. 2014;25(5):845–69. doi:10.1109/tnnls.2013.2292894.
25. Servedio RA. Smooth boosting and learning with malicious noise. J Mach Learn Res. 2003;4:633–48. doi:10.1162/153244304773936072.
26. Kalai A, Kanade V. Potential-Based Agnostic Boosting In: Bengio Y, Schuurmans D, Lafferty JD, Williams CKI, Culotta A, editors. Advances in neural information processing systems 22. Red Hook, NY: Curran Associates, Inc.; 2009. p. 880–8.
27. Bouveyron C. Weakly-supervised classification with mixture models for cervical cancer detection. In: Bio-Inspired Systems: Computational and Ambient Intelligence 10th International Work-Conference on Artificial Neural Networks, IWANN 2009, Salamanca, Spain, June 10-12, 2009. Proceedings, Part I. Berlin Heidelberg, Germany: Springer; 2009. p. 1021–8. doi:10.1007/978-3-642-02478-_128, http://dx.doi.org/10.1007/978-3-642-02478-8_128.
28. Morgan A, Lu Z, Wang X, Cohen A, Fluck J, Ruch P, et al. Overview of BioCreative II gene normalization. Genome Biol. 2008;9(Suppl 2):3. doi:10.1186/gb-2008-9-s2-s3.
29. Arighi C, Roberts P, Agarwal S, Bhattacharya S, Cesareni G, Aryamontri AC, et al. BioCreative III interactive task: an overview. BMC Bioinformatics. 2011;12(Suppl 8):4. doi:10.1186/1471-2105-12-s8-s4.
30. Kuo CJ, Ling M, Hsu CN. Soft tagging of overlapping high confidence gene mention variants for cross-species full-text gene normalization. BMC Bioinformatics. 2011;12(Suppl 8):6. doi:10.1186/1471-2105-12-s8-s6.

31. Arighi C, Lu Z, Krallinger M, Cohen K, Wilbur W, Valencia A, et al. Overview of the BioCreative III Workshop. BMC Bioinformatics. 2011;12(Suppl 8):1. doi:10.1186/1471-2105-12-s8-s1.
32. Agarwal S, Liu F, Yu H. Simple and efficient machine learning frameworks for identifying protein-protein interaction relevant articles and experimental methods used to study the interactions. BMC Bioinformatics. 2011;12(Suppl 8):10.
33. Raykar VC, Yu S, Zhao LH, Valadez GH, Florin C, Bogoni L, et al. Learning from crowds. J Mach Learn Res. 2010;11:1297–322.
34. Whitehill J, Ruvolo PL, Wu T, Bergsma J, Movellan JR. Whose vote should count more: optimal integration of labels from labelers of unknown expertise In: Bengio Y, Schuurmans D, Lafferty JD, Williams CKI, Culotta A, editors. Advances in Neural Information Processing Systems 22. Red Hook, NY: Curran Associates, Inc.; 2009. p. 2035–43.
35. Brodley CE, Uiversity P, Friedl MA, Uiversity B, Edu BP. Identifying mislabeled training data. J Artif Intell Res. 1999;11:131–67.
36. Leaman R, Islamaj Dogan R, Lu Z. DNorm: disease name normalization with pairwise learning to rank. Bioinformatics (Oxford, England). 2013;29(22):2909–17. doi:10.1093/bioinformatics/btt474.
37. Leaman R, Lu Z. Automated disease normalization with low rank approximations. In: Proceedings of BioNLP 2014. Stroudsburg, PA: Association for Computational Linguistics; 2014. p. 24–8.
38. Doğan RII, Leaman R, Lu Z. NCBI disease corpus: a resource for disease name recognition and concept normalization. J Biomed Inform. 2014;47: 1–10. doi:10.1016/j.jbi.2013.12.006.
39. Hindorff LA, MacArthur JAL, Morales J, Bowler EH, Hall P, Klemm K, et al. Comprehensive curation and visualization of ethnicity information from published genome-wide association studies (GWAS): an improved GWAS Catalog. In: The 64th american society of human genetics annual meeting. abstract ID: 977M. San Diego, CA; 2014.
40. Ruan YX, Lin HT, Tsai MF. Improving ranking performance with cost-sensitive ordinal classification via regression. Inf Retr. 2014;14(1): 1–20. doi:10.1007/s10791-013-9219-2.
41. Chang X, Zheng Q, Lin P. Cost-sensitive supported vector learning to rank imbalanced dataset. In: Proceedings of the intelligent computing 5th international conference on emerging intelligent computing technology and applications, ICIC'09. Berlin, Heidelberg: Springer; 2009. p. 305–14. http://portal.acm.org/citation.cfm?id=1788195.
42. Lo HY, Wang JC, Wang HM, Lin SD. Cost-sensitive multi-label learning for audio tag annotation and retrieval. Multimedia IEEE Trans. 2011;13(3): 518–29. doi:10.1109/tmm.2011.2129498.
43. European Bioinformatics Institute, GWAS to EFO Mappings, XLSX. 2014. http://www.ebi.ac.uk/fgpt/gwas/ontology/GWAS-EFO-Mappings201405.xlsx. (accessed July 21, 2015).
44. Malone J, Holloway E, Adamusiak T, Kapushesky M, Zheng J, Kolesnikov N, et al. Modeling sample variables with an experimental factor ontology. Bioinformatics. 2010;26(8):1112–8.
45. National Center for Biotechnology Information NationalLibraryofMedicine. PubMed Central Open Access Subset. 2015. http://www.ncbi.nlm.nih.gov/pmc/tools/ftp/. (accessed July 21, 2015).
46. Bhargava S. Preparing PDF scientific articles for biomedical text mining. San Diego: Master's thesis, Department of Computer Science and Engineering, University of California; 2015.
47. Constantin A, Pettifer S, Voronkov A. PDFX: fully-automated PDF-to-XML conversion of scientific literature. In: Proceedings of the 2013 ACM symposium on document engineering, DocEng '13. New York, NY, USA: ACM; 2013. p. 177–80. doi:10.1145/2494266.2494271, http://dx.doi.org/10.1145/2494266.2494271.
48. National Human Genome Research Institute. A Catalog of Published Genome-Wide Association Studies. 2015. https://www.genome.gov/26525384. (accessed July 21, 2015).
49. Bodenreider O. The unified medical language system (umls): integrating biomedical terminology. Nucleic Acids Res. 2004;32(suppl 1):267–70.
50. Central Intelligence Agency. The World Fact Book. 2015. https://www.cia.gov/library/publications/resources/the-world-factbook/. (accessed July 21, 2015).
51. Porter MF. An algorithm for suffix stripping. Program Electron Libr Inf Syst. 1980;40(3):211–8. doi:10.1108/00330330610681286.
52. Kuo CJ, Ling MH, Lin KT, Hsu CN. Bioadi: a machine learning approach to identifying abbreviations and definitions in biological literature. BMC Bioinformatics. 2009;10(Suppl 15):7.
53. Chowdhury M, Faisal M, et al. Disease mention recognition with specific features. In: Proceedings of the 2010 workshop on biomedical natural language processing. Stroudsburg, PA: Association for Computational Linguistics; 2010. p. 83–90.
54. Png E, Thalamuthu A, Ong RTH, Snippe H, Boland GJ, Seielstad M. A genome-wide association study of hepatitis B vaccine response in an Indonesian population reveals multiple independent risk variants in the HLA region. Hum Mol Genet. 2011;20(19):3893–8. doi:10.1093/hmg/ddr302.

Huang *et al. BMC Bioinformatics* 2016, **17**(Suppl 1):2
DOI 10.1186/s12859-015-0845-0

BMC Bioinformatics

PROCEEDINGS

Open Access

Drug repositioning for non-small cell lung cancer by using machine learning algorithms and topological graph theory

Chien-Hung Huang[1], Peter Mu-Hsin Chang[2], Chia-Wei Hsu[1], Chi-Ying F. Huang[3] and Ka-Lok Ng[4,5*]

From The Fourteenth Asia Pacific Bioinformatics Conference (APBC 2016)
San Francisco, CA, USA. 11 - 13 January 2016

Abstract

Background: Non-small cell lung cancer (NSCLC) is one of the leading causes of death globally, and research into NSCLC has been accumulating steadily over several years. Drug repositioning is the current trend in the pharmaceutical industry for identifying potential new uses for existing drugs and accelerating the development process of drugs, as well as reducing side effects.

Results: This work integrates two approaches - machine learning algorithms and topological parameter-based classification - to develop a novel pipeline of drug repositioning to analyze four lung cancer microarray datasets, enriched biological processes, potential therapeutic drugs and targeted genes for NSCLC treatments. A total of 7 (8) and 11 (12) promising drugs (targeted genes) were discovered for treating early- and late-stage NSCLC, respectively. The effectiveness of these drugs is supported by the literature, experimentally determined in-vitro IC_{50} and clinical trials. This work provides better drug prediction accuracy than competitive research according to IC_{50} measurements.

Conclusions: With the novel pipeline of drug repositioning, the discovery of enriched pathways and potential drugs related to NSCLC can provide insight into the key regulators of tumorigenesis and the treatment of NSCLC. Based on the verified effectiveness of the targeted drugs predicted by this pipeline, we suggest that our drug-finding pipeline is effective for repositioning drugs.

Keywords: Non-small cell lung cancer, Drug repositioning, Microarray data analysis, Machine learning algorithm, Topological parameters, Protein-protein interactions, Enrichment analysis, Connectivity Map

Background

Lung cancer is the leading cause of death globally [1] and non-small cell lung cancer (NSCLC) accounts for more than 85 % of all lung cancer cases; adenocarcinoma is the most common subtype. Many efforts have been made to development treatments for NSCLC, and they depend on finding suitable drugs for treating NSCLC within an effective time and at reasonable cost.

* Correspondence: ppiddi@gmail.com
[4]Department of Bioinformatics and Medical Engineering, Asia University, Taichung 41354, Taiwan
[5]Department of Medical Research, China Medical University Hospital, China Medical University, Taichung 40402, Taiwan
Full list of author information is available at the end of the article

Drug repositioning by the Food and Drug Administration (FDA) involves approving drugs with known side effects; it has become a major trend and seen some success. Stachnik et al. [2] showed that bisphosphonates can potentially be repurposed for the prevention and adjunctive therapy of HER1-driven cancers (such as NSCLC and breast cancers). Having constructed a drug-disease bipartite network, Chen et al. [3] utilized two inference methods, ProbS and HeatS, to predict direct drug-disease associations based on node degree in the network. Lee et al. [4] integrated the shared neighborhood scoring algorithm with a database of disease indications, drug development, and associated proteins, to identify new indications for known FDA-approved drugs. In earlier studies [5, 6], based on PPI (protein-protein interaction) community, we established a systematic strategy for identifying potential drugs and target genes for treating NSCLC, which can be extended in several respects that are addressed in the present study. Those two previous studies did not use the four features of machine learning algorithms that are used herein, and were proposed in our work in 2015 on the prediction of cancer proteins [7].

The machine learning method and the topological properties of biological networks have been used separately to identify cancer-related genes. For example, Bull et al. [8] utilized proteins' hydrophobicities, in vivo half-lives, propensity for being membrane-bound and the fraction of non-polar amino acids as features in the Random Forest classifier to predict drug targets. Carson et al. [9] utilized topological metrics, such as betweenness centralities, neighborhood connectivity and radiality, as features and used an alternating decision tree (ADTree) classifier to identify disease-associated genes. Many works on identifying repositioned drugs have been based on various computational methods, such as mapping gene expression profiles using drug response profiles [10–14], the use of side-effect-based similarities [15–17], heterogeneous network clustering [18], and the graph-based inference method [19–22]. However, most of these methods are either disease-centric or drug-centric. To the best of the authors' knowledge, few works have addressed the problem of drug repositioning by integrating machine learning methods, graph theory and meta-analysis. This work integrates two state-of-art methods - machine learning [7] and the graphing of topological properties [23] - to develop a new pipeline to identify potential therapeutic drugs and targeted genes for treating NSCLC.

In solving the targeted drug problem, the following issues must be addressed. First, different individuals may correspond to different sets of differentially expressed genes. Second, cancer is a heterogeneous disease: different stages of cancer require different drug targets and involve stage-specific cancer-associated genes. Third, the results of microarray profiling vary from study to study and a rigorous method is required to solve this problem. Fourth, the reliability of drug finding remains to be verified.

This study deals with the above four issues. First, to reduce the effect of biological heterogeneity among different individuals, tumor/adjacent non-tumor pairwise arrays for NSCLC were used, allowing pairwise statistical testing. Second, the samples were grouped into early-stage and late-stage samples. Third, meta-analysis was carried out to integrate multiple microarray profiles and results. Finally, potential drugs were validated by performing biochemical assays and with reference to the literature.

Methods

Cancer is a multistage process that arises from mutations of genetic sequences; early- and late- stage cancer-associated genes potentially differ considerably. This work elucidates a strategy for identifying stage-specific potential drugs for treating NSCLC based on an integrated analysis based on microarray profiling. This work proposes an *in silico* strategy for narrowing down the search for lung cancer genes. Figure 1 presents the workflow.

Microarray data for lung cancer were firstly separated into the early- and late-stage data. Two-pair tests (based on normal and cancer tissues from the same patient) were performed to identify differentially expressed genes (DEGs). A Robust Multi-array Average (RMA) was utilized to normalize gene expression, and eBayes analysis was then performed on the results thereof. DEGs were predicted using an adjusted p-value of 0.05. The selected DEGs were divided into two groups - an up-regulated group and a down-regulated group - based on the fold-changes (FC) in gene expression. These selected DEGs are separately filtered using machine learning classifiers and graph theory, and two corresponding sets of key genes are then derived. Gene set enrichment analysis and pathway analysis were then conducted on the two sets of key genes, and drug-gene interaction databases and the Connectivity Map (cMap) were used to identify potential drugs (with cMap p-value <0.1 and enrichment score <0) for treating NSCLC. The common enriched pathways and drugs that were returned by both machine learning algorithms and the classification of topological parameters were further investigated. The predictions of targeted drugs were confirmed by IC_{50} experiments, a review of the literature and clinical trials. Finally, the targeted genes were prioritized for reference.

Input datasets

The input data herein were taken from previous work [6] to enable this study to be compared with [6]. The microarray data for lung cancer were downloaded from GEO [24] and summarized in Table 1. The microarray datasets consist of data from experiments GSE7670 [25] and GSE10072 [26], which were conducted using the HG-U133A array; and data from experiments GSE19804 [27] and GSE27262 [28], which were performed using HG-U133 plus 2.0 chip.

To reduce the effect on integrating of biological heterogeneity among individuals, normal and cancer tissues were taken from each patient. Two-pair tests (on these normal and cancer tissues are taken from the same patient) were performed to identify differentially expressed genes (DEGs). Samples were divided into early- and late-stage samples. Early-stage samples were taken from patients with stage I, IA or IB cancer, while late-stage data were obtained from patients with stages III or IV cancer [6].

Microarray data analysis

In this study, the publicly available microarray data analysis package *Bioconductor* was used to identify DEGs

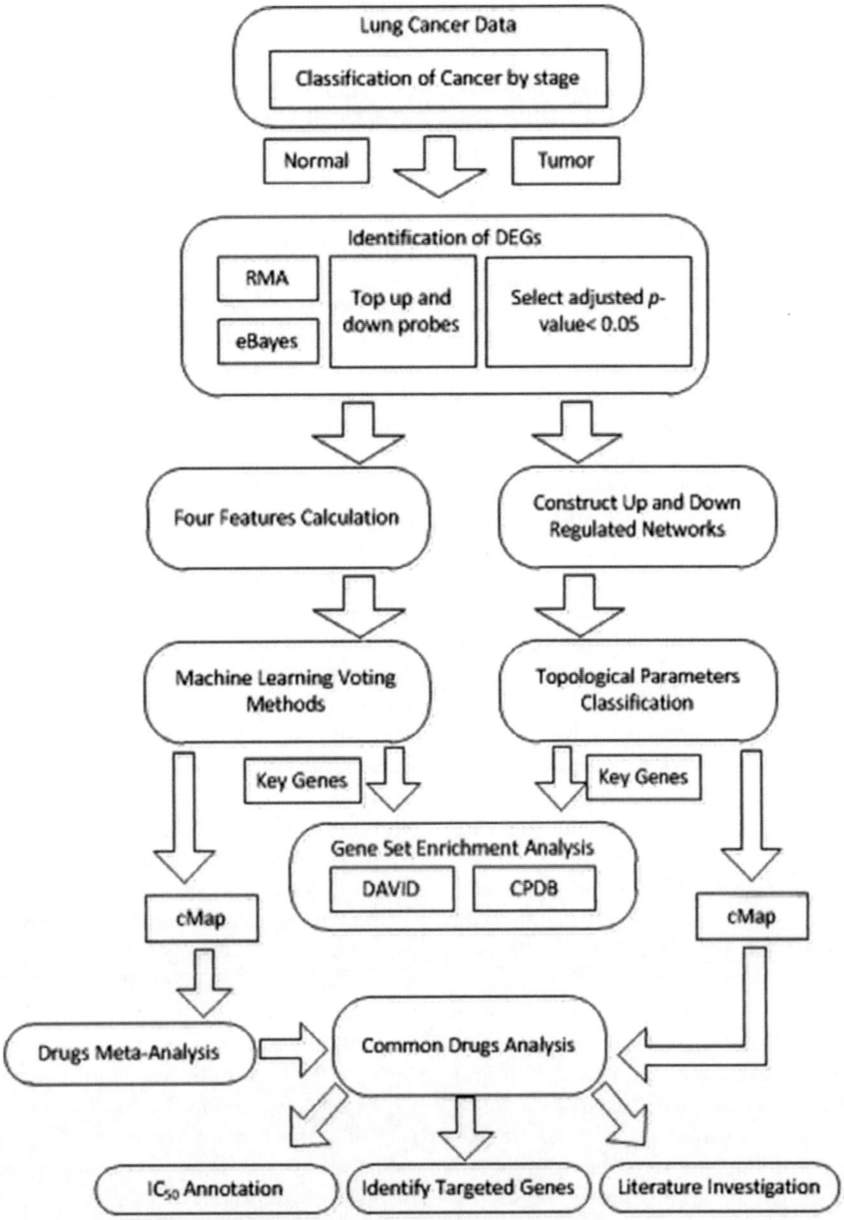

Fig. 1 Workflow of this study, which consists of (1) identification of DEGs, (2) machine learning approach, (3) topological parameter-based classification, (4) common pathway analysis, (5) common drug analysis and (6) effectiveness verification

Table 1 Summary of microarray datasets

GEO ID	Organization name	Number of samples (Early-stage)	Number of samples (Late-stage)
GSE7670	Taipei Veterans General Hospital	8	11
GSE10072	National Cancer Institute, NIH	15	9
GSE19804	National Taiwan University	35	13
GSE27262	National Yang Ming University	25	n/a

among a large number of gene expressions. Based on whether the log base 2 of the fold-change (FC) values for gene expression, $\log_2 FC$, was less than or greater than zero, the selected DEGs were divided into two groups - up-regulated (up probes in Fig. 1) and down-regulated (down probes in Fig. 1), respectively. The FC value of any gene expression level with a fold change value of less than 5.64 was set to 5.64 to facilitate the cMap [29] search.

Machine learning algorithms

In the previous study [7], we developed a simple and effective machine learning method, based on domain-domain interactions (DDI), weighted domain frequency score (DFS) and cancer linker degree data (CLD) to predict cancer proteins. We used the one-to-one interaction model to quantify the likelihood that was a cancer-specific DDI; the weighted DFS feature is used to measure the propensity of a domain to be present in cancer and non-cancer proteins, and the CLD feature is defined to identify the partners with which cancer and non-cancer proteins interact. The machine learning algorithms was implemented in the Weka software tool, and a ten-fold cross-validation test was used to train the supervised model. Based on our previous studies [30, 31], a balanced data set typically provides better performance than an unbalanced one, so, the machine learning algorithms were trained using positive and negative datasets that contained equal numbers of data.

Experimental results revealed that the proposed machine learning method identified cancer proteins with relatively high hit ratios (about 80 %). Five classifiers – three with the highest F1 values – the LMT, SimpleCart and J48 algorithms, and two with the highest AUC values – the LWL and Ridor algorithms, were used to identify potential cancer genes under strictly uniform voting, meaning that only a protein that was predicted by all five classifiers to be a cancer protein was considered. In the machine learning approach, the up- (down-) regulated DEGs in each microarray data are processed individually for each microarray.

Classification of topological parameters

The topological features provide valuable information for identifying crucial genes and clusters in a biological network. Recently, we proposed the identification of critical nodes for a network using topological parameters [23]. The five classified groups are: group 1: degree centrality; group 2: betweenness centrality; group 3: bridging centrality; group 4: closeness centrality and eccentricity centrality; group 5: clustering coefficient, brokering coefficient and local average connectivity. This classification enables nodes to be ranked by their topological importance in the networks. To apply topological parameter classification in this study, common up- (down-) regulated DEGs for the microarray datasets must be firstly extracted. Next, for early- and late-stage NSCLC, the corresponding up- (down-) regulated network was constructed by using the common DEGs for all microarray datasets and their neighbors in protein-protein interactions. The up- and down-regulated networks for early- and late-stage NSCLC are inputs for the topological parameter-based classification.

Enrichment analysis of gene set

Given a gene list, DAVID [32] performs batch annotation and GO [33] term enrichment analysis to highlight the most relevant GO terms. In contrast, the ConsensusPathDB (CPDB) [34] resource performs gene set analysis and metabolite set analysis. To find the enriched pathways of the proposed genetic signature for NSCLC, an over-representation pathway analysis was performed using both DAVID and CPDB using a p-value threshold of 0.05. Significant pathways were ranked by p-value. Both tools were utilized in this stage for cross-verification.

Potential target genes and drug discovery

The two sets of key genes that were obtained using machine learning algorithms and topological parameter-based classification were grouped up- and down-regulated genes to query the cMap database, which retained potential drugs with p-values of less than 0.05. Drugs that were output by cMap were mapped, and finally identified with known drug targets in the up- or down-regulated cancer PPI network.

Combining datasets raises some issues, such as the problem of data heterogeneity, varying sample sizes, and the problem of data dependence. In principle, these issues can be resolved using meta-analysis. Meta-analysis [35, 36] is a set of statistical methods for summarizing the results of several investigations as a single value. The advantage of meta-analysis is that it can identify relationships across many studies.

In this drug prediction study, a p-value and an enrichment score (ES) are obtained for each cMap drug. The Fisher summary statistical method [36] uses the p-values, defined as,

$$F_i = -2\sum_{j=1}^{N} \log\left(p_{ij}\right) \tag{1}$$

where F_i tests (χ^2 test with $2N$ degrees of freedom, where N is the sample size) the null hypothesis for gene i, and indices i and j indicate the ith gene in the jth dataset respectively.

The ES value lies between −1 and 1, and so can be treated as a sample correlation coefficient and an index of the size of effect in the meta-analysis [36]. In practice, the ES value is converted to a value on Fisher's z scale, and all analyses are performed using such converted values. After the analyses are completed, the z values are converted back to the original scale [6]. The ES value is transformed to a z value by,

$$z = \frac{1}{2}\ln\frac{1+ES}{1-ES} \tag{2}$$

and the variance of z is defined as $V_z = 1 / (N - 3)$, where N is the sample size. The variance of z is approximately proportional to N-3 (as proved by R. A. Fisher), which is independent of the value of the correlation among the population from which the sample drawn [37].

The weight that is assigned to each study in a fixed-effect model is given by,

$$W_i = \frac{1}{V_{Y_i}} \tag{3}$$

where W_i is the within-study variance in study i. The weighted mean (M) is computed as,

$$M = \frac{\sum_{i=1}^{k} W_i Y_i}{\sum_{i=1}^{k} W_i} \tag{4}$$

For unweighted calculations, W_i is unity. The variance of the summary effect (V_M) is given by,

$$V_M = \left(\sum_{i=1}^{k} W_i\right)^{-1} \tag{5}$$

For unweighted calculations, the Z-score for a normal distribution is defined as,

$$Z = \frac{M}{SE_M} \tag{6}$$

where SE_M is the standard error and equals $\sqrt{V_M}$.

For weighted calculations, the Z-score is defined as,

$$Z = \frac{\sum_{i=1}^{k} W_i Y_i}{\sqrt{\sum_{i=1}^{k} W_i^2}} \tag{7}$$

Equation (7) yields the one-tailed test p-value. The 95 % lower and upper limits on the summary effect are computed as,

$$LL_M = M - 1.96 \times SE_M$$
$$UL_M = M + 1.96 \times SE_M \tag{8}$$

The formula for the random-effects model can be found in a monograph that was written by Borenstein [36]. The above analyses enable the confidence interval of the ES to be determined.

The meta-analysis involves two models - the fixed-effect model and the random-effect mod-el [36]. In the fixed-effect model, only one true effect size is assumed to exist, and all differences among studies or batches are assumed to be caused by sampling errors only. In contrast, the random-effect model allows the effect size to vary among studies, and allows an effect size to be estimated for each study. This work considers both models.

A test for the homogeneity of the distribution of data was conducted. As the size of effect commonly found to vary among studies, the meta-analysis method is used herein. Q statistics and I^2 statistics are used to quantify the heterogeneity, to test it, and to incorporate it into the weighting scheme. The value of I^2 is defined as,

$$I^2 = \frac{Q - df}{Q} \times 100\% \tag{9}$$

where df is the number of degrees of freedom, and Q is given by,

$$Q = \sum_{i=1}^{k} W_i (Y_i - M) \tag{10}$$

where k, W, Y and M are the number of studies, the study weight, the size of the effect of interest in the study and the summary effect, respectively.

A p-value of 0.1 for I^2 statistics is used as the threshold for statistical significance. A p-value of larger than or equal to 0.1 indicates little variation among batches, and that a fixed-effect model may therefore be appropriate; otherwise, the random-effect model applies [36]. The I^2 value represents the degree of heterogeneity: an I^2 of less than 25 % implies no heterogeneity, whereas a value of larger than 75 % indicates extremely high heterogeneity.

If the studies are homogenous, then they are likely to have tested the same hypothesis. If estimates are heterogeneous, then the studies probably did not test the same hypothesis. Therefore, all of the study results may not be able to be combined in a single meta-analysis. In such a case, a separate meta-analysis, such as a meta-regression analysis, must be performed for various subsets of studies [36].

MTT™ cell viability test

To determine the effective cytotoxicity of screening drugs, MTT assay was used for cell viability and proliferation. In general, all incubated cancer cell lines (A549 and H460) were seeded in a 96-well microplate for up to 24 h dependent on the baseline growth rate. After incubation, candidate drugs were added into the plate and incubated together for 72 h. For performing the assay, 50 μl MTT solution (2 mg/ml) per well was added and incubated at 37 °C for 2 h. The 150 μl supernatant per well was then extracted and DMSO was filled to dissolve the recipe. The absorbance was set up at 570 nm and calculated by using ELISA reader (Infinite® M1000, TECAN, Switzerland). Ratio decrease comparing to the control group as 100 % viable was seemed as the inhibitory effect.

Clonogenic assay

We use two different high clonogenic lung cancer cell lines, A549 and H460 to perform the clonogenic assay. Cells were diluted to 500 cells per well and then seeded in 6-well plates up to 10 days according to the growth rate. Each well contained 1.5 ml RPMI medium as culture condition and screening compounds were added 24 h after the seeding. For the longer duration of incubation, medium and compounds were changed every 4 days. For performing the assay, cells were washed with PBS, and then the attached colonies were fixed with acetic acid (1: 3 diluted in methanol). The fixed colonies were stained with 0.5 % crystal violet. The colonies were then counted manually after removing the excess crystal violet and rinsing with tap water.

Results

Microarray data analysis

In this study, multiple microarray source data were used for analysis. The Robust Multi-array Average (RMA) was used to normalize gene expression. DEGs were predicted using an adjusted *p*-value of 0.005. Integrating DEGs data with the BioGrid [38] PPI data yielded a list of binary interactions among DEGs for both up and down groups.

The fact that that the use of various microarray platforms may raise the problem of heterogeneity is a concern, which can be tackled in the following two steps; (i) select common DEGs among all platforms for further analysis, and (ii) perform meta-analysis and test heterogeneity to determine whether the fixed-effect model or the random-effect model should be used.

Results of machine learning

In the machine learning method, every microarray dataset is processed individually. Before conducting the machine learning algorithms, the DEGs lacking of domain data or PPI data were excluded from the candidate DEGs. The input data concerned only the remaining DEGs. After the machine learning approach was implemented, only DEGs that were identified as cancer proteins by all five topological parameter-based classifiers were considered as key genes. Table 2 presents the statistical results in this stage.

Results of topological parameter-based classification

To identify key genes in the up- and down-regulated networks respectively the following process was implemented. For each group of DEGs that is classified by a topological parameter, a DEG that ranks in the top 20 % in that parameter will receive a score (S) of one. Clearly, a higher score for a DEG indicates greater importance in the network. DEGs with the highest scores in each group are selected for key genes. The key genes are the union of the two sets with the highest-scoring DEGs in the up- and down-regulated networks. In this work, this stage yielded 104 and 123 key genes for the early- and late-stage NSCLC, respectively. Focusing on the top 10 % rather than 20 % yields only 41 and 56 key genes for the early- and late-stage NSCLC. Relaxing the threshold to 30 % yields 170 and 200 key genes, respectively, which are too many; therefore, top 20 % of classified genes were chosen for key genes.

Enriched biological pathways

Pathways are annotated using DAVID and CPDB. Top-ranking pathways in REACTOME [39] and KEGG [40] with *p*-values of less than 0.05 are reported.

In the machine learning method, the selected DEGs are microarray-specific. Common DEGs were collected from all microarray datasets as the key genes for biological pathway analysis. The key genes that were selected by topological parameter-based classification of genes in up- and down-regulated networks are merged into a single set. The two sets of key genes from the different approaches are submitted to DAVID and CPDB to extract the common enriched biological pathways.

Table 3 presents the common enriched pathways for early NSCLC that are identified by the machine learning

Table 2 The number of DEGs derived from the machine learning method for each microarray dataset

Stage	GEO ID	Number of DEGs	Excluded	Net	Number of predicted key genes	Common genes
Early	GSE7670	801	350	451	259	136
	GSE10072	2835	890	1945	1173	
	GSE19804	4614	1924	2690	1697	
	GSE27262	8476	3161	5315	3310	
Late	GSE7670	1674	608	1066	511	182
	GSE10072	1656	574	1082	691	
	GSE19804	3391	1545	1846	1181	

Table 3 The common pathways by using DAVID and CPDB for early-stage NSCLC (the p_M-*value* and p_T-*value* represent the corresponding *p*-value obtained by machine learning algorithms and topological parameter-based classification)

DAVID					
KEGG			REACTOME		
pathname	p_M-*value*	p_T-*value*	pathname	p_M-*value*	p_T-*value*
NULL			NULL		
CPDB					
KEGG			REACTOME		
pathname	p_M-*value*	p_T-*value*	pathname	p_M-*value*	p_T-*value*
Endocytosis	0.01340	0.00045	Cell-Cell communication	0.02810	0.00432
Glycolysis/Gluconeogenesis	0.02330	0.00249	Glucose metabolism	0.02420	0.02000
Hematopoietic cell lineage	0.04060	0.03380	Regulation of PLK1 Activity at G2/M Transition	0.03700	0.03070
Gap junction	0.04700	0.03930	Metabolism of nucleotides	0.03940	0.03280
			Cell junction organization	0.00765	0.03590
			Platelet activation, signaling and aggregation	0.04720	0.03630

algorithms and topological parameter classification. According to Table 3, no common pathways were identified by DAVID, while some were found using CPDB. In KEGG, endocytosis, glycolysis/gluconeo-genesis, hematopoietic cell lineage and gap junction are the common enriched pathways for early-stage cancer. According to the literature, these common pathways are closely related to cancer. Among them, glycolysis/gluconeo-genesis has been identified as an enriched pathway for early-stage cancer [6]. Oncogenes and tumor suppressors are known to regulate metabolism. The mutations of oncogenes in the up-regulation of glucose transporters increase the consumption of glucose by cancer cells, increasing the rate of glucose metabolism [41, 42]. Endocytosis is closely related to cell regulation and is predicted to play an important role in human cancers [43]. Raf/MEK/ERK is typically associated with the proliferation and drug resistance of hematopoietic cells, while the activation of the Raf/MEK/ERK cascade is suppressed in some prostate cancer cell lines that have mutations at PTEN and express high levels of activated Akt [44]. Holder et al. claimed that persistent gap junction perturbation can have chronic effects, and various tumor promoters inhibit GJ intercellular communication [45]. Cancer cells typically have down-regulated levels of gap junctions, and many pieces of evidence suggest that loss of gap junctional intercellular communication is an important step in carcinogenesis [46].

In REACTOME, cell-cell communication, glucose metabolism, regulation of PLK1 activity at the G2/M transition, metabolism of nucleotides, organization of the cell junction and platelet activation, signaling and aggregation are enriched pathways for early NSCLC. Of them, glucose metabolism is like glycolysis/gluconeo-genesis and has been previously determined to be related to cancers. Tominaga et al. [47] demonstrated that cancer-derived extracellular vesicles (EVs), which are mediators of cell–cell communication, trigger the breakdown of the blood–brain barrier, which controls the migration of cancer cells. Arid and Zhang proposed that nucleotide metabolism causes tumor progression, and considered how this pathway can be targeted for cancer therapy by inducing the senescence of cancer cells [48]. Several cell junction components have functions that are associated with cell polarity and growth control and are specifically disrupted in cancerous cells [49]. PLK1 seems to be involved in the tumor suppressor p53-related pathways. Evidence suggests that PLK1 inhibits the transactivation and pro-apoptotic functions of p53 by physical interaction and phosphorylation [50]. Additionally, in cancer growth and dissemination, complex interactions between tumor cells and circulating platelets are critical. Evidence supports a role for physiological platelet receptors and platelet agonists in cancer metastases and angiogenesis [51].

Based on the pathway annotation database, REACTOME in DAVID, Table 4 presents the common enriched pathways for late-stage NSCLC that are identified by both methods. CPDB returns more pathways than DAVID. As noted in reference to Table 4, cell cycles are the common path that is identified using DAVID, and this finding is consistent with the results of our previous work [6]. Furthermore, many common paths were observed using CPDB, and these are dominated by the cell cycle. Notably, the endocytosis pathway appears in both Tables 3 and 4, indicating that this pathway is closely related to both early-stage and late-stage NSCLC.

DNA replication, repair and checkpoint activation pathways are highly regulated and coordinated. Defects in any of these functions cause genomic instability and may lead to cancer [52]. For example, BRCA2 participates in homologous recombination and regulating the S-phase checkpoint, and mutations of deficiencies in BRCA2 are strongly associated with tumorigenesis [53].

Table 4 The common paths using DAVID and CPDB for late-stage NSCLC (the p_M-*value* and p_T-*value* represent the corresponding *p*-value obtained by machine learning algorithms and topological parameter-based classification)

DAVID					
KEGG			REACTOME		
pathname	p_M-*value*	p_T-*value*	pathname	p_M-*value*	p_T-*value*
Cell cycle	0.03800	0.00140	Cell Cycle Checkpoints	0.00760	0.00872
			Cell Cycle, Mitotic	0.00100	0.02186
CPDB					
KEGG			REACTOME		
pathname	p_M-*value*	p_T-*value*	pathname	p_M-*value*	p_T-*value*
Cell cycle	0.00632	0.00004	Regulation of mitotic cell cycle	0.01500	0.00000
Inflammatory mediator regulation of TRP channels	0.03960	0.00048	APC/C:Cdc20 mediated degradation of mitotic proteins	0.01500	0.00000
Endocytosis	0.00051	0.00153	Activation of APC/C and APC/C: Cdc20 mediated degradation of mitotic proteins	0.00663	0.00001
Thyroid hormone synthesis	0.01560	0.00662	Cell Cycle Checkpoints	0.00729	0.00001
Salivary secretion	0.03120	0.01370	Cell Cycle	0.00007	0.00001
Long-term depression	0.04420	0.02330	Regulation of APC/C activators between G1/S and early anaphase	0.00177	0.00001
cGMP-PKG signaling pathway	0.02210	0.02740	G1/S Transition	0.01030	0.00002
Vascular smooth muscle contraction	0.00497	0.03440	Cdc20:Phospho-APC/C mediated degradation of Cyclin A	0.00837	0.00004
			Cell Cycle, Mitotic	0.00485	0.00014
			APC:Cdc20 mediated degradation of cell cycle proteins prior to satisfaction of the cell cycle checkpoint	0.00182	0.00014
			Mitotic G1-G1/S phases	0.00541	0.00016
			G2/M Checkpoints	0.02590	0.00023
			M Phase	0.00298	0.00119
			DNA Replication	0.00813	0.00153
			Resolution of Sister Chromatid Cohesion	0.00103	0.00297
			Mitotic Prometaphase	0.00230	0.00399
			Apoptotic cleavage of cellular proteins	0.00067	0.00572
			S Phase	0.01500	0.00762
			Mitotic Anaphase	0.02040	0.00875
			Mitotic Metaphase and Anaphase	0.00035	0.01140
			Apoptotic execution phase	0.00036	0.01170
			Synthesis of DNA	0.03370	0.01760
			Separation of Sister Chromatids	0.00540	0.01760

Table 4 agrees closely with the results of our previous work [6], which also identified cell-cycle, the mitotic anaphase, DNA replication, the sister-chromatid segregation process, the Cdc20:Phospho-APC/C-mediated degradation of Cy-clin A, the M-phase and mitotic G1-G1/S phases.

Although defective apoptosis is critical to the development and progression of cancer, apoptosis is important in the treatment of cancer as it is a popular target of many treatment strategies [54].

Wong et al. [55] noted that PKG-Iα kinase activity is necessary to maintaining high levels of cAMP response element binding (CREB) phosphorylation at ser133, and promotes the formation of colonies in NSCLC cells. The gene expression signature of the responses of vascular smooth muscle contraction to serum exposure is associated with a significantly poorer prognosis in cases of human cancer, and vascular injury response is therefore potentially linked to tumor progression [56].

According to Table 4, the mitotic process and CDC20 are involved in many enriched pathways. Mitotic progression and sister-chromatid segregation are controlled by the anaphase promoting complex/cyclosome (APC/C). APC/C forms a protein complex with its mitotic co-activator, CDC20, which controls mitotic progression. CDC20 protein level may directly influence the fate of cells during prolonged mitotic arrest and its turnover rate may critically affect the response of a cancer patient to anti-mitotic therapies [6].

In summary, combining machine learning methods with the classification of topological parameters reveals many cancer related pathways, which are well supported by the literature, providing insight into key regulators of the tumorigenesis of NSCLC.

Potential drugs for treating NSCLC and their targeted genes

Both sets of key genes that were identified by machine learning algorithms and topological parameter-based classification were analyzed using cMap to discover potential drugs. For the set of key genes from machine learning approach and individual microarray, meta-analysis was performed using the *p*-values that were obtained from cMap for an individual microarray. For example, in early-stage NSCLC, cMap outputs 1309 drugs for key genes from the microarray GSE7670. These 1309 drugs are then filtered to find those with cMap *p*-value <0.1, and 168 drugs are identified. For cMap *p*-value <0.1, the numbers of remaining drugs for the four microarray datasets of early-stage NSCLC are 168, 139, 149 and 85 respectively. A meta-analysis is then performed to integrate the four groups of remaining drugs, and nine drugs are finally extracted. IC_{50} experiments verified the therapeutic effectiveness of four of these drugs. The alternative method begins by extracting the drugs from cMap under the constraint ES < 0, yielding 597 drugs from the 1309 drugs for microarray GSE7670. Next, the meta-analysis is performed, and 383 drugs are filtered out of the 597 drugs. Finally, 60 drugs with a meta-analysis *p*-value (p_{MA}-value) of less than 0.1 are kept. Table 5 shows all of these results.

In Table 6, the first row presents the early- and late-stage ES and *p*-value that is used in the meta-analysis. The upper-diagonal includes the Jaccard Index (*JI*) score

Table 5 The number of potential drugs filtered by meta-analysis for early- and late- stage NSCLC using the enrichment score (ES) and cMap *p*-value (less than 0.1) for meta-analysis

	Early-stage		Late-stage	
	Potential drugs	IC_{50} verified	Potential drugs	IC_{50} verified
ES < 0 & cMap *p*-value <0.1	9	4	31	5
ES < 0 & cMap *p*-value <0.5	12	4	81	8
ES < 0 & p_{MA}-value <0.05	25	2	23	1
ES < 0 & p_{MA}-value <0.1	60	8	49	5

Table 6 The number of common drugs and *JI* score for early- and late-stage using the enrichment score (ES) and cMap *p*-value (less than 0.1) for meta-analysis

Effect size	Effect size	ES < 0 & p_{MA}-value <0.1		ES < 0 & cMap *p*-value < 0.1	
		Early-stage	Late-stage	Early-stage	Late-stage
ES < 0 & p_{MA}-value <0.1	Early-stage		0.557	0.078	0.152
	Late-stage	39		0.074	0.194
ES < 0 & cMap *p*-value < 0.1	Early-stage	5	4		0.143
	Late-stage	12	13	5	

of the corresponding effect size and NSCLC stage. Given two sets A and B, *JI*(A,B) is defined as |A∩B|/(|A∪B| - |A∩B |), where |A∩B|, |A| and |B| denote the cardinality of A ∩ B, A and B respectively. In contrast, entries in the lower diagonal are the number of common drugs for the corresponding effect size and NSCLC stage. For early-stage NSCLC, there are five common drugs (*JI* is 0.078) predicted under the two kinds of effect size, whereas, 13 common drugs (*JI* is 0.152) are identified for treating late-stage NSCLC.

The drugs that are predicted by the machine learning method are the union of the drugs that are predicted under the four conditions in Table 5. In contrast, the drugs that are predicted by the topological parameter-based classification are direct outputs of the cMap with a *p*-value of less than 0.05. Machine learning algorithms (topological parameter-based classification) identified 60 (17) potential drugs for treating early-stage NSCLC, among which eight (two) were validated as effective by MTT or clonogenic assay, and are presented in Table 7.

Machine learning algorithms (topological parameter-based classification) identified 49 (37) potential drugs for treating late-stage NSCLC, of which were five (five) were validated as effective by MTT or clonogenic assay, and are presented in Table 8.

Table 9 lists the common drugs that were identified by both machine learning algorithms and topological parameter-based classification. Of these, two (trichostatin A and vorinostat) were determined by IC_{50} to be effective against both early- and late-stage NSCLC, respectively. These common drugs are consistent with the findings of Ref. [5]. Seven drugs, including trichostatin A, vorinostat, MS-275, scriptaid, perhexiline, (–)-MK-801, and rifabutin, of the 18 predicted potential drugs for treating NSCLC had been reported in Ref. [5]. Interestingly, the first four drugs are HDAC inhibitors. Also, we found trichostatin A also among the 18 predicted potential drugs for treating NSCLC [6]. Notably, trichostatin A is the common drug

Table 7 IC_{50} values of potential drugs for early-stage NSCLC

Machine learning algorithms		
cMap drug name	MTT (μM)	Clonogenic (μM)
mebendazole	<1	
vorinostat	<1	
pyrvinium	<0.1	
niclosamide	>5	
nortriptyline		<10
piperlongumine	>5	
trichostatin A		
trioxysalen	>5	
Topological parameter-based classification		
cMap drug name	MTT (μM)	Clonogenic (μM)
trichostatin A	<1	
vorinostat	<1	

that was identified in all of the above studies. Five common drugs, including 4,5-dianilinophthalimide, perhexiline, puromycin, trichostatin A, and vorinostat, are identified for treating both early-stage and late-stage NSCLC in this study, suggesting that they may be stage-independent drugs.

Some of the above common drugs have been undergoing clinical trials for NSCLC treatment, including mepacrine (clinical trial NCT01839955), MS-275 (clinical trial NCT02437136) and Vorinostat (clinical trial NCT00667082). The results in this study are consistent with our previous work [5]; both studies identified nine drugs, of which had cytotoxic effects that were validated by IC_{50} experiments. These three drugs are trichostatin A, vorinosta and nortriptyline. The potential use in lung cancer

Table 8 IC_{50} values of potential drugs for late-stage NSCLC

Machine learning algorithms		
cMap drug name	MTT (μM)	Clonogenic (μM)
trichostatin A	<1	
Vorinostat	<1	
withaferin A	<1	
mebendazole	<1	
piperlongumine	>5	
Topological parameter classification		
cMap drug name	MTT (μM)	Clonogenic (μM)
acepromazine		<10
nortriptyline		<10
propafenone		<10
trichostatin A	<1	
vorinostat	<1	

Table 9 The common drugs identified by both two methods

Early-stage NSCLC			
4,5-dianilinophthalimide	mepacrine (quinacrine)	meptazinol	perhexiline
puromycin	trichostatin A	vorinostat	
Late-stage NSCLC			
(-)-MK-801	4,5-dianilinophthalimide	MS-275 (Entinostat)	perhexiline
puromycin	quinostatin	rifabutin	scriptaid
trichostatin A	vorinostat	Y-27632	

treatment warrants further exploration. Notably, Ref. [5] treated the early stage and late stage on the same footing, it is not stage-specific.

The machine learning approach has similar hit ratio to the topological parameter-based approach (early-stage: 8/60 vs. 2/17; late-stage: 5/49 vs. 5/37), as supported by in vitro IC_{50} measurements. Combining the machine learning approach with the topological parameter-based classification yielded the best hit ratio. The current method has a higher prediction accuracy (early-stage: 2/7 vs. 1/7; late-stage: 2/11 vs. 7/65) than the method of Ref. [6], consistent with the IC_{50} measurements.

The common drugs were submitted to DrugBank [57] and NCBI to search for their corresponding targeted genes. Among these targeted genes, we kept only those which are also key genes, finally yielding a total of 8 and 12 targeted genes for early- and late-stage NSCLC respectively, as shown in Table 10, which are the potential therapeutic targets for future lung cancer clinical trials. For each targeted gene in Table 10, the number in parentheses is the number of associated cMap drugs, and could be regarded as a metric for prioritizing the genes in the list. The *ADRB2* gene ranked top of the lists for both early- and late-stage NSCLC.

Whether a particular gene is related to most of the targeted genes in Table 10 is of interest. Therefore, networks of the targeted genes and their adjacent genes in PPI for early- and late- stage NSCLC were constructed. Figure 2a and b display the top three genes that exhibit the largest, the second largest and the third largest

Table 10 The targeted genes identified by the common drugs derived from both two methods (the parentheses represent the number of associated cMap drugs)

Early-stage NSCLC			
ADRB2 (4)	CASP1 (3)	KAT2A (2)	SNCA (1)
ARRB1(1)	PSIP1 (1)	PAFAH1B3 (1)	GAPDH (1)
Late-stage NSCLC			
ADRB2 (6)	ARRB1 (3)	NCOA1 (3)	PSIP1 (3)
SMARCA2 (3)	GAPDH (3)	CPT1A (1)	AURKB (1)
IRAK1 (1)	GRK5 (1)	SRPK1 (1)	AURKA (1)

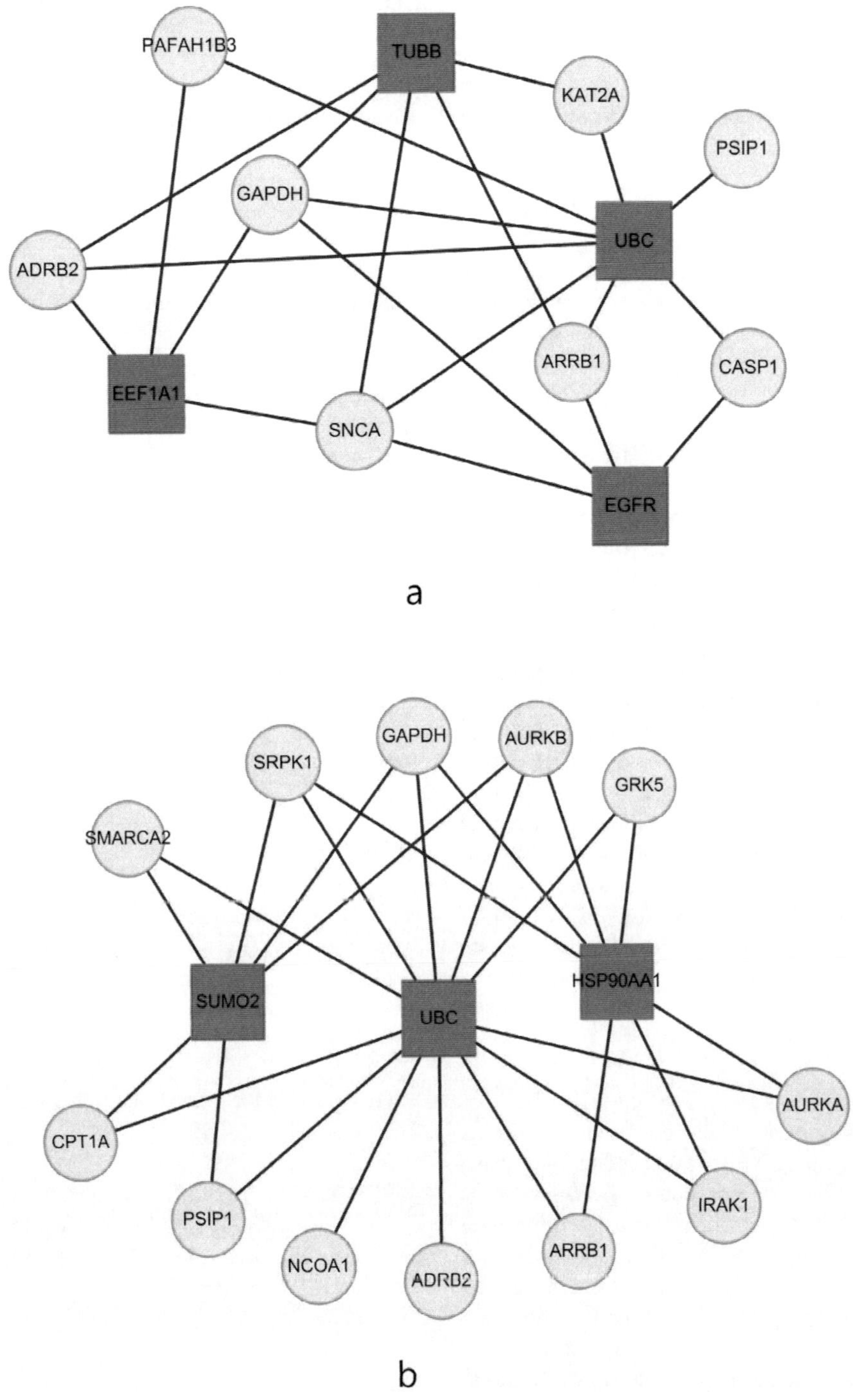

Fig. 2 The top three genes (*squares*) which connect to the largest, the second largest and the third largest degree of targeted genes (*circles*) for **a** early-stage; **b** late-stage network

degree in early- and late-stage networks, respectively. In the early-stage network, the *UBC* gene directly interacts with all of the 8 targeted genes, as shown in Fig. 2a, while the other genes (such as *TUBB*) connect to no more than 5 targeted genes. Similarly, in the late-stage network, the *UBC* gene connects to all of the 12 targeted genes as shown in Fig. 2b, while the others (such as *HSP90AA1*) connect to at most 7 targeted genes. This finding is in agreement with the findings of [5]. Although the *UBC* gene is neither a key gene nor a targeted gene, it dominates all of the targeted genes; whether this fact implies that the *UBC* gene acts as a master regulator in the cancer pathway deserves further experimental investigation.

A concern arises regarding how the *p*-values that are obtained by different methods are combined. In fact, only the *p*-values and the enrichment scores (ES) that were obtained from cMap are combined in meta-analysis. Please refer to the workflow in Fig. 1. Four *p*-values were obtained by (1) identification of DEGs, (2) gene set enrichment analysis, (3) cMap drug analysis and (4) meta-analysis of cMap drugs.

The *p*-values that were obtained in the DEG analysis are used to identify significant DEGs. Also, the *p*-values that were obtained in (2) and (3) are not related to each other, and they do not have to be combined. Since different microarray datasets yielded different drug predictions, meta-analysis was conducted using the cMap *p*-values and ES to achieve results in which confidence is high.

Some missense mutations and non-synonymous SNPs (nsSNPs) may damage protein functions, disrupting the drug actions. Our future work will account for this effect. Numerous web-based tools are available to facilitate such analysis. PolyPhen2 [58] is a tool that predicts the impact of an amino acid substitution on the function and structure of a protein using sequence-based and structure-based features. SNPdryad [59] is a web-based tool that elucidates the effect of nsSNPs based on multiple sequence alignments of orthologous proteins. MutationTaster [60] is another tool that uses NGS data to elucidate the effect of missense mutations on the expression and function of proteins.

Conclusion

In this study, two methods - machine learning algorithms and topological parameter-based classification - are compared and combined to identify potential reliable drugs for treating NSCLC, and meta-analysis is used to solve the problem of data heterogeneity. Since cancer is a multistage progressive disease, early- and late-stage cancer-related genes potentially differ substantially. Therefore, the proposed method was used to identify stage-specific DEGs, biological pathways and potential drugs. Some of the extracted biological pathways are supported by the literature, and some of the results herein concerning the identified drugs are supported by IC_{50} experiments. Seven and 11 potential drugs are discovered for treating early- and late-stage NSCLC, respectively, and warrant further investigation. Among them, perhexiline and trichostatin A are supported by the previous research. Interestingly, the *UBC* gene dominates all of the targeted genes associated with early- and late-stage NSCLC, so its role in the cancer pathway warrants further investigation.

Integrating machine learning algorithms and topological parameter-based classification herein increased drug prediction accuracy over that achieved in any previous research. This improvement is confirmed by IC_{50} experiments. The overlap of our discovered drug candidates with those that are undergoing clinical trials or are identified in the literature demonstrates the effectiveness of the proposed methods. The performance of the proposed methods can be further improved by incorporating more microarray datasets or verified gene-drug associations. In summary, many techniques were integrated to develop a novel pipeline of therapeutic drugs for NSCLC, and the efficiency of this pipeline was investigated. The approaches that were developed in this work are expected to inspire future studies, and the pipeline may be extended to the treatment of other diseases.

Competing interests

The authors declare that they have no competing interests.

Authors' contributions

CHH conducted the study design, analysis and drafted the manuscript. PMHC participated in discussion of the potental drugs. CWH conducted the related experiments. CYH provided the IC_{50} data. KLN participated in the interpretation and is the corresponding author of the article. All authors read and approved the manuscript.

Acknowledgements

The work of Chien-Hung Huang is supported by the Ministry of Science and Technology of Taiwan (MOST) under grant MOST 104-2221-E-150-039. The work of Ka-Lok Ng is supported by MOST under grants MOST 102-2632-E-468-001-MY3 and MOST 104-2221-E-468-012, and also supported by Asia University under the grants 103-asia-06. The work of Peter Mu-Hsin Chang is supported by MOST under grant MOST 103-2314-B-075-025 and Taipei Veterans General Hospital under grant V104C-090. The work of Chi-Ying Huang is supported by MOST under grant MOST104-2325-B-010-002 and by Center of Excellence for Cancer Research at Taipei Veterans General Hospital phase II: Integrated approach to reduce cancer incidence and mortality (MOHW104-TDU-B-211-124-001). Ted Knoy is appreciated for his editorial assistance.

Declarations

The source of funding for publication is supported by the Ministry of Science and Technology of Taiwan (MOST) under grant MOST 104-2221-E-150-039. This article has been published as part of *BMC Bioinformatics* Volume 17 Supplement 1, 2016: Selected articles from the Fourteenth Asia Pacific Bioinformatics Conference (APBC 2016). The full contents of the supplements are available online at http://www.biomedcentral.com/bmcbioinformatics/supplements/17/S1.

Author details

[1]Department of Computer Science and Information Engineering, National Formosa University, Hu-Wei 63205, Taiwan. [2]Division of Hematology and Oncology, Department of Medicine, Taipei Veterans General Hospital; Faculty of Medicine, National Yang Ming University, Taipei 112, Taiwan. [3]Institute of Biopharmaceutical Sciences, National Yang-Ming University, Taipei 112, Taiwan. [4]Department of Bioinformatics and Medical Engineering, Asia University, Taichung 41354, Taiwan. [5]Department of Medical Research, China Medical University Hospital, China Medical University, Taichung 40402, Taiwan.

Published: 11 January 2016

References

1. Berman AT, James SS, Rengan R. Structure, mechanism, and evolution of the mRNA capping apparatus. Cancers (Basel). 2015;7(3):1178–90.
2. Stachnik A, Yuen T, Iqbal J, Sgobba M, Gupta Y, Lu P, et al. Repurposing of bisphosphonates for the prevention and therapy of nonsmall cell lung and breast cancer. Proc Natl Acad Sci U S A. 2014;111(50):17995–8000.
3. Chen H, Zhang H, Zhang Z, Cao Y, Tang W. Network-based inference methods for drug repositioning. Comput Math Methods Med. 2015;2015:130620.
4. Lee HS, Bae T, Lee JH, Kim DG, Oh YS, Jang Y, et al. Rational drug repositioning guided by an integrated pharmacological network of protein, disease and drug. BMC Syst Biol. 2012;6:80.
5. Huang CH, Wu MY, Chang PM, Huang CY, Ng KL. In silico identification of potential targets and drugs for non-small cell lung cancer. IET Syst Biol. 2014;8(2):56–66.
6. Huang CH, Chang PM, Lin YJ, Wang CH, Huang CY, Ng KL. Drug repositioning discovery for early- and late-stage non-small-cell lung cancer. Biomed Res Int. 2014;2014:193817.
7. Huang CH, Peng HS, Ng KL. Prediction of cancer proteins by integrating protein interaction, domain frequency, and domain interaction data using machine learning algorithms. Biomed Res Int. 2015;2015:312047.
8. Bull SC, Doig AJ. Properties of protein drug target classes. PLoS One. 2015; 10(3), e0117955.
9. Carson MB, Lu H. Network-based prediction and knowledge mining of disease genes. BMC Med Genomics. 2015;8 Suppl 2:S9.
10. Iorio F, Bosotti R, Scacheri E, Belcastro V, Mithbaokar P, Ferriero R, et al. Discovery of drug mode of action and drug repositioning from transcriptional responses. Proc Natl Acad Sci U S A. 2010;107(33):14621–6.
11. Dudley JT, Sirota M, Shenoy M, Pai RK, Roedder S, Chiang AP, et al. Computational repositioning of the anticonvulsant topiramate for inflammatory bowel disease. Sci Transl Med. 2011;3(96):96ra76.
12. Sirota M, Dudley JT, Kim J, Chiang AP, Morgan AA, Sweet-Cordero A, et al. Discovery and preclinical validation of drug indications using compendia of public gene expression data. Sci Transl Med. 2011;3(96):96ra77.
13. Pacini C, Iorio F, Goncalves E, Iskar M, Klabunde T, Bork P, et al. DvD: An R/ Cytoscape pipeline for drug repurposing using public repositories of gene expression data. Bioinformatics. 2013;29(1):132–4.
14. Fortney K, Griesman J, Kotlyar M, Pastrello C, Angeli M, Sound-Tsao M, et al. Prioritizing therapeutics for lung cancer: an integrative meta-analysis of cancer gene signatures and chemogenomic data. PLoS Comput Biol. 2015;11(3), e1004068.
15. Campillos M, Kuhn M, Gavin AC, Jensen LJ, Bork P. Drug target identification using side-effect similarity. Science. 2008;321(5886):263–6.
16. Yang L, Agarwal P. Systematic drug repositioning based on clinical side-effects. PLoS One. 2011;6(12), e28025.
17. Duran-Frigola M, Aloy P. Recycling side-effects into clinical markers for drug repositioning. Genome Med. 2012;4(1):3.
18. Hurle MR, Yang L, Xie Q, Rajpal DK, Sanseau P, Agarwal P. Computational drug repositioning: from data to therapeutics. Clin Pharmacol Ther. 2013;93(4):335–41.
19. Keiser MJ, Setola V, Irwin JJ, Laggner C, Abbas AI, Hufeisen SJ, et al. Predicting new molecular targets for known drugs. Nature. 2009;462(7270):175–81.
20. Chiang AP, Butte AJ. Systematic evaluation of drug-disease relationships to identify leads for novel drug uses. Clin Pharmacol Ther. 2009;86(5):507–10.
21. Cheng F, Liu C, Jiang J, Lu W, Li W, Liu G, et al. Prediction of drug-target interactions and drug repositioning via network-based inference. PLoS Comput Biol. 2012;8(5), e1002503.
22. Fukuoka Y, Takei D, Ogawa H. A two-step drug repositioning method based on a protein-protein interaction network of genes shared by two diseases and the similarity of drugs. Bioinformation. 2013;9(2):89–93.
23. Huang CH, Peng HS, Ng KL. Graph theory and stability analysis of protein complex interaction networks. IET Syst Biol. 2015. doi:10.1049/iet-syb.2015.0007.
24. Barrett T, Wilhite SE, Ledoux P, Evangelista C, Kim IF, Tomashevsky M, et al. NCBI GEO: archive for functional genomics data sets—update. Nucleic Acids Res. 2013;41(Database issue):D991–5.
25. Su LJ, Chang CW, Wu YC, Chen KC, Lin CJ, Liang SC, et al. Selection of DDX5 as a novel internal control for Q-RT-PCR from microarray data using a block bootstrap re-sampling scheme. BMC Genomics. 2007;8:140.
26. Landi MT, Dracheva T, Rotunno M, Figueroa JD, Liu H, Dasgupta A, et al. Gene expression signature of cigarette smoking and its role in lung adenocarcinoma development and survival. PLoS One. 2008;3(2):e1651.
27. Lu TP, Tsai MH, Lee JM, Hsu CP, Chen PC, Lin CW, et al. Identification of a novel biomarker, SEMA5A, for non-small cell lung carcinoma in nonsmoking women. Cancer Epidemiol Biomarkers Prev. 2010;19(10):2590–7.
28. Wei TY, Juan CC, Hisa JY, Su LJ, Lee YC, Chou HY, et al. Protein arginine methyltransferase 5 is a potential oncoprotein that upregulates G1 cyclins/ cyclin-dependent kinases and the phosphoinositide 3-kinase/AKT signaling cascade. Cancer Sci. 2012;103(9):1640–50.
29. Lamb J. The Connectivity Map: a new tool for biomedical research. Nat Rev Cancer. 2007;7(1):54–60.
30. Huang CH, Chou SY, Ng KL. Improving protein complex classification accuracy using amino acid composition profile. Comput Biol Med. 2013; 43(9):1196–204.
31. Kurubanjerdjit N, Huang CH, Lee Y, Tsai JP, Ng KL. Prediction of microRNA-regulated protein interaction pathways in Arabidopsis using machine learning algorithms. Comput Biol Med. 2013;43(11):1645–52.
32. da Huang W, Sherman BT, Lempicki RA. Systematic and integrative analysis of large gene lists using DAVID bioinformatics resources. Nat Protoc. 2009;4(1):44–57.
33. Gene Ontology Consortium. The Gene Ontology (GO) project in 2006. Nucleic Acids Res. 2006;34(Database issue):D322–6.
34. Kamburov A, Wierling C, Lehrach H, Herwig R. ConsensusPathDB–a database for integrating human functional interaction networks. Nucleic Acids Res. 2009;37(Database issue):D623–8.
35. Wolf FM. Meta-Analysis: Quantitative Methods for Research Synthesis. California: Sage publications; 1986.
36. Borenstein M, Hedges LV, Higgins JPT, Rothstein HR. Introduction to meta-analysis. United Kingdom: Wiley press; 2009.
37. Erich L, Lehmann -Fisher, Neyman: the Creation of Classical Statistics. Ch 2, Fisher's Testing Methodology. Springer Science & Business Media; 2011. p. 24.
38. Breitkreutz BJ, Stark C, Reguly T, Boucher L, Breitkreutz A, Livstone M, et al. The BioGRID Interaction Database: 2008 update. Nucleic Acids Res. 2008; 36(Database issue):D637–40.
39. Croft D, O'Kelly G, Wu G, Haw R, Gillespie M, Matthews L, et al. Reactome: a database of reactions, pathways and biological processes. Nucleic Acids Res. 2011;39(Database issue):D691–7.
40. Kanehisa M, Goto S, Kawashima S, Okuno Y, Hattori M. The KEGG resource for deciphering the genome. Nucleic Acids Res. 2004;32(Database issue): D277–80.
41. Ganapathy-Kanniappan S, Geschwind JF. Tumor glycolysis as a target for cancer therapy: progress and prospects. Mol Cancer. 2013;12:152.
42. Gillies RJ, Robey I, Gatenby RA. Causes and consequences of increased glucose metabolism of cancers. J Nucl Med. 2008;49 Suppl 2:24S–42S.
43. Lanzetti L, Di Fiore PP. Endocytosis and cancer: an 'insider' network with dangerous liaisons. Traffic. 2008;9(12):2011–21.
44. McCubrey JA, Steelman LS, Chappell WH, Abrams SL, Wong EW, Chang F, et al. Roles of the Raf/MEK/ERK pathway in cell growth, malignant transformation and drug resistance. Biochim Biophys Acta. 2007;1773(8):1263–84.
45. Holder JW, Elmore E, Barrett JC. Gap junction function and cancer. Cancer Res. 1993;53(15):3475–85.
46. Leithe E, Sirnes S, Omori Y, Rivedal E. Downregulation of gap junctions in cancer cells. Crit Rev Oncog. 2006;12(3–4):225–56.
47. Tominaga N, Kosaka N, Ono M, Katsuda T, Yoshioka Y, Tamura K, et al. Brain metastatic cancer cells release microRNA-181c-containing extracellular vesicles capable of destructing blood–brain barrier. Nat Commun. 2015;6:6716.
48. Aird KM, Zhang R. Nucleotide metabolism, oncogene-induced senescence and cancer. Cancer Lett. 2015;356(2 Pt A):204–10.
49. Gates KL, Howell HA, Nair A, Vohwinkel CU, Welch LC, Beitel GJ, et al. Hypercapnia impairs lung neutrophil function and increases mortality in murine pseudomonas pneumonia. Am J Respir Cell Mol Biol. 2013;49(5):821–8.
50. Liu X, Erikson RL. Polo-like kinase (Plk)1 depletion induces apoptosis in cancer cells. Proc Natl Acad Sci U S A. 2003;100(10):5789–94.
51. Bambace NM, Holmes CE. The platelet contribution to cancer progression. J Thromb Haemost. 2011;9(2):237–49.
52. Mazouzi A, Velimezi G, Loizou JI. DNA replication stress: causes, resolution and disease. Exp Cell Res. 2014;329(1):85–93.
53. Venkitaraman AR. Cancer susceptibility and the functions of BRCA1 and BRCA2. Cell. 2002;108(2):171–82.
54. Wong RS. Apoptosis in cancer: from pathogenesis to treatment. J Exp Clin Cancer Res. 2011;30:87.
55. Wong JC, Bathina M, Fiscus RR. Cyclic GMP/protein kinase G type-Iα (PKG-Iα) signaling pathway promotes CREB phosphorylation and maintains higher c-IAP1, livin, survivin, and Mcl-1 expression and the inhibition of PKG-Iα kinase activity synergizes with cisplatin in non-small cell lung cancer cells. J Cell Biochem. 2012;113(11):3587–98.

56. Chi JT, Rodriguez EH, Wang Z, Nuyten DS, Mukherjee S, van de Rijn M, et al. Gene expression programs of human smooth muscle cells: tissue-specific differentiation and prognostic significance in breast cancers. PLoS Genet. 2007;3(9):1770–84.
57. Wishart DS, Knox C, Guo AC, Cheng D, Shrivastava S, Tzur D, et al. DrugBank: a knowledgebase for drugs, drug actions and drug targets. Nucleic Acids Res. 2008;36(Database issue):D901–6.
58. Adzhubei IA, Schmidt S, Peshkin L, Ramensky VE, Gerasimova A, Bork P, et al. A method and server for predicting damaging missense mutations. Nat Methods. 2010;7(4):248–9.
59. Wong KC, Zhang Z. SNPdryad: predicting deleterious non-synonymous human SNPs using only orthologous protein sequences. Bioinformatics. 2014;30(8):1112–9.
60. Schwarz JM, Cooper DN, Schuelke M, Seelow D. MutationTaster2: mutation prediction for the deep-sequencing age. Nat Methods. 2014;11(4):361–2.

Lee et al. BMC Bioinformatics 2016, 17(Suppl 1):3
DOI 10.1186/s12859-015-0848-x

BMC Bioinformatics

PROCEEDINGS Open Access

Gene expression profiling identifies candidate biomarkers for active and latent tuberculosis

Shih-Wei Lee[1,2†], Lawrence Shih-Hsin Wu[3†], Guan-Mau Huang[4], Kai-Yao Huang[4], Tzong-Yi Lee[4,5] and Julia Tzu-Ya Weng[4,5*]

From The Fourteenth Asia Pacific Bioinformatics Conference (APBC 2016)
San Francisco, CA, USA. 11 - 13 January 2016

Abstract

Background: Tuberculosis (TB) is a serious infectious disease in that 90 % of those latently infected with Mycobacterium tuberculosis present no symptoms, but possess a 10 % lifetime chance of developing active TB. To prevent the spread of the disease, early diagnosis is crucial. However, current methods of detection require improvement in sensitivity, efficiency or specificity. In the present study, we conducted a microarray experiment, comparing the gene expression profiles in the peripheral blood mononuclear cells among individuals with active TB, latent infection, and healthy conditions in a Taiwanese population.

Results: Bioinformatics analysis revealed that most of the differentially expressed genes belonged to immune responses, inflammation pathways, and cell cycle control. Subsequent RT-PCR validation identified four differentially expressed genes, *NEMF*, *ASUN*, *DHX29*, and *PTPRC*, as potential biomarkers for the detection of active and latent TB infections. Receiver operating characteristic analysis showed that the expression level of *PTPRC* may discriminate active TB patients from healthy individuals, while *ASUN* could differentiate between the latent state of TB infection and healthy condidtion. In contrast, *DHX29* may be used to identify latently infected individuals among active TB patients or healthy individuals. To test the concept of using these biomarkers as diagnostic support, we constructed classification models using these candidate biomarkers and found the Naïve Bayes-based model built with *ASUN*, *DHX29*, and *PTPRC* to yield the best performance.

Conclusions: Our study demonstrated that gene expression profiles in the blood can be used to identify not only active TB patients, but also to differentiate latently infected patients from their healthy counterparts. Validation of the constructed computational model in a larger sample size would confirm the reliability of the biomarkers and facilitate the development of a cost-effective and sensitive molecular diagnostic platform for TB.

Keywords: Tuberculosis, Latent infection, Gene expression, Biomarker

* Correspondence: julweng@saturn.yzu.edu.tw
†Equal contributors
[4]Department of Computer Science and Engineering, Yuan Ze University, Taoyuan, Taiwan
[5]Innovation Center for Big Data and Digital Convergence, Yuan Ze University, Taoyuan, Taiwan
Full list of author information is available at the end of the article

Background

Tuberculosis (TB) is an infectious disease caused by various strains of mycobacteria, with *Mycobacterium tuberculosis* (*Mtb*) being the most common causative agent [1]. It is a serious global health threat with one-third of the world's population estimated to be latently infected with *Mtb* [2]. Though about 90 % of those infected with *Mtb* are asymptomatic, possessing only a 10 % lifetime chance of developing active TB [3], even in developing countries with established healthcare systems, TB is still a deadly disease.

In 2006, the World Health Organization launched a "Global Plan to Stop Tuberculosis" that aims to save 14 million lives from TB by 2015. This objective is being hampered by the increase in HIV-associated tuberculosis and the emergence of multiple drug-resistant tuberculosis (MDR-TB) [4]. The only currently available vaccine is bacillus Calmette–Guérin (BCG) [5]. The vaccine is often administered to children, but the effectiveness of protection decreases after about 10 years.

With TB being one of the most common causes of death from infectious diseases, the current challenge is developing a sensitive and efficient method for the detection of latent TB infection (LTBI). The disease begins in the lungs via infection from the blood stream or aerosol droplets [6]. After TB bacteria enter the bloodstream, they can spread throughout the body and infect various tissues [7], such as the heart skeletal muscles, pancreas, or thyroid [8]. However, in LTBI, the bacteria remain dormant for several years before producing active TB. Even after treatment, the affected individual may still be susceptible to reactivation due to immunosuppression, or multiple-drug resistance in TB bacteria [4].

Substantial gene expression studies have revealed differences in the transcriptome between healthy controls and active TB or LTBI patients [9–11]. These findings not only uncovered important genetic signatures indicative of active TB and LTBI, but also identified transcriptionally regulated markers that are diverse in functions. In particular, these candidate genes are responsible for various key biological processes including inflammatory responses, immune defense, cell activation, homeostatic processes, regulation of cell proliferation and apoptosis. Moreover, these studies demonstrated the importance of cytokine and chemokine responses in the progression from latent infection to active disease [12–14]. However, the overall gene expression array results vary due to diverse genetic background of the study population and differences in the study design.

Early diagnosis of TB is crucial for preventing its spread, but the detection of LTBI is a major challenge as the carriers are often asymptomatic. Sputum smear acid-fast staining, though fast and inexpensive, is not the most sensitive and specific diagnostic test. While the tuberculin skin test represents a common diagnostic method, it has a tendency to produce false-positive results in individuals previously inoculated with BCG [15]. Culturing of TB bacteria usually takes time and diagnosis based on the test results is not always accurate. The interferon gamma release assays (IGRA) seem to have the potential of becoming the gold standard for TB test. The assays have been introduced into clinical practice to measure the amount of interferon-gamma (IFN-γ) released by blood cells infected wtih *Mtb* [16]. Unfortunately, this method is more expensive and requires blood samples with normal levels of viable leukocytes, which is not always possible in immunocompromised individuals. Consequently, an alternative quantitative polymerase chain reaction method was developed to detect the immune response to TB infection [17]. Yet, as most gene expression study results suggest, genetic background may influence the specificity and sensitivity of diagnosis.

Recently, Lu et al. (2011) conducted a gene expression microarray study to investigate the possibility of using mRNAs as biomarkers to differentiate active TB from LTBI [18]. Interestingly, in their study, the expression of IFN-γ, the biomarker used in IGRA, was not significantly different between the active TB and LTBI group [18]. Instead, the combination of three genes, *CXCL10* (chemokine C-X-C motif ligand 10), *ATP10A* (ATPase, class V, type 10A) and *TLR6* (toll-like receptor 6) appeared to be effective at distinguishing between active and latent TB infection. In contrast, *IL-8* (Interleukin 8), *FOXP3* (forkhead box P3), and *IL-12β* (interleukin 12 beta) were demonstrated to be the best discriminating biomarkers for TB and LTBI by Wu et al. [14]. Discrepancies between the two studies may be attributable to the differences in genetic background. At the same time, these findings suggest that not only are gene expression biomarkers more significant indicators of active TB, but they may also represent a more sensitive detection method for LTBI. Nevertheless, the same combination of genetic markers may not be applicable in another population.

For the present study, we attempted to compare the gene expression profiles in peripheral blood mononuclear cells among individuals with active TB, LTBI, and healthy conditions. We identified a panel of mRNAs that differed among these groups and subsequent validations with independent samples established the potential use of these gene expression biomarkers for the discrimination of LTBI from active TB in the Taiwanese population.

Results

Differentially expressed genes among TB, LTBI, and healthy controls

To identify candidate genes whose expression levels may differentiate among TB, LTBI, and healthy controls, we followed the workflow as illustrated in Fig. 1. The TB, LTBI, and healthy controls recruited for gene expression profiling did not differ significantly in age (One-way ANOVA: $F_{2,18} = 0.21$, $p = 0.81$; Additional file 1). Compared to healthy

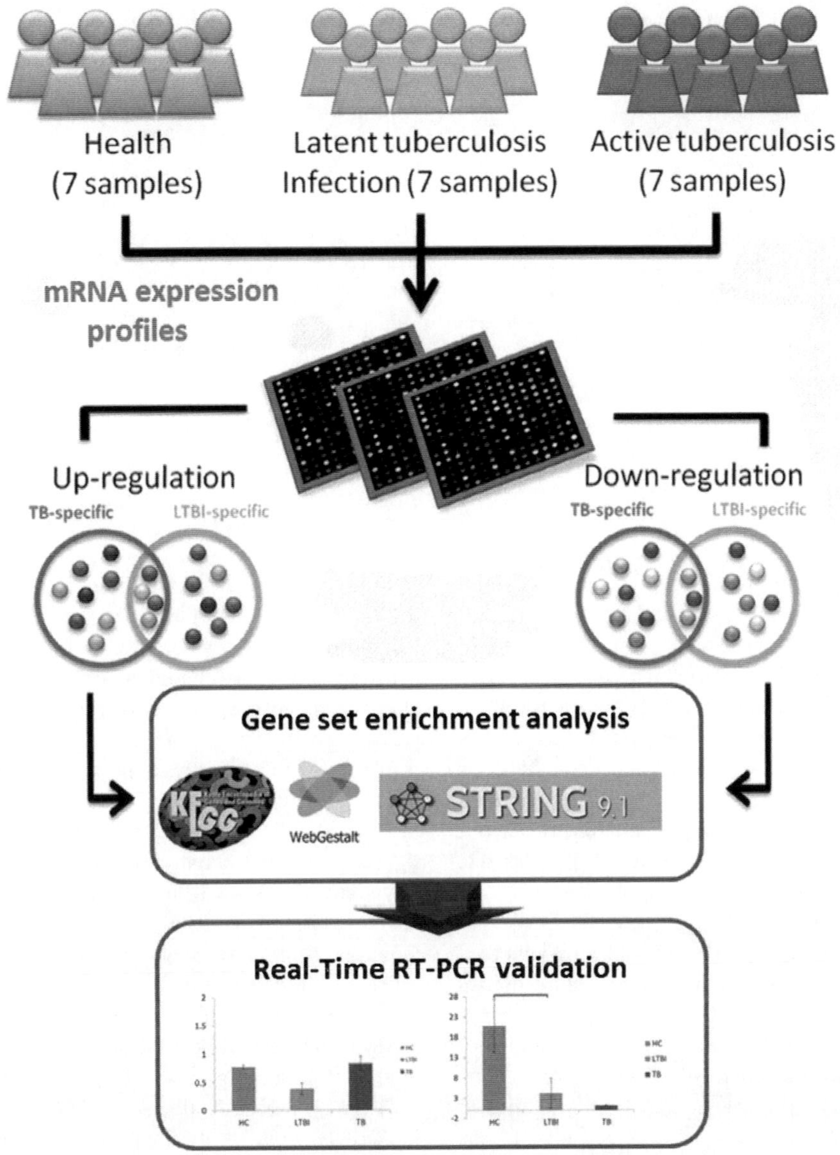

Fig. 1 System flow of our analysis

individuals, 31 and 16 genes were up-regulated in TB and LTBI, respectively (Fig. 2). While a total of 267 genes showed significantly reduced expression in TB patients relative to healthy controls, 111 genes appeared to be expressed at a lower level in those affected with LTBI compared with their healthy counterparts. Between TB and LTBI, 169 genes were differentially expressed, with 103 genes presenting increased abundance and 66 genes exhibiting decreased expression in LTBI relative to TB. Among these differentially expressed genes, three and 11 were also up-regulated and down-regulated, respectively, between LTBI and healthy controls. A list of the differentially expressed genes is provided in Additional file 2.

Functions, pathways, and interactions associated with the differentially expressed genes

According to the gene set enrichment analysis, genes differentially expressed among TB, LTBI, and healthy controls were over-represented in different GO categories (Table 1). The detailed lists of GO comparisons can be found in Additional files 3, 4, and 5 for LTBI reltiave to healthy controls, TB relative to healthy controls, and LTBI relative to TB, respectively. Compared with healthy controls, TB-specific gene expression profile appeared to be mostly related to leukocyte differentiation, lymphocyte activation, chemokine receptor activity, and regulation of immune response. In contrast, those latently

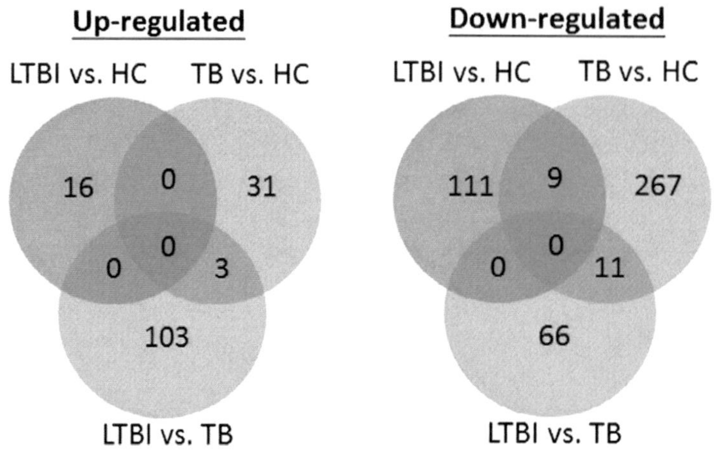

Fig. 2 Number of differentially expressed genes among TB, LTBI, and healthy control (HC) group. Significant differential expression is represented by an absolute $\log_2$ fold change ≥1, FDR < 0.05

infected with TB and healthy controls showed differing expression in genes belonging to regulation of metabolism, apoptosis, translation, and signal transduction pathways involving MAP kinase phosphatase and protein tyrosine/threonine phosphatase activities. Between the TB and LTBI group, the differentially expressed genes were not only enriched in immune system associated categories such as immune response activation and regulation, as well as natural killer cell and T-cell differentiation, but these genes were also involved in cellular processes like translation, transcription, and mRNA catabolism.

Pathway analysis revealed that relative to LTBI and healthy controls, most genes affected by active TB appeared to be involved in the regulation of immune responses (Table 2). For example, many differentially expressed genes between the TB and healthy control group were mapped to pathways associated with cytokine-cytokine receptor interaction, inflammatory responses such as rheumatoid arthritis, graft-versus-host disease and cancer. In addition, the transcriptional profiles that differed between LTBI and TB showed genes concentrated in apoptosis and signaling pathways involving chemokines, Toll-like receptors, and lymphocytes such as B- and T-cells. On the other hand, for genes differentially expressed between LTBI and healthy control, the most enriched pathway belonged to MAPK signaling cascade, followed by adipocytokine signaling modulated inflammatory response and Toll-like receptor signaling mediated innate immunity. The full lists of pathway comparisons for LTBI reltiave to healthy controls, TB relative to healthy controls, and LTBI relative to TB can be found in Additional files 6, 7, and 8, respectively.

Protein interaction analysis identified specific interaction network modules for active TB, LTBI, and healthy controls. The network modules were grouped according to their GO annotations and have been cross-validated with the STRING database [19, 20] (Fig. 3). Among the

Table 1 Gene Ontology categories enriched by differentially expressed genes among TB, LTBI, and healthy control (HC)

TB vs. HC	LTBI vs. HC	LTBI vs. TB
1. Immune response	Regulation of metabolic process	Response to cold
2. Leukocyte differentiation	Regulation of cellular Metabolic process	Immune Response-regulating signaling pathway
3. Immune system process	Regulation of biosynthetic process	Cellular process
4. B cell activation	Apoptotic process	Immune response- activating signal transduction
5. Lymphocyte differentiation	Death	Heterotypic cell-cell adhesion
6. Regulation of immune response	Regulation of gene expression	NK T cell differentiation
7. Positive regulation of response to stimulus	MAP kinase phosphatase activity	Regulation of mRNA catabolic process
8. Lymphocyte activation	Translation factor activity,	Translation regulator activity
9. Leukocyte activation	Translation initiation factor activity	NF-kappaB binding
10. Chemokine receptor activity	Protein tyrosine/threonine phosphatase activity	Translation repressor activity

WebGestalt setting: multiple test adjustment = Benjamini-Hochberg, significance level = top 10 (Benjamini-Hochberg adjusted $p < 0.05$); minimum number of genes for a category = 2

Table 2 Pathways enriched by differentially expressed genes among TB, LTBI, and healthy control (HC)

KEGG pathway	ID	Genes
TB vs. HC		
Cytokine-cytokine receptor interaction	04060	*CCR6, IL7R, CCR7, FLT1, CCR2, TNFSF13B, FASLG, IL2RB, CD27, IL23A, TGFBR2*
Rheumatoid arthritis	05323	*TLR4, IL23A, HLA-DOB, ATP6V0E2, FLT1, TNFSF13B*
Pathways in cancer	05200	*MYC, ITGA6, FASLG, RASSF5, HSP90AA1, CASP8, TRAF4, TRAF5, TGFBR2, ETS1*
Graft-versus-host disease	05332	*FASLG, KLRC1, PRF1, HLA-DOB*
MAPK signaling pathway	04010	*FASLG, RRAS2, TGFBR2, MYC, RASA2, PRKACB, MAP3K1*
LTBI vs. HC		
MAPK signaling pathway	04010	*TNF, JUN, DUSP2, ATF4, NR4A1, FOS, DUSP5, DDIT3, PDGFRB, DUSP1*
Adipocytokine signaling pathway	04920	*POMC, TNF, STK11, CAMKK2, NFKBIA, SOCS3*
Leishmaniasis	05140	*TNF, JUN, FOS, NFKBIA*
Toll-like receptor signaling pathway	04620	*TNF, JUN, FOS, NFKBIA*
Cytokine-cytokine receptor interaction	04060	*TNF, CXCL2, TNFSF14, PDGFRB, OSM*
LTBI vs. TB		
Chemokine signaling pathway	04062	*GSK3B, CCR6, CXCL16, ARRB2, PIK3CG, NFKBIA, PRKACB*
Apoptosis	04210	*CASP8, PIK3CG, NFKBIA, TNFRSF10B, PRKACB*
T cell receptor signaling pathway	04660	*GSK3B, PTPRC, FOS, PIK3CG, NFKBIA*
Toll-like receptor signaling pathway	04620	*LY96, CASP8, FOS, PIK3CG, NFKBIA*
B cell receptor signaling pathway	04662	*GSK3B, FOS, PIK3CG, NFKBIA*

Benjamini-Hochberg adjusted $p < 0.05$

genes differentially expressed between LTBI and healthy controls, protein interactions involved in transcriptional regulation (*ATF3, ATF4, JUNB, FOSB*, and *DDIT3*), as well as translation initiation (*EIF1* and *EIF5*) appeared to be the most important. Whereas proteins that regulate interferon-beta production (*LY96* and *TLR4*), apoptotic signaling (*HSP90AA1, LRRK2, TGFBR2*, FASLG, CASP8,), bacterial invasion (*SEPT1* and *SEPT6*), and Wnt signaling pathway (*HIC1* and *CTBP2*) seemed to represent the underlying variations between TB and healthy controls, the differences between TB and LTBI might be contributed by proteins that modulate transcription (*FOS* and *DDIT3*), phagosome formation (*TUBA1A* and *TUBB4B*), autophagy (*CASP8* and *TNFRSF10B*), and interferon-gamma signaling (*ARRB2*, PTAFR, *NFKBIA*). Additional files 9, 10, and 11 contain detailed lists of enriched protein interaction modules associated with genes differentially expressed in LTBI relative to healthy controls, TB relative to healthy controls, LTBI relative to TB, respectively.

Validation of differentially expressed candidate biomarkers

Many of the identified differentially expressed genes among the TB, LTBI, and healthy control group have also been implicated in TB pathology by other groups. However, as indicated by our analysis, several of these genes play roles in other infections, inflammatory diseases, cancers or even common cold. For real-time RT-PCR validation, we selected genes that are known to be expressed in the lungs and showed clear differences in transcript abundance (fold change ≥1) in at least one of the comparisons; that is, TB versus healthy controls, LTBI versus healthy controls, or LTBI versus TB. Additional volunteers were recruited for gene expression validation. To avoid overlaps with other respiratory tract infections, we chose three differentially expressed genes that may not be directly involved in mediating the immune and inflammatory responses against common respiratory infections. These genes were *NEMF* (nuclear export mediator factor), *ASUN* (asunder spermatogenesis regulator), and *DHX29* (DEAH (Asp-Glu-Ala-His) box polypeptide 29). Then, we selected *PTPRC* (protein tyrosine phosphatase, receptor type, C) or CD45, an estalished marker of active TB [21], as a reference standard.

Though the independently recruited participants differed significantly in age (Additional file 1), RT-PCR results successfully verified the array observations in the gene expression array experiment (Fig. 4), indicating that age might not have been a major factor. Subsequent ROC analyses confirmed that *PTPRC* expression may be able to detect active TB, while *ASUN* could discriminate TB or LTBI from healthy individuals (Fig. 5). Other than *PTPRC*, the transcript abundance of *DHX29* could also distinguish the differences between TB and healthy controls. In contrast, *NEMF* did not demonstrate to be a good discriminatory biomarker.

Finally, to assess the ability of *PTPRC, DHX29*, and *ASUN* in classifying TB, LTBI, and healthy individuals, as

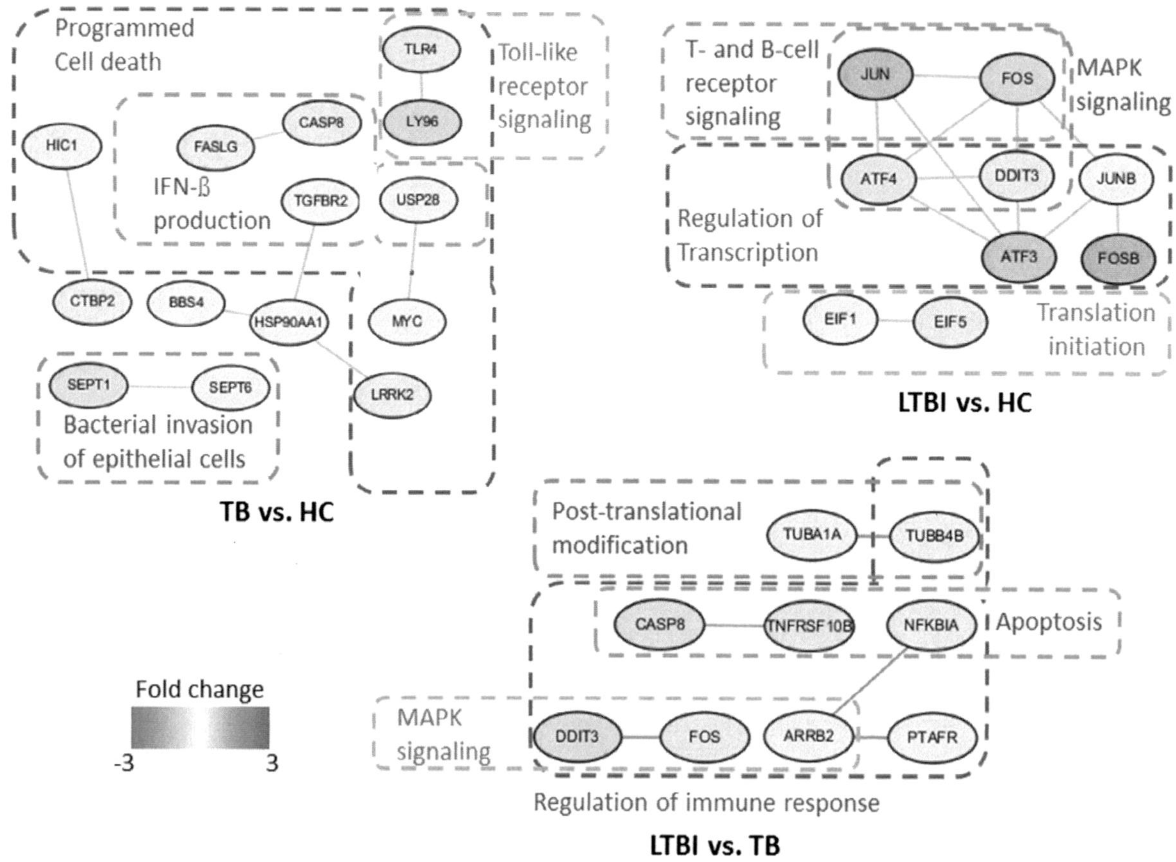

Fig. 3 Protein interaction networks of differentially expressed genes among TB, LTBI, and healthy control (HC). Genes are grouped according to their associated pathways and functions

a proof of concept experiment, we tested the performance of classification models built with the candidate biomarkers using a sample size of 17 LTBI, 15 TB, and 15 healthy individuals. We utilized four classifiers: decision tree, random forest, support vector machine (SVM), and Naïve Bayes. As evaluated by a 5-fold cross-validation approach, the accuracy, sensitivity, and specificity of the models constructed with single candidate genes were relatively low compared to those built using a combination of biomarkers (Additional file 12 for single gene models; Table 3 for hybrid models). The Naïve Bayes-based model, which was constructed with the expression levels of *PTPRC*, *DHX29*, and *ASUN* as the selected features, yielded the best performance (Table 3).

Discussion

TB is a serious health threat among the young, elderly, and immunocompromised. Variations in the transcriptional profile of human peripheral blood mononuclear cells in the presence of *Mtb* infection are complex and can be attributed to multiple factors, including age [22], genetic background [14, 23], and study designs. Therefore, identifying distinct genetic signatures of active TB, LTBI, and healthy individuals that is population-specific becomes a challenging task.

In the current study, we examined the transcriptomes of active TB, LTBI, and healthy individuals, and uncovered specific molecular markers and pathways associated with each group. We found that there were more genes showing differential expression between the TB group and healthy controls as compared with LTBI versus healthy controls. GO and pathway enrichment analyses revealed that the transcriptional profiles of TB individuals differed from those of healthy controls in immune system processes such as leukocyte and lymphocyte activation, differentiation, chemokine receptor activity. In contrast, although immune pathway alterations were indeed observed in individuals with LTBI at the transcript level, metabolic processes in these individuals also differed from the healthy controls. On the other hand, between TB and LTBI patients, the most important genes seemed to be mediators of inflammation, immune system responses, and apoptosis.

Our results support findings from other studies in that infection with *Mtb* triggers a relay of inflammatory signals

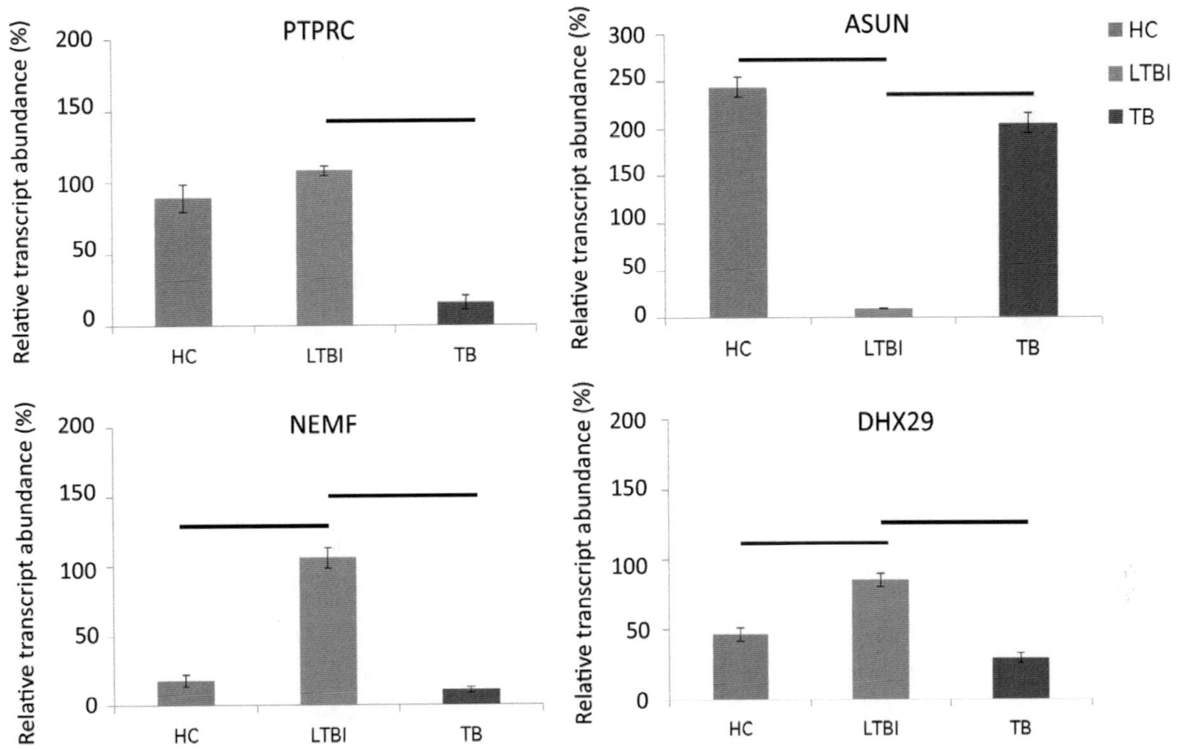

Fig. 4 Validation of four differentially expressed genes among TB, LTBI, and healthy controls (HC). Statistical significance ($p < 0.05$) is represented by a *horizontal bar*

and immune responses from the host. Upon entry, *Mtb* are recognized by various host receptors, including Toll-like receptors (TLRs) and nucleotide-binding oligomerization domain-like receptors that are expressed on immune cells [24]. This host-pathogen interaction initiates a cascade of inflammatory responses, whereby alveolar macrophages produce cytokines and chemokines to alert the host of the infection [25]. In response to this signal, T- and B-lymphocytes aggregate around the infected macrophages to form granulomas, a microenvironment to prevent bacterial spread and an isolated region for the lymphocytes to invoke apoptosis of the infected macrophages [26]. The bacteria have also evolved tactics to avoid this fate. After being engulfed by macrophages, the bacteria multiply in the phagosome, causing macrophage necrosis and allowing for their escape from the host defenses to infect other cells [25]. This is evident in our results in which a lot of the differences between TB and healthy controls were associated with mediators of immune responses.

In particular, our protein interaction analysis suggests that Toll-like receptor 4 (TLR4) may mediate the host defense against *Mtb*. It is known that TLR4 is important in modulating the balance between apoptosis and necrosis in *Mtb*-infected macrophages [27]. Our analysis showed that TLR4 may interact with LY96 (lymphocyte antigen 96) and S1PR1 (Sphingosine-1-phosphate receptor 1) in TB patients. LY96 is responsible for establishing the binding between the lipopolysaccharide on the bacterial cell membrane and TLR4 on the host cell surface, thus activating a series of immune responses in infected individuals [28].

The bacterial strain *Mtb* has quite a few tricks to blunt the bactericidal mechanisms of infected macrophages so as to promote their persistence in the host. In latent TB infection, the bacteria manipulate the host antigen presentation processes and enter a non-replicating state [29]. As a result, the bacteria remain dormant inside the phagosomes, and the granuloma became their safe hideout. This is evident in our observation that TB and LTBI differed not only in the regulation of immune responses, but also in the modulation of phagosome, autophagy and apoptosis. In fact, several apoptosis-associated molecules, such as decoy receptor 3, prostaglandin 2, and lipoxin, have been shown to correlate significantly with the status of *Mtb* infection [30, 31].

Compared with healthy controls, genes involved in cell cycle control seemed to be differentially regulated in individuals with LTBI. Our interaction analysis showed that this difference may be attributable to the interactions between mediators of the MAPK (mitogen-activated protein kinase) signaling pathway, such as JUN (Jun proto-oncogene), JUNB (Jun B proto-oncogene) and FOSB (FBJ

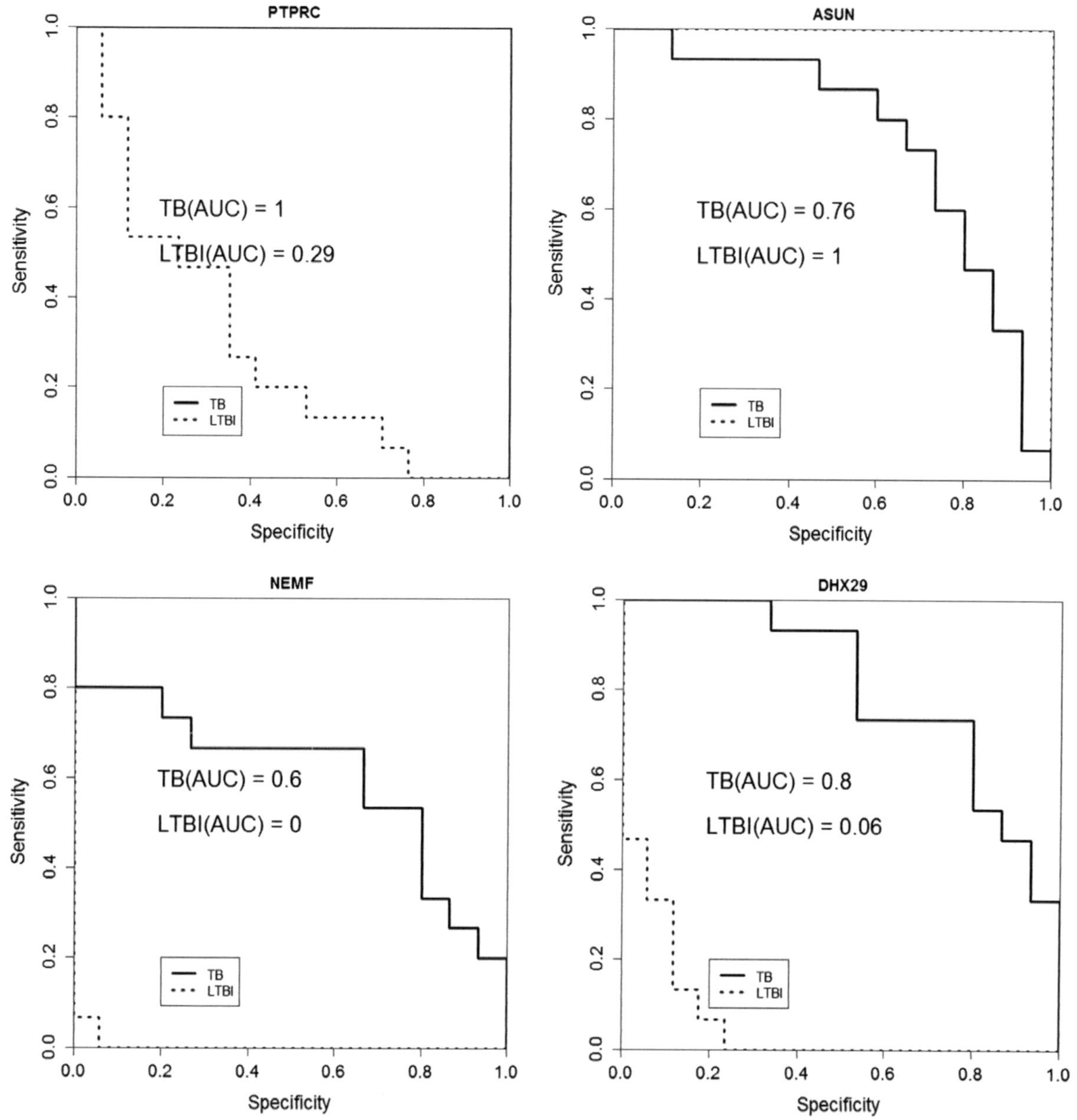

Fig. 5 ROC analysis of four differentially expressed genes among TB, LTBI, and healthy controls (HC). AUC represents the area under the curve

Table 3 Performance of diagnostic support models constructed using combinations of candidate biomarkers with various classifiers

Features	Classifier	Accuracy	Sensitivity	Precision	AUC
PTPRC+ASUN	Decision tree	91.49 %	91.5 %	97.7 %	0.943
PTPRC+ASUN+DHX29	Random Forest	93.62 %	93.6 %	93.6 %	0.982
PTPRC+ASUN+DHX29	SVM	95.74 %	95.7 %	96.2 %	0.969
PTPRC+ASUN+DHX29	Naïve Bayes	97.87 %	97.9 %	98 %	0.979

Sensitivity: TP/(TP+FN); Precision: TP/(TP+FP); performance was evaluated by 5-fold cross-validation murine osteosarcoma viral oncogene homolog B), as well as regulators of translation. MAPK signaling is one of the pathways responsible for modulating the host's innate immunity [32]. In fact, MAPK activation is important for inducing the expression of genes involved in inflammatory responses, but inactivation of MAPK activity is also important to prevent host tissue damage [32]. On the other hand, our interaction analysis also indicated that translational regulation represent important differences between LTBI and healthy individuals. It is known that bacteria can influence host tranlation in order to immobilize host defenses and promote their own survival [33]. Therefore, our finding supports that MAPK signaling and tranlsational control may underlie the differences between LTBI and healthy conditions.

In addition, alterations in the expression of genes involved in metabolic processes seemed to be a major difference between the LTBI group and healthy controls. Metabolism-related manifestations are known to be associated with TB [34]. It has been demonstrated that *Mtb*-infected macrophages would become lipid-loaded (foamy) in the granuloma, and the fatty acids accumulated as triacylglycerol represent the vital source of energy for dormant *Mtb* [35]. Also, certain immune-endocrine-metabolic alterations are thought to exist in TB patients, though molecular studies have yet to reach a consensus as to which molecules are the major players in this process [36–39].

To date, most identified biomarkers for active TB and LTBI are related to inflammatory and immune responses triggered by the pathogen infection [14, 18]. However, TB may be comorbid with various communicable diseases including influenza, bacterial pneumonia, HIV, syphilis, leishmaniasis, as well as non-communicable disorders such as diabetes, alcohol-related diseases, chronic obstructive pulmonary disease, coronary artery disease, cancer, etc. [40]. In fact, our gene set enrichment analysis mapped four genes, *TNF, JUN, FOS, NFKBIA*, that were differential expressed between LTBI and healthy controls to leshimaniasis-related pathways.

To avoid finding biomarkers that overlap with other diseases, we chose to verify differentially expressed genes that have not been identified as TB or LTBI biomarkers in independently recruited samples. We also included an established active TB marker, *PTPRC*, as a test reference. As a result, *ASUN, DHX29*, and *NEMF* were successfully confirmed to be differentially expressed among active TB, LTBI, and healthy individuals. Subsequent ROC analysis revealed the potential of *PTPRC, ASUN*, and *DHX29* in discriminating among TB, LTBI, and healthy conditions. Further classification experiments also indicated that combining the three canidate biomarkers may be more effective in achieving accurate identification of the different disease states.

To be a biomarker, the gene should to be related to the pathogenesis of the disease. However, since functional analysis was not performed in this study, we try to postulate how these candidate genes may be involved in TB pathology based on their known functions. Among the validated genes, PTPRC is an essential regulator of host immune response through the modulation of T- and B-cell receptor signal transduction [41]. In the guinea pigs, *PTPRC* expression appeared to increase after exposure to *Mtb* and decrease after the infection persisted for a longer period of time [42]. This is in line with our observation that *PTPRC* transcript level was decreased in active TB compared to LTBI and healthy state. *ASUN* is critical for the regulation of mitotic cell cycle [43]. As the number of T-cells can determine whether an individual is susceptible to *Mtb* infection or active TB [44], we suspect that the high level of *ASUN* expression in LTBI relative to the active TB and healthy controls may be associated with T-lymphocyte differentiation or proliferation. *NEFM* is a nuclear export mediator that have been implicated as a tumor suppressor in lung cancer [45]. Nuclear export is an important process for the regulation of autophagy [46]. A recent in vivo study performed in mice suggested that autophagy can suppress the progression towards active TB by inhibiting *Mtb* growth [47]. Finally DHX29 is a helicase protein that participates in translation initiation, and its down-regulation has an inhibitory effect on cancer growth [48], which may be related to the altered cellular processes in TB.

Note that our study is limited by the small sample size and lack of functional studies to determine the roles that the identified candidate biomarkers play in the pathogenesis of TB. This might have also been the reason that we did not observe more genes associated with the TLR signaling pathway, an established mechanism underlying *Mtb* recognition [49]. As a result, this pathway was not the most enriched in any of the comparisons. Equally likely is that the changes in *TLR* genes might be dependent on treatment duration. It has been demonstrated that the increase in *TLR4* expression level appeared to be more significant compared to *TLR2* after 1 month of treatment in TB patients when compared with healthy controls [50]. In our study, samples were collected at the time of diagnosis. More evident changes might be observed if we had examined the temporal expression profiles of TB and LTBI patients during treatment. Also, our experiments were focused on blood cells, instead of lung tissues. Therefore, the results are more relevant to biomarkers associated with TB and LTBI, and perhaps indirectly related to the pathology of the disease in the lungs. Moreover, BCG is known to have an effect on gene expression [51, 52]. In Taiwan, BCG vaccine should be administered to every newborn; therefore, all of our study participants have been inoculated with BCG. This may make our finding specific to not only the Taiwanese people, but also to those who have received this vaccine. Finally, despite the fact that evaluation of the biomarkers yielded relatively good results, due to the limited sample size and the variable prevalence of TB and LTBI in different seasons, we could not eliminate the possibility that other diseases or environmental factors may affect the effectiveness of the candidate biomarkers.

Conclusions

In conclusion, we have performed a genome-wide gene expression study comparing the transcriptional profiles among TB and LTBI patients, as well as healthy controls. Gene set enrichment analyses have identified specific biological processes and pathways associated with genes that are differentially expressed among these groups. We have uncovered novel discriminatory biomarkers for TB and LTBI. Moreover, we have demonstrated, as a proof of concept, that the expression levels of *PTPRC*, *ASUN*, and *DHX29* may be used as diagnostic biomarkers. Validation of the diagnostic support system in a larger sample size would help confirm the discriminative potential of these biomarkers and facilitate the development of a cost-effective and sensitive molecular diagnostic platform for TB.

Methods

Clinical sample collection

The analytical flow of our study is illustrated in Fig. 1. All procedures were reviewed and approved by the Institutional Review Board of Taoyuan General Hospital, Ministry of Health and Welfare, Taoyuan, Taiwan. Written informed consents were obtained from all participants. For those who had reduced ability to consent (including minors), the carers or guardians gave written informed consents on behalf of these participants. Eligibility for entry into the study was based on clinical signs and symptoms of Mtb infection. All participants were interviewed and examined by a physician. Each subject received the sputum smear test, T-SPOT TB test, and took a chest radiograph. TB subjects were those who showed clinical signs of TB, in addition to having been tested positive on all tests. LTBI subjects were recruited from close contacts (>8 h/day for a total of >40 h of close contact) with active TB patients, tested negative on the smear test, positive on the T-SPOT TB test, appeared normal on their chest radiographs, and exhibited no clinical evidence of active TB. Healthy controls were individuals who had not been in close contact with TB or LTBI patients, obtained negative results on all tests and showed no clinical signs of TB or LTBI. Individuals with allergic diseases, diabetes, cancer, immune-compromised conditions, and co-infections with any types of infectious diseases were excluded. In total, seven healthy individuals, seven patients with active TB, as well as seven subjects with LTBI, were included in the microarray experiment. Additional participants (15 TB, 17 LTBI, and 15 healthy individuals) were recruited as independent testing samples for validation of the expression profiling results. Age and gender information are provided in Additional file 1. In Taiwan, every newborn must be inoculated with BCG; therefore, the BCG inoculation status for every participant was positive.

RNA isolation

RNA was isolated from peripheral blood mononuclear cells (PMBC). The quality of RNA was determined by an optical density (OD) 260/280 ratio ≥1.8, and OD 260/230 ratio ≥1.5 on a spectrophotometer, in addition to the intensity of the 18S and 28S rRNA bands on a 1 % formaldehyde-agarose gel. RNA quantity was detected by a spectrophotometer. RNA integrity was examined on an Agilent Bioanalyzer. RNA with an RNA integrity number (RIN) ≥7.0 and 28S/18S >0.7 was subjected to microarray analysis.

Gene expression analysis

RNA samples were subjected to Human OneArray® v6 (Phalanx Biotech, Hsinchu, Taiwan). Data were analyzed with the Rosetta Resolver System software (Rosetta Biosoftware, USA). The criteria for identifying differentially expressed genes were: 1) an absolute log2 fold change ≥1; 2) a false discovery rate of <0.05; 3) an intensity difference of >1000 between two samples under comparison; 4) an individual intensity of >500. Genes showing significant differential expression were categorized into TB versus healthy controls, LTBI versus healthy controls, and LTBI versus TB. Our gene expression records can be found on the Gene Expression Omnibus under the accession number GSE62525.

Bioinformatics analysis

Differentially expressed genes were used as input for a series of bioinformatics analyses performed with the WEB-based GEne SeT AnaLysis Toolkit (WebGestalt) [53, 54]. WebGestalt is an open analytical platform that integrates gene ontology (GO) [55], KEGG [56], WikiPathway [57], protein interaction networks, microRNA binding sites and transcription factor targets [58], as well as cytogenetic band information, for a variety of enrichment analyses. The GO, KEGG, and protein interaction module tools were utilized to analyze the differentially expressed genes. Multiple testing bias was adjusted by a Benjamini-Hochberg threshold of $p < 0.05$. The enriched protein interaction network modules in each transcriptional profile were grouped according to their GO annotations or associated pathways. Experimentally confirmed interactions were cross-validated with the STRING database (v9.1) [19, 20] with a confidence level of 0.7 as the paratmeter setting. Visualization of the interaction networks was achieved using Cytoscape version 3.2.1 [59].

Real-time RT-PCR validation

The four differentially expressed genes selected for real-time RT-PCR validation were *PTPRC, ASUN, NEMF,* and *DHX29.* Total RNA from PBMC was extracted using TRIzol (Invitrogen, Carlsbad, CA, USA) from 17 TB, 15 LTBI, and 15 healthy individuals. Each extracted RNA sample was reversely transcribed using the First Strand cDNA Synthesis kit (Roche, Boulder, CO, USA) according to the manufacturer's instructions. Each cDNA sample was amplified with the FastStart Universal SYBR Green reagent (Roche, Mannheim, Germany) on the StepOnePlusTM Real-Time PCR system (Applied Biosystem, CA, USA). Briefly, the reaction conditions consisted of 10 ng of cDNA and 0.25 μM of primers in a final reaction volume of 20 μl in SYBR Green mixture. Each reaction was initiated with 10 min at 95 °C, followed by 40 cycles consisting of denaturation at 95 °C for 15 s, annealing at 60 °C for 1 min, and extension at 72 °C for 30 s. For each reaction, the β-actin gene was used as an endogenous control to normalize each sample. Primer sequences for each gene are listed in Additional file 13. The relative expression of each gene was compared using the 2-ΔΔ CT method and all experiments were run in triplicates and repeated three times. Differences in expression among TB, LTBI, and healthy controls were evaluated with one-way ANOVA followed by Tukey's post hoc test in SPSS 18.0 (IBM Corporation, NY, USA). A p-value of <0.05 was regarded as statistically significant. A receiver operating characteristic (ROC) curve analysis was performed in the R statistical environment (3.1.1) to assess the specificity and sensitivity of each validated biomarker.

Construction of a diagnostic support model

We wished to test if the candidate biomarkers could be integrated in a diagnostic support system. As a proof of concept, we used the expression levels of *PTPRC, ASUN,* and *DHX29* as features to build classification models based on the 57 volunteers who donated their blood samples for RT-PCR validation. Due to its low discriminating abiliy as evaluated by ROC analysis, *NEMF* was excluded. Experiments were conducted in LibSVM (version 3.12) [60] and WEKA, or Waikato Environment for Knowledge Analysis (version 3.6.5) [61]. Features were selected based on the previous ROC analysis results. The selected classifiers included the C4.5 decision tree algorithm [62], support vector machine (SVM) [63], Naïve Bayes [64, 65], and the random forest algorithm [66]. Models were built with single genes or a combination of the selected genes. Performance of each model was evaluated by 5-fold cross-validation.

Additional files

Additional file 1: Average age and gender distribution of the participants (XLSX 11 kb)

Additional file 2: List of differentially expressed genes among the TB, LTBI, and healthy individuals. (XLSX 19238 kb)

Additional file 3: GO enrichment of genes differentially expressed in LTBI relative to healthy controls. (XLSX 111 kb)

Additional file 4: GO enrichment of genes differentially expressed in TB relative to healthy controls. (XLSX 69 kb)

Additional file 5: GO enrichment of genes differentially expressed in LTBI relative to TB. (XLSX 66 kb)

Additional file 6: Pathway enrichment of genes differentially expressed in LTBI relative to healthy controls. (XLSX 19 kb)

Additional file 7: Pathway enrichment of genes differentially expressed in TB relative to healthy controls. (XLSX 20 kb)

Additional file 8: Pathway enrichment of genes differentially expressed in LTBI relative to TB. (XLSX 20 kb)

Additional file 9: Enriched protein interaction modules associated with genes differentially expressed in LTBI relative to healthy controls. (XLSX 14 kb)

Additional file 10: Enriched protein interaction modules associated with genes differentially expressed in TB relative to healthy controls. (XLSX 15 kb)

Additional file 11: Enriched protein interaction modules associated with genes differentially expressed in LTBI relative to TB. (XLSX 20 kb)

Additional file 12: Performance of diagnostic support models constructed using individual candidate biomarkers with various classifiers. (XLSX 11 kb)

Additional file 13: RT-PCR Primer for the selected differentially expressed genes. (XLSX 11 kb)

Competing interests

The authors have declared that no competing interests exist.

Authors' contributions

JTYW, SWL, and LSHW conceived and designed the experiments. SWL diagnosed the patients and collected the samples. LSHW was responsible for RNA isolation and RT-PCR validation. GMH conducted the ROC and classification experiments. KYH and JTYW performed the bioinformatics analyses and JTYW wrote the manuscript. TYL assisted with the bioinformatics analyses. All authors read and approved the final manuscript.

Acknowledgements

This study was supported by the Ministry of Science and Technology of the Republic of China, Taiwan, under the contract number of MOST 101-2628-E-155-002-MY2, as well as by Taoyuan General Hospital, Taiwan, under the contract number of 10409.

Declarations

The publication cost of this manuscript is supported by the Ministry of Science and Technology of the Republic of China, Taiwan, under the contract number of MOST 101-2628-E-155-002-MY2. This article has been published as part of *BMC Bioinformatics* Volume 17 Supplement 1, 2016: Selected articles from the Fourteenth Asia Pacific Bioinformatics Conference (APBC 2016). The full contents of the supplements are available online at http://www.biomedcentral.com/bmcbioinformatics/supplements/17/S1.

Author details

[1]Taoyuan General Hospital, Ministry of Health and Welfare, Taoyuan, Taiwan. [2]Department of Life Sciences, National Central University, Taoyuan, Taiwan. [3]Institute of Medical Sciences, Tzu Chi University, Hualien, Taiwan. [4]Department of Computer Science and Engineering, Yuan Ze University, Taoyuan, Taiwan. [5]Innovation Center for Big Data and Digital Convergence, Yuan Ze University, Taoyuan, Taiwan.

Published: 11 January 2016

References

1. Kumar V, Cotran RS, Robbins SL. Robbins basic pathology. 8th ed. New York: Saunders; 2007.
2. Dye C, Williams BG. The population dynamics and control of tuberculosis. Science. 2010;328(5980):856–61.
3. Skolnik RL. Global health 101. 2nd ed. Burlington, MA: Jones & Barlett Learning; 2012.
4. Lawn SD, Zumla AI. Tuberculosis. Lancet. 2011;378(9785):57–72.
5. McShane H. Tuberculosis vaccines: beyond bacille Calmette-Guerin. Philos Trans R Soc Lond Ser B Biol Sci. 2011;366(1579):2782–9.
6. Cole EC, Cook CE. Characterization of infectious aerosols in health care facilities: an aid to effective engineering controls and preventive strategies. Am J Infect Control. 1998;26(4):453–64.
7. Crowley L. An introduction to human disease: pathology and pathophysiology correlations. 9th ed. Burlington, MA: Jones & Bartlett Learning; 2010.
8. Agarwal R, Malhotra P, Awasthi A, Kakkar N, Gupta D. Tuberculous dilated cardiomyopathy: an under-recognized entity? BMC Infect Dis. 2005;5:29.
9. Berry MP, Graham CM, McNab FW, Xu Z, Bloch SA, Oni T, et al. An interferon-inducible neutrophil-driven blood transcriptional signature in human tuberculosis. Nature. 2010;466(7309):973–7.
10. Constantoulakis P, Filiou E, Rovina N, Chras G, Hamhougia A, Karabela S, et al. In vivo expression of innate immunity markers in patients with Mycobacterium tuberculosis infection. BMC Infect Dis. 2010;10:243.
11. Jacobsen M, Repsilber D, Gutschmidt A, Neher A, Feldmann K, Mollenkopf HJ, et al. Candidate biomarkers for discrimination between infection and disease caused by Mycobacterium tuberculosis. J Mol Med-Jmm. 2007;85(6):613–21.
12. Sutherland JS, de Jong BC, Jeffries DJ, Adetifa IM, Ota MO. Production of TNF-alpha, IL-12(p40) and IL-17 can discriminate between active TB disease and latent infection in a West African cohort. Plos One. 2010;5(8), e12365.
13. Chegou NN, Black GF, Kidd M, van Helden PD, Walzl G. Host markers in QuantiFERON supernatants differentiate active TB from latent TB infection: preliminary report. BMC pulmonary medicine. 2009;9:21.
14. Wu B, Huang C, Kato-Maeda M, Hopewell PC, Daley CL, Krensky AM, et al. Messenger RNA expression of IL-8, FOXP3, and IL-12beta differentiates latent tuberculosis infection from disease. J Immunol. 2007;178(6):3688–94.
15. Pai M. Alternatives to the tuberculin skin test: interferon-gamma assays in the diagnosis of mycobacterium tuberculosis infection. Indian J Med Microbiol. 2005;23(3):151–8.
16. Herrera V, Perry S, Parsonnet J, Banaei N. Clinical application and limitations of interferon-gamma release assays for the diagnosis of latent tuberculosis infection. Clinical infectious diseases : an official publication of the Infectious Diseases Society of America. 2011;52(8):1031–7.
17. Bibova I, Linhartova I, Stanek O, Rusnakova V, Kubista M, Suchanek M, et al. Detection of immune cell response to M. tuberculosis-specific antigens by quantitative polymerase chain reaction. Diagn Microbiol Infect Dis. 2012;72(1):68–78.
18. Lu C, Wu J, Wang H, Wang S, Diao N, Wang F, et al. Novel biomarkers distinguishing active tuberculosis from latent infection identified by gene expression profile of peripheral blood mononuclear cells. Plos One. 2011;6(8), e24290.
19. Franceschini A, Szklarczyk D, Frankild S, Kuhn M, Simonovic M, Roth A, et al. STRING v9.1: protein-protein interaction networks, with increased coverage and integration. Nucleic Acids Res. 2013;41(Database issue):D808–15.
20. Szklarczyk D, Franceschini A, Kuhn M, Simonovic M, Roth A, Minguez P, et al. The STRING database in 2011: functional interaction networks of proteins, globally integrated and scored. Nucleic Acids Res. 2011;39(Database issue):D561–8.
21. Nemeth J, Rumetshofer R, Winkler HM, Burghuber OC, Muller C, Winkler S. Active tuberculosis is characterized by an antigen specific and strictly localized expansion of effector T cells at the site of infection. Eur J Immunol. 2012;42(11):2844–50.
22. Verhagen LM, Zomer A, Maes M, Villalba JA, Del Nogal B, Eleveld M, et al. A predictive signature gene set for discriminating active from latent tuberculosis in Warao Amerindian children. BMC Genomics. 2013;14:74.
23. Wallis RS, Kim P, Cole S, Hanna D, Andrade BB, Maeurer M, Schito M, Zumla A: Tuberculosis biomarkers discovery: developments, needs, and challenges. Lancet Infect Dis. 2013; 13(4):362–372.
24. Jo EK. Mycobacterial interaction with innate receptors: TLRs, C-type lectins, and NLRs. Curr Opin Infect Dis. 2008;21(3):279–86.
25. Chen M, Gan H, Remold HG. A mechanism of virulence: virulent Mycobacterium tuberculosis strain H37Rv, but not attenuated H37Ra, causes significant mitochondrial inner membrane disruption in macrophages leading to necrosis. J Immunol. 2006;176(6):3707–16.
26. Grosset J. Mycobacterium tuberculosis in the extracellular compartment: an underestimated adversary. Antimicrob Agents Chemother. 2003;47(3):833–6.
27. Sanchez D, Rojas M, Hernandez I, Radzioch D, Garcia LF, Barrera LF. Role of TLR2- and TLR4-mediated signaling in Mycobacterium tuberculosis-induced macrophage death. Cell Immunol. 2010;260(2):128–36.
28. Palsson-McDermott EM, O'Neill LA. Signal transduction by the lipopolysaccharide receptor, Toll-like receptor-4. Immunology. 2004;113(2):153–62.
29. Ahmad S. Pathogenesis, immunology, and diagnosis of latent Mycobacterium tuberculosis infection. Clinical & developmental immunology. 2011;2011:814943.
30. Shu CC, Wu MF, Hsu CL, Huang CT, Wang JY, Hsieh SL, et al. Apoptosis-associated biomarkers in tuberculosis: promising for diagnosis and prognosis prediction. BMC Infect Dis. 2013;13:45.
31. Chen M, Divangahi M, Gan H, Shin DS, Hong S, Lee DM, et al. Lipid mediators in innate immunity against tuberculosis: opposing roles of PGE2 and LXA4 in the induction of macrophage death. The Journal of experimental medicine. 2008;205(12):2791–801.
32. Arthur JS, Ley SC. Mitogen-activated protein kinases in innate immunity. Nat Rev Immunol. 2013;13(9):679–92.
33. Mohr I, Sonenberg N. Host translation at the nexus of infection and immunity. Cell Host Microbe. 2012;12(4):470–83.
34. Chang SW, Pan WS, Lozano Beltran D, Oleyda Baldelomar L, Solano MA, Tuero I, et al. Gut hormones, appetite suppression and cachexia in patients with pulmonary TB. Plos One. 2013;8(1), e54564.
35. Daniel J, Maamar H, Deb C, Sirakova TD, Kolattukudy PE. Mycobacterium tuberculosis uses host triacylglycerol to accumulate lipid droplets and acquires a dormancy-like phenotype in lipid-loaded macrophages. PLoS Pathog. 2011;7(6), e1002093.
36. Santucci N, D'Attilio L, Kovalevski L, Bozza V, Besedovsky H, del Rey A, et al. A multifaceted analysis of immune-endocrine-metabolic alterations in patients with pulmonary tuberculosis. Plos One. 2011;6(10), e26363.
37. Yuksel I, Sencan M, Dokmetas HS, Dokmetas I, Ataseven H, Yonem O. The relation between serum leptin levels and body fat mass in patients with active lung tuberculosis. Endocr Res. 2003;29(3):257–64.
38. van Crevel R, Karyadi E, Netea MG, Verhoef H, Nelwan RH, West CE, et al. Decreased plasma leptin concentrations in tuberculosis patients are associated with wasting and inflammation. The Journal of clinical endocrinology and metabolism. 2002;87(2):758–63.
39. Kim JH, Lee CT, Yoon HI, Song J, Shin WG, Lee JH. Relation of ghrelin, leptin and inflammatory markers to nutritional status in active pulmonary tuberculosis. Clin Nutr. 2010;29(4):512–8.
40. Marais BJ, Lonnroth K, Lawn SD, Migliori GB, Mwaba P, Glaziou P, et al. Tuberculosis comorbidity with communicable and non-communicable diseases: integrating health services and control efforts. Lancet Infect Dis. 2013;13(5):436–48.
41. Hermiston ML, Xu Z, Weiss A. CD45: a critical regulator of signaling thresholds in immune cells. Annu Rev Immunol. 2003;21:107–37.
42. Ordway D, Palanisamy G, Henao-Tamayo M, Smith EE, Shanley C, Orme IM, et al. The cellular immune response to Mycobacterium tuberculosis infection in the guinea pig. J Immunol. 2007;179(4):2532–41.
43. Anderson MA, Jodoin JN, Lee E, Hales KG, Hays TS, Lee LA. Asunder is a critical regulator of dynein-dynactin localization during Drosophila spermatogenesis. Mol Biol Cell. 2009;20(11):2709–21.
44. Flynn JL, Ernst JD. Immune responses in tuberculosis. Curr Opin Immunol. 2000; 12(4):432–6.
45. Bi XL, Jones T, Abbasi F, Lee H, Stultz B, Hursh DA, et al. Drosophila caliban, a nuclear export mediator, can function as a tumor suppressor in human lung cancer cells. Oncogene. 2005;24(56):8229–39.
46. Funasaka T, Tsuka E, Wong RW. Regulation of autophagy by nucleoporin Tpr. Scientific reports. 2012;2:878.
47. Castillo EF, Dekonenko A, Arko-Mensah J, Mandell MA, Dupont N, Jiang S, et al. Autophagy protects against active tuberculosis by suppressing bacterial burden and inflammation. Proc Natl Acad Sci U S A. 2012;109(46):E3168–76.
48. Parsyan A, Shahbazian D, Martineau Y, Petroulakis E, Alain T, Larsson O, et al. The helicase protein DHX29 promotes translation initiation, cell proliferation, and tumorigenesis. Proc Natl Acad Sci U S A. 2009;106(52):22217–22.

49. Reiling N, Holscher C, Fehrenbach A, Kroger S, Kirschning CJ, Goyert S, et al. Cutting edge: Toll-like receptor (TLR)2- and TLR4-mediated pathogen recognition in resistance to airborne infection with Mycobacterium tuberculosis. J Immunol. 2002;169(7):3480–4.
50. de Oliveira LR, Peresi E, Golim Mde A, Gatto M, Araujo Junior JP, da Costa EA, et al. Analysis of Toll-like receptors, iNOS and cytokine profiles in patients with pulmonary tuberculosis during anti-tuberculosis treatment. Plos One. 2014;9(2), e88572.
51. Zarate-Blades CR, Silva CL, Passos GA. The impact of transcriptomics on the fight against tuberculosis: focus on biomarkers, BCG vaccination, and immunotherapy. Clinical & developmental immunology. 2011;2011:192630.
52. Aranday Cortes E, Kaveh D, Nunez-Garcia J, Hogarth PJ, Vordermeier HM. Mycobacterium bovis-BCG vaccination induces specific pulmonary transcriptome biosignatures in mice. Plos One. 2010;5(6), e11319.
53. Zhang B, Kirov S, Snoddy J. WebGestalt: an integrated system for exploring gene sets in various biological contexts. Nucleic Acids Res. 2005;33(Web Server issue):W741–8.
54. Wang J, Duncan D, Shi Z, Zhang B. WEB-based GEne SeT AnaLysis Toolkit (WebGestalt): update 2013. Nucleic Acids Res. 2013;41(Web Server issue):W77–83.
55. Ashburner M, Ball CA, Blake JA, Botstein D, Butler H, Cherry JM, et al. Gene ontology: tool for the unification of biology. The Gene Ontology Consortium. Nat Genet. 2000;25(1):25–9.
56. Kanehisa M, Goto S. KEGG: kyoto encyclopedia of genes and genomes. Nucleic Acids Res. 2000;28(1):27–30.
57. Pico AR, Kelder T, van Iersel MP, Hanspers K, Conklin BR, Evelo C. WikiPathways: pathway editing for the people. PLoS Biol. 2008;6(7), e184.
58. Liberzon A, Subramanian A, Pinchback R, Thorvaldsdottir H, Tamayo P, Mesirov JP. Molecular signatures database (MSigDB) 3.0. Bioinformatics. 2011;27(12):1739–40.
59. Shannon P, Markiel A, Ozier O, Baliga NS, Wang JT, Ramage D, et al. Cytoscape: a software environment for integrated models of biomolecular interaction networks. Genome Res. 2003;13(11):2498–504.
60. Chang C-C, Lin C-J. LIBSVM: A Library for Support Vector Machines. 2001.
61. Mark Hall EF, Holmes G, Pfahringer B, Reutemann P, Witten IH. The WEKA data mining software: an update. In: SIGKDD Explorations. 2009.
62. Quinlan JR. Improved use of continuous attributes in C4.5. J Artif Intell Res. 1996;4(1):77–90.
63. Corinna Cortes W. Support-vector networks. Mach Learn. 1995;20(3):273–97.
64. Chen J, Huang H. Feature selection for text classification with Naïve Bayes. Expert Systems with Applications. 2009;36(3):5432–5.
65. Guh R-S, Wu T-CJ, Weng S-P. Integrating genetic algorithm and decision tree learning for assistance in predicting in vitro fertilization outcomes. Expert Systems with Applications. 2011;38(4):4437–49.
66. Breiman L. Random forests. 2001.

Kumar and Bucher *BMC Bioinformatics* 2016, **17**(Suppl 1):4
DOI 10.1186/s12859-015-0846-z

BMC Bioinformatics

PROCEEDINGS Open Access

Predicting transcription factor site occupancy using DNA sequence intrinsic and cell-type specific chromatin features

Sunil Kumar[1,2] and Philipp Bucher[1,2*]

From The Fourteenth Asia Pacific Bioinformatics Conference (APBC 2016)
San Francisco, CA, USA. 11 - 13 January 2016

Abstract

Background: Understanding the mechanisms by which transcription factors (TF) are recruited to their physiological target sites is crucial for understanding gene regulation. DNA sequence intrinsic features such as predicted binding affinity are often not very effective in predicting in vivo site occupancy and in any case could not explain cell-type specific binding events. Recent reports show that chromatin accessibility, nucleosome occupancy and specific histone post-translational modifications greatly influence TF site occupancy in vivo. In this work, we use machine-learning methods to build predictive models and assess the relative importance of different sequence-intrinsic and chromatin features in the TF-to-target-site recruitment process.

Methods: Our study primarily relies on recent data published by the ENCODE consortium. Five dissimilar TFs assayed in multiple cell-types were selected as examples: CTCF, JunD, REST, GABP and USF2. We used two types of candidate target sites: (a) predicted sites obtained by scanning the whole genome with a position weight matrix, and (b) cell-type specific peak lists provided by ENCODE. Quantitative in vivo occupancy levels in different cell-types were based on ChIP-seq data for the corresponding TFs. In parallel, we computed a number of associated sequence-intrinsic and experimental features (histone modification, DNase I hypersensitivity, etc.) for each site. Machine learning algorithms were then used in a binary classification and regression framework to predict site occupancy and binding strength, for the purpose of assessing the relative importance of different contextual features.

Results: We observed striking differences in the feature importance rankings between the five factors tested. PWM-scores were amongst the most important features only for CTCF and REST but of little value for JunD and USF2. Chromatin accessibility and active histone marks are potent predictors for all factors except REST. Structural DNA parameters, repressive and gene body associated histone marks are generally of little or no predictive value.

Conclusions: We define a general and extensible computational framework for analyzing the importance of various DNA-intrinsic and chromatin-associated features in determining cell-type specific TF binding to target sites. The application of our methodology to ENCODE data has led to new insights on transcription regulatory processes and may serve as example for future studies encompassing even larger datasets.

* Correspondence: philipp.bucher@epfl.ch
[1]Swiss Institute for Experimental Cancer Research (ISREC), School of Life Sciences, EPFL, Station 15, Lausanne CH-1015, Switzerland
[2]Swiss Institute of Bioinformatics (SIB), EPFL, Station 15, Lausanne CH-1015, Switzerland

Background

Genes are regulated by transcription factors (TF) binding to physiological target sites in the genome. TFs may bind directly to target sites through sequence-specific protein-DNA interactions, or indirectly through protein-protein interactions with other TFs [1]. Understanding the mechanisms by which TFs are recruited to their target sites is essential for the understanding of gene regulation. For a long time, research in this area has been hampered by the lack of powerful assays to study TF binding events in vivo. This has drastically changed with the advent of the ChIP-seq technology which allows for comprehensive, genome-wide mapping of all in vivo bound sites of a given TF in a particular cell type at near base-pair resolution [2]. What has become clear from ChIP-seq experiments is that the intrinsic binding specificity of a TF can only partly explain the in vivo site occupancy patterns, which in addition were found to be tissue-specific [3]. The recruitment of TFs to target sites thus depends on both DNA-intrinsic properties and cell type specific covariates.

The intrinsic DNA binding specificity of a TF is commonly represented by a so-called position weight matrix (PWM) [4]. A PWM is a $N \times 4$ matrix whose elements define the TF's binding preferences for the four bases of the DNA alphabet along a binding site of length N. Base preferences may either be expressed as occurrence probabilities or as (additive) binding energies. PWMs are basic binding site models with known limitations. For instance, they cannot model nearest neighbor dependencies nor can they account for variable spacing between reverse-complementary half-sites of homodimeric TFs [5]. Nevertheless, it is generally agreed that at least some PWMs are good predictors of in vitro binding affinity of the corresponding TFs. Moreover, large collections of PWMs are available from public databases such as JASPAR [6]. More advanced modeling techniques have been proposed for describing more accurately the binding specificity of a TF [7] but corresponding factor-specific models are not yet available for more than a handful of TFs.

Other DNA sequence-intrinsic contextual features have been used to reduce false positive rates in PWM-based in vivo TF binding site (TFBS) prediction, for instance DNA structural properties [8, 9]. Double stranded DNA possesses anisotropic flexibility, which determines its stability and rigidity, properties that potentially interfere with DNA-protein binding. These structural properties, which are broadly classified into (i) DNA conformation (A-DNA philicity and Z-DNA stability energy), (ii) flexibility (B-DNA twist, protein DNA twist, propeller twist and bending stiffness) and (iii) stability (duplex disruption and stability free energy, stacking energy and denaturation) are sequence dependent, and at least partly predictable from structural characteristics of dinucleotides as revealed by crystal structures of double-stranded oligonucleotides. Cross-species conservation is another sequence-derived feature that has been successfully exploited for distinguishing biologically functional TFBS (evolving under purifying selection) from non-functional ones [10].

Several recent studies have reported that TF binding is influenced (and thus potentially predictable) by chromatin contextual features such as DNA accessibility, nucleosome occupancy, or the presence of specific histone post-translational modifications [11]. This is expected as these modifications were known to be associated with gene regulation processes long before. However, with the availability of high resolution ChIP-seq data for many histone marks in many cell types (e.g. ENCODE collection) it has become feasible to use these features in a systematic manner to build predictive models of in vivo TF binding. Proof of concept comes from a number of recent studies where chromatin features have successfully been exploited to predict TF site occupancy with high confidence [12–14].

In the current work, we have combined multiple sequence-intrinsic parameters with experimental chromatin feature types and used machine-learning methods to assess their relative importance in the TF-to-target-site recruitment process, in the hope to gain new insights into transcription regulatory mechanisms. We used five diverse TFs as examples: (i) the insulator protein CTCF featuring 11 zinc finger domains, which has also been attributed diverse functions including transcriptional repression, genomic imprinting and tumor suppressor [15], (ii) JunD, a leucine zipper protein and member of the activator protein 1 (AP1) family, (iii) the transcriptional activator GABPA, the only obligatory multimeric TF within the Ets family [16], (iv) the transcriptional repressor REST, also known as neuron-restrictive silencer factor (NRSF), and (v) USF2 a member of the evolutionary conserved basic helix-loop-helix leucine zipper TF family. These factors were chosen on the one hand because they represent diverse classes of transcription factors, and on the other hand because they were extensively assayed by ChIP-Seq and other chromatin profiling techniques (*e.g.* DNase I hypersensitivity assays) in multiple cell types by the ENCODE consortium.

Methods

Dataset definition and download

Binding sites were represented by a single genomic position corresponding to the center of the binding regions. Two different types of TFBS collections were used in this study.

- *Genome-wide predicted sites*: PWMs for CTCF (MA0139.1), JunD (MA0492.1), GABPA (MA0062.2), REST (MA0138.2) and USF2 (MA0526.1) were downloaded from the JASPAR database [6]. The human genome assembly hg19 was scanned with these PWMs using in-house tool PWMScan (http://ccg.vital-it.ch/pwmtools) with a P-value threshold of 0.0001. Predicted sites with no or only marginal ChIP-seq tag counts (<3) in any of the cell lines analyzed were removed. Final numbers of sites for each factors were: CTCF (77878), JunD (65526), GABP (51926), REST (48171) and USF2 (37138). The midpoint of the motif was used as the reference position in the peak lists. (For PWMs of even-numbered length, the position immediately upstream of the midpoint was used).

- *ENCODE peak lists* were downloaded from GEO (see Additional file 1 for sample ids). The "narrow peak" format version was used wherever available else the "broad peak" format was used. In either case, we used the midpoint of the peak region as reference point.

TFBS lists were annotated with cell-type specific and sequence-derived quantitative (numerical) features. These annotations were stored in a table with rows corresponding to binding sites and columns to annotation features. Cell-type specific experimental features were calculated as follows:

- ChIP-seq and DNase-seq tag coverage: Files containing mapped sequence tags from a single experiment were taken from the Mass Genome Annotation (MGA) repository of the Eukaryotic Promoter Database EPD [17]. Most of these datasets were originally downloaded in BAM format from the UCSC genome browser database (http://genome.ucsc.edu/ENCODE/downloads.html). Others were generated locally by mapping the sequence reads downloaded from the short read archive (SRA) [18] to the human genome assembly hg19 with the aid of Bowtie [19]. For determining the tag counts, we used only the 5' end positions of the mapped sequence reads. For non-histone ChIP-seq targets (TFs, cofactors, and PolII) these positions were shifted downstream or upstream towards the estimated fragment center by an appropriate distance (determined using cross correlation between + strand tags and – strand tags). The GEO sample accession numbers and specific shifting distances are given in Additional file 1. Multiple tags mapping exactly the same genome positions were counted only once per experiment. For transcription factors and corresponding cofactors (Rad21 for CTCF and FOSL for JunD) as well as for PolII, tags were counted in a window of ±100 bases relative to the TFBS center. A window of ±250 bases was used for DNase I hypersensitivity data. Histone modification tags were counted in windows of ±500 bases.
- Shape-based evaluation of DNase-seq (DNase I hypersensitivity data, also referred to as "digital genomic foot-printing" (DGF)) profiles. We essentially followed the probabilistic partitioning protocol used by Nair et al. [20] for ranking and refocusing CTCF sites. This method attempts to classify the input data into typical and atypical examples. Limited shifting of individual genomic regions is allowed to optimally match the aggregate profile of typical examples. Here, DNase-seq tags were extracted from a ±500 bp region around TFBS center positions and binned in 10 bp windows. Shifting was limited to ±50 bp (11 shift states). The method returns a probability p that a given genomic region constitutes a typical example plus an optimal shifting distance s. The output variable p was used as quantitative shape-based DGF feature in our analyses.

Sequence-derived cell type-independent features:

- *Position weight matrix (PWM) score:* For predicted TFBSs, the score was obtained as a by-product of scanning the whole-genome with the PWM. For ENCODE peaks, the PWM score of the best match to the JASPAR matrix within ±100 bp form the peak center was used. For computing the PWM scores, the base probability matrices from JASPAR were converted into log-odds matrices using the following base composition of the human genome as background frequencies: A - 29 %, C - 21 %, G - 21 %, T - 29 %.
- *Base (oligonucleotide) compositions:* mono-, di-, tri, tetra- and penta-nucleotide frequencies were calculated for ±100 base sequences around TFBS sites. The sequences were split into upstream and downstream parts relative to the TFBS center. Oligonucleotide frequencies were then determined separately for the two parts and for the complete sequence region, and the frequencies obtained in this way were used as three separate annotation features.
- Nucleosome occupancy prediction: average nucleosome occupancy per base was calculated for sequences around TFBS using the model built by Kaplan et al. (Version 3.0) [21]. Sequences ±1000 bases long were used in order to avoid border effects. The final feature consists of the average score of ±10 bases around the TFBS center.
- *Structural features*: These were calculated with the R script "dinucl.R" from the DnaFVP package (http://dnafvp.sourceforge.net/) . In essence, this script assigns a score taken from a table to each dinucleotide within a DNA sequence. The average value over the ±10 bp regions (20 dinculeotide positions) relative to the TFBS center was used as

quantitative structural feature. The following 10 features were used: A-DNA philicity, Z-DNA stability energy, duplex disruption free energy, duplex stability free energy, stacking energy, DNA denaturation, B-DNA twist, protein DNA twist, DNA propeller twist, and bending stiffness.

- *Conservation and polymorphism:* A compressed version of the phastCons46way track of about 10 bp resolution was used as input [22, 23]. The corresponding track file (available at ftp://ccg.vital-it.ch/mga/hg19/phastcons/) presents conservation profiles in a "counts per position" format similar to the ChIP-seq data files. The average PHASTCONS score within ±50 bases was used as conservation feature. SNP frequencies are based on dbSNP132 (~30 million SNPs) [24]. All variants of type SNP (excluding indels) were intersected with the genomic coordinates of extended TFBS regions (±100 bp) using Annovar [25]. The average SNP count within ±10 bases was used as final SNP feature.

Feature selection and model building

Data matrices consisting of features in columns and TF sites/peaks in rows were first normalized by centering columns on the mean and then scaled by dividing the centered values by the standard deviations of the corresponding columns.

Classifiers to discriminate strong from weak TF binding sites (top and bottom 20 %) were built and evaluated with the R functions *train* and *trainControl* from the *Caret* package. First, a given data table was randomly split into 90 % sites for model building and 10 % for testing. Support Vector Machines (SVM) and Random Forest (RF) were used to build a model under leave-one-group-out cross-validation (10 groups). Polynomial and radial basis function kernels were used with SVM. Area under the receiver operating curve (auROC) was computed with the R function *roc* from the *pROC* library and used as measure of accuracy. The *rfe* function from the *caret* library was used to carry out Recursive Feature Elimination (RFE) with SVM. In addition to RFE, which reduces the dimensionality of the parameter space in a way that takes into account correlations between features, an automatic selection of best performing SVM parameters (sigma was pre-estimated using *sigest* function form *kernlab* package and fixed for a range of cost C parameter which varied form 0.25 to 32) was used along with leave-one-group-out cross-validation to optimize models and reduce over-fitting.

Regression-based quantitative binding strength predictors were also built and cross validated with the R functions *train* and *trainControl.* The TF tag count coverage served as target variable. Cross-data set predictions were made with the R function *predict* and the relative importance of different features was extracted with *varImp* (all functions are part of the *caret* package). All sites from a given table were used in this case. SVM regression with radial basis function kernel was used as training method and a Pearson's correlation coefficient (PCC) between measured and predicted TF tags was used as performance indicator.

Results and discussions

The overall approach for prediction of TF site occupancy is outlined in Fig. 1. Two types of TFBS lists were used in this study: (i) genome wide-predicted sites resulting

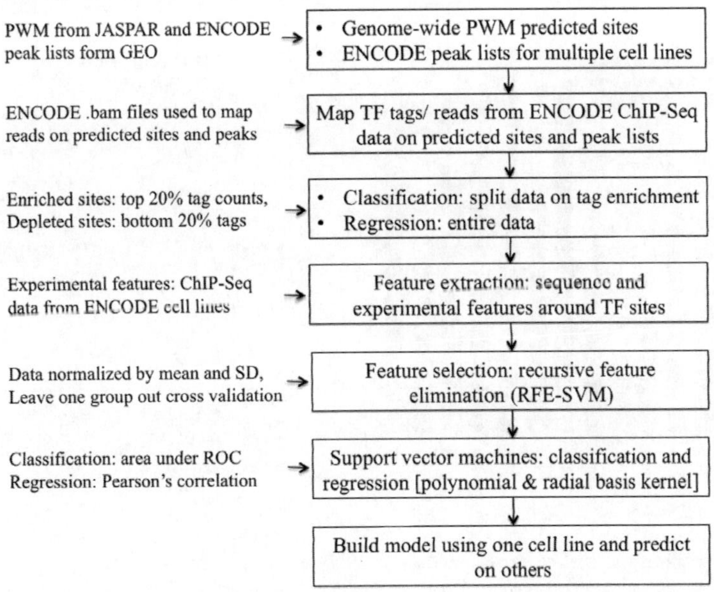

Fig. 1 Overall workflow. Overall approach for prediction of TF site occupancy

from a whole-genome PWM scan and (ii) cell-type specific ENCODE peak lists derived from ChIP-seq data. In both types of lists, we defined the genomic location of a TFBS by a single base position corresponding to the center of the PWM match or the peak region.

In the next step, these peak lists were annotated with the ChIP-seq tag counts from the corresponding experiments. Tags were counted in a ±100 base region relative to the TFBS center (see Methods for details). These tag counts represent binding strength estimates of an individual TF in different tissues. Note that both predicted binding sites and ENCODE peaks were retrospectively annotated with tag counts. For predicted sites, the same list was annotated with different ChIP-seq data representing different cell-types. In the case of ENCODE peak list, different lists were annotated with different ChIP-seq data pertaining to the corresponding cell types.

In two-class prediction experiments, the goal was to discriminate between strong and weak binding sites. To this end, only the top and bottom 20 % of TF binding sites (in terms of tag coverage) were used. For regression-based quantitative binding strength prediction, all TFBS of a given list were considered.

Each peak list was annotated with a large number of features that could be used as input variables for building predictors. Some of these features were directly derived from genome sequences or general annotation resources and thus invariable between different cell types. These features include a PWM-based TFBS score, base and oligo-nucleotide compositions of the binding regions (comprehensively referred to as "sequence" features, see Methods for more details), DNA structural parameters (10 in total), the distance to the nearest TSS, predicted nucleosome occupancy based on the model by Kaplan et al. [21], SNP density, and a cross-genome conservation score. Cell-type specific features included ChIP-seq derived relative abundances of 8 histone marks, PolII and co-factors known to be associated with the TFs under investigation (Rad21 for CTCF and FOSL1 for JunD). In addition, two measures of chromatin accessibility were computed from DGF. One of these measures is count-based whereas the other is shape-based (see Methods for details). The annotated peak lists were then used as data input for various machine-learning algorithms. For class prediction experiments, we used the area under the ROC curve as performance measure, for quantitative binding strength prediction, we used a Pearson correlation coefficient (PCC).

To gain preliminary insights into the relative importance and usefulness of the various features, we focused first on CTCF, a kind of model system for ChIP-seq data analysis. Specifically, we built binary classifiers to distinguish strong from weak binding sites using different features and feature classes. Classifiers were built for both predicted TFBS and ENCODE peaks. Data for the K562 cell line were used in this computational experiment.

The performance of the trained classifiers was evaluated in a 10-fold cross-validation setting. We tested three different machine-learning methods, RF and SVM with polynomial and radial basis function kernel (see Additional file 2 for a performance comparison of the three algorithms on a subset of data). The results obtained with the best performing method (SVM with radial kernel) are shown in Fig. 2a. Overall, a high auROC was achieved in binary classification for both datasets. The CTCF-interacting

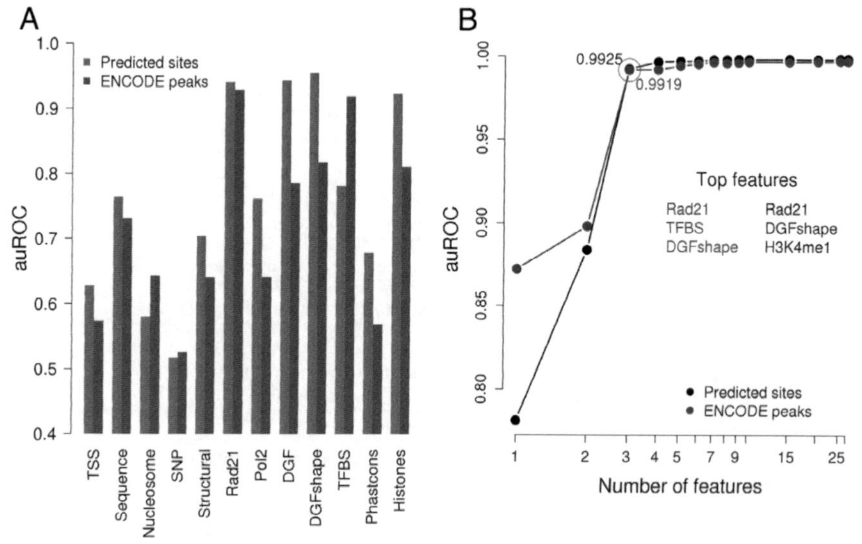

Fig. 2 Classification results in genome wide predicted sites and ENCODE peak lists. The performance in classifying strong versus weak binding sites is reported as area under ROC curve. **a** Performance of individual feature or feature classes on CTCF sites (predicted sites and ENCODE peak lists) in K562 cell-line. **b** Feature importance assessed by recursive feature elimination (RFE-SVM)

protein Rad21 was the pest performing feature, followed by chromatin accessibility (DGF) and the PWM-based binding score (TFBS). Histone marks together also performed quite well but may be partly redundant (correlated) with DGF. The good performance of Rad21 confirms previous reports that the majority of in vivo occupied CTCF sites in the genome are actually bound by the complete cohesin complex [26, 27].

Interestingly, experimental features such as DGF and histone mark perform better for predicted sites whereas the TFBS score is more useful for discriminating strong from weak binding sites in the ENCODE peak lists. The weaker performance of TFBS on predicted site could be explained by assuming that some of the high-scoring PWM matches reside in closed chromatin regions and thus are simply inaccessible to CTCF. Such "false positives" would be difficult to identify without experimental data, as chromatin accessibility varies substantially between cell types. Moreover, the diversity in TFBS is reduced compared to experimentally determined in vivo binding sites, as the predicted sites are selected for PWM matches with a high score. For analogous reasons, the weaker performance of DGF and histone marks on ENCODE peak lists is not surprising. In simplified terms, in vivo binding sites are sampled from accessible and active chromatin regions, resulting in reduced variability of these features as compared to predicted sites.

The predictive power of the numerous sequence features combined (over 4000 in total, see Methods) was equal or lower than the TFBS score alone. Furthermore, using that many sequence features was computationally expensive and time taking. Therefore we replaced these features in subsequent analyses by a simple G + C content feature, which we knew from parallel studies, was positively correlated with CTCF binding. Since count and shape-based DGF measures showed very similar performance, we kept only the simpler count-based measure. All other features were kept even though some showed very modest performance. The particularly weak performance of SNPs may be explained by the generally low SNP counts (often 0) in TFBS regions.

We carried out RFE, to identify the most informative individual features and to analyze the degree of redundancy between features. RFE works by backward elimination, starting with all the features and eliminating one at a time. At each step, the feature judged to be least useful for prediction is eliminated and the overall performance of the predictor is re-evaluated by cross-validation. The results of recursive feature elimination are shown in Fig. 2b. We note that for both predicted TFBS and ENCODE peaks maximal performance (0.99 accuracy) is reached with as few as three features. The cofactor Rad21 and DGF (shape-based) are part of the top features in both cases, complemented by the PWM score for ENCODE peaks and histone modification H3K4me1 for predicted sites.

In the next experiment, we used a quantitative (regression-based) prediction approach to address the following questions: (i) predict the binding strength of TFs (ii) do the previous observations concerning CTCF generalize to other TFs? (iii) Can a predictor trained on one cell type be applied to another cell type? We thus extended our study to another transcription factor, JunD, and to another cell type, H1hESC, an embryonic stem cell derived cell line. We divided the complete feature collections into two major groups: (i) chromatin-related (cell-type specific) features and (ii) sequence-derived (cell-type independent) features. Predictors were trained on each subset individually and on all the features together. As we found again that SVM with radial kernel performed better than the other two methods tested, we used this method here and in all subsequent analyses described in this paper.

Figure 3 shows the regression results for predictions within a cell line (10 fold cross-validation) and across a different cell line. Prediction accuracy of TF occupancy levels was higher for CTCF ($R = 0.85$ to 0.91) than for JunD ($R = 0.72$ to 0.82). The lower predictability of JunD binding strength is primarily due to the poor performance of the sequence and annotation-derived features ($R \sim 0.2$). We further noticed a higher cell type specificity of the trained predictors for JunD as compared to CTCF. The model built on the K562 CTCF peak list was a relatively good predictor of H1-hESC CTCF occupancy and vice versa. However, for JunD the models performed significantly better when tested on the same cell (drop of R from about 0.7 to 0.5). We also explored the role of structural features in regression analysis; even though we observed a slight improvement in cross-cell line prediction, the results obtained with structural features were not a significant improvement over those obtained with sequence plus annotation features alone.

The machine learning method used (from the R package caret, see Methods) automatically computes "importance values" for each feature. In simplified terms, these values correspond to the reduction in prediction accuracy resulting from removal of the features. Feature importance of the binding strength predictors for predicted sites and ENCODE peaks for K562 and H1hESC cells are shown in Fig. 4. For CTCF, the importance of Rad21, DGF and TFBS together remains very high for all TFBS lists, with the relative contribution of TFBS and DGF score changing between predicted sites and ENCODE peaks as observed in the binary classification assays (Fig. 2). Other features were assigned relatively low importance and vary considerably across the four data sets. Again the trends observed with binary classification are reproduced by regression analysis. Histone marks, especially H3K4me2/3, are of comparatively higher value for predicted TFBS than for ENCODE peaks.

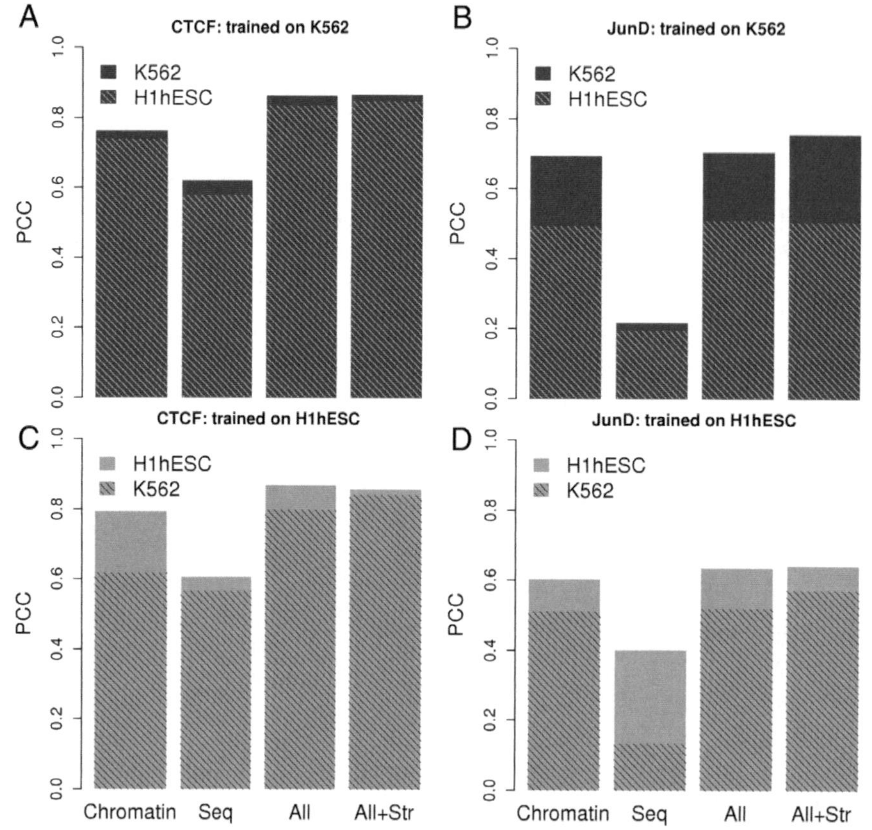

Fig. 3 Regression results with cross cell line prediction. The bar plots reflect the prediction ability (Pearson's correlation coefficient) of regression models trained on one cell line and tested on the same (cross-validation) or another cell line using four different feature sets on ENCODE peak lists. "Seq" consisted of sequence and annotation features; "Chromatin" features included DGF, various histone marks, PolII and co-factors (Rad21 for CTCF and FOSL1 for JunD); "All" included both of them, "All + Str" included structural features in addition to other features. Models for CTCF (**a**, **c**) and JunD (**b**, **d**) were alternatively trained on data from K562 (**a**, **b**) or H1hESC (**c**, **d**) cells

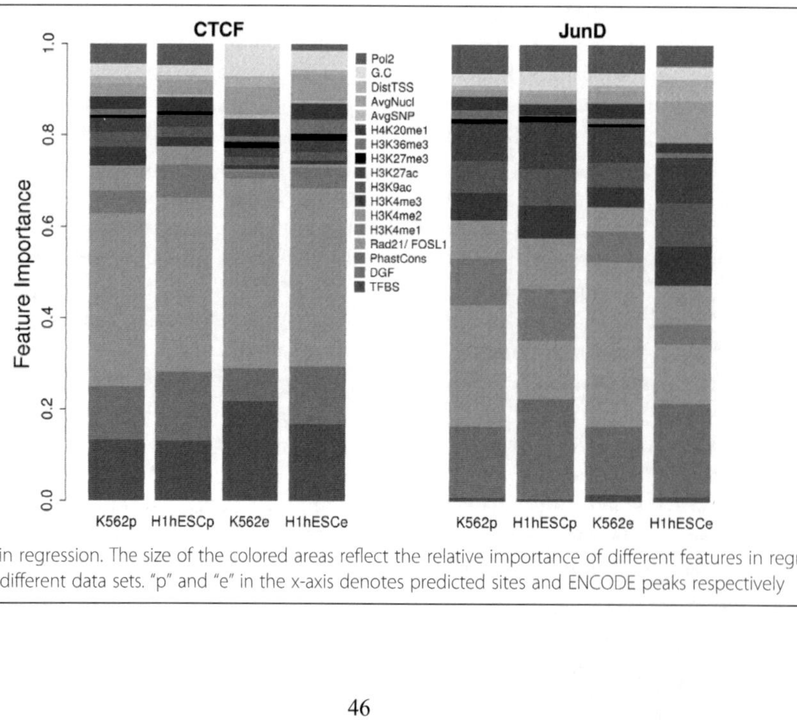

Fig. 4 Feature importance in regression. The size of the colored areas reflect the relative importance of different features in regression models built for CTCF and JunD in different data sets. "p" and "e" in the x-axis denotes predicted sites and ENCODE peaks respectively

A markedly different picture emerges for JunD. We first note the almost total insignificance of the PWM-based binding score, which largely explains the previously observed global ineffectiveness of sequence-derived features in JunD binding strength prediction. Interestingly, this holds for predicted as well as for ChIP-seq defined binding sites. The intrinsically low binding specificity of JunD reflected by the low information content of the corresponding JASPAR PWM may partly explain this observation. We also tested a secondary PWM for JunD from the Jaspar database (MA0491.2) and observed a marginally better (but still poor) performance. In biological terms, this probably means that sequence-specific protein-DNA interactions play only a minor role among the molecular processes that recruit JunD to its physiological target sites. Another interesting observation is that the cofactor FOSL1 is assigned lower feature importance in predictors built from H1hESC cells as compared to K562 cells. A hypothetical explanation for this unexpected difference could be that JunD preferentially associates with members of the AP1 family other than FOSL1 in H1hESC cell. Generally, active histone marks are of much greater importance for predicting JunD binding strength than CTCF binding strength. Repressive histone marks (H3K27me3) and gene-body marks (H3k36me3 and H4K20me1) are of little or no use for binding strength predictions, independently of cell type.

The results presented so far suggest that the intrinsic DNA binding specificity plays a much more important role for in vivo target site selection by CTCF as compared to JunD. At the same time, the selection process appears to be much more cell type-specific in the case of JunD. These two observations would imply a smaller overlap among the TFBS lists for JunD than for CTCF. To verify that this is indeed the case, we generated Venn diagrams for the three TFBS lists (predicted sites, ENCODE peaks for K562 and H1hESC) for the two factors (Fig. 5). Indeed, there is a much smaller overlap between the TFBS lists for JunD. There are only 1,680 JunD sites (0.9 %) common to all three lists from a combined total of 189,459 sites. In contrast, there are 30,073 common CTCF sites (18.6 %) out of 161,438 sites in total.

We then extended this type of regression-based machine modeling framework for predicting tag counts to three additional factors (REST, GABP and USF2) and three additional cell types (GM12878, HeLa and HepG2). The computational methods used were same as before, where we build models using 90 % of the data from each cell line/factor and prediction was carried out on the remaining 10 % test dataset. The purpose of this analysis was to obtain a more representative picture of the diversity of the TF-to-target-site recruitment processes operating in different cell types. We thus generated regression-based binding strength predictors with combinations of high performing features including histone marks, DGF, TFBS and PhastCons score using the predicted TFBS lists for each TF in all the cell lines. The cofactors Rad21 and FOSL1 were not used in this analysis to ensure fair conditions for all five TFs. The results are presented in a concise fashion in Fig. 6. Each subfigure compares two feature sets by means of a scatter plot. We first note that all factors (identified by different colors)

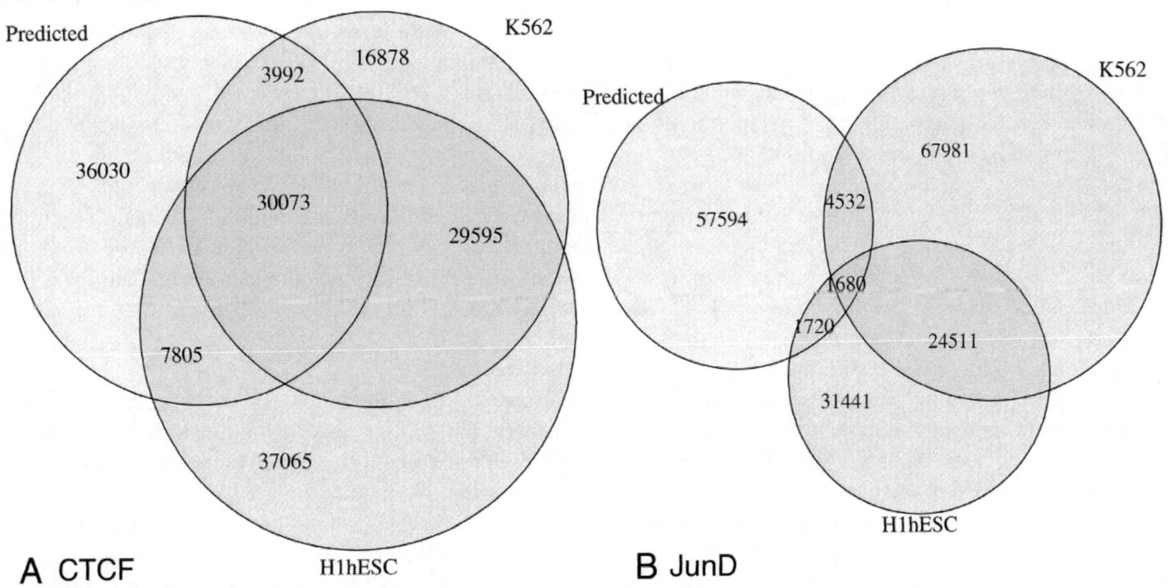

Fig. 5 Overlap between predicted sites and ENCODE peaks. The Venn diagrams show the overlap between predicted TFBS lists and ENCODE peak lists for two cell types (K562 and H1hESC) for (**a**) CTCF and (**b**) JunD

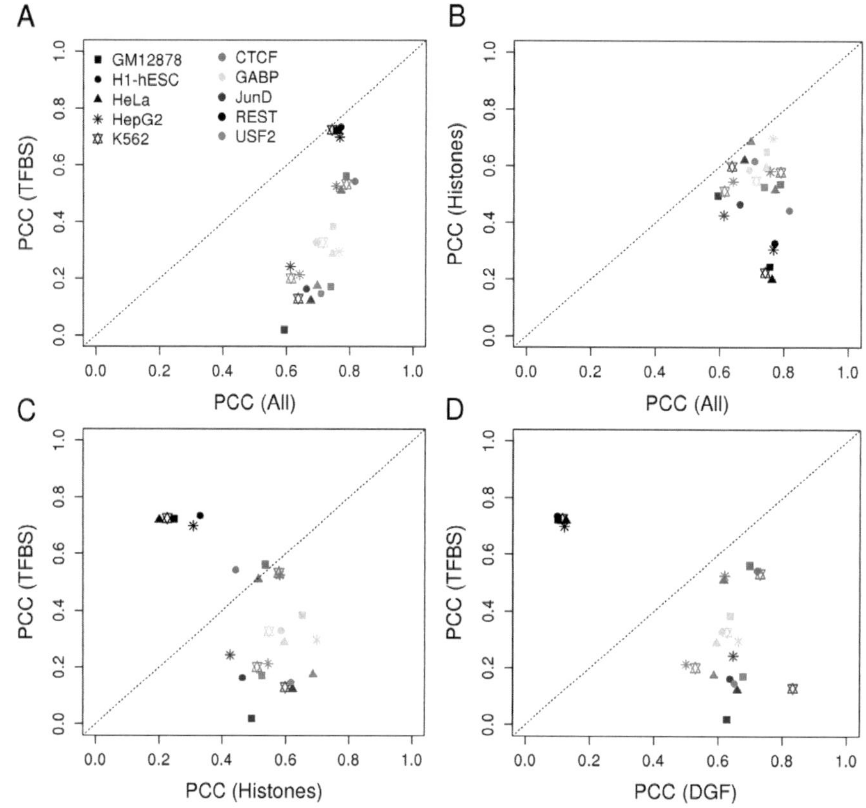

Fig. 6 Regression of multiple factors in different cell lines. Each scatter plot compares the prediction accuracy of regression models (SVM) trained with two different feature sets. (**a**) TFBS vs all features, (**b**) histone marks vs all features, (**c**) TFBS vs histone marks and (**d**) TFBS vs DGF feature. "Histones" includes the following seven marks: H3K4me2, H3K4me3, H4K20me1, H3K9ac, H3K27ac, H3K27me3, H3K36me3

form compact clouds in each plot, sometimes clearly separated from each other. The same is definitely not true for cell types (identified by different symbols). We conclude from this that the principles guiding the same transcription factors to its target sites are overall similar across cell types despite some undisputable cell-type specific effects for JunD, revealed by our results shown in Figs. 3 and 4.

On the other hand, the recruitment processes seem to be highly factor-specific. The most striking outlier among the five factors analyzed is REST whose binding strength is well predicted by its PWM score. However, all other features appear to be virtually useless for this factor. The fact that all dots lie on the main diagonal in Fig. 6a means that the prediction accuracy obtained with the PWM score cannot be increased by feeding more features to the learning algorithm even though histone modifications alone show a modest predictive value (Fig. 6b, $R \sim 0.25$). A dynamic interplay between REST binding and histone modifications was reported in a paper by Zheng et al. [28]. However, most of the histone marks selected in this study (except H3K27me3 and H3K36me3) did not show significant alterations with REST binding there. Perhaps a different selection of histone modifications would have shown better predictive ability of REST binding. This suggests that the histone modifications sometimes seen in the vicinity of a REST site are the consequence rather than the cause of REST binding to DNA. The results for REST are in sharp contrast to CTCF, where chromatin accessibility (DGF) and histone marks substantially increase prediction accuracy in all cell types. The other three factors can more or less be placed on a continuous line connecting CTCF to GABP, USF2 and JunD, in this order. The main discriminatory feature between these factors is the PWM score that varies in terms of its predictive value from high for CTFC to very low for JunD.

Conclusions

We presented a quantitative framework to investigate and compare the role of sequence-intrinsic (tissue-invariant) and cell-type specific chromatin features in the biological processes that recruit TFs to their physiological target sites. Surprisingly distinct feature importance landscapes were observed for different TFs. The PWM score reflecting the intrinsic affinity of a TF to DNA was a good predictor of in vivo binding strength only for the two factors possessing an information-rich recognition motif (CTCF and REST).

DGF came out high in terms of feature importance for all TFs studied except REST. Optimal performance was achieved with as few as three features suggesting high redundancy among the different features tested. DNA structural features, repressive and gene body associated histone marks were generally of little use in predicting in vivo binding site occupancy and strength. Our study highlights the immense value of large functional genomics data sets such as the ENCODE compendium for studying transcription regulation and illustrates the feasibility and effectiveness of pure *in silico* approaches in this field.

Additional files

Additional file 1: Samples used in the study. Details of samples used in the study. GEO and Jaspar ID are given. (PDF 84 kb)

Additional file 2: Performance of machine learning algorithms. Performance of various machine-learning algorithms to classify enriched and depleted CTCF sites (ENCODE peak list for K562 cell line) using multiple histone marks. (PDF 34 kb)

Competing interests

The authors declare that they have no competing interests.

Authors' contributions

SK and PB defined the research problem. Experiments were conducted by SK. PB and SK wrote the manuscript. Both authors read and approved the final manuscript.

Acknowledgements

Computationally intensive tasks were performed at the Vital-IT high-performance computing center of the Swiss Institute of Bioinformatics. SK was supported by Swiss National Science Foundation (grant number CRSII3_154500).

Declarations

Publication charges for this article have been funded by the Swiss Federal Institute of Technology Lausanne (EPFL). This article has been published as part of BMC Bioinformatics Volume 17 Supplement 1, 2016: Selected articles from the Fourteenth Asia Pacific Bioinformatics Conference (APBC 2016). The full contents of the supplements are available online at http://www.biomedcentral.com/bmcbioinformatics/supplements/17/S1.

Published: 11 January 2016

References

1. Gordan R, Hartemink AJ, Bulyk ML. Distinguishing direct versus indirect transcription factor-DNA interactions. Genome Res. 2009;19(11):2090–100.
2. Furey TS. ChIP-seq and beyond: new and improved methodologies to detect and characterize protein-DNA interactions. Nat Rev Genet. 2012;13(12):840–52.
3. Arvey A, Agius P, Noble WS, Leslie C. Sequence and chromatin determinants of cell-type-specific transcription factor binding. Genome Res. 2012;22(9):1723–34.
4. Stormo GD. Modeling the specificity of protein-DNA interactions. Quant Biol. 2013;1(2):115–30.
5. Badis G, Berger MF, Philippakis AA, Talukder S, Gehrke AR, Jaeger SA, et al. Diversity and complexity in DNA recognition by transcription factors. Science. 2009;324(5935):1720–3.
6. Mathelier A, Zhao X, Zhang AW, Parcy F, Worsley-Hunt R, Arenillas DJ, et al. JASPAR 2014: an extensively expanded and updated open-access database of transcription factor binding profiles. Nucleic Acids Res. 2014;42(Database issue):D142–7.
7. Sharon E, Lubliner S, Segal E. A feature-based approach to modeling protein-DNA interactions. PLoS Comput Biol. 2008;4(8):e1000154.
8. Florquin K, Saeys Y, Degroeve S, Rouze P, Van de Peer Y. Large-scale structural analysis of the core promoter in mammalian and plant genomes. Nucleic Acids Res. 2005;33(13):4255–64.
9. Tsai ZT, Shiu SH, Tsai HK. Contribution of sequence motif, chromatin state, and DNA structure features to predictive models of transcription factor binding in yeast. PLoS Comput Biol. 2015;11(8):e1004418.
10. Sinha S, Schroeder MD, Unnerstall U, Gaul U, Siggia ED. Cross-species comparison significantly improves genome-wide prediction of cis-regulatory modules in Drosophila. BMC Bioinformatics. 2004;5:129.
11. Wang J, Zhuang J, Iyer S, Lin X, Whitfield TW, Greven MC, et al. Sequence features and chromatin structure around the genomic regions bound by 119 human transcription factors. Genome Res. 2012;22(9):1798–812.
12. Luo K, Hartemink AJ. Using DNase digestion data to accurately identify transcription factor binding sites. Pac Symp Biocomput. 2013;80:91.
13. Liu L, Jin G, Zhou X. Modeling the relationship of epigenetic modifications to transcription factor binding. Nucleic Acids Res. 2015;43(8):3873–85.
14. Kahara J, Lahdesmaki H. BinDNase: a discriminatory approach for transcription factor binding prediction using DNase I hypersensitivity data. Bioinformatics. 2015;31(17):2852–9.
15. Kim TH, Abdullaev ZK, Smith AD, Ching KA, Loukinov DI, Green RD, et al. Analysis of the vertebrate insulator protein CTCF-binding sites in the human genome. Cell. 2007;128(6):1231–45.
16. Rosmarin AG, Resendes KK, Yang Z, McMillan JN, Fleming SL. GA-binding protein transcription factor: a review of GABP as an integrator of intracellular signaling and protein-protein interactions. Blood Cells Mol Dis. 2004;32(1):143–54.
17. Dreos R, Ambrosini G, Perier RC, Bucher P. The eukaryotic promoter database: expansion of EPDnew and new promoter analysis tools. Nucleic Acids Res. 2015;43(Database issue):D92–6.
18. Kodama Y, Shumway M, Leinonen R, International Nucleotide Sequence Database C. The sequence read archive: explosive growth of sequencing data. Nucleic Acids Res. 2012;40(Database issue):D54–6.
19. Langmead B, Trapnell C, Pop M, Salzberg SL. Ultrafast and memory-efficient alignment of short DNA sequences to the human genome. Genome Biol. 2009;10(3):R25.
20. Nair NU, Kumar S, Moret BM, Bucher P. Probabilistic partitioning methods to find significant patterns in ChIP-Seq data. Bioinformatics. 2014;30(17):2406–13.
21. Kaplan N, Moore IK, Fondufe-Mittendorf Y, Gossett AJ, Tillo D, Field Y, et al. The DNA-encoded nucleosome organization of a eukaryotic genome. Nature. 2009;458(7236):362–6.
22. Siepel A, Bejerano G, Pedersen JS, Hinrichs AS, Hou M, Rosenbloom K, et al. Evolutionarily conserved elements in vertebrate, insect, worm, and yeast genomes. Genome Res. 2005;15(8):1034–50.
23. Rosenbloom KR, Armstrong J, Barber GP, Casper J, Clawson H, Diekhans M, et al. The UCSC genome browser database: 2015 update. Nucleic Acids Res. 2015;43(Database issue):D670–81.
24. Sherry ST, Ward MH, Kholodov M, Baker J, Phan L, Smigielski EM, et al. dbSNP: the NCBI database of genetic variation. Nucleic Acids Res. 2001;29(1):308–11.
25. Wang K, Li M, Hakonarson H. ANNOVAR: functional annotation of genetic variants from high-throughput sequencing data. Nucleic Acids Res. 2010;38(16):e164.
26. Merkenschlager M, Odom DT. CTCF and cohesin: linking gene regulatory elements with their targets. Cell. 2013;152(6):1285–97.
27. Parelho V, Hadjur S, Spivakov M, Leleu M, Sauer S, Gregson HC, et al. Cohesins functionally associate with CTCF on mammalian chromosome arms. Cell. 2008;132(3):422–33.
28. Zheng D, Zhao K, Mehler MF. Profiling RE1/REST-mediated histone modifications in the human genome. Genome Biol. 2009;10(1):R9.

Lyu and Li *BMC Bioinformatics* 2016, **17**(Suppl 1):5
DOI 10.1186/s12859-015-0847-y

BMC Bioinformatics

PROCEEDINGS — Open Access

A semi-parametric statistical model for integrating gene expression profiles across different platforms

Yafei Lyu[1] and Qunhua Li[2*]

From The Fourteenth Asia Pacific Bioinformatics Conference (APBC 2016)
San Francisco, CA, USA. 11 - 13 January 2016

Abstract

Background: Determining differentially expressed genes (DEGs) between biological samples is the key to understand how genotype gives rise to phenotype. RNA-seq and microarray are two main technologies for profiling gene expression levels. However, considerable discrepancy has been found between DEGs detected using the two technologies. Integration data across these two platforms has the potential to improve the power and reliability of DEG detection.

Methods: We propose a rank-based semi-parametric model to determine DEGs using information across different sources and apply it to the integration of RNA-seq and microarray data. By incorporating both the significance of differential expression and the consistency across platforms, our method effectively detects DEGs with moderate but consistent signals. We demonstrate the effectiveness of our method using simulation studies, MAQC/SEQC data and a synthetic microRNA dataset.

Conclusions: Our integration method is not only robust to noise and heterogeneity in the data, but also adaptive to the structure of data. In our simulations and real data studies, our approach shows a higher discriminate power and identifies more biologically relevant DEGs than eBayes, DEseq and some commonly used meta-analysis methods.

Keywords: Data integration, Gene expression, Copula, Mixture model, Rank, Meta-analysis

Background

Detection of differentially expressed genes (DEGs) between biological samples is the key to understand how genotype gives rise to phenotype. With the rapid accumulation of consortium studies (e.g. ENCODE [1]) and public repositories (e.g. NCBI GEO [2]), a large number of RNA-seq and microarray data collected on similar samples from different sources have been made publicly available. Such collections make it possible to integrate similar studies from different sources and platforms in transcriptome analyses, potentially increasing statistical power and reliability in DEGs detection [3–6], while decreasing the cost of the analyses.

Despite these well-known benefits, combining gene expression data from different sources involves many intricate issues. For example, if data are collected from different platforms, the scales of measurements on individual studies may not be comparable. Even though many normalization methods have been developed, normalization across platforms still remains a challenge [7–10]. Furthermore, heterogeneity is often present in studies from different sources. Lab effects often are still retained among datasets produced by different laboratories even after normalization [11].

A convenient way to handle heterogeneity and noise in the data is to use rank-based approach, since rank is robust to outliers and is always comparable across platforms. It has been shown that ranking fold changes of differential gene expression produces better agreement of DEG lists across labs and platforms than using p-values in microarray

* Correspondence: qunhua.li@psu.edu
[2]Department of Statistics, Pennsylvania State University, University Park, PA 16802, USA
Full list of author information is available at the end of the article

gene expression studies [12]. Recently, several rank-based data integration methods have been developed, for example, RankProd [13], RankSum [13], product of ranks, [14] and sum of ranks [14]. They have been found effective in overcoming the heterogeneity among datasets. However, these methods typically are nonparametric methods with fixed rules for combining studies, such as, computing convolution of (transformed) ranks of fold change, hence are not adaptive to the structure of the data.

In this work, we develop a rank-based semi-parametric statistical method to integrate gene expression profiles from different sources. Our model emphasizes the biological intuition that true signals from samples measuring the same biological mechanism should have concordant signals. It builds in a strong preference for concordance of differential directionality and significance across sources, such that the genes that have moderate but consistent signals across studies can be effectively detected. Unlike meta-analysis methods based on fixed rules, our method explicitly models the structure of the data through a copula mixture model, making it both adaptive to the data and robust to noise.

We illustrate our method in the integration of microarray and RNA-seq data for DEG detection. Microarray has been the major experimental platform for gene expression study since mid-1990's [15, 16]. Despite its huge success, it is known to suffer from some limitations such as reliance on existing knowledge of transcript sequences, high background noise, and limited dynamic range of detection. Recently, RNA-Seq has emerged as a new experimental platform for transcriptome profiling and has been flourishing since then. Though it is believed that RNA-seq overcomes the major limitations of microarray [17], RNA-seq still demonstrates excessive variability [18], especially when sequencing depth is low or the gene expression level is low [1, 5]. Considerable differences have been reported between the DEGs detected in these two platforms [19–24]. In addition, due to cost constraints, many RNA-seq experiments nowadays still have no replicates, which limits the power and reliability of its inference. When both microarray and RNA-seq data are available for the same sample, it is natural to investigate whether integrating the data from these two platforms will combine the strengths of the platforms and improve the reliability of DEG identification.

We apply our method to microarray and RNA-seq data from the Microarray Quality Control (MAQC) [12] and Sequencing Quality Control (SEQC) [3] projects, as well as a synthetic microRNA dataset [25]. Our results show that our method substantially improved the accuracy for detecting DEGs.

Methods

Statistical model for gene expression profiles across platforms

Our goal is to develop a data integration method that is both robust to noise and heterogeneity in the data, and adaptive to the structure of data. We develop our method based on a copula mixture model in Li et al (2011) [26]. This model is originally developed for assessing the reproducibility of rank orders between two rank lists from high-throughput experiments. It has been successfully applied to the analysis of ChIP-seq data in ENCODE for comparing peak callers, identifying suboptimal experiments, and determining reporting thresholds for ChIP-seq peaks [27, 28].

Though this method was originally proposed for assessing reproducibility, it can also be viewed as a semi-parametric aggregation method that combines rank lists from different studies. The signals that are consistent across studies are weighed more favorably than those with similar significance but inconsistent across studies. However, this model only clusters the entries into two groups, with the top - ranked ones as interesting signals and the bottom - ranked ones as noise; whereas, for gene expression studies, DEGs reside on both ends of the rank lists, and both ends would be of interest.

Here we extend this model to the context of gene expression studies. We assume that the sample consists of non-DE, up-regulated, and down-regulated DE genes. We use the level of differential expression (e.g. fold change) as our data, such that the up-regulated, down-regulated, and non-DE genes are concentrated on the top, bottom, and middle part of the rank lists, respectively. For simplicity of discussion, we focus on the case of two studies in what follows, and provides the extension to the case with more studies in Additional file 1.

Suppose the level of differential expression for gene i on two studies are $(x_{i,1}, x_{i,2})$, we assume that $X_j = (x_{1,j}, x_{2,j}, \ldots, x_{n,j})$, $j = 1, 2$ is an independent and identically distributed sample with CDF F_j, where F_j is unknown and can vary across studies. Let K_i denote whether the ith gene is non-DE ($K_i = 0$), up-regulated ($K_i = 1$) or down-regulated ($K_i = 2$), and let π_0, π_1 and $\pi_2 = 1 - \pi_0 - \pi_1$ denote the corresponding proportions.

Because differentially expressed genes are expected to be concordant in both the direction of differentiation and the level of significance across studies, we expect the differential expression level of a gene to be positively correlated across studies for DEGs but not for non-DEGs. To model this dependence structure, we assume that, given $K_i = k$, the dependence across studies for the genes in the kth component is induced by a latent bivariate Gaussian random variable, $\boldsymbol{z_k} = (z_{i,1}, z_{i,2})|K_i = k$. The correlation coefficient between the two studies ρ_k is positive for $\boldsymbol{z}_1$ and $\boldsymbol{z}_2$, and 0 for $\boldsymbol{z}_0$. Though the marginal distribution of observed differential expression level, F_j, may be different across studies, it is natural to assume $z_{i,1}|K_i = k$ and $z_{i,2}|K_i = k$ have the same marginal distributions, as different studies are assumed to measure the same underlying biological process. To reflect up- and down-regulation, we assume z_1 has a higher mean than z_0, and z_0 has a higher mean than z_2. Finally, as the scales of the marginal distributions are unknown, only the difference in means between two latent variables and the ratio of their variances can be identified, but not their actual means and variances. Thus, we set z_0 to have mean 0 and variance 1. Putting the above together, the model that generates the dependence structure is

$$\begin{pmatrix} z_{i,1} \\ z_{i,2} \end{pmatrix} \Big| K_i = k \sim h_k = N\left(\begin{pmatrix} \mu_k \\ \mu_k \end{pmatrix}, \begin{pmatrix} \sigma_k^2 & \rho_k \sigma_k^2 \\ \rho_k \sigma_k^2 & \sigma_k^2 \end{pmatrix} \right) \quad (1)$$

where $\mu_0 = 0$, $\sigma_0^2 = 1$, $\rho_0 = 0$, $\mu_1 > 0 > \mu_2$, $1 > \rho_k > 0$ for $k = 1, 2$. Let $u_{i,j} \equiv G(z_{i,j}) = \sum_{k=0}^{2} \pi_k \Phi((z_{i,j} - \mu_k)/\sigma_k)$, where $\Phi(\cdot)$ is the CDF of the standard normal distribution. Then our actual observations $x_{i,j}$ are

$$x_{i,j} = F_j^{-1}(u_{i,j}).$$

Our model can be parameterized by $\theta = (\pi_0, \pi_1, \pi_2, \mu_1, \mu_2, \sigma_1, \sigma_1, \rho_1, \rho_2)$ and (F_1, F_2), where F_1 and F_2 will be substituted by the empirical distributions if they are unknown. Thus it is scale-free. The corresponding mixture likelihood for the data is

$$L(\theta) = \prod_{i=1}^{n} \left[\sum_{k=0}^{2} \pi_k h_k \left(G^{-1}(F_1(x_{i,1})), G^{-1}(F_2(x_{i,2})) \right) \right],$$

where h_k is the bivariate normal density function with parameters μ_k, σ_k^2 and ρ_k. The parameters θ can be estimated by maximizing the mixture likelihood using an estimation procedure similar to Li et al [26], with adaptation to three components. The detailed algorithm is provided in Additional file 1, Section 1.

Determination of differential expression

Given the parameter θ, the posterior probability that a gene i is in the kth group can be computed as

$$p_k(x_{i,1}, , x_{i,2}) = \frac{\pi_k h_k \left(G^{-1}(F_1(x_{i,1})), G^{-1}(F_2(x_{i,2})) \right)}{\sum_{k=0,1,2} \pi_k h_k \left(G^{-1}(F_1(x_{i,1})), G^{-1}(F_2(x_{i,2})) \right)}$$

The classification of a gene is determined by the component that possesses the highest posterior probability. To determine the cutoff for DEGs, we follow a selection procedure similar to the selection based on the IDR (irreproducible discovery rate) criterion in [26] with adaptation to three components, as follows:

1. Rank genes from low to high by $p_0(x_{i,1}, x_{i,2})$, $i = 1, \ldots n$
2. For the ith ordered gene, compute $Pr(nonDG|l \in I_i) = \sum_{l \in I_i} p_0(x_{l,1}, x_{l,2})/i$, where $I_i = \{(x_{l,1}, x_{l,2}) : p_0(x_{l,1}, x_{l,2}) < p_0(x_{(i),1}, x_{(i),2})\}$.
3. For a desired control level α, let $i_{max} = max\{i : Pr(nonDEG|l \in I_i) < \alpha\}$, then differentially expressed genes can be selected by selecting the genes $(x_{(l),1}, x_{(l),2})$ with $l = 1, \ldots, i_{max}$. This set of genes will have an expected rate of nonDEG discoveries no greater than α.

$Pr(nonDEG|l \in I_i)$ represents the expected proportion of nonDEG genes in the claimed DEGs, when the cutoff is set at the i th ordered gene.

Properties of this model

This model has several desirable properties as a data integration method. First, by modelling the two DEG components with positive correlation, the model builds in a strong preference for common directionality of significance across studies while not requiring the differential direction is known *a priori*. This is desirable, as the genes with concordant differentiation across studies are more likely to be real than the discordant ones. Second, the three-component clustering framework allows our method to estimate the two tails adaptively according to the data. When the proportions of up- and down-regulation are unequal, the asymmetry can be reflected in the clusters. In contrast, commonly-used meta-analysis methods, such as Fisher [29], Stouffer [30], and RankProd [13], implicitly assume that the rejection regions are symmetric on both sides, thus making it likely to lose power when asymmetry is present. Third, this model is scale-free, thus it is suitable for combining measurements on different scales or platforms. In our simulation and real data analyses, we will compare with both single-platform DEG detection methods and several commonly-used meta-analysis methods to illustrate the effectiveness of our method.

Results and discussion

Simulation studies with violation of model assumptions

We first examine the performance of our approach using a simple simulation study. In this simulation, the log fold changes are generated from a model similar to our model but with some violation of model assumptions. Our goal is to assess the robustness of our method against violation of model assumptions and to compare with commonly-used meta-analysis methods in this scenario.

Here we use model (1) as the basis to simulate the log fold changes on the two platforms. However, instead of assuming that the log fold change of all the up-regulated (or down-regulated) genes have the same distribution as in our model, we allow them to have different means and correlations, by letting μ_k and ρ_k $(k = 1, 2)$ drawn from uniform distributions. This setting is more flexible than model (1) and introduces mild violation to our model assumptions. Here we choose $\mu_1 \sim unif(0.58, 1.58)$, $\mu_2 \sim unif(-1.58, 0.58)$, $\rho_1 \sim unif(0.80, 0.88)$ and $\rho_2 \sim unif(0.80, 0.88)$. For each gene, we obtain a p-value on each platform from a two-sided z-test for $H_0 : \mu = 0$ vs $H_1 : \mu \neq 0$.

To compare our method with commonly-used meta-analysis methods, we combine the p-values using Fisher's method and Stouffer's method, and combine the log fold changes using our method and RankProd. RankProd is a non-parametric statistic that detects items that are consistently highly ranked in a number of lists [13]. Denote $r_{i,j}$ as the rank of the fold change of the ith gene in the jth platform and n_j as the total number of genes in the jth platform, RankProd is computed as $RP_{i,j} = \prod_{j=1,2} \frac{r_{i,j}}{n_j}$. A small value of RP indicates that a gene is consistently highly ranked across platforms.

We evaluate the performance in four scenarios: (S1) data with same proportion of up- and down-regulated DEGs, (S2) data with different proportions of up- and down-regulated DEGs, (S3) data with a small proportion of DEGs, and (4) data with low inter-platform consistency. The parameter setting is shown in Additional file 1: Table S1. For each parameter setting, we simulate 100 data sets, each of which consists of two replicates with 5,000 genes on each replicate.

Results of simulation studies with violation of model assumptions

In all simulations, our estimates for μ_k and ρ_k $(k = 1, 2)$ are close to the means of the corresponding uniform distributions, and the other estimated parameters are close to the true parameters (Additional file 1: Table S1). As a guide for the selection of the signals, the error rate of non-DEG discoveries estimated from our method should be well calibrated. To check the calibration, we compare the actual frequency of false calls (i.e. empirical FDR) with the estimated error rate, $Pr(nonDEG|l \in I_l)$. As shown in Fig. 1a and Additional file 1: Figure S1, our method is well-calibrated in all the scenarios. In addition, we also evaluate the trade-off between the numbers of correct and incorrect calls made at various thresholds for all methods. As shown in Fig. 1b and Additional file 1: Figure S2, our method shows the highest discriminative power among all the methods of comparison. These results indicate that our method is robust to mild violation of model assumptions.

Real data-based simulation studies

In this simulation, we simulate RNA-seq data and microarray data based on a real data set, following [31] and [32], respectively, and compare the performance of our method with single-platform DEG identification methods and some commonly-used meta-analysis methods in this more realistic setting.

To identify DEGs from a single-platform, we use DEseq for RNA-seq [33] and eBayes for microarray [34]. DEseq is one of the most commonly-used tools for identifying differentially expressed transcripts in RNA-seq data. It models the read counts based on a negative binomial model, with variance and mean linked by a data-driven local regression, and infers the significance of differentiation using an approach analogous to the Fisher's exact test. eBayes is a popular method for determining DEGs for microarray data. It estimates mean and variance of gene expression levels based on an empirical Bayes framework, and determines the significance of differentiation according to the empirical Bayes moderated t-statistics and their associated p-values. To compare our method with commonly-used meta-analysis methods, we combine the p-values from DEseq and eBayes using Fisher's method and Stouffer's method, and combine the fold changes using RankProd and our method.

In an attempt to simulate realistic data, we estimated the parameters for simulation from a real dataset in Marioni et al [23], which consists of microarray and RNA-seq measurements of the same biological samples from two cell types (kidney and liver), and then simulated the distribution of the gene expression levels on each platform based on the estimated parameters.

Simulation procedure for real data-based simulation studies

Here we provide a brief description of the simulation procedure. A detailed description and parameter settings can be found in Additional file 1, Section 2. Briefly, our simulation procedure consists of three parts, namely, simulation of the distribution of RNA-seq data, simulation of the expression levels of microarray data, and coupling of RNA-seq and microarray data. The counts of RNA-seq data are simulated from a negative binomial model following Kvam et al [31], with the mean parameter based on an estimate from the RNA-seq measurements of the kidney/liver samples in [23] and the over dispersion parameter drawn from a gamma distribution following Hardcastle and Kelly [35]. The microarray data is simulated following Xiao et al [32], where both the gene expression levels and the log-fold changes are simulated based on the estimates from the microarray measurements of the kidney/liver samples in [23]. After obtaining the distributions of RNA-seq and microarray data, the expression level of a gene on each platform then is generated by sampling the same quantile from the corresponding distribution.

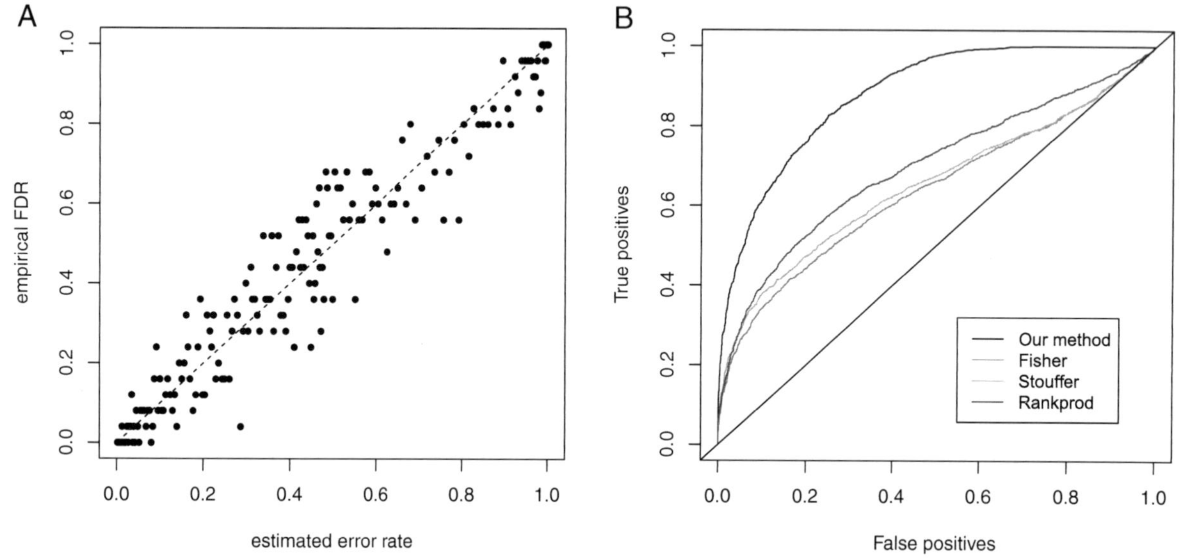

Fig. 1 Calibration and comparison of discriminative power in the simulation with violation of model assumption. **a**. The estimated error rate is compared with the empirical FDR for the simulation setting S1. Results from other simulation settings are similar (Additional file 1: Figure S1). **b**. The percentage of correct and incorrect calls at various thresholds for our method, Fisher's method, Stouffer's method and RankProd, for the simulation setting S1. Results from other simulation settings are similar (Additional file 1: Figure S2)

Here we evaluate the performance in several scenarios, in particular, the scenarios when two platforms have similar versus different data quality, when data quality is high versus low, and when the proportions of up- and down-regulated genes are equal versus unequal. For each scenario, three replicates are simulated under two conditions for each platform, with 10,000 genes for each replicate. Simulated expression levels then undergo the standard pre-processing procedure (Additional file 1, Section 3) prior to the application of DEG detection methods.

Results of real data-based simulation studies

As the significance levels from different methods may not be directly comparable, we evaluate the accuracy of DEG identifications for the top 10 %, 20 % and 30 % genes ranked by each method. At all three cutoffs, our method identifies more true DEGs than eBayes and DEseq. For example, when the cutoff is at 30 %, our method identifies 2626/3844 = 68.3 % true DEGs, whereas eBayes and DEseq identify 2508/3844 = 65.4 % and 2519/3844 = 65.9 %, respectively (Fig. 2). Among the true DEGs identified by our method, 28 were detected by our method exclusively. A close examination shows that the differential expression levels for these genes are moderate on both platforms; however, they are consistent across platforms (Fig. 3). Because our method not only takes account of the significance on individual platforms but also the consistency across platforms, these genes are ranked higher than the genes that are more significant in a single platform but inconsistent across platforms.

As shown in Fig. 4 and Additional file 1: Table S2, when the data consists of similar proportions of up- and down-regulated genes (Fig. 4a-c), our method and the other meta-analysis methods perform similarly. However, when the proportions of up- and down-regulated genes are considerably different (Fig. 4d-f), our method outperforms all the other methods and shows the highest area-under-the-curve (AUC) in the ROC curve (D: $AUC_{our} = 0.812$ vs $AUC_{other} = 0.756\text{-}0.796$; E: $AUC_{our} = 0.785$ vs $AUC_{other} = 0.675\text{-}0.766$; F: $AUC_{our} = 0.753$ vs $AUC_{other} = 0.678\text{-}0.733$; see Additional file 1: Table S2 for details). This is because our method is adaptive to the data and can effectively determine its rejection region according to the shapes of tails.

One concern in data integration is that integration may be deteriorated if data from one platform has poor quality. We therefore investigate how the quality of data from individual platforms affects the identification by simulating data with different quality. We consider two scenarios: the two platforms have similar data quality, and one platform has lower quality than the other. As shown in Fig. 4b (symmetric) and e (asymmetric), when one platform has apparently lower quality than the other, integrating the two platforms does not necessarily improve the discriminative power over using only the data from the platform with better quality. However, even in this scenario, our method (AUC = 0.790 for B and 0.785 for E) still shows a discriminative power that is as

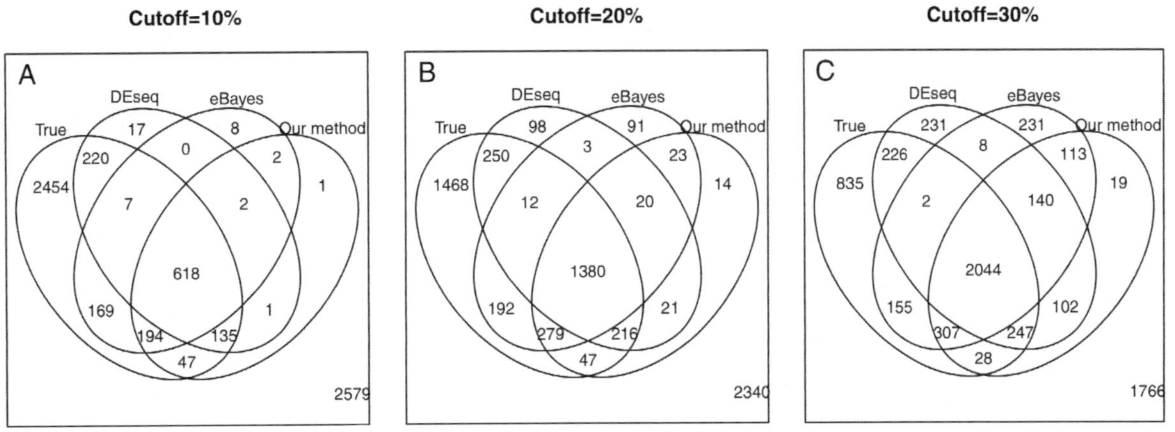

Fig. 2 Venn diagram for the detected DEGs reported by different methods in the real data-based simulation. Expression levels of 10000 genes were simulated, among them 30 % genes are up regulated, 30 % genes are down regulated and 40 % genes are non-DE genes. After removing genes with low counts, 6454 genes remain. The Venn diagrams (from left to right) show the intersection between true DEGs and top 10 %, 20 % and 30 % genes ranked by DEseq, eBayes and our method, respectively. At 10 % cutoff, the detected true positives are 994/3844 = 25.8 % (our), 980/3844 = 25.4 % (DEseq) and 988/3844 = 25.7 % (eBayes), respectively. At 20 % cutoff, the detected true positive are 1922/3844 = 50.0 % (our), 1858/3844 = 48.3 % (DEseq) and 1863/3844 = 48.5 % (eBayes), respectively. At 30 % cutoff, the detected true positive are 2626/3844 = 68.3 % (our), 2508/3844 = 65.4 % (DEseq) and 2519/3844 = 65.9 % (eBayes)

competitive as the better one of the single-platform methods (AUC = 0.799 for B and 0.766 for E). When the two platforms have similar quality (Fig. 4a, c, d and f, Additional file 1: Table S2), our method shows a more obvious gain over both single-platform methods regardless if the data quality is high or low.

Application to MAQC/SEQC project data

We apply our method to a dataset from MAQC/SEQC project [3, 12]. In this dataset, the mRNA samples were generated for universal human reference RNA (Stratagene) and human brain reference RNA (Ambion). The gene expression levels of each sample were measured by microarray and RNA-seq at multiple sites using multiple commercial platforms on multiple replicates, and were validated by qRT-PCR. Thus this dataset provides an ideal benchmark for objectively assessing the performance of our method. Here we use the microarray data generated using Affymetrix array platform at Affymetrix and the RNA-seq data generated using Illumina Hi-seq 2000 platform at BGI. Detailed information on the data can be found in [3, 12]

As the features measured by microarray data and RNA-seq data do not completely overlap, we only

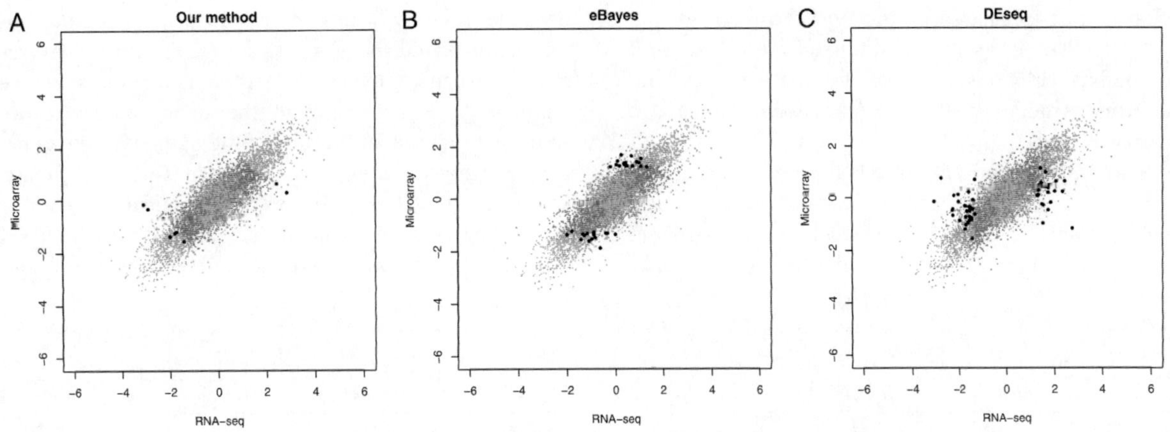

Fig. 3 Comparison of decision boundaries in the real data-based simulation study. The log2 fold changes of gene expression levels are shown for genes in the simulation in Fig. 2. The genes that are ranked in the top 20 % by each method are deemed as DEGs. Correctly detected DE genes, false positive, false negative and true negatives from each method are shown in green, black, red and grey, respectively. The difference in decision boundaries shows that genes with moderate but consistent signals are more likely to be identified by our method

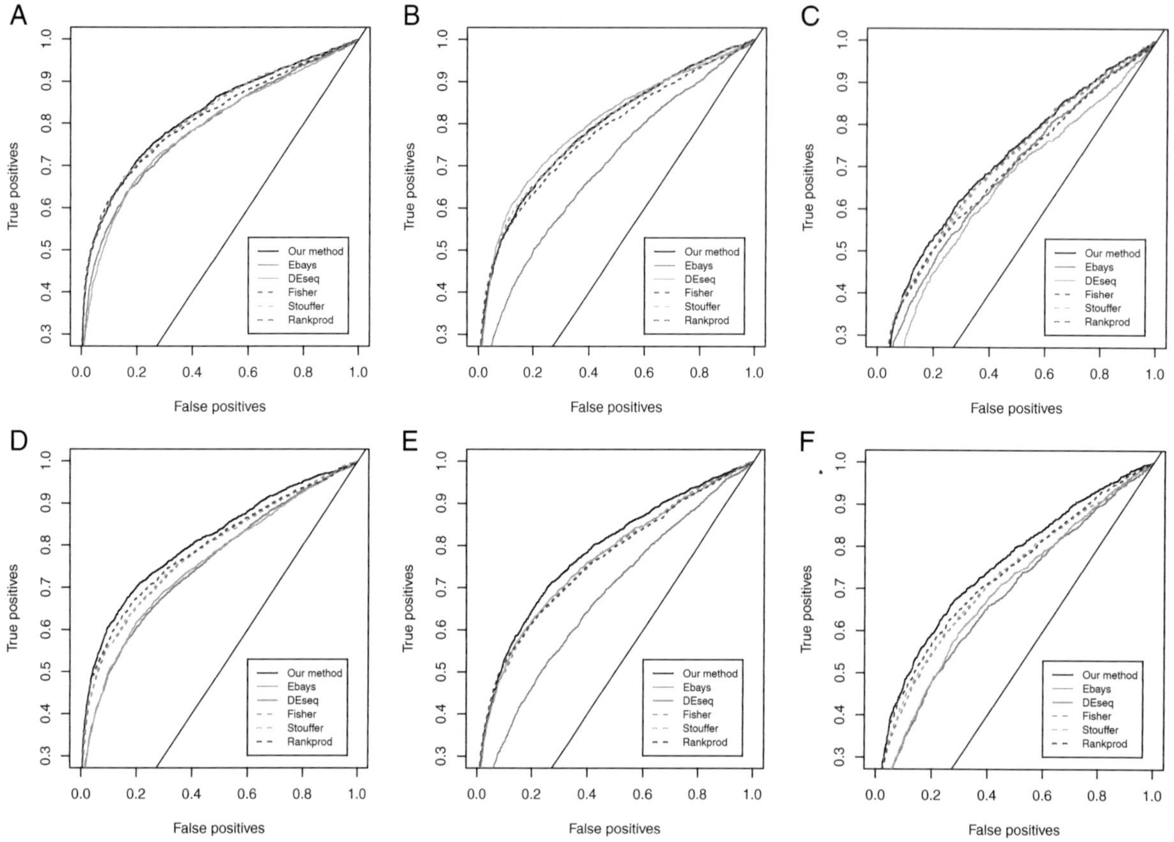

Fig. 4 Comparison of discriminative power in the real data-based simulation study. Figures show the percentage of correct and incorrect calls at various thresholds for six simulation settings. The AUCs are shown in Additional file 1: Table S2. **a**. 20 % genes are up-regulated and 20 % genes are down-regulated. Both platforms have high data quality. **b**. 20 % genes are up-regulated and 20 % genes are down-regulated. One platform has high data quality and the other platform has low data quality. **c**. 20 % genes are up-regulated and 20 % genes are down-regulated. Both platforms have low data quality. **d**. 10 % genes are up-regulated and 30 % genes are down-regulated. Both platforms have high data quality. **e**. 10 % genes are up-regulated and 30 % genes are down-regulated. One platform has high data quality and the other platform has low data quality. **f**. 10 % genes are up-regulated and 30 % genes are down-regulated. Both platforms have low data quality

included features shared by both platforms in our analysis. After data processing (Additional file 1, Section 3), we obtained 14546 genes that are measured on both platforms. Among them, 836 genes were validated by Taqman PCR.

We ran DEseq and eBayes using three replicates with their default parameter settings. We then integrated the p-values from DEseq and eBayes using Fisher's method and Stouffer's method, and integrated the fold change of the same three replicates across platforms using our approach and RankProd. Because the significance measures from these methods are not directly comparable, we rank the genes according to the significance measure from each method, and evaluate the Spearman rank correlation between these rankings and the fold change measured by PCR for the 836 PCR-validated genes in our comparison. As shown in Table 1, our method shows the highest rank correlation (0.872) with the fold

Table 1 Spearman correlation between PCR measured fold change /EST enrichment score and significance of differentiation for MAQC/SEQC analysis

Spearman correlation	Our method	DEseq	eBayes	Fisher	Stouffer[c]	Rankprod
Taqman PCR[a]	0.872	0.761	0.714	0.765	-	0.730
EST enrichment score[b]	0.276	0.111	0.169	0.106	0.201	0.281

[a]Correlation is calculated based on the 836 genes that are validated by Taqman PCR
[b]Correlation is calculated based on 118 brain specific genes obtained from TiGER database
[c]Correlation is not computed for Stouffer method as it generates many p-values at 0

Table 2 Average ranks of PCR measured fold change for top DEGs identified by different methods

DEGs	Our method	DEseq	eBayes	Fisher	Rankprod
Top 10	21.8	52.1	40.5	47.4	18.2
Top 50	45.7	57.2	71.3	55.2	56.9
Top 100	69.4	91.5	92.5	89.3	99.4

Small values represent top rankings

change measured by PCR, and is substantially higher than the correlations from either individual platforms or meta-analysis methods (0.714–0.765). Furthermore, we calculated the average ranks of the fold change measured by PCR for top 10, 50 and 100 differentially expressed genes identified by each method (Table 2). This quantity measures how well the significance of DEGs and the PCR fold change correspond to each other in ranks for top DEGs, which are often of primary scientific interests. A smaller value of the average rank (i.e. a top rank) indicates a better enrichment of the genes with high PCR measured fold change among the top DEGs. For the top 10 DEGs, RankProd (avg. rank = 18.2) and our method (avg. rank = 21.8) show a substantially better enrichment than all the other methods (avg. rank = 40.5–52.1). For the top 50 and 100 DEGs, our method (avg. rank = 45.7 and 69.4, respectively) shows the highest enrichment among all the methods of comparison in both cases (avg. rank of other methods = 55.2–71.3 and 89.3–99.4, respectively).

To evaluate the functional relevance of identified DE genes, we obtained the EST (Expressed sequence tag) enrichment score of brain preferentially expressed genes, which consists of 118 genes specifically expressed in brain, from the TiGER database (TiGER: http://bioinfo.wilmer.jhu.edu/tiger/db_tissue/est/brain-index.html). The EST enrichment score reflects the specificity of gene expression in a tissue and is expected to be correlated with the differential expression levels measured in the experimental data [36]. Here we calculated the Spearman correlation between the significance measures assigned by each method and the TiGER EST enrichment (Table 1). Though the correlation is low for all the methods (0.111–0.281), our method and RankProd show the highest correlation (our: 0.276, RankProd: 0.281, others: 0.111–0.201).

Application to synthetic microRNA data

We next illustrate the usefulness of our method for analyzing microRNA expression, for which effective sequencing technology is still under development [35]. Here we apply our method to a dataset consisting of synthetic microRNA samples with known concentrations, measured on both microarray and sequencing platforms [25]. As the amount of RNA is known in this dataset, it enables us to compare the detected DEGs with the true DEGs. Note that there is no replicate sample in this dataset, which is in fact quite common in practice for RNA-seq studies. It thus imposes challenges for the DEG detection methods based on a single platform to produce reliable statistical inference. Integrating information across platforms nevertheless may improve the reliability of DEG detection in this situation.

This dataset consists of two samples, A and B, each of which is a mixture of synthetic RNA oligos with various concentrations, including 11 differential gene expression levels, ranging from −4 to 4. In total, there are 281 genes with log2 fold change of ±4, ±3 or ±2, 278 genes with log2 fold change of ±1 or ±0.5, and 185 genes with log2 fold change 0. Detailed experimental design can be found in [32].

Since no replicates are available, eBayes cannot be applied, and DEseq can only be used with the single-replicate option. Consequently, Fisher's method and

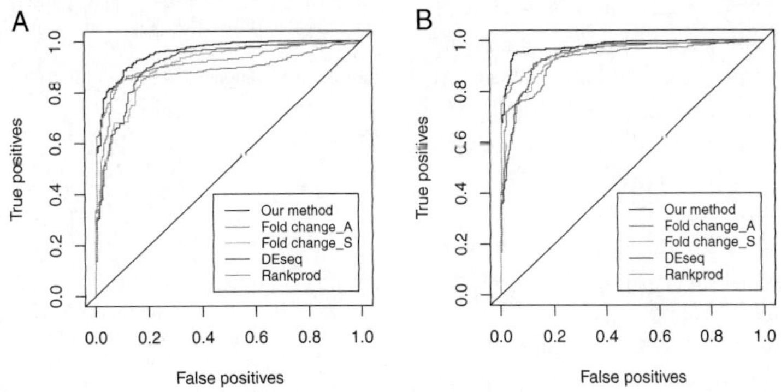

Fig. 5 Comparison of discriminative power in the analysis of synthetic microRNA data. Figures show the percentage of correct and incorrect calls at various thresholds for two levels of classification stringency. The AUCs are shown in Additional file 1: Table S3. **a**. Genes with no fold change as true non-DEGs and the rest as true DEGs. **b**. Genes with a log2 fold change less than ±0.5 as true non-DEGs and the rest as true DEGs. Fold changes measured by microarray and RNA-seq are denoted as Fold change_A and Fold change_S, respectively

Stouffer's method cannot be applied, due to lack of the p-values from eBayes. Only our method and RankProd can still be applied to combine the fold changes across platforms. Therefore, we only compare the performance of our method with RankProd but not with Fisher's method and Stouffer's method. To evaluate the gain over using a single platform, we also compare our method with the fold change of microarray, the fold change of RNA-seq, and the p-values from DEseq generated using the single-replicate option.

Similar to the MAQC/SEQC analysis, we use the rank correlation between the significance assigned by each method and true fold changes of RNA oligos, as an assessment of the performance. Our method shows the highest rank correlation (0.930) among all methods of comparison (DEseq: 0.888, fold change of RNA-seq: 0.888, fold change of microarray: 0.840, and RankProd: 0.894).

As true fold changes are known, the sensitivity and specificity of the identification of DEGs at various thresholds can be evaluated. Here we consider classification of DE and non-DE genes at two levels of stringency. One treats the genes with identical expression levels in both samples as true non-DEGs, and the rest as true DEGs; and the other treats the genes with log 2 fold change less than ±0.5 as true non-DEGs, and the rest as true DEGs. As shown in the ROC curve (Fig. 5), our method has the highest area under the curve (AUC = 0.957 for cutoff = 0; AUC = 0.978 for cutoff = ±0.5) among all methods of comparison in both cases (Other methods: AUC = 0.885–0.919 for cutoff = 0 and AUC = 0.927–0.958 for cutoff = ±0.5, see Additional file 1: Table S3).

Conclusions

In this paper, we present a semi-parametric statistical model for integrating gene expression profiles across studies. This method has several desirable properties as a data integration method. First, it is rank-based, thus is robust to noise in the data and offers a natural way to overcome the heterogeneity across datasets, especially datasets across different platforms. Second, it builds in a strong preference for common directionality of significance across samples, thus allowing genes that have moderate differential expression levels, but are consistent across studies, to be effectively identified. Third, comparing with the commonly-used nonparametric meta-analysis methods, it is adaptive and can reflect the asymmetry in its rejection regions, when the proportions of up- and down-regulation are asymmetric.

As shown in our application to integrate gene expression levels measured on microarray and RNA-seq platforms, our method effectively improved the biological relevance of the identified DEGs. Therefore, this method provides researchers a tool that can take advantage of the gene expression data on different platforms. Though we only illustrate the integration across microarray and RNA-seq platforms, this method is generic and can be applied to integrate rank lists from different sources in other high-throughput settings. The R code for this method is available upon request.

Additional file

Additional file 1: Supplementary materials. Section 1: Estimation algorithm for our model. Section 2: Simulation procedure for the real data-based simulation study. Section 3: Pre-processing procedure for MAQC/SEQC data. Section 4: Extension of our model to the case of more than two samples. Section 5: Tables S1-S3. Section 6: Figure S1-S2. (PDF 415 kb)

Abbreviations
DEG: differentially expressed genes.

Competing interests
The authors declare that they have no competing interests.

Authors' contributions
Algorithms were designed by QL. Algorithm code was implemented and tested by QL and YL. The manuscript was written by QL and YL. Both authors read and approved the final manuscript.

Acknowledgements
Y. L. is supported by NIH R01GM109453 and a Huck Institute Graduate Dissertation fellowship at Penn State University, and Q.L. is partially supported by NIH R01GM109453 and UL1 RR033184. Authors thank three anonymous referees for helpful comments. Authors also thank Feipeng Zhang for proofreading the algorithm in Additional file 1 Section 1, Rahul Vegesna for helpful discussions, and Claire Jin for editorial assistance.

Declarations
Publication charges for this article have been funded by a Huck Institute Graduate Dissertation fellowship at Penn State University awarded to Yafei Lyu. This article has been published as part of *BMC Bioinformatics* Volume 17 Supplement 1, 2016: Selected articles from the Fourteenth Asia Pacific Bioinformatics Conference (APBC 2016). The full contents of the supplements are available online at http://www.biomedcentral.com/bmcbioinformatics/supplements/17/S1.

Author details
[1]The Huck Institute of Life Science, Pennsylvania State University, University Park, PA 16802, USA. [2]Department of Statistics, Pennsylvania State University, University Park, PA 16802, USA.

Published: 11 January 2016

References
1. The ENCODE Consortium. The ENCODE (ENCyclopedia of DNA elements) project. Science. 2004;306(5696):636–40.
2. Barrett T, Troup DB, Wilhite SE, Ledoux P, Rudnev D, Evangelista C, et al. NCBI GEO: mining tens of millions of expression profiles—database and tools update. Nucleic Acids Res. 2007;35 suppl 1:D760–5.
3. Consortium SM-I. A comprehensive assessment of RNA-seq accuracy, reproducibility and information content by the Sequencing Quality Control Consortium. Nat Biotechnol. 2014;32(9):903–14.
4. Kim J, Patel K, Jung H, Kuo WP, Ohno-Machado L. AnyExpress: integrated toolkit for analysis of cross-platform gene expression data using a fast interval matching algorithm. BMC Bioinformatics. 2011;12(1):75.
5. Chavan SS, Bauer MA, Peterson EA, Heuck CJ, Johann DJ. Towards the integration, annotation and association of historical microarray experiments with RNA-seq. BMC Bioinformatics. 2013;14 Suppl 14:S4.

6. Bisognin A, Coppe A, Ferrari F, Risso D, Romualdi C, Bicciato S, et al. A-MADMAN: annotation-based microarray data meta-analysis tool. BMC Bioinformatics. 2009;10(1):201.
7. Johnson WE, Li C, Rabinovic A. Adjusting batch effects in microarray expression data using empirical Bayes methods. In: Biostatistics, vol. 8. 2007. p. 118–27.
8. Dillies M-A, Rau A, Aubert J, Hennequet-Antier C, Jeanmougin M, Servant N, et al. A comprehensive evaluation of normalization methods for Illumina high-throughput RNA sequencing data analysis. Brief Bioinform. 2013;14(6):671–83.
9. Hansen KD, Irizarry RA, Zhijin W. Removing technical variability in RNA-seq data using conditional quantile normalization. Biostatistics. 2012;13(2):204–16.
10. Roberts A, Trapnell C, Donaghey J, Rinn JL, Pachter L. Improving RNA-Seq expression estimates by correcting for fragment bias. Genome Biol. 2011;12(3):R22.
11. Vert G, Nemhauser JL, Geldner N, Hong F, Chory J. Molecular mechanisms of steroid hormone signaling in plants. Annu Rev Cell Dev Biol. 2005;21:177–201.
12. Shi L, Reid LH, Jones WD, Shippy R, Warrington JA, Baker SC, et al. The MicroArray Quality Control (MAQC) project shows inter-and intraplatform reproducibility of gene expression measurements. Nat Biotechnol. 2006;24(9):1151–61.
13. Hong F, Breitling R, McEntee CW, Wittner BS, Nemhauser JL, Chory J. RankProd: a bioconductor package for detecting differentially expressed genes in meta-analysis. Bioinformatics. 2006;22(22):2825–7.
14. Dreyfuss JM, Johnson MD, Park PJ. Meta-analysis of glioblastoma multiforme versus anaplastic astrocytoma identifies robust gene markers. Mol Cancer. 2009;8(1):71.
15. Schena M, Shalon D, Davis RW, Brown PO. Quantitative monitoring of gene expression patterns with a complementary DNA microarray. Science. 1995;270(5235):467–70.
16. Lockhart DJ, Dong H, Byrne MC, Follettie MT, Gallo MV, Chee MS, et al. Expression monitoring by hybridization to high-density oligonucleotide arrays. Nat Biotechnol. 1996;14(13):1675–80.
17. Wang Z, Gerstein M, Snyder M. RNA-Seq: a revolutionary tool for transcriptomics. Nat Rev Genet. 2009;10(1):57–63.
18. Schwartz S, Oren R, Ast G. Detection and removal of biases in the analysis of next-generation sequencing reads. PLoS One. 2011;6(1):e16685.
19. Liu S, Lin L, Jiang P, Wang D, Xing Y. A comparison of RNA-Seq and high-density exon array for detecting differential gene expression between closely related species. Nucleic Acids Res. 2011;39(2):578–88.
20. Tarazona S, García-Alcalde F, Dopazo J, Ferrer A, Conesa A. Differential expression in RNA-seq: a matter of depth. Genome Res. 2011;21(12):2213–23.
21. Zhao S, Fung-Leung W-P, Bittner A, Ngo K, Liu X. Comparison of RNA-Seq and microarray in transcriptome profiling of activated T cells. PloS one 2014;9(1):e78644.
22. Schmid MW, Schmidt A, Klostermeier UC, Barann M, Rosenstiel P, Grossniklaus U. A powerful method for transcriptional profiling of specific cell types in eukaryotes: laser-assisted microdissection and RNA sequencing. PLoS One. 2012;7(1):e29685.
23. Marioni JC, Mason CE, Mane SM, Stephens M, Gilad Y. RNA-seq: an assessment of technical reproducibility and comparison with gene expression arrays. Genome Res. 2008;18(9):1509–17.
24. Nookaew I, Papini M, Pornputtpong N, Scalcinati G, Fagerberg L, Uhlén M, et al. A comprehensive comparison of RNA-Seq-based transcriptome analysis from reads to differential gene expression and cross-comparison with microarrays: a case study in Saccharomyces cerevisiae. Nucleic acids research 2012;40(20):10084–10097.
25. Willenbrock H, Salomon J, Søkilde R, Barken KB, Hansen TN, Nielsen FC, et al. Quantitative miRNA expression analysis: comparing microarrays with next-generation sequencing. RNA. 2009;15(11):2028–34.
26. Li Q, Brown JB, Huang H, Bickel PJ. Measuring reproducibility of high-throughput experiments. The Annals of Applied Statistics 2011;5(3):1752–1779.
27. Consortium EP. An integrated encyclopedia of DNA elements in the human genome. Nature. 2012;489(7414):57–74.
28. Consortium EP. A user's guide to the encyclopedia of DNA elements (ENCODE). PLoS Biol. 2011;9(4):e1001046.
29. Fisher R. Statistical Methods for Research Workers. Edinburgh, UK: Oliver and Boyd; 1925.
30. Stouffer SA, Suchman EA, DeVinney LC, Star SA, Williams Jr RM. The American soldier: adjustment during army life.(Studies in social psychology in World War II, Vol. 1.). 1949.
31. Kvam VM, Liu P, Si Y. A comparison of statistical methods for detecting differentially expressed genes from RNA-seq data. Am J Bot. 2012;99(2):248–56.
32. Xu X, Zhang Y, Williams J, Antoniou E, McCombie WR, Wu S, et al. Parallel comparison of Illumina RNA-Seq and Affymetrix microarray platforms on transcriptomic profiles generated from 5-aza-deoxy-cytidine treated HT-29 colon cancer cells and simulated datasets. BMC Bioinformatics. 2013;14 Suppl 9:S1.
33. Anders S, Huber W. Differential expression analysis for sequence count data. Genome Biol. 2010;11(10):R106.
34. Berkeley C. Linear models and empirical Bayes methods for assessing differential expression in microarray experiments. 2004. E-book available at http://www.bepress.com/sagmb/vol3/iss1/art3 [PubMed].
35. Baker M. MicroRNA profiling: separating signal from noise. Nat Methods. 2010;7(9):687–92.
36. Liu X, Yu X, Zack DJ, Zhu H, Qian J. TiGER: a database for tissue-specific gene expression and regulation. BMC Bioinformatics. 2008;9(1):271.

Pagan et al. BMC Bioinformatics 2016, 17(Suppl 1):6
DOI 10.1186/s12859-015-0849-9

BMC Bioinformatics

PROCEEDINGS

Open Access

The diagnostic application of RNA sequencing in patients with thyroid cancer: an analysis of 851 variants and 133 fusions in 524 genes

Moraima Pagan, Richard T. Kloos, Chu-Fang Lin, Kevin J. Travers, Hajime Matsuzaki, Ed Y. Tom, Su Yeon Kim, Mei G. Wong, Andrew C. Stewart, Jing Huang, P. Sean Walsh, Robert J. Monroe and Giulia C. Kennedy*

From The Fourteenth Asia Pacific Bioinformatics Conference (APBC 2016)
San Francisco, CA, USA. 11 - 13 January 2016

Abstract

Background: Thyroid carcinomas are known to harbor oncogenic driver mutations and advances in sequencing technology now allow the detection of these in fine needle aspiration biopsies (FNA). Recent work by The Cancer Genome Atlas (TCGA) Research Network has expanded the number of genetic alterations detected in papillary thyroid carcinomas (PTC). We sought to investigate the prevalence of these and other genetic alterations in diverse subtypes of thyroid nodules beyond PTC, including a variety of samples with benign histopathology. This is the first clinical evaluation of a large panel of TCGA-reported genomic alterations in thyroid FNAs.

Results: In FNAs, genetic alterations were detected in 19/44 malignant samples (43 % sensitivity) and in 7/44 histopathology benign samples (84 % specificity). Overall, after adding a cohort of tissue samples, 38/76 (50 %) of histopathology malignant samples were found to harbor a genetic alteration, while 15/75 (20 %) of benign samples were also mutated. The most frequently mutated malignant subtypes were medullary thyroid carcinoma (9/12, 75 %) and PTC (14/30, 47 %). Additionally, follicular adenoma, a benign subtype of thyroid neoplasm, was also found to harbor mutations (12/29, 41 %). Frequently mutated genes in malignant samples included BRAF (20/76, 26 %) and RAS (9/76, 12 %). Of the TSHR variants detected, (6/7, 86 %) were in benign nodules. In a direct comparison of the same FNA also tested by an RNA-based gene expression classifier (GEC), the sensitivity of genetic alterations alone was 42 %, compared to the 91 % sensitivity achieved by the GEC. The specificity based only on genetic alterations was 84 %, compared to 77 % specificity with the GEC.

Conclusions: While the genomic landscape of all thyroid neoplasm subtypes will inevitably be elucidated, caution should be used in the early adoption of published mutations as the sole predictor of malignancy in thyroid. The largest set of such mutations known to date detects only a portion of thyroid carcinomas in preoperative FNAs in our cohort and thus is not sufficient to rule out cancer. Due to the finding that variants are also found in benign nodules, testing only GEC suspicious nodules may be helpful in avoiding false positives and altering the extent of treatment when selected mutations are found.

Keywords: Molecular testing, Variants, Fusions, RNA seq, Genomic alterations, Thyroid nodules, FNA

* Correspondence: giulia@veracyte.com
Veracyte, Inc., South San Francisco, CA, USA

Introduction

With the wide adoption of next-generation sequencing (NGS) technologies, the number of genome variants and fusions detected in disease tissues has increased dramatically. The Catalog of Somatic Mutations in Cancer (COSMIC) database contains >100,000 somatic mutations found across >400,000 tumors [1]. The publicly accessible ClinVar database contains >170,000 variant submissions, and >118,000 of these are with clinical interpretations [2]. The Cancer Gene Atlas (TCGA) project has generated extensive genomic data on >11,000 cases across 34 different cancers, including thyroid cancer [3]. With the discovery of so many variants in tumor tissues, expectations of their utility in diagnosing disease are high. In a recent statement by the American Thyroid Association [4], the suggestion is made that the expansion of gene sets informed by TCGA variant discovery "will likely improve the diagnostic accuracy of molecular analyses of thyroid cytology specimens and offer promise for personalizing surgical therapy, with the potential for cost and risk reduction in the diagnostic and therapeutic approaches to treating DTC (differentiated thyroid cancer)." However, the discovery of variants of uncertain significance (VUS), or variants found in benign conditions may temper these expectations.

We and others have studied the genomic landscape of thyroid neoplasms in the pursuit of diagnostic tools for managing patients with thyroid nodules. Several well-studied mutations in genes such as BRAF, N-, H-, K-RAS, and translocations of RET/PTC and PAX8/PPARG have been evaluated for use in the clinical decision-making of these patients. In recent years, a growing list of amino acid substitutions, deletions, insertions, frameshifts, truncations, extensions, gene-pair fusions, and other complex changes have been described in thyroid cancer [5–12]. The effort implemented by TCGA identified >500 novel somatic variants in papillary thyroid carcinoma (PTC), the most common type of thyroid cancer. While TCGA used malignant tumor vs. matched normal surgical tissue as the comparison for variant discovery and analysis, a comparison of malignant vs. benign thyroid neoplasms was not conducted, and an understanding of the frequency of these variants in benign thyroid nodules as well as their frequency in fine needle aspiration (FNA) biopsies is lacking. Here we used a deep RNA sequencing approach to test the hypothesis that diagnostic accuracy of thyroid cancer would be improved by testing a large number of TCGA and literature curated variants.

The management of thyroid nodules larger than 1 cm generally includes ultrasound-guided FNA for clinical decision-making. Evaluation of FNA by cytology leads to definitive benign or malignant results in 64 % of cases [13] with both high negative predictive value (NPV) >95 %, and high positive predictive value (PPV) >98 % [14–17]. Thirteen percent of cases are non-diagnostic [13]. Among asymptomatic thyroid nodules, a benign cytological diagnosis typically averts the need for diagnostic surgery. However, 23 % of nodules cannot be definitively diagnosed [13] and are instead grouped into one of three cytology indeterminate categories (Bethesda III, IV, V) [14] according to defined microscopic features. As most cytology indeterminate thyroid nodules are pathologically benign, consensus has arisen on the value of high sensitivity/high NPV value tests to "rule-out" cancer and safely avoid unnecessary diagnostic surgery [18].

The advent of molecular testing with an RNA gene expression classifier (GEC) to rule-out malignancy with high NPV has greatly reduced the number of diagnostic thyroid surgeries of cytology indeterminate nodules, as a majority of these are indeed benign [19–27]. Others have approached this problem using gene panels whose variants have been associated with malignancy [28, 29]. Whether these rule-in tests (associated with high specificity/high PPV) could be improved to have high sensitivity/high NPV for detecting malignancy (and thus become both a rule-in and rule-out test) has become a topic of interest. A panel measuring variants in seven genes failed to achieve the high sensitivity and NPV required to safely rule-out cancer, as more than half the cancers did not harbor an identified variant [29]. Panels measuring an increasing number of variants have been tested, but not validated in blinded or comprehensive studies, so their ability to safely rule-out cancer has yet to be determined.

The large-scale effort employed by TCGA, which generated exome sequencing, copy number and transcriptional profiling data on a large number of surgical tumor specimens, was a landmark study in elucidating the genomic landscape of many important cancers [3]. One of those studied was PTC, the most common thyroid malignancy [5]. Other types of thyroid cancer were not analyzed, therefore we extended the TCGA effort, in part, by evaluating other subtypes of thyroid cancer and by studying clinically relevant biopsies (thyroid FNA) collected pre-operatively. Here we report for the first time the prevalence of many somatic variants reported by TCGA and others across a wide spectrum of thyroid neoplasms, including benign lesions, in both preoperative FNA and surgical tissue. We studied FNA samples that were collected as part of a prospective, multicenter, blinded cohort, subjected to expert histopathology review [19]; these samples were subsequently blinded and then tested for known somatic genetic alterations using the largest panel of variants yet described. We demonstrate that this panel of 851 somatic variants and 133 somatic fusion gene-pairs has the power to detect only a portion of thyroid carcinomas in preoperative FNAs, raising doubt that such panels can safely rule-out cancer and avoid unnecessary diagnostic surgery, as has been suggested [4, 28]. Importantly, some of these genetic alterations were also commonly found in histopathology benign tumors, suggesting that as variant panels grow in size, an increasing fraction of benign nodules will be falsely identified as cancerous unless careful curation of variants based on their frequencies in benign and malignant nodules is incorporated into clinical decision-making.

Materials and methods

Sample cohorts

FNAs were prospectively collected from 88 patients as part of a previously reported, blinded multi-center study (VERA FNA) [19], or from a multi-center consecutive series of 110 de-identified, pre-operative specimens with remnant nucleic acids archived by Veracyte's CLIA laboratory (CLIA FNA). An additional 85 post-surgical snap-frozen tissues were procured from multiple clinical centers across the US, including samples obtained from Asterand (Detroit, MI), Cureline (South San Francisco, CA), Proteogenex (Culver City, CA) and CHTN http://www.chtn.nci.nih.gov/. All clinical protocols were approved by both central and intuition-specific investigational review boards, and when applicable patients provided written informed consent. All FNAs were blindly evaluated by a cytopathologist and were categorized according to the Bethesda System [14]. Blinded histopathology reference standard labels were determined by a panel of experts (VERA001 FNA) [19], or by blinded secondary review to confirm vendor labels (R. Monroe, tissue cohort) independent of all molecular testing. After secondary review 63/85 (74 %) of tissues had histopathology truth labels, while 22 were designated as follicular neoplasm (FN) with uncertain histopathology (Histo U). CLIA FNA were categorized as GEC Benign or GEC Suspicious, according to their original Afirma GEC test result. RNA was extracted using the AllPrep Micro kit (Qiagen) per the manufacturer's instructions. Sequencing libraries were prepared and pooled in groups of 16 using 10–20 ng of RNA per the manufacturer's instructions using the TruSeq RNA Access kit, followed by sequencing with a NextSeq 500 instrument (Illumina) to an average depth of 50 million (from 25 million fragments) using 75 bp paired-end reads.

Detection of genetic alterations

The genomic location of all variants and fusions was extracted directly from their source publications [5–7, 28, 29] or determined using reported gene ID, amino acid change and mutation annotation files (MAF) generated by Broad Institute, British Columbia Cancer Research Centre, or Baylor University and obtained through the TCGA public access portal. When this information was not available, variants were mapped to their genomic location using COSMIC, and/or cBioportal databases. The initial panel was comprised of 987 distinct genetic alterations, including 854 somatic variants and 133 somatic fusion gene-pairs comprising 524 genes, however three variants could not be captured by our platform. All genetic alterations included in the final panel ($n = 984$) are known to result in protein level changes, including amino acid substitutions, deletions, insertions, frameshifts, truncations, extensions, and other complex changes. Copy number alterations were not measured. While this study focused exclusively on curated somatic gene alteration reports [5–7, 28, 29], it is possible that some mutations are present in nodules due to germline variation. Samples were scored positive using the variant and fusion detection pipelines described below.

Reads were aligned to the reference genome, build 37, via STAR v2.4.1b [30] using state of the art best practices for variant and fusion detection as follows; splice-aware alignments used annotations of all known splice junctions from Ensembl 75. De-duplication was carried out with picard v1.123 MarkDuplicates and aligned reads were processed in GATK v3.3. Variants were detected using the GATK Haplotype Caller [31]. Low quality calls were filtered out with GATK VariantFiltration. Fusions were detected from STAR outputs using the Bioconductor package *chimera* [32] and all filtering settings turned on. Asymptotic standard logit intervals were used to calculate positive predictive values (PPV) and negative predictive values (NPV) along with their associated confidence intervals as previously described [33]. Whenever a zero occurs in a two-by-two contingency table, this method returns estimated adjusted logit intervals that are non-zero.

Results

Our analysis of genetic alterations using genome-wide transcriptome sequencing included 133 fusion pairs and 854 somatic variants. Of these, 133/133 (100 %) of fusions and 851/854 (99.6 %) of variants were detectable by our RNA sequencing effort. Only three variants could not be captured, because they are located at two genomic sites within the non-transcribed promoter region of the TERT gene. Hence, our results center on a final panel of 984 distinct genetic alterations spanning 524 genes. We measured sequencing quality metrics in this study, capturing a median of 48 million paired-end reads and 37 million aligned reads per sample, and computed the range of variants covered at various read depths (Additional file 1). We measured the median read-depth per variant across the pathology subtypes and found variation in coverage, consistent with biological variation in expression of transcripts harboring the genetic variants (Additional file 2). Given this biological variation among subtypes, it is important to assess individual level coverage. We found 93 % of the 851 variant locations have at least one sample with 20-fold coverage (Additional file 1). Variants discussed in this paper were identified using standard variant calling methods (GATK) and for the sake of comparison, additional variants identified at a more sensitive but less specific setting using another variant calling method (SamTools) are reported in Additional file 3.

We evaluated an FNA cohort with surgical truth and found that 19/44 of histopathology malignant samples scored positive for at least one of the genetic alterations, corresponding to 43 % sensitivity (Fig. 1, Table 1). Histopathology benign FNAs scored positive in 7/44 cases, corresponding to 84 % specificity (Table 1). The prevalence of malignant samples in this study cohort was 50 %, resulting in estimates of NPV of 60 % (53–66, CI), and PPV of 73 % (56–85, CI). The performance of the molecular panel was also evaluated for each Bethesda cytology category (Tables 2, 3, 4, 5 and 6) using two cancer prevalences (Table 1). While the number of samples was low in many groups, in cytology malignant (Bethesda VI) where the highest proportion of mutated samples was observed, 14/24 samples had a detectable variant or fusion, corresponding to 58 % sensitivity.

In the post-surgical tissue cohort, 19/32 histopathology malignant samples had a detectable genetic alteration (59 % sensitivity), while 8/31 histopathology benign samples scored positive, a specificity of 74 % (Fig. 2). A single GNAS variant was observed to occur in a follicular adenoma. As this variant has been considered to be a marker for benign nodules, we excluded it from performance calculations to avoid falsely lowering the estimate of specificity. The frequency of malignant samples in this small cohort was 51 % and its NPV was estimated at 64 % (53–74, CI), while PPV was 70 % (55–82, CI).

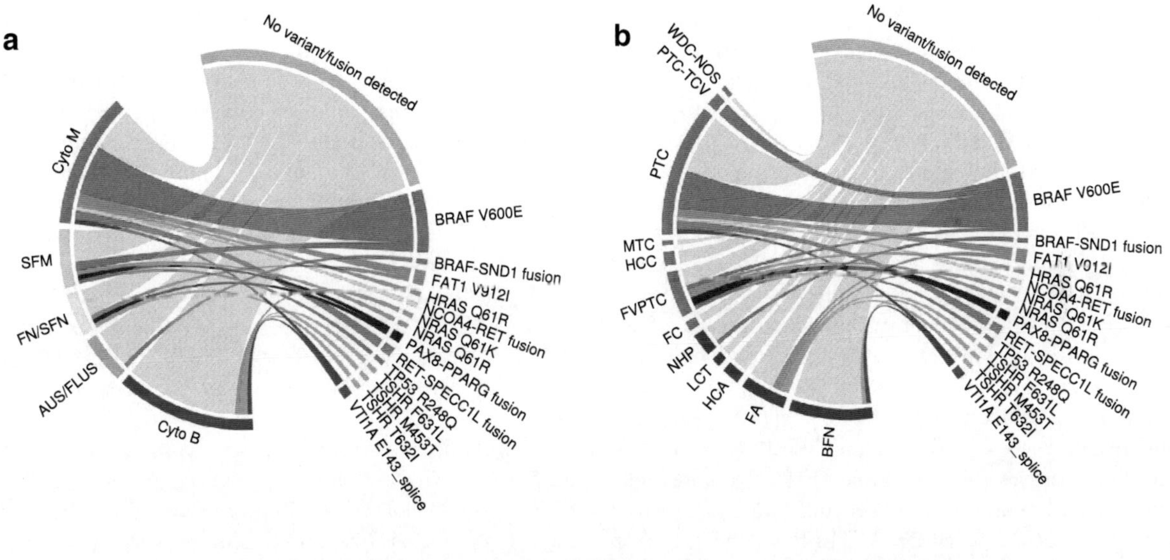

Genetic Alteration	Histology Malignant (n=44)	Histology Benign (n=44)
Positive	19	7
Negative	25	37

Sensitivity, 43% (28-59); Specificity, 84% (70-93); PPV, 73% (56-85); NPV, 60% (53-66)
Prevalence of malignant lesions, 50%. Estimated performance when prevalence equals 24%: PPV, 46% (29-65%); NPV, 82% (78-86).

Fig. 1 Performance across VERA FNA data set ($n = 88$) and genetic alterations observed by cytology (**a**) and histology subtype (**b**). The width of each band of the circos plots is proportional to the number of samples harboring a particular variant or fusion. A complete list of all variants and fusions detected can be found in Additional files 3 and 7. *Abbreviations*: *AUS/FLUS* atypia of undetermined significance/follicular lesion of undetermined significance, *BFN* benign follicular nodule, *Cyto B* cytology benign, *Cyto M* cytology malignant, *FA* follicular adenoma, *FC* follicular carcinoma, *FN/SFN* follicular neoplasm or suspicious for neoplasm, *FVPTC* follicular variant of papillary carcinoma, *HCA* Hurthle cell adenoma, *HCC* Hurthle cell carcinoma, *LCT* lymphocytic thyroiditis, *MTC* medullary thyroid carcinoma, *NHP* nodular hyperplasia, *PTC* papillary thyroid carcinoma, *PTC-TCV* papillary thyroid carcinoma-tall cell variant, *SFM* suspicious for malignancy, *WDC-NOS* well-differentiated carcinoma-not otherwise specified

Table 1 Performance comparison across sample cohorts

Performance of genetic alterations (Variants and Fusions)						
			Cancer prevalence used in calculation			
			PPV		NPV	
Sample cohort	Sensitivity	Specificity	24 %	40 %	24 %	40 %
All VERA FNA ($n = 88$)	43 % (28–59)	84 % (70–93)	46 % (29–65)	64 % (46–79)	82 % (78–86)	69 % (62–75)
AUS/FLUS & FN/SFN only VERA FNA ($n = 22$)	33 % (1–91)	84 % (60–97)	40 % (9–82)	58 % (17–90)	80 % (64–90)	65 % (45–81)
Tissue ($n = 63$)	59 % (41–76)	74 % (55–88)	42 % (27–58)	61 % (44–75)	85 % (78–90)	73 % (63–81)
Combined VERA FNA and tissue ($n = 151$)	50 % (38–62)	80 % (69–88)	44 % (32–57)	63 % (50–73)	84 % (80–87)	71 % (65–76)

We also studied a consecutive FNA series ($n = 110$) processed through our CLIA laboratory. This clinically representative cohort was comprised primarily of cytology indeterminate samples, and was evaluated to characterize the frequency of somatic genetic alterations in a broad, multicenter, US population (Fig. 3). In CLIA FNA with AUS/FLUS cytology, (Bethesda III) 9/52 (17 %) of samples scored positive for a genetic alteration, similar to FNAs with FN/SFN cytology (Bethesda IV) where 4/24 samples harbored a variant (17 %). Since histopathology truth was not available for any CLIA FNA, we could not compute sensitivity, specificity, NPV or PPV.

We compared the results obtained from testing for genetic alterations with results obtained from testing with the Afirma GEC, an RNA-based molecular classifier designed to rule-out malignancy with 90 % sensitivity and 94–95 % NPV [19]. The test reports either a GEC Benign or GEC Suspicious result. In a subset of 86 VERA FNA samples with truth labels and both GEC and genetic alteration results available, testing for genetic alterations alone identified 42 % of malignant samples, compared to the GEC, which identified 91 % of the true malignant samples (Fig. 4). These results are consistent with previously published reports confirming high sensitivity of the GEC [19, 21–26, 34]. In this cohort, testing for genetic alterations resulted in 84 % specificity, compared to the GEC, which resulted in 77 % specificity, higher than the 52 % reported in a large, prospective, blinded multicenter study [19]. The proportion of detected genetic alterations was far higher in samples with GEC Suspicious results; 22/49 (45 %) than those with GEC Benign results (3/37, 8 %), resulting in an overall PPV of potentially >60 % when cancer prevalence is 40 %, as is the case for Afirma Suspicious nodules. This suggests that variant testing would have greater clinical utility on GEC Suspicious FNAs where the detection of genetic alterations may alter the extent of treatment when selected mutations are found. Conversely, no cancer missed by the GEC had a genetic alteration, suggesting that mutational testing on this GEC Benign cohort would not have identified true malignant nodules and would therefore lack clinical utility.

We wanted to get a global view of genetic alterations in both of our cohorts. After combining FNA ($n = 88$) and tissue ($n = 63$) samples with truth labels, 38/76 (50 %) of histopathology malignant samples were found to harbor a genetic alteration, and 15/75 (20 %) of the histopathology benign samples were also mutated (Additional file 4). The most frequently mutated subtypes were MTC (9/12, 75 %), PTC (14/30, 47 %), and FA (12/29, 41 %). Frequently mutated genes in malignant samples included BRAF 21/76 (28 %), found predominantly in cytology malignant samples (Fig. 1). In FNAs with a histopathology diagnoses of PTC, FVPTC or PTC-TCV, 13/37 (35 %) were positive for the BRAF V600E mutation, and in surgical tissues with these same diagnoses, 7/12 (54 %) harbored this mutation (Additional file 5). Also frequently mutated was TSHR, found in 6/75 (8 %) of histology benign samples and only once in a malignant follicular carcinoma (Figs. 1 and 2).

Table 2 FNA performance per cytology group

Performance on Cytology Benign ($n = 30$)		
Genetic alteration	Histology malignant ($n = 6$)	Histology benign ($n = 24$)
Positive	0	4
Negative	6	20

Sensitivity, 0 % (0–46); Specificity, 83 % (63–95); PPV, 16 % (5–50); NPV, 79 % (73–85), prevalence of malignant lesions, 20 %. Estimated performance when malignancy is prevalent at 6 %: PPV, 5 % (1–20); NPV, 94 % (91–96)

Table 3 FNA performance per cytology group

Performance on AUS/FLUS ($n = 12$)		
Genetic alteration	Histology malignant ($n = 1$)	Histology benign ($n = 11$)
Positive	0	1
Negative	1	10

Sensitivity, 0 % (0–98); Specificity, 91 % (59–100); PPV, 16 % (4–45); NPV, 93 % (87–97), prevalence of malignant lesions, 8 %. Estimated performance when malignancy is prevalent at 24 %: PPV, 40 % (12–74); NPV, 81 % (66–90)

Table 4 FNA performance per cytology group

Performance on FN/SFN ($n = 10$) Genetic alteration	Histology malignant ($n = 2$)	Histology benign ($n = 8$)
Positive	1	2
Negative	1	6

Sensitivity, 50 % (1–99); Specificity, 75 % (35–97); PPV, 33 % (7–76); NPV, 86 % (59–96), prevalence of malignant lesions, 20 %. Estimated performance when malignancy is prevalent at 24 %: PPV, 39 % (9–80); NPV, 83 % (53–95)

Table 6 FNA performance per cytology group

Performance on Cytology Malignant ($n = 24$) Genetic alteration	Histology malignant ($n = 24$)	Histology benign ($n = 0$)
Positive	14	0
Negative	10	0

Sensitivity, 58 % (37–78); Specificity, NA% (NA); PPV, NA% (NA); NPV, NA% (NA), prevalence of malignant lesions, 100 %. Estimated performance when malignancy is prevalent at 95 %: PPV, NA% (NA); NPV, NA% (NA)

Given the rarity of most variants measured in this cohort, we investigated the effect of sample size and its impact on mutation rate estimation. At a true underlying mutation rate of 3 %, 95 % confidence intervals are large for sample sizes below 100 (Additional file 6), suggesting that larger cohorts would be required to achieve more precise estimates. However, the large proportion of malignant samples with no variant detected (50 %), along with the reasonably high rate of benign nodules with a variant detected (20 %) cannot be explained solely by an under-representation of these variants in a small sample size cohort. The variant detection rate observed in benign nodules limits the performance of the panel, in particular confers a specificity penalty that limits the panel's ability to attain both high sensitivity and high specificity.

Discussion

In this study we demonstrate that interrogation of a large number of somatic genomic alterations, as reported by TCGA [5, 6] and others [7, 28], is limited in its ability to detect cancer with high sensitivity when applied to a diverse set of thyroid carcinomas. Published data from TCGA shows 59.7 % of their PTC tissue cohort harbored a BRAF mutation, similar to the 54 % we found for the PTC tissues in our study. Despite this similarity, extension of our study to other subtypes and to FNAs reveals that of the histopathology malignant samples, 41 % of tissues and 57 % of FNAs scored negative for any of the 984 genetic alterations tested. These results are consistent with the observations of others [4] noting that a panel of the 17 most common genetic alterations [29] failed to detect 53 % of carcinomas with Bethesda III or IV cytology. Despite significant overlap with the mutations evaluated here and in those of other smaller panels [35, 36], a recent single-center study that was not blinded, reported both high sensitivity and high specificity measuring similar genetic alterations on FN/SFN [28] and/or AUS/FLUS samples [37]. While the difference in sensitivity may be partially explained by the measurement of very rare variants in these small cohorts, which is expected statistically to result in a wide range in performance, this would not explain the difference in specificity, as we detect many of the same variants in benign nodules. In addition, the impact of blinded versus unblinded histopathology truth labels, or local histopathology diagnostic trends among challenging neoplasms, may account for differing performance among similar genomic panels. Collectively, our results and those of other groups [29, 35, 36, 38, 39] do not provide sufficient evidence to support the notion that

Table 5 FNA performance per cytology group

Performance on SFM ($n = 12$) Genetic alteration	Histology malignant ($n = 11$)	Histology benign ($n = 1$)
Positive	4	0
Negative	7	1

Sensitivity, 36 % (11–69); Specificity, 100 % (3–100); PPV, 92 % (76–98); NPV, 9 % (4–17), prevalence of malignant lesions, 92 %. Estimated performance when malignancy is prevalent at 24 %: PPV, 25 % (8–53); NPV, 77 % (58–88). Estimated performance when malignancy is prevalent at 62 %: PPV, 64 % (32–85); NPV, 39 % (21–59)

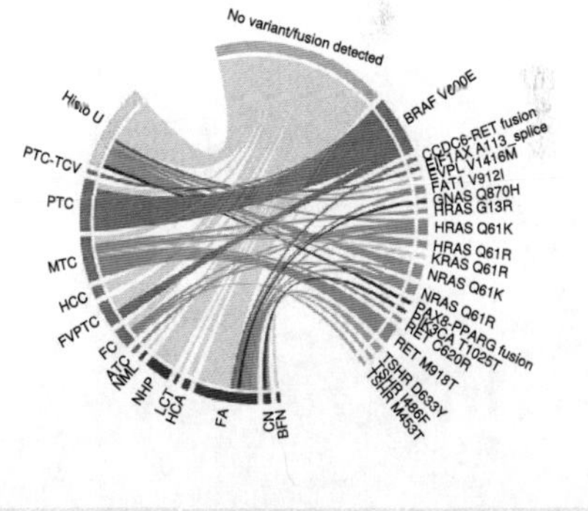

Genetic Alteration	Histology Malignant (n=32)	Histology Benign (n=31)
Positive	19	8
Negative	13	23

Sensitivity, 60% (41-70), Specificity, 74% (55-88); PPV, 70% (55-82); NPV, 64% (53-74), prevalence of malignant lesions, 51%.
Estimated performance when prevalence equals 24%: PPV, 42% (27-58); NPV, 85% (78-90).

Fig. 2 Performance across tissue data set ($n = 63$) showing genetic alterations observed per subtype. *Abbreviations*: *ATC* anaplastic thyroid carcinoma, *BFN* benign follicular nodule, *CN* colloid nodule, *FA* follicular adenoma, *FC* follicular carcinoma, *FVPTC* follicular variant of papillary carcinoma, *HCA* Hurthle cell adenoma, *HCC* Hurthle cell carcinoma, *Histo U* histology uncertain, *LCT* lymphocytic thyroiditis, *MTC* medullary thyroid carcinoma, *NHP* nodular hyperplasia, *NML* normal thyroid, *PTC* papillary thyroid carcinoma, *PTC-TCV* papillary thyroid carcinoma-tall cell variant, *SFM* suspicious for malignancy, *WDC-NOS* well-differentiated carcinoma-not otherwise specified

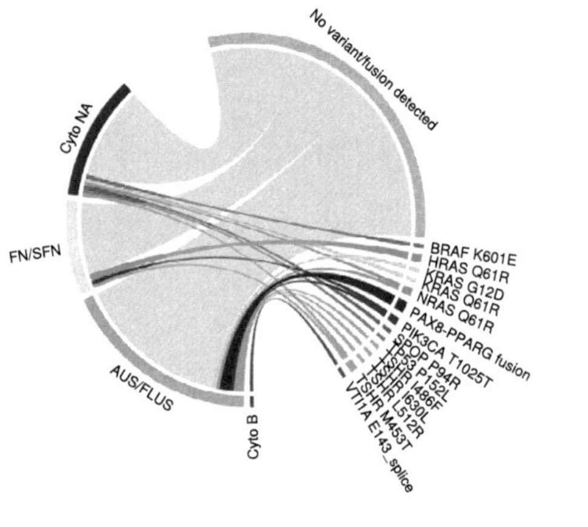

Genetic Alteration	Afirma Suspicious (n=50)	Afirma Benign (n=60)
Positive	15	5
Negative	35	55

PPA 30% (18-45); NPA, 92% (82-97).

Fig. 3 Variants detected in CLIA FNA data set (*n* = 110) showing genetic alterations observed per subtype. *Abbreviations*: *AUS/FLUS* atypia of undetermined significance/follicular lesion of undetermined significance, *Cyto B* cytology benign, *Cyto NA* cytology category not available, *FN/SFN* follicular neoplasm or suspicious for neoplasm, *NPA* negative percent agreement, *PPA* positive percent agreement

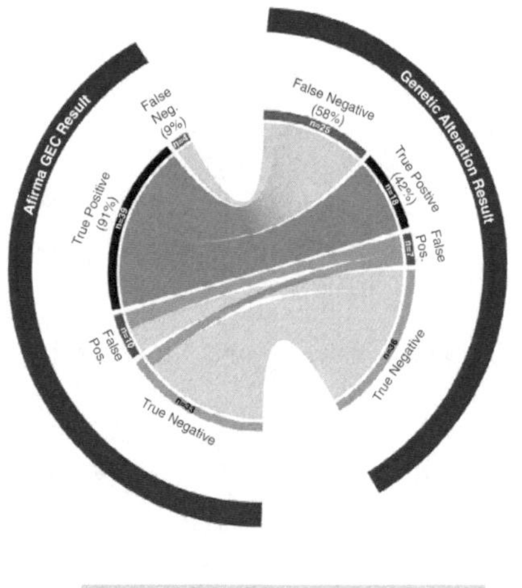

Genetic Alteration Test Results	GEC Test Results: True Neg.	False Pos.	True Pos.	False Neg.	Genetic Alteration Totals
True Neg.	30	6			36
False Pos.	3	4			7
True Pos.			18		18
False Neg.			21	4	25
GEC Totals	**33**	**10**	**39**	**4**	**86**

Fig. 4 Comparison of genetic alteration test results versus an RNA gene expression classifier (GEC) in FNA samples tested with both methods (*n* = 86). *Abbreviations*: *False Pos* false positive, *False Neg* false negative, *True Pos* true positive, *True Neg* true negative

detection of known genetic alterations alone has sufficient power at this time to safely rule out malignancy in all thyroid subtypes. Additional studies in larger FNA cohorts are clearly needed to evaluate further the utility of mutational testing in ruling out malignancy.

This analysis of a broad cohort of expertly diagnosed thyroid neoplasms establishes for the first time that recent TCGA-reported oncogenic driver alterations are commonly found in nodules with benign histopathology. While we recognize that our sample cohorts are of modest size and that more accurate mutation-detection methods may be developed and applied in the future, the overall mutation frequencies observed in histopathology benign FNA (16 %) and benign tissues (26 %) suggest that the use of panels with increasingly larger numbers of discovered variants may not help rule out malignancy in thyroid nodules. Without proper curation, such panels may increase the frequency of variant detection in benign nodules and generate false positive results. Many markers detected in the current study have been reported to vary widely in their individual specificity to detect malignancy [29, 39, 40]. Unfortunately, variants with high specificity such as BRAF V600E or RET/PTC fusions are the exception rather than the rule. Somatic genetic alterations such as those found in NRAS, HRAS, and PAX/PPARG are known to occur in benign nodules [35, 41–43]. The long-term significance of these mutations in benign nodules has not been fully studied [44], and some have suggested that these represent pre-malignant lesions [29, 35, 36, 45, 46]. Despite this suggestion, no clinical evidence exists to demonstrate that today's mutation-positive, but histopathology-benign tumors have a clinically meaningful rate of developing into cancer, or a higher rate of malignant transformation than mutation-negative benign lesions [44]. Thus, no evidence exists that patients benefit from the identification and treatment of mutation-positive benign nodules. In contrast, there may be a role in identifying variants in GEC Suspicious nodules, as these genomic alterations may alter the extent of treatment when selected mutations are found.

Because many gene variants are also detected in our histologically benign nodules, an approach whereby any mutation-positive nodule is designated as "test-positive" has the risk of artificially and erroneously raising the frequency of suspected thyroid carcinoma, thus negatively impacting patients who could have benefited from a watch-and-wait approach. An additional concern is that a number of studies in the literature use mutation status to assign a diagnosis, bypassing blinded expert review to confirm that a mutation positive nodule was indeed histopathology malignant [36, 47]. This chicken-egg conundrum can result in artificially high reported sensitivity and specificity, as truly benign nodules are declared malignant based on mutation status [29, 36, 47]. A recent report indicates that a diagnosis changes from benign to malignant in 30 % of RAS mutated cases when the pathologist is aware that a mutation is present [47].

Our approach measures gene variants found only in expressed genes, and therefore has the advantage of capturing biological processes such as allele-specific gene expression and imprinting. For example, variants found in DNA may have little relevance to a disease if the gene harboring the variant is not expressed in the tissue being sampled. This could be due to general lack of expression in a particular tissue, or could be due to a parental allele-of-origin effect. Our finding that gene transcripts harboring these variants shows expression patterns that differ amongst thyroid subtypes is consistent with the power of this approach to capture potential biological processes (Additional file 2). Thus in contrast to measuring variants in DNA, our RNA transcriptome approach allows the detection of variants and fusions that have the potential to be biologically relevant and clinically meaningful.

As some genomic variants are modestly or highly associated with thyroid malignancy, there may be value to their detection when their presence would increase the extent of initial surgery from a hemi-thyroidectomy to a total thyroidectomy to avoid the need for a second completion thyroidectomy surgery amongst those with cancer. Alternatively, in seven studies evaluating a total of 730 patients whose thyroid nodules were tested with the highly sensitive RNA-based GEC, 49 % were scored as benign [21–25, 34]. These patients are considered for clinical follow-up in lieu of diagnostic surgery with high accuracy [26], a data-driven approach that spares unnecessary morbidity. The implementation of a genomic panel in patients destined for surgery may reduce the need for this second (completion) surgery [48], especially when positive for highly specific variants.

Conclusions

Our studies in this cohort show that evaluating thyroid nodules for genetic variants increases the risk of false positives without rescuing any GEC false negatives. Our data suggest that even when testing with a large variant panel alone, a negative mutation result does not reasonably exclude cancer. However, our data also indicates that when malignancy cannot be ruled out with variant panels, a GEC benign result can first be implemented to safely exclude patients without malignancy and patients with suspicious GEC results may then benefit from evaluation with a genomic variant panel, if this new information results in a change to clinical management. A majority of nodules with indeterminate cytology prove benign on post-operative histopathology. Hence, tests that use genomic profiling to rule-out cancer with high sensitivity have demonstrated clinical utility in the management of cytology indeterminate nodules [19–26].

Given the extremely low frequency of many of the variants and fusions tested here, accurate estimates of test performance will require the testing of many more patient samples and cohorts. Hopefully, carefully conducted multi-center and blinded studies, with expert pathology annotation, will be forthcoming in the near future to shed light on the mutational spectrum of both benign and malignant thyroid nodules.

Additional files

Additional file 1: Sequencing quality metrics for 851 variants. (PDF 49 kb)

Additional file 2: Read-depths obtained per variant and per subtype. (PDF 393 kb)

Additional file 3: Variants included in the panel (*n* = 851). (PDF 596 kb)

Additional file 4: Contingency table combining FNA and tissue results. (PDF 44 kb)

Additional file 5: Variants detected by GATK and SamTools per sample. (PDF 231 kb)

Additional file 6: Estimated mutation rates as a function of cohort size. (JPG 96 kb)

Additional file 7: Fusions tested (n=133). (PDF 51 kb)

Abbreviations

ATA: American Thyroid Association; ATC: Anaplastic thyroid carcinoma; AUS/FLUS: Atypia of Undetermined Significance/Follicular Lesion of Undetermined Significance; BFN: Benign follicular nodule; CHTN: Cooperative Human Tissue Network; CLIA: Clinical Laboratory Improvement Amendments; CN: Colloid nodule; COSMIC: Catalogue of Somatic Mutations in Cancer; FA: Follicular adenoma; FC: Follicular carcinoma; FN: Follicular neoplasm; FN/SFN: Follicular neoplasm or suspicious for neoplasm; FNA: Fine needle aspirate; FVPTC: Follicular variant of papillary carcinoma; GEC: Gene expression classifier; HCA: Hurthle cell adenoma; HCC: Hurthle cell carcinoma; HTA: Hyalinizing trabecular adenoma; LCT: Lymphocytic thyroiditis; mPTC: Micro papillary thyroid carcinoma; MTC: Medullary thyroid carcinoma; NHP: Nodular hyperplasia; NML: Normal thyroid tissue; PTC: Papillary thyroid carcinoma; PTC-TCV: Papillary thyroid carcinoma-tall cell variant; SFM: Suspicious for malignancy; TCGA: The Cancer Genome Atlas; WDC-NOS: Well-differentiated carcinoma-not otherwise specified.

Competing interests

All authors are employed by Veracyte, Inc.

Authors' contributions

MP, GCK, PSW, RTK, RJM, and JH conceived of the study, participated in its design and helped draft the manuscript. KJT, YET, MGW carried out the molecular sequencing studies. CFH, HM, SYK, and ACS participated in sequence alignment and statistical analysis. All authors read and approved the final manuscript.

Acknowledgements

This study and the publication costs for this article were funded by Veracyte, Inc.

Declarations

This article has been published as part of *BMC Bioinformatics* Volume 17 Supplement 1, 2016: Selected articles from the Fourteenth Asia Pacific Bioinformatics Conference (APBC 2016). The full contents of the supplements are available online at http://www.biomedcentral.com/bmcbioinformatics/supplements/17/S1.

Published: 11 January 2016

References

1. Forbes SA, Tang G, Bindal N, Bamford S, Dawson E, Cole C, et al. COSMIC (the Catalogue of Somatic Mutations in Cancer): a resource to investigate acquired mutations in human cancer. Nucleic Acids Res. 2010;38(Database issue):D652–7. doi:10.1093/nar/gkp995.
2. Rehm HL, Berg JS, Brooks LD, Bustamante CD, Evans JP, Landrum MJ, et al. ClinGen–the clinical genome resource. N Engl J Med. 2015;372(23):2235–42. doi:10.1056/NEJMsr1406261.
3. The Cancer Genome Atlas Research Network. https://tcga-data.nci.nih.gov/tcga/tcgaHome2.jsp.
4. Ferris RL, Baloch Z, Bernet V, Chen A, Fahey 3rd TJ, Ganly I, et al. American thyroid association statement on surgical application of molecular profiling for thyroid nodules: current impact on perioperative decision making. Thyroid. 2015;25(7):760–8. doi:10.1089/thy.2014.0502.
5. Cancer Genome Atlas Research N. Integrated genomic characterization of papillary thyroid carcinoma. Cell. 2014;159(3):676–90. doi:10.1016/j.cell.2014.09.050.
6. Yoshihara K, Wang Q, Torres-Garcia W, Zheng S, Vegesna R, Kim H, et al. The landscape and therapeutic relevance of cancer-associated transcript fusions. Oncogene. 2014. doi:10.1038/onc.2014.406.
7. Liu D, Yang C, Bojdani E, Murugan AK, Xing M. Identification of RASAL1 as a major tumor suppressor gene in thyroid cancer. J Natl Cancer Inst. 2013; 105(21):1617–27. doi:10.1093/jnci/djt249.
8. Landa I, Ganly I, Chan TA, Mitsutake N, Matsuse M, Ibrahimpasic T, et al. Frequent somatic TERT promoter mutations in thyroid cancer: higher prevalence in advanced forms of the disease. J Clin Endocrinol Metab. 2013; 98(9):E1562–6. doi:10.1210/jc.2013-2383.
9. Giordano TJ, Beaudenon-Huibregtse S, Shinde R, Langfield L, Vinco M, Laosinchai-Wolf W, et al. Molecular testing for oncogenic gene mutations in thyroid lesions: a case-control validation study in 413 postsurgical specimens. Hum Pathol. 2014;45(7):1339–47. doi:10.1016/j.humpath.2014.03.010.
10. Kunstman JW, Juhlin CC, Goh G, Brown TC, Stenman A, Healy JM, et al. Characterization of the mutational landscape of anaplastic thyroid cancer via whole-exome sequencing. Hum Mol Genet. 2015;24(8):2318–29. doi:10.1093/hmg/ddu749.
11. Ganly I, Ricarte Filho J, Eng S, Ghossein R, Morris LG, Liang Y, et al. Genomic dissection of Hurthle cell carcinoma reveals a unique class of thyroid malignancy. J Clin Endocrinol Metab. 2013;98(5):E962–72. doi:10.1210/jc.2012-3539.
12. Marques AR, Espadinha C, Catarino AL, Moniz S, Pereira T, Sobrinho LG, et al. Expression of PAX8-PPAR gamma 1 rearrangements in both follicular thyroid carcinomas and adenomas. J Clin Endocrinol Metab. 2002;87(8): 3947–52. doi:10.1210/jcem.87.8.8756.
13. Bongiovanni M, Spitale A, Faquin WC, Mazzucchelli L, Baloch ZW. The Bethesda System for reporting thyroid cytopathology: a meta-analysis. Acta Cytol. 2012;56(4):333–9. doi:10.1159/000339959.
14. Cibas ES, Ali SZ. The Bethesda System for reporting thyroid cytopathology. Am J Clin Pathol. 2009;132(5):658–65. doi:10.1309/ajcpphlwmi3jv4la.
15. Cooper DS, Doherty GM, Haugen BR, Kloos RT, Lee SL, Mandel SJ, et al. Revised American Thyroid Association management guidelines for patients with thyroid nodules and differentiated thyroid cancer. Thyroid. 2009;19(11): 1167–214. doi:10.1089/thy.2009.0110.
16. Yassa L, Cibas ES, Benson CB, Frates MC, Doubilet PM, Gawande AA, et al. Long-term assessment of a multidisciplinary approach to thyroid nodule diagnostic evaluation. Cancer. 2007;111(6):508–16.
17. Baloch ZW, LiVolsi VA, Asa SL, Rosai J, Merino MJ, Randolph G, et al. Diagnostic terminology and morphologic criteria for cytologic diagnosis of thyroid lesions: a synopsis of the National Cancer Institute Thyroid Fine-Needle Aspiration State of the Science Conference. Diagn Cytopathol. 2008; 36(6):425–37. doi:10.1002/dc.20830.
18. Xing M, Haugen BR, Schlumberger M. Progress in molecular-based management of differentiated thyroid cancer. Lancet. 2013;381(9871): 1058–69. doi:10.1016/S0140-6736(13)60109-9.
19. Alexander EK, Kennedy GC, Baloch ZW, Cibas ES, Chudova D, Diggans J, et al. Preoperative diagnosis of benign thyroid nodules with indeterminate cytology. N Engl J Med. 2012;387(8):705–15. doi:10.1056/NEJMoa1203208.
20. Duick DS, Klopper J, Diggans JC, Friedman L, Kennedy GC, Lanman RB, et al. The impact of benign gene expression classifier test results on the endocrinologist-patient decision to operate on patients with thyroid nodules with indeterminate FNA cytopathology. Thyroid. 2012;22(10): 996–1001. doi:10.1089/thy.2012-0180.
21. Lastra RR, Pramick MR, Crammer CJ, LiVolsi VA, Baloch ZW. Implications of a suspicious afirma test result in thyroid fine-needle aspiration cytology: an institutional experience. Cancer Cytopathol. 2014;122(10):737–44. doi:10.1002/cncy.21455.
22. McIver B, Castro MR, Morris JC, Bernet V, Smallridge R, Henry M, et al. An independent study of a gene expression classifier (Afirma) in the evaluation of cytologically indeterminate thyroid nodules. J Clin Endocrinol Metab. 2014;99(11):4069–77. doi:10.1210/jc.2013-3584.
23. Harrell RM, Bimston DN. Surgical utility of Afirma: effects of high cancer prevalence and oncocytic cell types in patients with indeterminate thyroid cytology. Endocr Pract. 2014;20(4):364–9. doi:10.4158/EP13330.OR.
24. Marti JL, Avadhani V, Donatelli LA, Niyogi S, Wang B, Wong RJ, et al. Wide Inter-institutional variation in performance of a molecular classifier for indeterminate thyroid nodules. Ann Surg Oncol. 2015. doi:10.1245/s10434-015-4486-3.
25. Sullivan PS, Hirschowitz SL, Fung PC, Apple SK. The impact of atypia/follicular lesion of undetermined significance and repeat fine-needle aspiration: 5 years before and after implementation of the Bethesda System. Cancer Cytopathol. 2014;122(12):866–72. doi:10.1002/cncy.21468.
26. Witt RL. Outcome of thyroid gene expression classifier testing in clinical practice. Laryngoscope. 2015. doi:10.1002/lary.25607.
27. Angell TE, Frates MC, Medici M, Liu X, Kwong N, Cibas ES, et al. Afirma benign thyroid nodules show similar growth to cytologically benign nodules during follow up. J Clin Endocrinol Metab. 2015;jc20152658. doi:10.1210/jc.2015-2658.
28. Nikiforov YE, Carty SE, Chiosea SI, Coyne C, Duvvuri U, Ferris RL, et al. Highly accurate diagnosis of cancer in thyroid nodules with follicular neoplasm/suspicious for a follicular neoplasm cytology by ThyroSeq v2 next-generation sequencing assay. Cancer. 2014;120(23):3627–34. doi:10.1002/cncr.29038.
29. Beaudenon-Huibregtse S, Alexander EK, Guttler RB, Hershman JM, Babu V, Blevins TC, et al. Centralized molecular testing for oncogenic gene mutations complements the local cytopathologic diagnosis of thyroid nodules. Thyroid. 2014;24(10):1479–87. doi:10.1089/thy.2013.0640.
30. Dobin A, Davis CA, Schlesinger F, Drenkow J, Zaleski C, Jha S, et al. STAR: ultrafast universal RNA-seq aligner. Bioinformatics (Oxford, England). 2013; 29(1):15–21. doi:10.1093/bioinformatics/bts635.
31. McKenna A, Hanna M, Banks E, Sivachenko A, Cibulskis K, Kernytsky A, et al. The Genome Analysis Toolkit: a MapReduce framework for analyzing next-generation DNA sequencing data. Genome Res. 2010;20(9):1297–303. doi:10.1101/gr.107524.110.
32. Beccuti M, Carrara M, Cordero F, Lazzarato F, Donatelli S, Nadalin F, et al. Chimera: a bioconductor package for secondary analysis of fusion products. Bioinformatics (Oxford, England). 2014;30(24):3556–7. doi:10.1093/bioinformatics/btu662.
33. Mercaldo ND, Lau KF, Zhou XH. Confidence intervals for predictive values with an emphasis to case-control studies. Stat Med. 2007;26(10):2170–83. doi:10.1002/sim.2677.
34. Alexander EK, Schorr M, Klopper J, Kim C, Sipos J, Nabhan F, et al. Multicenter clinical experience with the Afirma gene expression classifier. J Clin Endocrinol Metab. 2014;99(1):119–25. doi:10.1210/jc.2013-2482.
35. Nikiforov YE, Ohori NP, Hodak SP, Carty SE, LeBeau SO, Ferris RL, et al. Impact of mutational testing on the diagnosis and management of patients with cytologically indeterminate thyroid nodules: a prospective analysis of 1056 FNA samples. J Clin Endocrinol Metab. 2011;96(11):3390–7. doi:10.1210/jc.2011-1469.
36. Eszlinger M, Krogdahl A, Munz S, Rehfeld C, Precht Jensen EM, Ferraz C, et al. Impact of molecular screening for point mutations and rearrangements in routine air-dried fine-needle aspiration samples of thyroid nodules. Thyroid. 2014;24(2):305–13. doi:10.1089/thy.2013.0278.
37. Nikiforov YE, Carty SE, Chiosea SI, Coyne C, Duvvuri U, Ferris RL, et al. Impact of the Multi-Gene ThyroSeq next-generation sequencing assay on cancer diagnosis in thyroid nodules with atypia of undetermined significance/follicular lesion of undetermined significance cytology. Thyroid. 2015. doi:10.1089/thy.2015.0305.
38. Labourier E, Shifrin A, Busseniers AE, Lupo MA, Manganelli ML, Andruss B, et al. Molecular testing for miRNA, mRNA, and DNA on fine-needle aspiration improves the preoperative diagnosis of thyroid nodules with indeterminate cytology. J Clin Endocrinol Metab. 2015;100(7):2743–50. doi:10.1210/jc.2015-1158.

39. Krane JF, Cibas ES, Alexander EK, Paschke R, Eszlinger M. Molecular analysis of residual ThinPrep material from thyroid FNAs increases diagnostic sensitivity. Cancer Cytopathol. 2015;123(6):356–61. doi:10.1002/cncy.21546.
40. Nikiforov YE, Nikiforova MN, Gnepp DR, Fagin JA. Prevalence of mutations of ras and p53 in benign and malignant thyroid tumors from children exposed to radiation after the Chernobyl nuclear accident. Oncogene. 1996;13(4):687–93.
41. Esapa CT, Johnson SJ, Kendall-Taylor P, Lennard TW, Harris PE. Prevalence of Ras mutations in thyroid neoplasia. Clin Endocrinol (Oxf). 1999;50(4):529–35.
42. Raman P, Koenig RJ. Pax-8-PPAR-gamma fusion protein in thyroid carcinoma. Nat Rev Endocrinol. 2014;10(10):616–23. doi:10.1038/nrendo.2014.115.
43. Gupta N, Dasyam AK, Carty SE, Nikiforova MN, Ohori NP, Armstrong M, et al. RAS mutations in thyroid FNA specimens are highly predictive of predominantly low-risk follicular-pattern cancers. J Clin Endocrinol Metab. 2013;98(5):E914–22. doi:10.1210/jc.2012-3396.
44. Medici M, Kwong N, Angell TE, Marqusee E, Kim MI, Frates MC, et al. The variable phenotype and low-risk nature of RAS-positive thyroid nodules. BMC Med. 2015;13(1):184. doi:10.1186/s12916-015-0419-z.
45. Nikiforova MN, Nikiforov YE. Molecular genetics of thyroid cancer: implications for diagnosis, treatment and prognosis. Expert Rev Mol Diagn. 2008;8(1):83–95. doi:10.1586/14737159.8.1.83.
46. Nikiforova MN, Nikiforov YE. Molecular diagnostics and predictors in thyroid cancer. Thyroid. 2009;19(12):1351–61. doi:10.1089/thy.2009.0240.
47. Eszlinger M, Piana S, Moll A, Bosenberg E, Bisagni A, Ciarrocchi A, et al. Molecular testing of thyroid fine-needle aspirations improves presurgical diagnosis and supports the histologic identification of minimally invasive follicular thyroid carcinomas. Thyroid. 2015;25(4):401–9. doi:10.1089/thy.2014.0362.
48. Yip L, Wharry LI, Armstrong MJ, Silbermann A, McCoy KL, Stang MT, et al. A clinical algorithm for fine-needle aspiration molecular testing effectively guides the appropriate extent of initial thyroidectomy. Ann Surg. 2014; 260(1):163–8. doi:10.1097/SLA.0000000000000215.

Wang *et al. BMC Bioinformatics* 2015, **17**(Suppl 1):7
DOI 10.1186/s12859-015-0850-3

BMC Bioinformatics

PROCEEDINGS

Open Access

A full Bayesian partition model for identifying hypo- and hyper-methylated loci from single nucleotide resolution sequencing data

Henan Wang[1], Chong He[1], Garima Kushwaha[2], Dong Xu[2] and Jing Qiu[3*]

From The Fourteenth Asia Pacific Bioinformatics Conference (APBC 2016)
San Francisco, CA, USA. 11 - 13 January 2016

Abstract

Backgroud: DNA methylation is an epigenetic modification that plays important roles on gene regulation. Study of whole-genome bisulfite sequencing and reduced representation bisulfite sequencing brings the availability of DNA methylation at single CpG resolution. The main interest of study on DNA methylation data is to test the methylation difference under two conditions of biological samples. However, the high cost and complexity of this sequencing experiment limits the number of biological replicates, which brings challenges to the development of statistical methods.

Results: Bayesian modeling is well known to be able to borrow strength across the genome, and hence is a powerful tool for high-dimensional- low-sample- size data. In order to provide accurate identification of methylation loci, especially for low coverage data, we propose a full Bayesian partition model to detect differentially methylated loci under two conditions of scientific study. Since hypo-methylation and hyper-methylation have distinct biological implication, it is desirable to differentiate these two types of differential methylation. The advantage of our Bayesian model is that it can produce one-step output of each locus being either equal-, hypo- or hyper-methylated locus without further post-hoc analysis. An R package named as MethyBayes implementing the proposed full Bayesian partition model will be submitted to the bioconductor website upon publication of the manuscript.

Conclusions: The proposed full Bayesian partition model outperforms existing methods in terms of power while maintaining a low false discovery rate based on simulation studies and real data analysis including bioinformatics analysis.

Keywords: DNA methylation, Full Bayesian partition model, Hypo-methylation, Hyper-methylation

Background

DNA methylation is methylation of cytosine residues at CpG dinucleotides in a DNA sequence and affects 70–80 % of all CpG dinucleotides in mammals [1]. It is the most widely studied epigenetic modification and is known to have profound effects on gene expression. It is involved in embryogensis, genomic imprinting [2], X-chromosome inactivation [3], and many diseases [4], particularly various types of cancers [5]. DNA methylation change in cancer tissue compared to normal tissue can be both increased, called hyper-methylation, or decreased, called hypo-methylation. DNA hyper-methylation is shown to be present at more specific locations, like CpG Islands, compared to the more diffused hypo-methylation all over the genome. CpG island and their methylation are not only present in promoters and upstream regions of genes, but also within gene bodies known as gene-body methylation.

Given the influence of methylation on the gene expression, there are a lot of studies aiming to identify differentially methylated loci in diseased tissue samples compared

*Correspondence: qiujing@udel.edu
[3]Department of Applied Economics and Statistics, University of Delaware, Newark, DE, USA
Full list of author information is available at the end of the article

to their respective normal samples. Among the methods developed to quantify the (relative) levels of CpG methylation in the whole genome, bisulfite sequencing is a common technique and has its advantages. This method involves treating DNA with sodium bisulfite [6], which converts un-methylated cytosines to uracil and leaves methylated cytosines unchanged. Treated DNA is then used to generate high-throughtput readouts by DNA sequencing technique. It can provide methylation level at a single nucleotide resolution.

The availability of this bisulfite sequencing (BS-seq), together with the influence of the DNA methylation on human disease, has led to extensive studies in detecting differentially methylated loci (DML) based on the case and control study. Several statistical methods have been applied to test DML. Fisher exact test [7] is a commonly used statistical approach for testing DML by pooling sequencing reads among the individuals in each condition. BSmooth [8] considers the variation among biological replicates, and uses a signal-to-noise statistics similar to the t statistics to discern differential methylation region via a smoothing approach across genome for each individual under two conditions. When considering the binary feature of each locus being methylated or not, Methylkit [9] utilizes a logistic regression model where a condition effect is incorporated to identify DML between the normal and cancer condition. Since logistic regression can be sensitive to small sample size, a filtering of the data is recommended before analysis so that only loci with at least 10 reads coverage for each sample are included for the analysis which will guarantee the overall sample size, the summation of read coverage over all samples in each condition, is large enough for the logistic regression to have good power. The DSS package [10] proposes an empirical Bayes Wald test to identify DML for single nucleotide resolution sequencing data. They consider a beta-binomial model to take into account of the biological variation that might exist for the methylation proportions and use an empirical Bayes approach to estimate the associated dispersion parameter. However the null distribution of the derived test statistics is unknown, although a normal distribution is recommended in their paper purely based on simulation studies.

All the above statistical approaches face a common problem that they only produce binary output of a locus being differentially methylated or not. However, for DNA methylation data, it is important to differentiate hypo- and hyper-methylation because they have very different biological implication. For instance, DNA hyper-methylation are usually associated with transcriptional inactivation of cancer-related genes by increased methylation in CpG island (regions with high CpG density) in their promoter region [11–13]. In contrast, DNA hypo-methylation is shown to be present within repeated DNA elements [14] and is linked to chromosomal instability, loss of imprinting and oncogene activation (eg. c-Myc). A common practice to further identify hypo- or hyper-methylated loci is based on the sign of the test statistics, which is an ad-hoc approach and ignores the uncertainty associated with the test statistics. In this article, we propose a full Bayesian partition model to identify hypo- and hyper-methylated loci simultaneously without further post-hoc analysis. In the proposed method, we introduce a latent variable representing the methylation group membership at each locus and the statistical inference is based on the posterior distribution of this latent variable with the final outcome of the analysis being whether a locus is an equal-, hypo- or hyper-methylated locus. In addition to producing one-step outputs, the Bayesian approach is also well-known to be able to borrow strength across the genome, and hence can be more powerful for small sample size study, which is fairly common for the DNA methylation studies due to the high cost of BS-seq experiments [10].

The rest of this article is organized as follows. In the Methods section, we introduce the proposed Bayesian model and implement a Metropolis-Hastings algorithm to obtain the posterior samples of genome-wide group membership for posterior inference. The section of simulation studies and the section of real data analysis present simulation studies and real data analysis including bioinformatics analysis respectively to evaluate the performance of the proposed Bayesian approach. The advantage of the proposed approach is shown by comparing with several competing methods. The discussion of the proposed method and future work is presented in the discussion section.

Methods

Model

We propose a full Bayesian partition model to identify DML based on a case-control study. Suppose there are a total of L CpG loci, with a number of n_1 samples in the case study and n_2 samples in the control study. Let C_{ijk} and M_{ijk} denote the read coverage and the number of methylated reads respectively at the ith locus ($i = 1, 2, \ldots, L$) of the jth ($j = 1, 2, \ldots, n_k$) replicate of the kth condition (k=1 for case, k=2 for control). Let p_{ik} be the true methylation proportion for the ith locus of the kth condition. For each locus i, since there are two kinds of reads, methylated and non-methylated, we model the distribution of the number of methylated reads by the following binomial distribution: $M_{ijk}|C_{ijk}, p_{ik} \sim \text{Binomial}(C_{ijk}, p_{ik})$.

In this study, since we are interested in identifying both hypo- and hyper-methylated loci, the L CpG loci are partitioned into three groups: group 0 containing equal-methylated loci where $p_{i1} = p_{i2}$; group 1 containing hypo-methylated loci where $p_{i1} < p_{i2}$ and group 2 containing hyper-methylated loci where $p_{i1} > p_{i2}$. We introduce a L-dimensional latent indicator variable $\boldsymbol{I} = (I_1, \ldots, I_L)$ with $I_i = 0, 1, 2$ to indicate the three possible group memberships of different loci. Let l_0, l_1 and l_2 denote the number of loci in each of the three groups ($l_0 + l_1 + l_2 = L$). Let the data matrix of read coverages for group 0 be $\boldsymbol{C}_0 = \left[\boldsymbol{c}_{i_1}, \boldsymbol{c}_{i_2}, \ldots, \boldsymbol{c}_{i_{l_0}}\right]$ where $\boldsymbol{c}_i = \left[C_{i11}, \ldots, C_{in_11}, C_{i12}, \ldots, C_{in_22}\right]'$ representing the vector of read coverages for the ith locus, consisting of observations for both case and control conditons, and $i_1, \ldots, i_{l_0}$ are indices for loci in group 0. Similarly, let $\boldsymbol{C}_1, \boldsymbol{C}_2$ denote the data matrices of read coverage for group 1 and 2 respectively, and $\boldsymbol{M}_0, \boldsymbol{M}_1, \boldsymbol{M}_2$ denote the data matrices of numbers of methylated reads for group 0, 1, and 2 respectively.

Assuming independence among loci, we describe the likelihood model as follows. Let $\boldsymbol{\Theta}_0 = \left\{(p_{i1}, p_{i2}) : I_i = 0\right\}$ be the methylation proportions for loci in group 0. Then the likelihood of $\boldsymbol{M}_0$ can be expressed as

$$p(\boldsymbol{M}_0|\boldsymbol{C}_0, \boldsymbol{\Theta}_0) = \prod_{i:I_i=0} \prod_{k=1}^{2} \prod_{j=1}^{n_k} \binom{C_{ijk}}{M_{ijk}} p_{ik}^{M_{ijk}} (1-p_{ik})^{C_{ijk}-M_{ijk}}. \tag{1}$$

We consider a conjugate beta prior distribution for the true methylation proportions p_{ik}. Note for loci in group 0, $p_{i1} = p_{i2}$. Hence we consider the joint prior density function for (p_{i1}, p_{i2}) in $\boldsymbol{\Theta}_0$ as follows:

$$p\left(p_{i1}, p_{i2}|\alpha_1, \beta_1, I_i = 0\right) = f\left(p_{i1}; \alpha_1, \beta_1\right) 1_{(p_{i1}=p_{i2})} \tag{2}$$

where $1_{(.)}$ is an indicator function and $f(p; \alpha, \beta) = \frac{1}{B(\alpha,\beta)} p^{\alpha-1}(1-p)^{\beta-1}$ with $B(\alpha, \beta) = \frac{\Gamma(\alpha)\Gamma(\beta)}{\Gamma(\alpha+\beta)}$ and $\Gamma(\alpha) = \int_0^\infty x^{\alpha-1} e^{-x} dx$. After integrating out $\boldsymbol{\Theta}_0$, we obtain the marginal probability

$$p(\boldsymbol{M}_0|\boldsymbol{C}_0, \boldsymbol{I}) = \prod_{\{i:I_i=0\}} \left[\prod_{k=1}^{2} \prod_{j=1}^{n_k} \frac{C_{ijk}!}{M_{ijk}! N_{ijk}!}\right] \frac{B\left(\alpha_1 + \sum_{k=1}^{2} \sum_{j=1}^{n_k} M_{ijk}, \beta_1 + \sum_{k=1}^{2} \sum_{j=1}^{n_k} N_{ijk}\right)}{B(\alpha_1, \beta_1)} \tag{3}$$

where $N_{ijk} = C_{ijk} - M_{ijk}$.

Loci in group 1 are hypo-methylated with $p_{i1} < p_{i2}$. Therefore, we consider a joint truncated beta distribution by adjusting the prior distribution proposed by [15] for two ordered means in the setting of microarray data analysis. Let $\boldsymbol{\Theta}_1 = \{(p_{i1}, p_{i2}) : I_i = 1\}$. Then the joint prior density function for (p_{i1}, p_{i2}) in $\boldsymbol{\Theta}_1$ is

$$p(p_{i1}, p_{i2}|\alpha_2, \beta_2, I_i = 1) = 2f(p_{i1}; \alpha_2, \beta_2) f(p_{i2}; \alpha_2, \beta_2) 1_{(p_{i1}<p_{i2})}. \tag{4}$$

Note that the likelihood function of $\boldsymbol{M}_1$ given $\boldsymbol{\Theta}_1$ and $\boldsymbol{C}_1$ is similar to Eq. (1). After integrating out $\boldsymbol{\Theta}_1$ using the prior density (4), we obtain the marginal probability

$$p(\boldsymbol{M}_1|\boldsymbol{C}_1, \boldsymbol{I}) = \prod_{\{i:I_i=1\}} \left\{\prod_{k=1}^{2} \left[\prod_{j=1}^{n_k} \frac{C_{ijk}!}{N_{ijk}! M_{ijk}!}\right] \frac{B\left(\sum_{j=1}^{n_k} M_{ijk} + \alpha_2, \sum_{j=1}^{n_k} N_{ijk} + \beta_2\right)}{B(\alpha_2, \beta_2)}\right\} \times 2P(X_i < Y_i) \tag{5}$$

where $X_i \sim Beta\left(\alpha_2 + \sum_{j=1}^{n_1} M_{ij1}, \beta_2 + \sum_{j=1}^{n_1} N_{ij1}\right)$ and is independent of $Y_i \sim Beta\left(\alpha_2 + \sum_{j=1}^{n_2} M_{ij2}, \beta_2 + \sum_{j=1}^{n_2} N_{ij2}\right)$.

For loci in group 2, the direction of the methylation proportion comparison between the case and control conditions is reversed. Let $\boldsymbol{\Theta}_2 = \{(p_{i1}, p_{i2}) : I_i = 2\}$. Then the joint prior density function for (p_{i1}, p_{i2}) in $\boldsymbol{\Theta}_2$ is

$$p(p_{i1}, p_{i2}|\alpha_3, \beta_3, I_i = 2) = 2f(p_{i1}; \alpha_3, \beta_3) f(p_{i2}; \alpha_3, \beta_3) 1_{(p_{i1}>p_{i2})}. \tag{6}$$

By integrating out $\boldsymbol{\Theta}_2$, we can obtain the marginal probability

$$p(\boldsymbol{M}_2|\boldsymbol{C}_2, \boldsymbol{I}) = \prod_{\{i:I_i=2\}} \left\{\prod_{k=1}^{2} \left[\prod_{j=1}^{n_k} \frac{C_{ijk}!}{N_{ijk}! M_{ijk}!}\right] \frac{B\left(\sum_{j=1}^{n_k} M_{ijk} + \alpha_3, \sum_{j=1}^{n_k} N_{ijk} + \beta_3\right)}{B(\alpha_3, \beta_3)}\right\} \times 2P(X_i > Y_i) \tag{7}$$

where $X_i \sim Beta\left(\alpha_3 + \sum_{j=1}^{n_1} M_{ij1}, \beta_3 + \sum_{j=1}^{n_1} N_{ij1}\right)$ and is independent of $Y_i \sim Beta\left(\alpha_3 + \sum_{j=1}^{n_2} M_{ij2}, \beta_3 + \sum_{j=1}^{n_2} N_{ij2}\right)$.

To make inference about the membership of each locus along the whole genome, the posterior distribution of $\boldsymbol{I}$ can be obtained by

$$p(\boldsymbol{I}|\boldsymbol{C}, \boldsymbol{M}) \propto p(\boldsymbol{M}_0|\boldsymbol{C}_0, \boldsymbol{I}) p(\boldsymbol{M}_1|\boldsymbol{C}_1, \boldsymbol{I}) p(\boldsymbol{M}_2|\boldsymbol{C}_2, \boldsymbol{I}) p(\boldsymbol{I}). \tag{8}$$

(See Additional file 1 for the derivation). Since I_i for $i = 1, \ldots, L$ has three possible entries, the prior distribution of I_i is modeled by multinomial $(1; \pi_0, \pi_1, \pi_2)$ with $\boldsymbol{\pi} = (\pi_0, \pi_1, \pi_2)$ to be the vector of probabilities of belonging to the three groups. Then L-vector $\boldsymbol{I}$ has density $p(\boldsymbol{I}|\boldsymbol{\pi}) = \pi_0^{l_0} \pi_1^{l_1} \pi_2^{l_2}$, where $l_j = \sum_{i=1}^{L} 1_{\{I_i=j\}}$ with $j = 0, 1, 2$. We further set a Dirichlet prior for parameters

$\boldsymbol{\pi}$: $p(\pi_0, \pi_1|k_0, k_1, k_2) = \frac{1}{B(k_0,k_1,k_2)} \prod_{i=0}^{2} \pi_i^{(k_i-1)}$, where $\pi_2 = 1 - \pi_0 - \pi_1$ and obtain that

$$p(\boldsymbol{I}) = \int p(\boldsymbol{I}|\boldsymbol{\pi})p(\boldsymbol{\pi}|k_0, k_1, k_2)d\boldsymbol{\pi} = \frac{B(k_0 + l_0, k_1 + l_1, k_2 + l_2)}{B(k_0, k_1, k_2)}. \quad (9)$$

where $B(k_0, k_1, k_2) = \frac{\Gamma(k_0)\Gamma(k_1)\Gamma(k_2)}{\Gamma(k_0+k_1+k_2)}$. Thus, the posterior distribution of $\boldsymbol{I}$ in (8) can be obtained by combining the formulas (3), (5), (7) and (9).

Model fitting via MCMC

We investigate the posterior distribution of L dimensional classification variable $\boldsymbol{I}$ using Markov chain Monte Carlo (MCMC) techniques [16], and in lack of conjugacy, a Metropolis Hastings sampling algorithm is applied to sample posterior draws of $\boldsymbol{I}$ in (8). Before we implement MCMC procedures, hyperparameter values in priors need to be specified. Beta priors in three groups are chosen to be non-informative, i.e., $\alpha_1, \alpha_2, \alpha_3, \beta_1, \beta_2$ and β_3 are all set to be ones. Dirichlet prior is also chosen to be non-informative with k_0, k_1 and k_2 being ones. We first randomly assign an initial state for each locus and then iteratively generate samples of $\boldsymbol{I}$ from its full conditional posterior distribution by the following steps. We define $\boldsymbol{I}^{old}$ to be current membership vector at the previous MCMC iteration, and $\boldsymbol{I}^{new}$ to be the proposed sample of new membership vector. First, we randomly choose one of the following two proposals for new value of $\boldsymbol{I}$: *(a)* Randomly pick one methylation group and choose a locus from this group, then change its membership value to one of other memberships; or *(b)* Randomly pick two methylation groups and choose a locus from each of these two groups from the $\boldsymbol{I}^{old}$ and exchange their membership values. Second, the proposed value $\boldsymbol{I}^{new}$ will be accepted based on the Metropolis Hastings probability:

$$\min\left\{1, \frac{p\left(\boldsymbol{I}^{new}|\boldsymbol{C},\boldsymbol{M}\right)}{p\left(\boldsymbol{I}^{old}|\boldsymbol{C},\boldsymbol{M}\right)} \times \frac{p\left(\boldsymbol{I}^{new} \rightarrow \boldsymbol{I}^{old}\right)}{p\left(\boldsymbol{I}^{old} \rightarrow \boldsymbol{I}^{new}\right)}\right\} \quad (10)$$

where $p(\boldsymbol{I}|\boldsymbol{C},\boldsymbol{M})$ is the posterior density of $\boldsymbol{I}$ in (8), and $p(\boldsymbol{I}^{old} \rightarrow \boldsymbol{I}^{new})$ is a transition probability from $\boldsymbol{I}^{old}$ to $\boldsymbol{I}^{new}$, i.e., the probability of generating $\boldsymbol{I}^{new}$ from $\boldsymbol{I}^{old}$ based on the above two proposals.

Post-burn-in MCMC samples are used to draw posterior inference, and the classification of each locus into equal-, hypo-, or hyper-methylated group can be made by using marginal posterior distribution. We calculate the empirical frequency of each locus belonging to each group by summarizing the Markov chain Monte Carlo output, and divide this empirical frequency by the total post-burn-in sample size to obtain the estimates of posterior membership probabilities for i=1,. . . , L. The methylation status of each locus is chosen as the membership group with highest posterior probability estimates. We tried different initial values of I and our simulation studies show that the posterior inference is robust to the choice of initial values.

Simulation studies

Simulation settings

We conducted two sets of simulation studies to evaluate the accuracy of the proposed method in identification of methylated loci, and compare the proposed procedure with some existing ones such as logistic regression model [9], z-test comparing two proportions, and empirical Bayes Wald test [10]. Here we didn't compare with the BSmooth method [8] because it was proposed for detecting differentially methylated region, not for DML. As stated in the user's guide of their package, the BSmooth algorithm depends heavily on smoothing and requires the methylation levels to be measured in bigger region of the genome, instead of single loci. Hence it was excluded from our comparison in this paper.

In order to best mimic the structure of real methylation data, we generated data based on a real dataset [17] where a genome-wide DNA methylation analysis for 11 $CD19^+$ B-cells from chronic lymphocytic leukemia (CLL) patients and 3 normal control samples were conducted to identify DML. For all case and control samples, methylation data were generated using reduced representation bisulfite sequencing (RRBS) at single-base resolution. For our simulation study, we selected two case and two control samples out of all samples in their real dataset. We selected only two samples for each condition to represent the common situation of small sample size often associated with DNA methylation data due to the high cost of BS-seq experiments. Also, as it is known that the variation can differ significantly among cancer samples [18], to minimize the sample variation among the samples in the case study so that it is justifiable to ignore the subject effect, we selected two CLL samples with matched percent identity, CD38 percent and IGHV mutation status. We arbitrary choose two control samples out of the three. Total 384,890 loci are commonly mapped by these two case samples and two control samples.

We first identified the three groups (equal, hypo- and hyper- methylated groups) of loci based on the real data using different cutoff values and then generate parameters for simulated data from the empirical distributions of the parameter estimates of each identified group. We calculated the maximum likelihood estimates of methylation proportions ($\hat{p}_{ik} = \sum_j M_{ijk}/\sum_j C_{ijk}$) under each condition (case and control), and computed the difference $\hat{d}_i = \hat{p}_{i1} - \hat{p}_{i2}$. Loci with $|\hat{d}_i| < 0.005$ are classified as equal-methylated group, loci with $\hat{d}_i > \gamma$ are classified as hyper-methylated group and loci with $\hat{d}_i < -\gamma$ are classified as hypo-methylated group, where the positive value γ is a tuning parameter to control the effect sizes for different simulation settings. The empirical distributions of the parameter estimators $\hat{p}_{ik}$ for each of the three identified groups were used to generate true parameters p_{ik} for the simulated data. When γ is small, the empirical distribution of $\hat{d}_i$ for the hypo- and hyper-methylated groups will cover more smaller values and hence the corresponding simulated data will have more smaller differential methylation levels for the hyper- and hypo- methylated groups. On the other hand, when γ is large, the empirical distribution of $\hat{d}_i$ for the hypo and hyper methylated groups will cover more larger values and hence the resultant simulated data have more larger signals. Therefore, we expect the power of various procedures is larger for large γ and smaller for small γ. (Note that γ is a tuning parameter to control the effect sizes for different simulation settings. We set different values of γ for simulation studies to see how our procedure performs in various situations. There is no such value for real data analysis. The only input for our procedure in real data analysis is the real data. There is no tuning parameter that need to be specified by the users).

The value of γ also determines the proportions of the three groups for the simulated data. For instance, when $\gamma = 0.01$, there are 43.28 % hypo-methylated loci and 29.44 % hyper-methylated loci based on the real data. While when $\gamma = 0.2$, there are 5.9 % hypo-methylated loci and 4.6 % hyper-methylated loci. For our simulated data, we generated three groups of loci according to these proportions. We considered different values of γ, $\gamma = 0.01, 0.05, 0.1, 0.15, 0.2$, to study the effects of the signal strength and percentage of signals in various simulation settings. Additional file 1: Table S1 gives the proportion of hypo- and hyper-methylated loci for the simulated data at different values of γ.

In the first set of simulation studies, we assumed p_{ik} does not depend on the replicates. For equal-methylated loci, we used the pooled estimates of the common methylation proportion of the real data as the true proportion: $\sum_{j,k} M_{ijk}/\sum_{j,k} C_{ijk}$. For hypo- (hyper-) methylated loci, the methylation proportion of the real data is estimated by $\sum_{j=1}^{n_1} M_{ij1}/\sum_{j=1}^{n_1} C_{ij1}$ and $\sum_{j=1}^{n_2} M_{ij2}/\sum_{j=1}^{n_2} C_{ij2}$ separately for each of the case and control conditions, and is then used as the true methylation proportion p_{ik} for the simulated data.

Our model assumed that the methylation proportion is the same for biological replicates within the case or control condition. However, in reality, biological variation might exist for the methylation proportions [10]. Therefore, we conducted further simulation studies to test the robustness of our model when such biological variation exists. We refer to the second set of simulation studies as *simulation studies with subject effect* (as a contrast, the first simulation setting is called the *simulation studies without subject effect*). To be specific, we calculated the estimated methylation proportion for each patient of the real data $\hat{p}_{ijk} = M_{ijk}/C_{ijk}$ and used it as the true parameter to generate the methylated counts from the following binomial distribution $M_{ijk} \sim \text{binomial}(C_{ijk}, p_{ijk})$, where C_{ijk} is the observed read coverage of the real data for the selected loci. Even though we allow the methylation proportion to vary across replicates within the case or control condition, the CpG loci can be still classified into equal- or hypo- or hyper-methylated groups based on the observed difference in methylation proportions between the case and control conditions as described in the first simulation setting.

Although our Bayesian model assumes beta prior distributions for the methylation proportions, our simulation studies try to mimic the real data as much as possible without making any parametric distribution for the methylation proportions. As described earlier in this section, the parameter p_{ik} is generated from the empirical distribution of the methylation proportion estimates of the real data. Here we consider the simulation results when there are two replicates per condition since it is quite common for this type of data to have small sample sizes [10]. For each simulation set, we generated a total of 20,000 loci, and the proportion of hypo- (or hyper-) methylated loci in the simulation data is determined by the tuning parameter γ (see Additional file 1: Table S1 for the detailed numbers).

Simulation results

We compared our Bayesian partition model to three existing methods: (1) the logistic regression model of Methylkit package [9]; (2) the standard z-test comparing two proportions; and (3) an empirical Bayes Wald test [10]. Note that a standard practice is to apply a two-sample t-test to the individual methylation proportion estimates M_{ijk}/C_{ijk} [10]. However, with two replicates per condition, it works very poorly. Hence we replaced it with the standard z-test for comparing two proportions which pooled sequencing reads among individuals under each condition. Note that all the three existing methods produce a p-value for each CpG locus and hence need to be adjusted for multiplicity. We applied the q-value approach in [19] to control false discovery rate at a nominal level. Since our Bayesian model makes inference based on the posterior probability and doesn't produce p-values, there is no FDR control. In order to make a fair comparison, we evaluated the actual FDR level of the Bayesian inference in the simulation studies and then used it as the nominal level of the FDR control for the other three methods.

Our Bayesian model produces three types of loci for each simulated data while the other three methods only produce binary results: DML or non-DML. Therefore, for these approaches, the direction of differential methylation is decided based on the signs of the estimates of the methylation proportion difference between the case and control conditions. Since for simulated data, it is known whether a locus is equal-, hypo- or hyper-methylated, we can evaluate different rates of false positives and true positives. We summarized our simulation results averaging over 100 simulation runs using five measures: FDR, mdFDR, TPR, TPR_hypo and TPR_hyper. Here the false discovery rate (FDR) is calculated without considering the misclassification between hypo- and hyper-methylated loci and defined as the proportion of truly non-DML among those identified as differentially methylated, while the mixed directional FDR (mdFDR) [20] considers the misspecification between the hypo- and hyper-methylated loci as false discoveries in addition to the mistakes of classifying non-DML as DML. The true positive rate (TPR) is defined as the proportion of truly DML that are detected by the method without considering the misspecification between the hypo- and hyper-methylated loci, while TPR for hypo-methylated loci (TPR_hypo) is defined as the proportion of truly hypo-methylated loci that are detected by the method. Similarly, TPR for hyper-methylated loci (TPR_hyper) is defined as the proportion of truly hyper-methylated loci that are detected by the method.

The results of the simulation studies without subject effect are reported in Table 1. Although our Bayesian inference does not control the FDR, Table 1 shows that the actual FDR level of our approach is very small and stable (ranges from 0.01 to 0.02 for different values of γ). Even when directional mistakes (namely hyper-methylated loci are declared as hypo-methylated or vice versa) are taken into account, the mdFDR of the proposed Bayesian method is still reasonably small, ranges from 0.01 to 0.03, and often equals to the FDR (in other words, there are zero directional mistakes, see Additional file 1: Table S2).

When the FDR of other procedures is controlled at the actual FDR level of the Bayesian approach using the q-value approach [19], our approach always outperforms other approaches in terms of power for different values of γ, whether the power is in terms of TPR, TPR_hypo or TPR_hyper. In terms of power, the ranking of the four approaches is very clear: the Bayesian approach works the best, the second one is the logistic regression approach, the third one is z-test and the last one is the empirical Bayes Wald test of the DSS package [10]. The power improvement of the proposed Bayesian approach over the logistic regression model, z test and the DSS procedure can be as high as 20 %, 30 % and 250 % respectively for small γ and can still be more than 7 %, 11 % and 46 % respectively for large γ. (See Additional file 1: Table S3 for the relative power improvement of the proposed approach over other methods for different γ). An interesting observation is that the power of the proposed approach to detect hypo-methylated loci (TPR_hypo) is always larger than its power to detect the hyper-methylated loci (TPR_hyper). This is likely due to the fact that there are more hypo-methylated loci than the hyper-methylated loci in the simulated data, which has the same proportion of differential methylation as the real data. (See Additional file 1: Table S1 for the proportions of hypo- and hyper-methylated loci of the real data for different values of γ).

As expected, the power of all procedures increases with the value γ since the magnitude of signal is larger and hence easier to detect even though the proportion

Table 1 Comparison of performance in terms of FDR, mdFDR, TPR, TPR_hypo and TPR_hyper of four procedures (logistic regression, DSS, z-test and the proposed Bayesian) at different γ values for the first set of simulation studies (without subject effect). Results are averaged over 100 replications of 20,000 CpG loci with two samples under each condition

		Logistic regression	DSS	z-test	Bayesian
$\gamma = 0.01$	FDR	0.0043	0.0001	0.0055	0.0091
	mdFDR	0.0048	0.0001	0.0059	0.0109
	TPR	0.2156	0.0837	0.2027	0.2548
	TPR_hypo	0.2162	0.0752	0.2129	0.2596
	TPR_hyper	0.2145	0.0952	0.1889	0.2471
$\gamma = 0.05$	FDR	0.0158	0.0003	0.0182	0.0184
	mdFDR	0.0159	0.0003	0.0183	0.0188
	TPR	0.3861	0.1672	0.3528	0.4099
	TPR_hypo	0.3921	0.159	0.3694	0.4251
	TPR_hyper	0.3787	0.1772	0.3326	0.3912
$\gamma = 0.1$	FDR	0.0207	0.0005	0.0244	0.0216
	mdFDR	0.0207	0.0005	0.0244	0.0216
	TPR	0.5398	0.2934	0.4964	0.5775
	TPR_hypo	0.5379	0.2851	0.5011	0.5879
	TPR_hyper	0.542	0.3032	0.4909	0.5652
$\gamma = 0.15$	FDR	0.0208	0.0005	0.0251	0.0217
	mdFDR	0.0208	0.0005	0.0251	0.0217
	TPR	0.6691	0.4345	0.6307	0.7161
	TPR_hypo	0.6619	0.422	0.627	0.7201
	TPR_hyper	0.6779	0.4497	0.6353	0.7113
$\gamma = 0.2$	FDR	0.02	0.0006	0.0257	0.0214
	mdFDR	0.02	0.0006	0.0257	0.0214
	TPR	0.7769	0.5751	0.7476	0.8214
	TPR_hypo	0.7708	0.5628	0.7422	0.8253
	TPR_hyper	0.7848	0.591	0.7546	0.8163

of signal in the data are decreasing. For instance, when $\gamma = 0.01$, the power of the proposed method is about 25 %, while when $\gamma = 0.2$, the power of the proposed method is about 82 %. This pattern is observed for all other procedures as well. However, a big surprise to us is the poor performance of the empirical Bayes Wald test (also called DSS) since it claims to borrow strength for better dispersion parameter estimates in their beta-binomial setting and should work well. A closer look at the actual FDR of this procedure tells us that this procedure is very conservative in its FDR control. The actual FDRs of the logistic regression model and z-test are close to their nominal levels, which is the actual FDR of the proposed Bayesian method. However, the actual FDR of the DSS procedure is significantly below the nominal level. A possible explanation might be that the null distribution of the DSS test statistics is no longer approximately normal in our simulation setting, which can be seen from the normal quantile quantile (QQ) plot of the DSS test statistics in Additional file 1: Figure S1, where all the CpG loci belong to the equal-methylated group in one simulated data and hence the graph represents the null distribution of the DSS test statistics. The QQ plot shows that the two tails of the null distribution of the DSS test not only deviate from the normal distribution but they are not very symmetric. Even if one thinks the normal distribution is a good approximation to the null distribution of the DSS test, one needs to be careful about what values to use for the mean and the standard deviation of this normal distribution. DSS [10] did not address this issue in their paper, but in their package (version 2.5.3.), the p-value of the DSS test was calculated assuming that the null distribution of the test statistics was a standard normal distribution (the authors also confirmed it through personal communication). However, the histogram of the DSS test statistics in Additional file 1: Figure S1 clearly shows that the null distribution of the DSS test deviates from the standard normal with variance smaller than one. Since we calculate the p-value of the DSS test based on the standard normal distribution as implemented in their package, it is not surprising that the null distribution of the p-values in Additional file 1: Figure S2 is skewed to the left and the test result is very conservative.

The results of the simulation studies with subject effect are reported in Table 2, Additional file 1: Table S4 and Additional file 1: Table S5. These numbers tell the same story as Table 1, Additional file 1: Table S2 and Additional file 1: Table S3. The actual FDR of the Bayesian approach is still reasonable small and stable for different values of γ and it ranges from 0.01 to 0.02 and the mdFDR is almost always equal to FDR, implying zero directional mistakes most of the time. The power of the proposed method to detect hypo-methylated loci is always higher than the one to detect hyper-methylated loci because there are more hypo-methylated loci in the data. The power of the proposed method ranges from around 25 % to 82 % for various values of γ. And for all values of γ, the proposed Bayesian approach has the best power among the four procedures under consideration with the power improvement over the other three approaches as high as over 30 %, 40 % and 1425 % respectively. Similar to the case without subject effect, the DSS procedure still performs poorly under the simulation setting with subject effect. The histogram in Additional file 1: Figure S3 shows that the null distribution of the DSS test statistics deviates even more from the standard normal distribution(with variance even smaller than one) than in the case without subject effect. Therefore, the p-values calculated assuming the standard normal distribution for the null distribution as implemented in the DSS package (version 2.5.3) tend to be very large (see the histogram of the p-values under the null hypothesis in right panels of Additional file 1: Figure S2)

Table 2 Similar to Table 1, but this comparison is based on the second set of simulation studies (with subject effect)

		Logistic regression	DSS	z-test	Bayesian
$\gamma = 0.01$	FDR	0.0027	0	0.0034	0.0069
	mdFDR	0.0029	0	0.0036	0.008
	TPR	0.1953	0.0171	0.183	0.2494
	TPR_hypo	0.1948	0.0167	0.1918	0.2545
	TPR_hyper	0.1958	0.0176	0.171	0.2418
$\gamma = 0.05$	FDR	0.0097	0	0.0113	0.014
	mdFDR	0.0097	0	0.0113	0.0141
	TPR	0.3554	0.0364	0.3235	0.4038
	TPR_hypo	0.3604	0.0375	0.3392	0.419
	TPR_hyper	0.3494	0.035	0.3044	0.3853
$\gamma = 0.1$	FDR	0.0119	0	0.0142	0.0156
	mdFDR	0.0119	0	0.0142	0.0156
	TPR	0.5053	0.0668	0.4588	0.5745
	TPR_hypo	0.5021	0.0692	0.4637	0.5836
	TPR_hyper	0.509	0.0641	0.453	0.5637
$\gamma = 0.15$	FDR	0.012	0	0.0147	0.0156
	mdFDR	0.012	0	0.0147	0.0156
	TPR	0.6385	0.1074	0.5939	0.7158
	TPR_hypo	0.6309	0.112	0.5905	0.7205
	TPR_hyper	0.6477	0.1016	0.5981	0.7101
$\gamma = 0.2$	FDR	0.0116	0.0001	0.0146	0.0153
	mdFDR	0.0116	0.0001	0.0146	0.0153
	TPR	0.7546	0.1576	0.7203	0.8254
	TPR_hypo	0.7459	0.1612	0.7129	0.8287
	TPR_hyper	0.7659	0.1529	0.7298	0.821

and lead to very conservative results. In summary, the results produced for simulation studies with subject effect are similar to those for simulation studies without subject effect. This might indicate that the two case samples we have chosen from the real data with matched covariates have ignorable subject effect as we have hoped for. On the other hand, it might also indicate that the proposed Bayesian model is not very sensitive to the presence of small subject effect in the data structure.

Real data analysis

In this section, we conduct both statistical analysis and bioinformatics analysis on a real dataset to compare the performance of difference procedures. The real dataset we analyze is the same CLL data [17] introduced at the beginning of the simulation section. Two CLL samples with matched covariates are selected to be compared with two control samples. There are 384,890 loci commonly mapped by these four samples. To reduce the dimension, we first remove loci that are either fully or non-methylated for all observed samples because there is no differential methylation associated with these loci. This reduces the dimension of the data to 324,126 loci. In other words, 60,764 loci are filtered out before we apply various procedures to detect DML.

Statistical analysis

We apply to the real dataset the same four statistical procedures studied in the simulation studies. However, since the simulation studies show that the logistic regression approach always performs better than the *z* test and the DSS procedure, we focus on the comparison of the proposed Bayesian method with the logistic regression for the real data analysis in the main text while reporting the results for the *z* test and the DSS procedures in the Additional file 1. Since we do not know the actual FDR level of the proposed Bayesian method for the real data, we cannot control the FDR level of the other three procedures at the same level as the proposed Bayesian method. However, from the simulation studies we learn that the actual FDR level for the proposed method ranges from around 0.01 to 0.03. Therefore we consider various nominal FDR levels for the logistic regression method (and the *z* test and the DSS procedures) to match with this range. To be specific, we consider nominal FDR levels of 0.01, 0.02, 0.03 and 0.05. Table 3 gives the numbers of hypo- and hyper-methylated loci identified by the proposed method and by the logistic regression approach at the four different nominal FDR levels (the results for the *z* test and DSS procedure are given in Additional file 1: Table S6). As expected, the power of the logistic regression method increases with the nominal FDR level. Since the actual FDR of the proposed Bayesian approach never exceeds 0.022 for all our simulations settings, it is reasonable to compare its result with that of the logistic regression method at nominal level 0.02 or 0.03. Table 3 shows that the proposed Bayesian approach detects more DML than the logistic regression approach with nominal FDR levels no greater than 0.03. Specifically, the improvement is 28.9 % for hypo-methylated loci and 19.8 % for hyper-methylated loci when compared to the logistic regression model at level 0.02 and the improvement is 15.3 % and 9.2 % for hypo- and hyper-methylated loci respectively when the FDR level is set to be 0.03. This is consistent with our simulation studies: the proposed Bayesian method is more powerful than the logistic regression model when their FDR level is matched. When the nominal FDR level increases to 0.05 for the logistic regression approach, it detects slightly more DML than the proposed Bayesian method with only 2 % more hypo-methylated loci and 5 % more hyper-methylated loci. However, the cost is its increased false discovery rate. To examine this, we look at the detected DML disagreed by these two methods in Table 4.

Table 3 Numbers of hypo- and hyper-methylated loci identified by the proposed Bayesian method and logistic regression method for the real data. FDR is controlled at levels 0.01, 0.02, 0.03 and 0.05 for logistic regression

	Hypo-methylation	Hyper-methylation
The proposed Bayesian method	31,328	21,270
Logistic Regression: FDR controlled at 0.01	20,527	15,166
Logistic Regression: FDR controlled at 0.02	24,309	17,758
Logistic Regression: FDR controlled at 0.03	27,172	19,483
Logistic Regression: FDR controlled at 0.05	31,863	22,419

Table 4 shows that there are more DML (both hypo- and hyper-methylated loci) uniquely identified by the proposed Bayesian method than the logistic regression approach when the nominal FDR level of the logistic regression approach is no greater than 0.03. When this nominal level increases to 0.05, there are slightly more uniquely identified DML by the logistic regression method. We also notice that as the nominal FDR level of the logistic regression method increases, the number of the DML uniquely identified by the proposed Bayesian method is smaller, which may imply that the DML uniquely detected by the proposed Bayesian method can be verified by the logistic regression method at slightly higher nominal FDR level and hence are more likely to be true signals than false positives. On the other hand, the increasing number of DML uniquely identified by the logistic regression approach might just be the result of including more false positives. To see this better, we compare the histograms of the observed methylation proportion differences of the uniquely identified DML by either

Table 4 The numbers of DML disagreed by the proposed Bayesian method and the logistic regression method for the real data analysis. The nominal FDR levels of the logistic regression model are 0.01,0.02, 0.03 and 0.05. In the table, 0, 1 and 2 represent equal-, hypo- and hyper-methylated loci respectively

		Logistic regression											
		FDR 0.01			FDR 0.02			FDR 0.03			FDR 0.05		
		0	1	2	0	1	2	0	1	2	0	1	2
	0		42	41		512	368		1,397	852		3,795	2,245
Bayesian method	1	10,843		0	7,531		0	5,553		0	3,260		0
	2	6,145	0		3,880	0		2,639	0		1,096	0	

the proposed Bayesian approach or the logistic regression approach at various nominal FDR levels in Fig. 1 and Additional file 1: Figures S4, S5 and S6. The right panels of Fig. 1 and Additional file 1: Figures S4, S5 and S6 show that the DML uniquely identified by the logistic regression method at various nominal FDR levels mostly have small observed methylation proportion difference. If one follows the common practice of filtering out DML with methylation proportion difference smaller than 0.2 or 0.25 [9], then the majority of the DML uniquely identified by the logistic regression method will be filtered out for further study. On the other hand, the DML uniquely identified by the proposed Bayesian method tends to have large effect size with half of the loci having difference larger than 0.2 (see left panels of the above mentioned figures) and hence these DML are not only statistically significant but also biologically significant.

When it comes to statistically significant DML, it is more likely for a small-effect-size locus to be a false positive than one with large effect size. Therefore, we suspect that the DML uniquely identified by the proposed Bayesian method are more likely to be true signals while

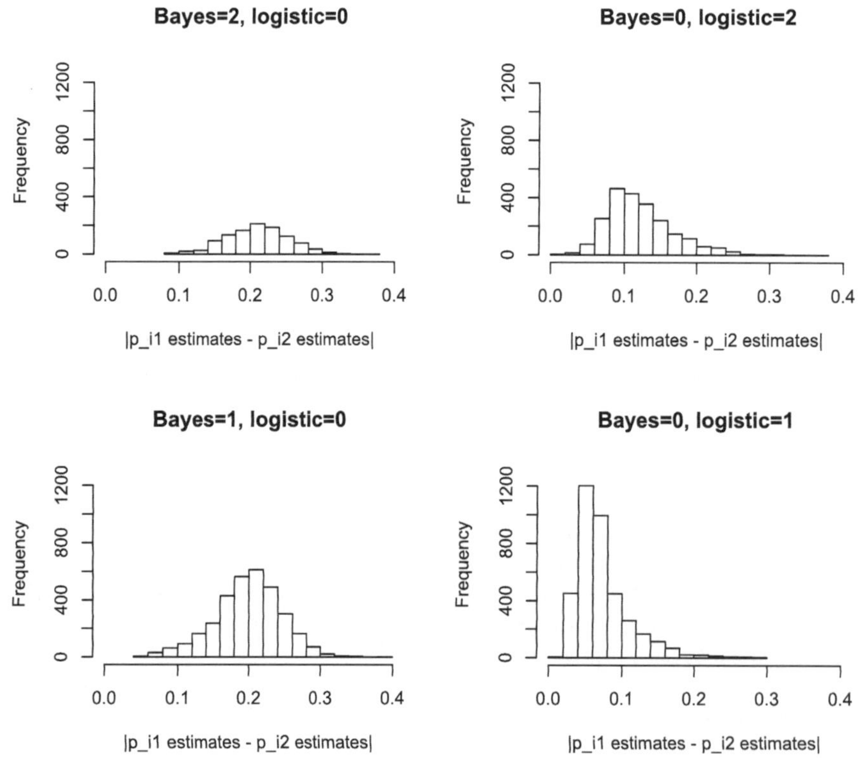

Fig. 1 For the loci uniquely identified to be hypo-methylated loci (*group 1*) or hyper-methylated loci (*group 2*) by either the proposed Bayesian method or logistic regression when applied to the real data, four panels are histograms of absolute difference for methylation proportion estimates between the case and control samples. These four panels correspond to the counts in Table 4 when FDR level for logistic regression is controlled at 0.05 level

the ones uniquely identified by the logistic regression method at FDR level 0.05 are more likely to be false positives. A simulation study is conducted to verify this conjecture and the results are presented in Additional file 1: Table S7. This table shows that on average, there are more DML uniquely identified by the logistic regression method at level 0.05 than the Bayesian method. However, among these uniquely identified DML by the logistic regression method, 57.42 % are false positives compared to 13.86 % false positives among the uniquely identified DML by the proposed Bayesian method. On the other hand, 88.25 % of the unique DML by the Bayesian method have proportion difference larger than 0.2 while only 26.84 % of the extra DML detected by the logistic regression method have difference larger than 0.2. Interestingly, almost 40 % of the uniquely identified DML by the logistic regression at level 0.05 have effect size less than 0.1 and among these small-effect-size loci, 99.48 % are false positives. On the other hand, all the DML uniquely identified by the proposed Bayesian method have methylation proportion difference larger than 0.1. Therefore, we can conclude that the proposed Bayesian method detects more large-effect-size DML with lower proportion of false positives than by the logistic regression method at level 0.05 although we did not specify a nominal FDR level for our procedure. This comparison helps us to realize that our Bayesian approach has another advantage over the other three approaches in that it doesn't depend on a pre-specified nominal FDR level while maintaining a small false discovery rate. In conclusion, the proposed Bayesian method is not only a powerful tool for identifying DML but it is also very reliable procedure and tends to pick up biologically significant DML missed by other procedures.

We also present the results of the z test and the DSS procedures in the Additional file 1. Comparing Additional file 1: Table S6 with Table 3, we can see that the ranking of the four procedures is about the same as seen in the simulation studies. The DSS is the most conservative procedure and the z test detects slightly fewer DML than the logistic regression model. Additional file 1: Figures S7, S8, S9 and S10 draw the Venn diagrams of the data analysis results of the four procedures at different nominal FDR levels (note that the result for the Bayesian approach doesn't depend on the specification of the nominal FDR level) and for hypo- and hyper-methylated loci separately. By examining these Venn diagrams, we can see that all DML identified by the DSS are also identified by the proposed Bayesian method and the logistic regression method, which might imply that the DSS procedure is good for ranking the DML but it needs the correct null distribution to produce the right p-value for appropriate FDR control. The current default assumption of standard normal distribution produces very conservative p-values for the DSS procedure and need to be improved. The z test does not overlap with the proposed Bayesian method as well as the logistic regression method. The histograms of the observed methylation differences in Additional file 1: Figures S11, S12, S13 and S14 show that the majority of the DML identified by the z test but not by the proposed Bayesian method have observed methylation proportion differences smaller than 0.1.

Bioinformatics analysis

To further compare the performance of the proposed Bayesian method and the logistic regression approach, we conduct some bioinformatics analysis. As we discuss in the introduction session, the differential methylation can influence the gene expression regulation. Therefore, a common next step after identifying the DML is to find the genes associated with the detected DML. Since multiple DML can associate with the same gene, identifying a larger number of DML does not guarantee identification of more associated genes. Therefore it is of interest to compare the number of genes uniquely identified by different methods. Table 5 gives the number of genes uniquely identified by either the proposed Bayesian method or the logistic regression approach or by both methods. It is very clear to see even in terms of genes, the proposed Bayesian method can detect more than the logistic regression methods at FDR level no greater than 0.03. We also notice that the number of commonly identified genes by both methods are increasing with the nominal FDR level of the logistic regression, which implies that the genes uniquely identified by the proposed Bayesian method can be verified

Table 5 Number of genes associated with hypo- and hyper-methylated loci uniquely identified by the proposed Bayesian method or the logistic regression method, or identified by both methods. The nominal FDR levels of the logistic regression model are 0.01,0.02, 0.03 and 0.05. In the table, 0, 1 and 2 represent equal-, hypo- and hyper-methylated loci respectively

		Logistic regression											
		FDR 0.01			FDR 0.02			FDR 0.03			FDR 0.05		
		0	1	2	0	1	2	0	1	2	0	1	2
	0		1	3		36	52		82	132		224	391
Bayesian method	1	745	2,293	0	475	2,563	0	329	2,709	0	129	2,909	0
	2	1,379	0	4,043	920	0	4,502	648	0	4,774	365	0	5,057

by the logistic regression method at higher nominal FDR level. Here we want to focus on the genes that are uniquely identified by our proposed Bayesian method but missed by the logistic regression method even when we increase its nominal level to 0.05 (recall it is unlikely that the proposed Bayesian method has actual FDR level greater than 0.05). There were 129 genes with hyper-methylated CpGs and 365 genes with hypo-methylated CpGs that were found only by the proposed Bayesian method and not by the logistic regression method at FDR cut-off 0.05. A list of these genes are provided in Additional file 1: Tables S8 and S9. Further investigation of these genes indicates their biological relevance towards cancer regulation and proliferation. Uniquely identified genes with hypo-methylated CpGs consisted of genes associated with important functions like negative regulation cell death/apoptosis (genes include *BCL6, BAG1, CD27, G2E3, POU3F3, RASA1, AZU1, EGFR, EDNRB, MAPK81P1, MALT1, MCL1 and TGFBR1*), cell-cycle regulation (genes include *CDKN1B, CDKN2A and RUNX1*) and leukocyte activation (genes include *BCL6, BCL11A, BST2, RAB27A, AZU1, CBFB, MALT1 and NCR1*). Uniquely identified genes with hyper-methylated CpGs mostly consisted of homeobox genes (eg *HOXA1, HMBOX1*) and genes associated with regulation of transcription (eg. *NFATC4, TCF12, HMBOX1, HOXA1, SOX5, SIX4, ESR1*) or transcription silencing (eg. *YY1*). The proposed method identified hyper-methylated CpG 37 bp upstream to the TSS (Transcription Start loci) of *CASP7*, which is one of the key regulator genes in apoptosis execution. Overall, 37 genes were uniquely identified with hyper-methylated CpG and 102 genes with hypo-methylated CpGs within 1000bp proximity to their TSS.

Interesting, if we compare the proposed Bayesian method with the logistic regression method at FDR 0.03 for the bioinformatics analysis, it uniquely detects 200 more genes associated with hypo-methylated CpGs and 283 more genes associated with hyper-methylated CpGs. Among these extra uniquely identified genes by the proposed Bayesian method, many are directly associated with B-cells (or B-lymphocytes/leukocyte) and CLL cancer type. GO annotations like "Leukocyte/lymphocyte differentiation and activatio" (for *CD1D, CARD11, FOXP1, HDAC4, IRF4, NTRK1, SPN, SYK, SNCA, YWHAZ*) and "Wnt signalling" (for *AXIN2, FZD2, MACF1, NXN, SLC9A3R1, WNT3*) and "regulation cell proliferation" (for *PINX1, TGIF1, AXIN2, FOXJ1, IGFBP7, LDOC1, PTH1R, RUNX3, B4GALT7*) were enriched for genes uniquely identified by the proposed Bayesian method for hyper-methylation. Similarly, B-lymphocyte specific biological process like "Leukocyte activation" (for *BCL6, BCL11A, BST2, RAB27A, ULBP1, AZU1CBFB, HSH2D, HSH2D, ILI2B, MALT1, NCR1, YWHAZ*), and cancer development related like "negative regulation of cell death/apoptosi" (*BCL6, BAG1, CD27, G2E3, IHH, POU3F3, IHH, RASA1, AZU1, EDNRB, EGFR1, MCL1, TGFBR1 etc.*), negative regulation of cell differentiation (*BCL6, IHH, LMX1A, THY1, BMP4, RUNX1, NRP1, FOXA2, etc*), G1/S transition of cell cycle (*E2F6, CDKN1B, CDKN2A, EGFR, GFI1, SPDYA*), Cell migration (*LMX1A, EDNRB, KIF5C, IL12B, TGFR1*) were enriched for genes uniquely identified by the proposed Bayesian method for hypo-methylation. Note that these biologically relevant genes, although were not identified by the logistic regression method at FDR level 0.03, were verified when the FDR level increases to 0.05. This is in consistent with our earlier conclusion that the DML identified uniquely by the proposed Bayesian method is more likely to be true positives and can often be verified by the logistic regression method at the cost of increased FDR level.

Conclusion and discussion

We propose a full Bayesian partition model for identifying differentially methylated loci under two conditions. It is well known that hypo-methylation and hyper-methylation plays different roles in gene regulations and have distinct biological meanings. Therefore it is important to differentiate these two types of differential methylation in the data analysis. Many existing methods including the logistic regression approach of Methylkit [9] and the empirical Bayes Wald test [10] only produce the binary output of a locus being differentially methylated or not. A common practice to further identify hypo- or hyper-methylated loci is based on the sign of the test statistics, which is an ad-hoc approach and ignores the uncertainty associated with the sign of the test statistics. Our proposed Bayesian partition model addresses the issue systematically by partitioning all loci into three groups: equal-, hypo- and hyper-methylated groups so that the final output of our analysis is more informative than just being differentially methylated or not. Compared to frequentist approaches such as logistic regression model of Methylkit [9], our Bayesian approach also has the advantage of borrowing strength across the loci and hence is more powerful for analyzing DNA methylation data with small sample sizes. Another advantage of the proposed Bayesian method is that one do not need to specify a nominal FDR level ahead of time, which sometimes can be a challenging job. Too small FDR level means shorter list of DML and too large FDR level can mean longer list of DML with more false positives. The proposed Bayesian model is shown by both simulation studies and real data analysis to have achieved higher power than other procedures while maintaining a low false discovery rate. It also tends to pick up biologically significant DML with large effect size missed by other methods. Therefore it is a both powerful and reliable procedure for identifying DML for single nucleotide resolution sequencing data.

Note our Bayesian model makes several assumptions such as parametric prior distributions and no subject effect for the methylation proportions within each condition. However our simulation studies try to mimic the real data as much as possible by generating the parameters from the empirical distribution of parameter estimates based on real data, which is not the prior distribution we assumed for the model. The good performance of our model show that our model is robust to the misspecification of the prior distribution. Furthermore, even when we generate data with subject effect on the methylation proportions, the proposed Bayesian model still performs similarly to the case without subject effect, which shows that our model is not very sensitive to the presence of small subject effect. However, it is important to address the issue of subject effect or biological variation on the methylation proportions for different biological samples within the same condition by modeling it correctly. This is one of our ongoing projects.

It is known that the methylation levels of adjacent CpG loci are correlated and hence it is important to consider the correlation structure of adjacent CpG loci. In this paper, our model assumes that the loci are conditional independent given parameters. Marginally, they are dependent because they share parameters. However,the correlation structure can also be modelled explicitly and this is one of our future work.

Although compared to other existing methods, our proposed method provides more accurate and reliable results, it is computationally more intensive. In terms of computing time, the proposed method takes approximately 42 hours to analyze 384,890 loci in a 2.80 GHz 4-core CPU, 16 GB memory environment while the logistic regression method, the DSS method and the z test take approximately 24, 5 and 3 minutes respectively in the same environment. Even though the computing time of our method is not infeasible in practice, it is desirable to improve the computing efficiency of our method to be closer to those of other methods, which is another future research direction for us.

Additional file

Additional file 1: Supplementary document. This is a PDF document with three parts not shown in the paper. The first part is the derivation for the posterior distribution of $\boldsymbol{I}$, and the second and third parts are additional tables and figures. Tables contains the proportion of hypo- and hyper-methylated loci in the real data at different γ values, comparison in identification of the four procedures for simulation study with and without subject effect, relative power improvement of the proposed Bayesian method compared to three other methods for simulation study with and without subject effect, number of hypo- and hyper-methylated loci identified by DSS and z test, simulation study of the proportion of false positives in the list of uniquely identified loci by either the proposed Bayesian method or the logistic regression method at 0.05 FDR and list of genes with hypo- and hyper-methylated CpGs uniquely identified by the proposed Bayesian. Figures include that Figures S1 and S3 are histograms and normal QQ plots of the empirical Bayes Wald test statistics for one simulated data with and without subject effect, Figure S2 is histograms of p values for the empirical Bayes Wald test under the null hypothesis on one simulated data with and without subject effect, Figures S4, S5 and S6 are histograms of absolute difference for methylation proportion estimates between the case and control samples in the real data for the loci uniquely identified to be hypo- and hyper-methylated loci by either the proposed Bayesian and logistic regression at different FDR levels 0.01, 0.02, 0.03, Figure S7, S8, S9 and S10 are Venn diagrams of detected hypo- and hyper-methylated loci in real data analysis for four procedures with FDR controlled at 0.01, 0.02, 0.03 and 0.05, and Figures S11, S12, S13 and S14 are similar histograms with Figures S4, S5 and S6 but different comparison of the methods by either the proposed Bayesian and z test at different FDR levels 0.01, 0.02, 0.03 and 0.05. (PDF 3160 kb)

Competing interests
The authors declare that they have no competing interests.

Authors' contributions
HW, CH and JQ participated in the method design, simulation study and results analysis. HW drafted the manuscript and developed an R package implementing the method. JQ revised the manuscript. GK and DX collected and processed the real data. GK conducted bioinformatics analysis in the paper. All authors have read and approved the final version of the manuscript.

Acknowledgements
This work was supported in part by the National Institute of Health grant DA025779.

Declaration
This article has been published as part of BMC Bioinformatics Volume 17 Supplement 1, 2016: Selected articles from the Fourteenth Asia Pacific Bioinformatics Conference (APBC 2016). The full contents of the supplements are available online at http://www.biomedcentral.com/bmcbioinformatics/supplements/17/S1.

Author details
[1]Department of Statistics, University of Missouri, Columbia, Missouri, USA. [2]Department of Computer Science and Informatics Institute, University of Missouri, Columbia, USA. [3]Department of Applied Economics and Statistics, University of Delaware, Newark, DE, USA.

Published: 11 January 2016

References
1. Ehrlich M, Gama-Sosa MA, Huang LH, Midgett RM, Kuo KC, McCune RA, Gehrke C. Amount and distribution of 5-methylcytosine in human dna from different types of tissues or cells. Nucleic Acids Res. 1982;10(8): 2709–21.
2. Razin A, Cedar H. Dna methylation and genomic imprinting. Cell. 1994;77(4):473–6.
3. Jaenisch R, Bird A. Epigenetic regulation of gene expression: how the genome integrates intrinsic and environmental signals. Nat Genet. 2003;33:245–54.
4. Li E. Chromatin modification and epigenetic reprogramming in mammalian development. Nat Rev Genet. 2002;3(9):662–73.
5. Jones PA, Takai D. The role of dna methylation in mammalian epigenetics. Science. 2001;293(5532):1068–70.
6. Clark SJ, Statham A, Stirzaker C, Molloy PL, Frommer M. Dna methylation: bisulphite modification and analysis. Nat Protoc. 2006;1(5):2353–64.
7. Lister R, Pelizzola M, Dowen RH, Hawkins RD, Hon G, Tonti-Filippini J, et al. Human dna methylomes at base resolution show widespread epigenomic differences. Nature. 2009;462(7271):315–22.
8. Hansen KD, Langmead B, Irizarry RA. Bsmooth: from whole genome bisulfite sequencing reads to differentially methylated regions. Genome Biol. 2012;13(10):83.
9. Akalin A, Kormaksson M, Li S, Garrett-Bakelman FE, Figueroa ME, Melnick A, et al. methylkit: a comprehensive r package for the analysis of genome-wide dna methylation profiles. Genome Biol. 2012;13(10):87.

10. Feng H, Conneely KN, Wu H. A bayesian hierarchical model to detect differentially methylated loci from single nucleotide resolution sequencing data. Nucleic Acids Res. 2014;42(8):69–9.
11. Das PM, Singal R. Dna methylation and cancer. J Clin Oncol. 2004;22(22): 4632–42.
12. Fujikane T, Nishikawa N, Toyota M, Suzuki H, Nojima M, Maruyama R, et al. Genomic screening for genes upregulated by demethylation revealed novel targets of epigenetic silencing in breast cancer. Breast Cancer Res Treat. 2010;122(3):699–710.
13. Kristensen LS, Nielsen HM, Hansen LL. Epigenetics and cancer treatment. Eur J Pharmacol. 2009;625(1):131–42.
14. Ehrlich M. Dna hypomethylation in cancer cells. Epigenomics. 2009;1(2): 239–59.
15. Newton MA, Noueiry A, Sarkar D, Ahlquist P. Detecting differential gene expression with a semiparametric hierarchical mixture method. Biostatistics. 2004;5(2):155–76.
16. Gamerman D, Lopes HF. Markov Chain Monte Carlo: Stochastic Simulation for Bayesian Inference. Boca Raton, FL: CRC Press; 2006.
17. Pei L, Choi JH, Liu J, Lee EJ, McCarthy B, Wilson JM, et al. Genome-wide dna methylation analysis reveals novel epigenetic changes in chronic lymphocytic leukemia. Epigenetics. 2012;7(6):567–78.
18. Hansen KD, Timp W, Bravo HC, Sabunciyan S, Langmead B, McDonald OG, et al. Increased methylation variation in epigenetic domains across cancer types. Nat Genet. 2011;43(8):768–75.
19. Storey JD. J R Stat Soc Ser B (Stat Methodol). 2002;64(3):479–98.
20. Guo W, Sarkar SK, Peddada SD. Controlling false discoveries in multidimensional directional decisions, with applications to gene expression data on ordered categories. Biometrics. 2010;66(2):485–92.

Fan et al. BMC Bioinformatics 2015, 17(Suppl 1):8
DOI 10.1186/s12859-015-0851-2

BMC Bioinformatics

PROCEEDINGS Open Access

PredRSA: a gradient boosted regression trees approach for predicting protein solvent accessibility

Chao Fan[1], Diwei Liu[1], Rui Huang[1], Zhigang Chen[1] and Lei Deng[1,2*]

From The Fourteenth Asia Pacific Bioinformatics Conference (APBC 2016)
San Francisco, CA, USA. 11 - 13 January 2016

Abstract

Background: Protein solvent accessibility prediction is a pivotal intermediate step towards modeling protein tertiary structures directly from one-dimensional sequences. It also plays an important part in identifying protein folds and domains. Although some methods have been presented to the protein solvent accessibility prediction in recent years, the performance is far from satisfactory. In this work, we propose PredRSA, a computational method that can accurately predict relative solvent accessible surface area (RSA) of residues by exploring various local and global sequence features which have been observed to be associated with solvent accessibility. Based on these features, a novel and efficient approach, Gradient Boosted Regression Trees (GBRT), is first adopted to predict RSA.

Results: Experimental results obtained from 5-fold cross-validation based on the Manesh-215 dataset show that the mean absolute error (MAE) and the Pearson correlation coefficient (PCC) of PredRSA are 9.0 % and 0.75, respectively, which are better than that of the existing methods. Moreover, we evaluate the performance of PredRSA using an independent test set of 68 proteins. Compared with the state-of-the-art approaches (SPINE-X and ASAquick), PredRSA achieves a significant improvement on the prediction quality.

Conclusions: Our experimental results show that the Gradient Boosted Regression Trees algorithm and the novel feature combination are quite effective in relative solvent accessibility prediction. The proposed PredRSA method could be useful in assisting the prediction of protein structures by applying the predicted RSA as useful restraints.

Keywords: Solvent accessibility, Sequence features, Gradient boosted regression trees

Background

Since the concept of solvent accessibility was first introduced by Lee and Richards [1], defined as the surface area of a protein that is accessible to a spherical solvent while probing the surface of that molecule, it has been considered as a key factor for understanding protein structure and function [2]. Predicting the three-dimensional (3D) structures of proteins from their one-dimensional sequences is a challenging issue because of the increasing gap between the enormous number of protein sequences and the number of known structures. Studies of solvent accessibility in proteins have provided many useful insights into the 3D structures of proteins [3]. Furthermore, knowledge of solvent accessibility has proved useful for structural domains identification [4], fold recognition [5], binding region identification [6–8] and protein intrinsic disorder [9]. The solvent accessibility is particularly important because it is associated with the spatial arrangement and packing of amino acids during the process of protein folding. It also plays an important role in predicting the active sites of protein-protein or protein-ligand binding [10].

*Correspondence: leideng@csu.edu.cn
[1] School of Software, Central South University, No.22 Shaoshan South Road, 410075 Changsha, China
[2] Shanghai Key Laboratory of Intelligent Information Processing, No.220 Handan Road, 200433 Shanghai, China

In many earlier studies, the solvent accessibility prediction was taken as a classification problem with varying thresholds, two-state (exposed or buried) or three-state (exposed, intermediate or buried) [11–15]. However, there is no standard definition for the thresholds of solvent accessibility states. For instance, a residue may be predicted to be exposed state based on a relative solvent accessibility threshold of 10 %, but the same residue may be predicted to be buried state based on a threshold of 20 %. In view of this, it is necessary to predict the real values of solvent accessibility. Some representative machine learning techniques have been proposed to predict the real values of solvent accessibility, including multiple linear regression [16], support vector regression [17–19], neural network [20, 21], energy optimization [22] and nearest neighbor method [23].

For the real-valued solvent accessibility prediction, Ahmad et al. [20] proposed a neural network method with only single sequence information as the input features. The result showed that this method achieved a MAE of 18.0–19.5 % on different data sets. Adamczak et al. [21] employed evolutionary information in the form of position-specific scoring matrix (PSSM) profile to train a neural network-based regression for the prediction. Compared with the single sequence based neural network [20], the prediction performance was improved and the MAE decreased by about 5 % on the PFAM database [24]. Subsequently, Lee et al. [16] applied PSSM profile by constructing a correlation matrix different window positions to train a multiple linear regression method. The result showed a performance of 16.6 % MAE and 0.63 PCC on the Barton-502 dataset. Garg et al. [25] took multiple sequence alignment and secondary structure as input features to predict RSA based on a feed-forward neural network. The result indicated that a lower MAE achieved on CASP6 was 15.9 % and a higher PCC was 0.68.

Although these methods for surface accessibility prediction were developed, several issues still exist and make surface accessibility prediction a very challenging task. Mainly, there are three reasons: (1) specific biological properties for precisely predicting surface accessibility are not fully exploited, and no single parameter can definitely estimate the accessible surface area, various combinations of different feature types, including PSSM profiles, secondary structure features, native disorder features as well as other global sequence features [26], need to be investigated comprehensively; (2) the performance of the existing methods is still unsatisfactory, especially in terms of independent testing and (3) high-performance ensemble learning algorithms such as boosted regression trees haven't been intensively used in this area.

In this article, we propose a new and efficient approach, PredRSA (Prediction of Relative Solvent Accessible surface area), that integrates gradient boosted regression trees (GBRT) algorithm with multiple sequence-based features (position-specific scoring matrix, secondary structure, conservation score, native disorder) and a global feature (side-chain environment) to predict RSA. We have benchmarked PredRSA using the Manesh training dataset and an independent dataset. Results show that PredRSA significantly outperforms the state-of-the-art methods and indicate that the GBRT algorithm and the novel feature combination are important determinants in the prediction of RSA.

Methods

The GBRT algorithm

Our approach utilizes an ensemble regression algorithm for predicting RSA values of amino acid residues in a protein sequence. Generally, a target residue in sequence can be described as an n-dimension vector. Let us denote an amino acid residue by $\boldsymbol{x} = (x_1, x_2, \cdots, x_n)$ where $x_i \in \boldsymbol{R}$ and the corresponding real-valued RSA by y. The goal of predicting RSA real value of the amino acid residue in sequence is to find a function $F^*(\boldsymbol{x})$ that maps $\boldsymbol{x}$ to y, such that over the joint distribution of all $(y, \boldsymbol{x})$-values, the expected value of some specified loss function $\Psi(y, F(\boldsymbol{x}))$ is minimized as follows:

$$\begin{aligned} F^*(\boldsymbol{x}) &= \underset{F(\boldsymbol{x})}{arg\ min}\ E_{y,\boldsymbol{x}}\Psi(y, F(\boldsymbol{x})) \\ &= \underset{F(\boldsymbol{x})}{arg\ min}\ E_{\boldsymbol{x}}[E_y(\Psi(y, F(\boldsymbol{x}))|\boldsymbol{x}] \end{aligned} \tag{1}$$

Let $\{y_i, x_i\}_1^N$ be a set of training data, N is the number of all amino acid residues in the training set. The GBRT algorithm iteratively constructs M different weak learners $h(\boldsymbol{x}, \Theta_1), \cdots, h(\boldsymbol{x}, \Theta_M)$ which consist of regression trees of fixed size from training set and constructs the following additive function $F(\boldsymbol{x})$:

$$F(\boldsymbol{x}) = \beta_0 + \sum_{m=1}^{M} h(\boldsymbol{x}, \Theta_m) \tag{2}$$

where β_m and Θ_m are a weight and vector of parameters for the mth weak regression tree $h(\boldsymbol{x}, \Theta_m)$, respectively, and β_0 is an initial value. Both the weight β_m and the parameters Θ_m are iteratively determined from $m = 1$ to $m = M$ so that a loss function $\Psi(y, F(\boldsymbol{x}))$ is minimized. That is, β_m and Θ_m for the mth regression tree are determined as follows:

$$(\beta_m, \Theta_m) = \underset{\beta,\Theta}{arg\ min} \sum_{i=1}^{N} \Psi(y_i, F_{m-1}(\boldsymbol{x}_i) + \beta h(\boldsymbol{x}_i, \Theta)) \tag{3}$$

where $F_0(\boldsymbol{x})$ is an initial value and given by $F_0(\boldsymbol{x}) = \beta_0 = arg\ min_\beta \sum_{i=1}^{N} \Psi(y_i, \beta)$, $F_{m-1}(\boldsymbol{x})$ is the $(m-1)$th additive function combined from the first to the $(m-1)$th weak regression tree.

However, in general, it is not straightforward to solve Eq. (3). Therefore, GBRT separately and approximately estimates (β_m, Θ_m) with a simple two-step fashion [27]. For the estimation of the parameters Θ_m, we determine them so that the function defined by the regression tree approximates a gradient with respect to the current function $F_{m-1}(\boldsymbol{x})$ in the sense of least-square error as follows:

$$\Theta_m = \underset{\Theta}{arg\ min} \sum_{i=1}^{N} (\tilde{y}_{im} - h(\boldsymbol{x}_i, \Theta))^2 \tag{4}$$

where $\tilde{y}_{im}$ is the gradient and is given by

$$\tilde{y}_{im} = -\left[\frac{\partial \Psi(y_i, F(\boldsymbol{x}_i))}{\partial F(\boldsymbol{x}_i)}\right]_{F(\boldsymbol{x})=F_{m-1}(\boldsymbol{x})} \tag{5}$$

When the mth regression tree using the Θ_m has L_m leaf nodes, the regression tree is given by

$$h\left(\boldsymbol{x}, \{R_{lm}\}_{l=1}^{L_m}\right) = \sum_{l=1}^{L_m} \bar{y}_{lm} l(\boldsymbol{x} \in R_{lm}) \tag{6}$$

where R_{lm} is a disjoint region that the lth leaf node of the mth regression tree defines. $l(.)$ is a Boolean function that outputs 1 in case the argument of the function is true. $\bar{y}_{lm}$ is a constant for the R_{lm}th region, defined as the mean of training data that belongs to the lth leaf node of the mth regression tree. The weight β_m can be straightforwardly chosen using line search:

$$\beta_m = \underset{\beta}{arg\ min} \sum_{i=1}^{N} \Psi\left(y_i, F_{m-1}(\boldsymbol{x}_i) - \beta \frac{\partial \Psi\left(y_i, F_{m-1}(\boldsymbol{x}_i)\right)}{\partial F_{m-1}(\boldsymbol{x}_i)}\right) \tag{7}$$

Then, a new additive function $F_m(\boldsymbol{x})$ is updated as follows:

$$F_m(\boldsymbol{x}) = F_{m-1}(\boldsymbol{x}) + \nu \sum_{l=1}^{L_m} \beta_m \bar{y}_{lm} l(\boldsymbol{x} \in R_{lm}) \tag{8}$$

where $0 < \nu < 1$ is a shrinkage parameter, also called the learning rate to scale the step length the the gradient descent procedure. In this work, we take Huber loss function [28] as the loss function given by

$$\Psi(y, F) = \begin{cases} \frac{1}{2}(y - F)^2 & if\ |y - F| \leqslant \delta \\ \delta(|y - F| - \delta/2) & if\ |y - F| > \delta \end{cases} \tag{9}$$

Hence, in Eq. (5), $\tilde{y}_{im}$ becomes:

$$\tilde{y}_{im} = \begin{cases} y_i - F_{m-1}(\boldsymbol{x}_i) & if\ |y_i - F_{m-1}(\boldsymbol{x}_i)| \leqslant \delta \\ \delta.sign(y_i - F_{m-1}(\boldsymbol{x}_i)) & if\ |y_i - F_{m-1}(\boldsymbol{x}_i)| > \delta \end{cases} \tag{10}$$

The value of the transition point δ depends on the iteration number m.

Finally, the resulting RSA value y corresponding to the amino acid residue $\boldsymbol{x}$ is given by: $y = F_M(\boldsymbol{x})$.

Sequence encoding schemes

Selecting appropriate features is a crucial step because it directly determines the prediction performance. In this article, we explore various sequenced-based features which have been shown to be related to the solvent accessibility or ever applied in the similar issues. These features include PSSM profiles [29–31], PSIPRED-predicted secondary structure [32], DISOPRED-predicted native disorder [33], conservation score and side-chain environment compositions [34]. In this section, a more detail description about how to extract and encode these different sequence-based features as follows.

PSI-BLAST-based profiles

Position-specific scoring matrix (PSSM) of a residue which is achieved by the PSI-BLAST program contains important evolutionary information that determines whether this residue is conserved in its family of related proteins. Each element in the PSSM represents the probability of each residue position in the multiple sequence alignment. Plenty of previous studies have shown that multiple sequence alignments in the form of PSSM can substantially improve overall prediction performance [35–38]. In this article, the PSSM profile for each protein sequence is generated with default parameters (3 iterations and 0.001 of E-value cutoff) against the non-redundant (nr) dataset obtained from the NCBI. We encode each residue using a local sliding window approach based on the PSSM profiles. The PSSM profile generated by PSI-BLAST consists of the likelihood of a particular residue substitution at a specific position. These likelihood values are normalized to [0,1] by standard logistic function:

$$x' = \frac{1}{1 + exp(-x)} \tag{11}$$

where x is the score derived from the PSSM profile and x' is the standardized value of x. For a given residue, its local sequence fragment is extracted and encoded as a $20 \times (2l + 1)$-dimensional vector by using a sliding window scheme where l denotes the half window size and $L = 2l + 1$ is the whole window length. Furthermore, the predictive performance of a variety of different local window sizes L (from 3–17) has been evaluated to select the optimal local window size L for the RSA prediction. Finally, in this encoding scheme, a residue is encoded by a $20 \times L = 20 \times (2l + 1)$-dimensional vector.

In addition, we try to introduce residue conservation score for the solvent accessibility prediction. The value of sequence conservation for residue is a measure of how often a given residue is seen at an equivalent position in an equivalent protein across different species. Generally, residue conservation score is proportional to its buried degree. The conservation score is obtained by PSI-BLAST search as well [39, 40].

PSIPRED-predicted secondary structure information

In this work, we use the PSIPRED program to predict the secondary structure information. PSIPRED provides highly accurate prediction for protein secondary structures by applying a feed-forward neural network. The outputs of PSIPRED are encoded by the probability profiles of three secondary structures (C for coil, H for helix and E for strand). Some previous works have shown that incorporation of PSIPRED-predicted secondary structure information can significantly improve the prediction performance [25, 41].

Analogously, for a given residue, its three-state secondary structure profiles are extracted and encoded using a sliding window of $L = 2l + 1$ consecutive residues. Therefore, in this encoding scheme, a residue is composed of a $3 \times L = 3 \times (2l + 1)$-dimensional vector.

DISOPRED-predicted native disorder information

In the past decade, protein disorder or unstructured regions have received considerable attention in that they are commonly responsible for important protein function. As such, there has been an increasing interest in studying such regions in proteins. Unstructured regions are found to be associated with molecular assembly, protein modification and molecular recognition [42–44]. Research shows unstructured regions have a large solvent accessible area, which explains why polar and charged residues which favorably interact with water are prevalent in these regions [45]. The conclusion is that disordered regions are strongly correlated with local solvent accessibility areas. Local solvent accessibility values are often used to find the disordered regions as well [46, 47].

In order to further improve the performance, in this study, we use DISOPRED program to output the predicted possibility of each residue being natively disordered or ordered. Similarly, a residue is encoded by a $2 \times L = 2 \times (2l + 1)$-dimensional vector in this encoding scheme.

Side-chain environment

The concept of side-chain environment was first purposed by Eisenberg et al. [34] and used to identify protein sequences that fold into a known three-dimensional structure. Then Li et al. [39] utilized it for prediction of protein-protein binding site.

The side-chain environment of a residue is typically defined as buried, partially buried, or exposed based on its solvent accessible surface area. The buried and partially buried residue environments can be further subdivided according to the fraction of side-chain area covered by polar atoms. Based on this, we divide the side-chain environment of a residue into six classes (see Fig. 1). The detailed definition of the side-chain environment were described in the work of Eisenberg et al. [34].

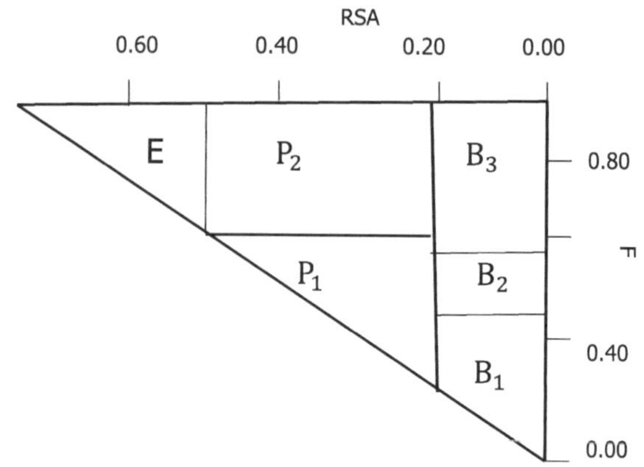

Fig. 1 The definition of the six side-chain environment categories. This figure shows the classification method of side-chain environment. *RSA* represents the relative accessible surface areas and *F* represents the fraction of the whole side-chain area covered by polar atoms. If $RSA < 0.09$, the residue will be placed into class B(buried). If $0.09 \leqslant RSA < 0.36$, the residue will be placed into class P(partial buried). If $RSA \geqslant 0.36$, the residue will be placed into class E(exposed). Within class B, if $F < 0.45$, the residue will be placed into B_1, if $0.45 \leqslant F < 0.58$, the residue will be placed into B_2, and if $F \geqslant 0.58$, the residue will be divided into class B_3. In class P, if $F < 0.67$, the reside will be divided into class P_1, and if $F \geqslant 0.67$, the reside will placed into class P_2

Framework of PredRSA

In this subsection, we describe the PredRSA framework that uses an accurate and effective ensemble computational approach for real values of relative solvent accessibility prediction from protein primary sequences. We are interested in investigating the influence of various sequence-based features and their combinations on the prediction performance of solvent accessibility. In order to fully exploit the sequence-derived features for RSA prediction, we propose a novel PredRSA approach which incorporates five different types of sequence-derived features as inputs. They are four local features (position-specific scoring matrix, secondary structure, conservation score, native disorder) and a global feature (side-chain environment). Figure 2 illustrates the flowchart of our proposed approach.

To determine the optimal local sliding window size L and the iterative tree number M, we calculate the prediction performance for L in the range of 3–17 with a step of 2 and M in the range of 100–1500 with a step of 50 using a grid search method. With $L = 7$ and $M = 800$, the PredRSA approach achieves the best performance for the RSA prediction.

Results and discussion

Datasets

Two non-homologous datasets of proteins chains with pair-wise sequence similarity less than 25 % have been

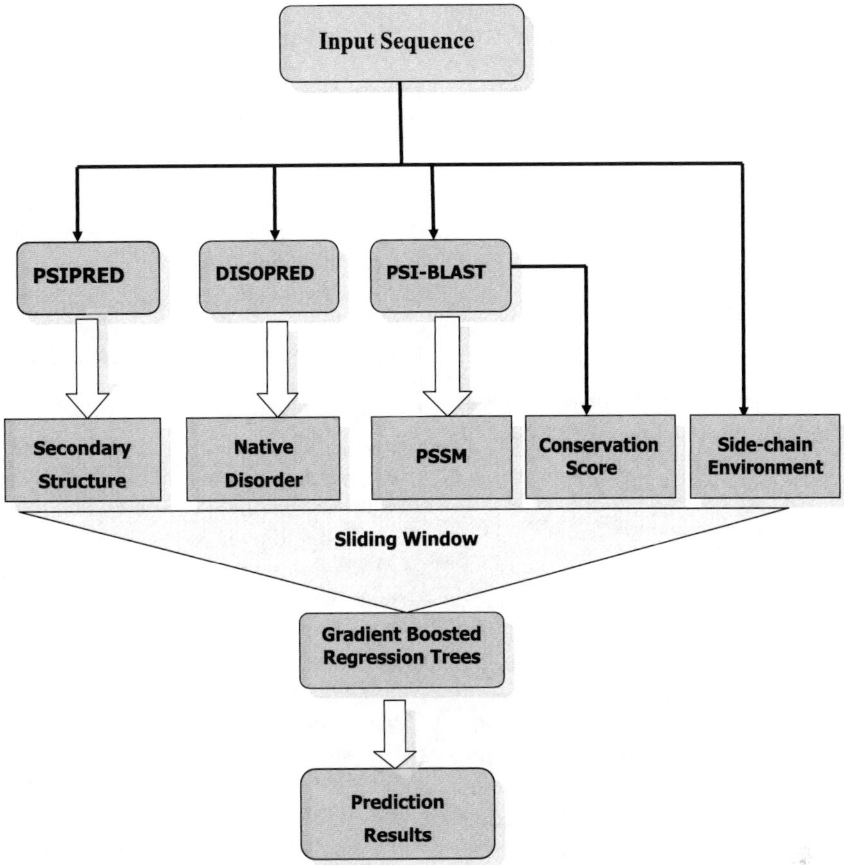

Fig. 2 The framework of PredRSA for protein relative solvent accessibility prediction. Five different types of sequence-derived features are generated and used as input to build the GBRT model. These features consist of PSSM in the form of PSI-BLAST profiles, predicted secondary structure information by PSIPRED, predicted native disorder information by DISOPRED, conservation score and side-chain environment

used in order to objectively compare our approach with other available methods developed previously. One dataset is consisted of 215 proteins, which was also used earlier by Manesh et al. [11] for solvent-accessible surface area of residues prediction. The other dataset is consisted of 502 proteins, obtained from the Cuff and Barton [48] dataset of 513 proteins , selected by removing those sequences, which have less than 30 residues. These two datasets have been referred to as Manesh-215 and CB-502, respectively. However, since the Manesh-215 dataset was widely used by researchers to benchmark prediction methods [18, 20, 25, 49], taking into account comparative purposes, we use Manesh-215 as the main data set for evaluation and analysis.

To further evaluate the performance of existing methods and the method developed in the present study, we also generate an independent dataset of CASP10 proteins. Originally, it contains 85 proteins [50], and we have removed 17 structures (containing chains) by using PISCES culling sever [51] with 25 % sequence similarity cutoff including X-ray (less than 3.0 Å resolution and 0.3 of R-factor) and NMR structures which contain more than 50 residues. Finally, the remaining 68 proteins are used for independent test.

Calculation of RSA

In this work, we take relative solvent accessibility, also called relative solvent accessible surface area (RSA) as the prediction of solvent accessibility. The RSA of a residue in a protein chain is a normalized value from 0–1. It is calculated as the ratio by dividing the solvent accessible surface area (ASA) by the maximum solvent accessibility according to Manesh's work [11] which uses Gly-X-Gly extended tripeptides. The values of ASA are calculated using DSSP [52] for all considered protein structures.

Evaluation measures

To measure the performance of real-valued RSA predictions, three widely used measures for real value RSA prediction are adopted in this study.

The first measure, mean absolute error (MAE), is defined as the average difference between the predicted and experimental RSA values of all residues:

$$MAE = \frac{\sum \left| RSA_{predicted} - RSA_{experimental} \right|}{N} \quad (12)$$

The second measure is the root mean square error (RMSE), which is defined as follows:

$$RMSE = \sqrt{\frac{1}{N} \sum_{N}^{i=1} \left(RSA_{predicted} - RSA_{experimental} \right)^2} \quad (13)$$

The third measure, Pearson correlation coefficient (PCC), the ratio of the covariance between the predicted and experimental RSA values which is given by:

$$PCC = \frac{\sum_{i=1}^{N} (x_i - \bar{x})(y_i - \bar{y})}{\sqrt{\sum_{i=1}^{N} (x_i - \bar{x})^2 \sum_{i=1}^{N} (y_i - \bar{y})^2}} \quad (14)$$

where N is the total number of residues in a protein sequence to predict; x_i and y_i are the experimental and predicted RSA values of the i-th residue, respectively; $\bar{x}$ and $\bar{y}$ are their corresponding means. $PCC = 1$ indicates that the two sets of values are fully correlated, while $PCC = 0$ indicates that they are completely uncorrelated.

Two-state (buried or exposed) predictions are evaluated according to various thresholds of RSA. Prediction accuracy which is defined by the percentage of correctly predicted residues divided by the total number of residues and Matthews correlation coefficient (MCC) are given as follows:

$$ACC = \frac{N_b + N_e}{N} \quad (15)$$

$$MCC = \frac{TP \times TN - FP \times FN}{\sqrt{(TP + FP)(TP + FN)(TN + FP)(TN + FN)}} \quad (16)$$

where N is the total number of residues in a chain, N_b and N_e represent the number of residues correctly predicted as buried and exposed, respectively. TP, TN, FP and FN are the numbers of the true positives, true negatives, false positives and false negatives, respectively.

Effect of different sequence encoding schemes on the prediction performance

We analyze the importance or contribution for each individual feature, which is useful to identify those features that have the most significant influence on overall prediction performance. The performance of each individual predictive is shown in Fig. 3. The feature of side-chain environment is first introduced to predict RSA and it is strongly related to solvent-accessible surface areas.

Table 1 compares the prediction performance of five different combinations of sequence-based features on Manesh-215 with 5-fold cross-validation. As shown in Table 1, the prediction performance of combining all five types of features is the best. It suggests that comprehensive sequence encoding schemes can improve the predictive performance. More importantly, incorporating side-chain environment into the model can significantly increase the prediction performance.

Performance comparison with other regression approaches

In this section, we compare the performance of PredRSA with that of other five existing real value RSA predictors,

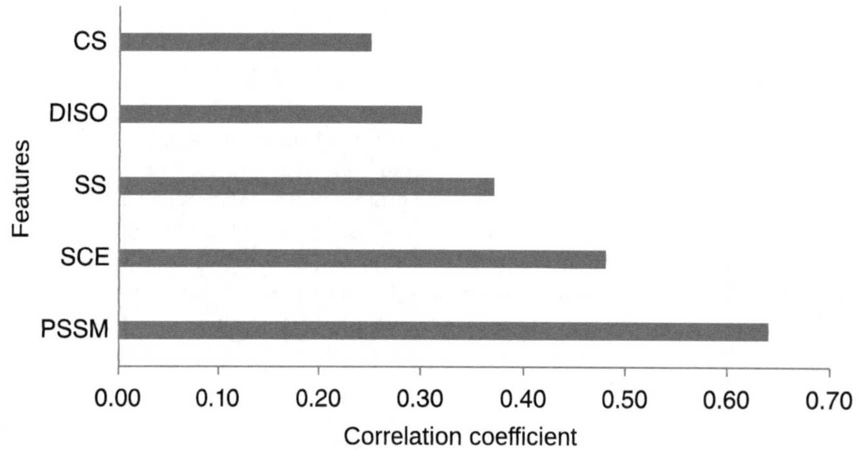

Fig. 3 The importance of the five relevant features used in PredRSA. PSSM, SS, DISO, SCE and CS stand for position specific scoring matrix, protein secondary structure, protein native disorder, side-chain environment and conservation score, respectively

Table 1 Prediction of real-valued RSA using the GBRT algorithm based on five different sequence encoding schemes that incorporate various combinations of sequence features

Feature	RMSE(%)	MAE(%)	PCC
PSSM	13.87	10.26	0.67
PSSM+DISO	13.38	9.93	0.69
PSSM+DISO+SS	13.04	9.64	0.71
PSSM+DISO+SS+SCE	12.23	8.99	0.74
PSSM+DISO+SS+SCE+CS	12.07	8.86	0.75

including a quadratic programming and buriability energy function for solvent accessibility prediction (QBES) [22], a neural network-based method using multiple sequence alignment and secondary structure (SARpred) [25], an improved two-layer neural network (Real-SPINE) [53], a support vector regression using enhanced PSSM features (SVR) [54] and an ensemble of artificial neural networks method (NetSurfP) [55]. Table 2 summarizes the results of these methods. We observe that our method achieves a significantly better performance over the compared predictors. Particularly, the PCC value of PredRSA is approximately 5 % higher than that of the previous predictors on Manesh-215. It is worth to point out that experimental maximum solvent accessibility scores are varied based on different references [11, 56, 57]. A higher maximum solvent accessibility score will lead to a lower RSA value, and thus a relatively lower MAE is obtained according to the definition of MAE. One reason for the differences of MAE between PredRSA and the other methods is that these methods may use different maximum solvent accessibility scores. On the other hand, the prediction precision of PredRSA is higher than that of the other methods and yields a lower MAE.

Performance comparison for two-state prediction

In the past, a plenty of approaches have been proposed for predicting the states (exposed or buried) of residues. Here we examine the performance of our method in terms of two-state prediction. We assign the label of a residue based on its predicted RSA value and a chosen threshold. Table 3 shows the performance of the two-state classification prediction.

Table 2 Performances comparison in predicting real values: PredRSA vs. other existing methods

	Manesh-215		CB-502	
Method	MAE(%)	PCC	MAE(%)	PCC
QBES	-	0.52	-	0.49
SARpred	14.9	0.68	15.9	0.66
SVR	14.2	0.69	14.8	0.68
Real-SPINE	13.8	0.70	14.5	0.68
NetSurfP	13.6	0.70	14.3	0.71
PredRSA	9.0	0.75	9.4	0.73

Table 3 Prediction performance of two-state classification based on different thresholds

	Manesh-215		CB-502		CASP10	
Threshold(%)	ACC(%)	MCC	ACC(%)	MCC	ACC(%)	MCC
5	80.1	0.54	77.9	0.50	78.5	0.48
10	81.7	0.63	79.0	0.58	79.1	0.57
20	81.0	0.61	80.5	0.60	78.3	0.56
25	81.2	0.58	81.0	0.57	79.7	0.56
30	82.4	0.54	82.1	0.52	80.5	0.51
40	87.1	0.42	86.8	0.39	85.0	0.40
50	93.2	0.25	93.0	0.23	91.2	0.30

We also compare the classification accuracy of PredRSA with that of other approaches by different thresholds. The threshold is used to determine the state (exposed or buried) of a predicted real value. For example, a 5 % threshold means a residue is defined as buried if its RSA value is less than 5 %. The methods for comparison include SARpred [25], pace regression algorithm (PR) [58], two-stage SVR [19] and SVR [54]. The prediction accuracy is showed in Table 4. Our method yields more than 80 % classification accuracy at any thresholds and obtains almost the highest accuracy across all the thresholds.

Independent test on the CASP10 dataset

An independent test (CASP10) is constructed to further validate the usability of our PredRSA method. We train the classifiers based on the Manesh-215 dataset and test against the CASP10 dataset which contains 68 proteins. Other state-of-the-art methods including SPINE-X [59] and ASAquick [60] are also evaluated. SPINE-X uses a multistep neural-network algorithm by coupling secondary structure prediction with prediction of solvent accessibility and backbone torsion angles in an iterative manner, while ASAquick utlizes solely sequential widow information and global features with a general neural network method. The Pearson correlation coefficient of PredRSA is 0.71, which outperform the results of SPINE-X and ASAquick by a rate of 2 % (0.69) and 4 % (0.67).

Table 4 Performance comparison of two-state classification: PredRSA vs. other existing predictors

	Accuracy for two-states prediction(%)						
Method	5 %	10 %	20 %	25 %	30 %	40 %	50 %
SARpred	74.9	77.2	77.7	-	77.8	78.1	80.5
PR	76.8	74.8	75.3	76.7	77.7	79.8	86.3
SVR	80.9	80.1	78.7	-	-	-	80.8
Two-stageSVR	81.1	78.7	77.6	77.3	-	-	79.5
PredRSA	80.0	81.6	80.9	81.1	82.2	87.1	93.2

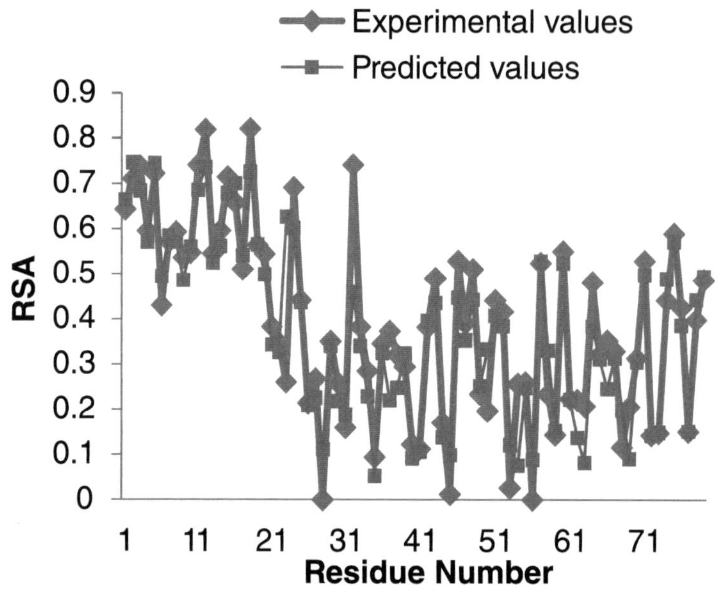

Fig. 4 Predicted and experimental values (%) of RSA for each residue of CASP10 T0675

Case study

For a better understanding of the power of our proposed PredRSA approach and illustrating the significance of PCC, RMSE and MAE measures used in this work, an example of the real-valued RSA for T0675 (Insulinoma-associated protein) from CASP10 is shown in Fig. 4. For this protein, our method gives a MAE of 5.31 %, a RMSE of 7.91 % and a PCC of 0.92. From Fig. 4, we can see that the majority of its predicted RSA values are in good agreement with the corresponding experimental RSA values calculated by DSSP, except for several separate positions.

In Fig. 5, the continuous real-value prediction of RSA and the actual continuous values are shown. Significant correlation between the true values and the predicted values is obtained.

Residue-specific variation in prediction error

In order to assess the prediction performance of various types of residues, we further calculate the average RSA values on the Manesh-215 dataset for all 20 amino acids (Fig. 6) from the PredRSA predictor. As can be seen from Fig. 6, an overwhelming majority of types of amino acids

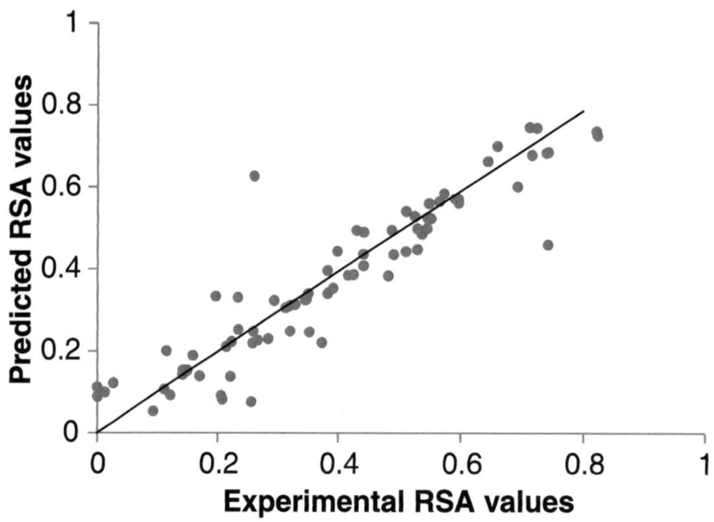

Fig. 5 Correlation between experimental RSA values and predicted RSA values of CASP10 T0675. The Pearson correlation coefficient score is 0.92 and the most buried residues are well predicted with the RSA values near zero

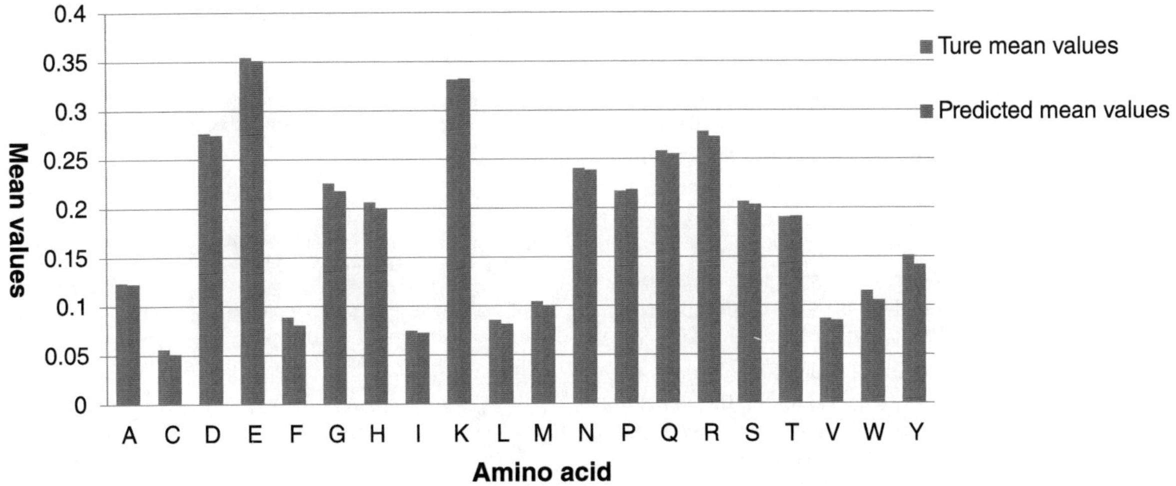

Fig. 6 Comparison between true mean values and predicted mean values for 20 amino acids on the Manesh-215 dataset

are predicted with <1 % mean error. All types of amino acids are predicted with <2 % mean error in our method. In particularly, we find that the true mean RSA values are in highly accord with the predicted mean RSA values for these amino acids, such as A (Ala), K (Lys), N (Asn), T (Thr).

Furthermore, we calculate the prediction errors of 20 amino acids on the Manesh-215 dataset. Figure 7 shows the mean absolute error (MAE) and the standard root mean square error (RMSE) of 20 amino acids. As expected, G (Gly) shows the highest MAE and RMSE due to its flexibility, and other polar residues show similar behavior. Hydrophobic amino acids including C (Cys), F (Phe), M (Met) and W (Trp) are better predicted than less hydrophobic amino acids. These results are also in good agreement with our PredRSA method.

Conclusions

Knowledge of residue solvent accessibility gives useful insights into protein structure and function prediction. In this work, we have presented PredRSA to predict real-valued relative solvent accessibility as well as classification state (buried or exposed) of a target residue. The method is based on a gradient boosted regression trees (GBRT)

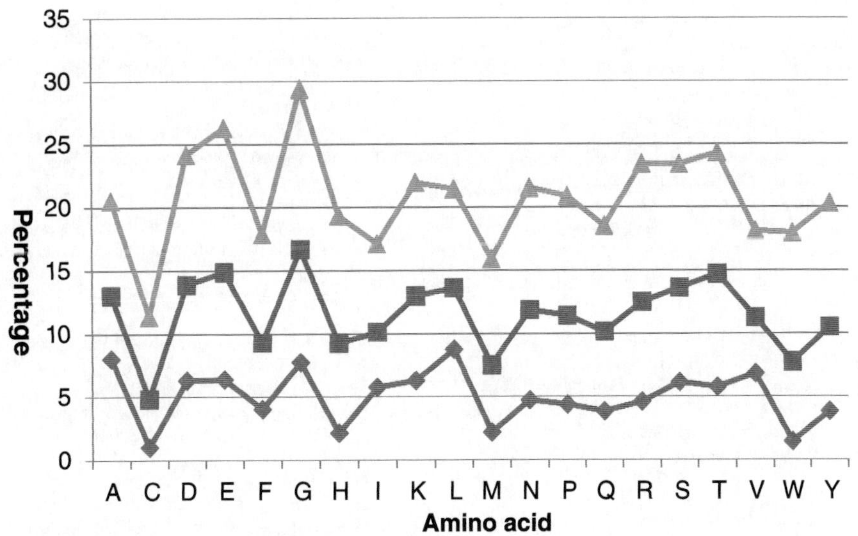

Fig. 7 Mean predicted errors of 20 amino acids on the Manesh-215 dataset. The green line represents standard root mean square error, the red line represents mean absolute error and the blue line represents the corresponding data distribution of 20 amino acids

algorithm combined with a novel set of features. The 5-fold cross-validated correlation coefficient between predicted and experimental RSA (0.75) is significantly better than existing methods on the Manesh-215 dataset. We also performed additional independent benchmark tests of PredRSA on the CASP10 set containing 68 proteins where we find that the proposed method outperforms existing methods. Furthermore, for prediction of discrete state, our method is able to achieve an accuracy of 79.7 % with an MCC value of 0.56 using two states classifications at a threshold of 25 %, which defines an approximately balanced division into the two classes.

Experimental results show GBRT is an efficient machine learning approach for continuous values of the solvent accessibility of a target residue. Compared with other traditional techniques, GBRT has several obvious advantages such as high prediction accuracy and stronger generalization capability.

On the other hand, PredRSA utilizes a variety of multiple sequence-derived features, including the position-specific scoring matrices and conservation score in the form of PSI-BLAST profiles, predicted secondary structure, predicted natively disordered region and side-chain environment. We have comprehensively assessed the effects of different sequence encoding schemes on the prediction performance of RSA, and the results show the prediction performance of RSA outperforms previous methods. Our work provides a complementary and useful approach towards the more accurate prediction of protein solvent accessibility.

Competing interests
The authors declare that they have no competing interests.

Authors' contributions
CF carried out the literature study, developed the new method and drafted the manuscript. CF and DL participated in several independent tests. LD participated in its design and coordination. LD, RH and ZC helped to draft and revise the manuscript. All authors read and approved the final manuscript.

Acknowledgements
This work was supported by National Natural Science Foundation of China under grants No. 61309010 and No. 61379057, China Postdoctoral Science Foundation under grant no. 2015T80886, Specialized Research Fund for the Doctoral Program of Higher Education of China under grant no. 20130162120073 and Shanghai Key Laboratory of Intelligent Information Processing under grant no. IIPL-2014-002.

Declarations
The publication fee of this article is funded by National Natural Science Foundation of China under grant No.61309010. This article has been published as part of BMC Bioinformatics Volume 17 Supplement 1, 2016: Selected articles from the Fourteenth Asia Pacific Bioinformatics Conference (APBC 2016). The full contents of the supplements are available online at http://www.biomedcentral.com/bmcbioinformatics/supplements/17/S1.

Published: 11 January 2016

References
1. Lee B, Richards FM. The interpretation of protein structures: estimation of static accessibility. J mole biol. 1971;55(3):379–4.
2. Eyal E, Najmanovich R, Mcconkey BJ, Edelman M, Sobolev V. Importance of solvent accessibility and contact surfaces in modeling side-chain conformations in proteins. J comput chem. 2004;25(5):712–24.
3. Rost B, Sander C. Conservation and prediction of solvent accessibility in protein families. Proteins Struct Funct Genet. 1994;20(3):216–26.
4. Wodak SJ, Janin J. Location of structural domains in proteins. Biochem. 1981;20(23):6544–52.
5. Liu S, Zhang C, Liang S, Zhou Y. Fold recognition by concurrent use of solvent accessibility and residue depth. Proteins Struct Funct Genet. 2007;68(3):636–45.
6. Eisenberg D, McLachlan AD. Solvation energy in protein folding and binding. Nature. 1986;319(6050):199–203.
7. Mooney C, Pollastri G, Shields DC, Haslam NJ. Prediction of short linear protein binding regions. J mol biol. 2012;415(1):193–204.
8. Zhang QC, Deng L, Fisher M, Guan J, Honig B, Petrey D. Predus: a web server for predicting protein interfaces using structural neighbors. Nucleic acids res. 2011;39(suppl 2):283–7.
9. He B, Wang K, Liu Y, Xue B, Uversky VN, Dunker AK. Predicting intrinsic disorder in proteins: an overview. Cell res. 2009;19(8):929–49.
10. Huang B, Schroeder M. Ligsitecsc: predicting ligand binding sites using the connolly surface and degree of conservation. BMC structural biol. 2006;6(1):19.
11. Naderi-Manesh H, Sadeghi M, Arab S, Moosavi Movahedi AA. Prediction of protein surface accessibility with information theory. Proteins Struct Funct Bioinforma. 2001;42(4):452–9.
12. Ahmad S, Gromiha MM. Netasa: neural network based prediction of solvent accessibility. Bioinforma. 2002;18(6):819–24.
13. Yuan Z, Burrage K, Mattick JS. Prediction of protein solvent accessibility using support vector machines. Proteins Struct Funct Bioinforma. 2002;48(3):566–70.
14. Kim H, Park H. Prediction of protein relative solvent accessibility with support vector machines and long-range interaction 3d local descriptor. Proteins Struct Funct Bioinforma. 2004;54(3):557–62.
15. Sim J, Kim SY, Lee J. Prediction of protein solvent accessibility using fuzzy k-nearest neighbor method. Bioinforma. 2005;21(12):2844–9.
16. Wang JY, Lee HM, Ahmad S. Prediction and evolutionary information analysis of protein solvent accessibility using multiple linear regression. Proteins Struct Funct Bioinforma. 2005;61(3):481–91.
17. Yuan Z, Huang B. Prediction of protein accessible surface areas by support vector regression. Proteins Struct Funct Bioinforma. 2004;57(3):558–64.
18. Xu W, Li A, Wang X, Jiang Z, Feng H. Improving prediction of residue solvent accessibility with svr and multiple sequence alignment profile. Conf Proc IEEE Eng Med Biol Soc. 2005;3:2595–8.
19. Nguyen MN, Rajapakse JC. Two-stage support vector regression approach for predicting accessible surface areas of amino acids. Proteins Struct Funct Bioinforma. 2006;63(3):542–50.
20. Ahmad S, Gromiha MM, Sarai A. Real value prediction of solvent accessibility from amino acid sequence. Proteins Struct Funct Bioinforma. 2003;50(4):629–35.
21. Adamczak R, Porollo A, Meller J. Accurate prediction of solvent accessibility using neural networks–based regression. Proteins Struct Funct Bioinforma. 2004;56(4):753–67.
22. Xu Z, Zhang C, Liu S, Zhou Y. Qbes: predicting real values of solvent accessibility from sequences by efficient, constrained energy optimization. Proteins Struct Funct Bioinforma. 2006;63(4):961–6.
23. Joo K, Lee SJ, Lee J. Sann: solvent accessibility prediction of proteins by nearest neighbor method. Proteins Struct Funct Bioinforma. 2012;80(7):1791–7.
24. Bateman A, Birney E, Cerruti L, Durbin R, Etwiller L, Eddy SR, et al. The pfam protein families database. Nucleic acids res. 2002;30(1):276–80.
25. Garg A, Kaur H, Raghava G. Real value prediction of solvent accessibility in proteins using multiple sequence alignment and secondary structure. Proteins Struct Funct Bioinforma. 2005;61(2):318–24.
26. Song J, Tan H, Wang M, Webb GI, Akutsu T. Tangle: two-level support vector regression approach for protein backbone torsion angle prediction from primary sequences. PloS ONE. 2012;7(2):30361.
27. Friedman JH. Greedy function approximation: a gradient boosting machine. Ann Stat. 2001;29:1189–232.
28. Huber PJ. Robust estimation of a location parameter. Ann Math Stat. 1964;35(1):73–101.

29. Altschul SF, Madden TL, Schäffer AA, Zhang J, Zhang Z, Miller W, et al. Gapped blast and psi-blast: a new generation of protein database search programs. Nucleic acids res. 1997;25(17):3389–402.
30. Deng L, Guan J, Wei X, Yi Y, Zhang QC, Zhou S. Boosting prediction performance of protein–protein interaction hot spots by using structural neighborhood properties. J Comput Biol. 2013;20(11):878–91.
31. Deng L, Zhang QC, Chen Z, Meng Y, Guan J, Zhou S. PredHS: a web server for predicting protein–protein interaction hot spots by using structural neighborhood properties. Nucleic acids res. 2014;42:W290–295.
32. Jones DT. Protein secondary structure prediction based on position-specific scoring matrices. J mol biol. 1999;292(2):195–202.
33. Ward JJ, Sodhi JS, McGuffin LJ, Buxton BF, Jones DT. Prediction and functional analysis of native disorder in proteins from the three kingdoms of life. J mol biol. 2004;337(3):635–45.
34. Bowie JU, Luthy R, Eisenberg D. A method to identify protein sequences that fold into a known three-dimensional structure. Science. 1991;253(5016):164–70.
35. Zhang J, Zhao X, Sun P, Ma Z. Psno: predicting cysteine s-nitrosylation sites by incorporating various sequence-derived features into the general form of chous pseaac. Int J Mol Sci. 2014;15(7):11204–19.
36. Song J, Burrage K, Yuan Z, Huber T. Prediction of cis/trans isomerization in proteins using psi-blast profiles and secondary structure information. BMC bioinforma. 2006;7(1):124.
37. Chen K, Kurgan L. Pfres: protein fold classification by using evolutionary information and predicted secondary structure. Bioinforma. 2007;23(21): 2843–50.
38. Mizianty MJ, Kurgan L. Improved identification of outer membrane beta barrel proteins using primary sequence, predicted secondary structure, and evolutionary information. Proteins Struct Funct Bioinforma. 2011;79(1):294–303.
39. Li N, Sun Z, Jiang F. Prediction of protein-protein binding site by using core interface residue and support vector machine. BMC bioinforma. 2008;9(1):553.
40. Deng L, Guan J, Dong Q, Zhou S. Prediction of protein-protein interaction sites using an ensemble method. BMC bioinforma. 2009;10(1):426.
41. Pugalenthi G, Kumar Kandaswamy K, Chou KC, Vivekanandan S, Kolatkar P. Rsarf: prediction of residue solvent accessibility from protein sequence using random forest method. Protein and peptide letters. 2012;19(1):50–6.
42. Dyson HJ, Wright PE. Intrinsically unstructured proteins and their functions. Nat Rev Mol Cell Biol. 2005;6(3):197–208.
43. Haynes C, Oldfield CJ, Ji F, Klitgord N, Cusick ME, Radivojac P, et al. Intrinsic disorder is a common feature of hub proteins from four eukaryotic interactomes. PLoS Comput Biol. 2006;2(8):100.
44. Gsponer J, Futschik ME, Teichmann SA, Babu MM. Tight regulation of unstructured proteins: from transcript synthesis to protein degradation. Science. 2008;322(5906):1365–8.
45. Schlessinger A, Punta M, Yachdav G, Kajan L, Rost B. Improved disorder prediction by combination of orthogonal approaches. PLoS ONE. 2009;4(2):4433.
46. Zhang H, Zhang T, Chen K, Shen S, Ruan J, Kurgan L. On the relation between residue flexibility and local solvent accessibility in proteins. Proteins Struct Funct Bioinforma. 2009;76(3):617–36.
47. Marsh JA. Buried and accessible surface area control intrinsic protein flexibility. J mol biol. 2013;425(17):3250–63.
48. Cuff JA, Barton GJ. Application of multiple sequence alignment profiles to improve protein secondary structure prediction. Proteins Struct Funct Bioinforma. 2000;40(3):502–11.
49. Wang JY, Ahmad S, Gromiha MM, Sarai A. Look-up tables for protein solvent accessibility prediction and nearest neighbor effect analysis. Biopolymers. 2004;75(3):209–16.
50. The CASP10 Database. http://predictioncenter.org/casp10/groups_analysis.cgi. Accessed 2012.
51. Wang G, Dunbrack RL. Pisces: a protein sequence culling server. Bioinforma. 2003;19(12):1589–91.
52. Kabsch W, Sander C. Dictionary of protein secondary structure: pattern recognition of hydrogen-bonded and geometrical features. Biopolymers. 1983;22(12):2577–637.
53. Faraggi E, Xue B, Zhou Y. Improving the prediction accuracy of residue solvent accessibility and real-value backbone torsion angles of proteins by guided-learning through a two-layer neural network. Proteins Struct Funct Bioinforma. 2009;74(4):847–56.
54. Chang DT, Huang HY, Syu YT, Wu CP. Real value prediction of protein solvent accessibility using enhanced pssm features. BMC bioinforma. 2008;9(Suppl 12):12.
55. Petersen B, Petersen TN, Andersen P, Nielsen M, Lundegaard C. A generic method for assignment of reliability scores applied to solvent accessibility predictions. BMC Struct Biol. 2009;9(1):51.
56. Chothia C. The nature of the accessible and buried surfaces in proteins. J mol biol. 1976;105(1):1–12.
57. Oobatake M, Ooi T. Hydration and heat stability effects on protein unfolding. Prog Biophys Mol Biol. 1993;59(3):237–84.
58. Meshkin A, Sadeghi M, Ghasem-Aghaee N. Prediction of relative solvent accessibility using pace regression. EXCLI J. 2009;8:211–7.
59. Faraggi E, Zhang T, Yang Y, Kurgan L, Zhou Y. Spine x: improving protein secondary structure prediction by multistep learning coupled with prediction of solvent accessible surface area and backbone torsion angles. J comput chem. 2012;33(3):259–67.
60. Faraggi E, Zhou Y, Kloczkowski A. Accurate single-sequence prediction of solvent accessible surface area using local and global features. Proteins Struct Funct Bioinforma. 2014;82(11):3170–6.

Chen et al. BMC Bioinformatics 2016, 17(Suppl 1):9
DOI 10.1186/s12859-015-0852-1

BMC Bioinformatics

PROCEEDINGS Open Access

Learning a hierarchical representation of the yeast transcriptomic machinery using an autoencoder model

Lujia Chen, Chunhui Cai, Vicky Chen and Xinghua Lu*

From The Fourteenth Asia Pacific Bioinformatics Conference (APBC 2016)
San Francisco, CA, USA. 11 -13 January 2016

Abstract

Background: A living cell has a complex, hierarchically organized signaling system that encodes and assimilates diverse environmental and intracellular signals, and it further transmits signals that control cellular responses, including a tightly controlled transcriptional program. An important and yet challenging task in systems biology is to reconstruct cellular signaling system in a data-driven manner. In this study, we investigate the utility of deep hierarchical neural networks in learning and representing the hierarchical organization of yeast transcriptomic machinery.

Results: We have designed a sparse autoencoder model consisting of a layer of observed variables and four layers of hidden variables. We applied the model to over a thousand of yeast microarrays to learn the encoding system of yeast transcriptomic machinery. After model selection, we evaluated whether the trained models captured biologically sensible information. We show that the latent variables in the first hidden layer correctly captured the signals of yeast transcription factors (TFs), obtaining a close to one-to-one mapping between latent variables and TFs. We further show that genes regulated by latent variables at higher hidden layers are often involved in a common biological process, and the hierarchical relationships between latent variables conform to existing knowledge. Finally, we show that information captured by the latent variables provide more abstract and concise representations of each microarray, enabling the identification of better separated clusters in comparison to gene-based representation.

Conclusions: Contemporary deep hierarchical latent variable models, such as the autoencoder, can be used to partially recover the organization of transcriptomic machinery.

Keywords: Yeast, Transcription, Gene expression, Transcriptomic machinery, Signal transduction, Deep learning, Deep hierarchical neural network, Unsupervised learning, Data mining

Background

A cell constantly responds to its changing environment and intracellular homeostasis. This is achieved by a signal transduction system that detects the signals, assimilates the information of diverse signals, and finally transmits its own signals to orchestra cellular responses. Many of such cellular responses involve tightly regulated transcriptomic activities, which can be measured by microarray or RNA-seq technology and used as readouts reflecting the state of the cellular signaling system.

Reverse engineering the signaling system controlling gene expression has been a focus area of bioinformatics and systems biology. However, this task is significantly hindered by the following difficulties: 1) a transcriptomic profile of a cell (with contemporary technology, often a population of cells) at a given time represents a convolution of all active signaling pathways regulating transcription in the cells, and 2) the states of the majority of these signaling pathway are not observed, making it a challenging task

* Correspondence: xinghua@pitt.edu
Department of Biomedical Informatics, University of Pittsburgh, 5607 Baum Blvd, 15237 Pittsburgh, PA, USA

to infer which genes are regulated by a common signal pathway, and it is even more challenging to reveal the relationships among signaling pathways.

Different latent variable models, such as principle component analysis [1], independent component analysis [2], Bayesian vector quantizer model [3], network component analysis [4–6], and non-negative matrix factorization [5, 6] models have been applied to analyze transcriptomic data, with an aim to represent the states of latent pathways using latent variables. Despite the different strengths and limitations of these models, they share a common drawback: the latent variables in these models are assumed to be independent, i.e., the latent variables are organized in single "flat" layer without any connection among them; as such the models lack the capability of representing the hierarchical organization of cellular signaling system.

Figure 1 illustrates the task of reverse-engineering a transcriptomic regulation system. Figure 1a illustrates the well-appreciated hierarchical organization of signaling molecules in cells and how the information encoded by signaling molecules are compositionally organized. It also shows that the convoluted signals eventually are emitted as changed gene expression. At this stage, all the hierarchical information of the signaling system is embedded in the data, a vector of gene expression value, in the form of context-specific and compositional covariance structures. When given a collection of transcriptomic profiles collected under different cellular conditions (Fig. 1b), the ultimate task is to recover the structure of the signaling systems shown in Fig. 1a, but the goal remains unattainable with current methodologies. In this study, we hypothesize that the hierarchical organization of cellular signals can be partially reconstructed by models capable of discovering and representing the context-specific and compositional covariance structure embedded in transcriptomic data. To this end, recent development in deep hierarchical models, commonly referred to as "deep learning" models, e.g., the autoencoder (deep belief network) shown in Fig. 1c, afford us the tools to reverse engineer the signaling systems of cells by mining systematic perturbation data.

In this family of deep hierarchical models, multiple layers of hidden (latent) variables are organized as a hierarchy, which can be used to capture the compositional relationships embedded in the transcriptomic data in a distributed fashion, i.e., different layers can capture different degrees of detail. For example, the relationships between TFs and their target genes can be captured by a hidden variable layer (hereafter referred to as hidden layer) immediately above the observed the layer of observed gene expression variables, whereas the function of pathways regulating TFs can be represented by higher hidden layers. Therefore, deep hierarchical models provide an abstract representation of the statistical structure of the transcriptomic data with flexibility and different degrees of granularity. We hypothesize that, if accurately trained, a deep hierarchical

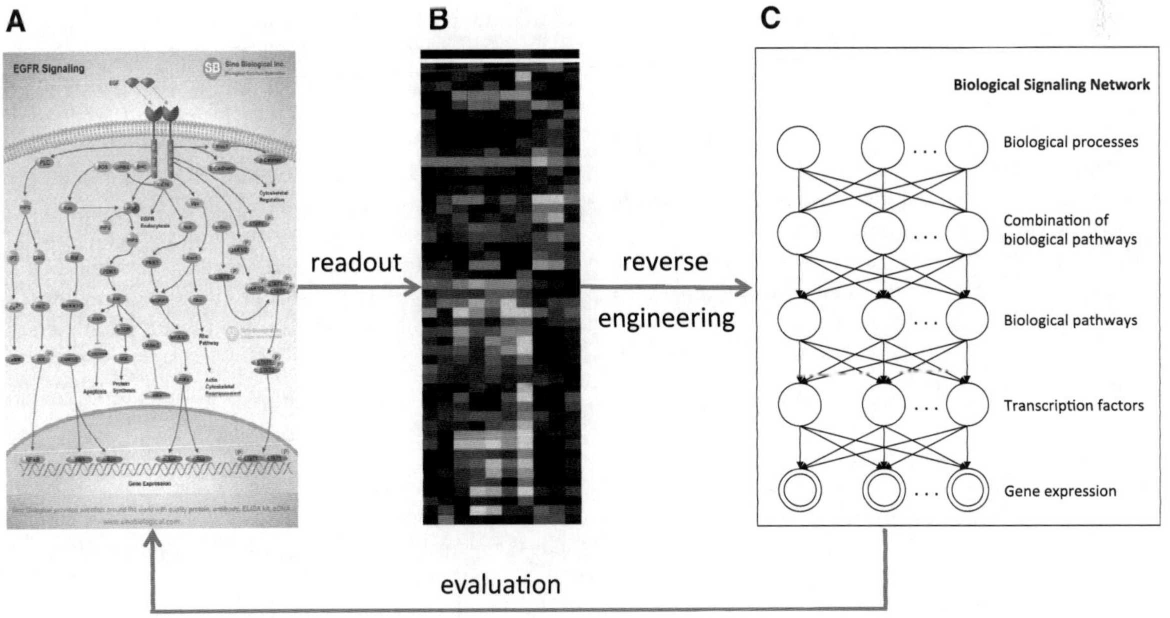

Fig. 1 An overview of studying molecular signaling transduction using an autoencoder. **a** An example of molecular signaling transduction. **b** An example of the heatmap of gene expression microarrays. **c** An autoencoder model consisting of hierarchically organized hidden variables. After the model was trained, we evaluated the information learnt from the autoencoder model by testing whether the information carried by hidden variables in the autoencoder has real biological entities

model can potentially represent the information of real biological entities and further reveal the relationships among them.

In this study, we designed and trained a sparse deep autoencoder model to learn how the information is encoded in yeast cells when subjected to diverse perturbations. Our results indicate that deep learning models can reveal biologically sensible information, thus learning a better representation of the transcriptomic machinery of yeast, and we believe that the approach is applicable to more complex organisms.

Methods

In this study, we investigated using the autoencoder model [7] and sparse autoencoder model [8] to represent the encoding system of the signal transduction systems of yeast cells. Before introducing the autoencoder model and sparse autoencoder model, we will first briefly review restricted Boltzmann machines (RBMs) as building blocks for the autoencoder.

Restricted Boltzmann Machines (RBMs)

A RBM is an undirected probabilistic graphical model that consists of two layers of stochastic binary variables (represented as nodes in the graph): a visible layer $v \in \{0, 1\}^D$ and a hidden layer $h \in \{0, 1\}^F$. The energy function E of the state $\{v, h\}$ of the RBM is:

$$\begin{aligned} \mathrm{E}(v, h; \theta) &= -a^{\mathrm{T}} v - b^{\mathrm{T}} h - v^{\mathrm{T}} W h \\ &= -\sum_{i=1}^{D} a_i v_i - \sum_{j=1}^{F} b_j h_j - \sum_{i=1}^{D}\sum_{j=1}^{F} v_i h_j w_{ij} \end{aligned}$$

In this equation, the binary state of visible variable i is represented by v_i, the binary state of hidden variable j is by h_j and the model parameters are $\theta = \{a, b, W\}$. The bias for visible variable i is a_i the bias for hidden variable j is b_j and the weight between visible variable i and hidden variable j is w_{ij}.

The joint distribution of the hidden and visible variables is defined using a Boltzmann distribution, and the conditional probability of the states of hidden variables and visible variables are as follows:

$$Pr(v, h; \theta) = \frac{1}{Z(\theta)} \exp(-\mathrm{E}(v, h; \theta))$$

$$Z(\theta) = \sum_{v,h} \exp(-\mathrm{E}(v, h; \theta))$$

$$Pr(\mathrm{h}_j = 1|v) = \sigma\left(\mathrm{b}_j + \sum_{i=1}^{m} \mathrm{W}_{ij} \mathrm{v}_i\right)$$

$$Pr(\mathrm{v}_i = 1|h) = \sigma\left(\mathrm{a}_i + \sum_{j=1}^{n} \mathrm{W}_{ij} \mathrm{h}_j\right)$$

where $\sigma(x)$ is the logistic function $1/(1 + \exp(-x))$, m is the total number of visible variables and n is the total number of hidden variables.

The efficient algorithm for learning parameters of the RBM model was introduced in detail in literature and our previous work [7, 9, 10].

Autoencoder

Unlike a RBM, which captures the statistical structure of data using a single layer of hidden nodes, an autoencoder uses multiple layers in a distributed manner, such that each layer captures the structure of different degrees of abstraction. As shown in Fig. 1c, an autoencoder contains one visible (input) layer and one or more hidden layers. To efficiently train the autoencoder, we treat it as a series of two-layered restricted Boltzmann machines (RBM) stacked on top of each other [7, 9]. The inference of the hidden node states and learning of model parameters are performed by learning the RBM stacks bottom-up, which is followed by a global optimization of generative parameters using the back-propagation algorithm. More details of the algorithm and pseudo code for training an autoencoder were discussed in both literature and our previous work [7, 9, 10].

Sparse autoencoder

In a conventional RBM model, each hidden unit is fully connected to the observed variables. After training, there is usually a non-zero weight between each pair of visible and hidden nodes. Based on the assumption that the change in gene expression due to a specific perturbation—most microarrays in this study are experiment-vs-control—is likely meditated by a small number of TFs or pathways, we adopted the sparse autoencoder model [8, 11] to simulate the cellular response to perturbations. The sparse autoencoder model enables one to specify that only a certain percent of hidden nodes have a high probability to be set to 1 ("on") by adding a penalization term to the optimization function. Optimization of the traditional RBM is performed by minimizing the negative log-likelihood of the data during RBM training within an autoencoder:

$$\text{minimize}_{\{\theta\}} - \sum\nolimits_{l=1}^{s} \log \sum\nolimits_{j=1}^{n} Pr\left(\mathrm{v}^l,\ \mathrm{h}_j^l|\theta\right)$$

Where s is the total number of samples, n is the total number of hidden units and $\theta = \{a, b, W\}$. The sparse RBM adds the regularization term [8] into the optimization:

$$\text{minimize}_{\{\theta\}} - \sum_{l=1}^{s} \log \sum_{j=1}^{n} Pr\left(\mathrm{v}^l, \mathrm{h}_j^l|\theta\right) + \lambda \sum_{j=1}^{n} \left| \mathrm{p} - \frac{1}{\mathrm{s}} \sum_{l=1}^{s} \mathrm{E}\left[\mathrm{h}_j^l|\mathrm{v}^l\right] \right|^2$$

Where λ is the regularization constant and p is a constant (usually representing the percent of nodes desired to be on) controlling the sparseness of the hidden units h_j. For the traditional RBM, the parameters are updated just based on the gradient of the log-likelihood term. But for the sparse RBM, the parameters are updated not only based on the gradient of the log-likelihood term but also the gradient of the regularization term.

Non-negative matrix factorization

Non-negative matrix factorization (NMF) has been applied to reduce the dimension of expression data from thousands of genes to a handful of hidden representations (ex. metagenes) [5]. NMF is an algorithm based on decomposition by parts that can reduce the dimension of a matrix V [6].

$$V = W * H$$

Given that the gene expression data is represented as matrix V, NMF factorizes it into a basis matrix (W) and a coefficient matrix (H). All three matrices should have non-negative elements. The number of hidden regulators is pre-defined, and is usually much smaller then the number of genes. In this study, we used the Matlab function nonnegative matrix factorization "nnmf" to perform NMF analysis.

Model selection of autoencoder and sparse autoencoder

We performed a series of cross-validation experiments to search for an "optimal" structure for autoencoders and sparse autoencoders. We adopted a four-layered autoencoder to represent the hierarchical structure of biological processes shown in Fig. 1c. We then explored models with different numbers of hidden units in each hidden layer. We set the initial structure of both autoencoder and sparse autoencoder to the following ranges: $h^{(1)}$: 100–428; $h^{(2)}$: 50–100; $h^{(3)}$: 50; and $h^{(4)}$: 25. We iteratively modified the structure of the model by changing the number of hidden nodes within a layer using a step size of 50 for the first and second hidden layer. Then we explored all combinations in the range stated above. In this case, the total number of models tested is 14 (7*2) for both autoencoder and sparse autoencoder. For the sparse autoencoder, we chose three sparsity constants that are 0.05, 0.1 and 0.15. Under each particular setting, we performed ten fold cross-validation to assess the performance of a model.

We used two criteria of evaluating the performance of the models. One is the reconstruction error, which is the difference between the original input data and the reconstructed data after training the model [12]. Due to the sparse features of the sparse autoencoder, we used Bayesian information criterion (BIC) [13] as another criteria for comparing models. BIC combines the factors of likelihood and number of free parameters to be estimated. The model with the lowest BIC is preferred.

$$BIC = -2 \cdot \ln\hat{L} + k \cdot \ln(n)$$
$$\hat{L} = \prod_{i=1}^{N} p^{m}(1-p)^{1-m}$$

where $\hat{L}$ is the maximized value of the likelihood function of the model, k is the number of free parameters to be estimated, n is the number of samples, p is the probability predicted from the model for a gene to be active in an experiment, and m is the true binary state of a unit in the input data.

Mapping between the hidden units and known biological components

Based on the weights between each hidden unit in the first hidden layer and all the visible units (genes), we used a threshold (top 15 % of the absolute values of weights) to cut the edges between a hidden node and the observed genes, such that an edge indicates that the hidden node regulates the gene. We then identified all genes predicted to be regulated by a hidden node as a gene set. Based on the DNA-Protein interaction table [14, 15], we also identified the gene set regulated by a known TF. We then assessed the significance of overlapping of gene sets regulated by hidden nodes and TFs using hypergeometric testing.

Consensus clustering of experiment samples

Consensus cluster clustering [16] was used to cluster the experiment samples using different datasets as input. The R implementation of ClusterCons [17] was downloaded from CRAN (https://cran.r-project.org/src/contrib/Archive/clusterCons/). The inputs for consensus clustering are the samples represented using original gene expression values, NMF megagenes values and the states of hidden variables under all experiment samples respectively. The partition around medoids (PAM) and *K*-means algorithms were used as base clustering algorithms. The inputs for cluster by cluster consensus clustering are the samples represented using samples clusters derived from the nodes from different hidden layers as features. If one sample belongs to a sample cluster, its input value is 1. Otherwise, its input value is 0.

Finding pheromone related hidden units

We calculated the significance between the state of a hidden node and the state of proteins related to pheromone signaling pathway by using the chi-square test. First, we used a threshold (top 15 %) to designate the state of a hidden unit as active or inactive based on its activation probability. Then, for each hidden unit, we created a contingency table to collect the counts of the joint state of the hidden node and whether any member of the pheromone pathway is perturbed in a specific experiment. We used the contingency table to perform the chi-square test. We used a p-value of 0.01 as the significance threshold.

Gene ontology analysis

GO [18] provides a standard description of what gene products do and where they are located. One of the frequently used databases that provide GO information for yeast is Saccharomyces Genome Database SGD. We first used the combination of weights [19] between neighboring hidden layers to get the weights between the hidden units in a particular hidden layer and the genes. A gene is regarded as being regulated by a hidden unit if their weight is in the top 15 % of all weights. When a gene set of interest associated with a hidden unit is available, we used the method mentioned in [20] to summarize the GO terms capturing as much as semantic information associated with those genes [21]. We identified the GO terms that could summarize the largest number of genes, while undergoing a minimal information loss.

Results and discussion

Training different models for representing yeast transcriptomic machinery

We collected a compendium of 1609 yeast cDNA microarrays from the Princeton University Microarray Database (puma.princeton.edu), and we combined them with 300 microarrays from the study by Hughes et al. [22], which was used in a previous study of the yeast signaling system [23]. The combined dataset is ideal for studying the yeast signaling system because it represents a large collection of perturbation experiments that are of biological interest. For example, the data from the study by Hughes et al. [22] were collected from yeast cells with genetic perturbations (deletion of genes) or chemical treatments, and similarly the microarrays from the database were collected from specific conditions and contrasted with "normal" growth condition. Taking advantage of the experiment-vs-control design of cDNA microarrays, we identified differentially expressed genes (3-fold change) in each array and retained 2228 genes that were changed in at least 5 % of the microarrays. We then represented the results of each microarray experiment as a binary vector, in which an element represented a gene, and its value was set to 1 if the gene was differentially expressed, and 0 otherwise. Thus, each microarray represented the *transcriptomic changes* in response to a certain condition, presumably regulated by certain specific signaling components, which is unknown to us.

We investigated the utility of the autoencoder model (also known as deep belief network) [9], with one observed layer representing the microarray results and 4 hidden variable layers (hereafter referred to as hidden layers) representing the yeast signaling components in yeast transcriptomic machinery. In this model, a hidden node is a binary variable, which may reflect the state of a collection of signaling molecules or a pathway, such that the switching of the node state between 1 and 0 can reflect the changing state of a pathway.

The probabilistic distribution of the state of a node in a given layer is dependent on the nodes in the adjacent parent layers, defined by a logistic function. The directed edges between nodes of adjacent layers indicate that, conditioning on the state of nodes in parent layer, the nodes in a child layer are independent. In other words, the statistical structure (patterns of joint probability of nodes) among the nodes in a child layer is captured by the nodes in the parent layer. For example, in our case, if the nodes in the 1st hidden layer (directly above the gene expression layer) represent the states of transcription factors, then co-differential expression (covariance) of a set of genes is solely dependent on (or explained by) the TFs that regulate the genes. Similarly, the co-regulation of TFs is determined by its parent layer, which may reflect the state of signaling pathways. Thus, this model is suited to capture the context-specific changes and compositional relationship among signaling components in a distributed manner. The model is referred to as autoencoder because, when given a collection of observed data, it learns to use hidden nodes to encode the statistical structure of observed data, and it is capable of probabilistically reconstructing the observed data.

Since the autoencoder model in our study is biologically motivated, we hypothesize that the nodes in the first hidden layer would likely capture the signal of TFs. Thus the number of nodes in this layer should be close to the number of known TFs for yeast, of which there are around 200 well-characterized yeast TFs [24]. However, for a given microarray from a perturbation experiment, genes that respond to a specific perturbation are likely regulated by a few transcription factors. Thus we also investigated a model referred to as *sparse autoencoder* [8, 25], which performs regularized learning of model parameters and allows a user to constrain the percent of nodes in a layer that can be set to the "on" state, see Methods for details. In our experiment, we constrained that, in the first hidden layer, around 10 % of hidden nodes should be used to encode the changes in a microarray.

We first evaluated how adding a sparse regularization term influenced the state of hidden units (the probability of hidden units to be active/on). We trained a conventional autoencoder and a sparse autoencoder (setting the sparsity constraint to 10 %) using the microarrays. For each microarray, the models probabilistically inferred the state of each hidden node (the probability of a node to take a value of 1). Figure 2 shows the histogram of

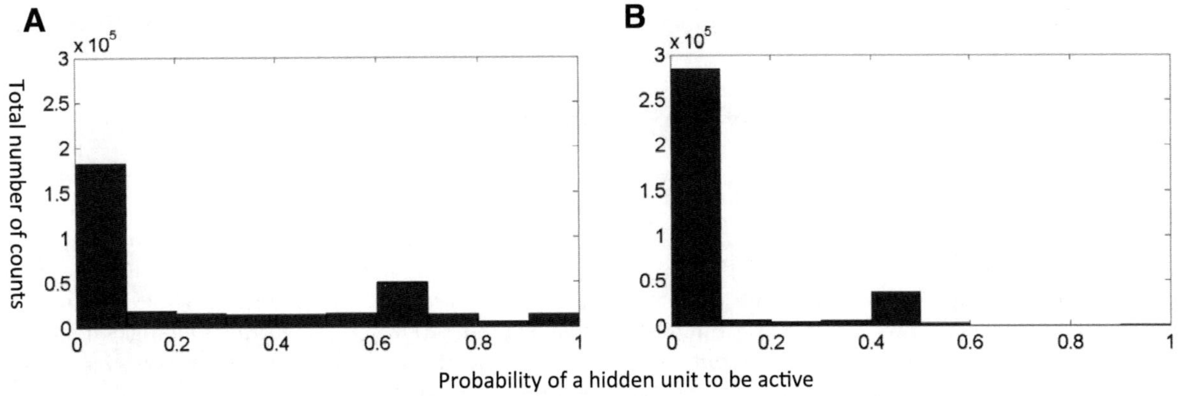

Fig. 2 The histogram of the expected states of hidden units (probability of hidden units to be on) in the first hidden layer for the conventional autoencoder (**a**) and sparse autoencoder (**b**) respectively. For both models, the number of hidden units from the first hidden layer to the fourth hidden layer is 214, 100, 50, and 25 respectively. The sparsity threshold for the sparse autoencoder is 0.1. A hidden unit has a state under each experiment condition. Therefore, the total number of states for all hidden units is the number of experiment condition (1609) * the number of hidden units (214). The x-axis is the probability of a hidden unit to be on ranging from 0 to 1. The y-axis is the count of states

the expected states of the nodes in the first hidden layer associated with all microarray samples. In the conventional autoencoder, a relatively larger number of the nodes in the first hidden layer had a non-zero probability to be 1 ("on state") (Fig. 2a), whereas the majority of the hidden nodes in the sparse autoencoder model were expected to take a value of 0 ("off state") (Fig. 2b). Thus, the sparse autoencoder strives to use less hidden nodes to encode the same statistical structure in the observed data, instead of using every hidden node, with each contributing a little to the expression of genes. This is a desired property conforming to our assumption that the response to a specific perturbation should be encoded by a relatively small number of TFs.

We further evaluated how well models with different architectures (mainly concentrating on the number of hidden nodes in the first hidden layer) can be used to represent the transcriptomic machinery of yeast. Table 1 shows the results of a limited model selection experiment based on our biological assumptions (note that an exhaustive search of possible combinations of architecture and parameter settings is intractable). We calculated the reconstruction error, which is the sum of the differences across all microarrays between the observed expression state (0 or 1) of genes in microarrays and the expected states of genes reconstructed by the autoencoder. The results indicate that models of the same type (conventional or sparse autoencoder) could learn to encode data with a similar accuracy across the range of the architectures studied here, although the sparse autoencoder had higher reconstruction errors.

While the results indicate that the reconstruction errors of sparse autoencoder models were a bit higher than the ones of the traditional autoencoder, it should be noted that the sparse autoencoder reconstructed the same data with a much smaller number of hidden variables. From the perspective of the minimum description length (MDL) principle [26], a model is preferred if it can encode the information of a dataset with a minimal description length while achieving a similar or better reconstruction of data. In information theory, the description length is measured as the number of bits needed to encode the data, and in our case each bit is encoded by a hidden node. Thus, the sparse autoencoder potentially is a more desirable model even if it suffers a higher reconstruction error. To quantify and compare the utility

Table 1 Reconstruction error of models with different architectures

Reconstruction error	Architecture 1 (100:100:50:25)	Architecture 2 (150:100:50:25)	Architecture 3 (214:100:50:25)	Architecture 4 (428:100:50:25)
Autoencoder training	150.20	150.23	148.94	150.81
Autoencoder test	188.57	189.12	190.76	189.63
Sparse autoencoder training (0.1)	170.19	170.07	170.41	171.94
Sparse autoencoder test (0.1)	206.58	208.40	203.99	203.27

Table 2 BIC scores of different models

	Arch 1 (100,100,50,25)	**Arch 2** (214, 100, 50, 25)
Autoencoder	**3.25e + 006** = 1.99e + 006 + 1.26e + 006;	**4.41e + 006** = 1.71e + 006 + 2.70e + 006;
Sparse autoencoder	**2.06e + 006** = 1.93e + 006 + 1.26e + 005;	**1.96e + 006** = 1.69e + 006 + 2.70e + 005;

The cells show the BIC score (bold) and the individual terms of the BIC (see Methods). The numbers in the parentheses associated with each architecture (Arch) indicate the number of hidden nodes in 1st – 4th hidden layers

of conventional and sparse autoencoders, we calculated the Bayesian information criteria (BIC) of the models, and the results are shown in Table 2. The results indicate that the BIC of the sparse autoencoder with an architecture consisting of hidden layers with 214, 100, 50, 25 hidden nodes (1st to 4th) respectively is the lowest (the best) among the compared models. Since the number of hidden nodes in the first hidden layer of this model agrees better with the knowledge of the number of transcription factors, we chose to investigate the results derived from this model in the following sections.

Distributed representation enhances discovery of signals of TFs

The motivation of using a hierarchical model is to allow latent variables in different hidden layers to capture information with different degree of abstraction in a distributed manner. When modeling transcriptomic data, one goal is to discover the signals of TFs. In a sparse autoencoder, it is natural to expect that the 1st hidden layer should capture the signals encoded by TFs. We test this hypothesis by evaluating the overlap of the genes predicted to be regulated by a hidden node in the 1st hidden layer and those known to be regulated by a TF. A statistically significant overlapping between them is shown in Fig. 3.

Indeed, the results indicate that sparse autoencoder is capable of capturing and representing the information of TFs, in that there is an almost one-to-one mapping between hidden nodes and known TFs as shown in Fig. 3a. For a few hidden variables that are significantly mapped to multiple TFs, we further investigated if these TFs are members of known TF complexes [27]. As an example, we found that a hidden node is significantly mapped to *AFT1* and *PUT3*, which are two yeast TFs known to cooperatively regulate genes involved in ion transportation [28]. As another example, a hidden node is mapped to both *MSN2* and *MSN4* [29], which belong to the same family and form hetero-dimers to regulate genes involved in general stress response.

To demonstrate the advantage of hierarchical and distributed representations, we compare our results with another latent variable model commonly used to represent microarray data, the non-negative matrix factorization model (NMF) [6]. The NMF can be thought of as a model consisting of an observed layer (gene expression) and a single hidden layer (hidden variables/metagenes [5]), which is used to capture all signals in embedded in microarrays, whereas the same information is distributed to multiple layers of hidden variables in the sparse

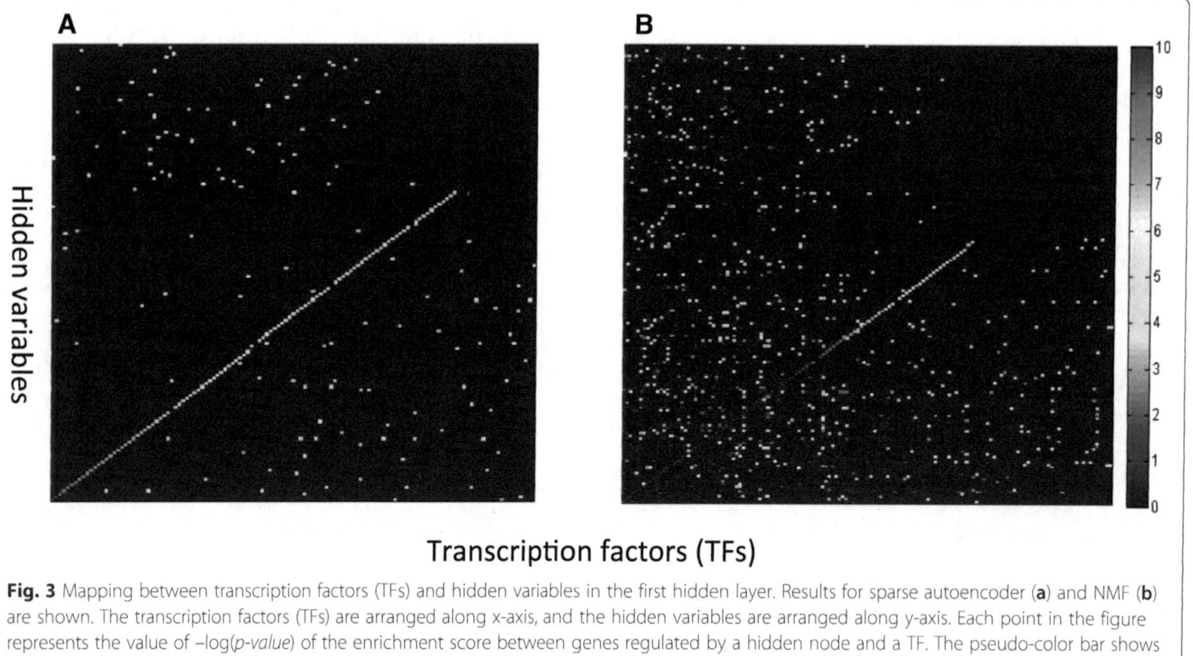

Fig. 3 Mapping between transcription factors (TFs) and hidden variables in the first hidden layer. Results for sparse autoencoder (**a**) and NMF (**b**) are shown. The transcription factors (TFs) are arranged along x-axis, and the hidden variables are arranged along y-axis. Each point in the figure represents the value of −log(*p-value*) of the enrichment score between genes regulated by a hidden node and a TF. The pseudo-color bar shows the scale of the −log(*p-value*)

autoencoder. We trained a NMF model with 214 "metagenes", which is the same as the number of hidden nodes in the 1st hidden layer of the sparse autoencoder, and the results of mapping between latent factors and TFs are shown in Fig. 3b.

Indeed, the results clearly demonstrate the expected difference between the two models. With the capability of capturing the context-specific and compositional relationship of signals regulating expression in a distributed manner, the hidden nodes in the 1st hidden layer clearly capture the specific signals of TFs, whereas the signals regulated TFs are delegated to the higher level hidden nodes. In contrast, with only a single layer of latent variables, all signals in the data need to be "squeezed" into these latent variables, such that a latent factor (a "metagene") has to represent the signal of multiple TFs. Therefore, the results support our hypothesis that, benefitting from the distributed representation of the statistical structures across multiple hidden layers, the sparse encoder can concisely learn and represent the information of biological entities, in this case the TFs.

Latent variables can capture the information of signaling pathways

We further investigated whether certain hidden nodes can represent the states of well-known yeast signaling pathways, i.e., whether the state of a hidden node can be mapped to the state of a collection of proteins in a pathway. In a previous study [23], we were able to recover the pheromone signaling pathway and a set of target genes whose transcription were regulated by the pheromone pathway, by mining the systematic perturbation data from the study by Hughes et al. [22] in which 14 genes involved in yeast pheromone signaling were perturbed by gene deletion. In the current study, we identified the microarray experiments in which the aforementioned 14 pheromone-pathway genes were perturbed, and we examined if the state of any hidden node was statistically associated with perturbation of pheromone pathway, using the chi-square test (see Methods). Interestingly, we found 2 hidden nodes in the 1st hidden layer that are significantly associated with the perturbation experiments, one with a chi-square test $p \leq 2.47e\text{-}05$, and the other with a $p \leq 3.82e\text{-}02$. Further inspecting the genes predicted to be regulated by these hidden nodes, we found a significant overlap between the pheromone target genes from our previous study and the genes regulated by these hidden nodes (data not shown). These results indicate that the hidden nodes of the sparse autoencoder model are capable of capturing the signals of specific yeast pathways. However, it should be noted that, by design, a hidden node in the high level layers of the sparse autoencoder might encode the signals of multiple pathways that share strong covariance.

The hierarchical structure captures signals of different degrees of abstraction

One advantage of the hierarchical structure of an autoencoder is to represent information with different degrees of abstraction. Intuitively, the lower level hidden layers should capture the highly specific signaling pathways or signaling molecules, such as TFs, whereas the high level hidden layers may encode more general information. To test this hypothesis, we identified the genes regulated by each hidden node by performing a linear weight combination experiment [19] (multiplication of weights between different hidden layers). We then applied a semantic analysis method previously developed by our group [20], which identifies the most appropriate concept from the Gene Ontology (GO) to summarize the genes. Interestingly, we found that genes regulated by AFT1 and PUT3 are significantly enriched among the genes regulated by a hidden node in the 2nd hidden layer, and the genes regulated by this hidden node is summarized by the GO term GO:0006826 (*iron ion transportation*), shown in Fig. 4. Using the same method, we found another hidden node whose related genes were annotated with GO:0006357 (*regulation of transcription from RNA polymerase II promoter*). Thus, the results indicate that the distributed representation of information enables the hidden nodes at the different level of hierarchy to capture information of different degrees of abstraction.

Concise representation enhances the discovery of global patterns

Given a large comprehensive dataset, it is often interesting to learn if distinct perturbations affect common biological processes of the cell [22, 23]. A common approach to discover such patterns is to perform clustering analysis and examine if certain samples (thereby experimental perturbations) are clustered together. In general, the result of a clustering analysis is significantly influenced by whether the features representing each sample are informative. Non-informative features may not reveal any real information, whereas features concisely reflecting the states of cellular signaling system may provide insights regarding the system. To examine if the signals represented by the latent variables are more informative than original gene expression values and NMF metagene values, we represented the samples in our dataset using the original gene expression values, NMF metagene values and the expected states of the hidden nodes in a hidden layer respectively, and we then compared the results of consensus clustering (Fig. 5).

The results clearly demonstrate that, if samples are represented in the high-dimensional gene expression space, the majority of the samples cannot be grouped into distinct clusters. When we use the low-dimensional NMF metagenes space, it performs slightly better than

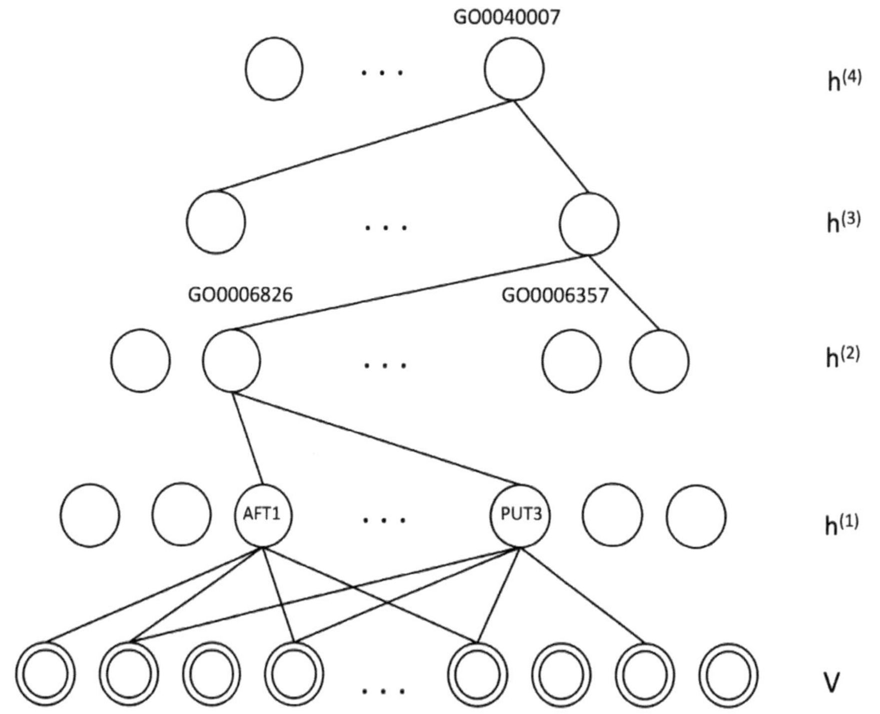

Fig. 4 Example of hierarchical Gene Ontology (GO) map for hierarchical hidden structure

using original gene expression space. But, the clusters derived are still not well separated. In contrast, when samples were represented using the expected states of hidden nodes of the 1st hidden layer, the samples can be consistently separated into distinct clusters. Representing samples using the expected states of the nodes from other hidden layers also produced clearly separated clusters (data not shown). The results indicate that the states of hidden nodes are much more informative in terms of representing distinct characteristics of individual samples, thus enabling clean separation of samples by the clustering algorithms. Although it would be interesting to systematically inspect the common characteristics of the samples in terms of whether the perturbation experiments affect common signaling pathways, such an analysis requires broad and in depth

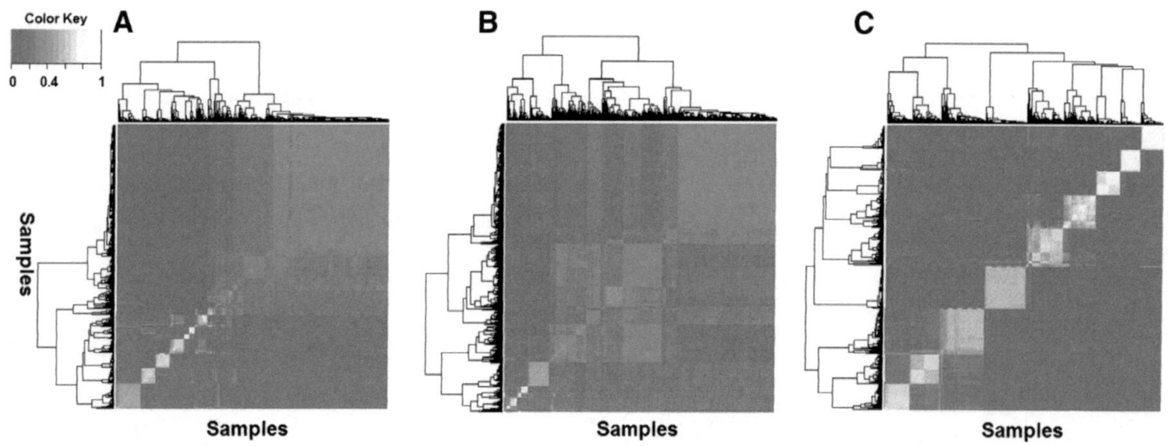

Fig. 5 Clustering of experiment samples represented using original gene expression data (**a**), NMF metagenes (**b**) and expected state of hidden nodes in the first hidden layer (**c**). Consensus clustering results show how consistently a set of samples is assigned into a common cluster during repeated clustering experiments using samples with replacement from a dataset. A yellow box indicates a set of samples that are consistently assigned to a common cluster, and the brightness of yellow reflects the consistency

knowledge of yeast genetics, which is beyond the expertise of our group.

Information embedded in data is consistently represented in different hidden layers

We hypothesized that, in a successful hierarchical representation of a dataset, the information embedded in data should be consistently encoded by different hidden layers, even though two different layers are of different dimensionality and identity of the nodes are totally different. In other words, when a sample is represented by the state of the nodes from different hidden layers, the characteristics that distinguish this sample from others (or make it similar to others) should be retained despite being represented by the nodes from different layers. To test this hypothesis, we first performed consensus clustering using nodes from different hidden layers as features to get sample clusters, and then we compare if members within a cluster derived using one representation significantly overlap with the members from another cluster derived using a different representation. Figure 6 shows the results of assessing the overlaps of the samples clusters derived using the nodes from the 1st and 2nd hidden layers as features respectively. The results indicate that the majority of the clusters derived with different representations agree. Interestingly, a cluster derived from the 2nd hidden layer as features subsumes (maps to) two clusters derived using the 1st hidden layer as features, indicating that the 2nd layer captures more general information. The results indicate that, even though the dimensionality and identity of features of each hidden layer are significantly different, the information encoded by the hidden nodes in a sparse autoencoder is preserved across different layers.

Conclusion

In this study, we investigated the utility of contemporary deep hierarchical models to learn a distributed representation of statistical structures embedded in transcriptomic

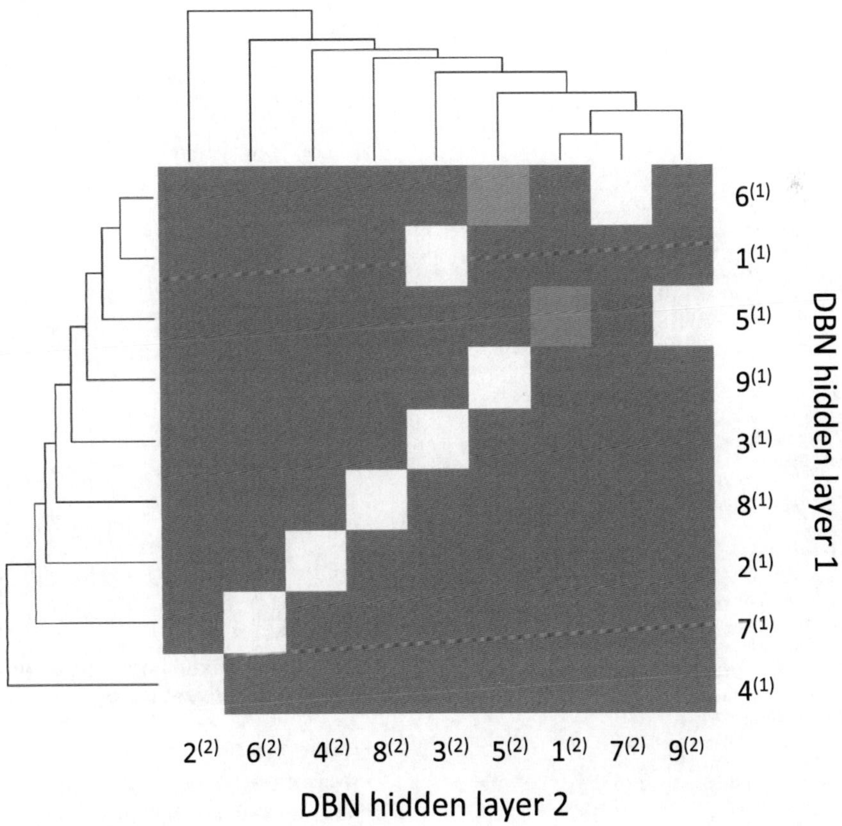

Fig. 6 Cluster by cluster clustering for clusters in the 1st hidden layer and the 2nd hidden layer. The x-axis is the 9 sample clusters using the states of hidden units in the 2nd hidden layer indexed by the superscript (2). The y-axis is the 9 samples clusters using the state of hidden units in the 1st hidden layer indexed by the superscript (1). For example, $1^{(1)}$ is the first sample cluster using the states of hidden units in the 1st hidden layer and $1^{(2)}$ is the first sample cluster using the states of hidden units in the 2nd hidden layer. A yellow box indicates that the members of two clusters (derived from different representation) significantly overlap, with the brightness of yellow reflecting the degree of overlap

data. We show that such a model is capable of learning biologically sensible representations of the data and revealing novel insights regarding the machinery regulating gene expression. We anticipate that such a model can be used to model more complex systems, such as perturbed signaling systems in cancer cells, thus directly contributing to the understanding of disease mechanisms in translational medicine.

Competing interests
The authors declare that they have no competing interests.

Authors' contributions
LC and XL designed the study, performed the computational experiments, interpreted the results and wrote the manuscript. CC participated in the design of the study, collected and preprocessed the experiment data. VC performed the GO analysis and participated in the manuscript writing. All authors read and approved the final manuscript.

Acknowledgements
The authors would like to thank Dr. Ruslan Salakhutdinov for discussion. This work is partially supported by the following NIH grants: R01LM012011, R01LM010144, R01LM011155. The publication cost is covered by the NIH grant R01LM011155.

Declarations
This article has been published as part of *BMC Bioinformatics* Volume 17 Supplement 1, 2016: Selected articles from the Fourteenth Asia Pacific Bioinformatics Conference (APBC 2016). The full contents of the supplements are available online at http://www.biomedcentral.com/bmcbioinformatics/supplements/17/S1.

Published: 11 January 2016

References
1. Raychaudhuri S, Stuart JM, Altman RB. Principal components analysis to summarize microarray experiments: application to sporulation time series. Pac Symp Biocomput. 2000:455–466.
2. Liebermeister W. Linear modes of gene expression determined by independent component analysis. Bioinformatics. 2002;18(1):51–60.
3. Lu X, Hauskrecht M, Day RS. Modeling cellular processes with variational Bayesian cooperative vector quantizer. Pac Symp Biocomput. 2004:533–544.
4. Liao JC, Boscolo R, Yang YL, Tran LM, Sabatti C, Roychowdhury VP. Network component analysis: reconstruction of regulatory signals in biological systems. Proc Natl Acad Sci U S A. 2003;100(26):15522–7.
5. Brunet JP, Tamayo P, Golub TR, Mesirov JP. Metagenes and molecular pattern discovery using matrix factorization. Proc Natl Acad Sci U S A. 2004;101(12):4164–9.
6. Devarajan K. Nonnegative Matrix Factorization: An Analytical and Interpretive Tool in Computational Biology. Plos Comput Biol. 2008;4(7):e1000029.
7. Hinton GE, Osindero S, Teh YW. A fast learning algorithm for deep belief nets. Neural Comput. 2006;18(7):1527–54.
8. Lee HE, Ekanadham, C, Ng, A.Y. Sparse deep belief net model for visual area V2. Advances in Neural Information Processing Systems 2008.
9. Hinton GE, Salakhutdinov RR. Reducing the dimensionality of data with neural networks. Science. 2006;313(5786):504–7.
10. Chen L, Cai C, Chen V, Lu X. Trans-species learning of cellular signaling systems with bimodal deep belief networks. Bioinformatics. 2015.
11. Goh HT, Thome, N, Cord M. Biasing Restricted Boltzmann Machines to Manipulate Latent Selectivity and Sparsity. NIPS. 2010.
12. Vincent P, Larochelle H, Bengio, Y.. Extracting and composing robust features with denoising autoencoders. Proceedings of the 25th international conference on Machine learning 2008.
13. Posada D, Buckley TR. Model selection and model averaging in phylogenetics: Advantages of akaike information criterion and Bayesian approaches over likelihood ratio tests. Syst Biol. 2004;53(5):793–808.
14. Huang SS, Fraenkel E. Integrating proteomic, transcriptional, and interactome data reveals hidden components of signaling and regulatory networks. Sci Signal. 2009;2(81):ra40.
15. Yeger-Lotem E, Riva L, Su LJ, Gitler AD, Cashikar AG, King OD, et al. Bridging high-throughput genetic and transcriptional data reveals cellular responses to alpha-synuclein toxicity. Nat Genet. 2009;41(3):316–23.
16. Monti S, Pablo T, Mesirov J, Golub T. Consensus Clustering: A Resampling-Based Method for Class Discovery and Visualization of Gene Expression Microarray Data. Mach Learn. 2003;52(1–2):91–118.
17. Simpson TI, Armstrong JD, Jarman AP. Merged consensus clustering to assess and improve class discovery with microarray data. BMC bioinformatics. 2010;11:590.
18. Ashburner M, Ball CA, Blake JA, Botstein D, Butler H, Cherry JM, et al. Gene Ontology: tool for the unification of biology. Nat Genet. 2000;25(1):25–9.
19. Dumitru EC, A., Bengio,Y. Understanding representations learned in deep architectures. Technical report,1355, Universite de Montreal/DIRO. 2010.
20. Chen V, Lu X. Conceptualization of molecular findings by mining gene annotations. BMC Proc. 2013;7 Suppl 7:S2.
21. Lu S, Lu X. Integrating genome and functional genomics data to reveal perturbed signaling pathways in ovarian cancers. AMIA Joint Summits on Translational Science proceedings AMIA Summit on Translational Science. 2012;2012:72–8.
22. Hughes TR, Marton MJ, Jones AR, Roberts CJ, Stoughton R, Armour CD, et al. Functional discovery via a compendium of expression profiles. Cell. 2000;102(1):109–26.
23. Lu S, Jin B, Cowart LA, Lu X. From data towards knowledge: revealing the architecture of signaling systems by unifying knowledge mining and data mining of systematic perturbation data. PLoS ONE. 2013;8(4):e61134.
24. Harbison CT, Gordon DB, Lee TI, Rinaldi NJ, Macisaac KD, Danford TW, et al. Transcriptional regulatory code of a eukaryotic genome. Nature. 2004;431(7004):99–104.
25. Ng A. Sparse autoencoder. CS294A Lecture notes. 2011.
26. Barron AR, Rissanen J, Yu B. The Minimum Description Length Principle in Coding and Modeling. Information Theory, IEEE Transactions on. 1998;44(6):2743–60.
27. Hill CS, Marais R, John S, Wynne J, Dalton S, Treisman R. Functional analysis of a growth factor-responsive transcription factor complex. Cell. 1993;73(2):395–406.
28. Alkim C, Benbadis L, Yilmaz U, Cakar ZP, Francois JM. Mechanisms other than activation of the iron regulon account for the hyper-resistance to cobalt of a Saccharomyces cerevisiae strain obtained by evolutionary engineering. Metallomics. 2013;5(8):1043–60.
29. Fabrizio P, Pozza F, Pletcher SD, Gendron CM, Longo VD. Regulation of longevity and stress resistance by Sch9 in yeast. Science. 2001;292(5515):288–90.

Yang *et al. BMC Bioinformatics* 2015, **17**(Suppl 1):10
DOI 10.1186/s12859-015-0853-0

BMC Bioinformatics

PROCEEDINGS Open Access

Missing value imputation for microRNA expression data by using a GO-based similarity measure

Yang Yang[1,2*], Zhuangdi Xu[1] and Dandan Song[3]

From The Fourteenth Asia Pacific Bioinformatics Conference (APBC 2016)
San Francisco, CA, USA. 11 - 13 January 2016

Abstract

Background: Missing values are commonly present in microarray data profiles. Instead of discarding genes or samples with incomplete expression level, missing values need to be properly imputed for accurate data analysis. The imputation methods can be roughly categorized as expression level-based and domain knowledge-based. The first type of methods only rely on expression data without the help of external data sources, while the second type incorporates available domain knowledge into expression data to improve imputation accuracy.
In recent years, microRNA (miRNA) microarray has been largely developed and used for identifying miRNA biomarkers in complex human disease studies. Similar to mRNA profiles, miRNA expression profiles with missing values can be treated with the existing imputation methods. However, the domain knowledge-based methods are hard to be applied due to the lack of direct functional annotation for miRNAs. With the rapid accumulation of miRNA microarray data, it is increasingly needed to develop domain knowledge-based imputation algorithms specific to miRNA expression profiles to improve the quality of miRNA data analysis.

Results: We connect miRNAs with domain knowledge of Gene Ontology (GO) via their target genes, and define miRNA functional similarity based on the semantic similarity of GO terms in GO graphs. A new measure combining miRNA functional similarity and expression similarity is used in the imputation of missing values. The new measure is tested on two miRNA microarray datasets from breast cancer research and achieves improved performance compared with the expression-based method on both datasets.

Conclusions: The experimental results demonstrate that the biological domain knowledge can benefit the estimation of missing values in miRNA profiles as well as mRNA profiles. Especially, functional similarity defined by GO terms annotated for the target genes of miRNAs can be useful complementary information for the expression-based method to improve the imputation accuracy of miRNA array data. Our method and data are available to the public upon request.

Background

Missing values are commonly present in microarray data due to various reasons, such as the limitation on detection sensitivity, contamination or error induced in experimental operations, and inappropriate data preprocessing. For the sake of accurate analysis on expression profiles, the estimation of missing values has been a basic and key issue in microarray data analysis.

Up to now, a variety of methods have been proposed for imputing missing values in gene expression profiles, which mainly fall into two categories. The first type is based solely on expression data, such as KNNImpute [1], LLSimpute [2] and Bayesian method [3]. The second type is based on domain knowledge in addition to expression data [4, 5].

*Correspondence: yangyang@cs.sjtu.edu.cn
[1]Department of Computer Science and Engineering, Shanghai Jiao Tong University, 800 Dongchuan Rd., Shanghai 200240, China
[2]Key Laboratory of Shanghai Education Commission for Intelligent Interaction and Cognitive Engineering, Shanghai 200240, China
Full list of author information is available at the end of the article

K-nearest neighbor (KNN) is the most widely used algorithm for missing value imputation, and many other methods implement various modified versions of KNN process. For example, assume that gene g_1 whose value on the first sample is missing, then the Euclidean distances of expression levels between g_1 and other genes are measured as Eq. 1.

$$d_{g_1,g_j} = \sqrt{\frac{\sum_{i=s_1}^{s_n} (G_{1,i} - G_{j,i})^2}{n}}, \quad (1)$$

where d_{g_1,g_j} denotes the distance between g_1 and g_j. $G_{1,i}$ and $G_{j,i}$ are the expression values of g_1 and g_j at the ith sample, respectively. $\{s_1, \cdots, s_n\}$ are the indexes of n samples where both g_1 and g_j have valid values. It should be noted that g_1 and g_j may have more than one missing values at different samples, and only the samples that both genes have valid values are considered. Thus the squared differences of expression values are averaged over n, i.e. the number of samples considered.

Once the distances between g_1 and all other genes have been computed, K nearest genes are selected and a weighted average of their expression values on the first sample is assigned to $G_{1,1}$.

In many cases, gene expression level alone cannot guarantee good imputation result because of noise, bias or a large proportion of missing values. Actually, the microarray experimental results are not simply numerical matrices. Instead, these expression data should be treated with the consideration of biological background, such as functional similarity, biological pathways that genes involved and interactions between gene products. If the external annotation data sources can be effectively incorporated into the algorithms, the imputation quality would be improved.

Common annotation data include Gene Ontology (GO) [6], KEGG pathways [7], and GenMAPP [8] etc. Since GO annotation covers most of human gene products, GO is most widely used for measuring functional similarity between genes. GO provides a controlled and standardized vocabulary, which has become an important tool and knowledge base in bioinformatics research. GO consists of three organizing principles, i.e. cellular component, biological process and molecular function. It represents gene characteristics of the three categories in terms, which are the basic elements of the GO database organized in directed acyclic graphs (DAG) including nodes (terms) and relationships (edges). The semantic similarity between GO terms regarding the locations in the DAG and common ancesters/descendands directly reflects the association between gene products [9, 10].

In the past two decades, many efforts have been put on studying the semantic similarity of GO terms. Early methods refer to related studies in natural language [11–13], and mainly consider information content of GO terms, as defined in Eq. 2. The drawback of these methods is that they care little about structural information, and treat the GO structure as a tree instead of a DAG.

$$IC(x) = -\log p(x) = -\log\left(\frac{|G_x|}{|G_{root}|}\right), \quad (2)$$

where $p(x)$ is the probability of an attribute described by a GO term x, G_x is the set of genes that are associated with GO term x (i.e. genes annotated by x and all of their descendants), and G_{root} is the set of genes that are associated with the root term (i.e. all genes that the DAG can annotate). Here, we use 'associated with' instead of 'annotated with' because in GO DAGs, parent nodes are generalized concept of children nodes, thus a gene annotated with some terms is also associated with their parent nodes. Recent methods [14–17] consider more structural information, such as G-SESAME [15] (http://bioinformatics.clemson.edu/G-SESAME/), which considers both common ancestors of two GO terms and their relative location in the GO graph.

Tuikkala et al. [4] investigated whether semantic similarity could improve the performance of missing value imputation. Their experimental results demonstrated that even a small proportion of annotated genes can provide improvements in data quality helpful for imputation. Besides GO, other biological knowledge has also been applied to missing value imputation. Xiang et al. [18] proposed a histone acetylation information aided imputation method, called HAIimpute, which incorporates histone acetylation information into conventional KNN and LLS algorithms. Ni et al. used protein-protein interaction (PPI) annotation data, and proposed PPI-KNNimpute and PPI-LLSimpute methods [5]. All of these three studies tested their methods on yeast cDNA microarray data, benefitting by the comprehensive annotation available for yeast.

In recent years, miRNA microarrays have become more and more common because miRNAs play important roles in many biological processes and the development of complex diseases. Especially, the correlation of miRNA expression profiles with clinical-pathological characteristics of cancer patients has been largely investigated. MiRNA expression data is of great help for tumor classification and can be effective biomarkers for prognosis even more precisely than protein-coding genes [19, 20]. The miRNAs expression profiles also have missing values due to the low expression intensity of miRNAs. Basically, the estimation of missing values in miRNA arrays can follow the same strategies for DNA arrays. However, the domain knowledge is hard to be utilized because there is no direct

GO annotation or similar standardized functional annotation for miRNAs. Therefore, the imputation of missing values for miRNA arrays with domain knowledge is a relatively tough job and seldom studied.

Considering that miRNAs play their roles mainly by regulating the expression of target genes at post-transcription level, the functional similarity of miRNAs can be inferred based on GO semantic similarities of their target genes [21]. Identification of target gene is an important issue and also a well-studied bioinformatics task in miRNA study. Some miRNA databases record known targets, such as TarBase [22], which collects manually curated target genes with experimental support. Widely used computational tools for predicting targets of mammalian miRNAs include picTar [23], TargetScan [24], miRanda [25, 26], etc. The GO annotation of target genes can be searched in GO database and pairwise similarity can be defined. Then the functional similarity between miRNAs can be computed according to their target genes' similarities.

We combine miRNA functional similarity with Euclidean distance of expression profile and develop a new imputation method. The new method is evaluated on two miRNA datasets from breast cancer research, on which the new method achieves improved performance compared with the expression-based method. The experimental results demonstrate that the functional knowledge retrieved from target genes of miRNAs can help to improve the estimation of missing values for miRNA expression profiles.

Results and discussion

Data source

In this study, we used two public miRNA data sets from NCBI [27], namely GSE26659 [28] and GSE40525 [29]. Both profiles were measured by Agilent-019118 Human miRNA Microarray 2.0 G4470B containing probes for 723 human and 76 human viral microRNAs from the Sanger database v.10.1.

In order to evaluate the performance of the new method, the original data sets were preprocessed by removing miRNAs with missing values. Then new data sets with different percentages of missing values were simulated from the preprocessed complete data. Considering that missing values are generally not evenly distributed over the whole profile, we did not generate the data sets by randomly selecting missing elements. Instead, we randomly selected 50 % of the miRNAs which were assumed to have incomplete profiles, and then randomly assigned missing elements in each of these miRNAs' profiles. The missing rate was set to be a random integer percentage between 1 and p % (maximum missing rate). Obviously, the bigger p is, the more missing values are present in the generated data sets. When p is equal to 20, it can be calculated that the expected value of missing rate of the whole simulated data set is 5 %. By changing the value of p, we can get simulated sets with different missing rates. The generation of data sets was repeated 100 times. Different imputation methods were tested and evaluated on all the simulated data set, and the accuracies were averaged to obtain the final result.

Evaluation criteria

There are multiple criteria for evaluating imputation algorithms. A commonly used one is the normalized root mean square error (*NRMSE*). Let G_{est} be the estimated data matrix by the imputation algorithm, and G_{true} be the matrix of the original complete data matrix. The *NRMSE* is defined as:

$$NRMSE = \frac{\sqrt{mean[\,(G_{est} - G_{true})^2]}}{\sigma[\,G_{true}]}, \tag{3}$$

where σ denotes standard deviation.

Experimental results

Based on semantic similarity retrieved from GO graph and BMA (best match average) strategy for calculating similarity of two sets, we obtain the GO-based functional similarity matrix H of miRNAs. Before proceeding to the evaluation of the new method, we conduct a hierarchical clustering and draw a heat map of H. Figure 1(a) and (b) show clustering results for the data sets extracted from GSE26659 and GSE40525, respectively. It can be observed that the miRNAs are grouped into four large clusters distinguished by depths of color. Different from conventional imputation method, given the new similarity measure, the nearest neighbors of a certain miRNA are more likely chosen within its function cluster. We use the new similarity measure in KNN imputation method. Figure 2(a) and (b) compare the *NRMSE* of conventional KNN and GO-based KNN with different values of K for GSE26659 and GSE40525, respectively, where maximum missing rate is 40 % (the expected value of missing rate for the whole simulated data set is 10 %). GO-based KNN has better performance than conventional KNN no matter what the value of K is. The results also suggest that K has great impact on the accuracy of imputation. The two data sets have a similar pattern of varying accuracies, and the optimal value of K is 4 or 5, which is much smaller than that used in the imputation for mRNA expression profiles.

Figure 3(a) and (b) compare expression-based and GO-based methods at different maximum missing rates, by using KNN and LLS (local least squares). The maximum missing rate, i.e. p, is an upper bound of the percentage of missing values. For example, when p equals 40 %, the missing rate is randomly assigned between 1 and 40 %. In GSE40525 data set which has 120 samples, a miRNA which is selected to have missing values can

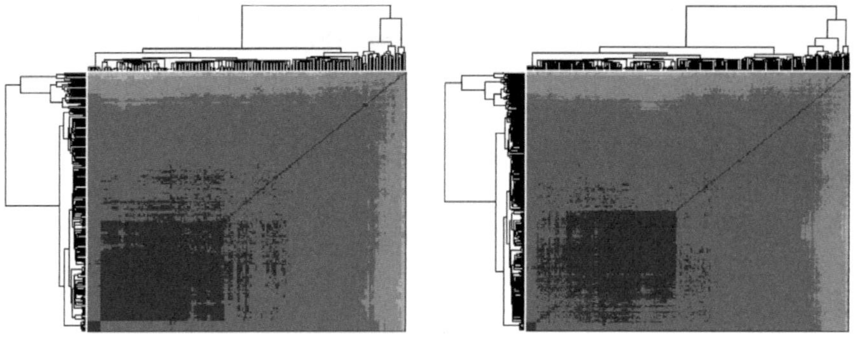

(a) GSE26659 (b) GSE40525

Fig. 1 Hierarchical clustering of functional similarity matrix of miRNAs

have 48 missing values at most and one missing value at least with equal probability. Such generation of missing values is more practical than completely random model with regard to the real distribution in microarray data. Here K is set to be 5. Solid lines denote conventional methods, and dotted lines denotes GO-based methods. These two figures show that incorporating GO knowledge can improve the imputation accuracy for both KNN and LLS. The improvement in LLS is not as obvious as in KNN, since we only use GO-based measure to select neighbors in LLS, thus the GO information has less impact on the performance of LLS than that of KNN. Generally, the LLS method has much less *NRMSE* while takes more computation time compared with KNN.

Compared with previous studies which utilize GO information for the imputation of missing values on mRNA expression data, the improvement on miRNA data is not that significant. Actually, we have found that the *NRMSE* of miRNA data sets using conventional imputation methods is much less than that of mRNA data. The possible reasons are: a) miRNA profiles with hundreds of miRNA have much smaller scale than mRNA profile with tens of thousands of probes; b) the expression levels of the same miRNA vary slightly across samples. In addition, we observe that higher maximum missing rate does not necessarily results in bigger *NRMSE*. Similar cases are also present in mRNA expression data sets [4]. And the non-completely random generation of missing values may also increase the probability of such result.

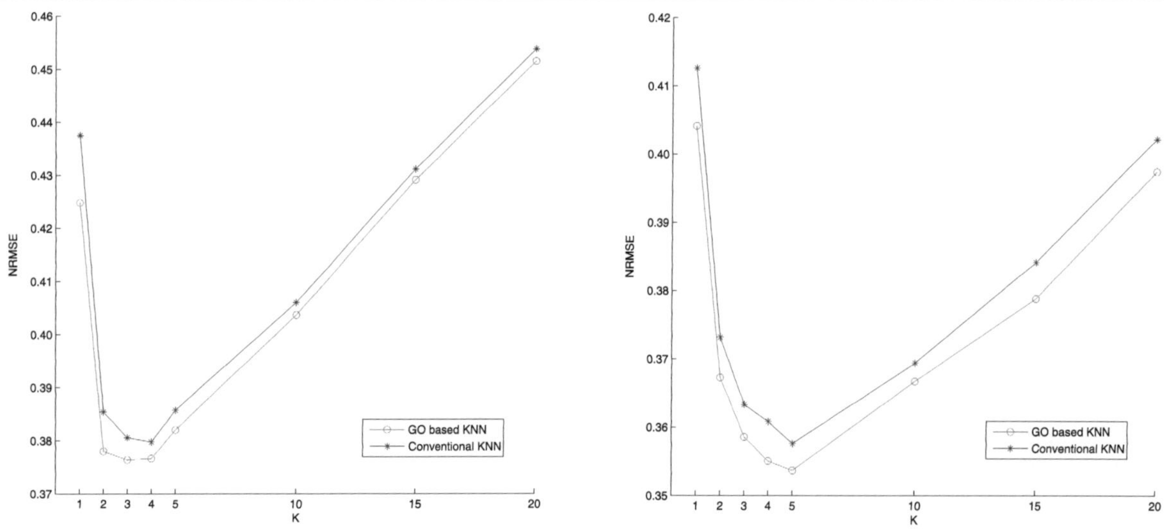

(a) GSE26659 (b) GSE40525

Fig. 2 Comparison of NRMSE at different values of K

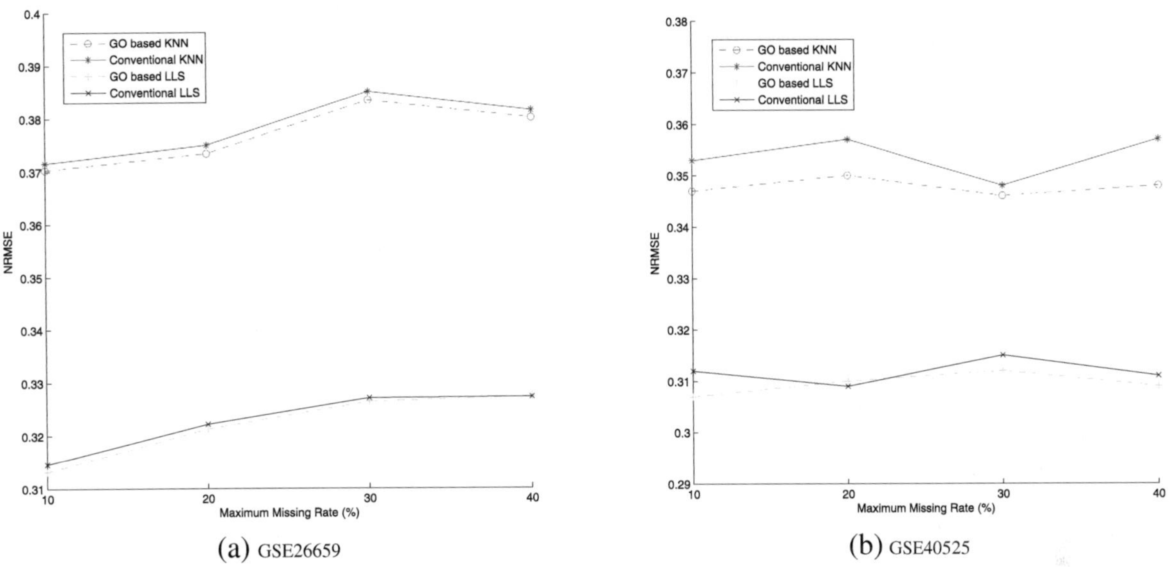

Fig. 3 Comparison of NRMSE at different maximum missing rates

Discussions

This study focuses on effective incorporation of biological knowledge into the imputation of missing values in miRNA microarray data. Unlike protein-coding genes, the functional annotation of miRNAs cannot be utilized directly because no such database is available. Since miRNAs play their roles mainly by suppressing or silencing their target genes' expression, two miRNAs may have strong correlation on their function and expression pattern if their targets are involved in the same biological process or have similar function. By transferring GO-based pairwise similarity of genes to the similarity of two target sets, the functional similarity of two miRNAs can be indirectly inferred.

Although the experiments have demonstrated the performance improvement, it is obvious that biological knowledge can only be treated as a data source complementary to expression levels. On one hand, function similarity does not directly correlate with expression similarity [30]. Similar function only indicates that two miRNAs have similar expression pattern with high probability. On the other hand, the process of calculating GO-based similarity for miRNAs has some drawbacks.

Firstly, the GO-based semantic similarity does not accurately reflect gene function similarity. Although we retrieve semantic information in the Molecular Function DAG, and the G-SESAME method has considered structural information, there is still a lot of information lost here. For example, the method only considers common ancestors but neglects common descendants. Moreover, even though we have a perfect method which comprehensively utilizes GO graph information, this semantic similarity is not necessarily corresponds to real functional similarity because the annotation is incomplete.

Secondly, in order to cover as many miRNAs in our data set as possible, we adopt computational tools for target gene identification, which may have unreliable results although we have set a threshold to control false positives and false negatives.

In addition, some process steps within the whole workflow, which have great impact on the final results, need further studies. The first issue is the simulation strategy for generating the evaluation set with artificial missing values. Although the partial-random model used in this study seems more reasonable than complete random model, how to simulate missing values according to their real distribution in microarray profiles needs careful studies. The second issue is the integration strategy from pair-wise GO similarity to miRNA similarity. Here a two-step best match average method is conducted, while other strategies may be examined. For example, each miRNA can be directly regarded as a GO set with redundant GO terms from its targets. A pairwise similarity for such redundant set can be defined.

As more and more miRNA functions have been revealed, available annotation for miRNAs increase rapidly, and other biological knowledge with regard to miRNAs can also be utilized. For example, Wang et al. [30] used microRNA-associated diseases to define miRNA functional similarity. Undoubtedly, the miRNA functional similarity provides complementary but important information to ensure the quality of data analysis on miRNA expression profiles.

Conclusions

Biological domain knowledge has been demonstrated to be very helpful to improve the imputation of missing values in DNA microarray data profiles. However, because of the lack of function annotation for miRNAs, the domain knowledge-based imputation methods have rarely been studied for miRNA expression data. In this paper, we propose to use the pairwise functional similarity of miRNAs based on the semantic similarity of target genes in GO graph as a complementary information for expression data and define a new similarity measure. This new measure can work together with conventional *K*NN and other imputation methods. The experiments have been conducted on two breast cancer miRNA expression profiles, and the normalized root mean square error (*NRMSE*) is compared under different values of *K* and maximum missing values. The GO-based new imputation method shows decreased *NRMSE* on both data sets.

Our method is applicable to not only human miRNAs but also other species as long as the target genes can be obtained. With the rapid accumulation of miRNA microarray data, the studies on domain knowledge-based function similarity for miRNAs are of great need to improve expression data quality and eventually benefit for the identification of novel miRNA biomarkers and novel potential functions of miRNAs.

Methods

As shown in Fig. 4, our method consists of four major parts. The first part builds a connection between miRNAs and target genes. The second part builds another connection between genes and GO terms. The third part focuses on the computation of pairwise similarity of GO terms, and the last part implements *K*NN (or other imputation algorithm) with the new similarity measure. The whole workflow can be summarized in the following steps:

1. Identify target genes of miRNAs via computational prediction tools or public databases
2. Search GO terms annotated for each target gene obtained in Step 1
3. Compute similarity for each pair of GO terms obtained in Step 2
4. Compute similarity for each pair of target genes from the similarity of their GO terms
5. Compute similarity of each pair of miRNAs from the similairty of their target genes
6. Define new similarity measure by combining GO-based similarity and expression level-based distance
7. Impute missing values with the new measure

GO similarity

In order to compute the functional similarity of two genes g_1 and g_2 from their GO annotation, we firstly retrieve GO term list of each target gene for miRNAs. Two genes correspond to two GO sets, i.e., $T_{g_1} = \{t_1, t_2, \cdots, t_k\}$, $T_{g_2} = \{t'_1, t'_2, \cdots, t'_s\}$, and the computation of gene pairwise similarity turns out to be the similarity between two GO sets. There are two issues here. The first issue is

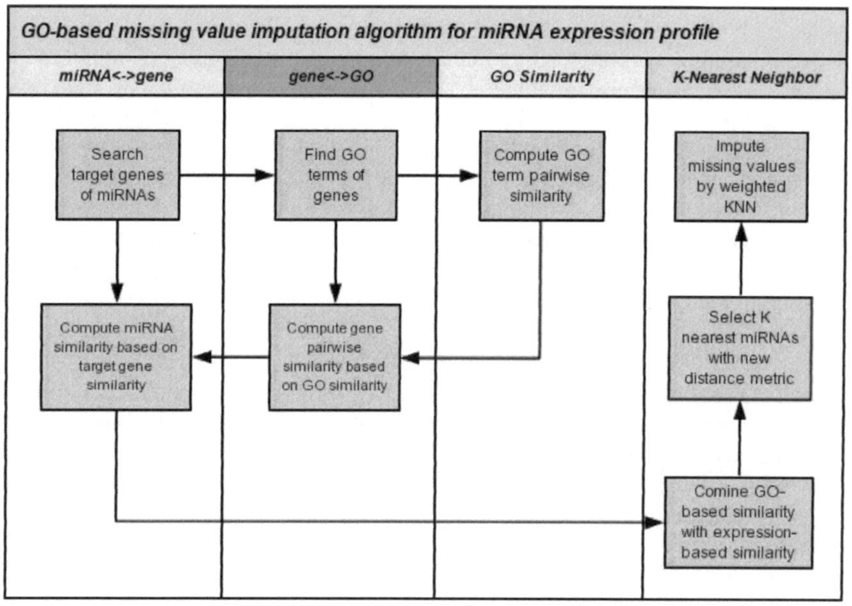

Fig. 4 Flowchart of GO-based imputation algorithm for miRNA expression data

to define the similarity between two GO terms according to their correlation in the GO DAGs. We adopt G-SESAME implemented in R package GOSemSim [31], and compute GO similarity in the DAG of Molecular Function (MF), which is a widely used knowledge base of gene function. The second issue is how to integrate the pairwise similarities to the similarity between two sets.

Given two sets of GO terms T_{g_1} and T_{g_2}, which have k and s terms respectively, and similarity of each pair of GO terms, there are four basic methods to obtain gene similarity, i.e. the similarity of these two sets, namely *max*, *avg*, *rcmax* and *BMA* [31]. The former two methods compute similarity for each pair of (t_1, t_2), where $t_1 \in T_{g_1}$ and $t_2 \in T_{g_2}$, and use the maximum (max) or average (avg) value to represent the similarity of the two genes. These two methods have obvious disadvantages. The *max* strategy fails to consider the overall similarity of two sets. For example, it regards the similarity of two sets as 1 even if they have only one term in common. As for the *ave* strategy, even though the two sets are exactly the same, the similarity between two sets is not 1. The *rcmax* method works like following:

1. For each term $t_i \in T_{g_1}$, find the maximum similarity between t_i and the terms in T_{g_2}
2. Get average value of the k maximum values obtained for all terms in T_{g_1} from 1), and name it as max_1
3. For each term $t_j \in T_{g_2}$, find the maximum similarity between t_j and the terms in T_{g_1}
4. Get average value of the s maximum values obtained for all terms in T_{g_2} from 3), and name it as max_2
5. Assign $max(max_1, max_2)$ to the gene similarity

rcmax is often biased when the sizes of GO sets differ considerably. The last method *BMA* (best match average) has the same steps 1) and 3) of rcmax. The difference is that it calculates the overall average value of k maximum values of terms in T_{g_1} and s maximum values of terms in T_{g_2}. The gene similarity is defined in Eq. 4.

$$sim_{BMA}(g_1, g_2) = \frac{\sum_{i=1}^{k} \max_{1 \le j \le s} sim(t_i, t'_j) + \sum_{j=1}^{s} \max_{1 \le i \le k} sim(t_i, t'_j)}{k+s} \tag{4}$$

BMA has been used in many studies [15, 21]. Here we also adopt *BMA* to compute gene similarity based on GO pairwise similarity. Since each miRNA corresponds to a set of target genes, miRNA similarity can be computed by gene similarity in the same manner.

In this study, we use computational prediction tools to identify targets of miRNAs, because the experimental-supported target databases, like TarBase, only house a small portion of miRNAs in the data sets. We have found that miRanda provides prediction results for the most miRNAs in our date sets than other tools. In order to avoid too many false positives, we set a threshold of mirSVR score ≤ -1 to ensure the prediction accuracy. It should be noted that the naming of some miRNAs is inconsistent between databases and miRNA array. We refer to miRBase [32] to deal with the inconsistency.

GO-based imputation algorithm

In our method, we combine the GO-based functional similarity with the Euclidean distance obtained from expression profiles as a new measure. Here we adopt a similar strategy used in Tuikkala's work [4]. Given the functional similarity matrix H and Euclidean distance matrix D, each element of the new distance matrix C is defined as Eq. 5.

$$C_{i,j} = (1 - H_{i,j})^{\alpha} D_{i,j}, \tag{5}$$

where $i, j = 1, \cdots, n$, and n is the number of miRNAs in the data set. α is a positive parameter of weight, denoting the impact of GO-based similarity in the combined measure. If $\alpha = 0$, C turns out to be D. The bigger α is, the more impact that GO information has on the selection of nearest neighbor and performance of imputation. The value of α is determined on an evaluation set extracted from the original complete data set.

Given the new distance matrix C, the following steps are the same as conventional expression-based imputation algorithm. Assume that we want to estimate miRNA g_i's missing value at the tth sample, and the nearest K miRNAs according to the new measure are $g_{s_1}, \cdots, g_{s_K}$. The estimated value $\bar{G}_{i,t} = w_{s_1} G_{s_1,t} + w_{s_2} G_{s_2,t} + \cdots + w_{s_K} G_{s_K,t}$. For KNN, $w_{s_j} = \frac{1/C_{g_i g_{s_j}}}{\sum 1/C_{g_i g_{s_j}}}, 1 \le j \le K$, and for LLS, w_{s_j} are the coefficients of the linear combination, obtained by solving the least squares optimization problem.

Competing interests

The authors declare that they have no competing interests.

Authors' contributions

YY and ZX designed the system, performed the computational tasks and analyzed the results. YY drafted the manuscript. DS participated in the target gene identification and the implementation of some algorithms. All authors read and approved the final manuscript.

Acknowledgements

Y. Yang is supported by the start-up fund of Shanghai Jiao Tong University, and the Scientific Research Foundation for the Returned Overseas Chinese Scholars, State Education Ministry.

Declarations

The publication costs for this article were partly funded by the start-up fund of Shanghai Jiao Tong University, and the Scientific Research Foundation for the Returned Overseas Chinese Scholars, State Education Ministry, grant number 15Z102050022.

This article has been published as part of BMC Bioinformatics Volume 17 Supplement 1, 2016: Selected articles from the Fourteenth Asia Pacific Bioinformatics Conference (APBC 2016). The full contents of the supplement are available online at http://www.biomedcentral.com/bmcbioinformatics/supplements/17/S1.

Author details

[1]Department of Computer Science and Engineering, Shanghai Jiao Tong University, 800 Dongchuan Rd., Shanghai 200240, China. [2]Key Laboratory of Shanghai Education Commission for Intelligent Interaction and Cognitive Engineering, Shanghai 200240, China. [3]School of Computer Science and Technology, Beijing Institute of Technology, Beijing 100081, China.

Published: 11 January 2016

References

1. Troyanskaya O, Cantor M, Sherlock G, Brown P, Hastie T, Tibshirani R, et al. Missing value estimation methods for dna microarrays. Bioinforma. 2001;17(6):520–5.
2. Kim H, Golub GH, Park H. Missing value estimation for dna microarray gene expression data: local least squares imputation. Bioinforma. 2005;21(2):187–98.
3. Zhou X, Wang X, Dougherty ER. Missing-value estimation using linear and non-linear regression with bayesian gene selection. Bioinforma. 2003;19(17):2302–7.
4. Tuikkala J, Elo L, Nevalainen OS, Aittokallio T. Improving missing value estimation in microarray data with gene ontology. Bioinforma. 2006;22(5): 566–72.
5. Qing-shan N, Guang-yun W, Lang-bo Q, Bo Q, Zheng-zhi W. Improving missing value of dna microarray data by using protein-protein interactions. Life Sci Res (in Chinese). 2008;12(2):104–9.
6. Ashburner M, Ball CA, Blake JA, Botstein D, Butler H, Cherry JM, et al. Gene ontology: tool for the unification of biology. Nature genet. 2000;25(1):25–9.
7. Kanehisa M, Goto S. Kegg: kyoto encyclopedia of genes and genomes. Nucleic Acids Res. 2000;28(1):27–30.
8. Schuler G, Boguski M, Stewart E, Stein L, Gyapay G, Rice K, et al. A gene map of the human genome. Sci. 1996;274(5287):540–6.
9. Lord PW, Stevens RD, Brass A, Goble CA. Investigating semantic similarity measures across the gene ontology: the relationship between sequence and annotation. Bioinforma. 2003;19(10):1275–83.
10. Sevilla JL, Segura V, Podhorski A, Guruceaga E, Mato JM, Martínez-Cruz L, et al. Correlation between gene expression and go semantic similarity. Comput Biol Bioinforma IEEE/ACM Transac. 2005;2(4):330–8.
11. Resnik P. Semantic similarity in a taxonomy: An information-based measure and its application to problems of ambiguity in natural language. J Artif Intell Res (JAIR). 1999;11:95–130.
12. Lin D. An information-theoretic definition of similarity. In: Proceedings of the International Conference on Machine Learning (ICML), 1998. p. 296–304.
13. Jiang JJ, Conrath DW. Semantic similarity based on corpus statistics and lexical taxonomy. In: Proceedings of International Conference Research on Computational Linguistics (ROCLING X). Taiwan; 1997.
14. Wu H, Su Z, Mao F, Olman V, Xu Y. Prediction of functional modules based on comparative genome analysis and gene ontology application. Nucleic acids res. 2005;33(9):2822–37.
15. Wang JZ, Du Z, Payattakool R, Philip SY, Chen CF. A new method to measure the semantic similarity of go terms. Bioinforma. 2007;23(10): 1274–81.
16. Zhang P, Zhang J, Sheng H, Russo JJ, Osborne B, Buetow K. Gene functional similarity search tool (gfsst). BMC bioinforma. 2006;7(1):135.
17. Peng J, Wang Y, Chen J. Towards integrative gene functional similarity measurement. BMC bioinforma. 2014;15(Suppl 2):5.
18. Xiang Q, Dai X, Deng Y, He C, Wang J, Feng J, et al. Missing value imputation for microarray gene expression data using histone acetylation information. BMC bioinforma. 2008;9(1):252.
19. Volinia S, Calin GA, Liu CG, Ambs S, Cimmino A, Petrocca F, et al. A microrna expression signature of human solid tumors defines cancer gene targets. Proc Natl Acad Sci USA. 2006;103(7):2257–61.
20. Lu J, Getz G, Miska EA, Alvarez-Saavedra E, Lamb J, Peck D, et al. Microrna expression profiles classify human cancers. nature. 2005;435(7043):834–8.
21. Yu G, Xiao CL, Bo X, Lu CH, Qin Y, Zhan S, et al. A new method for measuring functional similarity of micrornas. J integr omics. 2010;1(1): 49–54.
22. Sethupathy P, Corda B, Hatzigeorgiou AG. Tarbase: A comprehensive database of experimentally supported animal microrna targets. Rna. 2006;12(2):192–7.
23. Krek A, Grün D, Poy MN, Wolf R, Rosenberg L, Epstein EJ, et al. Combinatorial microrna target predictions. Nature gen. 2005;37(5): 495–500.
24. Lewis BP, Shih I-h, Jones-Rhoades MW, Bartel DP, Burge CB. Prediction of mammalian microrna targets. Cell. 2003;115(7):787–98.
25. John B, Enright AJ, Aravin A, Tuschl T, Sander C, Marks DS, et al. Human microrna targets. PLoS Biol. 2004;2(11):363.
26. Enright AJ, John B, Gaul U, Tuschl T, Sander C, Marks DS, et al. Microrna targets in drosophila. Genome biology. 2004;5(1):1–1.
27. Edgar R, Domrachev M, Lash AE. Gene expression omnibus: Ncbi gene expression and hybridization array data repository. Nucleic acids res. 2002;30(1):207–10.
28. Cimino D, De Pittà C, Orso F, Zampini M, Casara S, Penna E, et al. mir148b is a major coordinator of breast cancer progression in a relapse-associated microrna signature by targeting itga5, rock1, pik3ca, nras, and csf1. The FASEB J. 2013;27(3):1223–35.
29. Biagioni F, Ben-Moshe NB, Fontemaggi G, Canu V, Mori F, Antoniani B, et al. mir-10b*, a master inhibitor of the cell cycle, is down-regulated in human breast tumours. EMBO molecular med. 2012;4(11):1214–29.
30. Wang D, Wang J, Lu M, Song F, Cui Q. Inferring the human microrna functional similarity and functional network based on microrna-associated diseases. Bioinforma. 2010;26(13):1644–50.
31. Yu G, Li F, Qin Y, Bo X, Wu Y, Wang S. Gosemsim: an r package for measuring semantic similarity among go terms and gene products. Bioinforma. 2010;26(7):976–8.
32. Griffiths-Jones S, Grocock RJ, Van Dongen S, Bateman A, Enright AJ. mirbase: microrna sequences, targets and gene nomenclature. Nucleic acids res. 2006;34(suppl 1):140–4.

Chang et al. BMC Systems Biology 2016, 10(Suppl 1):1
DOI 10.1186/s12918-015-0242-3

BMC Systems Biology

PROCEEDINGS Open Access

c-Myc and viral cofactor Kaposin B co-operate to elicit angiogenesis through modulating miRNome traits of endothelial cells

Hsin-Chuan Chang[1], Tsung-Han Hsieh[1], Yi-Wei Lee[2], Cheng-Fong Tsai[2], Ya-Ni Tsai[1], Cheng-Chung Cheng[3*] and Hsei-Wei Wang[1,2,4]

From The Fourteenth Asia Pacific Bioinformatics Conference (APBC 2016)
San Francisco, CA, USA. 11 - 13 January 2016

Abstract

Background: MicroRNAs (miRNAs) have emerged as master regulators of angiogenesis and other cancer-related events. Discovering new angiogenesis-regulating microRNAs (angiomiRs) will eventually help in developing new therapeutic strategies for tumor angiogenesis and cardiovascular diseases. Kaposi's sarcoma (KS), which is induced by the etiological infectious agent KS-associated herpesvirus (KSHV), is a peculiar neoplasm that expresses both blood and lymphatic endothelial markers and possesses extensive neovasculature. Using KSHV and its proteins as baits will be an efficient way to discover new angiomiRs in endothelial cells. Kaposin B is one of the latent viral genes and is expressed in all KSHV tumor cells. Since Kaposin B is a nuclear protein with no DNA-binding domain, it may regulate gene expression by incorporating itself into a transcription complex.

Results: We demonstrated that c-Myc and Kaposin B form a transcription complex and bind to the miR-221/-222 promoter, thereby affecting their expression and anti-angiogenic ability. By small RNA sequencing (smRNA-Seq), we revealed that 72.1 % (173/240) of Kaposin B up-regulated and 46.5 % (113/243) of Kaposin B down-regulated known miRNAs were regulated by c-Myc. We also found that 77 novel miRNA were up-regulated and 28 novel miRNAs were down-regulated in cells expressing both c-Myc and Kaposin B compared with cells expressing Kaposin B only. The result was confirmed by RNA-IP-seq data.

Conclusions: Our study identifies known and novel c-Myc-regulated microRNAs and reveals that a c-Myc-oriented program is coordinated by Kaposin B in KSHV-infected cells.

Keywords: MicroRNAs, Angiogenesis, KS-associated herpesvirus, c-Myc, Kaposin B, Small RNA sequencing

Background

Kaposi's sarcoma-associated herpesvirus (KSHV), also known as human herpesvirus 8 (HHV8), is associated with Kaposi's sarcoma (KS) [1]. KS is the most common cancer in individuals with AIDS and one of the most common cancers in patients who have immunosuppression [2]. KS typically appears as colored lesions or blotches on the skin, although it can spread to internal organs [3]. This disease has been recognized as a peculiar neoplasm that has a diverse cellular makeup and an extensive neovasculature. In host endothelial cells, KSHV infection alters the expression of mRNAs and microRNAs required for many aspects of tumorigenesis, including angiogenesis [4]. Therefore, KS and KSHV are recognized as an excellent model to study angiogenesis and to discover new angiogenic mechanisms, including coding and non-coding RNAs.

Recently, the effect of KSHV infection on cellular microRNA (miRNAs) expression has been reported [5]. It was found that pre-miRNA signatures delineate the stages of KSHV-associated HUVEC transformation

* Correspondence: chengcc@mail.ndmctsgh.edu.tw
[3]Division of Cardiology, Department of Internal Medicine, Tri-Service General Hospital, National Defense Medical Center, Taipei, Taiwan
Full list of author information is available at the end of the article

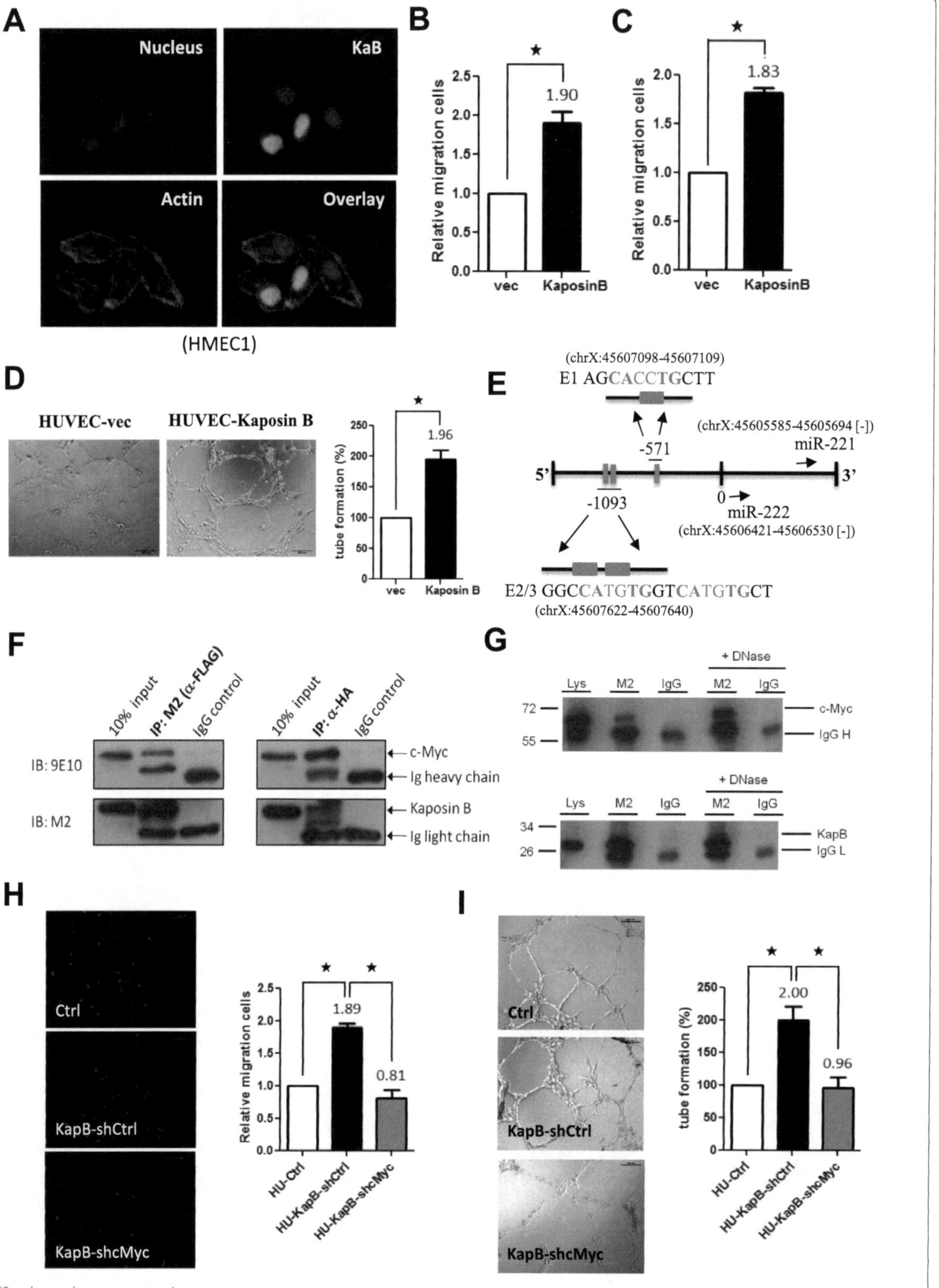

Fig. 1 (See legend on next page.)

(See figure on previous page.)

Fig. 1 Kaposin B binds to c-Myc for regulating endothelial cell angiogenic activities. **a** Nuclear distribution pattern of KSHV Kaposin B protein in cells. Immunofluorescence staining for Kaposin B proteins. HMEC1 cells with stable Kaposin B (KapB) expression were fixed, and Kaposin B proteins were detected using anti-Flag mAb, followed by anti–mouse IgG secondary antibody conjugated with FITC (*green*). Cell nuclei were counterstained with Hoechst 33342, while actin filaments with Texas Red phalloidin (Alexa Fluor 568). **b** HMEC1 cells stably expressing Kaposin B were subjected to the Transwell cell-migration assay ($n = 3$). **c** Kaposin B increases cell motility in HUVEC. Primary HUVEC stably transduced with Kaposin B or the vector control by lentivirus were used for Transwell cell-migration assays ($n = 3$). **d** Kaposin B enhances microvascular formation of HUVEC in an in vitro MatriGel angiogenesis assay. Pictures were taken after 6 h of incubation (*left*), and tube length was then measured and compared (*right*). **e** Schematic representation of miR-221/-222 proximal promoter. Three E-boxes (E1 and E2/3, in *red*) were found. **f** Co-immunoprecipitation assays show Kaposin B and c-Myc form a protein complex. Cell lysates were prepared from HMEC1 cells stably expressing Kaposin B. Five micrograms of anti-FLAG (clone M2; *left panel*), anti-HA (*right panel*) or isotype IgG control were incubated with 500 μl of cell extracts and then analyzed by western blotting with indicated mAbs. **g** The interaction between c-Myc and Kaposin B was independent of promoter DNA. Co-immunoprecipitation assays were performed with or without DNase pre-treatment on cell lysates prepared from HMEC1 cells stably expressing Kaposin B. Anti-FLAG (clone M2) or isotype IgG control were incubated with cell extracts, and pull-down products were analyzed by western blotting with anti-HA mAb for c-Myc (upper panel) or anti-FLAG M2 mAb for Kaposin B (*lower panel*). **h-i** Knockdown of endogenous c-Myc levels in Kaposin B(+) HUVECs inhibits Kaposin B-induced cell migration (**h**, *left*) and microvasculature formation (**i**, *right*) ($n = 3$)

[6, 7]. MiRNAs are small RNAs of 18–24 nucleotides in length that have emerged as master regulators of cancer and key modulators of angiogenesis [8–10]. An increasing number of studies have shown that some miRNAs, called angiomiRs (microRNAs that regulate angiogenesis), play a crucial role in regulating angiogenesis [11]. For example, miR-221 and miR-222, two closely related miRNAs encoded in cluster from a genomic region on chromosome X, can modulate the angiogenic properties of human umbilical vein endothelial cells (HUVECs) by targeting c-Kit [11] and endothelial nitric oxide synthase (eNOS) [12]. MiR-221 and miR-222 also regulate lymphatic endothelial cell (LEC) motility by targeting ETS2 and ETS1 transcription factors, respectively [4]. Loss of miR-221 expression demarks the transition from merely immortalized to fully tumorigenic HUVECs [6]. KSHV also induces endothelial cell motility by inducing miR-31 expression while suppressing that of miR-221 and miR-222 [4, 13].

Kaposin B, one of the three Kaposin proteins encoded from the KSHV genome K12 locus, is one of the viral genes expressed during latency and is therefore expressed in all KSHV tumor cells [4]. Kaposin B stabilizes various cytokine (such as IL6 and GM-CSF) mRNA containing AU-rich elements (AREs) *via* the p38/MK2 pathway. In response to LPS, Kaposin B and MK2 were shown to be exported to the cytoplasm, where mRNA stability is regulated [14]. Kaposin B also enhances the PROX1 mRNA stability during lymphatic reprogramming of vascular endothelial cells [15]. Kaposin B can influence cellular gene expression by regulating promoter activities of host genes: both Kaposin B and KSHV viral latency-associated nuclear antigen (LANA) proteins can down-regulate miR-221 and miR-222 levels by repressing the activity of miR-221/-222 cluster promoter [4]. Since there is no predictable DNA-binding domain on Kaposin B, how this nuclear protein can regulate mRNA and miRNA expression remains unclear.

c-Myc achieves its oncogenic effects by regulating transcription of protein-coding genes as well as microRNA genes such as miR-29b-1/miR-29a [16, 17]. c-Myc is also essential for vasculogenesis and angiogenesis during development and tumor progression [18] via inducing the expression of miR-17 ~ 92 angiogenic miRNA cluster [19]. Revealing the angiomiRs regulated by c-Myc and the underlying regulatory mechanisms will help to further understand c-Myc and endothelial cell biology.

Here, we showed that Kaposin B and c-Myc are in the same transcription complex that directly regulates the miR-221/-222 cluster promoter activity. A c-Myc-oriented circuit is therefore formed in the presence of Kaposin B in KSHV-infected endothelial cells. Furthermore, we also provide a global microRNA signature which is regulated by c-Myc and Kaposin B. We hope our roadmap will help the search and development of new therapeutic targets for virus- or cancer-induced angiogenesis, cancer formation and metastasis.

Methods

Cell culture and KSHV infection

Human primary umbilical vein endothelial cells (HUVECs) were purchased from Clonetics Inc. (Walkersville, Md.) and were cultured as described [4]. HMEC1, an immortalized human microvascular endothelial cell line, was cultured in endothelial cell growth medium MV (C-22020; PromoCell, Heidelberg, Germany). A recombinant virus, rKSHV.219, that expresses the red fluorescent protein (RFP) from the KSHV lytic PAN promoter, the green fluorescent protein (GFP) from the EF-1alpha promoter, and with a gene for puromycin resistance as a selectable marker, was constructed using JSC-1 cells as described previously [20].

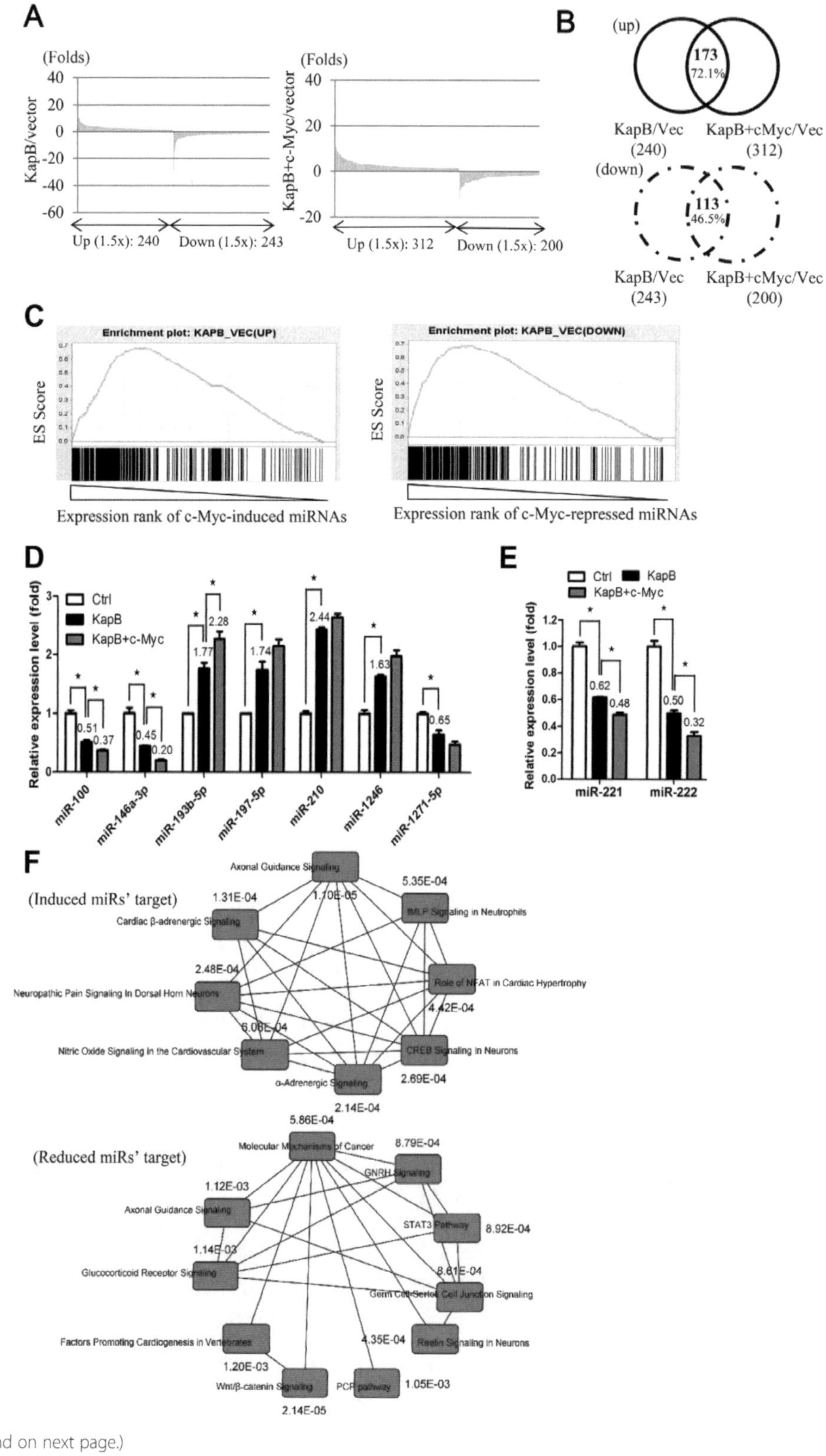

Fig. 2 (See legend on next page.)

(See figure on previous page.)
Fig. 2 c-Myc cooperates with Kaposin B to regulate cellular known miRNAs. **a** smRNA-Seq revealed differentially expressed miRNAs between Kaposin B(+) and HUVECs (≥1.5 folds, *left*), and between Kaposin B(+)c-Myc(+) and HUVECs (≥1.5 folds, *right*). **b** Venn diagrams summarizing significant overlap between c-Myc and Kaposin B known miRNAs signature. (*upper panel*) up-regulated miRNAs; (*lower panel*) down-regulated miRNAs. **c** Gene Set Enrichment Analysis (GSEA) verified the significant overlap between c-Myc and Kaposin B miRNAs signature. **d** RT-qPCR detected various known miRNAs expression. Mean expression levels of the target miRNAs are compared with the U6 control ($n = 3$). *: $P < .05$. **e** RT-qPCR detected miR-221/miR-222 expression in Kaposin B(+) and Kaposin B(+)c-Myc(+) cells. **f** The biological pathway of genes which were targeted by the Kaposin B(+)c-Myc(+) induced/reduced miRNAs. The *P* value were marked on the figure

Plasmid construction

Plasmids expressing KSHV Kaposin B and miR-221, miR-222, or miR-221/-222 were constructed as described previously [4]. c-Myc expression constructs (pcDNA3-HA-c-Myc and pHR-c-Myc) and knock down construct (pLKO.1) was kindly provided by Prof. Kenneth CW Wu [21]. The full-length miR-221/-222 promoter reporter plasmid was constructed as described previously [4], and primers for cloning miR-221/-222 promoter mutants are listed in Additional file 1.

Immunofluorescence assay (IFA)

Cells were fixed with 4 % paraformaldehyde (Sigma-Aldrich), permeabilized with 0.2 % Triton X-100 (Sigma-Aldrich), and then blocked with PBS containing 1 % bovine serum albumin (Sigma-Aldrich). For Kaposin B staining, cells were incubated with the monoclonal antibody anti-FLAG M2 at a 1:500 dilution for 60 to 120 min at 25 °C followed by incubation with FITC-conjugated goat anti-Mouse IgG (1:500, Jackson ImmunoResearch) for 60 to 120 min at 25 °C. Rhodamine-phalloidin (Molecular Probes, Invitrogen) was used to label actin cytoskeleton. Cell nuclei were counterstained with Hoechst 33342 (Sigma-Aldrich), and examined by fluorescence confocal microscopy (Olympus FV1000).

Transwell cell migration and endothelial cell tube formation assays

Cell migration ability was evaluated using Costar Transwell® Polycarbonate Permeable Supports (Corning, NY, USA) as described previously [13]. In brief, 5×10^4 cells in 500 μl of culture medium were applied to the upper chamber of the device, and 750 μl of medium containing 10 ng/ml human VEGF (R&D Systems, Minneapolis, MN, USA) was added to the lower chamber. A polycarbonate membrane with a pore size of 8 μm was placed in between the two chambers. After 6 h of incubation at 37 °C for HMEC1 and HUVEC cells, the membrane was fixed in 4 % paraformaldehyde (Sigma-Aldrich) for 20 min at room temperature and then stained with Hoechst 33342 solution (Sigma-Aldrich) for 30 min. On the upper side of membrane were identified un-migrated cells and removed. The cells under the membrane were counted under a microscope.

In vitro angiogenic activity of vascular endothelial cells was examined by tube formation assays: primary HUVECs were placed onto MatriGel® Basement Membrane Matrix gel (Becton-Dickinson, Franklin Lakes, NJ, USA) for 6 h at 37 °C. The formation of tubule structure was observed every hour under a microscope.

Small RNA sequencing (smRNA-Seq) and data analysis

Total RNA was collected and small RNA fractions were sequenced using Illumina HiSeq2000 sequencer (Illumina, San Diego, CA USA) according to the manufacturer's instructions. For data analysis, quality Fastq sequences, which were without poly-A, ambiguous nucleotides or a 5′ adapter, yet flanking 6–18 nt of 3′ adapter sequence, had the adapter sequences trimmed and the identical sequences were then collapsed to unique sequences. The resulting unique sequences that did not align to the mRNA database (UCSC genome browsers) but were aligned to known microRNA sequences (miRBase R20; http://www.mirbase.org/) were subjected to further quantification analysis. Sequencing reads were used to obtain a RPM (reads per million mapped reads) [22, 23] value calculated as $C/(MN) \times 10^9$, where C is "read numbers aligned to a given miRNA chromosomal region", M is "multiple mapping numbers across all miRNA regions" and N is "total read numbers that map to human genome sequence".

RNA extraction, real-time quantitative polymerase chain reaction (qPCR) and microRNA expression

RNA extraction and reverse transcription were performed as previously described [13]. The expression of mature human miR-221 and miR-222 was determined by a stem-loop real-time PCR system using the primer pairs previously published [4]. MicroRNA expression data were normalized to that of U6 snRNA. Details of qPCR primers are in Additional file 1.

Coimmunoprecipitation (co-IP), chromatin immunoprecipitation (ChIP) and Immunoblotting

Co-IP and ChIP assays were performed as previously described [24]. Briefly, Kaposin B or c-Myc expression plasmid was co-transfected into 293T cells using TurboFect™ (R0531; Fermentas, Glen Burnie, Maryland, USA). After 48 h, cells were lyzed in NET lysis buffer

Table 1 Top 50 miRNAs induced by both c-Myc and Kaposin B

Name	HU_KapB_cMyc (RPM)	HU_KaposinB (RPM)	cMyc + KapB/vec (Fold)	KapB/vec (Fold)
hsa-miR-5010-5p	2.56436	0.728448	34.99103515	9.93977038
hsa-miR-616-3p	1.08273	0.463558	14.77399565	6.325310904
hsa-miR-3121-3p	0.854786	0.596003	11.66366929	8.132540642
hsa-miR-6851-5p	0.683829	0.52978	9.330938158	7.228918951
hsa-miR-193b-5p	2.56436	1.05956	8.747752819	3.614457009
hsa-miR-6894-5p	1.25369	0.596003	8.553407199	4.066281418
hsa-miR-3163	0.626843	1.25823	8.553356566	17.16871662
hsa-miR-3168	0.626843	0.331113	8.553356566	4.518081167
hsa-miR-30b-3p	1.76656	1.05956	8.034967866	4.819270532
hsa-miR-3607-3p	0.569857	0.26489	7.775774975	3.614459475
hsa-miR-4804-5p	0.569857	0.26489	7.775774975	3.614459475
hsa-miR-483-5p	1.13971	0.993338	7.775768905	6.777133422
hsa-miR-4647	1.02574	0.927115	6.998198837	6.325321344
hsa-miR-7854-3p	1.02574	0.331113	6.998198837	2.259046748
hsa-miR-6735-5p	2.7923	1.58934	6.35022071	3.614461119
hsa-miR-3188	0.455886	0.198668	6.220625438	2.710851429
hsa-let-7e-5p	2636.49	1690.55	6.098609794	3.910504036
hsa-miR-197-5p	3.53312	2.58268	6.026232752	4.405123744
hsa-miR-210-5p	7.1802	6.15869	5.763201618	4.943284612
hsa-miR-3651	1.25369	0.860893	5.70224553	3.915659582
hsa-miR-1909-3p	0.3989	0.397335	5.443043847	5.421689213
hsa-miR-483-3p	1.9945	1.4569	5.443043847	3.975919068
hsa-miR-3173-5p	4.3879	1.78801	5.443038446	2.217964669
hsa-miR-3197	1.1967	0.79467	5.443033944	3.614452899
hsa-miR-139-3p	3.87503	2.31779	5.287530258	3.162655452
hsa-miR-423-5p	5484.82	3592.57	5.221947179	3.420387684
hsa-let-7f-5p	29803.4	20292.6	4.887141234	3.327566727
hsa-miR-1284	1.42464	0.52978	4.859847516	1.807228505
hsa-miR-3613-5p	3.47613	1.72179	4.743225873	2.349405482
hsa-miR-3120-3p	0.341914	0.132445	4.665462256	1.807229738
hsa-miR-4436b-3p	1.70957	1.12578	4.665462256	3.072283731
hsa-miR-4501	0.341914	0.728448	4.665462256	9.93977038
hsa-miR-497-3p	0.341914	0.331113	4.665462256	4.518081167
hsa-miR-544a	0.341914	0.132445	4.665462256	1.807229738
hsa-miR-664b-5p	0.341914	0.331113	4.665462256	4.518081167
hsa-miR-6862-5p	0.341914	0.26489	4.665462256	3.614459475
hsa-miR-7107-3p	0.341914	0.26489	4.665462256	3.614459475
hsa-miR-6816-3p	1.02574	0.596003	4.665444671	2.710841949
hsa-let-7a-5p	29230.6	20145.4	4.632835878	3.19289826
hsa-miR-210-3p	66.5023	47.8789	4.606251818	3.316310416
hsa-miR-3145-3p	0.626843	0.397335	4.276689954	2.710852004
hsa-miR-3675-5p	0.626843	0.596003	4.276689954	4.066281418
hsa-miR-6515-5p	1.1967	0.662225	4.0822801	2.259035631
hsa-miR-6797-5p	0.569857	0.397335	3.887898098	2.710852004

Table 1 Top 50 miRNAs induced by both c-Myc and Kaposin B *(Continued)*

hsa-miR-3183	0.284929	0.26489	3.88789431	3.614459475
hsa-miR-372-3p	0.284929	0.728448	3.88789431	9.93977038
hsa-miR-4488	0.284929	0.198668	3.88789431	2.710851429
hsa-miR-6797-3p	0.284929	0.198668	3.88789431	2.710851429
hsa-miR-548ar-3p	0.142464	0.132445	3.887880665	3.614459475
hsa-miR-6511b-3p	2.29842	1.2803	3.763465686	2.096381478

containing protease and phosphatase inhibitors and cell lysate (500 μl) was immunoprecipitated with anti-FLAG M2 (for detecting FLAG-tagged Kaposin B protein; Sigma-Aldrich), anti-HA or isotype IgG antibody. The immunoprecipitates were probed with anti-FLAG M2 (1:2000) and anti-c-Myc Tag (clone 9E10; MILLIPORE, 05–419, 1:2000) monoclonal antibodies, and visualized with horseradish peroxidase-conjugated secondary antibodies using the ECL reagent (Amersham Pharmacia Biosciences). For chromatin immunoprecipitation assay, chromosomal DNA fragments were prepared as described [25]. Briefly, lysates were incubated with isotype IgG or anti-FLAG antibody. The approximately 100-base pair products from miR-221/-222 proximal promoter (E1 region: −600 (45,607,128-45,607,214); E2/3 region: −1100 (45,607,635-45,607,764); NC region: −2600 (45,609,170-45,609,265)) were detected by real-time PCR.

Luciferase reporter assays

For reporter assays, 293T cells (5×10^4/well) in 24-well plates were transfected with either Kaposin B or c-Myc expression plasmid, along with the firefly luciferase reporter gene construct (500 ng) and 50 ng pRL-TK *Renilla* luciferase construct (for normalization) using TurboFect™. Cells were harvested 48 h after transfection, and the luciferase activity was measured using the Dual-Luciferase Reporter Assay System (E1910; Promega, Charbonnières, France).

Results

Kaposin B binds to c-Myc for regulating endothelial cell angiogenic activities

Kaposin B is a nuclear protein which can regulate miR-221 and miR-222 expressions without any predictable DNA-binding domain [4, 26]. Therefore, we hypothesized that Kaposin B may be a novel transcription cofactor which regulates promoter activities by binding to other transcription factor(s). To investigate, we overexpressed Kaposin B in HMEC1 and primary HUVEC cells and tested their migration and tube formation abilities. We verified expression of Kaposin B protein in lentivirus-transduced HUVEC using immunoblotting (Additional file 2-A), and immunofluorescence staining showed that localization of Kaposin B was in nucleus (Fig. 1a). Expression of Kaposin B induced an increase in cell migration rate in HMEC1 and primary HUVEC cells (Fig. 1b-c). Kaposin B also induced more microvascular structures formation by HUVEC in the in vitro tube formation assay (Fig. 1d).

To maintain KSHV latency and oncogenesis, KSHV-latency proteins, LANA and vIRF3, stabilized and activated c-Myc proteins [27–30] and c-Myc was shown to reduce miR-222 expression in mouse mammary tumors [31]. Therefore, we examined the proximal promoter region of miR-221/-222 and identified three putative c-Myc binding sites (E1, E2 and E3 E-boxes with a motif sequence CANNTG) (Fig. 1e). We hypothesized that Kaposin B may associate with c-Myc in regulating host gene expression. We therefore tested whether Kaposin B and c-Myc are in the same protein complex by co-transfecting FLAG-tagged Kaposin B and HA-tagged c-Myc into 293T cells. c-Myc was detected in immunoprecipitation (IP) assays using anti-FLAG monoclonal antibody (mAb) against Kaposin B. On the other hand, Kaposin B was detected in IP assay using anti-HA mAb against c-Myc (Fig. 1f). The interaction between c-Myc and Kaposin B was independent of promoter DNA since c-Myc was still detected in the anti-FLAG mAb pull-down products even when cell lysates were pre-treated with DNase digestion for removal of genomic DNA fragments (Fig. 1g, Additional file 2-B).

We further checked whether c-Myc is required for Kaposin B functions. Knockdown of endogenous c-Myc in Kaposin B-expressing HUVEC cells abolished Kaposin B-induced cell migration and microvasculature formation. (Fig. 1h-i, Additional file 2-C). We also knocked down c-Myc in Kaposin B-expressing HMEC1 cells and found that cell migration ability was down-regulated. (Additional file 2-D-E).

Small RNA sequencing reveals known miRNAs co-regulated by c-Myc and Kaposin B in primary endothelial cells

Previous study showed that Kaposin B can down-regulate miR-221 and miR-222 levels by repressing the

Table 2 Top 50 miRNAs reduced by both c-Myc and Kaposin B

Name	HU_KapB_cMyc (RPM)	HU_KaposinB (RPM)	cMyc + KapB/vec (Fold)	KapB/vec (Fold)
hsa-miR-3614-5p	25.7006	35.6277	0.117247798	0.162535869
hsa-miR-548f-5p	0.0121636	0.0245914	0.124207975	0.251113814
hsa-miR-4723-5p	0.0569857	0.0662225	0.129596309	0.150602547
hsa-miR-9-5p	2.10847	7.9467	0.139661522	0.526376101
hsa-miR-412-3p	0.170957	0.728448	0.145795596	0.621235225
hsa-miR-4484	0.0569857	0.0662225	0.1555155	0.180722974
hsa-miR-7844-5p	0.0569857	0.0662225	0.1555155	0.180722974
hsa-miR-3679-5p	0.0569857	0.198668	0.1555155	0.542170286
hsa-miR-4523	0.0569857	0.198668	0.1555155	0.542170286
hsa-miR-489-3p	0.170957	0.331113	0.155515833	0.301206233
hsa-miR-610	0.113971	0.198668	0.172794341	0.301205623
hsa-miR-548c-3p	0.07978	0.158934	0.209347972	0.417053279
hsa-miR-676-5p	0.170957	0.463558	0.212066256	0.575027693
hsa-miR-4485-3p	10.7703	26.3566	0.218044778	0.5335895
hsa-miR-570-3p	0.341914	0.596003	0.222164898	0.38726389
hsa-miR-891a-5p	0.455886	0.331113	0.222165584	0.161360325
hsa-miR-1271-5p	7.29417	17.2179	0.230927551	0.545104855
hsa-miR-3917	0.170957	0.397335	0.233273113	0.542168921
hsa-miR-5582-3p	0.455886	0.662225	0.239254975	0.347544399
hsa-miR-4284	2.84929	5.56269	0.241483672	0.471450365
hsa-miR-5091	0.113971	0.132445	0.259191707	0.301205093
hsa-miR-4720-5p	0.113971	0.198668	0.259191707	0.451808777
hsa-miR-144-3p	0.0569857	0.0662225	0.259192028	0.301204408
hsa-miR-3126-5p	0.0569857	0.132445	0.259192028	0.602408817
hsa-miR-3192-3p	0.0569857	0.132445	0.259192028	0.602408817
hsa-miR-4485-5p	0.0569857	0.132445	0.259192028	0.602408817
hsa-miR-1296-3p	0.227943	0.331113	0.25919255	0.376506507
hsa-miR-4521	3.47613	5.82758	0.269500868	0.451806426
hsa-miR-146a-3p	1.65259	2.78135	0.285440643	0.480403689
hsa-miR-7-5p	1.88053	2.45023	0.309157356	0.402815498
hsa-miR-4477b	0.0569857	0.0331113	0.31103015	0.180722753
hsa-miR-6733-5p	0.170957	0.132445	0.333246914	0.258175375
hsa-miR-548ab	0.372501	0.727659	0.335290464	0.654970387
hsa-miR-548 h-3p	0.487736	0.725844	0.337793045	0.502700344
hsa-miR-548z	0.487736	0.725844	0.337793045	0.502700344
hsa-miR-548b-5p	0.250673	0.433284	0.341223938	0.589799749
hsa-miR-200c-3p	0.227943	0.364224	0.345590197	0.552209298
hsa-miR-134-3p	0.7978	0.728448	0.351164459	0.320638065
hsa-miR-7705	10.0295	11.9863	0.372897929	0.445651972
hsa-miR-663a	0.113971	0.0662225	0.388787119	0.225903563
hsa-miR-301b-5p	0.170957	0.26489	0.388788698	0.602410187
hsa-miR-942-3p	0.284929	0.331113	0.388789431	0.451808117
hsa-miR-34b-3p	0.0569857	0.0662225	0.38878981	0.451808667
hsa-miR-6508-5p	0.0569857	0.0662225	0.38878981	0.451808667

Table 2 Top 50 miRNAs reduced by both c-Myc and Kaposin B *(Continued)*

hsa-miR-6510-3p	0.0569857	0.0662225	0.38878981	0.451808667
hsa-miR-548aj-5p	0.155689	0.149897	0.390962282	0.376417558
hsa-miR-548 g-5p	0.155689	0.149897	0.390962282	0.376417558
hsa-miR-548x-5p	0.155689	0.149897	0.390962282	0.376417558
hsa-miR-3944-3p	0.683829	0.728448	0.405693589	0.432164596
hsa-miR-153-3p	0.3989	0.0662225	0.418695505	0.069508807

activity of miR-221/-222 cluster promoter [4]. However, there are few studies exploring Kaposin B-induced global miRNAs alteration. Therefore, we overexpressed Kaposin B in HUVEC to examine its regulated miRNA expression. We have demonstrated that c-Myc binds to Kaposin B (Fig. 1f-g) and is required for its functions (Fig. 1h-i), hence we overexpressed c-Myc in Kaposin B(+) HUVEC to examine its ability to augment Kaposin B-regulated miRNA expression.

We analyzed small RNA sequencing (smRNA-Seq) data according to an in-house bioinformatics pipeline [32]. The Illumina next generation sequencing platform generated more than 29 million high-quality sequencing reads for HUVEC transduced with empty lentiviral vector alone, Kaposin B alone and Kaposin B with c-Myc. The initial operations included identifying sequence matches to the coding RNA database for eliminating reads of degraded mRNAs. Non-coding RNA reads that match previously annotated miRNAs deposited in the miRBase database (release 21) were subjected to normalization and quantitative profiling. Firstly, we explored known miRNAs expression. When we overexpressed Kaposin B alone (KapB/vec), we found 240 up-regulated and 243 down-regulated known miRNAs. In contrast, 312 up-regulated and 200 down-regulated miRNAs (≥1.5 fold change; Fig. 2a) were identified upon simultaneous overexpression of Kaposin B and c-Myc (KapB + c-Myc/vec). We further interrogated the contribution of c-Myc in Kaposin B-regulated miRNome changes. Comparison of the c-Myc(+)Kaposin B(+) miRNA signature with that triggered by Kaposin B alone in HUVEC indicated that c-Myc helped to further induced the expression of 72.1 % (173/240) of Kapsoin B up-regulated miRNAs (Fig. 2b, upper panel, fold change ≥1.5). Similarly, the miRNAs down-regulated by Kaposin B and c-Myc were 46.5 % (113/243); Fig. 2b, lower panel, fold change ≥1.5). c-Myc and Kaposin B-regulated miRNAs are enumerated in Tables 1 and 2 (only top 50 of c-Myc(+)Kaposin B(+) miRNA are listed). Gene Set Enrichment Analysis (GSEA) revealed that the expression of up- or down-regulated miRNAs by c-Myc correlated with those induced or repressed by Kaposin B in HUVEC (Fig. 2d). We also performed qPCR to detect miRNAs expression in Kaposin B(+) and Kaposin B(+)c-Myc(+) cells. We identified that miR-193b-5p, miR-197-5p, miR-210, and miR-1246 were up-regulated in Kaposin B(+) and Kaposin B(+)c-Myc(+) cells. In contrast, miR-100, miR-146a-3p and miR-1271-5p were down-regulated (Fig. 2d). RT-qPCR also demonstrated that miR-221 and miR-222 were down-regulated in KaposinB(+) and Kaposin B(+)c-Myc(+) cells (Fig. 2e). The miRNAs detected in RT-qPCR, except miR-1246 and miR-1271-5p, were similar to the ones detected in KSHV-infected HUVECs (Additional file 3). Moreover, we had used up to 3 different kind of tools (TargetScan (http://www.targetscan.org/vert_70/), mirDB (http://mirdb.org/miRDB/index.html) and TargetMiner (http://www.isical.ac.in/~bioinfo_miu/targetminer20.htm)) to analyze the targeted genes of c-Myc and Kaposin B induced/reduced top 5 miRNAs (Table 1). Based on our results, 806 genes were targeted by the up-regulated miRNAs (Additional file 4-A) and 818 genes were targeted by the down-regulated miRNAs (Additional file 4-B), respectively. The biological pathways were further analyzed, genes targeted by the up-regulated miRNAs contribute to axonal guidance, cardiac β-adrenergic signaling and α-adrenergic signaling; while the genes targeted by the down-regulated miRNAs contribute to Wnt/β signaling, reelin signaling and molecular mechanisms of cancer (Fig. 2f). This indicates Kaposin B + c-myc regulated miRNAs may trigger cells to more tumorigenic stage.

Small RNA sequencing identifies novel miRNAs co-regulated by c-Myc and Kaposin B in primary endothelial cells

To identify novel miRNAs, we first removed mRNA contamination and known miRNA loci. Using three independent bioinformatics algorithms (miRDeep2, mireap and miRanalyzer), 20–26 unique genomic loci yielded at the same time in Kaposin B(+)-, Kaposin B(+)c-Myc(+)-expressing cells and HUVEC control cells (Fig. 3a), and we listed all novel miRNA candidates in (Additional file 5). We found 325 and 374 novel miRNA candidates in Kaposin B-expressing cells and Kaposin B(+)c-Myc(+) cells (KapB + c-Myc/vec) that were up-regulated, respectively. In contrast,

A

	HUVEC	KaposinB	KaposinB + cMyc
Total fastq reads	29,456,450	29,977,991	30,220,181
Trimmed reads	24,560,718	26,779,115	27,806,647
Quality reads	18,695,451	19,985,073	23,025,222
Non-mRNA reads	16,960,215	18,223,162	20,999,919
Non-miRNA reads	6,155,363	6,687,898	7,743,383
Non-known reads	5,867,135	6,082,017	6,776,911
Genome mapping records	4,309,809	4,510,871	5,057,094
miRDeep2 miR prediction	137	146	167
mireap miR prediction	318	335	352
miRanalyzer miR prediction	223	221	233
Overlapped Locus	20	23	26

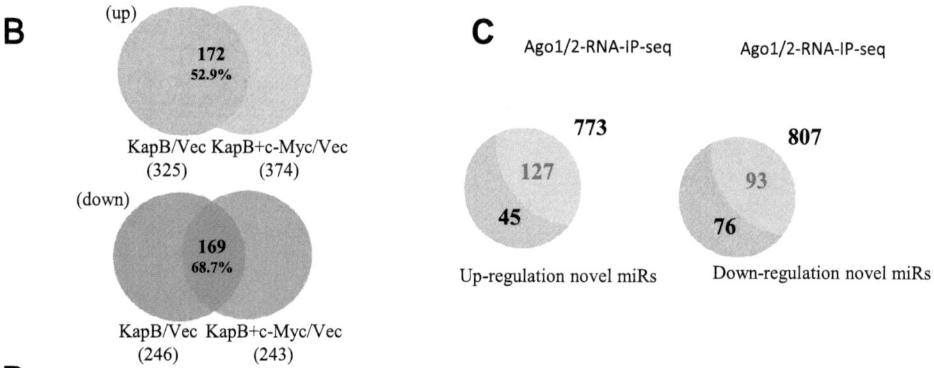

D

KaposinB+c-Myc < KaposinB

Mature_miR_ID	Precursor locus	strand	mature sequence
NM_hsa_4018	chr6:28626018-28626100	-	GAGGTGGGGATGTAGCTCAGTGGT
NM_hsa_4568	chr6:28795184-28795276	-	GGGGGTATAGCTCAGTGGTAGAGA
NM_hsa_3964	chr6:26328352-26328450	+	GTCAGAGGGCCAGTGGCGCAATGGA
NM_hsa_4564	chr6:28663697-28663809	-	GTCAGGGGGTGTAGATCAGTGGTA
NM_hsa_2539	chr13:42030048-42030148	-	GTTCGATTCCCGGTCAGGGAACAA

KaposinB+c-Myc > KaposinB

Mature_miR_ID	Precursor locus	strand	mature sequence
NM_hsa_5229	chr16:50903350-50903434	-	ATACCGGGTGCTGTAGGCTTAAAAA
NM_hsa_3403	chr20:54606257-54606348	-	TGAGGTAGTAGTTTGTGTCTT
NM_hsa_2154	chr1:70859275-70859325	+	TGAGGGAGTAGTTTGTGCAA
NM_hsa_4791	chr16:33964220-33964318	+	CGCTGAGAAGACAGTCGAACTTGACTA
NM_hsa_3949	chr6:393374-393467	+	AGCCGGCGGGGGCGCGCCGG

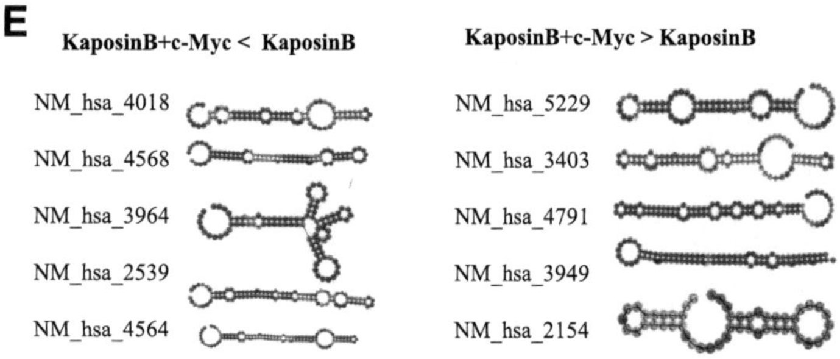

Fig. 3 (See legend on next page.)

(See figure on previous page.)
Fig. 3 c-Myc cooperates with Kaposin B to regulate cellular novel miRNAs. **a** The analysis pipeline for identification of known and novel miRNAs from smRNA-seq data. Reads or sequences pass each filtration process are indicated. **b** Venn diagrams summarizing significant overlap between c-Myc and Kaposin B signature novel miRNAs. (*upper panel*) up-regulated miRNAs; (*lower panel*) down-regulated miRNAs. **c** Venn diagrams summarizing significant overlap between up-regulated or down-regulated novel miRNA candidates and Ago1/2(+) RNA-IP-seq data. **d** Detailed information of 5 c-Myc-promoting and -reducing novel miRNAs that expressed higher reads in different groups. **e** Deduced RNA secondary structures of a set of newly discovered miRNAs

we identified 246 and 243 novel miRNA candidates in Kaposin B-expressing cells and down-regulated novel miRNA candidates in Kaposin B(+)c-Myc(+) cells (KapB + c-Myc/vec) that were down-regulated, respectively. To characterize c-Myc regulated novel miRNA candidates in Kaposin B cells, we compared the miRNA profiles between Kaposin B(+)c-Myc(+) cells and Kaposin B(+) cells. The miRNA profiles revealed that out of 325 miRNAs, 172 (52.9 %) were up-regulated and out of 246, 169 (68.7 %) were down-regulated novel mature miRNA candidates correlated with c-Myc expression in Kaposin B(+) cells (Fig. 3b). To investigate the RISC-binding properties of the novel miRNA candidates, we acquired nine public Ago1/2-mediated RNA-immunoprecipitation sequencing (Ago1/2-RNA-IP-seq) datasets from the GEO database and re-analyzed using our in-house pipelines [32]. The Ago1/2-RNA-IP-seq datasets verified 127 out of 172 up-regulated and 93 out of 169 down-regulated novel miRNA candidates (Fig. 3c).

To investigate c-Myc-promoting or -reducing novel miRNAs expression in Kaposin B(+) cells, we found that 77 out of 127 were up-regulated and 28 out of 93 were down-regulated novel mature miRNA candidates that were c-Myc-promoting or -reducing novel miRNAs expression in Kaposin B(+) cells (Additional files 6 and 7). We showed 5 up-regulated or down-regulated novel c-Myc-regulating miRNAs candidates (KaposinB + c-Myc > KapsoinB or KaposinB > KaposinB + c-Myc) that expressed higher read numbers in Kaposin B(+) and Kaposin B(+)c-Myc(+) cells (Fig. 3d) and their detailed information, including precursor locus, strand and mature sequence. Based on RNA-fold analysis (http://rna.tbi.univic.ac.at/cgi-bin/RNA-fold.cgi), we demonstrated that these novel miRNAs could fold into hairpin secondary structure, which is another miRNA property (Fig. 3e).

c-Myc enhances Kaposin B association with and regulation of miRNA promoters

Previous studies showed that miR-221/-222 promoted angiogenesis and metastasis in various cancers, including gliomas, colon cancers and breast cancers [33–37] and that c-Myc reduces miR-222 expression in mouse mammary tumors [31]. Furthermore, miR-221/-222 were crucial in KSHV-mediated endothelial cell motility [4]. Therefore, to investigate whether c-Myc can enhance Kaposin B association with and regulation of miR-221/-222 promoters, we knocked down c-Myc expression and we found that miR-221/-222 level were rescued (Fig. 4a).

Next, we performed chromatin immunoprecipitation (ChIP) assays using an anti-FLAG mAb to pull down Kaposin B and its associated genomic DNA fragments. ChIP data identified the presence of c-Myc-interacting miR-221/-222 promoter fragments containing the E1 (−600) or E2/E3 (−1100) E-boxes in HMEC1 cells stably expressing Kaposin B (Fig. 4b). Overexpressing c-Myc further enhanced the association between Kaposin B and the miR-221/-222 promoter (Fig. 4b). A negative control PCR containing a region with no putative E-box motif (the "−2600" group) showed no off-target enrichment from the same ChIP set (Fig. 4b). Direct involvement of E1 and E2/3 motifs in Kaposin B-c-Myc complex binding was verified by ChIP assays using the site-directed mutagenesis of all the E1 and E2/E3 E-boxes plasmids. A wild type −1600 promoter construct, double mutant construct and Kaposin B (empty pLenti4/V5 vector) were co-transfected into 293T cells, and ChIP assays were performed using anti-FLAG mAb. qPCR on IP-purified DNA fragments showed that these E-box motifs were essential for Kaposin B to associate with miR-221/-222 promoter (Fig. 4c). We next examined the critical role of endogenous c-Myc in Kaposin B function. Knockdown of endogenous c-Myc in Kaposin B(+) HMEC1 cells significantly reduced Kaposin B and miR-221/-222 promoter association (Fig. 4d).

Repression of miR-221/-222 promoter activity was observed after transient co-transfection of Kaposin B- or c-Myc-expressing plasmids with the full-length reporter construct into 293T cells (Fig. 4f, the "−1600 bp" group). Ectopic expression of both Kaposin B and c-Myc repressed miR-221/-222 promoter activity significantly more than expression of Kaposin B or c-Myc alone (Fig. 4e), suggesting c-Myc enhances the capability of Kaposin B to regulate the miR-221/-222 promoter. Deleting the region containing E2/3 E-boxes did not dramatically influence Kaposin B or c-Myc function (Fig. 4e, the "−600" group), but additional deletion of the E1 E-box significantly abolished the repression of the miR-221/-222 promoter by Kaposin B, c-Myc or both

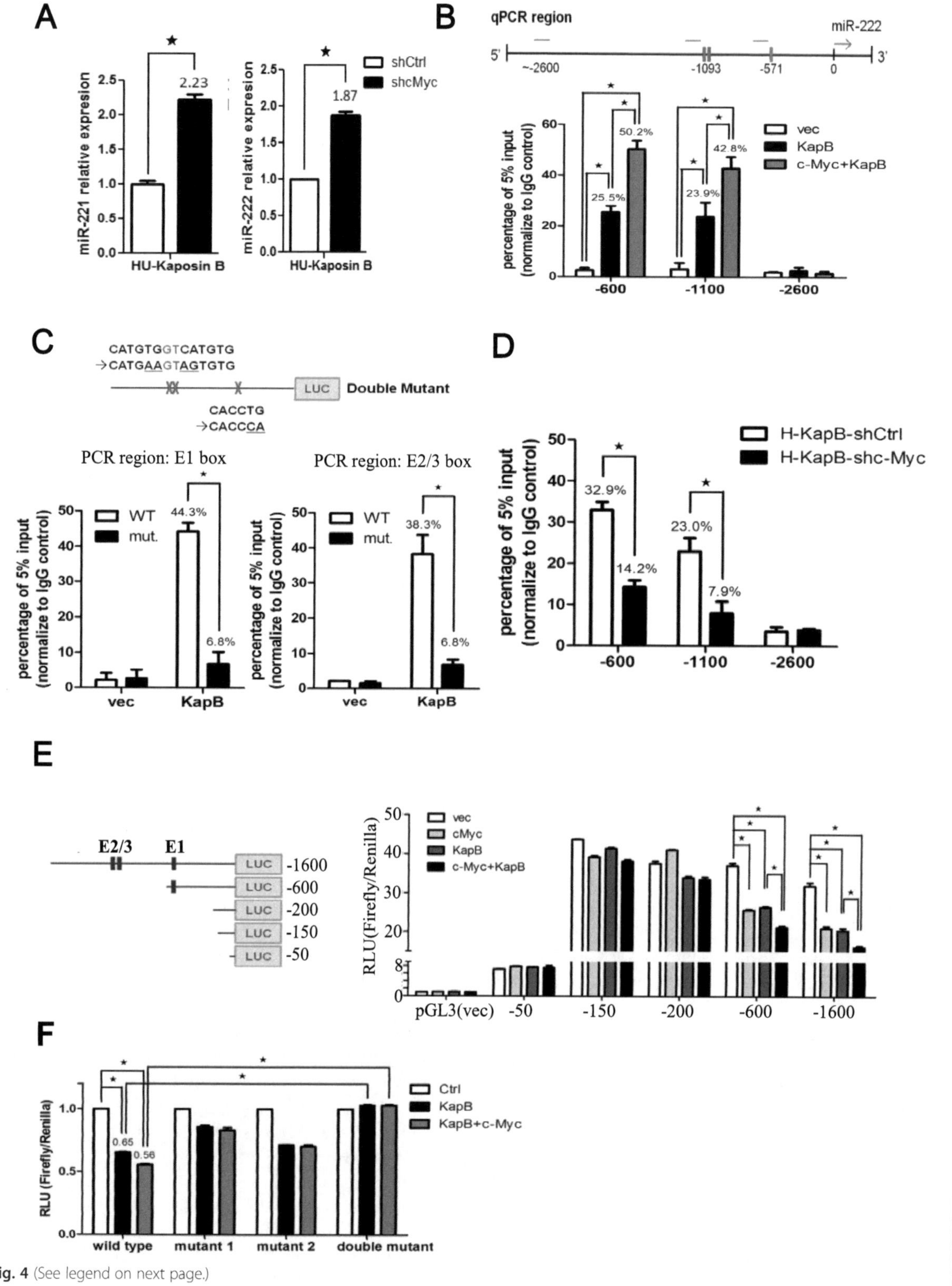

Fig. 4 (See legend on next page.)

(See figure on previous page.)
Fig. 4 c-Myc enhances Kaposin B to regulate microRNA promoter activity. **a** Knockdown of endogenous c-Myc levels in Kaposin B(+) HUVECs rescues miR-221 (*left*) and miR-222 (*right*) expression. **b** ChIP analysis with immunoglobulin G control or anti-FLAG antibody (detecting Kaposin B). (*upper*) Schematic representation of miR-221/-222 promoter region and PCR fragments. (*lower*) HMEC1 cells stably expressing empty vector, Kaposin B, or Kaposin B + c-Myc were subjected into ChIP assays. The −2600 promoter region PCR product was used as a negative control. **c** Wild type (−1600) promoter construct or double mutant construct (*upper*) was transfected into a HMEC1 cell line stably expressing Kaposin B (HMEC1-Kaposin B), and ChIP-qPCR assays were performed on transfectants using anti-FLAG mAb. **d** Endogenous c-Myc was stably knocked down using shRNA in a HMEC1-Kaposin B stable cell line, and cells with or without c-Myc knockdown were subjected to ChIP assays and qPCR for indicated regions ($n = 3$). **e** Reporter activity of tested reporter plasmids after co-transfection with Kaposin B, c-Myc, or c-Myc + Kaposin B. Schematic representation of reporter constructs (*left*) and reporter assay results ($n = 3$) (*right*). *: $P < .05$. **f** Reporter assays on E1 or E2/3 single mutants, and on the E1 and E2/E3 E-boxes double mutant. Only the double mutant showed significant promoter activity restoration. ($n = 3$) *: $P < .05$

Kaposin B and c-Myc together (Fig. 4e, the "−200", "−150" and "−50" group). Site-directed mutagenesis of all the E1 and E2/E3 E-boxes, but not E1 or E2/3 alone, hampered Kaposin B- and/or c-Myc-mediated repression, indicating c-Myc could bind to all the 3 E-boxes to direct Kaposin B to the miR-221/-222 promoter (Fig. 4f). Figure 5 is an illustration of the mechanism whereby Kaposin B and c-Myc in the same transcription complex can directly regulate the miR-221/-222 cluster promoter activity.

Discussion

Kaposin B is known to stabilize various AU-rich elements (ARE)-containing cytokine mRNAs, such as IL6, *via* the p38/MK2 pathway [14, 38]. AREs are usually 50 to 150 nucleotides in length and located in the 3′ untranslated region (UTR) of short half-life mRNAs. AREs serve as an mRNA-destabilizing determinant by promoting degradation of mRNAs. AREs are now grouped into three classes based on the number and distribution of the core AUUUA pentamers: cluster I ARE genes, such as cell cycle regulators c-Myc and cyclin D1, contain several dispersed copies of the AUUUA motif within the AU-rich region [39]. Cluster II ARE genes (such as cytokines IL1B, IL6, IL8 and granulocyte/macrophage colony-stimulating factor (GM-CSF)) have at least 2 overlapping UUAUUUA(U/A)(U/A) motifs, and cluster III ARE genes (including c-JUN and p53) are less characterized and do not contain the canonical AUUUA motif [39]. Database search revealed that there is no ARE on miR-221-222 pri-miRNA. Whether Kaposin B collaborates with c-Myc to regulate promoter activities of these ARE(−) genes is an interesting direction to follow and can be answered by global genomics approaches, such as ChIP-Seq.

In this study, we use miR-221/-222 as a model to elucidate how Kaposin B regulates promoter activity. We demonstrate that c-Myc together with Kaposin B regulate the activity of miR-221/-222 cluster promoter. c-Myc is known to maintain KSHV latency [40]. KSHV latency proteins, such as LANA and vIRF3, have been demonstrated to regulate c-Myc. LANA stabilizes and activates c-Myc through inhibition of GSK-3 phosphorylation and c-Myc ubiquitination [27, 28]. LANA protein that lacks a known DNA-binding domain, also represses the expression of miR-221/-222 cluster [4]. Nevertheless, pilot studies showed that the interaction of Kaposin B and c-Myc is independent of LANA (not shown). vIRF3 associated with c-Myc suppressor, MM-1α, to stimulate c-Myc function [29]. A c-Myc-oriented angiogenesis program is therefore activated and plays a crucial role in KSHV-infected cells and even in KS. Since KS is a neoplasm with an extensive neovasculature, the induction of endothelial cell migration and microvascular formation by c-Myc and Kaposin B may eventually benefit KS formation and progression.

MYC protein, including c-Myc, N-Myc and L-Myc, have been shown to be involved in pathogenesis of most cancers, such as burkitt's lymphoma, breast cancer, prostate cancer, gastrointestinal cancer etc. [41]. c-Myc affects most important oncogenic programs through regulating several genes and miRNAs expression [16, 42, 43]. In c-Myc-induced mouse mammary tumors, 50 and 59 miRNAs show increased and decreased expression, respectively. The expression of miR-20a, −20b, and miR-9 were increased, while that of miR-222 was decreased [16, 31]. In angiogenesis, previous report showed that c-Myc activated miR-17 cluster expression to attenuate clusterin expression, thereby promoting angiogenesis [44]. In this report, we provide a comprehensive c-Myc-regulated miRome trait in primary human endothelial cells. Most of the c-Myc and/or Kaposin B-affected miRNAs have not been linked to angiogenesis, cell motility or tumorigenesis. Among miRNAs induced by both c-Myc and Kaposin B (Table 1), miR-210, the most prominent hypoxia-induced microRNA, modulates cellular response to hypoxia through MNT, a MAX-interacting transcriptional repressor that functions as a c-Myc antagonist [45]. miR-483-5p, a microRNA embedded in the intron of insulin-like growth factor 2 (Igf2), inhibits angiogenesis *via* targeting serum response factor [46]. miR-372 is involved in cell migration, proliferation, apoptosis, and globally affect the regulation circuits centered on MAPK/ERK signaling in colon cancer cells

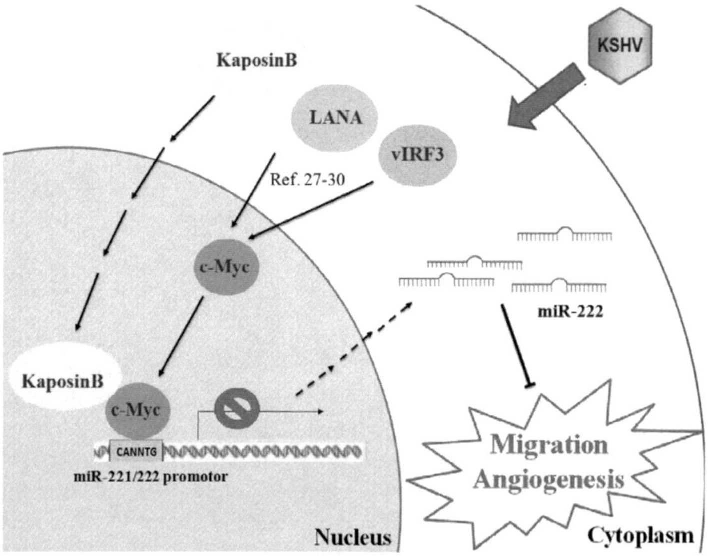

Fig. 5 A proposed model of regulation of endothelial cell activities by the Kaposin B-c-Myc circuit

[47]. Knockdown of miR-139 in HUVEC significantly suppressed cell viability [48]. miR-423 promotes cell growth and regulates G(1)/S transition by targeting p21Cip1/Waf1 in hepatocellular carcinoma [49].

Conclusions

In this study, we provide a global picture of KSHV- and/or Kaposin B-induced microRNAs, especially those co-regulated by c-Myc. We have demonstrated that c-Myc is essential for Kaposin B to stimulate endothelial cell activity and to activate miRNA promoters. In addition to providing a global picture, this study also benefits the development of new therapeutic approaches to treat KSHV-induced tumor progression in future.

Additional files

Additional file 1: List of primers used in the study. (PDF 72 kb)

Additional file 2: (A) Primary HUVEC stably transduced with Kaposin B or the vector control by lentivirus. Immunoblotting of Kaposin B proteins with anti-Flag mAb (bottom panel). β-actin was used as an internal control. (**B**) DNase treatment effects were verified by agarose gel electrophoresis. (**C-D**) knockdown of endogenous c-Myc levels with shRNA in Kaposin B(+) HUVEC (**C**) and Kaposin B(+) HMEC1 cells (**D**) were validated by immunoblotting and qRT-PCR. (**E**) knockdown of endogenous c-Myc levels with shRNA in Kaposin B(+) HMEC1 cells repressed Kaposin B-induced cellular migration ($n = 3$). (PDF 311 kb)

Additional file 3: Validation of smRNA-Seq data by RT-qPCR in KSHV-infected cells. Mean expression levels of the target miRNAs are compared with the U6 control ($n = 3$). *: $P < .05$. (PDF 76 kb)

Additional file 4: Genes were targeted by the up-regulated miRNAs (A) and down-regulated miRNAs (B). The genes interactions were analyzed by STRING (http://string-db.org/newstring_cgi/show_input_page.pl?UserId=PZAhu49PCqxs&sessionId=A8hUzxbU17ST). (PDF 8481 kb)

Additional file 5: Detail information of all novel miRNAs. (PDF 477 kb)

Additional file 6: List of Kaposin B and c-Myc down-regulated novel miRNAs which exist in Ago1/2-IP-seq data. (PDF 79 kb)

Additional file 7: List of Kaposin B and c-Myc up-regulated novel miRNAs which exist in Ago1/2-IP-seq data. (PDF 83 kb)

Competing interests
The authors declare that they no competing interests.

Authors' contributions
HWW and CCC designed the analysis approach; HCC, THH, CFT and YNT performed the research and analyzed the data; HWW provided biologic guidance during the analysis process; HWW and HCC wrote the manuscript; and all authors read and approved the final manuscript.

Acknowledgements
The authors acknowledge the technical services provided by the Microarray & Gene Expression Analysis Core Facility, which is supported by the Ministry of Science and Technology (MOST), of the National Yang-Ming University VGH-YM Genome Research Center. This work is funded by the Ministry of Science and Technology (103-2911-I-010-506, 102-2314-B-010-045, and 101-2320-B-010-059-MY3), Mackay Memorial Hospital (MMH-YM-10304), Veterans General Hospitals University System of Taiwan (VGHUST) Joint Research Program, Tsou's Foundation (VGHUST102-G7-3-2), National Health Research Institutes (10401-62-014), National Yang-Ming University (Ministry of Education, Aim for the Top University Plan) and Tri-Service General Hospital (TSGH-C104-028). This work is also funded in part by the UST-UCSD International Center for Excellence in Advanced Bioengineering sponsored by the Taiwan MOST I-RiCE Program (102-2911-I-009-101). The authors would like to dedicate this paper to the memory of Dr. Hsei-Wei Wang (Apr. 1963-Sep. 2015), who passed away during the period of this research. Dr. Hsei-Wei Wang focused on bioinformatics and biology studying; without his long-term devotion to KSHV research, this paper could not have been completed.

Declarations
The publication cost for this article was funded by Tri-Service General Hospital, grant number (TSGH-C104-028).

Author details
[1]Institute of Microbiology and Immunology, National Yang-Ming University, Taipei, Taiwan. [2]Institute of Biomedical Informatics, National Yang-Ming

University, Taipei, Taiwan. [3]Division of Cardiology, Department of Internal Medicine, Tri-Service General Hospital, National Defense Medical Center, Taipei, Taiwan. [4]VGH-YM Genome Research Center, National Yang-Ming University, Taipei, Taiwan.

Published: 11 January 2016

References

1. Mesri EA, Cesarman E, Boshoff C. Kaposi's sarcoma and its associated herpesvirus. Nat Rev Cancer. 2010;10(10):707–19.
2. Hengge UR, Ruzicka T, Tyring SK, Stuschke M, Roggendorf M, Schwartz RA, et al. Update on Kaposi's sarcoma and other HHV8 associated diseases. Part 2: pathogenesis, Castleman's disease, and pleural effusion lymphoma. Lancet Infect Dis. 2002;2(6):344–52.
3. Hengge UR, Ruzicka T, Tyring SK, Stuschke M, Roggendorf M, Schwartz RA, et al. Update on Kaposi's sarcoma and other HHV8 associated diseases. Part 1: epidemiology, environmental predispositions, clinical manifestations, and therapy. Lancet Infect Dis. 2002;2(5):281–92.
4. Wu YH, Hu TF, Chen YC, Tsai YN, Tsai YH, Cheng CC, et al. The manipulation of miRNA-gene regulatory networks by KSHV induces endothelial cell motility. Blood. 2011;118(10):2896–905.
5. Lagos D, Pollara G, Henderson S, Gratrix F, Fabani M, Milne RS, et al. miR-132 regulates antiviral innate immunity through suppression of the p300 transcriptional co-activator. Nat Cell Biol. 2010;12(5):513–9.
6. O'Hara AJ, Chugh P, Wang L, Netto EM, Luz E, Harrington WJ, et al. Pre-micro RNA signatures delineate stages of endothelial cell transformation in Kaposi sarcoma. PLoS Pathog. 2009;5(4):e1000389.
7. O'Hara AJ, Wang L, Dezube BJ, Harrington Jr WJ, Damania B, Dittmer DP. Tumor suppressor microRNAs are underrepresented in primary effusion lymphoma and Kaposi sarcoma. Blood. 2009;113(23):5938–41.
8. Croce CM, Calin GA. miRNAs, cancer, and stem cell division. Cell. 2005;122(1):6–7.
9. Schickel R, Boyerinas B, Park SM, Peter ME. MicroRNAs: key players in the immune system, differentiation, tumorigenesis and cell death. Oncogene. 2008;27(45):5959–74.
10. Zimmerman AL, Wu S. MicroRNAs, cancer and cancer stem cells. Cancer Lett. 2011;300(1):10–9.
11. Poliseno L, Tuccoli A, Mariani L, Evangelista M, Citti L, Woods K, et al. MicroRNAs modulate the angiogenic properties of HUVECs. Blood. 2006;108(9):3068–71.
12. Suarez Y, Fernandez-Hernando C, Pober JS, Sessa WC. Dicer dependent microRNAs regulate gene expression and functions in human endothelial cells. Circ Res. 2007;100(8):1164–73.
13. Tsai YH, Wu MF, Wu YH, Chang SJ, Lin SF, Sharp TV, et al. The M type K15 protein of Kaposi's sarcoma-associated herpesvirus regulates microRNA expression via its SH2-binding motif to induce cell migration and invasion. J Virol. 2009;83(2):622–32.
14. McCormick C, Ganem D. The kaposin B protein of KSHV activates the p38/MK2 pathway and stabilizes cytokine mRNAs. Science. 2005;307(5710):739–41.
15. Yoo J, Kang J, Lee HN, Aguilar B, Kafka D, Lee S, et al. Kaposin-B enhances the PROX1 mRNA stability during lymphatic reprogramming of vascular endothelial cells by Kaposi's sarcoma herpes virus. PLoS Pathog. 2010;6(8):e1001046.
16. Bui TV, Mendell JT. Myc: Maestro of MicroRNAs. Genes Cancer. 2010;1(6):568–75.
17. Mott JL, Kurita S, Cazanave SC, Bronk SF, Werneburg NW, Fernandez-Zapico ME. Transcriptional suppression of mir-29b-1/mir-29a promoter by c-Myc, hedgehog, and NF-kappaB. J Cell Biochem. 2010;110(5):1155–64.
18. Baudino TA, McKay C, Pendeville-Samain H, Nilsson JA, Maclean KH, White EL, et al. c-Myc is essential for vasculogenesis and angiogenesis during development and tumor progression. Genes Dev. 2002;16(19):2530–43.
19. Dews M, Homayouni A, Yu D, Murphy D, Sevignani C, Wentzel E, et al. Augmentation of tumor angiogenesis by a Myc-activated microRNA cluster. Nat Genet. 2006;38(9):1060–5.
20. Vieira J, O'Hearn PM. Use of the red fluorescent protein as a marker of Kaposi's sarcoma-associated herpesvirus lytic gene expression. Virology. 2004;325(2):225–40.
21. Chou YT, Lin HH, Lien YC, Wang YH, Hong CF, Kao YR, et al. EGFR promotes lung tumorigenesis by activating miR-7 through a Ras/ERK/Myc pathway that targets the Ets2 transcriptional repressor ERF. Cancer Res. 2010;70(21):8822–31.
22. Mortazavi A, Williams BA, McCue K, Schaeffer L, Wold B. Mapping and quantifying mammalian transcriptomes by RNA-Seq. Nat Methods. 2008;5(7):621–8.
23. Hsieh JY, Huang TS, Cheng SM, Lin WS, Tsai TN, Lee OK, et al. miR-146a-5p circuitry uncouples cell proliferation and migration, but not differentiation, in human mesenchymal stem cells. Nucleic Acids Res. 2013;41(21):9753–63.
24. Hwang WL, Yang MH, Tsai ML, Lan HY, Su SH, Chang SC, et al. SNAIL regulates interleukin-8 expression, stem cell-like activity, and tumorigenicity of human colorectal carcinoma cells. Gastroenterology. 2011;141(1):279–91. 291 e271-275.
25. Yang MH, Wu MZ, Chiou SH, Chen PM, Chang SY, Liu CJ, et al. Direct regulation of TWIST by HIF-1alpha promotes metastasis. Nat Cell Biol. 2008;10(3):295–305.
26. Sadler R, Wu L, Forghani B, Renne R, Zhong W, Herndier B, et al. A complex translational program generates multiple novel proteins from the latently expressed kaposin (K12) locus of Kaposi's sarcoma-associated herpesvirus. J Virol. 1999;73(7):5722–30.
27. Bubman D, Guasparri I, Cesarman E. Deregulation of c-Myc in primary effusion lymphoma by Kaposi's sarcoma herpesvirus latency-associated nuclear antigen. Oncogene. 2007;26(34):4979–86.
28. Liu J, Martin HJ, Liao G, Hayward SD. The Kaposi's sarcoma-associated herpesvirus LANA protein stabilizes and activates c-Myc. J Virol. 2007;81(19):10451–9.
29. Lubyova B, Kellum MJ, Frisancho JA, Pitha PM. Stimulation of c-Myc transcriptional activity by vIRF-3 of Kaposi sarcoma-associated herpesvirus. J Biol Chem. 2007;282(44):31944–53.
30. Nador RG, Cesarman E, Chadburn A, Dawson DB, Ansari MQ, Sald J, et al. Primary effusion lymphoma: a distinct clinicopathologic entity associated with the Kaposi's sarcoma-associated herpes virus. Blood. 1996;88(2):645–56.
31. Sun Y, Wu J, Wu SH, Thakur A, Bollig A, Huang Y, et al. Expression profile of microRNAs in c-Myc induced mouse mammary tumors. Breast Cancer Res Treat. 2009;118(1):185–96.
32. Cheng WC, Chung IF, Tsai CF, Huang TS, Chen CY, Wang SC, et al. YM500v2: a small RNA sequencing (smRNA-seq) database for human cancer miRNome research. Nucleic Acids Res. 2015;43(Database issue):D862–867.
33. Shah MY, Calin GA. MicroRNAs miR-221 and miR-222: a new level of regulation in aggressive breast cancer. Genome Med. 2011;3(8):56.
34. Ye X, Bai W, Zhu H, Zhang X, Chen Y, Wang L, et al. MiR-221 promotes trastuzumab-resistance and metastasis in HER2-positive breast cancers by targeting PTEN. BMB Rep. 2014;47(5):268–73.
35. Qin J, Luo M. MicroRNA-221 promotes colorectal cancer cell invasion and metastasis by targeting RECK. FEBS Lett. 2014;588(1):99–104.
36. Yang F, Wang W, Zhou C, Xi W, Yuan L, Chen X, et al. MiR-221/-222 promote human glioma cell invasion and angiogenesis by targeting TIMP2. Tumour Biol. 2015;36(5):3763–73.
37. Hur K, Toiyama Y, Schetter AJ, Okugawa Y, Harris CC, Boland CR, et al. Identification of a metastasis-specific MicroRNA signature in human colorectal cancer. J Natl Cancer Inst. 2015;107(3):dju492.
38. Wang HW, Boshoff C. Linking Kaposi virus to cancer-associated cytokines. Trends Mol Med. 2005;11(7):309–12.
39. Barreau C, Paillard L, Osborne HB. AU-rich elements and associated factors: are there unifying principles? Nucleic Acids Res. 2005;33(22):7138–50.
40. Li X, Chen S, Feng J, Deng H, Sun R. Myc is required for the maintenance of Kaposi's sarcoma-associated herpesvirus latency. J Virol. 2010;84(17):8945–8.
41. Nesbit CE, Tersak JM, Prochownik EV. MYC oncogenes and human neoplastic disease. Oncogene. 1999;18(19):3004–16.
42. Dang CV. MYC on the path to cancer. Cell. 2012;149(1):22–35.
43. Meyer N, Penn LZ. Reflecting on 25 years with MYC. Nat Rev Cancer. 2008;8(12):976–90.
44. Dews M, Fox JL, Hultine S, Sundaram P, Wang W, Liu YY, et al. The myc-miR-17 ~ 92 axis blunts TGF{beta} signaling and production of multiple TGF{beta}-dependent antiangiogenic factors. Cancer Res. 2010;70(20):8233–46.
45. Zhang Z, Sun H, Dai H, Walsh R, Imakura M, Schelter J, et al. MicroRNA miR-210 modulates cellular response to hypoxia through the MYC antagonist MNT. Cell Cycle. 2009;8(17):2756–68.
46. Qiao Y, Ma N, Wang X, Hui Y, Li F, Xiang Y, et al. MiR-483-5p controls angiogenesis in vitro and targets serum response factor. FEBS Lett. 2011;585(19):3095–100.

47. Ragusa M, Statello L, Maugeri M, Majorana A, Barbagallo D, Salito L, et al. Specific alterations of the microRNA transcriptome and global network structure in colorectal cancer after treatment with MAPK/ERK inhibitors. J Mol Med. 2012;90(12):1421–38.
48. Li YX, Liu DQ, Zheng C, Zheng SQ, Liu M, Li X, et al. miR-200a modulate HUVECs viability and migration. IUBMB Life. 2011;63(7):553–9.
49. Lin J, Huang S, Wu S, Ding J, Zhao Y, Liang L, et al. MicroRNA-423 promotes cell growth and regulates G(1)/S transition by targeting p21Cip1/Waf1 in hepatocellular carcinoma. Carcinogenesis. 2011;32(11):1641–7.

Yu et al. BMC Systems Biology 2016, 10(Suppl 1):2
DOI 10.1186/s12918-015-0243-2

BMC Systems Biology

PROCEEDINGS Open Access

Prediction of drugs having opposite effects on disease genes in a directed network

Hasun Yu[1,2], Sungji Choo[1,2], Junseok Park[1,2], Jinmyung Jung[1,2], Yeeok Kang[1,2] and Doheon Lee[1,2*]

From The Fourteenth Asia Pacific Bioinformatics Conference (APBC 2016)
San Francisco, CA, USA. 11 - 13 January 2016

Abstract

Background: Developing novel uses of approved drugs, called drug repositioning, can reduce costs and times in traditional drug development. Network-based approaches have presented promising results in this field. However, even though various types of interactions such as activation or inhibition exist in drug-target interactions and molecular pathways, most of previous network-based studies disregarded this information.

Methods: We developed a novel computational method, Prediction of Drugs having Opposite effects on Disease genes (PDOD), for identifying drugs having opposite effects on altered states of disease genes. PDOD utilized drug-drug target interactions with 'effect type', an integrated directed molecular network with 'effect type' and 'effect direction', and disease genes with regulated states in disease patients. With this information, we proposed a scoring function to discover drugs likely to restore altered states of disease genes using the path from a drug to a disease through the drug-drug target interactions, shortest paths from drug targets to disease genes in molecular pathways, and disease gene-disease associations.

Results: We collected drug-drug target interactions, molecular pathways, and disease genes with their regulated states in the diseases. PDOD is applied to 898 drugs with known drug-drug target interactions and nine diseases. We compared performance of PDOD for predicting known therapeutic drug-disease associations with the previous methods. PDOD outperformed other previous approaches which do not exploit directional information in molecular network. In addition, we provide a simple web service that researchers can submit genes of interest with their altered states and will obtain drugs seeming to have opposite effects on altered states of input genes at http://gto.kaist.ac.kr/pdod/index.php/main.

Conclusions: Our results showed that 'effect type' and 'effect direction' information in the network based approaches can be utilized to identify drugs having opposite effects on diseases. Our study can offer a novel insight into the field of network-based drug repositioning.

Keywords: Drug repositioning, Network biology, Directed network

Background

A drug is a chemical compound used to treat diseases and it usually has opposite effects to disease states. The traditional drug development remains expensive and time consuming process with low success rate. Developing a new drug to market takes about 15 years and 1 billion US dollars [1] and more than 85 % of drugs are failed to be released [2]. The identification of new therapeutic indications for existing drugs suggests an alternative for this situation [3]. The repurposing of approved drugs reduces costs and times involved in early phase of drug discovery.

In silico methods have provided candidates for novel uses of existing drugs. With the emerging technologies generating gene expression profiles, one of prevalent computational approaches to identify novel indications for existing drugs is based on drug and disease gene expression signatures from public databases [4–7]. These

* Correspondence: dhlee@biosoft.kaist.ac.kr
[1]Department of Bio and Brain Engineering, KAIST, 291 Daehak-ro, Yuseong-gu, Daejeon 305-701, Republic of Korea
[2]Bio–Synergy Research Center, 291 Daehak–ro, Yuseong–gu, Daejeon 305-701, Republic of Korea

methods predicted drug and disease pairs showing therapeutic relationships through comparison of signatures of drugs effect and diseases pathophysiology. They discovered some previously uncharacterized uses of known drugs to diseases. However, these studies may have low precision since gene expression profiles, especially for drugs, can be generated under various conditions such as different tissue, cell line/types, and different doses of drugs [5].

Other recent popular *in silico* approaches in drug repositioning are network-based approaches including: 1. prediction of interactions between drugs and target proteins and using them for drug repositioning [8–11]; 2. prediction of drug effects to disease genes on biological networks [9, 12, 13]. These network-based studies have revealed promising results. However, even though there are diverse types of drug-drug target interactions such as activation or inhibition, the first type of network-based methods using drug-target interactions (DTI) regarded drug-target protein interactions as binary interactions, except for one [11], which did not illustrate how this method could be used in drug repositioning. The second type of network-based methods using molecular biological networks did not fully take into account the way drugs affect to disease genes because these methods did not consider 'effect types', 'effect directions', and 'altered disease genes' in diseases (Fig. 1). Thus, candidate drugs for the treatment of diseases from previous studies could activate over expressed genes in the disease state or inhibit under-expressed genes and could accelerate progression of the disease. In addition, candidate drugs from previous studies could not exert influence on diseases genes due to 'effect direction' (Fig. 1).

Here, we proposed a novel approach to identify novel indications of existing drugs by predicting drugs having opposite effects on altered disease genes. Our hypothesis is that if a drug compensates differentially regulated states of disease genes in patient, then that drug has the potential to be a therapy of the disease. Entire routes from drugs to disease genes are considered with the 'effect types', 'effect directions', and 'altered disease genes' information: interactions between drugs and drug targets; directed pathway between drug targets and disease genes; regulated states of disease genes in patients. This approach is termed as Predicting Drugs having Opposite effects Disease genes (PDOD).

In our PDOD approach, we integrated directed pathways having 'effect type' and 'effect direction', mapped drug-drug target interactions with their 'effect types', and mapped disease genes with their differentially regulated states. Based on this entire path, we predicted a drug whose targets are close to disease genes and activate down-regulated disease genes or inhibit up-regulated disease genes in patients as a therapy for the diseases (Fig. 2 and Additional file 1: Figure S1). We also showed that our method would be able to verify known uses of drugs and predict novel indications for existing drugs.

Methods

Datasets

To identify how drugs influenced on disease genes, we considered the shortest paths from drugs to diseases. We collected directed molecular pathways, drug-target protein interactions, and disease related genes with their regulated states from various data sources. The datasets used in our method are described here.

Integrated directed pathways

We constructed a directed backbone network having 'effect type' and 'effect direction' by integrating Kyoto Encyclopedia of Genes and Genomes (KEGG) pathway [14]. Although KEGG pathway does not cover entire molecular entries and pathways in human, it provides

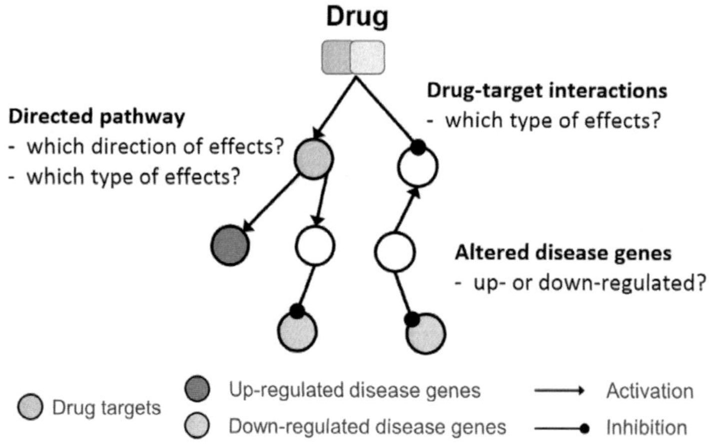

Fig. 1 The way a drug has effects on disease genes. The entire path from a drug to disease genes consists of 'effect type' such as activation or inhibition, 'effect direction', and 'altered states of disease genes' like up- or down-regulated states in disease patients

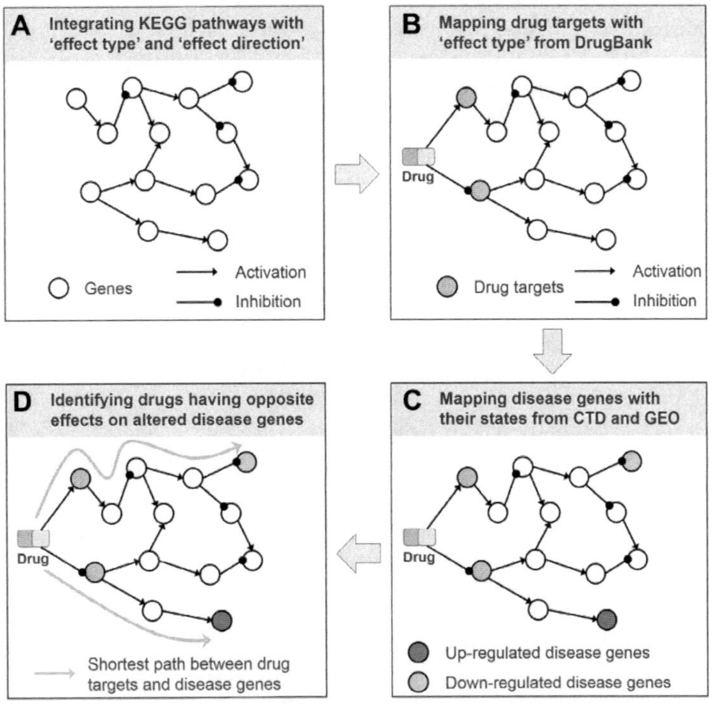

Fig. 2 Overview of PDOD. **a** Integrating pathways whose edges have 'effect type' and 'effect direction' information. **b** Mapping drug-drug target interactions on an integrated pathway with their 'effect type' and 'effect direction'. **c** Mapping disease genes with their regulated states in patients. **d** Predicting drugs having opposite effects on states of disease genes in disease states based on shortest paths

manually curated pathways with highly reliable relations having 'effect type' and 'effect direction' information. We downloaded KEGG Markup Language (KGML) files of 286 human pathways using KEGGgraph package [15]. We only used gene entries and relations with obvious 'effect direction' like activation, inhibition, expression, and repression. Then, the 166 human pathways including signal transduction and transcription pathways were merged into a Directed KEGG pathways (DKEGG) consisting of 3,307 nodes and 28,808 directed edges. Each node in DKEGG is a gene and each edge represents a relation with 'effect type' and 'effect direction' from gene A to gene B, for example, gene A activates (inhibits) gene B. In DKEGG, if the effect type from gene A to gene B was 'activation-like effect types' like activation or expression, we assigned the edge weight from node A to B as '+1' and if the effect type from node A to B was inhibition or repression, the edge weight from node A to B was set to '-1'.

Drug-drug target interactions

Drug-drug target interactions were obtained from DrugBank database (version 4.0) [16]. We collected drugs which possess known 'activation-like' or 'inhibition-like' drug-drug target interactions. We regarded activation, agonist, activator, simulator, and partial agonist interactions as 'activation-like drug-drug target interactions' and inhibition, inhibitor, antagonist, negative modulator, inverse agonist, suppressor, inhibitor (competitive), partial antagonist, reducer, and blocker as 'inhibition-like drug-drug target interactions'. We filtered out drugs whose targets were not included in DKEGG. Finally, we extracted 2,434 drug-drug target interactions for 364 drug targets (genes) and 898 drugs.

Selection of diseases

We selected diseases satisfying several conditions: 1. Known disease-gene associations with direct evidence are in Comparative Toxicogenomics Database (CTD) [17]; 2. Diseases related tissues are clear; 3. Gene expression profiles of disease cases and controls obtained from related tissues exist in Gene Expression Omnibus (GEO) database [18]. Based on these conditions, we selected original 16 disease. Gene expression data information and related tissues about 16 diseases we firstly chose were shown in Additional file 2: Table S1.

Disease genes with their altered states

Disease-gene associations for 16 diseases were collected from CTD. We only used disease-gene associations having direct evidences. MeSH IDs [19] of diseases and the number of the associated disease genes are described in Additional file 2: Table S1.

We assigned altered states of disease genes using gene expression data. We downloaded microarray gene expression datasets for each disease from GEO. The altered states of disease genes were assigned to up-regulated or down-regulated if disease genes are significantly differentially expressed genes between disease and control samples. We identified significantly differentially expressed genes using linear models and empirical Bayes methods (limma) [20] representing several practical advantages compared to other methods [21]. We found altered disease genes of 16 diseases whose FDR corrected p-values were less than 0.05. Five diseases, Alzheimer's disease, bipolar disorder, chronic obstructive pulmonary disease, idiopathic pulmonary fibrosis, and major depressive disorder, were excluded because none of their disease genes was significantly differentially expressed in disease states. Then, we checked whether differentially expressed disease genes exist in DKEGG or not. Multiple sclerosis was removed because its differentially expressed disease genes do not exist in DKEGG. Pulmonary hypertension was also filtered out because a differentially expressed disease gene is an isolated node in DKEGG. Finally, nine diseases were used in our further analysis.

Collecting therapeutic drug-disease associations

The chemical-disease association file was downloaded from CTD. MeSH IDs, presented in Additional file 2: Table S1, were utilized to match nine diseases in our study to diseases in the chemical-disease association field. We collected answer drugs for each disease which mean the drugs could be used to treat the diseases. The chemicals were employed as the answer sets in the further analysis according to two criteria: 1. Chemicals have therapeutic associations for diseases; 2. Chemicals have DrugBank ID and drug-drug target interactions with their effect types. The entire list of answer drugs with DrugBank ID for each disease is shown in Additional file 3: Table S2.

Resolving conflicts in DKEGG

We hypothesized drugs whose targets have short distances to disease genes and activate (inhibit) down (up)-regulated genes in disease states can be used to treat the diseases. To calculate distances between drug targets and disease genes, we inferred all possible shortest paths from each drug target to each disease gene using Breadth-first-search algorithm in DKEGG with 'effect type' and 'effect direction'. In our approach, a distance from gene A to gene B can be a negative value if gene A inhibits gene B.

In directed biological pathways, there exist conflicts. Conflict refers to a situation that two or more contradictory relations coexist in directed biological pathways [22]. Conflicts usually occur in biological pathways due to different biological contexts [23], for example, a dual way of TAK1 working corresponding to different cellular contexts [24]. In addition, conflicts were also observed in shortest paths from node A and node B in the directed pathway. For example, there were cases that gene A activated gene B in one shortest path while gene A inhibited gene B in other shortest paths: gene A activates gene C and gene C activates B; gene A activates gene D and gene D inhibits B.

To resolve this problem, we searched all possible shortest paths and set a distance $d_c(r, g)$ from a drug target r to a disease gene g with the consideration of the conflict problem by

$$d_c(r,g) = \frac{n_a + n_i}{n_a - n_i}|d(r,g)| \quad (1)$$

where n_a is the number of 'activation-like shortest paths', n_i is the number of 'inhibition-like shortest paths', and $|d(\cdot)|$ is an absolute value of distances of shortest paths between a drug target and a disease gene. Whether a shortest path is an activation or inhibition-like path is determined by the sum of the number of inhibition and repression edges. If the sum of the number of inhibition and repression edges is an even number including zero in the shortest path, it is regarded as an activation-like path and if an odd number, an inhibition-like path. The sign of d_C indicates whether the drug target would activate or inhibit the disease gene. When the number of activation-like shortest paths is greater than inhibition-like shortest paths, the sign of d_C is positive, or vice versa. The absolute value of d_C reveals how reliable the effect from a drug target to a disease gene is. As the number of activation and inhibition-like paths are similar, the absolute value of d_C increases than an original absolute value of distances $|d(\cdot)|$ and it represents the drug target would give little influence on the disease gene. If the number of activation and inhibition-like paths are equal or a drug target r and disease gene g are not connected in DKEGG, $d_C(r, g)$ will be a value of infinity. The distribution of d_C between all genes in DKEGG is presented in Additional file 4: Figure S2.

A score between a drug and a disease

For each drug and disease pair, we calculated a score using drug-target interactions, distances from drug targets disease genes with consideration of conflicts, and the regulated states of disease genes. In our PDOD approach, a PDOD score between a drug R having drug targets r_i and a disease G with its associated disease genes g_j was computed as below:

$$\text{PDOD score}(R,G) = \frac{1}{n_g n_r}\sum_{i=1}^{n_r}\sum_{j=1}^{n_g} sgn\left(r_i, g_j\right) \times \frac{1}{1+\left|\frac{d_c\left(r_i,g_j\right)}{\alpha}\right|^2} \quad (2)$$

$$sgn\left(r_i, g_j\right) = i(r_i) \times sign\left(d_c\left(r_i,g_j\right)\right) \times s\left(g_j\right) \quad (3)$$

where α is the parameter of a bell shaped function meaning the half width of the bell shaped function, n_r is the number of drug targets of the drug R, n_g is the number of disease genes of the disease G, $i(r_i)$ is the corresponding value for the predefined 'effect type' of drug-target interactions and it is assigned to '+1' for 'activation-like drug-drug target effect type' and '-1' for 'inhibition-like drug-drug target effect type', $s(g_j)$ is '+1' in case that g_j is a down-regulated gene in disease states and '-1' in case of an up-regulated-gene, and $sign(\cdot)$ is the sign function. Thus, the function $sgn(r_i, g_j)$ has a positive value when a drug has on opposite effect on a disease gene g_j through a drug target r_i. We set our scoring function as a bell shaped function since we thought that if the distances from drug targets to a disease genes were longer than the specific distance, α, the possibility that the drug had an effect on the disease genes would decrease rapidly. Thus, an absolute value of our PDOD score is in inverse proportion to the norm of $d_c(r_i, g_j)$. We divided our scores by the number of drug targets of the drug because we thought target proteins which did not affect disease genes might act as 'off-targets' for treatment of diseases and these targets could lead to unexpected effects such as side effects [25]. Our scores were also divided by the number of disease genes for normalization between diseases and thus our scores had values between −1 and 1. Finally, in our scoring function, a drug will receive a high score in case that drug targets of the drug are close to disease genes, drugs targets activate down-regulated disease genes and inhibit up-regulated disease genes in patients, and the number of off-targets is small.

Results and discussions

We applied our PDOD approach to nine diseases, adenocarcinoma of lung, acute myelocytic leukemia, astrocytoma, asthma, glioblastoma, oligodendroglioma, Parkinson disease, schizophrenia, and thyroid carcinoma. For each disease, we estimated the possibility of 898 drugs for the treatment of each disease based on PDOD scores which use drug-drug target interactions with 'effect type' from DrugBank, shortest paths from drug targets to disease genes with the consideration of conflicts in DKEGG, and changed states of disease genes in the disease with disease genes from CTD and expression profiles of GEO. The distribution of PDOD scores between all drugs and nine diseases we used presented in in Additional file 4: Figure S2–S3.

Performance of PDOD and previous methods

Comparison to a network-based method without directional information

To verify the importance of directional information used in PDOD, we make a Predicting Drugs having effects on Disease genes (PDD) score which uses drug-drug target interactions without 'effect type', paths from drug targets to disease genes on DKEGG without 'effect type' and 'effect direction', and disease genes without their regulated states.

We calculated PDOD scores and PDD scores between all 898 drugs and each disease respectively and compared performance of PDOD to PDD for predicting known drugs of each disease from CTD. Fig. 3 shows the area under receiver operating characteristic curves (AUC) values of PDOD and PDD approaches for nine diseases. As shown in Fig. 3, PDOD shows better performance than PDD in the prediction of answer drugs for diseases except for Parkinson disease. This illustrates the importance of directional information used in PDOD. We set a value

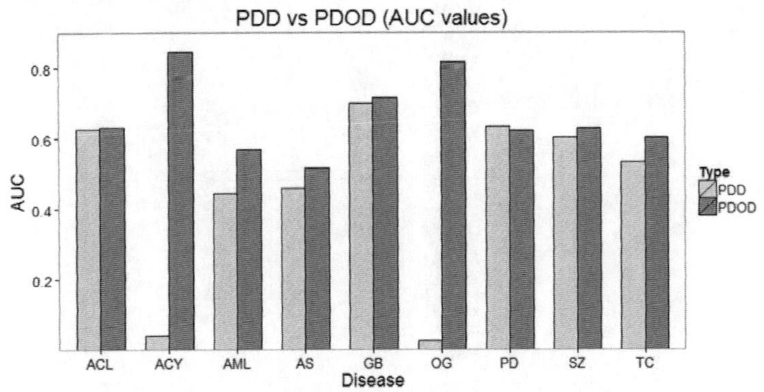

Fig. 3 Comparison of PDOD with PDD. The AUC values of PDOD and PDD for nine diseases

of parameter α to three for AUC values of both PDOD and PDD in Fig. 3. AUC values through various α values are given in Additional file 5: Figure S4–S12. As shown in figures of Additional file 5: Figure S4–S12, some diseases like adenocarcinoma of lung and schizophrenia revealed slightly different AUC values according to values of parameter α and they have high AUC values when α is bigger than three. It may be related to results from the previous work [26] which shows that the averages of distances between drug targets and disease genes are dependent on the types of diseases.

Comparison to a previous method without directed network

Our approach was also compared with a previous method using gene expression profiles of both drugs and diseases [6, 7]. These previous approaches are analogous to PDOD in that they intended to identify drugs likely to have complementary effects on a disease state as therapies of the disease. Thus, we compared performance of PDOD, which is a network–based approach, to the previous method using only gene expression profiles for prediction of known therapeutic relationship between drugs and diseases.

Disease gene expression signatures of the previous method were obtained identical datasets from GEO (Additional file 3: Table S1). We assigned that the numbers of up- and down-regulated disease gene expression signatures for each disease are equal to the numbers of up- and down-regulated disease genes used in PDOD. With disease gene expression signatures, the enrichment score (ES) introduced in Connectivity Map [6] was used for determining the potentiality of a drug for the treatment of the disease. A drug having high negative enrichment score was expected to repress the disease. Since some of answer drugs did not have drug exposure expression data in Connectivity Map and some drugs existing in Connectivity Map did not have drug-target interactions used in our approach. Thus, the intersection of two drug sets, 317 drugs, was used in this evaluation. In this process, the answer drug of astrocytoma and oligodendroglioma are filtered out. Thus, we compared performances of PDOD and the previous method for seven diseases (Fig. 4). Figure 4 indicates PDOD using directed network usually outperformed a previous approach using only gene expression signatures.

Potentiality of high ranked drugs in PDOD as therapies

Top 10 ranked drugs receiving high scores in our PDOD for whole diseases are revealed in Table 1. All of the ten drugs getting high scores are not included in answer drugs for each disease. We investigated published literature which can support the possibility of uses of these drugs for the treatment of the disease. Six drugs have supporting evidence in literature for their expected usages [27–36]. These results show the potential of our PDOD approach for identification of novel drug indications.

Specially, tetracycline [DrugBank:DB00759], which is the drug receiving the highest PDOD score and is not contained in our answer drug sets for oligodendroglioma, has already a supporting literature evidence for the treatment of oligodendroglioma [27]. This drug inhibits its target [Entrez ID:5621]. The distance d_c from its target [Entrez ID:5621] to up-regulated disease gene [Entrez ID:3280] is '+3' which is obtained by two 'activation like paths' from DKEGG and thus its target activates up-regulated disease genes. Therefore, this drug may inhibit an up-regulated disease gene [Entrez ID:3280] of oligodendroglioma not directly but through its target [Entrez ID:5621] and DKEGG.

Five drugs, fulvestrant [DrugBank:DB00947], cetuximab [DrugBank:DB00002], gefitinib [DrugBank:DB00317], erlotinib [DrugBank:DB00530], and panitumumab [DrugBank:DB01269], do not exist in our answer drug sets but they also have a potential to treat thyroid carcinoma

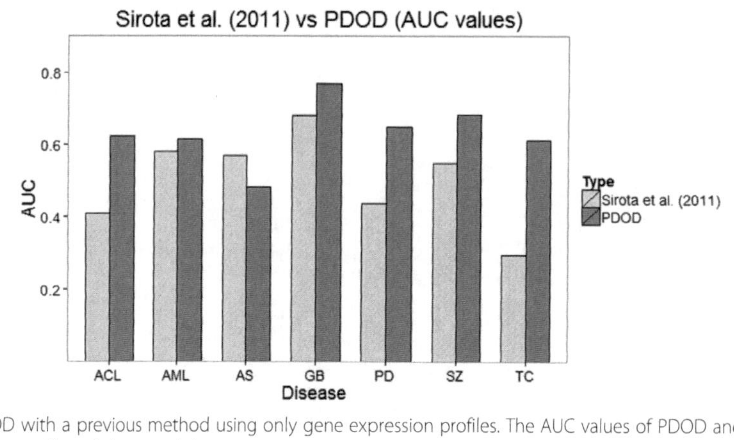

Fig. 4 Comparison of PDOD with a previous method using only gene expression profiles. The AUC values of PDOD and a previous method which uses gene expression profiles of drugs and diseases

Table 1 Top 10 ranked drugs in PDD and their potentiality for the therapy of diseases

Disease (MeSH ID)	Drug (DrugBank ID)	Description in DrugBank	Supporting evidence from literature
OG (D009837)	Tetracycline (DB00759)	Bacterial infections	[27]
TC (D013964)	Clomifene (DB00882)	Female infertility due to anovulation	n/a
TC (D013964)	Fulvestrant (DB00947)	Metastatic breast cancer	[28]
TC (D013964)	Ospemifene (DB04938)	Dyspareunia	n/a
TC (D013964)	Cetuximab (DB00002)	Metastatic colorectal cancer	[29, 30]
TC (D013964)	Gefitinib (DB00317)	Certain types of cancer	[31–33]
TC (D013964)	Erlotinib (DB00530)	Non-small cell lung cancer	[34–36]
TC (D013964)	Panitumumab (DB01269)	Antineoplastic agent	[36]
AML (D015470)	Mecasermin (DB01277)	Primary IGF 1 deficiency	n/a
TC (D013964)	Purvalanol (DB02733)	n/a	n/a

according to literature. All of five drugs also inhibit an up-regulated disease gene [Entrez ID:595] and activate a down-regulated disease gene [Entrez ID:83439] through their targets [Entrez ID:1956, 2099] and DKEGG. For example, fulvestrant have an antagonistic effect on its target [Entrez ID:2099]. The distance from its target [Entrez ID:2099] to up-regulated disease gene [Entrez ID:595] is '+1' by one 'activation like path'. The distance from its target [Entrez ID:2099] to down-regulated disease gene [Entrez ID:83439] is '-3' by one 'inhibition like path'. Thus, fulvestrant would inhibit an up-regulated disease genes and activate a down-regulated disease gene. For another example, gefitinib have an antagonistic effect on its target [Entrez ID:1956]. The distance from its target [Entrez ID:1956] to up-regulated disease gene [Entrez ID:595] is about '2.57' by eight 'activation like paths' and one 'inhibition like path'. The distance from its target [Entrez ID:2099] to down-regulated disease gene [Entrez ID:83439] is '-2' by one 'inhibition like path'. Therefore, gefitinib also would inhibit an up-regulated disease genes and activate a down-regulated disease gene and have an opposite effects on altered states of disease genes through drug-drug target interactions and paths from drug target to disease genes in DKEGG. These results not only are consistent with the previous study [26] but also indicate the importance of the consideration of biological networks in predicting indications of drugs.

Implementation in online

For researchers to infer drugs having opposite effects on altered genes they interested, we implemented PDOD in online (http://gto.kaist.ac.kr/pdod/index.php/main). Users can submit a gene list with their Entrez gene IDs [37] and altered states which they gained from their experiments. Users can select and submit alpha values according to the disease of interest [26] or the distribution of d_C in Additional file 4: Figure S2. Then, users will get a drug list likely to have reverse effects on their genes if more than one gene existed in DKEGG. In addition, users can simply access our web page via mobile.

Conclusions

In this paper, we developed a novel network-based method, PDOD, to identify novel indications of existing drugs. We inferred drugs having reverse effects on altered states of disease genes with the consideration of 'effect type' and 'effect direction' in directed network. For a given disease gene set with altered states, we predicted a drug as a medicine of the disease when the drug satisfies two conditions: 1. Target proteins of a drug are close to disease genes; 2. A drug restores altered states of disease genes through paths from a drug to disease genes. We applied our method to infer candidate therapies for nine diseases. Our PDOD is compared to two previous approaches, the network-based method without 'effect type','effect direction', and'altered states of disease gene' information and the method which did not utilize any biological network. This comparison showed the importance of this kind of information. In addition, we represented the potentialities of some of high ranked drugs in PDOD as novel treatments of the diseases.

We constructed a web based tool implementing PDOD. Our online tool may help researchers to predict drugs having opposite effects on disease genes and generate a hypothesis easily by using PDOD. Since average values of distances between drug targets and disease genes are various according to categories of diseases [26], users can set parameters depending on the disease of interest. In addition, even though we identified altered states of disease genes using expression data, researchers can input genes with changed states from any kind of experiment.

In spite of our promising results for a case study, our works have several drawbacks. One of the major drawbacks in our approach is that the number of 'effect types' used in our method is limited to four interactions. Relations having ambiguous 'effect type' and 'effect direction', for example, binding, association, and phosphorylation were excluded at this present; thus, the scale of backbone network is relatively small compared to a real molecular network. We did not include these relations because they can bring out false positives in prediction of drugs; however, it causes that some diseases like multiple sclerosis cannot be applied PDOD to since genes related to these diseases do not exist in our constructed network. Another drawback of our approach is that we considered only the shortest paths although there would be longer paths from drug targets to disease genes. Even though we only used the shortest paths based on the assumption that closer genes may be more interactive, which is commonly used in other papers using biological network [38, 39], the consideration of longer paths could give us more comprehensive results.

In further study, we plan to enlarge scales of the backbone network not only by including other interactions like phosphorylation, methylation, binding/association, and metabolic reaction but also by combining other databases containing 'effect type' and 'effect direction information' such as pathway interaction database (PID) and small molecule pathway database (SMPDB). We will apply our method to more disease cases and validate promising drugs from our PDOD with wet lab experiment through collaboration. In addition, we expect that a negative score between a drug and disease in PDOD may explain the drug that is likely to have same effects on altered states of disease genes. In other words, the disease may be regarded as side effects of the drug. Based on this idea, the proposed approach may be feasible to predict side effects of drugs and drug-drug interactions.

Homeostasis is one of the most fundamental features of mankind. If a disease occurred and homeostasis could not be maintained, a drug having opposite effects on the disease could restore the altered states in patients. Our approach would be appropriate to predict these kinds of drugs.

Additional files

Additional file 1: Figure S1. Flow diagram of PDOD. The figure that illustrates how the databases are used to generate the inputs and expected outputs. (PDF 166 kb)

Additional file 2: Table S1. 16 diseases considered in this work. Statistics and information of data for 16 diseases are described in this file: Disease name; MeSH ID of diseases; GSE numbers we used; platforms of the GSE; tissues where GSE are generated; the number of disease genes from CTD; the number of differentially expressed disease genes; the number of differentially expressed disease genes in DKEGG. (XLSX 12 kb)

Additional file 3: Table S2. Answer drugs used in this work for nine diseases. Information about answer drugs used in this work are described in this file: Disease name; MeSH ID of diseases; the number of answer drugs for each disease; the list of answer drugs (DrugBank ID). (XLSX 11 kb)

Additional file 4: Figure S2–S3. The distribution of dc and PDOD scores. Each figure presents the distribution of dc between all genes in DKEGG and PDOD scores between all drugs and nine diseases we used. (PDF 236 kb)

Additional file 5: Figure S4–S12. AUC values for nine diseases with different values of α. AUC values of PDD and PDOD for nine diseases according to different values of parameter α are enclosed in this file. (PDF 131 kb)

Abbreviations
PDOD: Prediction of Drugs having opposite effects on Disease genes; DKEGG: Directed KEGG pathways, which we constructed, containing only directed relations such as 'activation' and 'inhibition'; PDD: Prediction of Drugs having effects on Disease genes; ACL: Adenocarcinoma of lung; AML: Acute myelocytic leukemia; ACY: Astrocytoma; AS: Asthma; GB: Glioblastoma; OG: Oligodendroglioma; PD: Parkinson disease; PH: Pulmonary hypertension; SZ: Schizophrenia; TC: Thyroid carcinoma.

Competing interests
All authors declare that they have no conflict of interest to the publication of this article, and approved the manuscript and this submission.

Authors' contributions
HY suggested the main idea and DH supervised this work. HY and SC not only resolved conflict problems in directed network but also extracted drug-drug target interaction and drug-disease association information. SC and JP developed a web service. HY and JJ integrated directed pathways. SC and YK found literature evidences of high scored drugs in PDOD. All authors read, wrote, and approved the manuscript.

Acknowledgements
We thank to the distributors of well-established databases, R project, and R libraries used in this research, including CTD, GEO, KEGG, KEGGgraph, DrugBank and Connectivity Map. We thank to DJ Jang and JH Lee for being involved in drafting the manuscript. We also thank all of other researchers giving us inspiration. This work was supported by the Bio-Synergy Research Project (2012M3A9C4048758) of the Ministry of Science, ICT and Future Planning through the National Research Foundation. The publication costs for this article were also funded by the Bio-Synergy Research Project (2012M3A9C4048758).

Published: 11 January 2016

References
1. DiMasi JA, Hansen RW, Grabowski HG. The price of innovation: new estimates of drug development costs. J Health Econ. 2003;22(2):151–85.
2. Hay M, Thomas DW, Craighead JL, Economides C, Rosenthal J. Clinical development success rates for investigational drugs. Nat Biotechnol. 2014;32(1):40–51.
3. Ashburn TT, Thor KB. Drug repositioning: identifying and developing new uses for existing drugs. Nat Rev Drug Discov. 2004;3(8):673–83.
4. Dudley JT, Sirota M, Shenoy M, Pai RK, Roedder S, Chiang AP. Computational repositioning of the anticonvulsant topiramate for inflammatory bowel disease. Sci Transl Med. 2011;3(96):96ra76.
5. Hu G, Agarwal P. Human disease-drug network based on genomic expression profiles. PLoS One. 2009;4(8):e6536.
6. Lamb J, Crawford ED, Peck D, Modell JW, Blat IC, Wrobel MJ. The Connectivity Map: using gene-expression signatures to connect small molecules, genes, and disease. Science. 2006;313(5795):1929–35.
7. Sirota M, Dudley JT, Kim J, Chiang AP, Morgan AA, Sweet-Cordero A. Discovery and preclinical validation of drug indications using compendia of public gene expression data. Sci Transl Med. 2011;3(96):96ra77.
8. Cheng F, Liu C, Jiang J, Lu W, Li W, Liu G, et al. Prediction of drug-target interactions and drug repositioning via network-based inference. PLoS Comput Biol. 2012;8(5):e1002503.
9. Emig D, Ivliev A, Pustovalova O, Lancashire L, Bureeva S, Nikolsky Y, et al. Drug target prediction and repositioning using an integrated network-based approach. PLoS One. 2013;8(4):e60618.
10. Wang W, Yang S, Zhang X, Li J. Drug repositioning by integrating target information through a heterogeneous network model. Bioinformatics. 2014;30(20):2923–30.

11. Wang Y, Zeng J. Predicting drug-target interactions using restricted Boltzmann machines. Bioinformatics. 2013;29(13):i126–i34.
12. Qabaja A, Alshalalfa M, Alanazi E, Alhajj R. Prediction of novel drug indications using network driven biological data prioritization and integration. J Cheminformatics. 2014;6(1):1.
13. Fukuoka Y, Takei D, Ogawa H. A two-step drug repositioning method based on a protein-protein interaction network of genes shared by two diseases and the similarity of drugs. Bioinformation. 2013;9(2):89–93. doi:10.6026/97320630009089.
14. Kanehisa M, Goto S. KEGG: kyoto encyclopedia of genes and genomes. Nucleic Acids Res. 2000;28(1):27–30.
15. Zhang JD, Wiemann S. KEGGgraph: a graph approach to KEGG PATHWAY in R and bioconductor. Bioinformatics. 2009;25(11):1470–1.
16. Wishart DS, Knox C, Guo AC, Shrivastava S, Hassanali M, Stothard P, et al. DrugBank: a comprehensive resource for in silico drug discovery and exploration. Nucleic Acids Res. 2006;34 suppl 1:D668–D72.
17. Davis AP, Murphy CG, Saraceni-Richards CA, Rosenstein MC, Wiegers TC, Mattingly CJ. Comparative Toxicogenomics Database: a knowledgebase and discovery tool for chemical–gene–disease networks. Nucleic Acids Res. 2009;37 suppl 1:D786–D92.
18. Edgar R, Domrachev M, Lash AE. Gene Expression Omnibus: NCBI gene expression and hybridization array data repository. Nucleic Acids Res. 2002;30(1):207–10.
19. Lowe HJ, Barnett GO. Understanding and using the medical subject headings (MeSH) vocabulary to perform literature searches. Jama. 1994;271(14):1103–8.
20. Smyth GK. Limma: linear models for microarray data. Bioinformatics and computational biology solutions using R and Bioconductor. New York: Springer; 2005. p. 397–420.
21. Jeanmougin M, De Reynies A, Marisa L, Paccard C, Nuel G, Guedj M. Should we abandon the *t*-test in the analysis of gene expression microarray data: a comparison of variance modeling strategies. PLoS One. 2010;5(9):e12336.
22. Stobbe MD, Swertz MA, Thiele I, Rengaw T, Van Kampen AH, Moerland PD. Consensus and conflict cards for metabolic pathway databases. BMC Syst Biol. 2013;7(1):50.
23. Yoon S, Jung J, Yu H, Kwon M, Choo S, Park K et al. Context-based resolution of semantic conflicts in biological pathways. BMC medical informatics and decision making. 2015;15 Suppl 1:S3. doi:10.1186/1472-6947-15-S1-S3.
24. Ajibade AA, Wang HY, Wang R-F. Cell type-specific function of TAK1 in innate immune signaling. Trends Immunol. 2013;34(7):307–16.
25. Scheiber J, Chen B, Milik M, Sukuru SCK, Bender A, Mikhailov D, et al. Gaining insight into off-target mediated effects of drug candidates with a comprehensive systems chemical biology analysis. J Chem Inf Model. 2009;49(2):308–17.
26. Yıldırım MA, Goh K-I, Cusick ME, Barabási A-L, Vidal M. Drug—target network. Nat Biotechnol. 2007;25(10):1119–26.
27. Domercq M, Matute C. Neuroprotection by tetracyclines. Trends Pharmacol Sci. 2004;25(12):609–12. doi:10.1016/j.tips.2004.10.001.
28. Rajoria S, Suriano R, George A, Shanmugam A, Schantz SP, Geliebter J, et al. Estrogen induced metastatic modulators MMP-2 and MMP-9 are targets of 3,3′-diindolylmethane in thyroid cancer. PLoS One. 2011;6(1):e15879. doi:10.1371/journal.pone.0015879.
29. Prichard CN, Kim S, Yazici YD, Doan DD, Jasser SA, Mandal M, et al. Concurrent cetuximab and bevacizumab therapy in a murine orthotopic model of anaplastic thyroid carcinoma. Laryngoscope. 2007;117(4):674–9. doi:10.1097/MLG.0b013e318031055e.
30. Kim S, Prichard CN, Younes MN, Yazici YD, Jasser SA, Bekele BN, et al. Cetuximab and irinotecan interact synergistically to inhibit the growth of orthotopic anaplastic thyroid carcinoma xenografts in nude mice. Clin Cancer Res. 2006;12(2):600–7. doi:10.1158/1078-0432.CCR-05-1325.
31. Pennell NA, Daniels GH, Haddad RI, Ross DS, Evans T, Wirth LJ, et al. A phase II study of gefitinib in patients with advanced thyroid cancer. Thyroid. 2008; 18(3):317–23. doi:10.1089/thy.2007.0120.
32. Frasca F, Vella V, Nicolosi ML, Messina RL, Giani F, Lotta S, et al. Thyroid cancer cell resistance to gefitinib depends on the constitutive oncogenic activation of the ERK pathway. J Clin Endocrinol Metab. 2013;98(6):2502–12. doi:10.1210/jc.2012-3623.
33. Kurebayashi J, Okubo S, Yamamoto Y, Ikeda M, Tanaka K, Otsuki T, et al. Additive antitumor effects of gefitinib and imatinib on anaplastic thyroid cancer cells. Cancer Chemother Pharmacol. 2006;58(4):460–70. doi:10.1007/s00280-006-0185-x.
34. Hogan T, Jing Jie Y, Williams HJ, Altaha R, Xiaobing L, Qi H. Oncocytic, focally anaplastic, thyroid cancer responding to erlotinib. J Oncol Pharm Pract. 2009;15(2):111–7. doi:10.1177/1078155208101212.
35. Masago K, Miura M, Toyama Y, Togashi Y, Mishima M. Good clinical response to erlotinib in a patient with anaplastic thyroid carcinoma harboring an epidermal growth factor somatic mutation, L858R, in exon 21. J Clin Oncol. 2011;29(16):e465–7. doi:10.1200/JCO.2010.34.0216.
36. Landriscina M, Maddalena F, Fabiano A, Piscazzi A, La Macchia O, Cignarelli M. Erlotinib enhances the proapoptotic activity of cytotoxic agents and synergizes with paclitaxel in poorly-differentiated thyroid carcinoma cells. Anticancer Res. 2010;30(2):473–80.
37. Maglott D, Ostell J, Pruitt KD, Tatusova T. Entrez Gene: gene-centered information at NCBI. Nucleic Acids Res. 2005;33 suppl 1:D54–D8.
38. Li S, Zhang B, Zhang N. Network target for screening synergistic drug combinations with application to traditional Chinese medicine. BMC Syst Biol. 2011;5 Suppl 1:S10. doi:10.1186/1752-0509-5-S1-S10.
39. Zou J, Ji P, Zhao YL, Li LL, Wei YQ, Chen YZ, et al. Neighbor communities in drug combination networks characterize synergistic effect. Mol BioSyst. 2012;8(12):3185–96. doi:10.1039/c2mb25267h.

Huang *et al. BMC Systems Biology* 2016, **10**(Suppl 1):3
DOI 10.1186/s12918-015-0244-1

BMC Systems Biology

PROCEEDINGS

Open Access

A new scheme to discover functional associations and regulatory networks of E3 ubiquitin ligases

Kai-Yao Huang[1], Julia Tzu-Ya Weng[1,2], Tzong-Yi Lee[1,2*] and Shun-Long Weng[3,4,5*]

From The Fourteenth Asia Pacific Bioinformatics Conference (APBC 2016)
San Francisco, CA, USA. 11 - 13 January 2016

Abstract

Background: Protein ubiquitination catalyzed by E3 ubiquitin ligases play important modulatory roles in various biological processes. With the emergence of high-throughput mass spectrometry technology, the proteomics research community embraced the development of numerous experimental methods for the determination of ubiquitination sites. The result is an accumulation of ubiquitinome data, coupled with a lack of available resources for investigating the regulatory networks among E3 ligases and ubiquitinated proteins. In this study, by integrating existing ubiquitinome data, experimentally validated E3 ligases and established protein-protein interactions, we have devised a strategy to construct a comprehensive map of protein ubiquitination networks.

Results: In total, 41,392 experimentally verified ubiquitination sites from 12,786 ubiquitinated proteins of humans have been obtained for this study. Additional 494 E3 ligases along with 1220 functional annotations and 28588 protein domains were manually curated. To characterize the regulatory networks among E3 ligases and ubiquitinated proteins, a well-established network viewer was utilized for the exploration of ubiquitination networks from 40892 protein-protein interactions. The effectiveness of the proposed approach was demonstrated in a case study examining E3 ligases involved in the ubiquitination of tumor suppressor p53. In addition to Mdm2, a known regulator of p53, the investigation also revealed other potential E3 ligases that may participate in the ubiquitination of p53.

Conclusion: Aside from the ability to facilitate comprehensive investigations of protein ubiquitination networks, by integrating information regarding protein-protein interactions and substrate specificities, the proposed method could discover potential E3 ligases for ubiquitinated proteins. Our strategy presents an efficient means for the preliminary screen of ubiquitination networks and overcomes the challenge as a result of limited knowledge about E3 ligase-regulated ubiquitination.

Keywords: Ubiquitination, Ubiquitin, E3 ubiquitin ligase, Protein-protein interaction, Ubiquitination network

* Correspondence: francis@saturn.yzu.edu.tw; 4467@mmh.org.tw
[1]Department of Computer Science and Engineering, Yuan Ze University, Taoyuan 320, Taiwan
[3]Department of Obstetrics and Gynecology, Hsinchu Mackay Memorial Hospital, Hsin-Chu 300, Taiwan
Full list of author information is available at the end of the article

Introduction

Protein ubiquitination involves a series of enzymatic reactions such as E1 activation, E2 conjugation, and E3 ligation, resulting in the conjugation of single or multiple ubiquitin proteins at a target lysine residue [1]. Numerous substrate proteins with ubiquitination sites have been characterized to date, owing to the emergence of high-throughput mass spectrometry-based proteomics approaches [2–4]. Identified to play key roles in transcriptional regulation, signal transduction, development, apoptosis, endocytosis, cell proliferation and cancers, ubiquitination of the lysine residue has been regarded as an essential mediator of various biological processes [5–7]. Among the enzymes that catalyze protein ubiquitination, E3 ligases are particularly important for the recognition of substrate sites to facilitate ubiquitin-mediated protein degradation [8]. The relationships between E3 ligase and substrates are complex. Multiple substrates could be targeted by a single E3 ligase; alternatively, multiple E3 ligases could catalyze the ubiquitination of a single substrate [9]. These substrate-enzyme correlations could be used to construct E3-specific regulatory networks and map to the associated cellular pathways, making possible the characterization of complex cellular processes and functional analysis of E3-sbustrate relationships [9]. This approach has allowed the discovery of the role that anaphase-promoting complex (APC)/cyclosome plays in modulating key targets of the cell cycle, such as cyclins and their related E3 ligases [10–12].

To date, a significant amount of research efforts have been invested towards the characterization of E3 structures and examination of the mechanisms underlying E3-mediated regulatory networks, as well as E3-related diseases [13–21]. Based on their catalytic mechanisms in the ubiquitination process, E3 ligases can be classified into three major types: the HECT (homologous to E6-AP C-terminus), the RING (really interesting new gene), and U-box domain types [22]. The HECT-type is responsible for catalyzing the attachment of ubiquitin to substrate proteins. In contrast, the RING-type and U-box-type, similar in both structure and function, facilitate the interaction between an E2 enzyme and the target proteins. Regardless of the types, the significance of E3-mediated ubiquitination is obvious from their association with diseases [23]. Several studies have suggested that the inhibition of E3 ligases may cause growth suppression or cell death, as evidenced by the overexpression of Mdm2/Hdm2, IAPs, and SCF in various human cancers [24]. Therefore, regulation of E3 ligase activities and functions may be a promising approach for cancer treatments.

Many databases and tools have been developed to aid in the study of E3 ligases. For example, E3Miner [25] offers a text mining approach to identify ubiquitin-protein ligases, whereas E3Net [9] allows users to search through a collection of 1671 E3-substrate relationships among 493 E3s and 1277 substrates in 42 organisms. In contrast, by analyzing protein sequence similarities, domains, and distributions across different species, Sakiyama et al. [26] constructed a useful database for the exploration of proteins involved in the ubiquitin signaling cascade. Unfortunately, the present accumulation of large-scale ubiquitinome data demands for the development of tools that investigate the regulatory networks of E3 ligases and their substrates. Here, we present a new strategy that utilizes an interactive network viewer to assist with the discovery of novel protein ubiquitination networks. Furthermore, to effectively investigate the relationships between E3 ligases and their substrates, metabolic pathways and protein-protein interactions (PPIs) were integrated to construct comprehensive protein ubiquitination networks. The ability of the proposed method to identify E3 ligase-mediated ubiquitination networks and their biological significance was demonstrated by case studies. The results indicated that, despite the current limited knowledge about regulatory relationship between E3 ligases and ubiquitinated proteins, our approach could uncover potential E3 ligase-substrate relationships based on based on protein-protein interaction information and substrate site specificities.

Materials and method

Construction of the protein ubiquitination networks involved collection of E3 ligase and ubiquitinated protein data, integration of ubiquitinated proteins' functional data, computational identification of ubiquitination sites based on substrate motifs, as well as network construction using protein-protein interactions and metabolic pathways (Fig. 1). A network viewer was employed to provide a visualization of the ubiquitination regulatory network, with implemented functional information, for a group of proteins of interest. The detailed workflow is described as follows.

Data collection of E3 ubiquitin ligases and ubiquitinated proteins

Experimentally validated E1 activating, E2 conjugating, and E3 ligating enzyme data were obtained from various sources. From UUCD-Version 1.0 [27], seven distinct E1 activating enzymes were collected. From E3Net, UUCD [27], hUbiquitome [28], and UniProtKB [29], 494 non-redundant E3 ubiquitin ligases and their biological functions were extracted. In addition, a total of 46 non-redundant E2 conjugating enzymes were collected from UUCD [27], hUbiquitome [28] and UniProtKB [29]. Experimentally verified ubiquitination sites from dbPTM [30–32] were also included. Next, search keywords, such as "ubiquitinated", "ubiquitination", "ubiquitylated", or "ubiquitylation", were entered on the PubMed database to

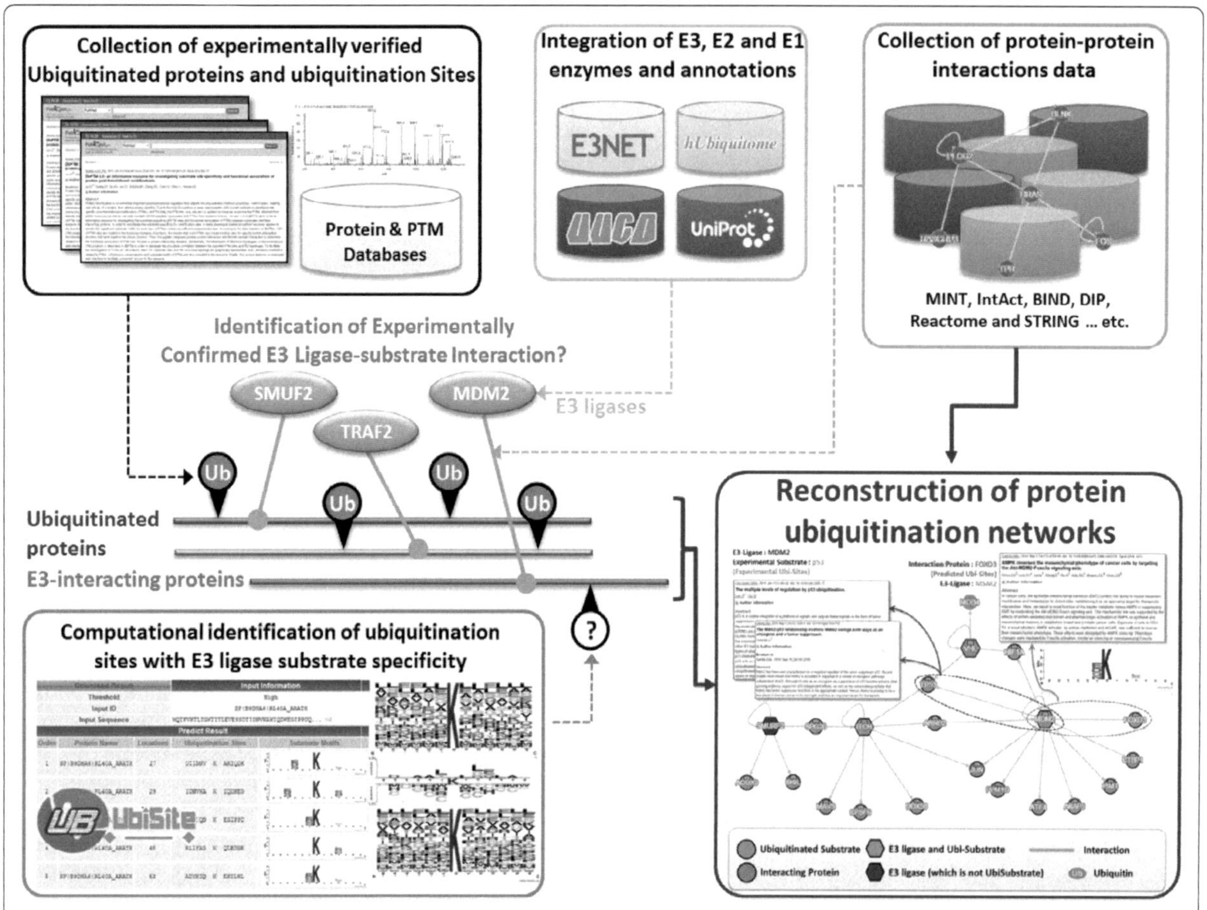

Fig. 1 System flow of protein ubiquitination network construction

extract ubiquitinated protein data from research articles. Specifically, full texts of the matched research articles were manually reviewed to ensure that the exact ubiquitinated peptide and modified lysine residue information were extracted. Finally, redundant data were removed, generating a total of 41,392 ubiquitinated lysines from 12,786 ubiquitinated human proteins.

Characterization of protein ubiquitination sites

To characterize the amino acid composition of protein ubiquitination sites, WebLogo [33, 34] was utilized to generate the relative frequency of the corresponding amino acid at each position around the ubiquitination sites as represented by the graphical sequence logo. As well, to further discriminate the amino acid composition of ubiquitinated sites from their non-ubiquitinated counterparts, TwoSampleLogo [35] was adopted to display statistically significant differences in position-specific symbol compositions. The inherent complexity of large-scale ubiquitinome data may make it difficult to uncover conserved motifs. To overcome this problem, MDDLogo [36] was applied to identify potential motifs for the curated protein ubiquitination sites. MDDLogo is a program that uses the maximal dependence decomposition (MDD) approach to discover conserved motifs from groups of aligned signal sequences through a recursive process that divides the data sets into tree-like subgroups. The effectiveness of MDDLogo has been demonstrated in the identification of substrate motifs for phosphorylation [37–40], *S*-nitrosylation [41], O-GlcNAcylation sites [42], *S*-glutathionylation [43], as well as ubiquitin conjugation sites [2].

Data integration for functional investigation of ubiquitinated proteins

To investigate the biological significance of ubiquitinated proteins, various biological databases, such as Gene Ontology (GO) [44], InterPro [45], as well as KEGG Diseases and Pathways [46], were incorporated. To provide comprehensive functional annotations of proteins associated with ubiquitination, the ubiquitinated proteins were classified according to their molecular functions, biological processes, and cellular components. Since ubiquitination is known to regulate the cellular localization, interactions, and degradation of proteins [47–49], the biological roles of ubiquitination sites within a specific protein domain could be inferred from the functional annotation of the domain. For this purpose, essential protein family, domain, and functional site information was obtained from InterPro [45], a database which integrates data from various sources such as the PROSITE [50], PRINTS [51], Pfam [52], and ProDom [53].

Network construction using protein-protein interactions and metabolic pathways

Substantial evidence supports the role that protein ubiquitination plays in the regulation of cellular processes. Thus, by integrating experimentally validated mammalian E3 ubiquitin ligases and their functional information, we hoped to provide a foundation for navigating ubiquitination regulatory networks in mammals. To facilitate the exploration of regulatory relationships between E3 ligases and their ubiquitinated substrates, associated metabolic pathways and protein-protein interactions (PPIs) were included for the comprehensive construction of protein ubiquitination networks. The human metabolic pathways were extracted from KEGG [54]. Experimentally verified PPIs were obtained from over ten PPI databases (Additional file 1: Table S1). Potential PPIs predicted based on co-regulation, co-occurrence in the literature, co-expression, and genomic context were curated from the STRING database [55]; each interaction included a confidence score calculated by the STRING built-in function.

Next, a graph theory [56, 57] approach has been adopted to illustrate the relationships between E3 ligases and substrates. Specifically, we use a directed and cyclic graph $G = (V, E)$ to symbolize a protein ubiquitination network, where $x, y \in V$ and $(x, y) \in E$. The E3 ligases and substrate proteins were represented by x and y, respectively, and protein ubiquitination was denoted by $(x, y) \in E$ to indicate the recognition of a specific substrate y by E3 ligase x (Additional file 2: Figure S1). Due to limited knowledge about ubiquitinated substrates that are recognized by E3 ligases, (x, y) could also represent a type of protein-protein interaction between E3 ligase x and ubiquitinated protein y. We used V to refer to all human proteins and E, to all experimentally confirmed PPIs. Cytoscape [58], a publicly available network viewer, was employed for the visualization of regulatory networks among E3 ligases and ubiquitinated substrates.

Results and discussion

Data statistics in this investigation

Data used for building the protein ubiquitination networks in this study were experimentally validated and supported with 39,814 research articles. Over 500 research articles were manually reviewed via a text mining method. In total, 41,392 ubiquitination sites from 12,786 ubiquitinated proteins in humans were extracted from 406 literatures. After removing redundant data among heterogeneous online resources, 494 experimentally verified human E3 ubiquitin ligases remained in the resulting data. PPIs between E3 ligases and ubiquitinated proteins were retrieved to deduce potential regulatory relationships between E3 ligases and ubiquitinated substrates to compensate for the limited information about E3 ligase targets. As shown in Table 1, 9,271 physical PPIs between 426 E3 ligases and 2,649 ubiquitinated proteins were curated. In particular, by incorporating the substrate motifs identified by the MDDLogo ubiquitination site prediction method [36], potential substrates of E3 ligases could be inferred from the 27,227 PPIs between E3 ligases and other proteins. Moreover, 377,117 PPIs that appeared to involve ubiquitinated proteins could be integrated for the investigation of their functional associations in the context of ubiquitination.

Table 1 Data statistics in this work

Data content	Number of records
Ubiquitinated protein (potential E3 substrates)	12,786
Ubiquitination sites	41,392
E3 ubiquitin ligases	494
PPIs between E3 ligases and other proteins	27,227
PPIs between E3 ligases and ubiquitinated proteins	9,271
E3 ligases interacting with ubiquitinated proteins	436
Ubiquitinated proteins interacting with E3 ligases	2,649
Supported articles	39,814

Substrate specificities of human ubiquitination sites

The entropy plots generated by the sequence logo was used to graphically visualize the amino acid sequences flanking the substrate sites (at position 0). This allows for the easy observation of amino acid conservation surrounding the ubiquitination sites. Figure 2a shows Leu (L), Glu (E), and Ala (A) to be the most conserved amino acid residues as indicated by the position-specific amino aicd composition around the ubiquitinated lysines. Furthermore, using TwoSampleLogo, the differences in position-specific amino acid composition between ubiquitinated and non-ubiquitinated sites were revealed (Fig. 2b). The residues surrounding the ubiquitination sites were significantly enriched with Ala (A), Asp (D), Glu (E), Leu (L), Gly (G) and Thr (T), and depleted in Cys (C), His (H), Arg (R), Trp (W) and Met (M) ($p < 0.005$).

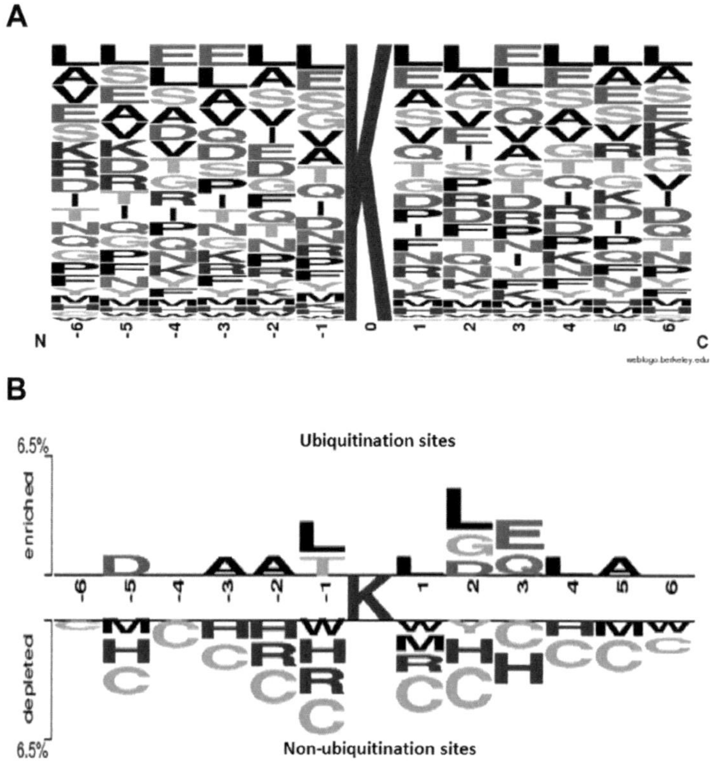

Fig. 2 Amino acid composition of protein ubiquitination sites. **a** The frequency plot of ubiquitinated sites. **b** The compositional biases of amino acids around ubiquitination sites compared to the non-ubiquitination sites

To overcome the difficulty of discovering conserved motifs from large-scale ubiquitinome data, the MDDLogo clustering method was adopted to search for substrate motifs from the curated human ubiquitination sites using a 13-mer window length. MDDLogo identified a total of nine subgroups containing conserved motifs from non-homologous human ubiquitination sites (Additional file 3: Table S2). While subgroup 1 (241 ubiquitination sites) contained the conserved amino acid composition at positions +3 and +5, the conserved motif of subgroup 2 included Arginine (R), Lysine (K), Phenylalanine (F), Tyrosine (Y) and Tryptophan (W) residues at position +5. The conserved motifs of Subgroups 3 and 8 comprised Glutamic acid (E), Aspartic acid (D), Glutamine (Q) and Asparagine (N) residues at positions +3 and -2, respectively. In contrast, the remaining subgroups consisted of Phenylalanine (F), Tyrosine (Y) and Tryptophan (W) residues at various specific positions in their conserved motifs. Thus, substrate motifs for ubiquitination sites may be determined by the position-specific conservation of Phenylalanine (F), Tyrosine (Y) and Tryptophan (W) residues. Furthermore, MDDLogo could be utilized to identify putative ubiquitination sites and potential interaction between E3 ligase and ubiquitinated proteins based on substrate motif conservation.

Functional associations of E3 ligases and ubiquitinated proteins

Distributions of GO annotations for E3 ligases and ubiquitinated proteins categorized according to their corresponding biological processes, molecular functions and cellular components are provided in (Additional file 4: Table S3 and Additional file 5: Table S4), respectively. Following the InterPro annotation, the most abundant protein domain for E3 ligases appeared to belong to the "Zinc finger, RING-type RNA" (Table 2). In a genome-wide study of E3 ligases, it was suggested that the mammalian genomes encode more than 600 potential RING finger E3s [59]. E3 ligases containing the RING finger domain facilitate the interaction between an E2 enzyme and a substrate to mediate the transfer of ubiquitin from E2 to the target [60, 61]. On the other hand, those with the HECT domain are involved in the regulation of cellular trafficking, immune response, cellular growth and proliferation [62]. The HECT domain containing E3 ligases form a catalytic intermediate with ubiquitin and is responsible for the catalysis of two reactions: 1) transesterification reaction, in which ubiquitin from the cysteine residue at the E2 active site is transferred to another cysteine residue in the HECT domain [60]; 2) the subsequent attack of a substrate lysine on the thioester of the ubiquitin-bound HECT domain [63] (Additional file

Table 2 The distribution of top 20 functional domains for human E3 ligases

No.	InterPro ID	Domain terms	Number of proteins	% Total
1	IPR001680	WD40 repeat	287	57.5150 %
2	IPR006652	Kelch repeat type 1	68	13.6273 %
3	IPR000408	Regulator of chromosome condensation, RCC1	55	11.0220 %
4	IPR001841	Zinc finger, RING-type	54	10.8216 %
5	IPR000315	Zinc finger, B-box	48	9.6192 %
6	IPR003877	SPla/RYanodine receptor SPRY	38	7.6152 %
7	IPR013069	BTB/POZ	28	5.6112 %
8	IPR000569	HECT	28	5.6112 %
9	IPR001202	WW/Rsp5/WWP	28	5.6112 %
10	IPR020683	Ankyrin repeat-containing domain	27	5.4108 %
11	IPR001496	SOCS protein, C-terminal	25	5.0100 %
12	IPR002867	Zinc finger, C6HC-type	23	4.6092 %
13	IPR018957	Zinc finger, C3HC4 RING-type	22	4.4088 %
14	IPR001258	NHL repeat	21	4.2084 %
15	IPR011705	BTB/Kelch-associated	15	3.0060 %
16	IPR002110	Ankyrin repeat	14	2.8056 %
17	IPR000571	Zinc finger, CCCH-type	11	2.2044 %
18	IPR001876	Zinc finger, RanBP2-type	11	2.2044 %
19	IPR011016	Zinc finger, RING-CH-type	11	2.2044 %
20	IPR001452	Src homology-3 domain	10	2.0040 %

6: Figure S2). Whereas the C-terminus of the HECT domain is more conserved, the N-terminus, the part that mediates substrate targeting, is more diverse [62].

According to the annotation information on InterPro, approximately 70 % of established ubiquitination sites are mapped to specific functional domains, suggesting that ubiquitination may modulate a variety of biological functions. The top 50 InterPro functional domains containing ubiquitinated sites in humans are given in Table 3. It appeared that most ubiquitination sites could be found in the MHC class I (alpha chain) protein domains. It has been reported that viral proteins could induce the degradation of the histocompatibility complex (MHC) class I protein in the endoplasmic reticulum and at the cell surface by ubiquitinating the MHC class I domain [64]. The immunoglobulin C1-set domain, or classical Ig-like domains that resemble the antibody constant domain, is another domain found to be enriched with ubiquitinated sites. Interestingly, these domains were found exclusively in mediators of immune response, including various T-cell receptors, MHC class I and II complexes [65].

Network analysis for a group of interested proteins

To allow users to efficiently search for the proteins of their interest, a convenient interactive network viewer was implemented in the proposed method implemented an interactive network viewer. An example of constructing a protein ubiquitination network using our approach is illustrated in (Additional file 7: Figure S3). The network was built with four E3 ligases, 14 ubiquitinated proteins and three other proteins. While the established interactions between the four E3 ligases and 14 ubiquitinated proteins were immediately recognized, three other proteins interacting with two of the E3 ligases were also revealed as potential ubiquitinated substrates. For instance, E3 ligase MDM2 was predicted to target Forkhead box protein O3 (FOXO3) for ubiquitination. This is consistent with a recent study supporting MDM2 to be an E3 ligase responsible for the ubiquitin-mediated degradation of FOXO3 [66]. As well, our approach could provide the potential ubiquitination sites and the corresponding substrate motifs for a specific protein. Furthermore, for a specific E3 ligase and their interacting ubiquitinated proteins, the analysis could even be extended to exploring their functional associations and creating a comprehensive ubiquitin regulatory network.

A case study of the discovered E3 ligases associated with the regulation of p53

In cases where information is limited with respect to the interaction between an E3 ligase and its corresponding substrates, our strategy could still identify the potential

Table 3 Distribution of the top 50 functional domains covering ubiquitination sites

No.	Domain (InterPro) ID	Domain (InterPro) terms	Number of sites	% Total
1	IPR001039	MHC class I, alpha chain, alpha1/alpha2	1625	3.6574 %
2	IPR003597	Immunoglobulin C1-set	1321	2.9732 %
3	IPR001680	WD40 repeat	1043	2.3475 %
4	IPR002017	Spectrin repeat	1012	2.2777 %
5	IPR010579	MHC class I, alpha chain, C-terminal	729	1.6408 %
6	IPR003961	Fibronectin, type III	469	1.0556 %
7	IPR000504	RNA recognition motif domain	419	0.9431 %
8	IPR000719	Protein kinase, catalytic domain	338	0.7607 %
9	IPR013098	Immunoglobulin I-set	232	0.5222 %
10	IPR020683	Ankyrin repeat-containing domain	230	0.5177 %
11	IPR017868	Filamin/ABP280 repeat-like	212	0.4772 %
12	IPR006209	EGF-like domain	179	0.4029 %
13	IPR001715	Calponin homology domain	177	0.3984 %
14	IPR010630	Neuroblastoma breakpoint family (NBPF) domain	168	0.3781 %
15	IPR001452	Src homology-3 domain	167	0.3759 %
16	IPR006652	Kelch repeat type 1	158	0.3556 %
17	IPR000408	Regulator of chromosome condensation, RCC1	157	0.3534 %
18	IPR000225	Armadillo	156	0.3511 %
19	IPR004088	K Homology domain, type 1	154	0.3466 %
20	IPR001650	Helicase, C-terminal	146	0.3286 %
21	IPR002126	Cadherin	145	0.3264 %
22	IPR000048	IQ motif, EF-hand binding site	143	0.3219 %
23	IPR000083	Fibronectin, type I	143	0.3219 %
24	IPR000626	Ubiquitin	142	0.3196 %
25	IPR000008	C2 calcium-dependent membrane targeting	135	0.3038 %
26	IPR001806	Small GTPase superfamily	125	0.2813 %
27	IPR001245	Serine-threonine/tyrosine-protein kinase catalytic domain	125	0.2813 %
28	IPR000433	Zinc finger, ZZ-type	123	0.2768 %
29	IPR001478	PDZ domain	122	0.2746 %
30	IPR001849	Pleckstrin homology domain	120	0.2701 %
31	IPR011545	DNA/RNA helicase, DEAD/DEAH box type, N-terminal	112	0.2521 %
32	IPR019787	Zinc finger, PHD-finger	111	0.2498 %
33	IPR001440	Tetratricopeptide TPR-1	111	0.2498 %
34	IPR001202	WW/Rsp5/WWP	110	0.2476 %
35	IPR018108	Mitochondrial substrate/solute carrier	110	0.2476 %
36	IPR001781	Zinc finger, LIM-type	106	0.2386 %
37	IPR001101	Plectin repeat	104	0.2341 %
38	IPR002049	EGF-like, laminin	100	0.2251 %
39	IPR002429	Cytochrome c oxidase subunit II C-terminal	96	0.2161 %
40	IPR011759	Cytochrome C oxidase subunit II, transmembrane domain	96	0.2161 %
41	IPR003439	ABC transporter-like	91	0.2048 %
42	IPR001487	Bromodomain	91	0.2048 %
43	IPR018502	Annexin repeat	88	0.1981 %
44	IPR000980	SH2 domain	86	0.1936 %

Table 3 Distribution of the top 50 functional domains covering ubiquitination sites *(Continued)*

45	IPR003008	Tubulin/FtsZ, GTPase domain	77	0.1733 %
46	IPR001421	ATPase, F0 complex, subunit 8, mitochondrial, Metazoan	75	0.1688 %
47	IPR013069	BTB/POZ	75	0.1688 %
48	IPR003959	ATPase, AAA-type, core	74	0.1666 %
49	IPR007087	Zinc finger, C2H2	74	0.1666 %
50	IPR000197	Zinc finger, TAZ-type	74	0.1666 %

E3 ligases that may target a specific ubiquitinated protein. A case study is shown in Fig. 3, demonstrating the ability of the proposed method to construct an interaction map for the ubiquitination of tumor suppressor p53 (TP53). The resulting network is consistent with the literature. As a transcription factor, the tumor suppressor protein p53 responds to stress such as DNA damage by inducing cell cycle arrest and apoptosis [67]. Recent evidence has established that MDM2, a RING oncoprotein and a negative regulator of p53 [24], modulates the proteasomal degradation of p53 via a RING-finger-dependent manner [68–72]. Yet, our approach discovered other E3 ligases that may also regulate the ubiquitination of p53. Thus, the proposed strategy has the ability to uncover potential substrates for a specific E3 ligase, as well as potential E3 ligases for ubiquitinated proteins.

Conclusion

In an attempt to characterize the regulatory role protein ubiquitination plays in a variety of biological processes, we combined the information of E3 ligases, ubiquitinated proteins, and protein-protein interactions to construct a comprehensive network of E3 ligases and their

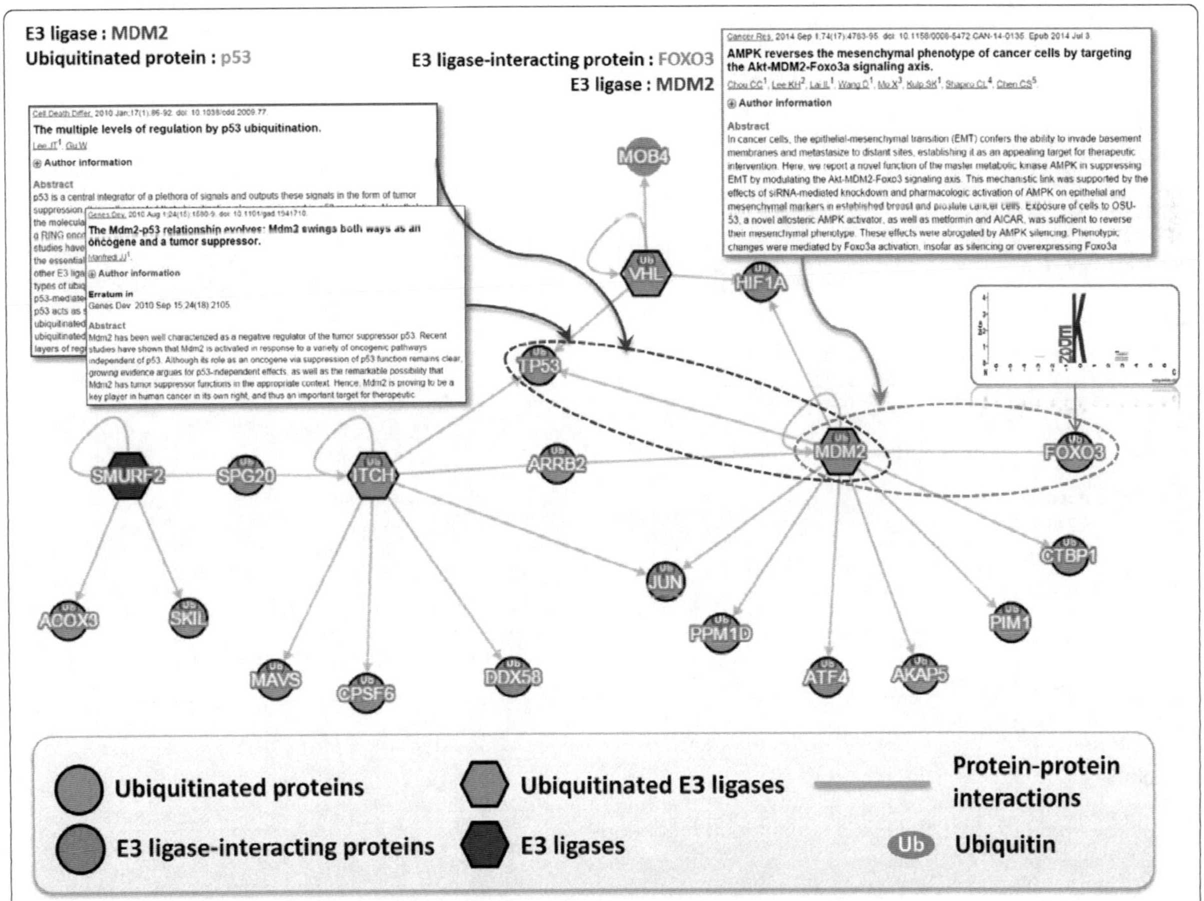

Fig. 3 A case study of the discovered E3 ligases associated with the regulation of *tumor suppressor p53* (TP53)

ubiquitinated substrates. Designed to serve as not only a meaningful platform for investigating E3-substrate regulatory networks but also a new strategy to uncover potential E3 ligases for ubiquitinated substrates, the proposed approach allows for the efficient characterization of protein ubiquitination networks from large-scale ubiquitinome data. With access to more updated data, the proposed scheme can be further refined for the study of E1 activating enzymes, E2 conjugating enzymes, and E3 ubiquitin ligases. Also, recent publications regarding the structural environment of experimentally validated ubiquitination sites based on protein tertiary structures [73–76] could be incorporated to infer the functional interactions between the enzymes and substrates. Finally, confirmed functional annotations of ubiquitination sites could be extracted from the literature via a more advanced information retrieval system to collect more adequate information required for further functional analyses.

Additional files

Additional file 1: Table S1. Data statistics of protein-protein interactions obtained from public resources. (PDF 8 kb)

Additional file 2: Figure S1. A conceptual diagram of exploiting Cytoscape software to construct protein ubiquitination networks based on graph theory. (PDF 86 kb)

Additional file 3: Table S2. The entropy plot of sequence logos for MDDLogo-identified motifs obtained from non-homologous ubiquitination sites in humans. (PDF 86 kb)

Additional file 4: Table S3. Distribution of the top 15 GO annotations for human E3 ligases. (PDF 31 kb)

Additional file 5: Table S4. Distribution of the top 20 GO annotations for human ubiquitinated proteins. (PDF 52 kb)

Additional file 6: Figure S2. The HECT-type structure of E3 ligases. (PDF 133 kb)

Additional file 7: Figure S3. An example to construct protein ubiquitination network for 21 proteins (containing four E3 ligases and 14 ubiquitinated proteins). (PDF 345 kb)

Competing interests
The authors have declared that no competing interests exist.

Authors' contributions
TYL and SLW conceived and supervised the project. KYH was responsible for the project design and computational analyses. KYH and JTYW drafted the manuscript with revisions by TYL and SLW. All authors read and approved the final manuscript.

Acknowledgements
The authors sincerely appreciate the Ministry of Science and Technology of Taiwan for financially supporting this research under Contract Number of MOST 103-2221-E-155-020-MY3, MOST 103-2633-E-155-002 and MOST104-2221-E-155-036-MY2.

Declarations
Publication charge for this work was funded by MOST grant under contract number of MOST 103-2221-E-155-020-MY3 and MOST 104-2221-E-155-036-MY2 to TYL.

Author details
[1]Department of Computer Science and Engineering, Yuan Ze University, Taoyuan 320, Taiwan. [2]Innovation Center for Big Data and Digital Convergence, Yuan Ze University, Taoyuan 320, Taiwan. [3]Department of Obstetrics and Gynecology, Hsinchu Mackay Memorial Hospital, Hsin-Chu 300, Taiwan. [4]Mackay Junior College of Medicine, Nursing and Management, Taipei 112, Taiwan. [5]Department of Medicine, Mackay Medical College, New Taipei City 252, Taiwan.

Published: 11 January 2016

References
1. Hershko A, Ciechanover A. The ubiquitin system. Annu Rev Biochem. 1998;67:425–79.
2. Nguyen VN, Huang KY, Huang CH, Chang TH, Bretana N, Lai K, et al. Characterization and identification of ubiquitin conjugation sites with E3 ligase recognition specificities. BMC bioinformatics. 2015;16 Suppl 1:S1.
3. Wagner SA, Beli P, Weinert BT, Nielsen ML, Cox J, Mann M, et al. A proteome-wide, quantitative survey of in vivo ubiquitylation sites reveals widespread regulatory roles. Molecular & cellular proteomics: MCP. 2011;10(10):M111 013284.
4. Lee TY, Chen SA, Hung HY, Ou YY. Incorporating distant sequence features and radial basis function networks to identify ubiquitin conjugation sites. Plos One. 2011;6(3):e17331.
5. Hurley JH, Lee S, Prag G. Ubiquitin-binding domains. Biochem J. 2006;399:361–72.
6. Hicke L, Schubert HL, Hill CP. Ubiquitin-binding domains. Nat Rev Mol Cell Bio. 2005;6(8):610–21.
7. Peng JM, Schwartz D, Elias JE, Thoreen CC, Cheng DM, Marsischky G, et al. A proteomics approach to understanding protein ubiquitination. Nat Biotechnol. 2003;21(8):921–6.
8. Wilkinson KD. The discovery of ubiquitin-dependent proteolysis. Proc Natl Acad Sci U S A. 2005;102(43):15280–2.
9. Han Y, Lee H, Park JC, Yi GS. E3Net: a system for exploring E3-mediated regulatory networks of cellular functions. Mol Cell Proteomics. 2012;11(4):O111.014076.
10. Zhang J, Wan L, Dai X, Sun Y, Wei W. Functional characterization of anaphase promoting complex/cyclosome (APC/C) E3 ubiquitin ligases in tumorigenesis. Biochim Biophys Acta. 2014;1845(2):277–93.
11. Manchado E, Eguren M, Malumbres M. The anaphase-promoting complex/cyclosome (APC/C): cell-cycle-dependent and -independent functions. Biochem Soc Trans. 2010;38(Pt 1):65–71.
12. Acquaviva C, Pines J. The anaphase-promoting complex/cyclosome: APC/C. J Cell Sci. 2006;119(Pt 12):2401–4.
13. Spratt DE, Walden H, Shaw GS. RBR E3 ubiquitin ligases: new structures, new insights, new questions. Biochem J. 2014;458(3):421–37.
14. Paul I, Ghosh MK. The E3 ligase CHIP: insights into its structure and regulation. Biomed Res Int. 2014.
15. Duplan V, Rivas S. E3 ubiquitin-ligases and their target proteins during the regulation of plant innate immunity. Front Plant Sci. 2014;5.
16. Snoek BC, de Wilt LH, Jansen G, Peters GJ. Role of E3 ubiquitin ligases in lung cancer. World journal of clinical oncology. 2013;4(3):58–69.
17. Plechanovova A, Jaffray EG, Tatham MH, Naismith JH, Hay RT. Structure of a RING E3 ligase and ubiquitin-loaded E2 primed for catalysis. Nature. 2012;489(7414):115–20.
18. Metzger MB, Hristova VA, Weissman AM. HECT and RING finger families of E3 ubiquitin ligases at a glance. J Cell Sci. 2012;125(3):531–7.
19. Bernassola F, Karin M, Ciechanover A, Melino G. The HECT family of E3 ubiquitin ligases: Multiple players in cancer development. Cancer Cell. 2008;14(1):10–21.
20. Scheffner M, Staub O. HECT E3s and human disease. BMC Biochem. 2007;8.
21. Mazzucotelli E, Belloni S, Marone D, De Leonardis AM, Guerra D, Fonzo N, et al. The E3 ubiquitin ligase gene family in plants: regulation by degradation. Curr Genomics. 2006;7(8):509–22.
22. Robinson PA, Ardley HC. Ubiquitin-protein ligases. J Cell Sci. 2004;117(22):5191–4.
23. Sun Y. Targeting E3 ubiquitin ligases for cancer therapy. Cancer Biol Ther. 2003;2(6):623–9.
24. Manfredi JJ. The Mdm2-p53 relationship evolves: Mdm2 swings both ways as an oncogene and a tumor suppressor. Genes Dev. 2010;24(15):1580–9.
25. Lee H, Yi GS, Park JC. E3Miner: a text mining tool for ubiquitin-protein ligases. Nucleic Acids Res. 2008;36(Web Server issue):W416–22.
26. Sakiyama T, Kawashima S, Yoshizawa AC, Kanehisa M. The construction of a database for ubiquitin signaling cascade. Genome Informatics. 2003;14:653–4.

27. Gao T, Liu Z, Wang Y, Cheng H, Yang Q, Guo A, et al. UUCD: a family-based database of ubiquitin and ubiquitin-like conjugation. Nucleic Acids Res. 2013;41(Database issue):D445–51.
28. Du YP, Xu NF, Lu M, Li TT. hUbiquitome: a database of experimentally verified ubiquitination cascades in humans. Database-Oxford. 2011.
29. Boeckmann B, Bairoch A, Apweiler R, Blatter MC, Estreicher A, Gasteiger E, et al. The SWISS-PROT protein knowledgebase and its supplement TrEMBL in 2003. Nucleic Acids Res. 2003;31(1):365–70.
30. Lee TY, Huang HD, Hung JH, Huang HY, Yang YS, Wang TH. dbPTM: an information repository of protein post-translational modification. Nucleic Acids Res. 2006;34(Database issue):D622–7.
31. Lu CT, Huang KY, Su MG, Lee TY, Bretana NA, Chang WC, et al. dbPTM 3.0: an informative resource for investigating substrate site specificity and functional association of protein post-translational modifications. Nucleic Acids Res. 2013;41(D1):D295–305.
32. Su MG, Huang KY, Lu CT, Kao HJ, Chang YH, Lee TY. topPTM: a new module of dbPTM for identifying functional post-translational modifications in transmembrane proteins. Nucleic Acids Res. 2014;42(Database issue):D537–45.
33. Crooks GE, Hon G, Chandonia JM, Brenner SE. WebLogo: a sequence logo generator. Genome Res. 2004;14(6):1188–90.
34. Schneider TD, Stephens RM. Sequence logos: a new way to display consensus sequences. Nucleic Acids Res. 1990;18(20):6097–100.
35. Vacic V, Iakoucheva LM, Radivojac P. Two sample logo: a graphical representation of the differences between two sets of sequence alignments. Bioinformatics. 2006;22(12):1536–7.
36. Lee TY, Lin ZQ, Hsieh SJ, Bretana NA, Lu CT. Exploiting maximal dependence decomposition to identify conserved motifs from a group of aligned signal sequences. Bioinformatics. 2011;27(13):1780–7.
37. Wong YH, Lee TY, Liang HK, Huang CM, Wang TY, Yang YH, et al. KinasePhos 2.0: a web server for identifying protein kinase-specific phosphorylation sites based on sequences and coupling patterns. Nucleic Acids Res. 2007;35(Web Server issue):W588–94.
38. Huang HD, Lee TY, Tzeng SW, Horng JT. KinasePhos: a web tool for identifying protein kinase-specific phosphorylation sites. Nucleic Acids Res. 2005;33(Web Server issue):W226–9.
39. Lee TY, Bretana NA, Lu CT. PlantPhos: using maximal dependence decomposition to identify plant phosphorylation sites with substrate site specificity. BMC Bioinformatics. 2011;12:261.
40. Bretana NA, Lu CT, Chiang CY, Su MG, Huang KY, Lee TY, et al. Identifying protein phosphorylation sites with kinase substrate specificity on human viruses. PLoS One. 2012;7(7):e40694.
41. Lee TY, Chen YJ, Lu TC, Huang HD. SNOSite: exploiting maximal dependence decomposition to identify cysteine S-nitrosylation with substrate site specificity. PLoS One. 2011;6(7):e21849.
42. Wu HY, Lu CT, Kao HJ, Chen YJ, Chen YJ, Lee TY. Characterization and identification of protein O-GlcNAcylation sites with substrate specificity. BMC bioinformatics. 2014;15 Suppl 16:S1.
43. Chen YJ, Lu CT, Huang KY, Wu HY, Chen YJ, Lee TY. GSHSite: exploiting an iteratively statistical method to identify s-glutathionylation sites with substrate specificity. PLoS One. 2015;10(4):e0118752.
44. Gene Ontology C, Blake JA, Dolan M, Drabkin H, Hill DP, Li N, et al. Gene ontology annotations and resources. Nucleic Acids Res. 2013;41(Database issue):D530–5.
45. Hunter S, Jones P, Mitchell A, Apweiler R, Attwood TK, Bateman A, et al. InterPro in 2011: new developments in the family and domain prediction database. Nucleic Acids Res. 2011;40(Database issue):D306–12.
46. Kanehisa M, Goto S, Sato Y, Furumichi M, Tanabe M. KEGG for integration and interpretation of large-scale molecular data sets. Nucleic Acids Res. 2012;40(Database issue):D109–14.
47. Mukhopadhyay D, Riezman H. Proteasome-independent functions of ubiquitin in endocytosis and signaling. Science. 2007;315(5809):201–5.
48. Schnell JD, Hicke L. Non-traditional functions of ubiquitin and ubiquitin-binding proteins. The Journal of biological chemistry. 2003;278(38):35857–60.
49. Glickman MH, Ciechanover A. The ubiquitin-proteasome proteolytic pathway: destruction for the sake of construction. Physiol Rev. 2002;82(2):373–428.
50. Bairoch A. PROSITE: a dictionary of sites and patterns in proteins. Nucleic Acids Res. 1991;19(Suppl):2241–5.
51. Attwood TK, Beck ME, Bleasby AJ, Parry-Smith DJ. PRINTS–a database of protein motif fingerprints. Nucleic Acids Res. 1994;22(17):3590–6.
52. Sonnhammer EL, Eddy SR, Durbin R. Pfam: a comprehensive database of protein domain families based on seed alignments. Proteins. 1997;28(3):405–20.
53. Corpet F, Gouzy J, Kahn D. The ProDom database of protein domain families. Nucleic Acids Res. 1998;26(1):323–6.
54. Ogata H, Goto S, Sato K, Fujibuchi W, Bono H, Kanehisa M. KEGG: Kyoto encyclopedia of genes and genomes. Nucleic Acids Res. 1999;27(1):29–34.
55. von Mering C, Huynen M, Jaeggi D, Schmidt S, Bork P, Snel B. STRING: a database of predicted functional associations between proteins. Nucleic Acids Res. 2003;31(1):258–61.
56. Lee TY, Bo-Kai Hsu J, Chang WC, Huang HD. RegPhos: a system to explore the protein kinase-substrate phosphorylation network in humans. Nucleic Acids Res. 2011;39(Database issue):D777–87.
57. Huang KY, Wu HY, Chen YJ, Lu CT, Su MG, Hsieh YC, et al. RegPhos 2.0: an updated resource to explore protein kinase-substrate phosphorylation networks in mammals. Database : the journal of biological databases and curation. 2014;2014(0):bau034.
58. Kohl M, Wiese S, Warscheid B. Cytoscape: software for visualization and analysis of biological networks. Methods Mol Biol. 2010;696:291–303.
59. Li W, Bengtson MH, Ulbrich A, Matsuda A, Reddy VA, Orth A, et al. Genome-wide and functional annotation of human E3 ubiquitin ligases identifies MULAN, a mitochondrial E3 that regulates the organelle's dynamics and signaling. Plos One. 2008;3(1):e1487.
60. Berndsen CE, Wolberger C. New insights into ubiquitin E3 ligase mechanism. Nat Struct Mol Biol. 2014;21(4):301–7.
61. Zheng N, Wang P, Jeffrey PD, Pavletich NP. Structure of a c-Cbl-UbcH7 complex: RING domain function in ubiquitin-protein ligases. Cell. 2000;102(4):533–9.
62. Rotin D, Kumar S. Physiological functions of the HECT family of ubiquitin ligases. Nat Rev Mol Cell Biol. 2009;10(6):398–409.
63. Huibregtse JM, Scheffner M, Beaudenon S, Howley PM. A family of proteins structurally and functionally related to the E6-AP ubiquitin-protein ligase. Proc Natl Acad Sci U S A. 1995;92(11):5249.
64. Burr ML, Boname JM, Lehner PJ. Studying ubiquitination of MHC class I molecules. Methods Mol Biol. 2013;960:109–25.
65. Cresswell P, Ackerman AL, Giodini A, Peaper DR, Wearsch PA. Mechanisms of MHC class I-restricted antigen processing and cross-presentation. Immunol Rev. 2005;207:145–57.
66. Chou CC, Lee KH, Lai IL, Wang D, Mo X, Kulp SK, et al. AMPK reverses the mesenchymal phenotype of cancer cells by targeting the Akt-MDM2-Foxo3a signaling axis. Cancer Res. 2014;74(17):4783–95.
67. Lee JT, Gu W. The multiple levels of regulation by p53 ubiquitination. Cell Death Differ. 2010;17(1):86–92.
68. Haupt Y, Maya R, Kazaz A, Oren M. Mdm2 promotes the rapid degradation of p53. Nature. 1997;387(6630):296–9.
69. Kubbutat MHG, Jones SN, Vousden KH. Regulation of p53 stability by Mdm2. Nature. 1997;387(6630):299–303.
70. Honda R, Tanaka H, Yasuda H. Oncoprotein MDM2 is a ubiquitin ligase E3 for tumor suppressor p53. Febs Lett. 1997;420(1):25–7.
71. Fang SY, Jensen JP, Ludwig RL, Vousden KH, Weissman AM. Mdm2 is a RING finger-dependent ubiquitin protein ligase for itself and p53. J Biol Chem. 2000;275(12):8945–51.
72. Honda R, Yasuda H. Activity of MDM2, a ubiquitin Ligase, toward p53 or itself is dependent on the RING finger domain of the ligase. Oncogene. 2000;19(11):1473–6.
73. Su MG, Lee TY. Incorporating substrate sequence motifs and spatial amino acid composition to identify kinase-specific phosphorylation sites on protein three-dimensional structures. BMC bioinformatics. 2013;14 Suppl 16:S2.
74. Lee TY, Chen YJ, Lu CT, Ching WC, Teng YC, Huang HD. dbSNO: a database of cysteine S-nitrosylation. Bioinformatics. 2012;28(17):2293–5.
75. Chen YJ, Lu CT, Su MG, Huang KY, Ching WC, Yang HH, et al. dbSNO 2.0: a resource for exploring structural environment, functional and disease association and regulatory network of protein S-nitrosylation. Nucleic Acids Res. 2015;43(Database issue):D503–11.
76. Chen YJ, Lu CT, Lee TY, Chen YJ. dbGSH: a database of S-glutathionylation. Bioinformatics. 2014;30(16):2386–8.

Yuan *et al. BMC Systems Biology* 2016, **10**(Suppl 1):8
DOI 10.1186/s12918-015-0245-0

BMC Systems Biology

PROCEEDINGS Open Access

A network based covariance test for detecting multivariate eQTL in saccharomyces cerevisiae

Huili Yuan[1], Zhenye Li[1], Nelson L. S. Tang[2] and Minghua Deng[1,3,4*]

From The Fourteenth Asia Pacific Bioinformatics Conference (APBC 2016)
San Francisco, CA, USA. 11 - 13 January 2016

Abstract

Background: Expression quantitative trait locus (eQTL) analysis has been widely used to understand how genetic variations affect gene expressions in the biological systems. Traditional eQTL is investigated in a pair-wise manner in which one SNP affects the expression of one gene. In this way, some associated markers found in GWAS have been related to disease mechanism by eQTL study. However, in real life, biological process is usually performed by a group of genes. Although some methods have been proposed to identify a group of SNPs that affect the mean of gene expressions in the network, the change of co-expression pattern has not been considered. So we propose a process and algorithm to identify the marker which affects the co-expression pattern of a pathway. Considering two genes may have different correlations under different isoforms which is hard to detect by the linear test, we also consider the nonlinear test.

Results: When we applied our method to yeast eQTL dataset profiled under both the glucose and ethanol conditions, we identified a total of 166 modules, with each module consisting of a group of genes and one eQTL where the eQTL regulate the co-expression patterns of the group of genes. We found that many of these modules have biological significance.

Conclusions: We propose a network based covariance test to identify the SNP which affects the structure of a pathway. We also consider the nonlinear test as considering two genes may have different correlations under different isoforms which is hard to detect by linear test.

Keywords: eQTL, Pathway, Isoform

Background

GWAS aims to detect the association between genetic variation and complex diseases. Recent years, GWAS has found 2000 loci associated to complex diseases by statistical methods [1]. As the development of the next-generation sequencing and other high-throughput technology, various types of genome-scale datasets have been collected, providing opportunity to find the mechanism of genetic variation leading to complex diseases by connect the high-throughout data to GWAS. The eQTL study is one of them, which aims to uncover the genetic effects to gene expression and have been conducted in many organisms [2–5]. A common approach in eQTL data analysis is to consider association between each expression trait and each genetic marker through regression analysis. Despite great success with this approach, some regulatory signals may not be detected due to complex interaction between SNPs like epistasis.

Although most eQTL studies considered the expression levels of individual genes as response (single outcome variable), the change of correlation between genes under different genetic status still contains some biological information. For example, post-transcriptional

*Correspondence: dengmh@math.pku.edu.cn
[1]LMAM, School of Mathematical Sciences, Peking University, Yiheyuan Road, 100871 Beijing, China
[3]Center for Quantitative Biology, Peking University, Yiheyuan Road, 100871 Beijing, China
Full list of author information is available at the end of the article

regulations such as phosphorylations and dephosphorylations often affect the activities of transcriptional factors (TFs), which further affect the correlation among TF genes and TF target genes, also the co-expression patterns of the targets of TFs. However, such regulations are hard to be detected if only individual gene considered because there may be little change at the expression levels of individual TF genes. The approach considering "liquid association" (LA) between a pair of genes proposed by [6] is a method to identify such loci, which is later introduced into eQTL study [7]. Subsequently, conditional bi-variate normal model has been developed to capture the change of correlation between a pair of genes [8–10].

However, a biological process is usually performed by a group of genes (more than two genes as in the bivariate model). Network approaches should be used to study these interactions [11–13]. If we want to see the effects of a cellular change to the organism, it is better for us to consider the change in a functional gene-set such as a pathway. Therefore, some papers has considered the multivariate circumstances by applying CCA to gene expressions and SNP (or CNV) data [14–16]. However, these methods do not consider the network structure when finding the association between gene sets and genetic variant, which will miss the information contained in the network. Li et al. [17], Kim and Xing [18], Zhang and Kim [19], Casale et al. [20] have considered pathway structure when studied the association between genetic variation and gene expression. However, they assume the network structure is the same (static) under different genetic variant. In fact, network structure may be dynamic and biologists have realized that differential network analysis will become a standard mode in network analysis and insightful discoveries could be made with differential network analysis [21]. For example, [22] identified a cancer point mutation in the kinase domain of RET, which causes multiple endocrine neoplasia type 2B by leading to a switch in peptide specificity and then altering the network structure.

So we propose a method to test whether the co-expression pattern in a pathway is affected by a SNP. Our goal is to test for a global change in covariance structure in each pathway, which is different from other network-based methods, which tries to detect non-zero edges from all pairs of genes. When we applied our method to a yeast eQTL dataset, we were able to find some pathway-SNP modules that have biological significance.

Methods

Let $(X_1, X_2, \ldots, X_p)$ be the expression levels of a group of genes and $(Z_1, Z_2, \ldots, Z_m)$ be the set of SNPs. Suppose that there are n independent samples and let $(x_{1i}, x_{2i}, \ldots, x_{pi})_{i=1,\ldots,n}$ denote the expression level of $(X_1, X_2, \ldots, X_p)$ in the ith sample and $(Z_{1i}, Z_{2i}, \ldots, Z_{mi})$ denote the SNP types of the SNP set in the ith sample. Since the mean expression levels of $(X_1, X_2, \ldots, X_p)$ are also possibly affected by some SNPs in $(Z_1, Z_2, \ldots, Z_m)$, we can imitate the procedure in [9] that we first perform regression analysis or penalized regression analysis such as Lasso [23] or SCAD [24] to adjust the effects of $(Z_1, Z_2, \ldots, Z_m)$ on the means and then model the residuals. We assume that the covariate-adjusted expression levels are appropriately centered to have mean values of zero and our interest is to test whether the covariate-adjusted covariance of expression levels is changed under each SNP. In our analysis, the group of genes are a pathway in KEGG [25]. Figure 1 describes our strategy to detect pathway-SNP associations. In this manuscript, we define a module as the collection of a SNP and a pathway, and our objective is to find pathway-SNP modules where the SNP

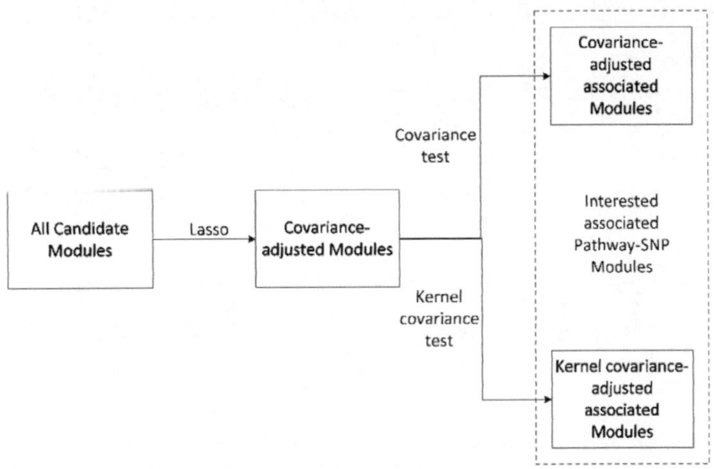

Fig. 1 Flowchart of our strategy. Flowchart of our strategy for detecting pathway-associated SNP. We first perform Lasso to adjust the effects of SNPs on the means of gene expression. Then we use covariance test and kernel covariance test to select candidate pathway-SNP modules

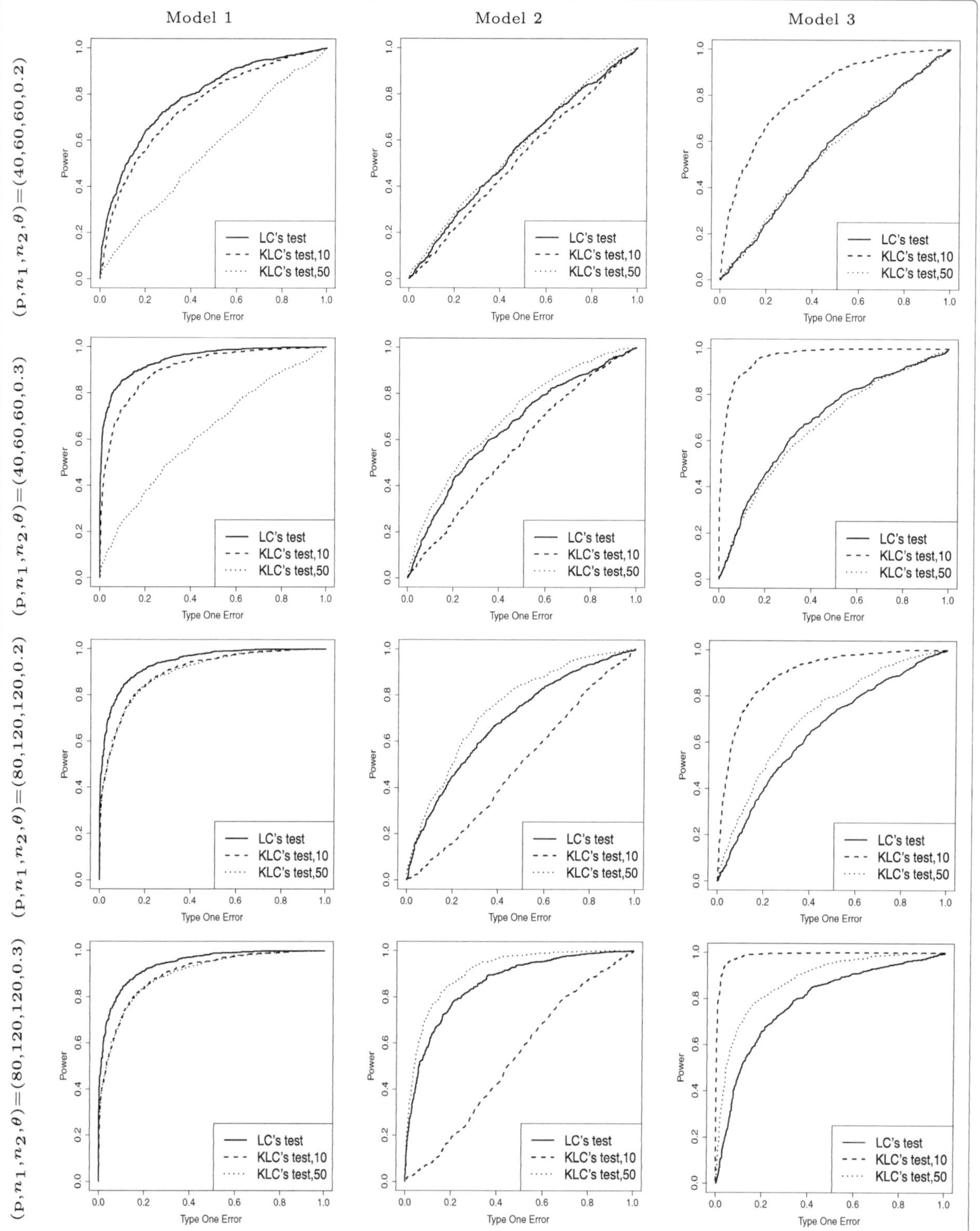

Fig. 2 Comparison between linear method and kernel method. Simulations under different setups. Setup of the first column is under model 1, the second column is under model 2 and the third column is under model 3. First row: (p, n_1, n_2, θ) = (40, 60, 60, 0.2); Second row: (p, n_1, n_2, θ) = (40, 60, 60, 0.3); Third row: (p, n_1, n_2, θ) = (80, 120, 120, 0.2); Fourth row: (p, n_1, n_2, θ) = (80,120,120, 0.3)

affect the co-expression patterns among the genes in the pathway.

Model

We use covariance test to find the pathway-SNP modules. There are three key elements of covariance test for a given gene set S. We consider the strategy similar to [26].

- **Calculation of T statistics.** We calculate a T statistics that reflects the difference of the covariance matrix of the two classes of samples. The statistics is calculated by estimating the Frobenius norm of the difference of the covariance matrix. We first perform the method by [27] to do the test:

$$H_0 : \Sigma_1 = \Sigma_2, \;\; H_1 : \Sigma_1 \neq \Sigma_2 \tag{1}$$

where Σ_1 is the covariance matrix of gene expression under one genotype and Σ_2 is that of gene expression under the other genotype. Then we consider the nonlinear relationship between gene expressions by applying kernel method.

- **Estimation of significance level of T statistics.** We estimate the statistical significance (nominal P value) of the T statistics by using an empirical SNP-based permutation test procedure that preserves the complex correlation structure of the gene expression data. Specifically, we permute the SNP labels and recompute the T statistics of the gene set for the permuted data, which generates a null distribution for the T statistics. The empirical, nominal P value of the observed T statistics is then calculated relative to this null distribution. Importantly, the permutation of class labels preserves gene-gene correlations and, thus, provides a more biologically reasonable assessment of significance than would be obtained by permuting genes.

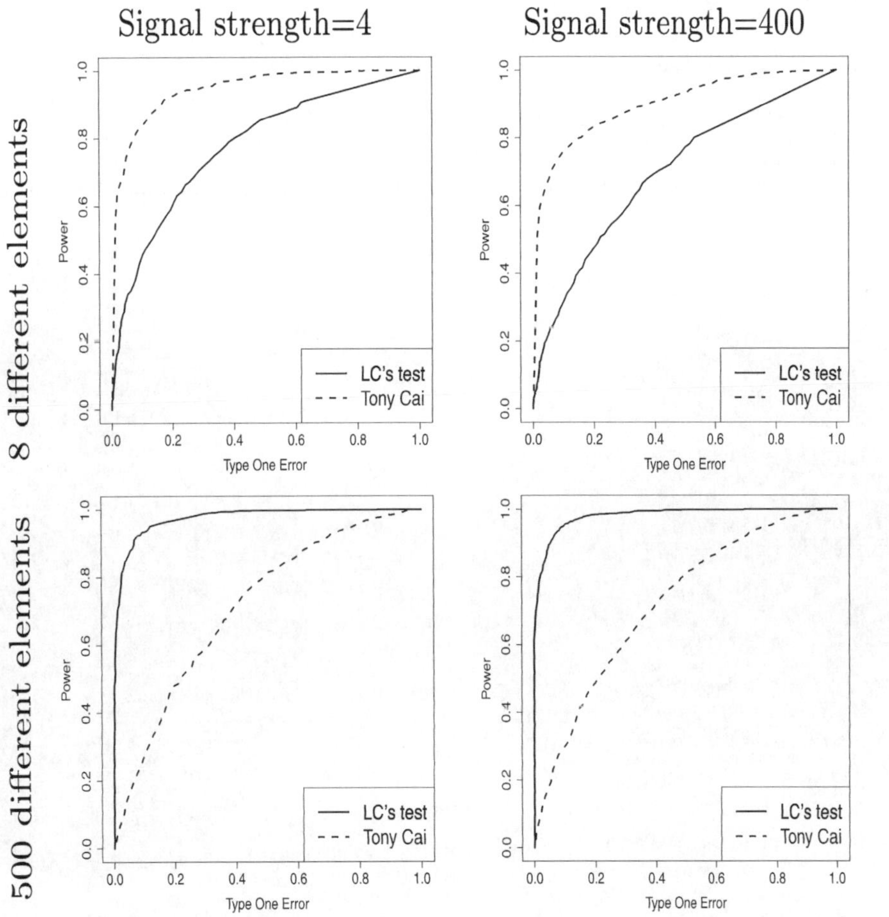

Fig. 3 Comparison between Chen's linear method and other method. *Topleft*: The two covariance matrices have eight different elements, each with a magnitude generated from $Unif(0,4) * \max_{1\leq j\leq p} \sigma_{jj}$; *Topright*: The two covariance matrices have eight different elements, each with a magnitude generated from $Unif(0,400) * \max_{1\leq j\leq p} \sigma_{jj}$; *Bottomleft*: The two covariance matrices have 500t different elements, each with a magnitude generated from $Unif(0,4) * \max_{1\leq j\leq p} \sigma_{jj}$; *Bottomright*: The two covariance matrices have 500 different elements, each with a magnitude generated from $Unif(0,400) * \max_{1\leq j\leq p} \sigma_{jj}$

- **Adjustment for multiple hypothesis testing.** When an entire database of gene sets is evaluated, we adjust the estimated significance level to account for multiple hypothesis testing. We first normalize the T statistics for each gene set to account for the size of the set, yielding a normalized T statistics. We then control the proportion of false positives by calculating the false discovery rate (FDR) corresponding to each NT statistics. The FDR is the estimated probability that a set with a given NT statistics represents a false positive finding; it is computed by comparing the tails of the observed and null distributions for the NT statistics. To capture the change of the structure of the gene network, we consider the covariance of the gene expression.

Test for high-dimensional covariance matrices

To simplify the problem, we just consider there are two possible values of each SNP. Covariance matrices under two genotypes of the SNP are denoted as Σ_1 and Σ_2, respectively. The primary interest is to test

$$H_0 : \Sigma_1 = \Sigma_2,\ H_1 : \Sigma_1 \neq \Sigma_2$$

which is a nontrivial statistical problem because the number of genes is greater than the number of samples sometimes. The test statistic for the hypothesis is formulated by targeting on $\mathrm{tr}(\Sigma_1 - \Sigma_2)^2$, the squared Frobenius norm of $\Sigma_1 - \Sigma_2$ [27]. Specifically, the test statistic is

$$T_{n_1,n_2} = A_{n_1} + A_{n_2} - 2C_{n_1 n_2}$$

$$A_{n_h} = \frac{1}{n_h(n_h-1)} \sum_{i \neq j} (X'_{hi} X_{hj})^2 - \frac{2}{n_h(n_h-1)(n_h-2)} \sum_{i,j,k}^{*} X'_{hi} X_{hj} X'_{hj} X_{hk} + \frac{1}{n_h(n_h-1)(n_h-2)(n_h-3)} \sum_{i,j,k,l}^{*} X'_{hi} X_{hj} X'_{hk} X_{hl}$$

$$C_{n_1 n_2} = \frac{1}{n_1(n_2)} \sum_i \sum_j (X'_{1i} X_{2j})^2 - \frac{1}{n_1 n_2 (n_1-1)} \sum_{i,k}^{*} \sum_j X'_{1i} X_{2j} X'_{2j} X_{1k} - \frac{1}{n_1 n_2 (n_2-1)} \sum_{i,k}^{*} \sum_j X'_{2i} X_{1j} X'_{1j} X_{2k} + \frac{1}{n_1 n_2 (n_1-1)(n_2-1)} \sum_{i,k}^{*} \sum_{j,l}^{*} X'_{1i} X_{2j} X'_{1k} X_{2l}$$

where h refers to a subpopulation with a particular SNP.

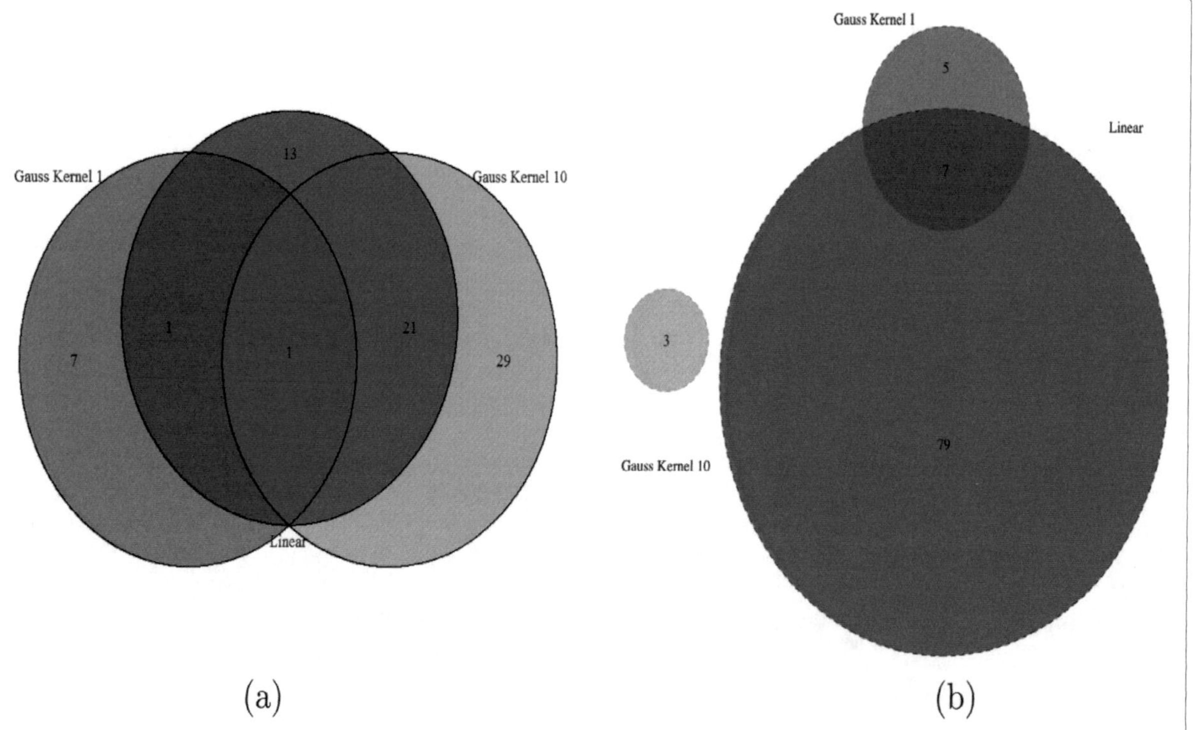

Fig. 4 Different modules detected under different conditions by different method. **a** Detected modules under ethanol condition. We found 36 modules by covariance test, 9 modules by kernel covariance test with parameter 1 and 51 modules by kernel covariance test with parameter 10; **b** Detected modules under glucose condition. We found 86 modules by covariance test, 3 modules by kernel covariance test with parameter 10 and 12 modules by kernel covariance test with parameter 1

Table 1 New associated pathways and SNPs under ethanol condition

Pathways	Associated markers
Glycolysis/Gluconeogenesis	*gOL02*$^{(10)}$
Synthesis and degradation of ketone bodies	*YLR257W*$^{(10)}$, *YLR261C*$^{(10)}$
Steroid biosynthesis	*YEL021W*^{L}, *YFR035C*^{L}, *YJL001W*^{L}, *YJR006W*$^{L,(10)}$, *YJL007C*$^{L,(10)}$,
Valine, leucine and isoleucine degradation	*YOR006C*^{L}, *NOR005W*$^{L,(10)}$, *YOR051C*$^{L,(10)}$, *YOR076C*$^{L,(10)}$
Valine, leucine and isoleucine biosynthesis	*NLR116W*$^{L,(10)}$, *YOR076C*^{L}, *YCL023C*$^{(10)}$, *YLR257W*$^{(10)}$
Histidine metabolism	*gOL02*$^{L,(10)}$, *YOR025W*$^{(10)}$
Tyrosine metabolism	*YFL019C*^{L}, *gOL02*$^{L,(10)}$
Phenylalanine metabolism	*gOL02*$^{L,(10)}$
beta-Alanine metabolism	*gOL02*$^{L,(10)}$
Taurine and hypotaurine metabolism	*gOL02*$^{(10)}$
Selenocompound metabolism	*YOR006C*$^{L,(10)}$, *YOR019W*$^{L,(10)}$, *YOR025W*$^{L,(10)}$, *NOR005W*$^{L,(10)}$
Purine metabolism	*NNL035W*$^{(1)}$
Cyanoamino acid metabolism	*YLR027C*$^{(1)}$
Arachidonic acid metabolism	*gPL09*$^{(1)}$
Linoleic acid metabolism	*YFL029C*$^{L,(1),(10)}$, *YFL019C*$^{(1)}$
Glyoxylate and dicarboxylate metabolism	*NNL035W*$^{(1)}$, *YNL074C*$^{(1)}$
Porphyrin and chlorophyll metabolism	*NBR008W*$^{(1)}$
Sphingolipid metabolism	*YHL047C*^{L}
Pantothenate and CoA biosynthesis	*YGL053W*^{L}, *NLR116W*$^{L,(10)}$
Terpenoid backbone biosynthesis	*YJL007C*$^{L,(10)}$, *YJL001W*$^{L,(10)}$, *YJR006W*$^{L,(10)}$, *NJR006C*$^{L,(10)}$
Sesquiterpenoid and triterpenoid biosynthesis	*YOR334W*^{L}, *YOR343C*^{L}, *YLR261C*$^{(10)}$, *NLR116W*$^{(10)}$, *YLR257W*$^{(10)}$
Metabolic pathways	*YIL078W*$^{(10)}$, *YLR257W*$^{(10)}$, *YLR308W*$^{(10)}$, *NNL035W*$^{(10)}$, *gOL02*$^{(10)}$, *YOR006C*$^{(10)}$, *YOR051C*$^{(10)}$, *YOR019W*$^{(10)}$, *YNL066W*$^{(10)}$, *YLR261C*$^{(10)}$
Biosynthesis of secondary metabolites	*YOR025W*$^{(10)}$, *YOR063W*$^{(10)}$
Carbon metabolism	*YOR019W*$^{(10)}$
2-Oxocarboxylic acid metabolism	*YLR261C*$^{L,(10)}$, *YLR308W*$^{L,(10)}$, *YCL022C*$^{(10)}$, *YLR265C*$^{(10)}$, *NLR116W*$^{(10)}$, *YLR322W*^{L}

Table 1 New associated pathways and SNPs under ethanol condition *(Continued)*

mRNA surveillance pathway	*YOR072W*^{L}
Mismatch repair	*gKR08*^{L}
Non-homologous end	*YGR006W*^{L}
Biosynthesis of amino acids	*gOL02*$^{(10)}$
MAPK signaling pathway	*YDR164C*$^{(10)}$, *gDR10*$^{(10)}$

L means detected by covariance test, (1) means detected by kernel covariance test with parameter 1 and (10) means detected by kernel covariance test with parameter 10. The FDR of the covariance test, kernel covariance test with parameter 1 and kernel covariance test with parameter 10 are 0.25, 0.33 and 0.25 respectively. The FWER of the test by Tony Cai is 0.2

For test

$H_0 : \Sigma_1 = \Sigma_2 = \Sigma_3, H_1 : \Sigma_1 \neq \Sigma_2$ or $\Sigma_2 \neq \Sigma_3$

We consider $\mathrm{tr}(\Sigma_1 - \Sigma_2)^2 + \mathrm{tr}(\Sigma_2 - \Sigma_3)^2$. Specifically, the test statistic is

$$T_{n_1,n_2} + T_{n_2,n_3}$$

where T_{n_2,n_3} is defined similar to T_{n_1,n_2}.

Kernel method

We generalize the method of [27] to the kernel space inspired by the method of [28]. We give the similar definition of Frobenius norm and covariance matrix. Let p_x and p_y be Borel probability measures defined on a domain Ω. Given observations $X := \{x_1, \ldots, x_m\}$ and $Y := \{y_1, \ldots, y_n\}$, drawn independently and identically distributed(i.i.d.) from p_x and p_y, respectively.

Definition (HSDCC) Given separable reproducing kernel Hilbert space (RKHS) $\mathcal{F}$, and measures p_x, p_y over $(\mathcal{X}, \Gamma)$, we define the Hilbert-Schmidt Different Covariance Criterion(HSDCC) as the squared HS-norm of the difference of covariance Σ_{xx} and Σ_{yy}:

$$HSDCC(p_x, p_y, \mathcal{F}) := \| \Sigma_{xx} - \Sigma_{yy} \|_{HS}^2$$

The detailed computation of above norm can be found in text of the Additional file 1. We give the unbiased statistics to $HSDCC(P_x, P_y, \mathcal{F})$ like [27]

$$A_{n_h} = \frac{1}{n_h(n_h-1)} \sum_{i \neq j} k(X_{hi}, X_{hj})^2 - \frac{2}{n_h(n_h-1)(n_h-2)} \sum_{i,j,k}^{*} k(X_{hi}, X_{hj}) k(X_{hj}, X_{hk}) + \frac{1}{n_h(n_h-1)(n_h-2)(n_h-3)} \sum_{i,j,k,l}^{*} k(X_{hi}, X_{hj}) k(X_{hk}, X_{hl})$$

Table 2 New associated pathways and SNPs under glucose condition

Pathways	Associated markers
Synthesis and degradation of ketone bodies	*gJL07*$^{(10)}$
Inositol phosphate metabolism	*YBR259W*$^{(10)}$
Riboflavin metabolism	*YML056C*$^{(10)}$
Fatty acid degradation	*YBR045C*$^{(1)}$
Cysteine and methionine metabolism	*YGL195W*$^{(1)}$
Valine, leucine and isoleucine biosynthesis	*YCL025C*^{L}, *NGR093C*^{L}
	YOR253W^{L}, *YOR274W*^{L}
	YOR326W^{L}, *YOR334W*^{L}
	YOR343C^{L}, *YCL022C*$^{(1)}$
Phenylalanine metabolism	*YJR040W*^{L}, *YOL123W*^{L}
	YOL118C^{L}, *YOL106W*^{L}
	YOL093W^{L}, *YOL088C*^{L}
	gOL02$^{L,(1)}$
beta-Alanine metabolism	*YBR271W*^{L}, *gOL02*$^{L,(1)}$
	NJR007C^{L}, *YOL106W*^{L}
Arachidonic acid metabolism	*YIR022W*$^{L,(1)}$
Vitamin B6 metabolism	*YKL118W*$^{(1)}$
Porphyrin and chlorophyll metabolism	*YML071C*$^{(1)}$, *gFL02*^{L}
Degradation of aromatic compounds	*YMR316C*$^{L,(1)}$, *YMR316C*^{L}
ABC transporters	*YBR131W*$^{L,(1)}$, *YBR137W*^{L}
Glycolysis/Gluconeogenesis	*YJR071W*^{L}
Pentose phosphate pathway	*NOL043W*^{L}, *YOL151W*^{L}
	YOL123W^{L}, *YOL094C*^{L}
	YOL093W^{L}, *YOL088C*^{L}
	gOL02^{L}
Pentose and glucuronate interconversions	*YGL263W*^{L}
Purine metabolism	*YLR140W*^{L}
Pyrimidine metabolism	*YBL010C*^{L}, *YGL217C*^{L}
Glycine, serine and threonine metabolism	*YCL065W*^{L}, *YJR038C*^{L}
Lysine biosynthesis	*YBR087W*^{L}
Histidine metabolism	*YBR271W*^{L}, *NJR007C*^{L}
	YJR040W^{L}, *YJR057W*^{L}
	YOL106W^{L}, *YOL093W*^{L},
	gOL02$^{L,(1)}$
Tyrosine metabolism	*YOL123W*^{L}, *YOL106W*^{L}
	YOL094C^{L}, *gOL02*$^{L,(1)}$
Cyanoamino acid metabolism	*YDR351W*^{L}
Starch and sucrose metabolism	*YER095W*^{L}, *YER116C*^{L}
Linoleic acid metabolism	*NDR174C*^{L}
Butanoate metabolism	*YBR271W*^{L}
Pantothenate and CoA biosynthesis	*YOR274W*^{L}
Lipoic acid metabolism	*gLL01*^{L}, *YNL158W*^{L}
Folate biosynthesis	*NML013W*^{L}, *YNL066W*^{L}, *YNL050C*^{L}
Sesquiterpenoid and triterpenoid biosynthesis	*YMR084W*^{L}
Aminoacyl-tRNA biosynthesis	*YCL065W*^{L}, *YCL047C*^{L}
	YCL039W^{L}, *NJR007C*^{L}, *YNL010W*^{L}
Biosynthesis of unsaturated fatty acids	*YFL029C*^{L}
Metabolic pathways	*YCL065W*^{L}, *YJR071W*^{L}
Biosynthesis of secondary metabolites	*YJR038C6L*
Biosynthesis of amino acids	*YJR071W*^{L}, *YOL123W*^{L}
	YOL118C^{L}, *YOL106W*^{L}
	YOL094C^{L}, *YOL093W*^{L}
	YOL088C^{L}, *gOL02*^{L}
Ribosome	*YAR035W*^{L}, *YJL026W*^{L}
RNA transport	*YBL010C*^{L}
RNA polymerase	*YLR140W*^{L}
Proteasome	*YBL010C*^{L}
Phosphatidylinositol signaling system	*YBR045C*^{L}
Meiosis - yeast	*YOL106W*^{L}

L means detected by covariance test, (1) means detected by kernel covariance test with parameter 1 and (10) means detected by kernel covariance test with parameter 10. The FDR of the covariance test, kernel covariance test with parameter 1 and kernel covariance test with parameter 10 are 0.20, 0.24 and 0.33 respectively. The FWER of the test by Tony Cai is 0.2

$$C_{n_1 n_2} = \frac{1}{n_1(n_2)} \sum_i \sum_j k(X_{1i}, X_{2j})^2 - \frac{1}{n_1 n_2 (n_1 - 1)} \sum_{i,k}^{*} \sum_j k(X_{1i}, X_{2j}) k(X_{2j}, X_{1k}) - \frac{1}{n_1 n_2 (n_2 - 1)} \sum_{i,k}^{*} \sum_j k(X_{2i}, X_{1j}) k(X_{1j}, X_{2k}) + \frac{1}{n_1 n_2 (n_1 - 1)(n_2 - 1)} \sum_{i,k}^{*} \sum_{j,l}^{*} k(X_{1i}, X_{2j}) k(X_{1k}, X_{2l})$$

$$T_{n_1,n_2} = A_{n_1} + A_{n_2} - 2C_{n_1 n_2}$$

For test
$H_0 : \Sigma_{xx} = \Sigma_{yy} = \Sigma_{zz}, H_1 : \Sigma_{xx} \neq \Sigma_{yy}$ or $\Sigma_{yy} \neq \Sigma_{zz}$
We consider $\| \Sigma_{xx} - \Sigma_{yy} \|_{HS}^2 + \| \Sigma_{yy} - \Sigma_{zz} \|_{HS}^2$.
Specifically, the test statistic is

$$T_{n_1,n_2} + T_{n_2,n_3}$$

where T_{n_2,n_3} is defined similar to T_{n_1,n_2}.

Results

Simulation

Comparison between linear method and kernel method

We performed a simulation study to evaluate the power of the proposed kernel methods, and compared the results with the primary method by [27]. Three models have been considered, as below.

Model 1: $X_{ijk} = Z_{ijk} + \theta Z_{ijk+1}$, where Z_{ijk} were i.i.d. standard normally distributed, and $\theta = 0.5$ in the null hypothesis while 0.2 or 0.3 in the alternative hypothesis.

Model 2: $X_{ijk} = Z_{ijk}^3 + \theta Z_{ijk+1}^3$, where Z_{ijk} and θ were defined the same as that in Model 1.

Model 3: $X_{ijk} = e^{Z_{ijk}} + \theta e^{Z_{ijk+1}}$, where Z_{ijk} and θ were defined the same as that in Model 1.

The correlation between variables are linear in model 1, while the correlation between variables are nonlinear in model 2 and 3.

We chose (p, n_1, n_2)=(40, 60, 60) and (80, 120, 120) respectively. The power of the tests are shown by ROC curves (Fig. 2). All the simulation results reported were based on 1000 simulations. We can see from the simulation that kernel methods with some parameters have higher power than the linear test when the true relationships between variables are nonlinear (Model 2 and Model 3). A similar simulation results with different setup of parameters can be found in Additional file 1: Figure S3.

Comparison between Chen et al.'s linear method and other method

We conducted a simulation to compare the power of Chen et al.'s method [27] and Tony Cai et al.'s method [29]. We consider four simulation setups represented different signal quantities and strength, the first of which is the same as the model 2 in [29].

Model 1: Let

$$\Sigma^* = (\sigma_{ij}^*),\ where\ \omega_{ij}^* = 0.5^{|i-j|}\ for\ 1 \le i,j \le p.$$

$$\Sigma = D^{1/2}\Sigma^* D^{1/2},\ where\ D = (d_{ij}),\ d_{ii} = Unif(0.5, 2.5), 1 \le i \le p$$

$$\Sigma_1 = \Sigma + \delta I,\ \Sigma_2 = \Sigma + U + \delta I,\ where\ \delta$$

$$= |\min\{\lambda_{min}(\Sigma + U), \lambda_{min}(\Sigma)\}| + 0.05,$$

$U = (u_{kl})$ be a matrix with eight random nonzero entries, each with a magnitude generated from $Unif(0,4) * \max_{1 \le j \le p} \sigma_{jj}$. The number of each class samples is 50 and the number of variables is 50.

Model 2: $U = (u_{kl})$ be a matrix with eight random nonzero entries, each with a magnitude generated from $Unif(0,400) * \max_{1 \le j \le p} \sigma_{jj}$.

Model 3: $U = (u_{kl})$ be a matrix with 500 random nonzero entries, each with a magnitude generated from $Unif(0,4) * \max_{1 \le j \le p} \sigma_{jj}$.

Model 4: $U = (u_{kl})$ be a matrix with 500 random nonzero entries, each with a magnitude generated from $Unif(0,400) * \max_{1 \le j \le p} \sigma_{jj}$.

As shown in Fig. 3, under the sparse setups (Model 1 and 2), the results of Tony Cai et al.'s method is much better than those of Chen et al.'s method. Chen et al.'s method is better than Tony Cai et al.'s method when the setups are not sparse (Model 3 and 4). Since Tony Cai et al.'s method corresponds to testing each element in the covariance matrix by Hoteling's test and then give the judgement according to the maximum statistic of all of the Hoteling's tests, so Chen et al.'s

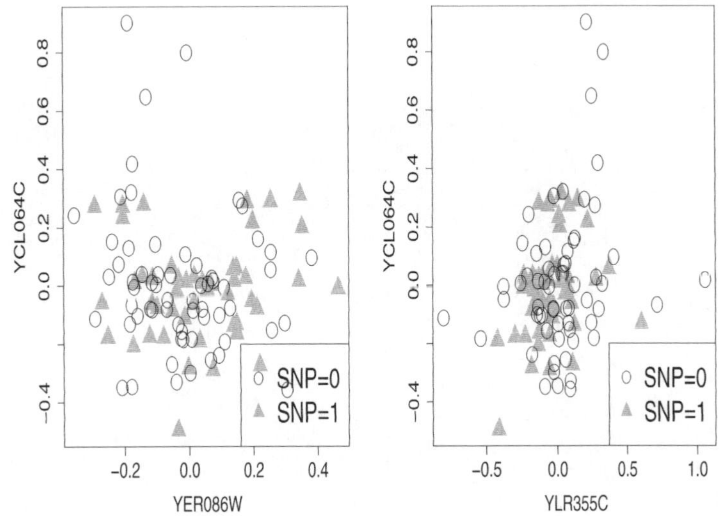

Fig. 5 Isoform-specific structure change. *Left*: Scatter plot of gene YER086W and YCL064C under two different genotypes of YCL023C; *Right*: Scatter plot of gene YLR355C and YCL064C under two different genotypes of YCL023C. We found the associated pathway-SNP modules only by kernel covariance test. The scatter figures show that YER086W-YCL064C and YLR355C-YCL064C were nonlinear correlated under genotypes of marker YCL023C

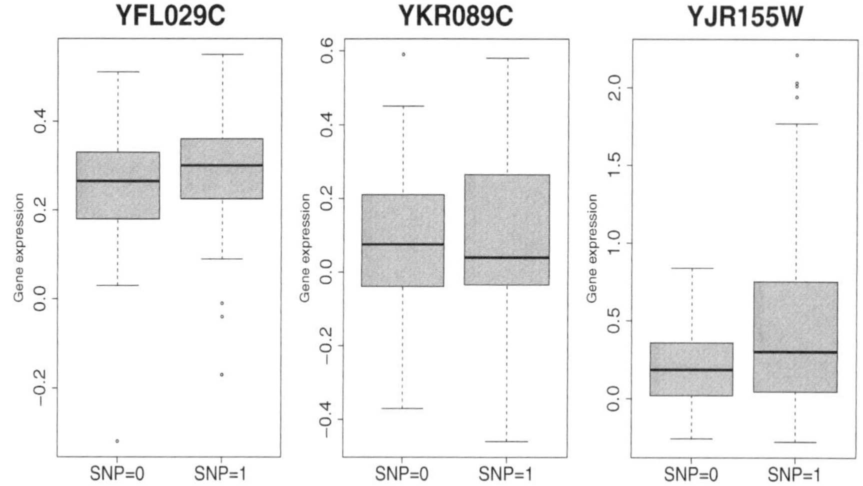

Fig. 6 Failure to detect correlation between single gene expression level and genotype of YFL029C. *Left*: Boxplot of expression level of YFL029C; *middle*, Boxplot of expression level of YKR089C; *right*: Boxplot of expression level of YJR155W. We can see that the means of YKR089C and YJR155W expressions do not change significantly

linear method has higher power than bi-variate model when the setups are not sparse. A similar simulation results with different number of samples can be found in Additional file 1: Figure S2.

Real data results

Associated SNP and pathways

We analyzed the yeast dataset collected by Kruglyak and colleagues [30]. The expression data were downloaded from http://journals.plos.org/plosbiology/article?id=10.1371/journal.pbio.0060083, with 4482 genes measured in 109 segregants derived from a cross between BY and RM. The experiments were performed under two conditions, glucose and ethanol. We did the pre-processing like [10], after which 4419 genes and 820 merged markers remained. We mapped 4419 genes to 103 pathways and analyzed the effect of each SNP to each pathway. Therefore, we tested 103*820 times. The algorithm was implemented in R, which can be found at http://www.math.pku.edu.cn/teachers/dengmh/NetworkBiomarker.

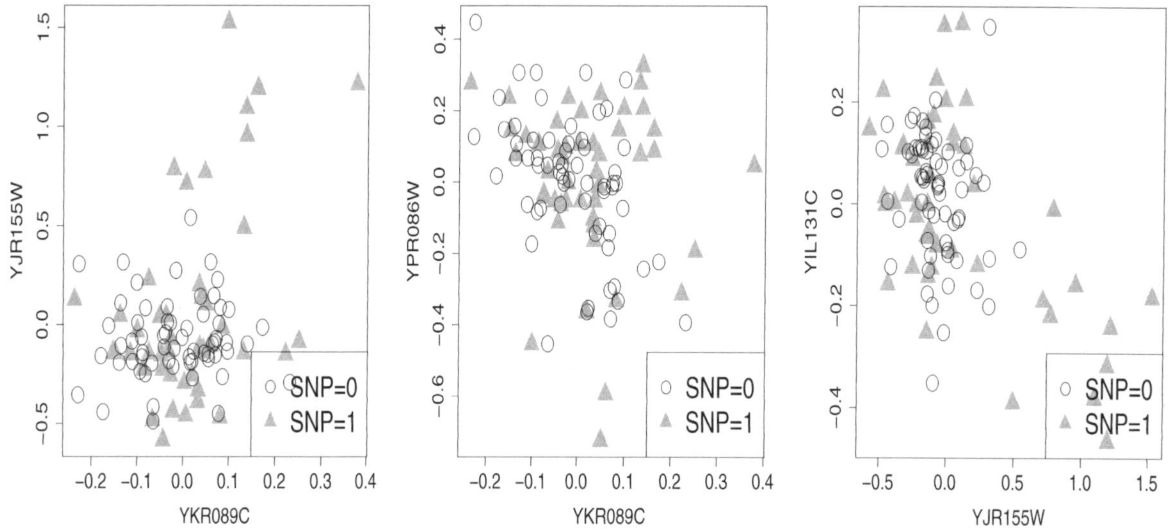

Fig. 7 Three examples of differential coexpression patterns of 2 genes due to genotype of YFL029C. *Left*: the co-expression patterns between the two genes YKR089C and YJR155W depend on the genotype of YFL029C. For samples with genotype 1, the co-expression correlation is different from the other samples. *Middle*: the co-expression patterns between the two genes YKR089C and YPR086W depend on the genotype of YFL029C. *Right*: the co-expression patterns between the two genes YJR155W and YIL131C depend on the genotype of YFL029C

We performed covariance test and kernel covariance with parameter 1 and 10 respectively to the pathway-SNP adjusted modules. We consider 103 pathways in KEGG [25] (the number of genes in each pathway can be found in the Additional file 1: Figure S1) and 820 merged markers under ethanol and glucose condition respectively. We

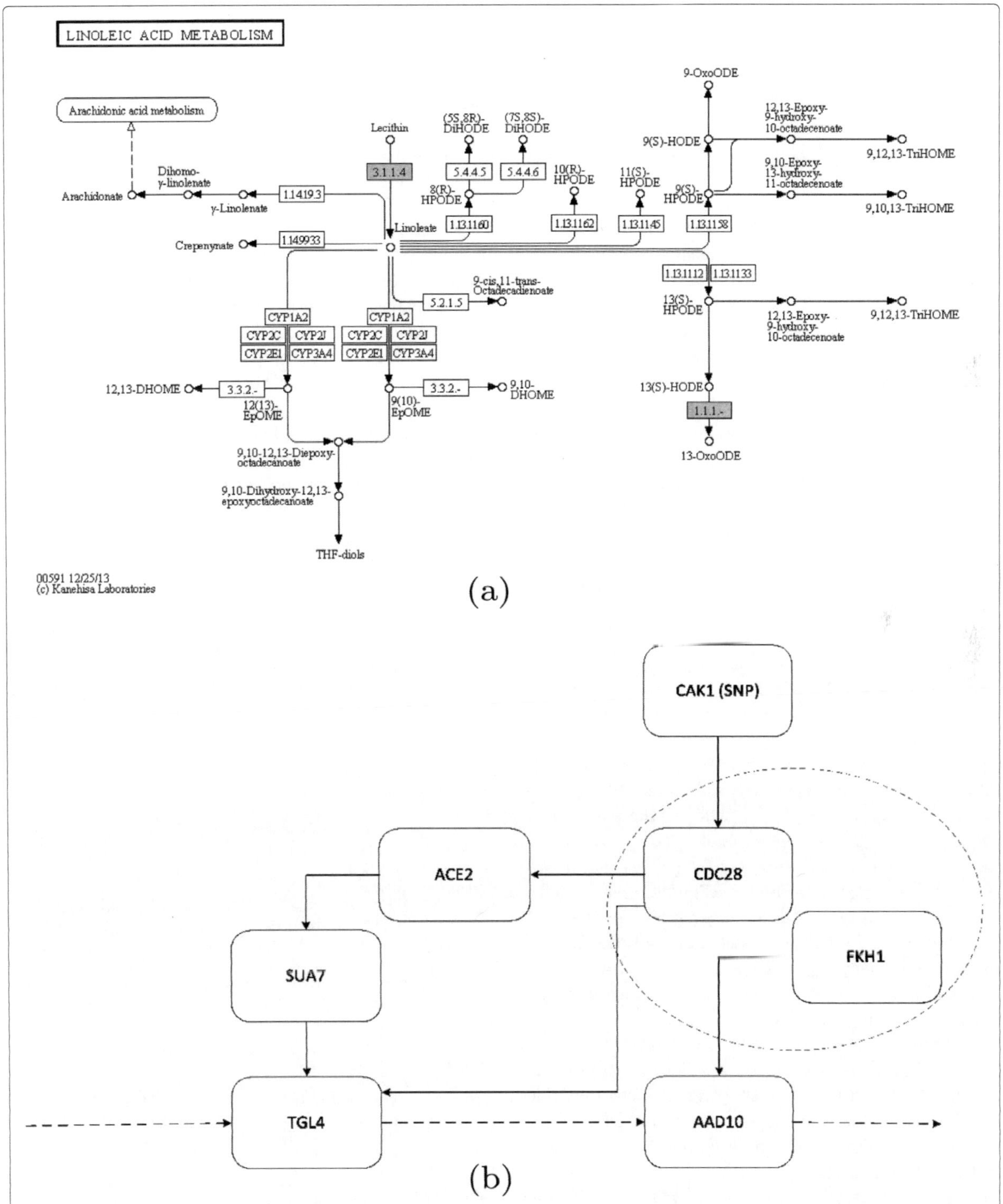

Fig. 8 The possible regulatory mechanism of the marker YFL029C to Linoleic acid metabolism pathway. **a** KEGG pathway of Linoleic acid metabolism pathway; **b** The potential regulatory relationship between marker YFL029C and Linoleic acid metabolism pathway

found 72 pathway-SNP modules under ethanol condition and 94 modules under glucose condition. Specifically, we found 36 modules by covariance test, 9 modules by kernel covariance test with parameter 1 and 51 modules by kernel covariance test with parameter 10 under ethanol condition, while 86 modules by covariance test, 3 modules by kernel covariance test with parameter 10 and 12 modules by kernel covariance test with parameter 1 under glucose condition (Fig. 4). Table 1 showed the associated pathways and SNPs under ethanol condition while Table 2 showed the associated pathways and SNPs under glucose condition.

Kernel Method found isoform-specific structure change

In our result, we found Valine, leucine and isoleucine biosynthesis pathway was associated with YCL023C marker only by kernel method under ethanol condition. Figure 5 shows the non-linear correlation between two pairs of genes, YER086W-YCL064C and YLR355C-YCL064C were nonlinear correlated with genotypes of YCL023C. And more than 10 isoforms of YER086W and 6 isoforms of YLR355C have been found (Saccharomyces Genome Database, http://www.yeastgenome.org/). The nonlinear correlation between two pairs of genes might be caused by samples in different isoforms. Specifically, two genes may be positive correlated under one isoform while negetive correlated under another isoform. However, the correlation of two genes might be missed if when we only considered linear correlation.

Linoleic acid metabolism is associated to cell cycle

Our method found YFL029C is associated with linoleic acid metabolism pathway under ethanol condition. With single gene correlation analysis, both of the mean of expression levels of YKR089C (TGL4) and YJR155W (AAD10) were not associated with YFL029C (Fig. 6 middle and right). Specifically, with p-value 0.5174 and 0.002804 (not significant for multiple test). However, the scatterplot after correction shows the correlation of two genes change apparently under YFL029C (Fig. 7 left). Under one status of SNP, the two genes are positive correlated while under the other status of SNP, the two are nearly independent. To understand this from the biological meaning which was showned in Fig. 8b, we found that marker locates in gene CAK1 (The expression of CAK1 is slightly different under two SNP status which was shown in Fig. 6 left.), the product of which can increase the activity of CDC28 [31]. CDC28 plays an important role in cell cycle. It can control the progress of cell cycle by phosphorylate different transcription factor. In our case, CDC28 phosphorylate ACE2 [32] which can increase the activity of transcription factor, SUA7 [33]. SUA7 is the transcription factor of TGL4, which is a lipase in linoleic acid metabolism pathway. Meanwhile, CDC28 and FKH1 can form complex [34] and FKH1 is the transcription factor of AAD10, which is another enzyme in linoleic acid metabolism pathway. The correlation between YKR089C and its TF was shown in Fig. 7 middle and the correlation between YJR155W and its TF was shown in Fig. 7 right. From the structure of the pathway in KEGG [25] as shown in Fig. 8a, the different status of the SNP YFL029C might lead to different amounts of intermediate product in the pathway.

Discussion and conclusion

We propose a network based covariance test to identify the marker which affects the structure of a pathway. It has an advantage that a static network structure is not assumed. The biomarker we defined is the SNP associated to the structure of genes in the pathway. Considering two genes may have different correlations under different isoforms which is hard to detect by linear test, so we also consider the nonlinear test. We identified a total of 166 modules, with each module consisting of a group of genes and one eQTL where the eQTL regulate the co-expression patterns of the group of genes. We found that many of these modules have biological interpretations. Till now, we consider the difference of two networks by covariance matrix and covariance operators. We will focus on difference of precision matrix in the future research.

Additional file

Additional file 1: Supplementary materials for the computation of HSDCC and additional figures Figures S1–S3. (PDF 1474 kb)

Competing interests

The authors declare that they have no competing interests.

Authors' contributions

MD and NT conceived of the project. The flowchart of the strategy was designed by MD and HY. HY and ZL developed the kernel method. HY conducted the simulation and real data analysis. HY wrote the first draft of the manuscript, which all authors revised and approved.

Acknowledgements

This work is supported by the National Natural Science Foundation of China (Nos. 31171262, 31428012, 31471246), and the National Key Basic Research Project of China (No. 2015CB910303). We thank Dr. Lin Wang for helpful feedback during the saccharomyces cerevisiae data pre-processing and Dr. Minping Qian and Dr. Kai Song's discussions. The publication costs for this article are from the National Natural Science Foundation of China (No. 31471246).

Declarations

This article has been published as part of BMC Systems Biology Volume 10 Supplement 1, 2016: Selected articles from the Fourteenth Asia Pacific Bioinformatics Conference (APBC 2016): Systems Biology. The full contents of the supplement are available online at http://www.biomedcentral.com/bmcsystbiol/supplements/10/S1.

Author details

[1]LMAM, School of Mathematical Sciences, Peking University, Yiheyuan Road, 100871 Beijing, China. [2]Department of Chemical Pathology, Prince of Wales Hospital, Faculty of Medicine, The Chinese University of Hong Kong, Shatin,

Hong Kong, China. [3]Center for Quantitative Biology, Peking University, Yiheyuan Road, 100871 Beijing, China. [4]Center for Statistical Sciences, Peking University, Yiheyuan Road, 100871 Beijing, China.

Published: 11 January 2016

References

1. Visscher PM, Brown MA, McCarthy MI, Yang J. Five years of gwas discovery. Am J Hum Genet. 2012;90(1):7–24.
2. Albert FW, Kruglyak L. The role of regulatory variation in complex traits and disease. Nat Rev Genet. 2015;16:197–212. doi:10.1038/nrg3891.
3. Cookson W, Liang L, Abecasis G, Moffatt M, Lathrop M. Mapping complex disease traits with global gene expression. Nat Rev Genet. 2009;10(3):184–94.
4. Nica AC, Dermitzakis ET. Expression quantitative trait loci: present and future. Phil Trans R Soc B: Biol Sci. 2013;368(1620):20120362.
5. Wang P, Dawson JA, Keller MP, Yandell BS, Thornberry NA, Zhang BB, et al. A model selection approach for expression quantitative trait loci (eqtl) mapping. Genetics. 2011;187(2):611–21.
6. Li KC. Genome-wide coexpression dynamics: theory and application. Proc Natl Acad Sci. 2002;99(26):16875–80.
7. Sun W, Yuan S, Li KC. Trait-trait dynamic interaction: 2d-trait eqtl mapping for genetic variation study. BMC Genomics. 2008;9(1):242.
8. Ho YY, Parmigiani G, Louis TA, Cope LM. Modeling liquid association. Biometrics. 2011;67(1):133–41.
9. Chen J, Xie J, Li H. A penalized likelihood approach for bivariate conditional normal models for dynamic co-expression analysis. Biometrics. 2011;67(1):299–308.
10. Wang L, Zheng W, Zhao H, Deng M. Statistical analysis reveals co-expression patterns of many pairs of genes in yeast are jointly regulated by interacting loci. PLoS Genet. 2013;9(3):1003414.
11. Basso K, Margolin AA, Stolovitzky G, Klein U, Dalla-Favera R, Califano A. Reverse engineering of regulatory networks in human b cells. Nat Genet. 2005;37(4):382–90.
12. Bonneau R, Facciotti MT, Reiss DJ, Schmid AK, Pan M, Kaur A, et al. A predictive model for transcriptional control of physiology in a free living cell. Cell. 2007;131(7):1354–65.
13. Pereira-Leal JB, Enright AJ, Ouzounis CA. Detection of functional modules from protein interaction networks. Proteins Struct Function Bioinforma. 2004;54(1):49–57.
14. Witten DM, Tibshirani R, Hastie T. A penalized matrix decomposition, with applications to sparse principal components and canonical correlation analysis. Biostatistics. 2009;10(3):515–34. doi:10.1093/biostatistics/kxp008.
15. Naylor MG, Lin X, Weiss ST, Raby BA, Lange C. Using canonical correlation analysis to discover genetic regulatory variants. PloS ONE. 2010;5(5):10395.
16. Lin D, Zhang J, Li J, Calhoun VD, Deng HW, Wang YP. Group sparse canonical correlation analysis for genomic data integration. BMC Bioinformatics. 2013;14(1):245.
17. Li Y, Nan B, Zhu J. Multivariate sparse group lasso for the multivariate multiple linear regression with an arbitrary group structure. Biometrics. 2015.
18. Kim S, Xing EP. Statistical estimation of correlated genome associations to a quantitative trait network. PLoS Genet. 2009;5(8):1000587.
19. Zhang L, Kim S. Learning gene networks under snp perturbations using eqtl datasets. PLoS Comput Biol. 2014;10(2):1003420.
20. Casale FP, Rakitsch B, Lippert C, Stegle O. Efficient set tests for the genetic analysis of correlated traits. Nat Meth. 2015.
21. Ideker T, Krogan NJ. Differential network biology. Mol Syst Biol. 2012;8.
22. Zhou S, Carraway KL, Eck MJ, Harrison SC, Feldman RA, Mohammadi M, et al. Catalytic specificity of protein-tyrosine kinases is critical for selective signalling. Nature. 1995;373(6514):536–9.
23. Tibshirani R. Regression shrinkage and selection via the lasso. J R Stat Soc. Series B (Methodological). 1996;58(1):267–88.
24. Fan J, Li R. Variable selection via nonconcave penalized likelihood and its oracle properties. J Am Stat Assoc. 2001;96(456):1348–60.
25. Ogata H, Goto S, Sato K, Fujibuchi W, Bono H, Kanehisa M. KEGG: Kyoto Encyclopedia of Genes and Genomes. Nucleic Acids Res. 1999;27:29–34.
26. Subramanian A, Tamayo P, Mootha VK, Mukherjee S, Ebert BL, Gillette MA, et al. Gene set enrichment analysis: a knowledge-based approach for interpreting genome-wide expression profiles. Proc Natl Acad Sci of the USA. 2005;102(43):15545–50.
27. Li J, Chen SX. Two sample tests for high-dimensional covariance matrices. Ann Stat. 2012;40(2):908–40.
28. Jain S, Simon HU, Tomita E. Algorithmic Learning Theory. 16th International Conference, ALT 2005, Singapore, October 8–11, 2005, Proceedings. Springer-Verlag Berlin Heidelberg 2005.
29. Cai T, Liu W, Xia Y. Two-sample covariance matrix testing and support recovery in high-dimensional and sparse settings. J Am Stat Assoc. 2013;108(501):265–77.
30. Smith EN, Kruglyak L. Gene–environment interaction in yeast gene expression. PLoS Biol. 2008;6(4):83.
31. Kurat CF, Wolinski H, Petschnigg J, Kaluarachchi S, Andrews B, Natter K, et al. Cdk1/cdc28-dependent activation of the major triacylglycerol lipase tgl4 in yeast links lipolysis to cell-cycle progression. Mol Cell. 2009;33(1):53–63.
32. Ubersax JA, Woodbury EL, Quang PN, Paraz M, Blethrow JD, Shah K, et al. Targets of the cyclin-dependent kinase cdk1. Nature. 2003;425(6960):859–64.
33. Zhang DY, Dorsey MJ, Voth WP, Carson DJ, Zeng X, Stillman DJ, et al. Intramolecular interaction of yeast tfiib in transcription control. Nucleic Acids Res. 2000;28(9):1913–20.
34. Venters BJ, Wachi S, Mavrich TN, Andersen BE, Jena P, Sinnamon AJ, et al. A comprehensive genomic binding map of gene and chromatin regulatory proteins in saccharomyces. Mol Cell. 2011;41(4):480–92.

Huang et al. BMC Systems Biology 2016, 10(Suppl 1):6
DOI 10.1186/s12918-015-0246-z

BMC Systems Biology

PROCEEDINGS Open Access

UbiSite: incorporating two-layered machine learning method with substrate motifs to predict ubiquitin-conjugation site on lysines

Chien-Hsun Huang[1,2], Min-Gang Su[1], Hui-Ju Kao[1], Jhih-Hua Jhong[1], Shun-Long Weng[3,4,5*] and Tzong-Yi Lee[1,6*]

From The Fourteenth Asia Pacific Bioinformatics Conference (APBC 2016)
San Francisco, CA, USA. 11 - 13 January 2016

Abstract

Background: The conjugation of ubiquitin to a substrate protein (protein ubiquitylation), which involves a sequential process – E1 activation, E2 conjugation and E3 ligation, is crucial to the regulation of protein function and activity in eukaryotes. This ubiquitin-conjugation process typically binds the last amino acid of ubiquitin (glycine 76) to a lysine residue of a target protein. The high-throughput of mass spectrometry-based proteomics has stimulated a large-scale identification of ubiquitin-conjugated peptides. Hence, a new web resource, UbiSite, was developed to identify ubiquitin-conjugation site on lysines based on large-scale proteome dataset.

Results: Given a total of 37,647 ubiquitin-conjugated proteins, including 128026 ubiquitylated peptides, obtained from various resources, this study carries out a large-scale investigation on ubiquitin-conjugation sites based on sequenced and structural characteristics. A TwoSampleLogo reveals that a significant depletion of histidine (H), arginine (R) and cysteine (C) residues around ubiquitylation sites may impact the conjugation of ubiquitins in closed three-dimensional environments. Based on the large-scale ubiquitylation dataset, a motif discovery tool, MDDLogo, has been adopted to characterize the potential substrate motifs for ubiquitin conjugation. Not only are single features such as amino acid composition (AAC), positional weighted matrix (PWM), position-specific scoring matrix (PSSM) and solvent-accessible surface area (SASA) considered, but also the effectiveness of incorporating MDDLogo-identified substrate motifs into a two-layered prediction model is taken into account. Evaluation by five-fold cross-validation showed that PSSM is the best feature in discriminating between ubiquitylation and non-ubiquitylation sites, based on support vector machine (SVM). Additionally, the two-layered SVM model integrating MDDLogo-identified substrate motifs could obtain a promising accuracy and the Matthews Correlation Coefficient (MCC) at 81.06 % and 0.586, respectively. Furthermore, the independent testing showed that the two-layered SVM model could outperform other prediction tools, reaching at 85.10 % sensitivity, 69.69 % specificity, 73.69 % accuracy and the 0.483 of MCC value.

Conclusion: The independent testing result indicated the effectiveness of incorporating MDDLogo-identified motifs into the prediction of ubiquitylation sites. In order to provide meaningful assistance to researchers interested in large-scale ubiquitinome data, the two-layered SVM model has been implemented onto a web-based system (UbiSite), which is freely available at http://csb.cse.yzu.edu.tw/UbiSite/. Two cases given in the UbiSite provide a demonstration of effective identification of ubiquitylation sites with reference to substrate motifs.

Keywords: Ubiquitin conjugation, Ubiquitylation, Substrate motif, Position-specific scoring matrix

* Correspondence: 4467@mmh.org.tw; francis@saturn.yzu.edu.tw
[3]Department of Obstetrics and Gynecology, Hsinchu Mackay Memorial Hospital, Hsin-Chu 300, Taiwan
[1]Department of Computer Science and Engineering, Yuan Ze University, Taoyuan 320, Taiwan
Full list of author information is available at the end of the article

Background

In 1975, Goldstein et al. [1] discovered ubiquitin, which is a small protein of approximately 8.5 kDa and is composed of 76 amino acids [2]. The attachment of mono-ubiquitin or poly-ubiquitin chains to proteins at particular lysines is a critical post-translational modification (PTM), which is called protein ubiquitylation or ubiquitination [3, 4]. Protein ubiquitylation, involving a sequential process – E1 activation, E2 conjugation and E3 ligation enzyme, is important for modulating various cellular functions, such as transcriptional regulation, signal transduction, development, apoptosis, endocytosis, and cell proliferation [5–7]. These three major classes of enzymes have critical roles in promoting and controlling the activities that are involved in the process of protein ubiquitylation system. In particular, the E3 ligases recognize a specific substrate protein and then catalyze the attachment of ubiquitin to the target lysine of a protein [8, 9]. The HECT domain of E3 ligases plays prominent roles in trafficking, immune response, and several signaling pathways that regulate cellular growth and proliferation [10]. As illustrated in Figure S1 (Additional file 1), the HECT domain consists of two major components: a N-terminal N-lobe that interacts with the E2, and a C-terminal C-lobe which contains the active-site cysteine forming the thioester linkage with ubiquitin [11–13]. The conserved HECT domain is located at the C-terminus of these enzymes, whereas their N-terminal domains are diverse and mediate substrate targeting.

With the biological importance of protein ubiquitylation, the high-throughput of mass spectrometry-based proteomics has stimulated a large-scale identification of ubiquitin-conjugated peptides [14–16]. Thus, numerous databases [9, 17–20] of protein ubiquitylation sites have been developed to date, owing to the regulatory significance of ubiquitin in cellular processes. In addition, the labor-intensive experiment in site-specific identification of ubiquitin-conjugated peptides motivated an increasing number of computational methods [14, 16, 21–25] developed for the identification of ubiquitylation sites. As the number of experimentally verified ubiquitylation sites is increasing, the development of new bioinformatics tools that can be applied to large scale proteome data is required. Recently, an ensemble classifier was developed to identify ubiquitylation sites based on lysines [26]. The WPNNA (weighted passive nearest neighbor algorithm) classifier for identifying ubiquitylation sites based on hybrid features has also been proposed [27]. The UbiProber [28], which integrates key position and amino acid residue features, was designed to predict both general and species-specific ubiquitylation sites. Additionally, independent testing has demonstrated that the combined model improves the area under the operating characteristic curve of the receiver (AUC) by ~15 %. Although the fact that the accuracy and stability of these prediction tools have been demonstrated, room typically exists to improve its predictive power. Moreover, as more experimentally confirmed ubiquitylation sites become available, the lack of methods for characterizing the substrate site specificities based on large-scale datasets is serious.

An increasing number of site-specific ubiquitylated peptides obtained from mass spectrometry-based proteomics has stimulated this work to characterize and identify substrate sites of ubiquitin conjugation based on large-scale dataset. Experimentally confirmed ubiquitin-conjugated peptides were mainly collected from several online resources, including hCKSAAP_UbiSite [22], dbPTM [19, 20] and mUbiSiDa [18]. Not only are sequence-based features, such as amino acid composition (AAC), amino acid pairwise composition (AAPC), positional weighted matrix (PWM) and position-specific scoring matrix (PSSM) assessed, but also the effectiveness of solvent-accessible surface area (SASA) or secondary structure is examined. A best feature for the prediction of ubiquitylation sites was determined based on five-fold cross-validation. Additionally, a motif discovery tool, MDDLogo [29], was adopted to characterize the potential substrate motifs of ubiquitylation sites and to generate a two-layered prediction model by integrating the MDDLogo-identified substrate motifs. The independent testing showed that the two-layered SVM model could provide a promising accuracy (73.69 %) and perform better than other prediction tools. Finally, the two-layered prediction model was adopted to implement a web-based system (UbiSite) for providing effective assistance to researchers who are interested in large-scale proteome data.

Methods

Collection and preprocessing of training dataset

Experimentally validated ubiquitin-conjugated peptides were mainly obtained from several online resources, including hCKSAAP_UbiSite [22], dbPTM [19, 20], and mUbiSiDa [18]. As presented in Table 1, totally 6118 ubiquitylated lysines were taken from 2500 ubiquitylated proteins. The dataset associated with protein ubiquitylation was also obtained from version 3.0 of dbPTM, which is a comprehensive database of protein post-translational modifications that has accumulated 23,949 ubiquitylated lysines from 6259 proteins. Additionally, a large-scale ubiquitin-conjugated peptides (110,695 ubiquitylated lysines in mammals) were obtained from mUbiSiDa, which is a comprehensive database containing 35,494 ubiquitylated proteins. After redundant data were removed, 37,647 ubiquitylated proteins, consisting of 128,026 ubiquitylated lysines, were regarded as the training dataset.

To construct positive data of training dataset, herein, a window of length $2n+1$ was used to extract sequence fragments that were centered at the ubiquitylated lysine

Table 1 Data statistics in the construction of training dataset and independent testing dataset

Data set	Data resource	Number of ubiquitylated proteins	Number of ubiquitylated lysines	Number of non-ubiquitylated lysines
Training set	hCKSAAP_UbiSite	2500	6118	6118
	dbPTM 3.0	6259	23,949	228,441
	mUbiSiDa	35,494	110,695	1,217,977
	Combined non-redundant data	37,647	128,026	1,317,734
	Non-homologous data (sequence identity ≦ 30 %)	4828	5438	12,663
Independent testing set	CPLM 2.0	32,429	139,950	1,109,432
	Non-homologous data (sequence identity ≦ 30 %)	2894	3732	10,664

(K) residue and contained n upstream and n downstream flanking amino acids. Given 37,647 experimentally verified ubiquitylated proteins, the sequence fragments that contained window length of $2n + 1$ amino acids and were centered at the lysine residue without an annotation of ubiquitylation were regarded as the negative data of training dataset (non-ubiquitylated lysines). According to another work [14] and the preliminary evaluation using windows of various lengths, a window size of 13 ($n = 6$) maximizes the accuracy in the prediction of ubiquitylation sites. Based on a window size of 13, the training dataset contains 128,026 positive data and 1,317,734 negative data.

As for binary classification, the performance of the predictive models may be overestimated or underestimated owing to the homologous sequences in positive and negative datasets. In order to obtain a reasonable prediction performance, the homologous sequences should be removed from the training dataset by using CD-HIT [30] program. Additionally, an analysis of sequence fragments indicates that, owing to the incompleteness of available information about experimentally validated ubiquitylation sites, some negative data may be homologous to positive data in the training dataset, potentially resulting in false positive or false negative predictions. Accordingly, CD-HIT was used again by running cd-hit-2d across positive and negative training data with 50 % sequence similarity to filter out the homologous sequences in negative data. Table S1 (Additional file 2) presents statistics concerning the data after the homologous sequences had been removed using CD-HIT, based on various values of sequence identity. After homologous sequences with 30 % sequence identity had been filtered out using cd-hit and psi-cd-hit, the training dataset comprised 5438 positive sequences and 12,663 negative sequences.

Features extraction and encoding

In this work, several sequence-based features, including amino acid composition (AAC), amino acid pairwise composition (AAPC), positional weighted matrix (PWM), and position-specific scoring matrix (PSSM), were examined. Additionally, the solvent-accessible surface area (SASA) and secondary structure around the ubiquitylation sites were also investigated. Amino acid sequences with a lysine in the center were individually extracted from positive and negative training sets using a window of length $2n + 1$, where n was set to six. To transform the amino acid sequence into a numeric vector for model construction, a basic encoding method, namely 20-dimensional binary coding (20D), was adopted. For instance, Alanine (A) was encoded as "10000000000000000000", and Cysteine (C) was encoded as "01000000000000000000". The number of dimensions for 20D feature was $(2n + 1) \times 20$ to represent the flanking amino acids surrounding the ubiquitylation sites. The training data contained k vectors $\{x_i, i = 1, 2 ..., k\}$ corresponding to the k fragment sequences. The labels +1 and −1 were used to mark the positive and negative data classes, respectively, for each numeric vector.

For the coding of amino acid composition, each sequence fragment was encoded as a 21-dimensional vector $\{x_i, i = 1,...,21\}$ that comprised 20 types of amino acid and an non-existing residue, which specify the number of occurrences of the 20 types of amino acid normalized with the total number of residues in a sequence fragment. For the coding of amino acid pair composition, a 21x21-dimensional matrix $\{x_{ij}, i, j = 1, ..., 21\}$ was used to encode each sequence fragment. Each element x_{ij} specifies the number of occurrences of amino acid pairs normalized with the total number of amino acid pairs in a sequence fragment. By referring to the SulfoSite method [31], PWM was determined by calculating the occurrence rate of 20 amino acids surrounding a substrate sites, and was utilized in encoding the fragment sequences. Each sequence fragment of the training dataset was represented by a matrix of $(2n + 1) \times w$ elements, where w stands for 21 elements including 20 types of amino acids and one for the non-existing residue.

In the viewpoint of protein sequence evolution, several amino acid residues of a protein can be mutated without changing its core structure or functional domain, and two proteins may have similar tertiary structures with different amino acid compositions. Position Specific Scoring Matrix (PSSM) profiles, which have been extensively utilized in protein domain finding, secondary structure prediction, subcellular localization and other biological problems [32–34], was adopted with evolutionary significance. As presented in Figure S2 (Additional file 3), the PSSM profiles were obtained by PSI-BLAST [35] against non-redundant set of ubiquitylated protein sequences. Then, matrices of $(2n + 1) \times 20$ elements, with rows that were centered at ubiquitylation sites and non-ubiquitylation sites, were extracted from PSSM profiles. Then, the $(2n + 1) \times 20$ matrix was transformed into a 20×20 matrix by summing up the rows that were associated with the same type of amino acid. Finally, every element in 20×20 matrix was divided by the window length $2n + 1$ and normalized using the expression: $\frac{1}{1+e^{-x}}$.

It has been reported that a side-chain of amino acid that undergoes post-translational modification prefers to be accessible on the surface of a protein [36]. Thus, the solvent-accessible surface area (SASA) was used to evaluate the ability of predicting ubiquitylation sites. In this investigation, the RVP-Net [37, 38] was applied to compute the ASA value from the full-length protein sequence. The computed ASA was the percentage of the solvent-accessible area of each amino acid on the protein. The ASA values of amino acids around the ubiquitylation and non-ubiquitylation sites were extracted and normalized to be between zero and one. In the investigation of secondary structure, version 2.0 of PSIPRED [39] was employed to compute the secondary structure from the protein sequence. The output of PSIPRED is given in terms of "H," "E" and "C" which represent helix, sheet and coil, respectively. The full-length protein sequences with ubiquitylated lysines were submitted to PSIPRED to obtain the secondary structure of all residues. Then, an orthogonal binary coding scheme was used to transform the three types of secondary structure into numeric vectors: "100" for helix, "010" for sheet, and "001" for coil structure.

Model building and performance evaluation

Based on binary classification, the positive and negative datasets were used to build a predictive model using support vector machine (SVM). We employed a public SVM library, LIBSVM [40], to implement the construction of predictive model and the evaluation of prediction performance. The radial basis function (RBF):

$$K(S_i, S_j) = \exp\left(-\gamma \left\| S_i - S_j \right\|^2\right) \quad (1)$$

was adopted as the kernel function for transforming the input samples into a higher dimensional space, and finding a hyper-plane to discriminate between the two classes with maximal margin and minimal error. The power of RBF kernel could be determined by tuning the gamma (g) parameter, while the cost (c) parameter controls the hyper-plane softness. In this work, each feature was adopted to generate a predictive model by LIBSVM library; then, the feature performing best was selected as the final feature to implement the prediction tool.

In the evaluation of predictive power, the five-fold cross-validation was carried out for each SVM model trained with different feature to test their predictive performances. The method of cross-validation could increase the reliability of performance evaluation, because it considers all original data in both the training and testing data sets; typically, each data was regarded as a test set only once [41]. To gauge the predictive performance of training model, the following measures were used: sensitivity (Sn), specificity (Sp), accuracy (Acc) and Matthews Correlation Coefficient (MCC):

$$\mathrm{Sn} = \frac{\mathrm{TP}}{\mathrm{TP} + \mathrm{FN}} \quad (2)$$

$$\mathrm{Sp} = \frac{\mathrm{TN}}{\mathrm{TN} = \mathrm{FP}} \quad (3)$$

$$\mathrm{Acc} = \frac{\mathrm{TP} + \mathrm{TN}}{\mathrm{TP} + \mathrm{FP} + \mathrm{TN} + \mathrm{FN}} \quad (4)$$

$$\mathrm{MCC} = \frac{(TP \times TN) - (FN \times FP)}{\sqrt{(TP + FN) \times (TN + FP) \times (TP + FP) \times (TN + FN)}} \quad (5)$$

where TP, TN, FP and FN represent the numbers of true positives, true negatives, false positives and false negatives, respectively. Sensitivity is the percentage of true positive predictions, while specificity represents that of true negative predictions. Accuracy reflects the overall proportion of true positive and true negative predictions. As for binary classifications, accuracy is sometimes not useful when the two classes are of very different sizes. Accordingly, the MCC is typically considered as a balanced measure, even if the two classes are of very different sizes [42]. The MCC value is ranging from −1 to +1: a coefficient value of +1 represents a perfect prediction, while the values 0 and −1 represent random and opposite predictions, respectively. In this work, the model containing a higher positive MCC value indicates a better prediction performance for classifying positive and negative data correctly.

Discovery of substrate motifs of ubiquitin-conjugation sites

Along the protein ubiquitylation pathway, E3 ubiquitin ligase has a critical role in catalyzing the attachment of ubiquitin to lysine residue of a protein by recognizing a specific substrate site. Owing to an increasing number of large-scale ubiquitylation data obtained from mass spectrometry-based proteomics, we are motivated to explore potential substrate motifs for ubiquitin-conjugated sites. Although the WebLogo [43] can display position-specific amino acid composition for a group of aligned signal sequences, it is difficult to explore conserved motifs for large-scale sequence data. For instance, a sequence logo for all phosphorylation data involved with various catalytic kinases fails to obviously present the kinase-specific substrate specificity. Hence, the MDDLogo [29] was applied to the training data in an attempt to discover substrate motif signatures at ubiquitylation sites. Previous works [44–55] have demonstrated the effectiveness of dividing a group of protein sequences into smaller subgroups before the computational identification of PTM sites. MDDLogo applies chi-square test to evaluate the dependence of the occurrence of amino acids between two positions, A_i and A_j, which are adjacent to the modification site (Figure S3 in Additional file 4). After the recursive chi-square test, MDDLogo divides a group of aligned sequences into subsets that capture the most significant dependencies of positions on each other. When applying MDDLogo, a parameter, i.e., the maximum-cluster-size, should be set. If the size of a subgroup is less than a specified value of maximum-cluster-size, the subgroup will not be divided any further. The MDDLogo terminates after all of the subgroup sizes are less than the value of the specified value of maximum-cluster-size.

Construction of two-layered predictive model

In this investigation MDDLogo was utilized to sub-divide 5438 ubiquitylated sequence fragments (positive training data) into 12 subgroups that contain significant substrate motifs. As presented in Fig. 1, the LIBSVM was employed to generate first-layered SVM model for each MDDLogo-identified substrate motifs. The negative data for each MDDLogo-clustered subgroup were selected from the negative training data (12,663 non-ubiquitylated sequences) with a ratio of approximately 1:2.33 (which almost equals the ratio of the number of positive data to the number of negative data, 5438:12663). In first layer, each SVM model will output a value of probability estimate ranging from 0 to 1 for each prediction. Thus, the values of probability estimates from each SVM classifier trained according to a specific motif were adopted to form an input vector for second-layered SVM classifier. The so-called two-layered prediction model could be used to identify ubiquitin-conjugation sites along with their corresponding substrate motifs.

Construction of independent testing dataset

In the prediction of PTM sites, after selecting the best predictive model with the highest MCC value in cross-validation process, an independent testing dataset, which was definitely blind to the training dataset, was generated to evaluate the real determinant of the predictive performance of the selected model. The dataset for independent testing was collected from version 2.0 of CPLM [17], which is a comprehensive database of protein lysine modifications. The positive and negative testing datasets were generated using the same method as was applied to the training dataset. The program cd-hit-2d was used with a sequence identity cutoff of 30 % to remove homologous sequences between independent testing set and training set, yielding the final independent testing dataset that contained 3732 positive and 10,664 negative data (Table 1). The testing dataset was also used to evaluate the predictive power of other prediction tools, which were compared with our method in terms of predictive performance.

Results and discussion

Investigation of amino acid composition and structural characteristics neighboring with ubiquitin-conjugation sites

Based on the analysis of amino acid composition, the frequency of occurrence of 20 amino acids surrounding the substrate site could be determined to find the potential consensus motifs. The flanking amino acid sequences of 5438 ubiquitylation sites (at position 0) can be graphically visualized in the entropy plots of the sequence logo by WebLogo; however, there is no conservation of amino acids around the ubiquitylation sites [14]. Hence, an effective tool, TwoSampleLogo [56], was applied to detect statistically noteworthy differences in position-specific amino acid composition between positive and negative training datasets. As presented in Fig. 2a, the lysine residue was placed in the middle of the fragment sequences, and positions of the flanking amino acids were described in range from −6 to +6. The comparison of position-specific amino acid composition between 5438 ubiquitylated and 12,663 non-ubiquitylated sites reveals that the nonpolar and aliphatic amino acids, such as alanine (A), leucine (L) and glycine (G), are enriched around ubiquitylation sites at positions −3, −2, −1, +1, +2, +4 and +5 (with $p\text{-}value < 0.01$). Additionally, negatively charged amino acids, aspartic acid (D) and glutamic acid (E), are found at positions −5, +2, and +3. By contrast, non-ubiquitylated sites have the enrichment of positively charged amino acids, histidine (H) and arginine (R) at positions −5, −3, −2, −1, +1, +2, +3

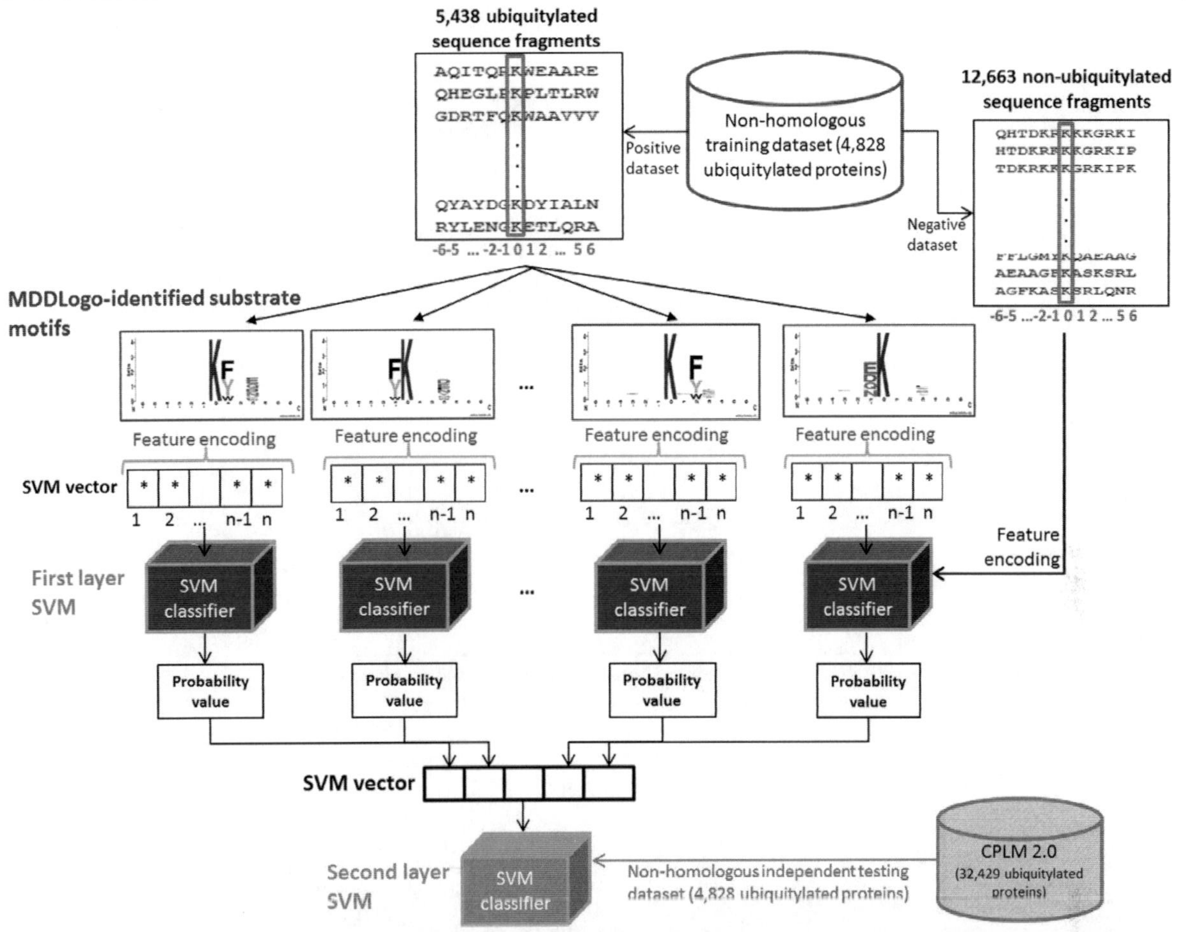

Fig. 1 Flowchart of constructing two-layered prediction model based on MDDLogo-identified substrate motifs

and +4. It would be noticed that a neutral amino acid, cysteine (C), is enriched around non-ubiquitylation sites; oppositely, cysteine is highly depleted in close to ubiquitylation sites. In summary, this investigation indicates a significant depletion of H, R and C residues for ubiquitylation sites.

To investigate the preference of the solvent-accessible surface area (SASA) that surrounds ubiquitin conjugation sites in protein tertiary structures, all experimentally identified ubiquitylation sites were mapped to the corresponding positions of the protein entries in the Protein Data Bank (PDB) [57]. However, only 302 ubiquitylation sites have the corresponding positions in PDB. Based on protein tertiary structures, Figure S4 (Additional file 5) indicates that the substrate sites of ubiquitin conjugation prefer to be accessible on the surface of proteins. Since most of the ubiquitylated proteins do not have corresponding protein tertiary structures in PDB, RVP-Net and PSIPRED have been adopted to compute the SASA value and secondary structure, respectively, from the protein sequence. Figure 2b presents the comparison of flanking SASA between ubiquitylation and non-ubiquitylation sites. This investigation reveals that most of the ubiquitylation sites prefer to locate in a region with higher SASA (≥35 %), especially at positions +1, +2 and +3. Overall, the mean SASA that surrounds the ubiquitylation sites slightly exceeds that around non-ubiquitylation sites. In the investigation of secondary structure, Fig. 2c indicates a higher preference of coil (loop) or helix structures around ubiquitylation sites, which is consistent with the analysis from Catic et al. [58].

Performance evaluation of the investigated features in identifying ubiquitylation sites

To examine the effectiveness of various features, namely 20D, AAC, AAPC, PWM, PSSM, SASA and SS, in discriminating between 5438 ubiquitylation sites and 12,663 non-ubiquitylation sites, the LIBSVM was employed to learn a predictive model for each feature. As shown in Table 2, the SVM model trained with PSSM profile has

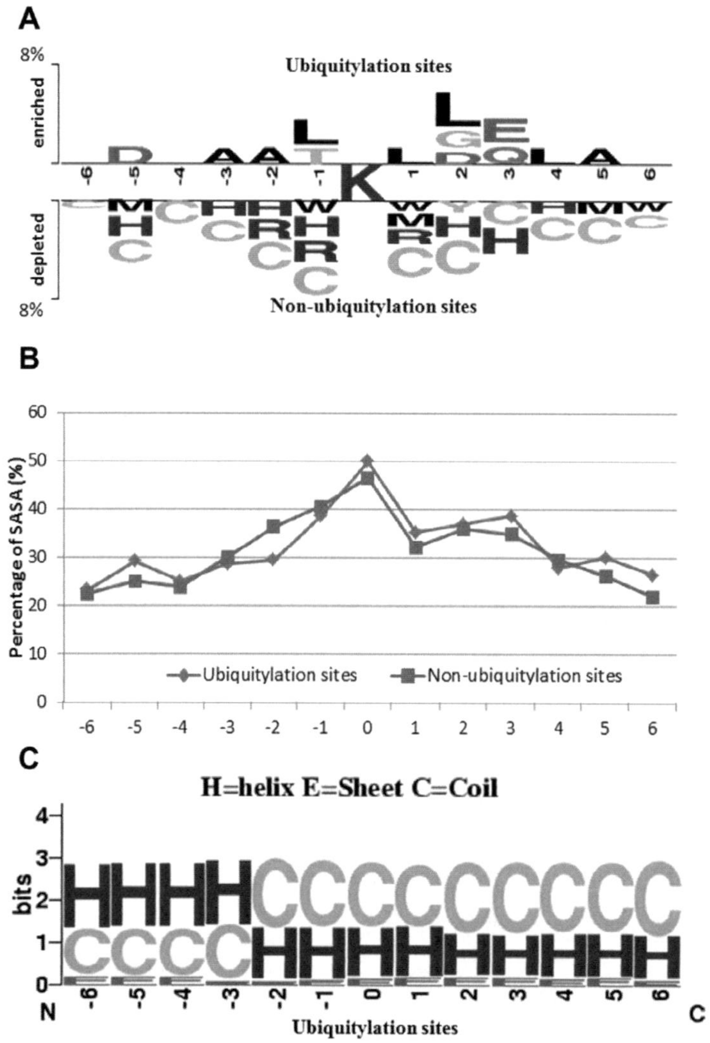

Fig. 2 Sequenced and structural characteristics of ubiquitin-conjugation sites. **a** Comparison of position-specific amino acid composition between ubiquitylation and non-ubiquitylation sites. **b** Comparison of solvent-accessible surface area between ubiquitylation and non-ubiquitylation sites. **c** Distribution of secondary structure around ubiquitylation sites

Table 2 Performance evaluation of the investigated features in identifying ubiquitylation sites based on five-fold cross-validation

Investigated features	Sensitivity	Specificity	Accuracy	MCC
20D binary coding (20D)	65.59 %	67.09 %	66.64 %	0.303
Amino Acid Composition (AAC)	64.34 %	65.44 %	65.11 %	0.275
Amino Acid Pair Composition (AAPC)	68.70 %	70.72 %	70.11 %	0.367
Position Weight Matrix (PWM)	68.08 %	67.99 %	68.01 %	0.334
Position-Specific Scoring Matrix (PSSM)	69.46 %	70.69 %	70.32 %	0.374
Solvent-Accessible Surface Area (SASA)	64.58 %	65.47 %	65.20 %	0.278
Secondary Structure (SS)	55.20 %	60.51 %	58.91 %	0.145

the highest MCC value at 0.0.374, and relatively high sensitivity, specificity and accuracy at 69.46 %, 70.69 % and 70.32 %, respectively. Based on the large-scale dataset, the SVM model, learned from AAPC with 441-dimensional vector, also performs as best as PSSM model. In the evaluation of structural characteristics, the SVM model trained with SASA yields an acceptable performance, including 64.58 % sensitivity, 65.47 % specificity, 65.20 % accuracy, and MCC value at 0.278. On the other hand, the secondary structure (SS) is found to be the worse feature for the prediction of ubiquitylation sites, with sensitivity at 55.20 %, specificity at 60.51 %, accuracy at 58.91 %, and MCC at 0.145. In addition, the best feature (PSSM) was considered to combine with other features for obtaining a better performance in identifying ubiquitylation sites. However, the hybrid features that incorporated PSSM with other single features provide similar prediction performance with using only PSSM feature. Overall, the SVM model learned from PSSM profiles could provide best predictive performance based on the large-scale ubiquitylation data. This investigation indicates that the PSSM profile, obtained by PSI-BLAST against non-redundant set of ubiquitylated protein sequences, could reflects the evolutionary conservation of amino acids which are prone to occur in the conserved domains for ubiquitin conjugation.

In order to assess the practicability of the constructed models, an independent testing dataset was generated from CPLM database. After the removal of homologous sequences with sequence identity ≤ 30 %, the independent testing dataset is composed of 3732 positive and 10,664 negative data. Table 3 reveals that the PSSM model could yield best MCC and accuracy at 0.369 and 69.68 %, respectively, in the independent testing dataset. Interestingly, the PWM model, reaching highest sensitivity at 73.90 %, has similar prediction performance with PSSM model; otherwise, the AAPC model has the best predictive specificity at 69.05 %. Overall, the SVM model learned from PSSM profiles outperforms all others based on the large-scale independent testing dataset.

Effectiveness of incorporating substrate motifs into the identification of ubiquitylation sites

In this investigation, the MDDLogo was adopted to explore the conserved motifs by dividing positive training dataset (5438 sites) into 12 subgroups. Each subgroup represents a potential ubiquitin-conjugated motif that contains statistically significant dependencies of amino acid composition between specific positions. Figure 3 provides a tree-like visualization of MDDLogo-clustered subgroups with statistically significant motifs for 5438 ubiquitylation sites. On the left subtree, six motifs (subgroups Ub1 to Ub6) out of all MDDLogo-clustered subgroups were detected based on the occurrence of acidic and amide amino acids (D, E, N and Q) at position +3, with maximal dependence value. At the same time, subgroups Ub1 (192 sites), Ub2 (266 sites) and Ub3 (248 sites) represented the occurrence of aromatic amino acids (F, Y and W) at positions +1, +2 and −1, respectively, with maximal dependence values. Subgroup Ub4 (378 sites) had another motif of acidic and amide amino acids at position −4. Additionally, subgroup Ub5 (467 sites) had the occurrence of polar amino acids at position +6, whereas subgroup Ub6 (792 sites) had no occurrence of polar amino acids at position +6. On the right subtree of Fig. 3, subgroups Ub7 (847 sites) and Ub8 (563 sites) had the motif of acidic and amide amino acids at positions −3 and −1, respectively. Subgroups Ub9 (672 sites), Ub10 (323 sites) and Ub11 (149 sites) represented the motif of aromatic amino acids at positions +1, +2 and −5, respectively, with maximal dependence values. Finally, the remaining 541 positive data resulted in the twelfth subgroup (Ub12), which contains a little conservation of amino acids at position +3.

In order to evaluate the effectiveness of MDDLogo-identified substrate motifs in identifying ubiquitylation sites, the LIBSVM was utilized to generate a predictive model for each subgroup based on PSSM feature. Table S2 (Additional file 6) presents the five-fold cross-validation performance for 12 MDDLogo-identified motifs obtained from 5438 non-homologous ubiquitylation sites. Overall, the subgroups containing motif of aromatic

Table 3 Performance evaluation of the SVM models trained with various features based on independent testing dataset (3732 ubiquitylation sites and 10,664 non-ubiquitylation sites)

Training features	Sensitivity	Specificity	Accuracy	MCC
20D binary coding (20D)	62.59 %	65.85 %	65.00 %	0.253
Amino Acid Composition (AAC)	66.37 %	64.63 %	65.08 %	0.274
Amino Acid Pair Composition (AAPC)	69.05 %	69.05 %	69.05 %	0.340
Position Weight Matrix (PWM)	73.90 %	67.29 %	69.01 %	0.364
Position-Specific Scoring Matrix (PSSM)	73.20 %	68.45 %	69.68 %	0.369
Solvent-Accessible Surface Area (SASA)	63.91 %	61.36 %	62.02 %	0.223
Secondary Structure (SS)	55.60 %	51.34 %	52.45 %	0.061

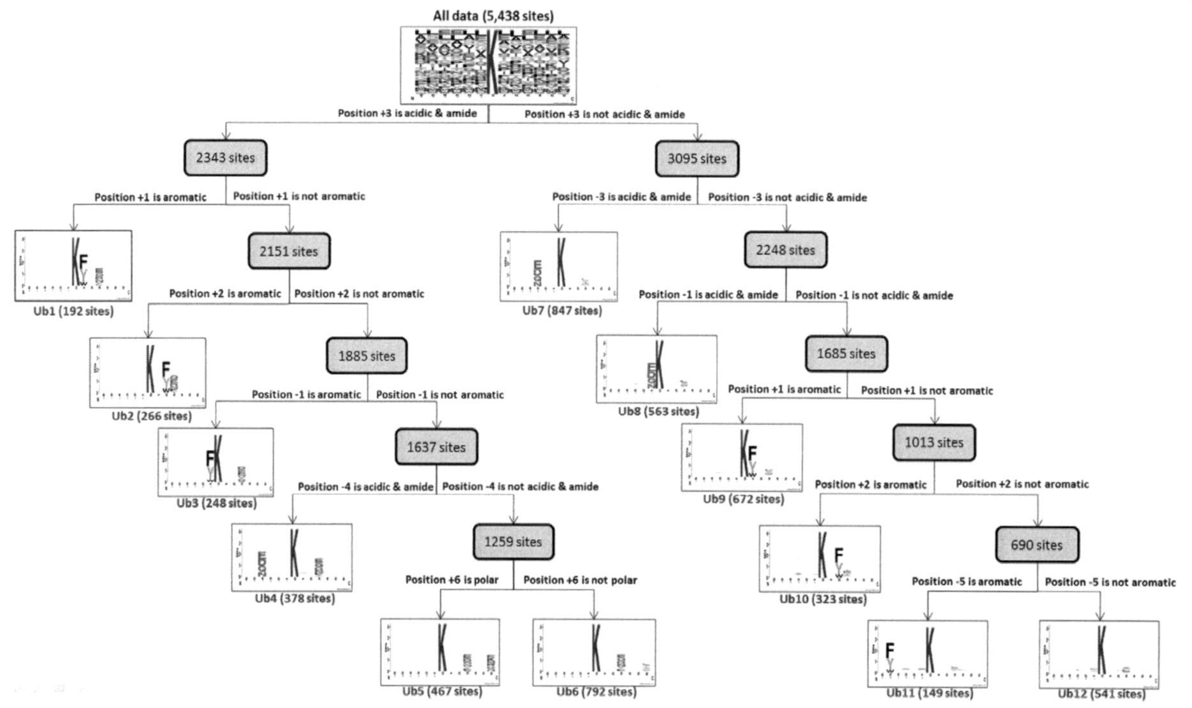

Fig. 3 Tree view of MDDLogo-clustered subgroups with statistically significant motifs for 5438 ubiquitylation sites

amino acids (F, Y and W) appeared to generate better predictive performances. For instance, the subgroup Ub2, containing the conserved motif of aromatic residues at position +2 as well as acidic and amide residues at position +3, yielded the highest MCC and accuracy at 0.881 and 94.70 %, respectively. Having similar substrate motifs with Ub2, subgroups Ub1 and Ub3 obtained predictive accuracies at 90.61 % and 89.70 %, respectively. Additionally, subgroups Ub9, Ub10 and Ub11, which contain the conserved motif of aromatic residues at specific positions, also obtained relatively high accuracies at 89.40, 91.16 and 86.04 %, respectively. In particular, subgroup Ub4, containing the conserved motif of acidic and amide residues at positions −4 and +3, yielded the second high accuracy at 92.84 %. The subgroups without clearly conserved motifs of aromatic residues generally showed relatively low accuracy (≤ 80.0 %). Subgroup Ub12, which has no conserved motif, yielded the lowest accuracy at 67.18 %.

In general, almost all subgroups, containing the conserved motif of aromatic amino acids at specific positions, could yield a promising accuracy. This investigation indicates that the substrate sites for ubiquitin conjugation may depend on the occurrence of aromatic residues. Table S2 (Additional file 6) shows that most of the SVM models trained with MDDLogo-identified motifs outperformed the SVM models trained with all training dataset. The mean accuracy of combining the 12 SVM models was 81.06 %, and the mean MCC was 0.586. Additionally, the independent testing dataset was used to examine the effectiveness of the two-layered SVM model, which integrates 12 MDDLogo-identified motifs, against that of single SVM model, based on the best feature (PSSM). As presented in Fig. 4, the testing results further support the effectiveness of the MDDLogo-identified motifs in the prediction of ubiquitylation sites. Moreover, the independent testing dataset was used to compare the two-layered SVM model with three available prediction tools, namely UbiPred [59], UbiProber [28] and hCKSAAP_UbiSite [22]. As shown in Table S3 (Additional file 7), the two-layered SVM model obtained highest performance when compared to others, reaching at 85.10 % sensitivity, 69.69 % specificity, 73.69 % accuracy and the 0.483 of MCC value.

Web resource for identifying ubiquitylated lysines with substrate motifs

Owing to the time-consuming and lab-intensive experimental identification of site-specific ubiquitinome, a biologist may only concluded that a protein can be ubiquitylated but the precise substrate sites for ubiquitin conjugation remains unknown. Hence, an effective prediction system can help focusing efficiently on potential substrate sites. After evaluation by cross-validation and independent testing, the two-layered SVM model incorporating 12 MDDLogo-identified substrate motifs has been adopted to develop a web resource, named UbiSite, for

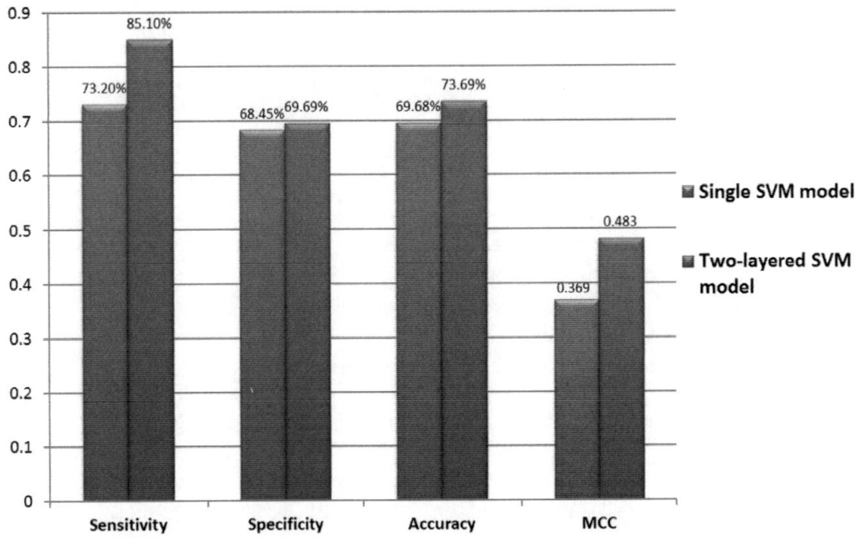

Fig. 4 Comparison of independent testing performance between single SVM model and two-layered SVM model

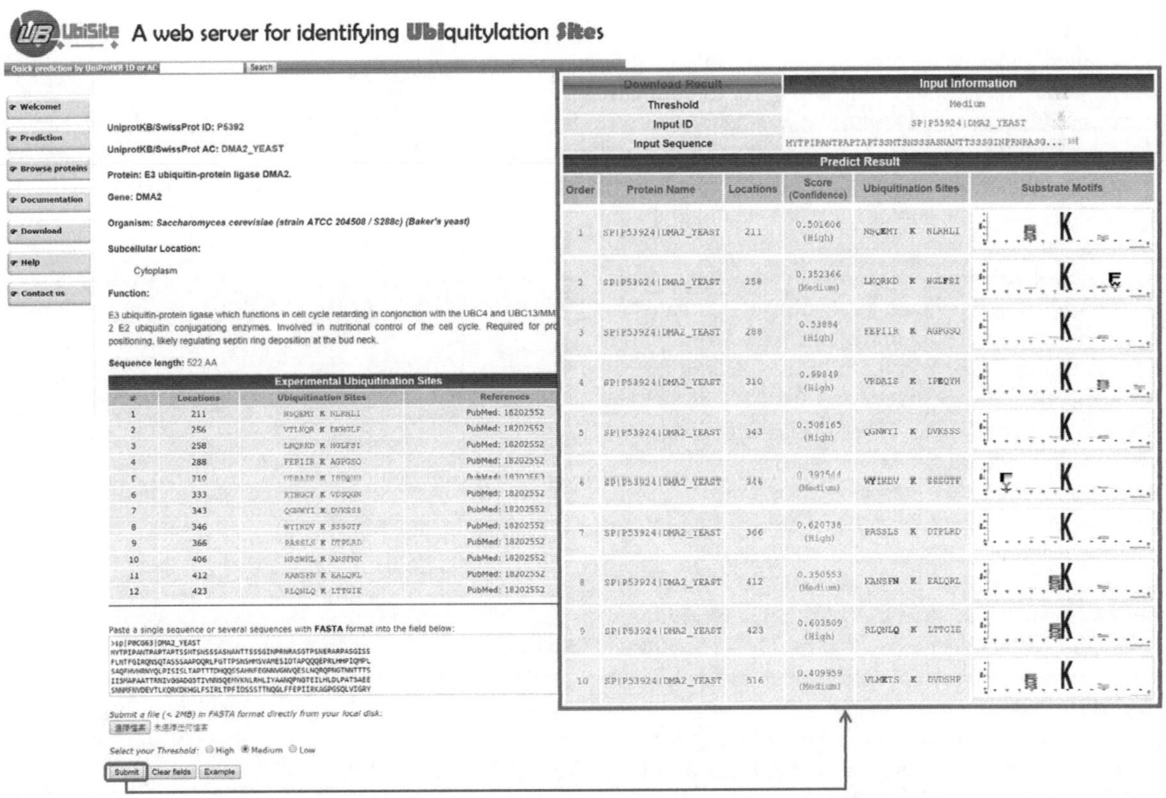

Fig. 5 Case study of identifying ubiquitylation sites on *E3 ubiquitin-protein ligase* DMA2 of *Saccharomyces cerevisiae*

identifying lysine ubiquitylation sites based on MDDLogo-identified substrate motifs. UbiSite allows users to submit their protein sequences in FASTA format; then, the system returns the prediction results, including ubiquitylated positions, flanking amino acids, and corresponding substrate motifs. A case study of *E3 ubiquitin-protein ligase* DMA2 from *Saccharomyces cerevisiae*, which is not included in the training dataset, was provided to demonstrate the effectiveness of UbiSite. As presented in Fig. 5, the DMA2 had 12 experimentally confirmed ubiquitylation sites located at positions 211, 256, 258, 288, 310, 333, 343, 346, 366, 406, 412 and 423 [60]. UbiSite predicted ten ubiquitylation sites for the protein sequence of DMA2, at positions 211, 258, 288, 310, 343, 346, 366, 412, 423 and 516, as well as their corresponding substrate motifs. The position 516 is a false positive prediction resulting in the predictive accuracy of 90.0 %. Another case study regards to the prediction on *tumor antigen p53* (TP53) in *Homo sapiens* (Human). TP53 has 11 ubiquitylation sites at positions 101, 120, 132, 164, 291, 292, 305, 320, 321, 357 and 370 [15, 61, 62]. As shown in Figure S5 (Additional file 8), UbiSite identified 11 ubiquitylation sites at positions 101, 120, 291, 292, 305, 319, 320, 321, 351, 357 and 382, with corresponding substrate motifs. Positions 319, 351 and 382 are false positive predictions, while positions 132, 164 and 370 are false negative predictions.

Conclusion

In proteomics, understanding the mechanisms that underlie ubiquitylation and ubiquitin binding specificity is both essential and challenging. The high throughput of mass spectrometry-based proteomics has provided an opportunity to explore the substrate motifs for ubiquitin conjugation, based on large-scale ubiquitinome dataset. Through Two-SampleLogo, the analysis of position-specific amino acids composition between ubiquitylation and non-ubiquitylation sites revealed the significant dependencies of flanking amino acids around the substrate sites. This investigation also found that the solvent-accessible surface area of amino acids surrounding ubiquitylation sites have a tendency to higher than that around non-ubiquitylation sites. Based on the results of five-fold cross-validation, the PSSM, which reflected the evolutionary conservation of amino acids in the domain of ubiquitin conjugation, was estimated as the best feature with the highest proportion of sensitivity, specificity, accuracy and MCC value. By using MDDLogo, all ubiquitylated sequences were clustered into 12 subgroups corresponding with 12 conserved motifs. The 12 identified motifs can thus be adopted to construct a two-layered SVM model to identify ubiquitylation sites based on large-scale ubiquitylation dataset. As expected, the two-layered SVM model, which combined 12 MDDLogo-identified substrate motifs, could yield the best performance in discriminating between ubiquitylation and non-ubiquitylation sites. Consequently, the two-layered SVM model was employed to set up a web-based resource, named UbiSite, to identify ubiquitylation sites and their corresponding substrate motifs.

Availability

The proposed method is implemented as a web-based resource, which is now freely available to all interested users at http://csb.cse.yzu.edu.tw/UbiSite/. All of the data set used in this work is also available for download in the website.

Additional files

Additional file 1: Figure S1. The HECT-type structure of E3 ligase. (DOCX 172 kb)

Additional file 2: Table S1. Data statistics of using CD-HIT with various parameters of sequence identity. (DOCX 15 kb)

Additional file 3: Figure S2. Detailed steps for obtaining the PSSM vector in this study. (DOCX 768 kb)

Additional file 4: Figure S3. Flowchart of MDDLogo analysis on ubiquitylated sequences. (DOCX 385 kb)

Additional file 5: Figure S4. Solvent-accessible surface area around ubiquitylated lysines based on protein tertiary structures. (DOCX 70 kb)

Additional file 6: Table S2. The five-fold cross-validation performance for 12 MDDLogo-identified motifs obtained from 5438 non-homologous ubiquitylation sites. (DOCX 93 kb)

Additional file 7: Table S3. Comparison of predictive performance between the presented models and other prediction tools based on independent testing dataset (3732 ubiquitylation sites and 10,664 non-ubiquitylation sites). (DOCX 16 kb)

Additional file 8: Figure S5. Case study of identifying ubiquitylation sites on *tumor antigen p53* (TP53) in *Homo sapiens* (Human). (DOCX 713 kb)

Competing interests

The authors have declared that no competing interests exist.

Authors' contributions

TYL and SLW conceived and supervised the project. CHH, MGS, HJK and JHJ were responsible for the design, computational analyses, and the implementation of web tool. CHH drafted the manuscript with revisions provided by SLW and TYL. All authors read and approved the final manuscript.

Acknowledgements

The authors sincerely appreciate the Ministry of Science and Technology (MOST) of Taiwan for financially supporting this research under contract number of MOST 103-2221-E-155-020-MY3, MOST 103-2633-E-155-002, and MOST 104-2221-E-155-036-MY2.

Declarations

Publication charge for this work was funded by MOST grant under contract number of MOST 103-2221-E-155-020-MY3 and MOST 104-2221-E-155-036-MY2 to TYL.

Author details

[1]Department of Computer Science and Engineering, Yuan Ze University, Taoyuan 320, Taiwan. [2]Ministry of Health & Welfare, Tao-Yuan Hospital, Taoyuan 320, Taiwan. [3]Department of Obstetrics and Gynecology, Hsinchu Mackay Memorial Hospital, Hsin-Chu 300, Taiwan. [4]Mackay Junior College of Medicine, Nursing and Management , Taipei 112, Taiwan. [5]Department of Medicine, Mackay Medical College, New Taipei City 252, Taiwan. [6]Innovation Center for Big Data and Digital Convergence, Yuan Ze University, Taoyuan 320, Taiwan.

Published: 11 January 2016

References

1. Goldstein G, Scheid M, Hammerling U, Schlesinger DH, Niall HD, Boyse EA. Isolation of a polypeptide that has lymphocyte-differentiating properties and is probably represented universally in living cells. Proc Natl Acad Sci U S A. 1975;72(1):11–5.
2. Wilkinson KD. The discovery of ubiquitin-dependent proteolysis. Proc Natl Acad Sci U S A. 2005;102(43):15280–2.
3. Pickart CM, Eddins MJ. Ubiquitin: structures, functions, mechanisms. Biochim Biophys Acta. 2004;1695(1–3):55–72.
4. Welchman RL, Gordon C, Mayer RJ. Ubiquitin and ubiquitin-like proteins as multifunctional signals. Nat Rev Mol Cell Biol. 2005;6(8):599–609.
5. Hurley JH, Lee S, Prag G. Ubiquitin-binding domains. Biochem J. 2006;399: 361–72.
6. Hicke L, Schubert HL, Hill CP. Ubiquitin-binding domains. Nat Rev Mol Cell Biol. 2005;6(8):610–21.
7. Peng JM, Schwartz D, Elias JE, Thoreen CC, Cheng DM, Marsischky G, et al. A proteomics approach to understanding protein ubiquitination. Nat Biotechnol. 2003;21(8):921–6.
8. Hershko A, Ciechanover A. The ubiquitin system. Annu Rev Biochem. 1998; 67:425–79.
9. Gao T, Liu Z, Wang Y, Cheng H, Yang Q, Guo A, et al. UUCD: a family-based database of ubiquitin and ubiquitin-like conjugation. Nucleic Acids Res. 2013;41(Database issue):D445–451.
10. Rotin D, Kumar S. Physiological functions of the HECT family of ubiquitin ligases. Nat Rev Mol Cell Biol. 2009;10(6):398–409.
11. Huang L, Kinnucan E, Wang GL, Beaudenon S, Howley PM, Huibregtse JM, et al. Structure of an E6AP-UbcH7 complex: Insights into ubiquitination by the E2-E3 enzyme cascade. Science. 1999;286(5443):1321–6.
12. Metzger MB, Hristova VA, Weissman AM. HECT and RING finger families of E3 ubiquitin ligases at a glance. J Cell Sci. 2012;125(3):531–7.
13. Lin DY, Diao J, Chen J. Crystal structures of two bacterial HECT-like E3 ligases in complex with a human E2 reveal atomic details of pathogen-host interactions. Proc Natl Acad Sci U S A. 2012;109(6):1925–30.
14. Nguyen VN, Huang KY, Huang CH, Chang TH, Bretana N, Lai K, et al. Characterization and identification of ubiquitin conjugation sites with E3 ligase recognition specificities. BMC Bioinformatics. 2015;16 Suppl 1:S1.
15. Wagner SA, Beli P, Weinert BT, Nielsen ML, Cox J, Mann M, et al. A proteome-wide, quantitative survey of in vivo ubiquitylation sites reveals widespread regulatory roles. Mol Cell Proteomics. 2011;10(10):M111 013284.
16. Lee TY, Chen SA, Hung HY, Ou YY. Incorporating distant sequence features and radial basis function networks to identify ubiquitin conjugation sites. PLoS One. 2011;6(3):e17331.
17. Liu Z, Wang Y, Gao T, Pan Z, Cheng H, Yang Q, et al. CPLM: a database of protein lysine modifications. Nucleic Acids Res. 2014;42(Database issue): D531–536.
18. Chen T, Zhou T, He B, Yu HY, Guo XJ, Song XF, et al. mUbiSiDa: a comprehensive database for protein ubiquitination sites in mammals. PLoS One. 2014;9(1):e85744.
19. Lu CT, Huang KY, Su MG, Lee TY, Bretana NA, Chang WC, et al. dbPTM 3.0: an informative resource for investigating substrate site specificity and functional association of protein post-translational modifications. Nucleic Acids Res. 2013;41(D1):D295–305.
20. Lee TY, Huang HD, Hung JH, Huang HY, Yang YS, Wang TH. dbPTM: an information repository of protein post-translational modification. Nucleic Acids Res. 2006;34(Database issue):D622–627.
21. Chen Z, Zhou Y, Zhang Z, Song J. Towards more accurate prediction of ubiquitination sites: a comprehensive review of current methods, tools and features. Brief Bioinform. 2014;16(4):640–57.
22. Chen Z, Zhou Y, Song JN, Zhang ZD. hCKSAAP_UbSite: improved prediction of human ubiquitination sites by exploiting amino acid pattern and properties. Biochimica et Biophysica Acta-Proteins Proteom. 2013;1834(8):1461–7.
23. Cai YD, Huang T, Hu LL, Shi XH, Xie L, Li YX. Prediction of lysine ubiquitination with mRMR feature selection and analysis. Amino Acids. 2012; 42(4):1387–95.
24. Chen Z, Chen YZ, Wang XF, Wang C, Yan RX, Zhang ZD. Prediction of ubiquitination sites by using the composition of k-spaced amino acid pairs. PLoS One. 2011;6(7):e22930.
25. Radivojac P, Vacic V, Haynes C, Cocklin RR, Mohan A, Heyen JW, et al. Identification, analysis, and prediction of protein ubiquitination sites. Proteins. 2010;78(2):365–80.
26. Zhao XW, Li XT, Ma ZQ, Yin MH. Prediction of lysine ubiquitylation with ensemble classifier and feature selection. Int J Mol Sci. 2011;12(12):8347–61.
27. Feng KY, Huang T, Feng KR, Liu XJ. Using WPNNA classifier in ubiquitination site prediction based on hybrid features. Protein Peptide Lett. 2013;20(3): 318–23.
28. Chen X, Qiu JD, Shi SP, Suo SB, Huang SY, Liang RP. Incorporating key position and amino acid residue features to identify general and species-specific Ubiquitin conjugation sites. Bioinformatics. 2013;29(13):1614–22.
29. Lee TY, Lin ZQ, Hsieh SJ, Bretana NA, Lu CT. Exploiting maximal dependence decomposition to identify conserved motifs from a group of aligned signal sequences. Bioinformatics. 2011;27(13):1780–7.
30. Huang Y, Niu BF, Gao Y, Fu LM, Li WZ. CD-HIT Suite: a web server for clustering and comparing biological sequences. Bioinformatics. 2010;26(5): 680–2.
31. Chang WC, Lee TY, Shien DM, Hsu JB, Horng JT, Hsu PC, et al. Incorporating support vector machine for identifying protein tyrosine sulfation sites. J Comput Chem. 2009;30(15):2526–37.
32. Hsu JB, Bretana NA, Lee TY, Huang HD. Incorporating evolutionary information and functional domains for identifying RNA splicing factors in humans. PLoS One. 2011;6(11):e27567.
33. Xie D, Li A, Wang M, Fan Z, Feng H. LOCSVMPSI: a web server for subcellular localization of eukaryotic proteins using SVM and profile of PSI-BLAST. Nucleic Acids Res. 2005;33(Web Server issue):W105–110.
34. Jones DT. Protein secondary structure prediction based on position-specific scoring matrices. J Mol Biol. 1999;292(2):195–202.
35. Altschul SF, Madden TL, Schaffer AA, Zhang J, Zhang Z, Miller W, et al. Gapped BLAST and PSI-BLAST: a new generation of protein database search programs. Nucleic Acids Res. 1997;25(17):3389–402.
36. Pang CN, Hayen A, Wilkins MR. Surface accessibility of protein post-translational modifications. J Proteome Res. 2007;6(5):1833–45.
37. Ahmad S, Gromiha MM, Sarai A. RVP-net: online prediction of real valued accessible surface area of proteins from single sequences. Bioinformatics. 2003;19(14):1849–51.
38. Ahmad S, Gromiha MM, Sarai A. Real value prediction of solvent accessibility from amino acid sequence. Proteins. 2003;50(4):629–35.
39. Bryson K, McGuffin LJ, Marsden RL, Ward JJ, Sodhi JS, Jones DT. Protein structure prediction servers at University College London. Nucleic Acids Res. 2005;33(Web Server issue):W36–38.
40. Chang C-C, Lin C-J. LIBSVM : a library for support vector machines. ACM Transactions on Intelligent Systems and Technology. 2011;2(27):1–27.
41. Lu CT, Chen SA, Bretana NA, Cheng TH, Lee TY. Carboxylator: incorporating solvent-accessible surface area for identifying protein carboxylation sites. J Comput Aided Mol Des. 2011;25(10):987–95.
42. Matthews BW. Comparison of the predicted and observed secondary structure of T4 phage lysozyme. Biochim Biophys Acta. 1975;405(2):442–51.
43. Crooks GE, Hon G, Chandonia JM, Brenner SE. WebLogo: a sequence logo generator. Genome Res. 2004;14(6):1188–90.
44. Lee TY, Chen YJ, Lu CT, Ching WC, Teng YC, Huang HD, et al. dbSNO: a database of cysteine S-nitrosylation. Bioinformatics. 2012;28(17):2293–5.
45. Bretana NA, Lu CT, Chiang CY, Su MG, Huang KY, Lee TY, et al. Identifying protein phosphorylation sites with kinase substrate specificity on human viruses. PLoS One. 2012;7(7):+.
46. Lee TY, Chen YJ, Lu TC, Huang HD. SNOSite: exploiting maximal dependence decomposition to identify cysteine S-nitrosylation with substrate site specificity. PLoS One. 2011;6(7):+.
47. Lee TY, Bretana NA, Lu CT. PlantPhos: using maximal dependence decomposition to identify plant phosphorylation sites with substrate site specificity. BMC Bioinformatics. 2011;12:261.
48. Wong YH, Lee TY, Liang HK, Huang CM, Wang TY, Yang YH, et al. KinasePhos 2.0: a web server for identifying protein kinase-specific phosphorylation sites based on sequences and coupling patterns. Nucleic Acids Res. 2007;35(Web Server issue):W588–594.
49. Huang HD, Lee TY, Tzeng SW, Wu LC, Horng JT, Tsou AP, et al. Incorporating hidden Markov models for identifying protein kinase-specific phosphorylation sites. J Comput Chem. 2005;26(10):1032–41.
50. Huang HD, Lee TY, Tzeng SW, Horng JT. KinasePhos: a web tool for identifying protein kinase-specific phosphorylation sites. Nucleic Acids Res. 2005;33(Web Server issue):W226–229.

51. Chen YJ, Lu CT, Su MG, Huang KY, Ching WC, Yang HH, et al. dbSNO 2.0: a resource for exploring structural environment, functional and disease association and regulatory network of protein S-nitrosylation. Nucleic Acids Res. 2015;43(Database issue):D503–511.
52. Chen YJ, Lu CT, Huang KY, Wu HY, Chen YJ, Lee TY. GSHSite: exploiting an iteratively statistical method to identify s-glutathionylation sites with substrate specificity. PLoS One. 2015;10(4):e0118752.
53. Wu HY, Lu CT, Kao HJ, Chen YJ, Chen YJ, Lee TY. Characterization and identification of protein O-GlcNAcylation sites with substrate specificity. BMC Bioinformatics. 2014;15 Suppl 16:S1.
54. Lu CT, Lee TY, Chen YJ, Chen YJ. An intelligent system for identifying acetylated lysine on histones and nonhistone proteins. Biomed Res Int. 2014;2014:528650.
55. Huang KY, Lu CT, Bretana N, Lee TY, Chang TH. ViralPhos: incorporating a recursively statistical method to predict phosphorylation sites on virus proteins. BMC Bioinformatics. 2013;14 Suppl 16:S10.
56. Vacic V, Iakoucheva LM, Radivojac P. Two sample logo: a graphical representation of the differences between two sets of sequence alignments. Bioinformatics. 2006;22(12):1536–7.
57. Berman HM, Westbrook J, Feng Z, Gilliland G, Bhat TN, Weissig H, et al. The Protein Data Bank. Nucleic Acids Res. 2000;28(1):235–42.
58. Catic A, Collins C, Church GM, Ploegh HL. Preferred in vivo ubiquitination sites. Bioinformatics. 2004;20(18):3302–7.
59. Tung CW, Ho SY. Computational identification of ubiquitylation sites from protein sequences. BMC Bioinformatics. 2008;9:310.
60. Loring GL, Christensen KC, Gerber SA, Brenner C. Yeast Chfr homologs retard cell cycle at G1 and G2/M via Ubc4 and Ubc13/Mms2-dependent ubiquitination. Cell Cycle. 2008;7(1):96–105.
61. Oshikawa K, Matsumoto M, Oyamada K, Nakayama KI. Proteome-wide identification of ubiquitylation sites by conjugation of engineered lysine-less ubiquitin. J Proteome Res. 2012;11(2):796–807.
62. Kim W, Bennett EJ, Huttlin EL, Guo A, Li J, Possemato A, et al. Systematic and quantitative assessment of the ubiquitin-modified proteome. Mol Cell. 2011;44(2):325–40.

Zhang *et al. BMC Systems Biology* 2016, **10**(Suppl 1):4
DOI 10.1186/s12918-015-0247-y

BMC Systems Biology

PROCEEDINGS Open Access

Inference of domain-disease associations from domain-protein, protein-disease and disease-disease relationships

Wangshu Zhang[1], Marcelo P. Coba[2,3] and Fengzhu Sun[1,4*]

From The Fourteenth Asia Pacific Bioinformatics Conference (APBC 2016)
San Francisco, CA, USA. 11 - 13 January 2016

Abstract

Background: Protein domains can be viewed as portable units of biological function that defines the functional properties of proteins. Therefore, if a protein is associated with a disease, protein domains might also be associated and define disease endophenotypes. However, knowledge about such domain-disease relationships is rarely available. Thus, identification of domains associated with human diseases would greatly improve our understandingof the mechanism of human complex diseases and further improve the prevention, diagnosis and treatment of these diseases.

Methods: Based on phenotypic similarities among diseases, we first group diseases into overlapping modules. We then develop a framework to infer associations between domains and diseases through known relationships between diseases and modules, domains and proteins, as well as proteins and disease modules. Different methods including Association, Maximum likelihood estimation (MLE), Domain-disease pair exclusion analysis (DPEA), Bayesian, and Parsimonious explanation (PE) approaches are developed to predict domain-disease associations.

Results: We demonstrate the effectiveness of all the five approaches via a series of validation experiments, and show the robustness of the MLE, Bayesian and PE approaches to the involved parameters. We also study the effects of disease modularization in inferring novel domain-disease associations. Through validation, the AUC (Area Under the operating characteristic Curve) scores for Bayesian, MLE, DPEA, PE, and Association approaches are 0.86, 0.84, 0.83, 0.83 and 0.79, respectively, indicating the usefulness of these approaches for predicting domain-disease relationships. Finally, we choose the Bayesian approach to infer domains associated with two common diseases, Crohn's disease and type 2 diabetes.

Conclusions: The Bayesian approach has the best performance for the inference of domain-disease relationships. The predicted landscape between domains and diseases provides a more detailed view about the disease mechanisms.

Background

Uncovering the mechanisms underlying human complex diseases is one of the central goals of human disease studies. Recent developments in human genetics and computational biology made it possible to identify a number of genes that are associated with complex diseases [1]. For example, recent genome-wide association studies have detected more than 2000 genetic loci associated with human complex diseases or traits [2, 3]. Most of the identified loci, however, represent novel discoveries with no obvious candidate genes and molecular mechanisms [4], rendering difficulties in medical treatment according to genes [5–7]. Even if particular disease associated genes are identified [8–12], narrowing down to particular domains can be challenging because genes may encode for proteins containing a variety of domains. Protein domains are structural units of proteins that can also function independently from other regions of the protein. If a gene product (protein) contains multiple domains [13] and the gene is associated with a disease, one

* Correspondence: fsun@usc.edu
[1]Molecular and Computational Biology Program, University of Southern California, 1050 Childs Way, Los Angeles, USA
[4]Centre for Computational Systems Biology, School of Mathematical Sciences, Fudan University, Shanghai, China
Full list of author information is available at the end of the article

of the domains might be associated with the disease. Narrowing down domains associated with complex diseases will greatly improve our understanding about the pathogenesis of the diseases and facilitate the discovery of drugs as well as personalized medicine.

Several pioneering studies have developed methods for large-scale inference of associations between domains and human diseases based on domain-domain interactions and disease phenotype similarities [14, 15]. These studies have two drawbacks. Firstly, both studies rely on a relatively small set of domain-disease associations compiled by bridging domains that contain known deleterious nsSNPs and human diseases with these nsSNPs [15]. To circumvent this problem we seek evidences of domain-disease associations at the gene level, and instead of considering inadequate number of disease mutations in the domains, we resort to highly abundant publicly available gene-disease associations [16–20]. Secondly, these studies depend on domain-domain interactions that are generally incomplete and contain many false positive and false negative domain interactions [14, 15]. In this study, we use the domain-protein, protein-disease and disease-disease relationships to infer domain-disease relationships without using domain interactions. The basic idea is that if a disease is associated with many genes with their corresponding products (proteins) containing common domains, the common domains are more likely to be associated with the disease.

We *surprisingly* noted that inferring domain-disease relationships based on domain-protein and protein-disease relationship is closely related to the problem of inferring domain-domain interactions based on protein-protein interactions, a problem that have been studied extensively over the past decade [21–32]. Therefore, we adopted some of the promising methods for protein domain interactions based on protein interactions to the inference of domain-disease relationships. These methods include the simple Association method, the Maximum likelihood estimation (MLE) approach studied in Deng et al. [33], Domain pair exclusion analysis (DPEA) approach proposed by Riley et al.[25], a Bayesian version of the MLE approach as developed by Kim et al. [34], and Parsimonious explanation (PE) approach proposed by Guimaraes et al. [26].

Since a particular disease/trait generally has a relatively small number of associated genes and inferring the domains related to the disease based on the small number of proteins can be unreliable, therefore we group single diseases into modules. Each module comprises several disease phenotypes that are highly similar to each other, and different modules may share diseases. This modularization process takes into consideration of the comorbidities of various diseases [35, 36], and overlaps among modules are allowed indicating that one disease may belong to more than one module since for complex disorders, different genetic mutations may lead to the same clinical outcome [37, 38]. For example, schizophrenia is a group of heritable disorders presenting distinct clinical syndromes, and schizophrenia has been shown to be associated with eight separate networks of genetic mutations [20]. Dubowitz syndrome is also a complex disease comprised of multiple, genetically distinct but phenotypically overlapping disorders [39]. As a result, we are able to define a larger number of proteins associated with a particular disease module, and the data insufficiency problem could be solved. In practice we group the diseases based on phenotype similarities obtained from a recent work of Jiang et al. [40] and then identify potentially overlapping disease modules. Then we identify domains associated with each disease contained in these modules.

We develop a framework to infer associations between domains and diseases through known relationships between diseases and modules, domains and proteins, as well as proteins and disease modules. We demonstrate the validity and robustness of these approaches, compare their performance, and predict domain-disease as well as gene-disease associations. We further illustrate the consistency between our inference results and the evidences from genome-wide association studies for two common diseases: Crohn's disease and type 2 diabetes.

Materials and methods

Data sources

The developed scheme to infer domain-disease relationships depends on the relationships between domains and proteins, proteins and diseases, and phenotypic similarities among diseases.

The relationships between domains and human proteins were obtained from the Pfam database, which provides a large collection of both high quality protein domain families (Pfam-A) and low quality protein domain families (Pfam-B) [41]. In version 27.0 of the Pfam-A collection (released in March 2013), there were 146,442 associations between 5561 domains and 100,977 human proteins. A domain is referred as associated with a protein if the protein contains the domain. The relationships between genes and human diseases were extracted from the OMIM database [16], from which 4951 associations between 3313 genes and 4151 diseases were established. Moreover, from a recent work of Jiang et al. [40], we obtained a pair-wise phenotypic similarity profile for 7719 diseases. The author applied text mining techniques to extract feature vectors from three vocabularies, UMLS, MeSH, and HPO, respectively, and a cosine value between two feature vectors for a pair of diseases was calculated to measure their similarity. Here we only selected the phenotype similarities obtained using the UMLS terms, since it has a full coverage of all 7719 disease phenotypes and also achieves the best performance in the leave-one-out cross-validation experiments of measuring the strength of association between a candidate gene and a query disease [40].

Identification of overlapping disease modules

Given the disease phenotypic similarity profile and a predefined threshold on disease similarity, we construct a disease network with diseases as nodes and two diseases are linked if their pair-wise similarity is above this threshold. Considering computational feasibility, in this paper we choose a threshold of 0.5, and obtain 39,506 linked pairs among 6033 diseases with similarity above this threshold, based on which we construct our disease network.

Then, we use the ClusterONE (Clustering with Overlapping Neighborhood Expansion) method [42] to detect overlapping modules in the disease network. This method was originally proposed to detect potentially overlapping protein complexes from protein-protein interaction data. Here we apply the algorithm to detect overlapping disease modules, with all default parameter values, and retrieve modules containing at least 5 diseases and having p-values below 0.05, a measure to evaluate the statistical significance of the module.

As a result, we obtain altogether 255 modules, with 3430 diseases included. We say a gene is associated with a module if the gene is associated with at least one of the diseases in the module. We only include modules containing at least one disease having known associated genes. Similarly, we only include genes associated with at least one disease module. A gene is referred as associated with a disease module if it is associated with at least one disease in the module. Finally, we obtain 2096 associations between 1106 domains and 1238 proteins, 4096 associations between 3430 diseases and 255 modules, and 1789 associations between 1238 proteins and 255 modules. Relationships among domains, proteins, diseases and modules are shown in Fig. 1.

As mentioned above we construct our disease network by choosing a threshold of 0.5. Here we also try other thresholds from 0.1 to 0.9 with a step of 0.1 to identify disease modules, and list the numbers of domains, proteins, diseases and modules obtained using the same method, in the Additional file 1: Table S1. From the result we see that when the threshold is selected as 0.5, we obtain the highest number of 255 modules, and when the threshold is selected as 0.4, the numbers of collected domains, diseases and proteins are the largest (1145 domains, 4090 diseases, and 1298 proteins). We also see that as the threshold increases from 0.5 to 0.9, or decreases from 0.4 to 0.1, the numbers of domains, diseases, proteins and modules are all decreasing, but drop relatively slightly when the threshold is between 0.3 and 0.6 (the numbers of domains, diseases, proteins, and modules are larger than 800, 2300, 900, 50, respectively). Since if a threshold larger than 0.6 is selected, there would be fewer diseases that have pair-wise similarities above this threshold, while if a threshold smaller than 0.3 is selected, there would be less disease modules being identified for the increasing insignificance of disease modules due to the excessive connections in the disease network. Therefore, we set the threshold at 0.5 to keep a reasonable number of domains, diseases, proteins and modules.

Methods to predict domain-disease associations

We develop a scheme to infer domain-disease associations from known relationships between domain-proteins, disease-modules and protein-modules as demonstrated in Fig. 2. As shown in Fig. 2, for a given disease/trait T_n, we first extract all module(s) containing this disease (in the figure M_j) and all the other diseases contained in M_j. Then we consider the modules that share at least one disease

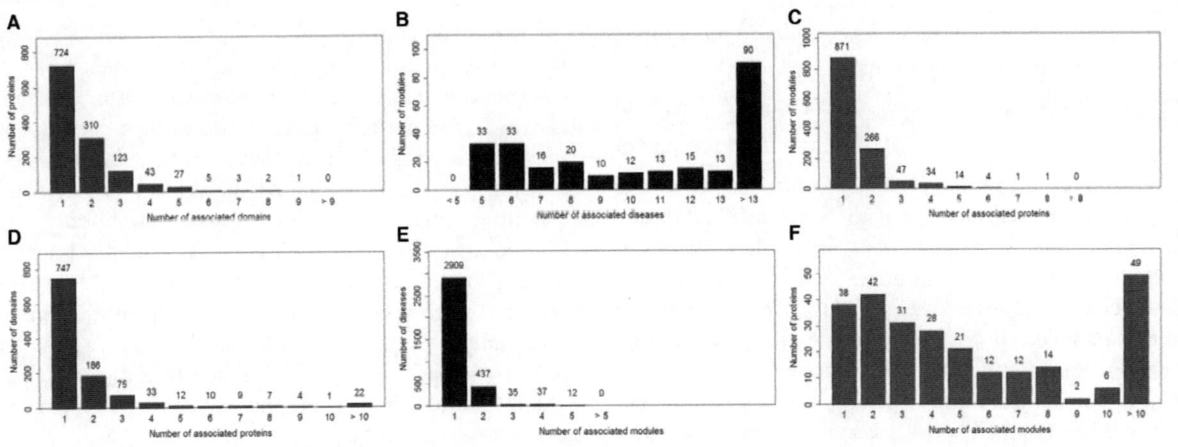

Fig. 1 The relationships between the different data types. The histograms of the number of **a** proteins with respect to the number of domains the protein contains, **b** disease modules with respect to the number of diseases the module contains, **c** disease modules with respect to the number of proteins the module associates, **d** domains with respect to the number of proteins the domain associates, **e** diseases with respect to the number of disease modules the disease associates, and **f** proteins with respect to the number of disease modules the protein associates

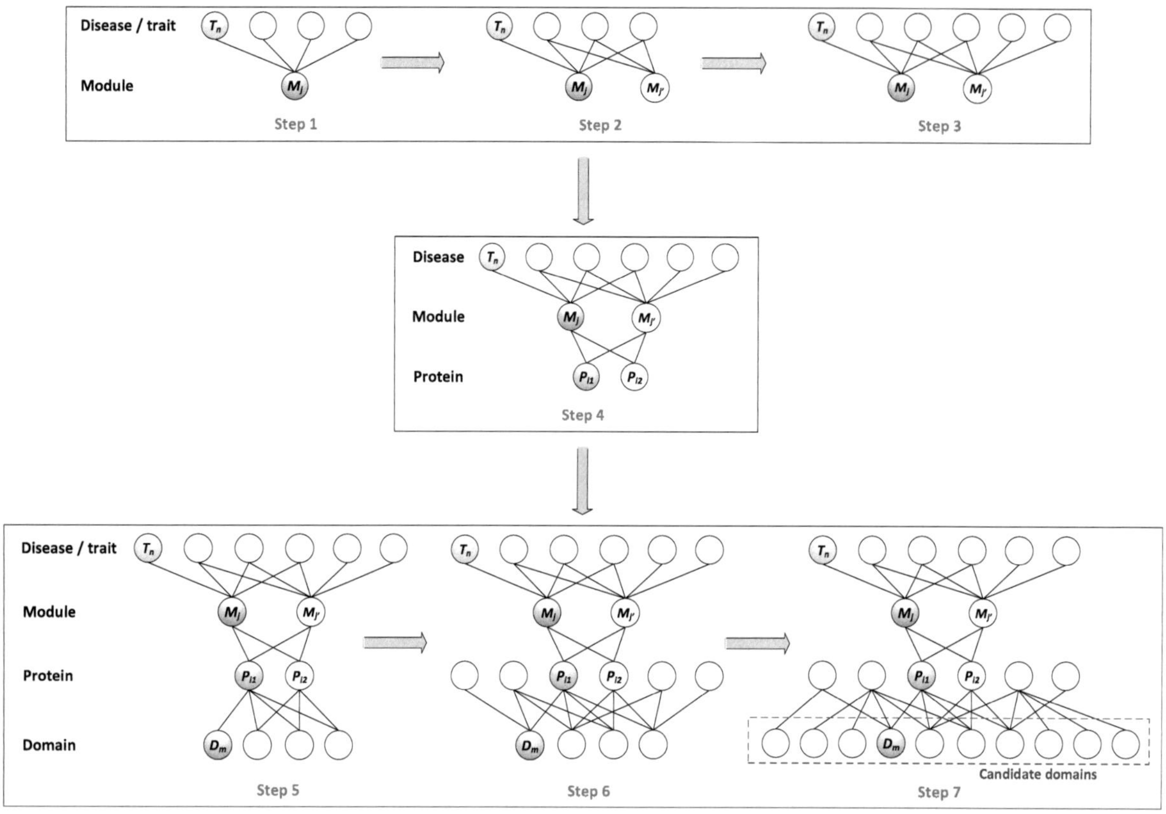

Fig. 2 Scheme for predicting domain-disease relationships. Nodes represent diseases/traits, modules, proteins and domains. An edge connecting two nodes represents a known association. Steps 1-7 demonstrate the procedure that, when predicting for a specific disease, how to obtain its candidate domains. Step 1: For a given disease T_n, all module(s) containing this disease (in the figure M_j) and all the other diseases/traits contained in M_j are extracted. Step 2: Module(s) sharing at least one disease with module M_j (in the figure $M_{j'}$) are extracted. Step 3: All the other diseases/traits in $M_{j'}$ are included in the prediction scheme. Step 4: All proteins associated with the set of $\left\{M_j, , M_{j'}\right\}$ (in the figure P_{i1} and P_{i2}) are extracted. Step 5: All domains contained in the set of $\{P_{i1}, P_{i2}\}$ are included in the prediction scheme. Step 6: All proteins sharing domains with proteins $\{P_{i1}, P_{i2}\}$ are included in the prediction scheme. Step 7: All the other domains in all proteins produced at Step 6 are included in the prediction scheme and the resulting set of domains are called candidate domains

with module M_j (in the figure $M_{j'}$), and also incorporate other diseases in $M_{j'}$. Next, from the protein-module associations we collect all proteins associated with the set of $\left\{M_j, , M_{j'}\right\}$ (in the figure P_{i1} and P_{i2}), and further find domains contained in these proteins. Similarly, we consider all proteins sharing domains with proteins $\{P_{i1}, P_{i2}\}$, and incorporate their domains as well. The resulting set of domains are called candidate domains. Finally, we predict the associations between each of the collected candidate domains and the disease/trait T_n. For different diseases, the numbers of candidate domains can vary, and only 247,112 domain-disease relationships will be predicted. For each disease, the number of associated domains is small and thus the standard machine learning based approach cannot be used to predict domain-disease relationships. For clarity of presentation, we summarize all the notations used in the five approaches in Additional file 2.

The association approach

We first propose a naive Association approach to rank the candidate domain-disease relationship. For a candidate domain $D_m$and a query disease/trait T_n, their potential of being associated is measured by the fraction of associated protein-module pairs among all the protein-module pairs containing this domain-disease pair. Specifically, let $A_{mn} = \{(P, M);\ D_m \in P,\ T_n \in M,\ P \text{ is associated with } M\}$, where P indicates proteins and M indicates disease modules, be the set of associated protein-module pairs containing the domain-disease pair (D_m, T_n). Similarly, let $N_{mn} = \{(P, M); D_m \in P, T_n \in M\}$, be the complete set of protein-module pairs containing (D_m, T_n). The association score is given by

$$Score(D_m, T_n) = \frac{|A_{mn}|}{|N_{mn}|}.$$

If the domain is not included in any proteins that are known to be associated with the disease module containing the disease, the corresponding score is defined as zero [33]. With this definition, we calculate the scores of all candidate domains for each disease.

The maximum likelihood estimation (MLE) approach

Deng et al. [33] developed a Maximum Likelihood Estimation (MLE) approach to predict domain-domain interactions from protein-protein interactions. Here we extend the approach to infer domain-disease associations, with parameters of this model estimated in terms of the Expectation Maximization (EM) algorithm [33]. Let D denote domains, T denote diseases/traits, P denote proteins and M denote disease modules. Let $\phi_{mn} = 1$ denote that domain D_m associates with disease T_n and $\phi_{mn} = 0$, otherwise. Similarly, let $\psi_{ij} = 1$ denote protein P_i associates with module M_j and $\psi_{ij} = 0$, otherwise.

It is assumed that a protein associates with a module if and only if at least one of domains inside this protein associates with one of diseases belonging to the module. Under this assumption, we have

$$\Pr\left(\psi_{ij} = 1\right) = 1 - \prod_{(D_m,T_n)\in(P_i,M_j)} (1-\lambda_{mn})$$

where $\lambda_{mn} = \Pr(\phi_{mn} = 1)$ denotes the probability that domain D_m associates with disease/trait T_n, and we thus define the association score as

$$Score(D_m, T_n) = \lambda_{mn}.$$

If we consider about two types of errors, false positive rate fp and false negative rate fn, and let O_{ij} be the indicator variable denoting if protein P_i and module M_j are observed to be associated, then

$$fp = \Pr\left(O_{ij} = 1|\psi_{ij} = 0\right)$$

$$fn = \Pr\left(O_{ij} = 0|\psi_{ij} = 1\right).$$

Thus, the probability for the observed protein-module association is

$\Pr(O_{ij} = 1) = \Pr(\psi_{ij} = 1)(1 - fn) + (1 - \Pr(\psi_{ij} = 1))fp$, and the likelihood function for all the observed protein-module relationships is:

$$L = \prod_{ij} \left(\Pr(O_{ij} = 1)\right)^{O_{ij}} \left(1-\Pr(O_{ij} = 1)\right)^{1-O_{ij}}$$

The likelihood L is a function of $\theta = \{\lambda, fp, fn\}$. We then apply an EM algorithm to estimate θ, as was developed by Deng et al. [33] for the inference of domain-domain interactions. The details of implementing the EM algorithm is described in Additional file 2. We initialize all $\lambda_{mn}^{\text{init}} = 1/|D^{(n)}|$, where $|D^{(n)}|$ denotes the total number of candidate domains for disease/trait n, meaning that each domain has an equal probability to be associated with a specific disease. Also, we perform a grid search for different combinations of $\{fp, fn\}$, and find results are quite robust when fp is close to 0 while fn is close to 1. Therefore we use $\{fp, fn\} = \{0, 0.9\}$ throughout this paper.

The domain-disease pair exclusion analysis (DPEA) approach

Riley et al. [25] developed the DPEA (Domain Pair Exclusion Analysis) approach for predicting domain interactions from protein interactions. Here we extend the approach to predict domain-disease associations based on protein-module relationships. This approach extends the MLE approach, by first estimating the probability for each potential domain-disease association through the MLE approach, and then measuring association strength in terms of the changes in likelihood ratios when a potential underlying domain-disease pair is assumed to be not associated [25].

The DPEA approach does not consider the false positive and negative rates inherent in protein-module associations, which means, $\{fp, fn\} = \{0, 0\}$. However, it identifies low-probable but high-specificity domain-disease associations that are not detected by the MLE approach.

The likelihood L is a function of $\theta = \lambda$, and we apply an EM algorithm to estimate θ. Then we use the log-likelihood ratio without restricting λ_{mn} versus that when $\lambda_{mn} = 0$ to measure the strength of association between the domain D_m and trait T_n. The score is thus defined as

$$Score(D_m, T_n) = \sum_{ij} \log \frac{\Pr\left(O_{ij} = 1|D_m, T_n \text{ can associate}\right)}{\Pr\left(O_{ij} = 1|D_m, T_n \text{ do not associate}\right)}$$

$$= \sum_{ij} \log \frac{1 - \prod_{(D_k,T_l)\in(P_i,M_j)} (1-\lambda_{kl})}{1 - \prod_{(D_k,T_l)\in(P_i,M_j)} (1-\bar{\lambda}_{kl}^{mn})}$$

where $\bar{\lambda}^{mn}$ is obtained from λ by setting the probability of domain D_m associated with disease T_n to be zero, and is also estimated by the EM algorithm. The details of deducing the score function of the DPEA approach is described in Additional file 2.

The Bayesian approach

Kim et al. [34] extended the MLE approach to a Bayesian approach for the prediction of domain-domain interactions based on protein interactions. We adopt the Bayesian approach for the prediction of domain-disease associations based on protein-module associations. We assume a uniform prior distributions of false positive rate fp and false negative rate fn: $fp \sim \text{Unif}[u_p, v_p]$ and $fn \sim \text{Unif}[u_n, v_n]$, and also assume the domain-disease association probability λ_{mn}

has a Beta prior distribution: $\lambda_{mn} \sim \text{Beta}(\alpha, \beta)$ as in Kim et al. [34]. Therefore, the posterior distribution of λ_{mn} is proportional to

$$
\begin{aligned}
[\lambda|rest] &\propto L(O|fn,fp,\lambda)f(\lambda|fn,fp) \\
&\propto \prod_{ij} (\Pr(O_{ij}=1))^{O_{ij}}(1-\Pr(O_{ij}=1))^{1-O_{ij}} \times f(\lambda|fn,fp) \\
&\propto \prod_{ij} [h_{ij}(\lambda)(1-fn)+\{1-h_{ij}(\lambda)\}fp]^{O_{ij}}[1-h_{ij}(\lambda)(1-fn) \\
&\qquad -\{1-h_{ij}(\lambda)\}fp]^{1-O_{ij}} \\
&\qquad \times f(\lambda|fn,fp)
\end{aligned}
$$

Here $\lambda = \{\lambda_{mn}; D_m \in P,\ T_n \in M\}$, where P indicates proteins and M indicates disease modules, and $h_{ij}(\lambda) = \Pr(\psi_{ij} = 1)$.

In addition, the posterior distributions of fp and fn are proportional to

$$
\begin{aligned}
[fp|rest] &\propto L(O|fn,fp,\lambda)f(fp|\lambda,fn)f(fn|\lambda) \\
&\propto \prod_{ij} (\Pr(O_{ij}=1))^{O_{ij}}(1-\Pr(O_{ij}=1))^{1-O_{ij}} \times f(fp|\lambda,fn)f(fn|\lambda) \\
&\propto \prod_{ij} (\Pr(O_{ij}=1))^{O_{ij}}(1-\Pr(O_{ij}=1))^{1-O_{ij}} \times f(fp|\lambda,fn) \\
&\propto \prod_{ij} [h_{ij}(\lambda)(1-fn)+\{1-h_{ij}(\lambda)\}fp]^{O_{ij}}[1-h_{ij}(\lambda)(1-fn) \\
&\qquad -\{1-h_{ij}(\lambda)\}fp]^{1-O_{ij}} \\
&\times f(fp|\lambda,fn)
\end{aligned}
$$

and

$$
\begin{aligned}
[fn|rest] &\propto L(O|fn,fp,\lambda)f(fn|\lambda,fp)f(fp|\lambda) \\
&\propto \prod_{ij} (\Pr(O_{ij}=1))^{O_{ij}}(1-\Pr(O_{ij}=1))^{1-O_{ij}} \times f(fn|\lambda,fp)f(fp|\lambda) \\
&\propto \prod_{ij} (\Pr(O_{ij}=1))^{O_{ij}}(1-\Pr(O_{ij}=1))^{1-O_{ij}} \times f(fn|\lambda,fp) \\
&\propto \prod_{ij} [h_{ij}(\lambda)(1-fn)+\{1-h_{ij}(\lambda)\}fp]^{O_{ij}}[1-h_{ij}(\lambda)(1-fn) \\
&\qquad -\{1-h_{ij}(\lambda)\}fp]^{1-O_{ij}} \\
&\times f(fn|\lambda,fp)
\end{aligned}
$$

An adaptive rejection sampling algorithm [43] is applied to sample from the posteriors, and parameters can be estimated using the posterior means.

The parsimony explanation (PE) approach

Guimaraes et al.[26] originally developed a PE approach for predicting domain interactions based on protein interactions. Here we extend the PE approach for predicting domain-disease associations from protein-module associations. The PE approach formulates the problem of predicting domain-disease associations into a linear programming framework. The optimization objective is to minimize the number of overall domain-disease associations necessary to justify each of the underlying protein-module associations. We define a set of domain-disease pairs $\Delta = \{\{D_m, T_n\} | D_m \in P,\ T_n \in M$, and P associates with $M\}$, where P indicates proteins and M indicates disease modules, underlying all associated protein-module pairs. Therefore, domain D_m and disease T_n are assumed to have an association variable x_{mn} if and only if there exists an associated protein-module pair P and M containing domain D_m and disease T_n, respectively. We propose to solve the following linear programming (LP) problem:

Minimize $\sum_{\{D_m, T_n\} \in \Delta} x_{mn}$,

Subject to: $\sum_{(D_m, T_n) \in (P_i, M_j)} x_{mn} \geq 1$, for every associated protein-module pair (P_i, M_j) corresponding to one linear constraint. The resulting value x_{mn} is then used to measure the strength of potential domain-disease association between domain D_m and disease T_n. To account for the noise in the protein-module associations, we define r as the reliability rate, which is, the probability that a protein-module association actually exists. For a pre-selected r, we include each protein-module association into the constraints with probability r, and perform the linear programming 1000 times. The average x_{mn} is defined as the LP-score and used to measure the strength of association between domain D_m and disease T_n.

To control for possible over-prediction of associations between frequently occurring domain-disease pairs, we assign a promiscuity versus witnesses (pw)-score to each domain-disease pair. Specifically, a pw-score between domain D_m and disease T_n is defined as

$$\text{pw-score}(D_m, T_n) = \min\left(p\text{-value}(D_m, T_n), (1-r)^{w(D_m, T_n)}\right)$$

where $w(D_m, T_n)$ is the number of occurrences (witnesses) for a given domain-disease pair (D_m, T_n) in each associated protein-module pair (P_i, M_j), and r is the reliability rate of the protein-module association. Thus, the value of $(1-r)^{w(D_m, T_n)}$ is the probability that all witnesses of domain-disease pairs in (P_i, M_j) are false positives. We then define $p\text{-value}(D_m, T_n)$ to measure the influence of the frequently occurring domain-disease pairs on the LP-score. To estimate this, we generate 1000 random protein-module associations by preserving the same protein-domain and module-disease compositions as well as the total number of protein-module associations, but selecting associations between proteins and modules at random. We perform the linear programming 1000 times, and the $p\text{-value}(D_m, T_n)$ is calculated as the frequency of obtaining the same or higher LP-score in the 1000 runs when the protein-module pair containing the domain-disease pair (D_m, T_n) is randomized.

Potential associations with LP-scores above a threshold and pw-scores below another threshold are predicted to be putative associations between domains and diseases.

Validation methods and evaluation criteria

We compile an independent validation set to test how well the Association, MLE, DPEA, Bayesian, and PE approaches perform in recovering known associations between domains and diseases obtained from the Ensembl database (Human Release 80, accessed in June 2015) [44]. By running the BioMart tool [45] within Ensembl, we collect 3004 associations between 869 domains and 1484 diseases in our study. We then consider these 3004 associations as positives, and all the other 244,108 possible domain-disease associations as negatives (247,112 candidate domain-disease associations minus 3004 known associations). For each of the approaches developed above, we rank the domain-disease pairs together in descending order of their scores. Given a predefined threshold of rank, we define the *sensitivity* (or called *recall*) as the percentage of positive domain-disease pairs that are ranked above the threshold, the *specificity* as the percentage of negative domain-disease pairs that are ranked below the threshold, and the *precision* as the percentage of domain-disease pairs ranked above the threshold that are truly positives. By varying the threshold values, we are able to plot a ROC (*R*eceiver *O*perating *C*haracteristic) curve, which demonstrates the relationships between *sensitivity* and 1-*specificity*, as shown in Fig. 3 (a), and a Precision-Recall curve, which demonstrates the relationships between *precision* and *recall*, as shown in Fig. 3 (b).

We use three criteria to measure the performance of each approach. The first criterion is the *Area Under the ROC curve (AUC)*, which provides an overall measure for the performance of the prioritization approach. The second criterion is the *Accuracy*, which is the fraction of positive domain-disease pairs with the corresponding domain being ranked at the top 10 of the disease. Since the average length of candidate domain lists for all 3430 diseases is 72 (247,112 candidate domain-disease associations divided by 3430), therefore by a random guess the probability to rank a known domain-disease association at the top 10 is $10/72 \approx 0.1389$. Obviously, a high value of accuracy that is much larger than 0.1389 represents an approach with high prediction power. The third criterion is the *Mean Rank Ratio (MRR)*. For each positive domain-disease pair, the rank ratio is calculated as the rank of corresponding domain for the disease divided by the total number of candidate domains, and the mean rank ratio is the average of rank ratios for all the positive associations in the validation experiment. This criterion provides a summary of the ranks of all positive domain-disease associations, and the smaller the mean rank ratio is, the better the approach performs.

Results

Validation and comparison of different approaches

On the basis of the compiled known associations between domains and diseases that comprise an independent test data set, we implement a series of large-scale validation experiments to demonstrate the effectiveness of all the approaches developed in this paper. As is described in the "Materials and Methods" section, in each run of the experiments, we prioritize candidate domains according to the scores generated by a selected approach under investigation, with performances being evaluated by AUC, accuracy and mean rank ratio of known domain-disease associations, and list the results in Table 1. Corresponding ROC curves together with the Precision-recall curves of the Association, MLE, DPEA, Bayesian, and PE approaches are shown in Fig. 3.

From the results in Table 1 we see that, all the approaches perform reasonably well recovering the known associations between domains and diseases. The AUC scores obtained using the five approaches are all above 0.79, the accuracies are all above 0.62, and the mean rank ratios are all below 0.25, suggesting the effectiveness of the approaches. The best performance is achieved using the Bayesian approach, with an AUC score of 0.8554, accuracy

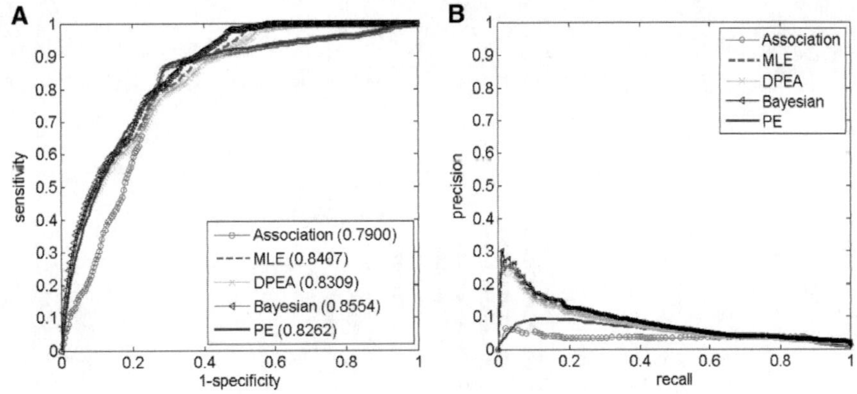

Fig. 3 Receiver Operating Characteristic (ROC) and Precision-Recall curves of the different approaches. The figure shows ROC curves (Subplot **a**) and precision-recall curves (Subplot **b**) of the Association, MLE ($fp = 0, fn = 0.9$), DPEA, Bayesian ($u_p, u_n = 0$, $v_p, v_n = 1$, and $\alpha = 2, \beta = 2$), and PE ($r = 100\,\%$, and pw threshold ≤ 0.01) approaches, respectively. Based on both ROC and precision-recall curves, the three MLE based approaches including DPEA, MLE and Bayesian outperform PE and Association. The Bayesian approach performs slightly better than DPEA and MLE

Table 1 Performance of the five approaches

Approach	AUC	Accuracy	Mean rank ratio
Association	0.7900	0.6256	0.2432
MLE	0.8407	0.7074	0.1914
DPEA	0.8309	0.6513	0.2177
Bayesian	0.8554	0.7289	0.1872
PE	0.8262	0.6525	0.2282

The AUC, accuracy and mean rank ratio of the Association, MLE ($fp = 0, fn = 0.9$), DPEA, Bayesian ($u_p, u_n = 0$, $v_p, v_n = 1$, and $\alpha = 2, \beta = 2$), and PE ($r = 100\,\%$, and pw threshold ≤ 0.01) approaches for predicting domain-disease associations, respectively

of 0.7289 and a mean rank ratio of 0.1872. The MLE approach is a little inferior to the Bayesian approach whose AUC score is 0.8407, accuracy is 0.7074, and the mean rank ratio is 0.1914. The DPEA and PE approaches are two less effective ones with the performances being quite close to each other (AUCs are 0.8309 and 0.8262, accuracies are 0.6513 and 0.6525, and mean rank ratios are 0.2177 and 0.2282), while the Association approach performs worst with a AUC score of 0.7900, accuracy of 0.6256, and a mean rank ratio of 0.2432. From Fig. 3, we observe that the ROC curve of the Bayesian approach is above those of the other approaches, suggesting that the performance of Bayesian approach is superior over that of the others. From the precision-recall curves we also see the same trend. Therefore, we see from the results that the predictive power of the five approaches follows an order of: Bayesian > MLE > DPEA > PE > Association.

Robustness of the approaches

We notice that some of these approaches contain free parameters such as the false positive rate fp and the false negative rate fn in the MLE approach, the hyper-parameters in prior distributions of the Bayesian approach, in addition to the reliability rate r and the pw-score threshold in the PE approach. In the above validation experiments, we only use predefined values of these parameters for simplicity, yet there is still necessity to show whether these approaches are sensitive to these parameters. Hence, for each approach we select several values across the range of corresponding parameters, perform the validation experiments, and see how the results change accordingly.

Specifically, for the MLE approach we implement a grid search of fp and fn at the same time, increasing from 0 to 1 with a step of 0.1, respectively. Performances in terms of AUC, accuracy, and mean rank ratio are shown in Fig. 4 (a-c) and Additional file 1: Table S2. From the figure we can see when both fp and fn are relatively large (the shading areas in Additional file 1: Table S2), the AUC is lower than 0.72, the accuracy is lower than 0.48, and the mean rank ratio is higher than 0.35. Otherwise as fp increases, performances become worse. For example, when $fn = 0.2$, the peak performance is obtained at $fp = 0$ (AUC = 0.8313, accuracy = 0.6935, and mean rank ratio = 0.2128), and the worst performance is obtained at $fp = 0.9$ (AUC = 0.7944, accuracy = 0.6069, and mean rank ratio = 0.2815). The best performance is obtained at $fp = 0$ combined with $fn = 0.9$ (AUC = 0.8407, accuracy = 0.7074, and mean rank ratio = 0.1914). From the results we conjecture that better performance is

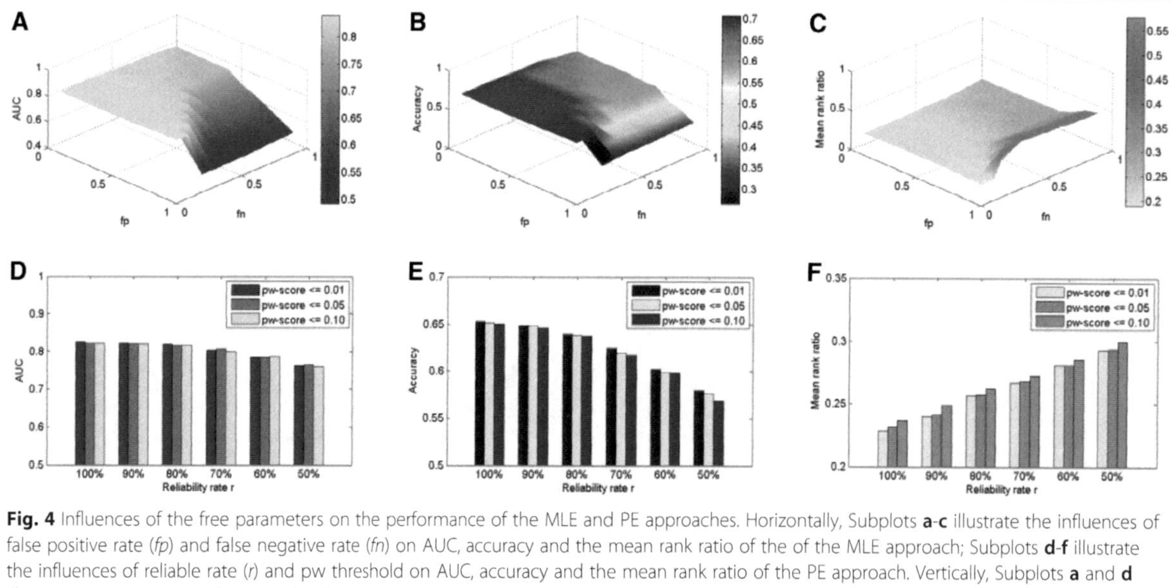

Fig. 4 Influences of the free parameters on the performance of the MLE and PE approaches. Horizontally, Subplots **a-c** illustrate the influences of false positive rate (*fp*) and false negative rate (*fn*) on AUC, accuracy and the mean rank ratio of the of the MLE approach; Subplots **d-f** illustrate the influences of reliable rate (*r*) and pw threshold on AUC, accuracy and the mean rank ratio of the PE approach. Vertically, Subplots **a** and **d** illustrate AUC scores; Subplots B and E illustrate accuracies; Subplots C and F illustrate mean rank ratios, respectively

usually obtained when fp is small and fn is large. This is reasonable since according to the experience of Deng et al. [33], fp usually takes a very small value while fn usually takes a large value close to 1. In the further analysis we just use $fp = 0$ and $fn = 0.9$ so as to acquire a higher predictive power.

Additionally, for the Bayesian approach, we have assigned a beta prior distribution for λ, and to eliminate the effect of shape hyper-parameters on the results, we choose to use a 0-1 uniform distribution instead. After re-run of the validation experiments, we obtain AUC = 0.8553, accuracy = 0.7289, and mean rank ratio = 0.1874, which brings tiny change to the inference results. Hence we conjecture that the Bayesian approach is quite robust when an appropriate flat prior distribution is used, and compared to the likelihood functions, priors are not factors playing decisive roles here.

Finally, for the PE approach we decrease the reliability rate r from 100 to 50 % with a step of 10 % while fixing the pw-score threshold as 0.01, 0.05, and 0.10, respectively, with the changes of AUC, accuracies and mean rank ratios shown in Fig. 4 (d-f) and Additional file 1: Table S3. First, as the pw-score threshold increases, performances in terms of the three criteria become worse. For example when r is 100 %, the performance obtained at pw threshold = 0.01 is the best (AUC = 0.8262, accuracy = 0.6525, and mean rank ratio = 0.2282), followed by the performance obtained at pw threshold = 0.05 (AUC = 0.8238, accuracy = 0.6509, and mean rank ratio = 0.2314), while the worst performance is obtained at pw threshold = 0.10 (AUC = 0.8235, accuracy = 0.6499, and mean rank ratio = 0.2365). The same trend can also be observed for other values of r. Second, as r decreases, performances become worse since both AUC and accuracy increase, and mean rank ratio decreases. For example, when the pw-score threshold is 0.01, the peak performance is obtained at $r = 100$ % (AUC = 0.8262, accuracy = 0.6525, and mean rank ratio = 0.2282), and the worst performance is obtained at $r = 50$ % (AUC = 0.7635, accuracy = 0.5793, and mean rank ratio = 0.2929). The change of AUC is not as obvious as the other two criteria, and all the values for the three criteria are generally stable. We then conjecture that the PE approach is not sensitive to this free parameter on the basis that pw threshold is smaller than 0.10 and r is greater than 50 %. In the further analysis we use pw threshold = 0.01 and $r = 100$ % to achieve maximal number of new predictions according to Guimaraes et al. [26].

Effects of disease modularization on the performance of different approaches

Beyond the parameters in the approaches, our proposed inference scheme also highly relies on the formulation of disease modules. To demonstrate it, we select the disease RENAL TUBULAR ACIDOSIS DISTAL AUTOSOMAL DOMINANT (OMIM: 179800) as an example. Using the scheme illustrated in Fig. 2, we extract all relevant relationships among modules, diseases, proteins and domains, as shown in Fig. 5, to infer the strength of associations among all domains and diseases shown in the figure. Each disease and its known associated gene (protein) are marked with the same color.

By clustering single disease phenotypes into modules in a disease network, known associations between proteins and diseases are actually expanded. Considering a disease that has no known associated proteins, if this disease belongs to a module whose other member diseases have known associated proteins, then as a result, relations between these proteins and the disease are actually bridged, the hidden effects of which would greatly benefit us to uncover more novel associated domains or genes for the disease. For example, the disease RENAL TUBULAR ACIDOSIS PROXIMAL (OMIM: 179830) does not have any known associated genes in the OMIM database. However, since it belongs to the same module as the disease RENAL TUBULAR ACIDOSIS DISTAL AUTOSOMAL DOMINANT (OMIM: 179800), we are able to predict its associated domains. We further map the predicted domains to all genes with corresponding proteins containing these domains, and therefore we are able to predict the genes associated with the disease.

Moreover, we compare the disease RENAL TUBULAR ACIDOSIS DISTAL AUTOSOMAL DOMINANT (OMIM: 179800) with another disease OSTEOPETROSIS, AUTOSOMAL RECESSIVE 3 (OPTB3; OMIM: 259730), since the former disease is only contained in module #127 and the latter one is contained in both module #127 and module #161. The results in terms of all five approaches are listed in Table 2. First, we observe that the Association and PE approaches generate zero scores for a domain-disease association if the domain is not contained in any proteins that are associated with the modules containing the disease, while the other three approaches usually provide such association with a non-zero score. This explains why the Association and the PE approach have relatively lower predictive power compared to the other three approaches. Second, if we only focus on disease RENAL TUBULAR ACIDOSIS DISTAL AUTOSOMAL DOMINANT (OMIM: 179800), we observe that comparatively the domains ATP-synt_ab (PF00006), ATP-synt_ab_N (PF02874), Carb_anhydrase (PF00194), HCO3_cotransp (PF00955), Band_3_cyto (PF07565) and ATP-synt_ab_C (PF00306) have more indirect associations with the disease, compared to the domains TNF (PF00229), IMPDH (PF00478), CBS (PF00571), Voltage_CLC (PF00654), and V_ATPase_I (PF01496), and therefore should have comparatively higher scores. This is generally the case when we see from the left half part of Table 2, except

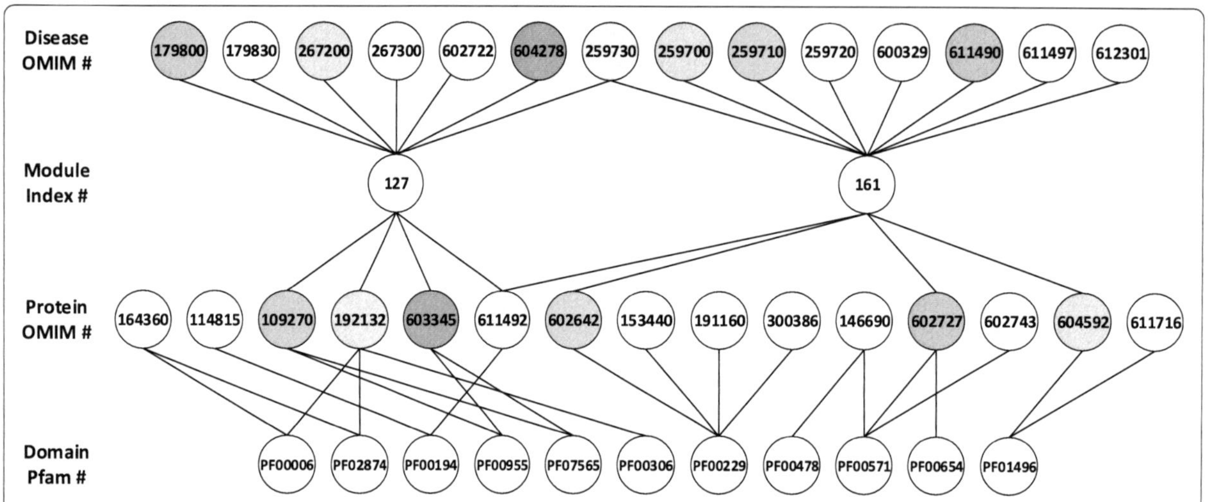

Fig. 5 Example for illustration of module effect. Nodes represent diseases with OMIM numbers, modules with index numbers, proteins with OMIM numbers and domains with Pfam numbers. Edges connecting two nodes represent a known association. Nodes with the same background colors represent 7 known associations between corresponding diseases and proteins. (i) Disease OMIM # corresponds to disease/trait names as: [179800]: RENAL TUBULAR ACIDOSIS, DISTAL, AUTOSOMAL DOMINANT. [179830]: RENAL TUBULAR ACIDOSIS, PROXIMAL. [267200]: RENAL TUBULAR ACIDOSIS III. [267300]: RENAL TUBULAR ACIDOSIS, DISTAL, WITH PROGRESSIVE NERVE DEAFNESS. [602722]: RENAL TUBULAR ACIDOSIS, DISTAL, AUTOSOMAL RECESSIVE; RTADR. [604278]: RENAL TUBULAR ACIDOSIS, PROXIMAL, WITH OCULAR ABNORMALITIES AND MENTAL RETARDATION. [259730]: OSTEOPETROSIS, AUTOSOMAL RECESSIVE 3; OPTB3. [259700]: OSTEOPETROSIS, AUTOSOMAL RECESSIVE 1; OPTB1. [259710]: OSTEOPETROSIS, AUTOSOMAL RECESSIVE 2; OPTB2. [259720]: OSTEOPETROSIS, AUTOSOMAL RECESSIVE 5; OPTB5. [600329]: OSTEOPETROSIS AND INFANTILE NEUROAXONAL DYSTROPHY. [611490]: OSTEOPETROSIS, AUTOSOMAL RECESSIVE 4; OPTB4. [611497]: OSTEOPETROSIS, AUTOSOMAL RECESSIVE 6; OPTB6. [612301]: OSTEOPETROSIS, AUTOSOMAL RECESSIVE 7; OPTB7. (ii) Protein OMIM # corresponds to gene name as: [164360]: ATP5A1; [114815]: CA8; [109270]: SLC4A1; [192132]: ATP6V1B1; [603345]: SLC4A4; [611492]: CA2; [602642]: TNFSF11; [153440]: LTA; [191160]: TNF; [300386]: CD40LG; [146690]: IMPDH1; [602727]: CLCN7; [602743]: PRKAG2; [604592]: TCIRG1; [611716]: ATP6V0A2. (iii) Domain Pfam # corresponds to domain name as: [PF00006]: ATP-synt_ab; [PF02874]: ATP-synt_ab_N; [PF00194]: Carb_anhydrase; [PF00955]: HCO3_cotransp; [PF07565]: Band_3_cyto; [PF00306]: ATP-synt_ab_C; [PF00229]: TNF; [PF00478]: IMPDH; [PF00571]: CBS; [PF00654]: Voltage_CLC; [PF01496]: V_ATPase_I

Table 2 Effect of disease modularization on the performances of different approaches

		Disease (OMIM ID)									
		RENAL TUBULAR ACIDOSIS, DISTAL, AUTOSOMAL DOMINANT (OMIM: 179800)					OSTEOPETROSIS, AUTOSOMAL RECESSIVE 3; OPTB3 (OMIM: 259730)				
		Association	MLE	DPEA	Bayesian	PE	Association	MLE	DPEA	Bayesian	PE
Domain(Pfam ID)	ATP-synt_ab(PF00006)	0.5	0.0971	0.0041	0.6256	0.0476	0.25	0.0930	0.0024	0.6084	0.0476
	ATP-synt_ab_N(PF02874)	0.5	0.0971	0.0041	0.6843	0.0476	0.25	0.0930	0.0024	0.6366	0.0476
	Carb_anhydrase(PF00194)	0.5	0.1566	0.062	0.8031	0.1646	0.5	0.1499	0.0541	0.7741	0.1210
	HCO3_cotransp(PF00955)	1	0.1300	0.0065	0.7569	0.0714	0.5	0.1086	0.0003	0.6438	0.0714
	Band_3_cyto(PF07565)	1	0.1300	0.0065	0.7078	0.0714	0.5	0.1086	0.0003	0.6859	0.0714
	ATP-synt_ab_C(PF00306)	1	0.1069	0.0002	0.6879	0.0476	0.5	0.0983	0.0001	0.6048	0.0476
	TNF(PF00229)	0	0.0806	0.1900	0.4526	0	0.125	0.0968	0.1906	0.5360	0.1250
	IMPDH(PF00478)	0	0.0874	0.007	0.4572	0	0	0.0878	0.0059	0.4301	0
	CBS(PF00571)	0	0.0851	0.0676	0.4381	0	0.1667	0.0910	0.0646	0.6035	0.06250
	Voltage_CLC(PF00654)	0	0.0874	0.0058	0.4547	0	0.5	0.1040	0.0061	0.6864	0.06250
	V_ATPase_I(PF01496)	0	0.0808	0.0823	0.4801	0	0.25	0.1103	0.0823	0.7057	0.1250

The scores of predicting the associations between two diseases RENAL TUBULAR ACIDOSIS, DISTAL, AUTOSOMAL DOMINANT (OMIM: 179800) as well as OSTEOPETROSIS, AUTOSOMAL RECESSIVE 3; OPTB3 (OMIM: 259730)), and their candidate domains ATP-synt_ab (PF00006), ATP-synt_ab_N (PF02874), Carb_anhydrase (PF00194), HCO3_cotransp (PF00955), Band_3_cyto (PF07565), ATP-synt_ab_C (PF00306), TNF (PF00229), IMPDH (PF00478), CBS (PF00571), Voltage_CLC (PF00654) as well as V_ATPase_I (PF01496), in terms of the Association, MLE ($fp = 0, fn = 0.9$), DPEA, Bayesian ($u_p, u_n = 0$, $v_p, v_n = 1$, and $\alpha = 2, \beta = 2$), and PE ($r = 100$ %, and pw threshold ≤ 0.01) approaches, respectively for the results of DPEA method, that due to the too many connection with proteins, the domain PF00229 has the highest score. Third, if focus on domains ATP-synt_ab (PF00006), ATP-synt_ab_N (PF02874), Carb_anhydrase (PF00194), HCO3_cotransp (PF00955), Band_3_-cyto (PF07565) and ATP-synt_ab_C (PF00306), we observe that their scores obtained for disease RENAL TUBULAR ACIDOSIS DISTAL AUTOSOMAL DOMINANT (OMIM: 179800) are generally higher than those obtained for disease OSTEOPETROSIS, AUTOSOMAL RECESSIVE 3 (OPTB3; OMIM: 259730). Since the disease RENAL TUBULAR ACIDOSIS DISTAL AUTOSOMAL DOMINANT (OMIM: 179800) is only contained in one module while disease OSTEOPETROSIS, AUTOSOMAL RECESSIVE 3 (OPTB3; OMIM: 259730) is contained in two, we conjecture from the results that, for a domain under investigation, if a disease is only contained in one module associated with the proteins containing this domain while another disease is also contained in other modules that are not associated with the corresponding proteins, usually the first disease would receive higher score to be associated with this domain since its relationship with the domain is more "specific".

Top predictions of domain-disease and gene-disease associations

We list the 50 highest-scoring domain-disease associations predicted using the Bayesian approach, as shown in Table 3. We then map the domains predicted for these diseases to all genes with corresponding proteins containing these domains, and mark the genes that are known as disease genes in the OMIM database using a bold italic style. Among these predictions, 8 are in the above 3004 known domain-disease associations compiled by the Ensembl BioMart tool and we mark them using a bold style in Table 3 (diseases with ranks 6, 7, 9, 19, 20, 25, 42, and 50, respectively). Four of these diseases, CORNEAL DYSTROPHY EPITHELIAL BASEMENT MEMBRANE (EBMD; OMIM: 121820), CORNEAL DYSTROPHY GROENOUW TYPE I (CDGG1; OMIM: 121900), CORNEAL DYSTROPHY LATTICE TYPE I (LCD1; OMIM: 122200) and LUBS X-LINKED MENTAL RETARDATION SYNDROME (MRXSL; OMIM: 300260) have their predicted domains exactly mapped to their only known genes in the OMIM database, while the other 4 diseases have extra predictions of genes that are unknown to be associated with these diseases before. Some of these predictions might be redundant, such as for the disease ALPHA-THALASSEMIA (OMIM: 604131), the gene HBB, which is short for HEMOGLOBIN–BETA LOCUS, is obviously a cause of another type of thalassemia called BETA-THALASSEMIA (OMIM: 613985) but not ALPHA-THALASSEMIA [46]. However, for diseases that without any known genes associated, the predictions might indicate novel genetic findings for the diseases. As listed in the table, all the rows without bold or italic correspond to diseases that have no known OMIM genes at all. But by our methods we can predict for each of them a highest scored domain and a short list of genes, which are highly possible to be the real causal ones since the top ranked diseases are mostly rare diseases and the majority of them are caused by altered functions of single genes [47].

Finally, there are still 7 domain-disease associations that are unknown in our study, but they have at least one known genes in the OMIM database, and we mark them using an italic style (diseases with ranks 5, 8, 15, 21, 28, 30, and 48, respectively). From the prediction result we see that only one of these diseases, QUESTION MARK EARS ISOLATED (QME; OMIM: 612798) successfully recover its known disease genes EDN1 (OMIM: 131240), while the other 6 diseases fail to prioritize their known associated genes at the top. This is one main drawback of the modularization, i.e., it will lose some information of known disease-gene associations when constructing protein-module association. A trade-off is called for to achieve the balance between the exploration of novel disease-gene predictions and the failure of recovering known associations.

Evidences in genome-wide association studies

Over the past decade, genome-wide association studies (GWAS) have led to the identification of susceptible single nucleotide polymorphisms (SNPs) conferring risk for common human diseases [48]. The NHGRI-EBI catalog of published genome-wide association studies [3] is a quality controlled, manually curated, literature-derived collection of all published genome-wide association studies assaying at least 100,000 SNPs and all SNP-trait associations with p-values $< 1.0 \times 10^{-5}$ [49]. With this resource, it is of interest to test the consistency of our inferred domain-disease associations with the GWAS results.

We choose the inference results obtained using the Bayesian approach for further analysis since it performs the best in the above validation experiments. Then for a given disease of interest, from the GWAS catalog we are able to collect a list of reported susceptible SNPs, and check how many of these SNPs appear within 5 Mbp around the domains that are ranked in top 10 in our inference results. We next select Crohn's disease (OMIM: 266600) and Type 2 diabetes (OMIM: 125853) as two typical examples of common polygenetic disorders with multiple genes contributing to the diseases, therefore GWAS studies provide a collection of risk factors for these two disorders.

Table 3 Novel predictions of domain-disease and gene-disease associations

Rank	Disease	$OMIM_d$	Module Index	Domain	Pfam	Gene	$OMIM_g$
1	APNEA, OBSTRUCTIVE SLEEP	107650	152	Acetyltransf_1	PF00583	NAA10	300013
						AANAT	600950
2	ARTERIES, ANOMALIES OF	108000	182	Sugar_tr	PF00083	SLC2A1	138140
						SLC2A2	138160
						SLC2A9	606142
						SLC2A10	606145
3	ATRESIA OF EXTERNAL AUDITORY CANAL AND CONDUCTIVE DEAFNESS	108760	133	zf-C2H2_2	PF12756	TSHZ1	614427
4	CELIAC ARTERY STENOSIS FROM COMPRESSION BY MEDIAN ARCUATE LIGAMENT OF DIAPHRAGM	116870	182	Sugar_tr	PF00083	SLC2A1	138140
						SLC2A2	138160
						SLC2A9	606142
						SLC2A10	606145
5	*SCHNYDER CORNEAL DYSTROPHY; SCCD*	*121800*	*77*	*Fasciclin*	*PF02469*	*TGFBI*	*601692*
6	CORNEAL DYSTROPHY, EPITHELIAL BASEMENT MEMBRANE; EBMD	121820	77	Fasciclin	PF02469	TGFBI	*601692*
7	CORNEAL DYSTROPHY, GROENOUW TYPE I; CDGG1	121900	77	Fasciclin	PF02469	TGFBI	*601692*
8	*CORNEAL DYSTROPHY, MEESMANN; MECD*	*122100*	*77*	*Fasciclin*	*PF02469*	*TGFBI*	*601692*
9	CORNEAL DYSTROPHY, LATTICE TYPE I; LCD1	122200	77	Fasciclin	PF02469	TGFBI	*601692*
10	CORONARY ARTERY DISSECTION, SPONTANEOUS	122455	182	Sugar_tr	PF00083	SLC2A1	138140
						SLC2A2	138160
						SLC2A9	606142
						SLC2A10	606145
11	DEAFNESS, CONDUCTIVE STAPEDIAL, WITH EAR MALFORMATION AND FACIAL PALSY	124490	137	Endothelin	PF00322	EDN1	131240
						EDN3	131242
12	EAR FOLDING	128500	137	Endothelin	PF00322	EDN1	131240
						EDN3	131242
13	PREAURICULAR FISTULAE, CONGENITAL	128700	137	Endothelin	PF00322	EDN1	131240
						EDN3	131242
14	EAR PITS, POSTERIOR HELICAL	128710	137	Endothelin	PF00322	EDN1	131240
						EDN3	131242
15	*EAR WITHOUT HELIX*	*128800*	*137*	*Endothelin*	*PF00322*	*EDN1*	*131240*
						EDN3	*131242*
16	EXTERNAL AUDITORY CANAL, BILATERAL ATRESIA OF, WITH CONGENITAL VERTICAL TALUS	133705	133	zf-C2H2_2	PF12756	TSHZ1	614427
17	FIBROMUSCULAR DYSPLASIA OF ARTERIES	135580	182	Sugar_tr	PF00083	SLC2A1	138140
						SLC2A2	138160
						SLC2A9	606142
						SLC2A10	606145
18	GLAUCOMA AND SLEEP APNEA	137763	152	Acetyltransf_1	PF00583	NAA10	300013
						AANAT	600950
19	LUBS X-LINKED MENTAL RETARDATION SYNDROME; MRXSL	300260	66	MBD	PF01429	MECP2	*300005*
20	ALPHA-THALASSEMIA	604131	188	Globin	PF00042	HBA1	*141800*
						HBA2	*141850*
						HBB	141900
21	*HOLT-ORAM SYNDROME; HOS*	*142900*	*140*	*LMBR1*	*PF04791*	*LMBR1*	*605522*
22	INTERNAL CAROTID ARTERY, SPONTANEOUS DISSECTION OF	147820	182	Sugar_tr	PF00083	SLC2A1	138140

Table 3 Novel predictions of domain-disease and gene-disease associations *(Continued)*

						SLC2A2	138160
						SLC2A9	606142
						SLC2A10	606145
23	LITHIUM TRANSPORT	152420	180	SNF	PF00209	SLC6A3	126455
						SLC6A19	608893
24	MACULAR DYSTROPHY, FENESTRATED SHEEN TYPE	153890	77	Fasciclin	PF02469	TGFBI	601692
25	MULLERIAN APLASIA AND HYPERANDROGENISM	158330	199	wnt	PF00110	WNT5A	164975
						WNT10B	601906
						WNT4	*603490*
						WNT10A	606268
26	OSSIFIED EAR CARTILAGES	165670	137	Endothelin	PF00322	EDN1	131240
						EDN3	131242
27	ENCHONDROMATOSIS, MULTIPLE, OLLIER TYPE	166000	166	Iso_dh	PF00180	IDH2	147650
						IDH1	147700
						IDH3B	604526
28	*MACULAR DYSTROPHY, PATTERNED, 1; MDPT1*	*169150*	*77*	*Fasciclin*	*PF02469*	*TGFBI*	*601692*
29	RADIAL RAY HYPOPLASIA WITH CHOANAL ATRESIA	179270	140	LMBR1	PF04791	LMBR1	605522
30	*QUESTION MARK EARS, ISOLATED; QME*	*612798*	*137*	*Endothelin*	*PF00322*	*EDN1*	*131240*
						EDN3	*131242*
31	THUMB DEFORMITY	188100	140	LMBR1	PF04791	LMBR1	605522
32	THYROID HORMONE PLASMA MEMBRANE TRANSPORT DEFECT	188560	180	SNF	PF00209	SLC6A3	126455
						SLC6A19	608893
33	TRACHEOESOPHAGEAL FISTULA WITH OR WITHOUT ESOPHAGEAL ATRESIA	189960	133, 181	zf-C2H2_2	PF12756	TSHZ1	614427
34	TRIGGER THUMB	190410	140	LMBR1	PF04791	LMBR1	605522
35	TRIPHALANGEAL THUMB WITH DOUBLE PHALANGES	190500	140	LMBR1	PF04791	LMBR1	605522
36	TRIPHALANGEAL THUMB, NONOPPOSABLE	190600	140	LMBR1	PF04791	LMBR1	605522
37	UTERINE ANOMALIES	192000	199	wnt	PF00110	WNT5A	164975
						WNT10B	601906
						WNT4	603490
						WNT10A	606268
38	UTERUS BICORNIS BICOLLIS WITH PARTIAL VAGINAL SEPTUM AND UNILATERAL HEMATOCOLPOS WITH IPSILATERAL RENAL AGENESIS	192050	199	wnt	PF00110	WNT5A	164975
						WNT10B	601906
						WNT4	603490
						WNT10A	606268
39	ACRORENAL-MANDIBULAR SYNDROME	200980	199	wnt	PF00110	WNT5A	164975
						WNT10B	601906
						WNT4	603490
						WNT10A	606268
40	ADDUCTED THUMBS SYNDROME	201550	140	LMBR1	PF04791	LMBR1	605522
41	APNEA, CENTRAL SLEEP	207720	152	Acetyltransf_1	PF00583	NAA10	300013
						AANAT	600950
42	ARTERIAL TORTUOSITY SYNDROME; ATS	208050	182	Sugar_tr	PF00083	SLC2A1	138140
						SLC2A2	138160
						SLC2A9	606142
						SLC2A10	*606145*

Table 3 Novel predictions of domain-disease and gene-disease associations *(Continued)*

43	AURAL ATRESIA, MULTIPLE CONGENITAL ANOMALIES, AND MENTAL RETARDATION	209770	133, 181	zf-C2H2_2	PF12756	TSHZ1	614427
44	BILIARY ATRESIA, EXTRAHEPATIC; EHBA	210500	133, 181	zf-C2H2_2	PF12756	TSHZ1	614427
45	CITRULLINE TRANSPORT DEFECT	215720	180	SNF	PF00209	SLC6A3	126455
						SLC6A19	608893
46	CENTRAL CLOUDY DYSTROPHY OF FRANCOIS; CCDF	217600	77	Fasciclin	PF02469	TGFBI	601692
47	DEAFNESS, CONDUCTIVE, WITH MALFORMED EXTERNAL EAR	221300	137	Endothelin	PF00322	EDN1	131240
						EDN3	131242
48	*DICARBOXYLIC AMINOACIDURIA; DCBXA*	*222730*	*180*	*SNF*	*PF00209*	*SLC6A3*	*126455*
						SLC6A19	*608893*
49	DUODENAL ATRESIA	223400	181	zf-C2H2_2	PF12756	TSHZ1	614427
50	HARTNUP DISORDER; HND	234500	180	SNF	PF00209	SLC6A3	126455
						SLC6A19	*608893*

"Rank" is the rank of predicted domain-disease associations in terms of Bayesian scores; "Disease" is the name of the disease phenotype, "$OMIM_d$" is ID of disease phenotype in the OMIM database; "Module Index" is the index of module including the disease; "Domain" is the name of domain; "Pfam" is the domain ID in the Pfam database; "Gene" is mapped gene from corresponding domain; and "$OMIM_g$" is the gene ID in the OMIM database. The bold rows represent known domain-disease associations compiled by the Ensembl BioMart tool. The italic rows represent domain-disease associations that are unknown in our study but have at least one known genes in the OMIM database. The bold italic elements in the "$OMIM_g$" column represent that the predicted genes are known as disease genes in the OMIM database

Crohn's disease

Crohn's disease is a type of chronic inflammatory bowel disease (IBD) of the gastrointestinal tract [50] affecting 26–200 per 100,000 in European populations [51], and is thought to be caused by a combination of environmental, immune and bacterial factors in genetically susceptible individuals [52]. Recently, genome-wide association studies (GWAS) have made remarkable progress in Crohn's disease identifying at least 140 genome-wide significant loci [53]. In our study, we compile from GWAS catalog 203 reported susceptible SNPs, and 54 of them are found to be within 5 Mbp regions of all domains that are ranked at the top 10. Then for each of the top 10 domains, we check regions of their corresponding genes, to see if any GWAS reported SNPs are inside, upstream, or downstream within 5Mbp of every gene region, and list the results in Additional file 1: Table S4.

From the table we observe 18 times that a susceptible SNP locates inside a domain with 10 times in ATG16 (PF08614), 4 times in NACHT (PF05729), and 4 times in CARD (PF00619). We also observe 24 times that a susceptible SNP locates within 1 Mbp upstream or downstream of a domain, 52 times that a susceptible SNP locates within 1 to 5 Mbp region of a domain, and only 33 times that a susceptible SNP locates beyond 5 Mbp region of a domain. From these results we conjecture that domains ranked among the top 10 indeed tends to be closer to, even include, known susceptible SNPs for this disease.

Type 2 diabetes

Type 2 diabetes, which is also called noninsulin-dependent diabetes mellitus (NIDDM) or adult-onset diabetes, is a metabolic disorder that is characterized by high blood sugar in the context of insulin resistance and relative lack of insulin [54]. It results from interaction between genetic and environmental risk factors [55–57]. To uncover the genetic basis of the disease, at least 88 SNPs have been reported being associated with increased risk for Type 2 diabetes [58]. In our study, we compile from GWAS catalog 221 reported susceptible SNPs, and 35 of them are found to be within 5 Mbp regions of 8 domains that are ranked at the top 10 (except domain Peptidase_M10 (PF00413) and LTD (PF00932). Similarly, for each of the top 10 domains, we check regions of their corresponding genes, to see if any GWAS reported SNPs are inside, upstream, or downstream within 5Mbp of every gene region, and list the results in Additional file 1: Table S5.

From the table we observe 3 times that a susceptible SNP locates inside a domain with once in PH (PF00169) and twice in IRK (PF01007). We also observe 19 times that a susceptible SNP locates within [13] 1 Mbp upstream or downstream of a domain, 32 times that a susceptible SNP locates within 1 to 5 Mbp region of a domain, and also 32 times that a susceptible SNP locates beyond 5 Mbp region of a domain. From these results we conjecture that domains ranked among the top 10 indeed tends to be closer to, and even include, known susceptible SNPs for this disease.

Conclusions and discussion

In this paper, we implemented a comparative study of inferring protein domains that are associated with human inherited diseases, by means of five approaches that are previously used to predict domain-domain interactions from protein-protein interactions. On the basis of several network data sources, we first constructed disease modules from the disease phenotype similarity network, and then proposed a framework to make the inference through known relations between diseases and modules, domains and proteins, as well as proteins and disease modules. We further demonstrate the effectiveness and robustness of these approaches, through a series of large-scale validation experiments, and discussed about the benefits brought by modularization, while comparing the performances of the five approaches in terms of three evaluation criteria (AUC score, Accuracy, and Mean rank ratio). We finally illustrate the consistency between our inference results and the evidences from genome-wide association studies for two common diseases: Crohn's disease and Type 2 diabetes.

Our main contribution therefore lies in the following parts: (1) effectively utilization of protein-domain, gene-disease and phenotype similarity sources to make large-scale inference of domain-disease associations and even novel gene-disease associations; (2) comparisons of the predictive powers of five well-known approaches in solving the domain-disease association inference problem; (3) comprehensive analysis of the robustness of the approaches to changing parameters; (4) illustration of the effects of disease modularization in predicting novel domain-disease and gene-disease associations; (5) analysis of top 50 predictions of domain-disease and corresponding gene-disease associations; (6) demonstration of the consistency of our inference results with the genome-wide association studies, providing genetic support to our domain-disease association inference results.

Notwithstanding so, our work can be further explored from the following aspects. First, the success of all the five approaches is mainly owing to the known gene-disease, protein-domain relationships and the construction of disease modules based on phenotype similarities. Therefore, the abundances and qualities of these relationships are of great importance to the inference results. However, current studies concerning the structures and functions protein domains are still too few, making the domain-centric approach instead of the gene-centric ones challenging [13]. To improve this situation one may resort to more network containing the domain information as an alternative, such as the domain-domain interactions from the DOMINE database [59]. As for the construction of disease module, although there are already several methods for calculating similarities between disease phenotypes, how to categorize the diseases remains a challenge. In our work we just simply construct the phenotype similarity network according to a threshold, through which the numerical information of similarities among diseases is not sufficiently utilized.

Table 4 Strengths and weaknesses of the approaches

	Strength	Weakness
Association	• Fast in implementation • No need to pre-determine parameters • Results can be easily validated by hand calculation	• Unsatisfactory in predictive power • Does not consider the structures all relevant protein-module associations as a whole • Do not have control for possible over-prediction of associations between frequently occurring domain-disease pairs
MLE	• Good in predictive power • Fast in implementation • Take into account the structures all relevant protein-module associations as a whole	• Need to pre-determine parameters • Do not have control for possible over-prediction of associations between frequently occurring domain-disease pairs
DPEA	• Satisfactory in predictive power • No need to pre-determine parameters • Have control for possible over-prediction of associations between frequently occurring domain-disease pairs	• Slow in implementation when the number of candidate domain-disease associations is large
Bayesian	• Excellent in predictive power • Take into account the structures all relevant protein-module associations as a whole	• Slow in implementation when the number of candidate domain-disease associations is large • Do not have control for possible over-prediction of associations between frequently occurring domain-disease pairs • Failure when the log-concave conditions of parameters are not satisfied
PE	• Satisfactory in predictive power • Only one pre-determined parameter • Have control for possible over-prediction of associations between frequently occurring domain-disease pairs	• Slow in implementation when the number of candidate domain-disease associations is large

The threshold itself needs to be pre-determined as well, and in this paper we select its value mainly for the consideration of computational feasibility. However, different values of the threshold would change the number of diseases, proteins, and domains involved, and different clustering approaches and their potential parameters used may also influence significantly. For clustering diseases into biological meaningful modules, more knowledge about disease categorization is needed.

Second, although we successfully developed the disease modules to circumvent the limitations of the insufficient protein-disease associations, the modularization does not differentiate the diseases in the same modules, and known protein-disease information is reduced to protein-module information instead. As a result, for the diseases that already have known genes associated, these disease genes are not guaranteed to be scored highest for the diseases. Therefore, to overcome this problem, we need to assign different weights to diseases in the same module during the disease modularization process.

Third, as shown in Table 4, we list the strengths and weaknesses for all the five approaches referred in this paper. Although the Bayesian approach performs the best in our comparative study, it also suffers from some mathematical and computational issues that need to be improved upon. Although the Adaptive Rejection Sampling method in the Bayesian approach only needs that the posterior distribution to be log-concave, we still need to determine the prior distributions for the parameters involved. As is known, the specification of prior is intrinsically complicated and subjective. The main consideration is that the posterior mean and variance should not depend on the units in which the disease similarities are measured and should also be invariant to the shift of the response variable. Therefore, one can consider the use of Jeffrey's prior instead of the conjugate prior. Also, the computational time of the Bayesian approach is much longer than the other four approaches, and as the scale of networks increase, the iteration processes becomes unbearable. More computational efficient approaches to estimate the posterior probability distributions are needed.

Finally, our study is mainly motivated by the hypothesis that protein domains can be viewed as functional units of proteins, and are considered important in disease development. However, in addition to domains, the linker sequences between domains could also be essential for the biological functions of the proteins, which has been demonstrated by a recent study [60]. Some linker sequences can be viewed as flexible "domains", by connecting various domains in a single protein without interfering with the function of each domain [61], and their relationships between linker sequences and diseases is a topic for a future research.

Additional files

Additional file 1: Table S1. Number of domains, diseases, proteins and modules collected based on different thresholds used in the disease module identification. **Table S2**: The performances as measured by AUC, accuracy and mean rank ratio of the MLE approach under different combinations of the false positive rate *fp* and the false negative rate *fn*. **Table S3**: The performances as measured by AUC, accuracy and mean rank ratio of the PE approach under different reliability rate and pw-score threshold. **Table S4**: GWAS evidence between domains and Crohn's disease. **Table S5**: GWAS evidence between domains and Type 2 Diabetes. (DOCX 47 kb)

Additional file 2: Supplemental methods. This file contains a) symbols used in the equations and formulations of the methods, b) details of implementing the EM algorithm in the maximum likelihood estimation (MLE) approach, and c) details of deducing the score function in the domain-disease pair exclusion analysis (DPEA) approach. (DOCX 197 kb)

Competing interests
No conflicts of interests declared.

Authors' contributions
WZ implemented the method, collected the results, and wrote the manuscript. MPC motivated the study, provided biological supports, and revised the manuscript. FS designed the research and wrote the manuscript. All authors read and approved the final manuscript.

Acknowledgements
The publication costs for this article were partly funded by U.S. National Institute of Health (NIH), grant number U01HL108634. WZ received the USC Provost Fellowship for graduate studies.

Author details
[1]Molecular and Computational Biology Program, University of Southern California, 1050 Childs Way, Los Angeles, USA. [2]Zilkha Neurogenetic Institute, Keck School of Medicine, University of Southern California, Los Angeles, CA, USA. [3]Department of Psychiatry and Behavioral Sciences, Keck School of Medicine, University of Southern California, Los Angeles, CA, USA. [4]Centre for Computational Systems Biology, School of Mathematical Sciences, Fudan University, Shanghai, China.

Published: 11 January 2016

References
1. Rehm HL. Disease-targeted sequencing: a cornerstone in the clinic. Nat Rev Genet. 2013;14(4):295–300. doi:10.1038/nrg3463.
2. Visscher PM, Brown MA, McCarthy MI, Yang J. Five years of GWAS discovery. Am J Hum Genet. 2012;90(1):7–24. doi:10.1016/j.ajhg.2011.11.029.
3. Welter D, MacArthur J, Morales J, Burdett T, Hall P, Junkins H, et al. The NHGRI GWAS Catalog, a curated resource of SNP-trait associations. Nucleic Acids Res. 2014;42(Database issue):D1001–6. doi:10.1093/nar/gkt1229.
4. McCarthy MI, Hirschhorn JN. Genome-wide association studies: potential next steps on a genetic journey. Hum Mol Genet. 2008;17(R2):R156–65. doi:10.1093/hmg/ddn289.
5. Arrowsmith CH, Bountra C, Fish PV, Lee K, Schapira M. Epigenetic protein families: a new frontier for drug discovery. Nat Rev Drug Discov. 2012;11(5): 384–400. doi:10.1038/nrd3674.
6. Dev KK. Making protein interactions druggable: targeting PDZ domains. Nat Rev Drug Discov. 2004;3(12):1047–56. doi:10.1038/nrd1578.
7. Cao C, Moult J. GWAS and drug targets. BMC Genomics. 2014;15 Suppl 4:S5. doi:10.1186/1471-2164-15-S4-S5.
8. Lage K, Hansen NT, Karlberg EO, Eklund AC, Roque FS, Donahoe PK, et al. A large-scale analysis of tissue-specific pathology and gene expression of human disease genes and complexes. Proc Natl Acad Sci U S A. 2008; 105(52):20870–5. doi:10.1073/pnas.0810772105.
9. Bornigen D, Pers TH, Thorrez L, Huttenhower C, Moreau Y, Brunak S. Concordance of gene expression in human protein complexes reveals tissue specificity and pathology. Nucleic Acids Res. 2013;41(18):e171. doi:10.1093/nar/gkt661.

10. Oellrich A, Sanger Mouse Genetics Project, Smedley D. Linking tissues to phenotypes using gene expression profiles. Database (Oxford). 2014;2014: bau017. doi:10.1093/database/bau017.
11. Farh KK, Marson A, Zhu J, Kleinewietfeld M, Housley WJ, Beik S, et al. Genetic and epigenetic fine mapping of causal autoimmune disease variants. Nature. 2015;518(7539):337–43. doi:10.1038/nature13835.
12. Yang H, Robinson PN, Wang K. Phenolyzer: phenotype-based prioritization of candidate genes for human diseases. Nat Methods. 2015;12(9):841–3. doi:10.1038/nmeth.3484.
13. Peterson TA, Park D, Kann MG. A protein domain-centric approach for the comparative analysis of human and yeast phenotypically relevant mutations. BMC Genomics. 2013;14 Suppl 3:S5. doi:10.1186/1471-2164-14-S3-S5.
14. Wang W, Zhang W, Jiang R, Luan Y. Prioritisation of associations between protein domains and complex diseases using domain-domain interaction networks. IET Syst Biol. 2010;4(3):212–22. doi:10.1049/iet-syb.2009.0037.
15. Zhang W, Chen Y, Sun F, Jiang R. DomainRBF: a Bayesian regression approach to the prioritization of candidate domains for complex diseases. BMC Syst Biol. 2011;5:55. doi:10.1186/1752-0509-5-55.
16. Amberger JS, Bocchini CA, Schiettecatte F, Scott AF, Hamosh A. OMIM.org: Online Mendelian Inheritance in Man (OMIM®), an online catalog of human genes and genetic disorders. Nucleic Acids Res. 2015;43(Database issue): D789–98. doi:10.1093/nar/gku1205.
17. Tryka KA, Hao L, Sturcke A, Jin Y, Wang ZY, Ziyabari L, et al. NCBI's Database of Genotypes and Phenotypes: dbGaP. Nucleic Acids Res. 2014;42(Database issue):D975–9. doi:10.1093/nar/gkt1211.
18. Kibbe WA, Arze C, Felix V, Mitraka E, Bolton E, Fu G, et al. Disease Ontology 2015 update: an expanded and updated database of human diseases for linking biomedical knowledge through disease data. Nucleic Acids Res. 2015;43(Database issue):D1071–8. doi:10.1093/nar/gku1011.
19. Pagon RA AM, Ardinger HH, et al. GeneReviews® [Internet]. Seattle (WA): University of Washington, Seattle; 1993-2015.
20. National Library of Medicine (US). Genetics Home Reference [Internet]. Bethesda (MD): The Library; 2015. [cited 2015 Sep 28]. Available from: http://ghr.nlm.nih.gov/.
21. Lee H, Deng M, Sun F, Chen T. An integrated approach to the prediction of domain-domain interactions. BMC Bioinformatics. 2006;7:269. doi:10.1186/1471-2105-7-269.
22. Jothi R, Cherukuri PF, Tasneem A, Przytycka TM. Co-evolutionary analysis of domains in interacting proteins reveals insights into domain-domain interactions mediating protein-protein interactions. J Mol Biol. 2006;362(4): 861–75. doi:10.1016/j.jmb.2006.07.072.
23. Nye TM, Berzuini C, Gilks WR, Babu MM, Teichmann SA. Statistical analysis of domains in interacting protein pairs. Bioinformatics. 2005;21(7):993–1001. doi:10.1093/bioinformatics/bti086.
24. Ng SK, Zhang Z, Tan SH, Lin K. InterDom: a database of putative interacting protein domains for validating predicted protein interactions and complexes. Nucleic Acids Res. 2003;31(1):251–4.
25. Riley R, Lee C, Sabatti C, Eisenberg D. Inferring protein domain interactions from databases of interacting proteins. Genome Biol. 2005; 6(10):R89. doi:10.1186/gb-2005-6-10-r89.
26. Guimaraes KS, Jothi R, Zotenko E, Przytycka TM. Predicting domain-domain interactions using a parsimony approach. Genome Biol. 2006; 7(11):R104. doi:10.1186/gb-2006-7-11-r104.
27. Guimaraes KS, Przytycka TM. Interrogating domain-domain interactions with parsimony based approaches. BMC Bioinformatics. 2008;9:171. doi:10.1186/1471-2105-9-171.
28. Zhao XM, Chen L, Aihara K. A discriminative approach for identifying domain-domain interactions from protein-protein interactions. Proteins. 2010;78(5):1243–53. doi:10.1002/prot.22643.
29. Chen XW, Liu M. Prediction of protein-protein interactions using random decision forest framework. Bioinformatics. 2005;21(24):4394–400. doi:10.1093/bioinformatics/bti721.
30. Liu M, Chen XW, Jothi R. Knowledge-guided inference of domain-domain interactions from incomplete protein-protein interaction networks. Bioinformatics. 2009;25(19):2492–9. doi:10.1093/bioinformatics/btp480.
31. Wang H, Segal E, Ben-Hur A, Li QR, Vidal M, Koller D. InSite: a computational method for identifying protein-protein interaction binding sites on a proteome-wide scale. Genome Biol. 2007;8(9):R192. doi:10.1186/gb-2007-8-9-r192.
32. Singhal M, Resat H. A domain-based approach to predict protein-protein interactions. BMC Bioinformatics. 2007;8:199. doi:10.1186/1471-2105-8-199.
33. Deng M, Mehta S, Sun F, Chen T. Inferring domain-domain interactions from protein-protein interactions. Genome Res. 2002;12(10):1540–8. doi:10.1101/gr.153002.
34. Kim I, Liu Y, Zhao H. Bayesian methods for predicting interacting protein pairs using domain information. Biometrics. 2007;63(3):824–33. doi:10.1111/j.1541-0420.2007.00755.x.
35. Barabasi AL, Gulbahce N, Loscalzo J. Network medicine: a network-based approach to human disease. Nat Rev Genet. 2011;12(1):56–68. doi:10.1038/nrg2918.
36. Menche J, Sharma A, Kitsak M, Ghiassian SD, Vidal M, Loscalzo J, et al. Disease networks. Uncovering disease-disease relationships through the incomplete interactome. Science. 2015;347(6224):1257601. doi:10.1126/science.1257601.
37. de Vries PJ. Genetics and neuropsychiatric disorders: genome-wide, yet narrow. Nat Med. 2009;15(8):850–1. doi:10.1038/nm0809-850.
38. Arnedo J, Svrakic DM, Del Val C, Romero-Zaliz R, Hernández-Cuervo H, Molecular Genetics of Schizophrenia Consortium, et al. Uncovering the hidden risk architecture of the schizophrenias: confirmation in three independent genome-wide association studies. Am J Psychiatry. 2015; 172(2):139–53. doi:10.1176/appi.ajp.2014.14040435.
39. Stewart DR, Pemov A, Johnston JJ, Sapp JC, Yeager M, He J, et al. Dubowitz syndrome is a complex comprised of multiple, genetically distinct and phenotypically overlapping disorders. PLoS One. 2014;9(6): e98686. doi:10.1371/journal.pone.0098686.
40. Jiang R. Walking on multiple disease-gene networks to prioritize candidate genes. J Mol Cell Biol. 2015. doi:10.1093/jmcb/mjv008.
41. Finn RD, Bateman A, Clements J, Coggill P, Eberhardt RY, Eddy SR, et al. Pfam: the protein families database. Nucleic Acids Res. 2014;42(Database issue):D222–30. doi:10.1093/nar/gkt1223.
42. Nepusz T, Yu H, Paccanaro A. Detecting overlapping protein complexes in protein-protein interaction networks. Nat Methods. 2012;9(5):471–2. doi:10.1038/nmeth.1938.
43. Gilks WR, Wild P. Adaptive Rejection Sampling for Gibbs Sampling. Applied Statistics-Journal of the Royal Statistical Society Series C. 1992;41(2):337–48. doi:10.2307/2347565.
44. Cunningham F, Amode MR, Barrell D, Beal K, Billis K, Brent S, et al. Ensembl 2015. Nucleic Acids Res. 2015;43(Database issue):D662–9. doi:10.1093/nar/gku1010.
45. Haider S, Ballester B, Smedley D, Zhang J, Rice P, Kasprzyk A. BioMart Central Portal–unified access to biological data. Nucleic Acids Res. 2009;37(Web Server issue):W23–7. doi:10.1093/nar/gkp265.
46. Cao A, Galanello R. Beta-thalassemia. Genet Med. 2010;12(2):61–76. doi:10.1097/GIM.0b013e3181cd68ed.
47. Boycott KM, Vanstone MR, Bulman DE, MacKenzie AE. Rare-disease genetics in the era of next-generation sequencing: discovery to translation. Nat Rev Genet. 2013;14(10):681–91. doi:10.1038/nrg3555.
48. Pal LR, Moult J. Genetic Basis of Common Human Disease: Insight into the Role of Missense SNPs from Genome-Wide Association Studies. J Mol Biol. 2015;427(13):2271–89. doi:10.1016/j.jmb.2015.04.014.
49. Hindorff LA, Sethupathy P, Junkins HA, Ramos EM, Mehta JP, Collins FS, et al. Potential etiologic and functional implications of genome-wide association loci for human diseases and traits. Proc Natl Acad Sci U S A. 2009;106(23):9362–7. doi:10.1073/pnas.0903103106.
50. Baumgart DC, Sandborn WJ. Crohn's disease. Lancet. 2012;380(9853): 1590–605. doi:10.1016/S0140-6736(12)60026-9.
51. Liu JZ, Anderson CA. Genetic studies of Crohn's disease: past, present and future. Best Pract Res Clin Gastroenterol. 2014;28(3):373–86. doi:10.1016/j.bpg.2014.04.009.
52. Khor B, Gardet A, Xavier RJ. Genetics and pathogenesis of inflammatory bowel disease. Nature. 2011;474(7351):307–17. doi:10.1038/nature10209.
53. Ning K, Gettler K, Zhang W, Ng SM, Bowen BM, Hyams J, et al. Improved integrative framework combining association data with gene expression features to prioritize Crohn's disease genes. Hum Mol Genet. 2015;24(14): 4147–57. doi:10.1093/hmg/ddv142.
54. Kahn CR. Banting Lecture. Insulin action, diabetogenes, and the cause of type II diabetes. Diabetes. 1994;43(8):1066–84.
55. Murea M, Ma L, Freedman BI. Genetic and environmental factors associated with type 2 diabetes and diabetic vascular complications. Rev Diabet Stud. 2012;9(1):6–22. doi:10.1900/RDS.2012.9.6.
56. Olokoba AB, Obateru OA, Olokoba LB. Type 2 diabetes mellitus: a review of current trends. Oman Med J. 2012;27(4):269–73. doi:10.5001/omj.2012.68.

57. Chen L, Magliano DJ, Zimmet PZ. The worldwide epidemiology of type 2 diabetes mellitus–present and future perspectives. Nat Rev Endocrinol. 2012; 8(4):228–36. doi:10.1038/nrendo.2011.183.
58. Mohlke KL, Boehnke M. Recent advances in understanding the genetic architecture of type 2 diabetes. Hum Mol Genet. 2015. doi:10.1093/hmg/ddv264.
59. Yellaboina S, Tasneem A, Zaykin DV, Raghavachari B, Jothi R. DOMINE: a comprehensive collection of known and predicted domain-domain interactions. Nucleic Acids Res. 2011;39(Database issue):D730–5. doi:10.1093/nar/gkq1229.
60. Gueroussov S, Gonatopoulos-Pournatzis T, Irimia M, Raj B, Lin ZY, Gingras AC, et al. RNA SPLICING. An alternative splicing event amplifies evolutionary differences between vertebrates. Science. 2015;349(6250): 868–73. doi:10.1126/science.aaa8381.
61. Reddy Chichili VP, Kumar V, Sivaraman J. Linkers in the structural biology of protein-protein interactions. Protein Sci. 2013;22(2):153–67. doi:10.1002/pro.2206.

Mangericao et al. BMC Systems Biology 2016, 10(Suppl 1):5
DOI 10.1186/s12918-015-0248-x

BMC Systems Biology

PROCEEDINGS Open Access

Computational prediction of CRISPR cassettes in gut metagenome samples from Chinese type-2 diabetic patients and healthy controls

Tatiana C. Mangericao[1,2], Zhanhao Peng[1] and Xuegong Zhang[1,3*]

From The Fourteenth Asia Pacific Bioinformatics Conference (APBC 2016)
San Francisco, CA, USA. 11 - 13 January 2016

Abstract

Background: CRISPR has been becoming a hot topic as a powerful technique for genome editing for human and other higher organisms. The original CRISPR-Cas (Clustered Regularly Interspaced Short Palindromic Repeats coupled with CRISPR-associated proteins) is an important adaptive defence system for prokaryotes that provides resistance against invading elements such as viruses and plasmids. A CRISPR cassette contains short nucleotide sequences called spacers. These unique regions retain a history of the interactions between prokaryotes and their invaders in individual strains and ecosystems. One important ecosystem in the human body is the human gut, a rich habitat populated by a great diversity of microorganisms. Gut microbiomes are important for human physiology and health. Metagenome sequencing has been widely applied for studying the gut microbiomes. Most efforts in metagenome study has been focused on profiling taxa compositions and gene catalogues and identifying their associations with human health. Less attention has been paid to the analysis of the ecosystems of microbiomes themselves especially their CRISPR composition.

Results: We conducted a preliminary analysis of CRISPR sequences in a human gut metagenomic data set of Chinese individuals of type-2 diabetes patients and healthy controls. Applying an available CRISPR-identification algorithm, PILER-CR, we identified 3169 CRISPR cassettes in the data, from which we constructed a set of 1302 unique repeat sequences and 36,709 spacers. A more extensive analysis was made for the CRISPR repeats: these repeats were submitted to a more comprehensive clustering and classification using the web server tool CRISPRmap. All repeats were compared with known CRISPRs in the database CRISPRdb. A total of 784 repeats had matches in the database, and the remaining 518 repeats from our set are potentially novel ones.

Conclusions: The computational analysis of CRISPR composition based contigs of metagenome sequencing data is feasible. It provides an efficient approach for finding potential novel CRISPR arrays and for analysing the ecosystem and history of human microbiomes.

Keywords: CRISPR, Human gut microbiome, Metagenome

* Correspondence: zhangxg@tsinghua.edu.cn
[1]MOE Key Lab of Bioinformatics/Bioinformatics Division, TNLIST/Center for Synthetic and Systems Biology, and Department of Automation, Tsinghua University, Beijing 100084, China
[3]School of Life Sciences, Tsinghua University, Beijing 100084, China
Full list of author information is available at the end of the article

Background

The human body is host to this complex community of symbiotic, pathogenic and commensal microorganisms (microbiome), whose abundance is estimated to exceed the number of human cells by at least an order of magnitude [1]. It has been estimated that the microbes in our bodies collectively make up to 100 trillion cells, tenfold the number of human cells [2]. To understand and exploit the impact of the microbes on human health and wellbeing, it is necessary to interpret the content, diversity and function of the microbial community. Metagenomic sequencing has been proven to be a powerful tool for analysing complex microbial communities. And in this context many projects have been developed all over the world to study human microbiomes of multiple body sites. For example, the US-based Human Microbiome Project (HMP) [3] and the EU-based MetaHIT project [4] have generated resources that can enable the comprehensive characterization of the human microbiome and analysis of its role in human health and disease.

Among all body sites, the diversity of microorganisms in the human gut is known to be among the highest [1]. This community has been discovered to be associated with human physiology through processes related to development, nutrition, immunity, and resistance to pathogens [5–9]. While bacteria are responsible for these functions and most research has focused on the interaction of bacteria with the host, bacteriophages, in turn, influence the composition and abundances of bacteria in the human gut [10, 11].

Bacteria and Archaea have evolved defence and regulatory mechanisms to cope with various environmental stressors especially virus attacks. The understanding on this arsenal has been expanded by the discovery of the versatile CRISPR-*Cas* system. Bacteria can remember their viral invaders by sampling short DNA sequences, known as protospacers, from the genetic materials of viruses or phage. These sequences become integrated into the bacterium's own DNA, specifically into an array of repeat sequences called clustered regularly interspaced short palindromic repeats (CRISPR). The integrated sequences are called spacers [12]. When these sequences are transcribed and processed into small RNAs, they guide a multifunctional protein complex (*Cas* proteins – CRISPR associated proteins) to recognize and cleave incoming foreign genetic material [13]. The diversity of *Cas* genes suggests that multiple pathways have been developed to use the basic information contained in the CRISPR cassettes in diverse defence mechanisms [14]. This adaptive immunity system was first observed in *Escherichia coli* in 1987, although its significance was not straightaway apparent. Since then, CRISPR arrays have been identified in approximately 40 % of Bacteria and 90 % of Archaea [15].

CRISPR cassettes were already characterized across human body sites in different individuals in independent projects [16–18] and as a part of the Human Microbiome Project (HMP) [14] with particular focus in the gut metagenome [14, 19, 20]. So far there have been two main approaches for the study of CRISPR in metagenomic samples: one focuses more on the analysis of spacers in raw reads, which are then used to search for CRISPR cassettes [19]; the other is based on the reconstruction of CRISPR arrays, where direct repeat consensus sequences from known CRISPR types identified in reference genomes are used to recruit reads from the dataset and then be assembled into CRISPR loci [14]. For the present work, and since we didn't have access to raw reads, we decided to follow a third approach, which is based on the identification of CRISPR cassettes in assembled contigs/scaffolds [21]. This approach differentiates from the previous ones because it allows finding novel CRISPR cassettes, as *de novo* prediction of CRISPRs relies on sequence features of CRISPRs that don't exist in short reads. Partially following the strategy of Gogleva et al [21], we analysed the CRISPR composition of the metagenomic data of a set of Chinese individuals [22]. The findings were compared with those on the 3 different datasets used in Gogleva et al [21].

Materials and methods

Metagenomic datasets

We used the gut metagenome data of a set of Chinese individuals published in [22]. The data contain 145 individuals, with their gut microbiota sequenced using whole genome sequencing [22]. The samples include 71 type-2 diabetic individuals and 74 non-diabetic individuals used as controls. Individual metagenomes were assembled in contigs with the average size of 10,687 bp. The the total length of contigs comprised 15.96 Gb. More information about the samples are provided in Additional file 1.

Identification and analysis of CRISPR cassettes

To construct a set of CRISPR cassettes from the metagenomic dataset, we used the software PILER-CR [23]. PILER-CR searches for CRISPR repeats by doing a multiple alignment between them in each array. It also shows flanking regions and spacers, making it visually obvious when PILER-CR mis-identifies the repeat boundary by one or a few positions. In the output, repeats are clustered by similarity to reveal relationships between different arrays and, where applicable, to provide confirmatory evidence of marginal arrays (i.e. arrays with only 3 or 4 repeats that may not be perfectly conserved). Sequences are considered similar, i.e., clustered in similar groups, if they have an estimated mutual identity of more than 50 %. The parameters used in this work followed the default parameters proposed by PILER-CR, (Table 1). With PILER-

Table 1 PILER-CR algorithm parameters

Parameter	Value	Description
Minrepeat	16	Minimum length of a repeat.
Maxrepeat	64	Maximum length of a repeat.
Minspacer	8	Minimum spacer length.
Minarray	3	Minimum number of repeats in an array/cassette.
Maxspacer	64	Maximum spacer length.
Minrepeat ratio	0,9	Repeat length ratio
Minspacer ratio	0,75	Spacer length ratio
Mincons	0,9	A run of minarray repeats with at least mincons similarity is required for an array to be reported. The value is specified as fractional identity 0.0... 1.0, with 1.0 meaning identical sequences.

CR it is possible to extract the repeat consensus sequences and the spacers' sequences. There is information about the array size, as well as information about where the CRISPR array is located (which scaffold/contig).

Repeat clustering and the CRISPRmap tool

Repeat consensus sequences were submitted to the standard BLASTClust procedure to form repeat clusters (parameters: -L 0.5 -S 50 -e F -p F -W 15; L is length coverage threshold, S is bit score coverage, default is 1.75) [24]. BLASTClust clusters nucleotide sequences based on pairwise matches found using the MegaBLAST algorithm for DNA/nucleotides. A single MegaBLAST search is performed for all the sequences combined against a database created from the same sequences. BLASTClust finds pairs of sequences that have statistically significant matches and clusters them using single-linkage clustering [24]. We used a first BLASTClust procedure as a filtering step to find identical repeats and construct a set of unique repeat sequences. And a second BLASTClust procedure was used to form clusters between unique repeat sequences previously found.

Another tool used for repeat clustering and analysis was CRISPRmap [25]. This tool allows CRISPR repeats to be divided into types as part of a comprehensive independent clustering analysis to determine conserved sequence families, potential structure motifs for endoribonucleases, and evolutionary relationships [25]. The repeats are divided among 6 superclasses, named A to F, that comprise 24 families and 18 structural motifs.

Repeat BLAST to CRISPR database

Unique repeat consensus sequences were matched against the CRISPR database CRISPRdb [26] (http://crispr.u-psud.fr/crispr/BLAST/CRISPRsBlast.php) using a standard BLAST with an *e-value* threshold of 0.01, in order to find hits for known repeats in our findings and to access the possibility of new repeat sequences. We added a step to remove the matches that have continuous indels of more than 5 nucleotides at either ends of the query or target sequences.

Results and discussions

Characteristics of CRISPR

We used data from the Chinese type-2 diabetes gut microbiota study [22] to search for CRISPR cassettes in gut metagenome sequences using PILER-CR with a similar pipeline as in Gogleva et al [21]. In the original pipeline by Gogleva et. al. [21], they also applied both the software CRT (CRISPR Recognition Tool) [27] and CRISPR Finder [28] to the dataset in order to find CRISPR cassettes. However, there are some issues in using these software: CRISPR Finder is an online tool and it's not available for download. Trying to upload large data to the website makes the processing time too long; for CRT, on the other hand, although it is available for offline usage, it also has some usage problems since it considers scaffolds of each individual as connected to each other as a single sequence, which is incorrect and can cause unreliable results. In further research, we should find a way for the program to consider each scaffold individually and make CRT results comparable to PILER-CR results. But for the current project, we chose only to use PILER-CR for this step.

Applying PILER-CR with default parameters, we found a total of 3169 candidate CRISPR cassettes. Given the said conditions, a filtering system was not applied, and all the CRISPR cassettes found were kept for further analysis. Among the total number of CRISPRs found, 1624 cassettes belong to the 74 individuals from the control group and the remaining 1545 cassettes belong to the 71 individuals with type-2 diabetes. CRISPR cassettes have been predicted in all single individuals. The full set of identified cassettes is shown and characterized in Additional file 2.

The identified 3169 CRISPR cassettes comprised a total of 36,709 spacers and 1302 unique repeats, meaning that only about 40 % of the repeats were unique. This could imply that the same CRISPR-containing bacteria were recurrent in multiple individuals. Within this dataset, there is an average of 22 CRISPR array per individual, and each cassette contains an average of 12 spacers, hence the average size of 767 bp per CRISPR. It should be referred that if we analyse both healthy and diabetic individuals separately the results are very similar, meaning that we couldn't find an obvious distinction between the number of CRISPR cassettes or number of spacers found (Table 2).

Comparing our findings with the ones from Gogleva et al. [21] on gut samples of from 124 European individuals, downloaded as an assembly of 1,889,651 contigs with a total size of 3732 Mb (http://public.genomics.org.cn/BGI/gutmeta/UniSet/), the most similar dataset to the one we used in terms of data size, we found that the number of CRISPR cassettes predicted by PILER-CR in our work was

Table 2 Summary of the results' comparison between the set of CRISPR cassetes found for type-2 diabetic individuals and healthy individuals

	Metagenomic dataset group	Type-2 diabetic individuals	Healthy (non-diabetic) individuals
	Cassettes		
Identified by PILER-CR		1624	1545
Average Length (in bp)		757	777
	Spacers		
Total number		16 949	17 547
	Repeats		
Average size (in bp)		31	32

larger. They initially predicted 121 CRISPR cassettes for 124 individuals. After the filtering process, their reliable set of CRISPR had a total of 78 cassettes including 378 spacers and 74 unique repeats. According to Gogleva et al.'s results from the 3 software they used, PILER-CR presented the least number of predicted CRISPR cassettes, which may be explained by its high specificity considering genomic repeats and low-complexity regions as a candidate CRISPR cassette, a process followed by several refinement steps [21, 23].

The relatively large number of predicted CRISPR cassettes in our work was not filtered with extra criteria. However, based on the results of filtering in the reference, we can predict that, even if we applied those filtering steps, the number of reliable CRISPR cassettes in our data would still be larger than that in Gogleva et al. [21]. This might lead us to suggest that the gut microbiota of the Chinese individuals, compared with European individuals, is composed by a larger number of bacteria "armed" with CRISPR adaptive defence mechanism. But the current data are insufficient to support such an observation. A more possible reason that might cause the difference in the number of identified CRISPR cassettes is that the two metagenome data were sequenced at least two years apart in time, which is a period when sequencing technology and protocols have developed very rapidly. Differences in sequencing protocol settings in the data may cause differences in predicting CRISPRs from the data. The data we used has 8,039,994 contigs with the total length of 16,345 Mb, which are much larger than the data used in Gogleva et al.. Besides data size, the number of CRISPRs found is conditioned by a series of other known or unknown factors such as types of bacteria present in the microbiota of the individual, the size of CRISPR systems which can dependent on the history of horizontal gene transfer and former phage invasions, the program used to search for CRISPR, protocols and settings in the sequencing experiment, and the quality of the sequencing data and assembly of the metagenomic sample, etc.. The observations in this study suggested that more studies on both real and simulated data are needed to build a better understanding of the distribution of CRISPRs in metagenome data and to develop better methods/strategies for predicting CRISPRs in metagenome data. The summary of the comparison of the results is given in Table 3.

Repeat clustering with CRISPRmap

After collecting the CRISPR cassettes that resulted from the PILER-CR software, we applied a standard BLASTClust procedure and also submitted our data to the CRISPRmap web server. BLASTClust was used in the first phase to find and construct a set of unique repeat sequences. In this work, an ID number or the denomination of "RepeatCluster" was attributed to the repeats. The latter refers to the repeat sequences that are identical and shared by more than two CRISPR arrays. As stated above, the set of unique repeats was composed by a total of 1302 repeat sequences. The set of unique repeats is available in Additional file 3A.

In the second phase, we constructed repeat clusters based on sequence similarity. See Additional file 3B. We consider groups of two or more repeats as a cluster. In this way 121 repeat clusters were formed. The majority of the clusters (68 out of 121) are composed by two repeats. Another 37 clusters contained three or four repeats. The biggest cluster contained 12 repeats. Since BLASTClust can only cluster sequences of high similarity (few mismatches) we consider it can be used as a reliable tool to find identical repeats from an "ocean" of sequences.

CRISPRmap was designed for independent and comprehensive classification of all known CRISPR cassettes based entirely on the repeat properties [25]. Their approach considers not only the pairwise similarities between repeats, but also how the binding affinity of *Cas* proteins is affected by the repeat structure, namely, a small hairpin structure (a key binding motif for *Cas* endoribonucleases in several systems) [25]. Their classification system provides a more complete insight into the diversity of CRISPR

Table 3 Summary of the results' comparison between the CRISPR analysis in this present work and Gogleva et al

	Study	In Gogleva et al., 2014	The present work
	Cassettes		
Identified by PILER-CR		121	3169
Average length		338 bp	767 bp
	Spacers		
Total number		378	36 709
	Repeats		
Total number		121	3169
Unique repeats		74	1305

systems. Applying CRISPRmap to our identified CRISPR repeats we found that 670 repeats were attributed superclasses, while the remaining 632 repeats were not assigned to any superclass. Additional file 4 provides the CRISPRmap results. Representatives for all superclasses were found. The most populated superclasses are: superclass A with 138 unique repeats, superclass C with 168 unique repeats, and superclass F with 182 unique repeats. To 365 out of 1302 repeats it was also attributed a structure motif. It's observed that for the majority of repeats belonging to the same superclass, the structure motif is similar. The summary on the superclasses is given in Table 4.

The superclasses were ordered according to generally decreasing conservation. Superclass A contains highly conserved CRISPRs repeats at the sequence level, but only a few small structure motifs. Superclasses B and C contain sequence families that roughly correspond to one structure motif each; the same is true for half of superclass D. The other half of superclass D and superclass E contain little sequence conservation, but many small conserved motifs. The bacterial repeats in superclass F are divergent [25]. For the repeats to which a superclass was attributed, 78 of those was also attributed to a sequence family. It's noted that families are associated with superclasses, and to no repeat was assigned a sequence family but not a superclass.

Repeat BLAST to CRISPRdb database

Our set of unique repeat sequences was then compared to the already known repeats from CRISPR cassettes collected in CRISPRdb [26] (e-value threshold of 0.01). CRISPRdb is a public database which is updated monthly from newly released genome sequences, to display CRISPRs and to generate dictionaries of spacers and repeats. This database is constructed using CRISPRFinder as a CRISPR identifying software. At the moment, they have in their database a total of 2612 bacterial genomes analysed (last updated 2015-08-05) where they found 1176 strains with convincing CRISPR(s) and 126 out of 150 Archae genomes with convincing CRISPR(s).

From our collected set of repeats, 932 out of the 1302 unique repeats had matches in the CRISPRdb. However, 148 of them have continuous indels at either ends of the query or target sequences. They may be overlapping repeats from different CRISPRs. We took them as unmatched repeats to keep higher confidence in the matched repeats. The repeat sequences that weren't matched to CRISPRdb can possibly be of novel CRISPR cassettes that have not been reported before. The detailed results are given in Additional file 5. In total, a number of 190 different strains were found. The majority of the matches belonged to *Megamonas hypermegale ART12/1*, although the number of matches isn't particularly relevant since it only ends up representing 3 % of the total matches. The second most matched strains, within our repeats, were *Bacteroides helcogenes P36-108*, *Eubacterium rectale M104/1* and *Faecalibacterium prausnitzii SL3/3*, all strains commonly found in the human gut microbiota [29, 30]. The last two strains are bacteria that are part of the colonic microbiomes responsible for the fermentation of hexose and pentose sugars [31]. The results obtained may not be completely accurate, but they indicated a large diversity of CRISPR-containing strains present in the human gut microbiome.

Table 4 CRISPR map clustering of the collection of unique repeat sequences from Chinese individuals' CRISPR cassettes

	Number of repeats	Sequence families associated	Motifs
Superclass A	138	2, 7	6
Superclass B	102	1,6	1,2,7
Superclass C	168	5	3,4
Superclass D	32	None	11
Superclass E	48	9,24	3,4,15,16
Superclass F	182	13,23	3,4,15,17,18

Comparison of CRISPR repeats found in different data

We compared the findings from the diabetics patient group and the control group. The diabetics group has 3,777,094 contigs. A total of 1624 repeats were identified, with 790 unique repeats. The control group has 4,262,900 contigs. A total of 1545 repeats were found, with only 600 unique repeats. There are 88 unique repeats shared between the two groups. Additional file 3 shows unique repeats in two groups.

We also compared using BLASTclust the repeats we found with those found by Gogleva et. al. [21] on the three datasets they used: the European data of 124 individuals, the JPN data of 13 healthy Japanese individuals, and the DG data of 2 healthy adults in the Distal Gut metagenomic project. The latter two data were sequenced with different sequencing platforms. In the DG results, 4 out of 14 unique repeats were clustered into the same group with the unique repeats we found in the Chinese dataset. In the results on the European data, 7 out of 74 unique repeats were clustered with those in our data. In the results on the JPN data, 119 out of 296 are clustered with unique repeats we found in the Chinese data, which means 40 % of the repeats are highly similar with the unique repeats we've analysed. This may suggest that the two Asian gut metagenome datasets share higher similarity in terms of CRISPR compositions, but further investigation is needed for a systematic comparison.

Conclusions

We analysed CRISPR content in a human gut metagenomic dataset of Chinese individuals of healthy and type-2 diabetes groups. With some newly released tools for CRISPR identification and post-processing, we were able to profile CRISPR cassettes and their corresponding repeats and spacers in the gut microbiome of this sample set. Comparison with the existing database show that the majority of the identified CRISPRs have been reported in the literature, while there are a quarter of the identifications indicate newly discovered CRISPRs that have not been reported before.

The human gut microbiota is one of the most complicated microbial ecosystems in the human body and has important associations with human health. CRISPR is a major system that microbes use to deal with phage. Therefore analysing the CRISPR composition of a microbiome and of a group of microbiomes can be very informative for understanding the history and function of the microbiome. Also, the microbiota is composed of highly diverse bacteria species that cannot be cultured. Identifying new CRISPRs from metagenome data might also provide an efficient approach for finding possible novel CRISPRs that may be used for genome editing applications.

The collection of spacers and repeats constructed in this work constitutes a base for further studies. The preliminary comparison between CRISPRs found in different groups of data and between different studies showed interesting observations. As multiple gut metagenome datasets have been published in the projects like HMP, MetaHIT and other projects that focus on specific groups of people or specific human diseases, it'll be interesting and promising to conduct systematic comparative study among different datasets, which may lead to better understanding of the forming, shaping, changing and function of gut microbiome populations in different people.

The current strategy for computational profiling of CRISPRs is still preliminary as there has been no stringent control on the possible false discoveries. Some filtering steps may be necessary to reduce false discoveries. For example, Gogleva et al. [21] suggested three steps to form a reliable set of CRISPR cassettes: (1) Cassettes predicted by more than two CRISPR predicting programs, should be kept. Using more than one algorithm is recommended as they have complementary strengths for precision and recall; (2) Cassettes predicted by less than two programs, i.e. candidate cassettes, which are adjacent to *cas* genes, are added to the set of cassettes constructed in the first step; (3) candidate cassettes whose repeat consensus sequence are part of a repeat cluster containing repeats from already established as reliable CRISPR cassettes, are added to the set. This strategy can be further calibrated and improved when more data and features are available.

More advanced studies could investigate more on the *Cas* proteins and their relationship with CRISPR array repeats, for better understanding of the defence mechanism. Methods like CRISPRmap are important for further studies regarding CRISPR-*Cas* proteins systems. We should also pay more attention to the spacer sequences and, for example, on discovering how this defence mechanism recognizes short sequence motifs, known to be adjacent to the spacer precursor in the invaders genomes.

Additional files

Additional file 1: Information of the metagenome samples. The Chinese dataset consisted in faecal samples belonging to 145 Chinese individuals living in the south of China, collected by Shenzhen Second People's Hospital, Peking University Shenzhen Hospital and Medical Research Center of Guangdong General Hospital. The set comprised 71 type-2 diabetic individuals - classified DLF, DLM, DOF and DOM - and 74 healthy controls - NLF, NLM, NOF and NOM. (XLSX 27 kb)

Additional file 2: PILER-CR predicted CRISPR cassettes. This file shows the detailed info about the predicted CRISPR cassettes with PILER-CR software. The number of cassettes found in each individual, cassette origin scaffold, length (in base pairs), number of spacers and repeat consensus sequences. (XLSX 257 kb)

Additional file 3: A: BLASTClust results. Table S4A includes the collection of unique repeat sequences and Table S4B the results of BLASTClust alignment. (XLSX 60 kb)

Additional file 4: CRISPRmap results. The results for CRISPRmap contain the distribution of the unique repeat sequences for the 6 superclasses (A-F), sequence families and structural motifs. There is a summary table of the results as well as a detailed table with data about the sequence families. (XLSX 90 kb)

Additional file 5: Repeat sequence's BLAST hits against CRISPRdb database. The results for the BLAST search against CRISPRdb of known repeats include a table showing for each unique repeat consensus the corresponding hit in the database, strain name and NCBI code, as well as the *e-value*. The excel file also includes a summary of the results (percentage of unique repeats that found a hit in the database), as well as the distribution of the known repeats for the different Bacteria *phylum*. Those removed repeat sequences that have continuous indels at either ends of the hits are shown in Table S5C. (XLSX 147 kb)

Competing interests

The authors have declared that no competing interests exist.

Authors' contributions

XZ initiated and designed the project. TCM implemented the project with ZP helping on the computational tools. TCM, ZP and XZ analysed the results. TCM, ZP and XZ wrote the paper. All authors read and approved the final manuscript.

Acknowledgements

We thank Hongfei Cui for her help in pre-processing of the data. This project is supported by the National Basic Research Program of China (Project 2012CB316504). The publication costs for this article were partly funded by the Basic Research Fund of Tsinghua National Laboratory for Information Science and Technology. Tatiana Mangericao is supported by the Student Exchange Program between Tsinghua University, China and Instituto Superior Técnico, Portugal.

Declarations

This article has been published as part of *BMC Systems Biology* Volume 10 Supplement 1, 2016: Selected articles from the Fourteenth Asia Pacific Bioinformatics Conference (APBC 2016): Systems Biology. The full contents of the supplement are available online at http://www.biomedcentral.com/bmcsystbiol/supplements/10/S1.

Author details
[1]MOE Key Lab of Bioinformatics/Bioinformatics Division, TNLIST/Center for Synthetic and Systems Biology, and Department of Automation, Tsinghua University, Beijing 100084, China. [2]Department of Bioengineering, Instituto Superior Técnico (IST), Lisbon, Portugal. [3]School of Life Sciences, Tsinghua University, Beijing 100084, China.

Published: 11 January 2016

References

1. Li K, Bihan M, Yooseph S, Methé BA. Analyses of the microbial diversity across the human microbiome. PLoS One. 2012;7(6):e32118.
2. Qin J, Li R, Raes J, Arumugam M, Burgdorf KS, Manichanh C, et al. A human gut microbial gene catalogue established by metagenomic sequencing. Nature. 2010;464:59–65.
3. The Human Microbiome Project (HMP). [http://hmpdacc.org/overview/about.php].
4. The MetaHit project. [http://www.metahit.eu/index.php?id=453].
5. Conly JM, Stein K, Worobetz L, Rutledge-Harding S. The contribution of vitamin K2 (menaquinones) produced by the intestinal microflora to human nutritional requirements for vitamin K. Am J Gastroenterol. 1994;89(6):915–23.
6. Hill MJ. Intestinal flora and endogenous vitamin synthesis. Eur J Cancer Prev. 1997;6(2):S43–5.
7. Cummings JH. Microbial digestion of complex carbohydrates in man. Proc Nutr Soc. 1984;43(01):35–44.
8. Black DD, Jianhui D, Heng W. Regulation of apolipoprotein secretion by long-chain polyunsaturated fatty acids in newborn swine intestinal epithelial cells. Pediatr Res. 1998;43:98.
9. Backhed F. Host-bacterial mutualism in the human intestine. Science. 2005; 307(5717):1915–920.
10. Paterson S, Vogwill T, Buckling A, Benmayor R, Spiers AJ, Thomson NR, et al. Antagonistic coevolution accelerates molecular evolution. Nature. 2010; 464(7286):275–8.
11. Riley PA. Bacteriophages in autoimmune disease and other inflammatory conditions. Med Hypotheses. 2004;62(4):493–8.
12. Yosef I, Qimron U. Microbiology: how bacteria get spacers from invaders. Nature. 2015;519(7542):166–67.
13. Bhaya D, Davison M, Barrangou R. CRISPR-Cas systems in bacteria and archaea: versatile small RNAs for adaptive defense and regulation. Annu Rev Genet. 2011;45(1):273–97.
14. Rho M, Wu Y-W, Tang H, Doak TG, Ye Y. Diverse CRISPRs evolving in human microbiomes. PLoS Genet. 2012;8(6):e1002441.
15. Sorek R, Kunin V, Hugenholtz P. CRISPR–a widespread system that provides acquired resistance against phages in bacteria and archaea. Nat Rev Microbiol. 2008;6:181–6.
16. Pride DT, Sun CL, Salzman J, Rao N, Loomer P, Armitage GC, et al. Analysis of streptococcal CRISPRs from human saliva reveals substantial sequence diversity within and between subjects over time. Genome Res. 2011;21(1): 126–36.
17. Pride DT, Salzman J, Relman DA. Comparisons of clustered regularly interspaced short palindromic repeats and viromes in human saliva reveal bacterial adaptations to salivary viruses. Environ Microbiol. 2012;14(9):2564–76.
18. Robles-Sikisaka R, Ly M, Boehm T, Naidu M, Salzman J, Pride DT. Association between living environment and human oral viral ecology. ISME J. 2013; 7(9):1710–24.
19. Stern A, Mick E, Tirosh I, Sagy O, Sorek R. CRISPR targeting reveals a reservoir of common phages associated with the human gut microbiome. Genome Res. 2012;22(10):1985–94.
20. Mick E, Stern A, Sorek R. Holding a grudge: persisting anti-phage CRISPR immunity in multiple human gut microbiomes. RNA Biol. 2013;10(5):900–6.
21. Gogleva AA, Gelfand MS, Artamonova II. Comparative analysis of CRISPR cassettes from the human gut metagenomic contigs. BMC Genomics. 2014;15:202.
22. Qin J, Li Y, Cai Z, Li S, Zhu J, Zhang F, et al. A metagenome-wide association study of gut microbiota in type 2 diabetes. Nature. 2012;490(7418):55–60.
23. Edgar RC, Myers EW. PILER: Identification and classification of genomic repeats. Bioinformatics. 2005;21 Suppl 1:152–8.
24. BLASTClust tool. [http://www.animalgenome.org/blast/doc/blastclust.html].
25. Lange SJ, Alkhnbashi OS, Rose D, Will S, Backofen R. CRISPRmap: an automated classification of repeat conservation in prokaryotic adaptive immune systems. Nucleic Acids Res. 2013;41(17):8034–044.
26. Grissa I, Vergnaud G, Pourcel C. The CRISPRdb database and tools to display CRISPRs and to generate dictionaries of spacers and repeats. BMC Bioinformatics. 2007;8(1):172.
27. Bland C, Ramsey TL, Sabree F, Lowe M, Brown K, Kyrpides NC, et al. CRISPR recognition tool (CRT): a tool for automatic detection of clustered regularly interspaced palindromic repeats. BMC Bioinformatics. 2007;8(1):209.
28. CRISPR Finder online tool. [http://crispr.u-psud.fr/Server/CRISPRfinder.php].
29. Sun J, Chang EB. Exploring gut microbes in human health and disease: pushing the envelope. Genes Dis. 2014;1(2):132–39.
30. The normal bacterial flora of humans. Online textbook. [http://textbookofbacteriology.net/normalflora.html].
31. Russell WR, Hoyles L, Flint HJ, Dumas M-E. Colonic bacterial metabolites and human health. Curr Opin Microbiol. 2013;16(3):246–54.

Zhang *et al. BMC Systems Biology* 2015, **10**(Suppl 1):7
DOI 10.1186/s12918-015-0249-9

BMC Systems Biology

PROCEEDINGS Open Access

Generalized logical model based on network topology to capture the dynamical trends of cellular signaling pathways

Fan Zhang[1], Haoting Chen[1,2], Li Na Zhao[3], Hui Liu[1,4], Teresa M. Przytycka[5] and Jie Zheng[1,6,7*]

From The Fourteenth Asia Pacific Bioinformatics Conference (APBC 2016)
San Francisco, CA, USA. 11 - 13 January 2016

Abstract

Background: Cellular responses to extracellular perturbations require signaling pathways to capture and transmit the signals. However, the underlying molecular mechanisms of signal transduction are not yet fully understood, thus detailed and comprehensive models may not be available for all the signaling pathways. In particular, insufficient knowledge of parameters, which is a long-standing hindrance for quantitative kinetic modeling necessitates the use of parameter-free methods for modeling and simulation to capture dynamic properties of signaling pathways.

Results: We present a computational model that is able to simulate the graded responses to degradations, the sigmoidal biological relationships between signaling molecules and the effects of scheduled perturbations to the cells. The simulation results are validated using experimental data of protein phosphorylation, demonstrating that the proposed model is capable of capturing the main trend of protein activities during the process of signal transduction. Compared with existing simulators, our model has better performance on predicting the state transitions of signaling networks.

Conclusion: The proposed simulation tool provides a valuable resource for modeling cellular signaling pathways using a knowledge-based method.

Keywords: Generalized logical model, Signaling pathways, Dynamical system, Cancer

Introduction

Signal transduction plays an essential role in the cellular processes in which cell responds to extracellular perturbations (e.g., the exposure to drugs or ligands). According to the signals, the cell adjust its metabolism, shape, gene expression, etc., to adapt to the environment. It is widely believed that the dysregulation of signal transduction is one of the most important pathogeneses of many human diseases including cancer. Although high-throughput experimental data show a great potential for uncovering unprecedented details of biological systems, it is still challenging to understand signaling networks at systems level. Therefore, computational simulation, which is a systems biology approach, is highly desirable for the analysis of the underlying mechanisms of how the signals are transmitted through signaling pathways.

Many existing models are able to simulate the process of signal transduction, such as Boolean network models, fuzzy logic models and kinetic models based on ordinary differential equations (ODEs). Boolean network is a simple and promising framework for the modeling of protein-protein interactions and signaling pathways. It has been used with some success in identifying stable states of a system [1, 2], simulating the influence of deletion/knockout of important nodes in a network [3], predicting carcinogenesis and targeted therapy outcomes [4], reproducing the dynamics of the yeast MAPK pathways [5], modeling the mammalian cell cycle [6] and analyzing the behaviors

*Correspondence: zhengjie@ntu.edu.sg
[1] Biomedical Informatics Graduate Lab, School of Computer Engineering, Nanyang Technological University, Singapore 639798, Singapore
[6] Complexity Institute, Nanyang Technological University, Singapore 637723, Singapore
Full list of author information is available at the end of the article

of the apoptosis pathways [7, 8]. However, its inability of encoding graded responses and the typically sigmoidal biological relationships becomes a significant limitation since it is able to handle only binary values, i.e., a simple on/off state which is over–simplified compared with a real signaling network. To overcome this limitation, fuzzy logic models, which generalize the on/off characteristic to a continuous range from 0 to 1, have been successfully applied to analyzing the crosstalk among the TNF/EGF/Insulin-induced signaling pathways [9] and the liver cell responses to inflammatory stimuli [10]. However, a large amount of prior knowledge is needed for the assembly of the membership functions and logical rules for the fuzzy logic models. On the other hand, ODEs have also been applied to modeling various biological processes, such as the simulation of physiological responses of mammalian cells to the control of cell cycle [11], mathematical modeling of the mechanisms for regulating the differentiation of hematopoietic stem cells [12], discovery of signaling pathway rewiring [13] and exploring the dynamics of the pathways controlling cell apoptosis [14, 15]. However, ODEs-based models require a relatively detailed knowledge of kinetic parameters which is hardly available for all the pathways. Previously, we proposed a simulation tool called SimBoolNet [16] which is based on an extended Boolean network model. Although the performance of SimBoolNet in predicting protein activities was promising [17], it has limited capability of dealing with blocking effects, degradations and sequenced perturbations.

Here, we present a generalized logical model, which is capable of revealing the process of degradation, the sigmoidal biological relationships between molecules and the effects of scheduled perturbations to signaling networks. Compared with SimBoolNet [16] and GINsim [1] (a Boolean network based simulation tool), the proposed simulator can not only predict the stable states of the signal transduction system but also dynamically simulate the effects induced by various timing and ordering of perturbations. The simulations are validated using experimental phosphoproteomics data of breast cancer cells perturbed by different combinations of drug additions [18]. The simulated time-series data of protein activity levels show significant correlations with the real time-course data, thereby demonstrating that the proposed model is able to capture the key features of the signaling pathways.

Methods

Computational model for dynamical simulation

Our model takes a directed graph as the input network to do simulation. In the network, each node denotes a molecular species (e.g., a protein) and each directed edge (u, v) represents signal transduction from node u to node v. One variable with a nonnegative value from 0 (fully inhibited) to 1 (fully activated) is associated with each node to represent the activity level of the protein. The edge weight is also a variable with the value between 0 and 1 to denote the strength of the interaction and a sign ('+' or '–') to denote the type of the interaction (i.e., positive means activation and negative blockage). Users can select input nodes and a virtual node is added upstream of all the input nodes. This virtual node is the abstraction of the extracellular environment which is able to generate signals that can stimulate or inhibit the input nodes (e.g., the receptors) of the signaling network. Given random initial activities for all the nodes in the network, their states will be updated synchronously based on their own previous states and the incoming signals from their parent nodes, according to Eq. (1). In this formula, X_t is the activity level of node x at time t, d is a pre-defined parameter denoting the degradation rate of the activated x from time $t-1$ to t, A_i (or B_j) is the signals (i.e., the activity level times the edge weight) transmitted from the ith activating (or jth inhibiting) parent node upstream of x; $\left[1-\prod(1-A_i)\right]$ is the overall activating effect generated by all the incoming activating signals and $\prod(1-B_j)$ is the probability that the incoming inhibiting signals do not affect x. Altogether, they act on the inactivated form of x at the $(t-1)$th iteration (i.e., $(1-X_{t-1})$). The blocking effect acting on the activated form of x follows the similar logic. Overall, the updated state is defined by the nondegraded part and the newly activated part minus the inhibited part. Given a user-defined number of simulation iterations, the discrete steps are employed to approximate the process that the activity levels of the nodes change over time. Figure 1 shows the workflow of the simulation using the proposed model.

$$X_t = (1-d)\times X_{t-1} + \left[1-\prod(1-A_i)\right]\times\prod(1-B_j) \times (1-X_{t-1}) - \prod(1-A_i)\times\left[1-\prod(1-B_j)\right]\times X_{t-1} \quad (1)$$

It is suggested that cells respond to external perturbations through a time–dependent (e.g., the schedule and duration of drug addition) process [18]. Wet–lab experiments have shown that different ordering and timing of drug additions have significantly different drug effects, such as inducing specific alterations of signaling pathways [10, 19, 20] and showing different efficiencies in killing cancer cells [18]. However, most existing simulation tools are not able to accommodate the time-staggered design of drug treatments in biological experiments. Therefore, our model introduces time-staggered perturbations to explore the effects of not only dosage, but also the schedule and duration of the perturbations to cellular systems with a knowledge-based model. The timing and the order of drug additions can be specified by users as parameters.

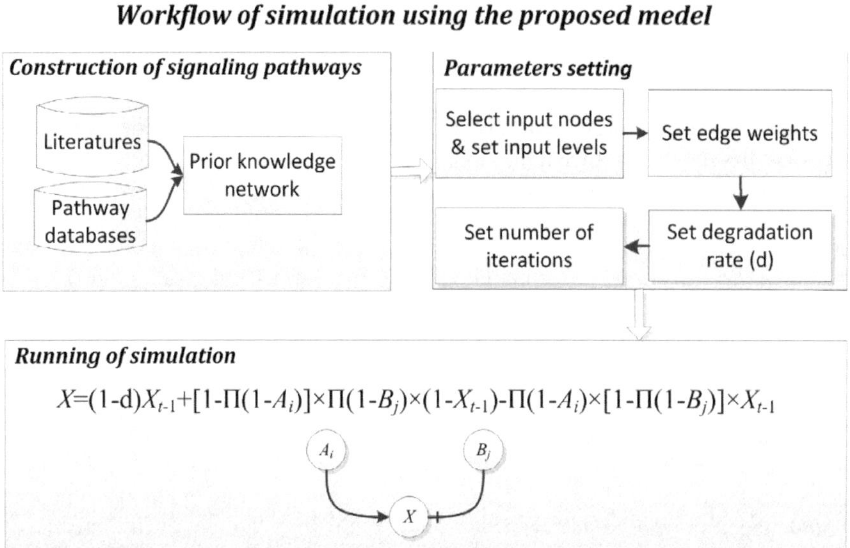

Fig. 1 The workflow of simulation using the proposed model. The activity level of X is calculated based on its own previous activity with a degradation rate, and the activation and inhibition effects produced by the signals transmitted from its parent nodes

For example, the drug can start to affect its target at the kth simulation iteration with a user-defined k. The target, input level, type of interaction (stimulation or inhibition) and schedule of the perturbations can all be specified according to user's design of experiment.

Network structure

A signaling network [21] (Fig. 2) is constructed according to well-known pathway databases (GeneGO MetaCore [22] and KEGG [23]). The network comprises 35 nodes, indicating 32 signaling proteins or stimuli and 3 cell fates, and 57 edges denoting signal transduction from the source nodes to the target nodes. Dark blue nodes in the pathways represent 21 signaling proteins that have been experimentally measured in [18]. Activation and inhibition interactions are denoted by green arrow and red flat-head edges, respectively.

Results

Performance comparison on simulating signaling responses to perturbations

For comparison, we run simulations using our program based on SimBoolNet [16], GINsim [1] and the proposed model on the same network in Fig. 2. Two different inputs are introduced: (1) the input levels of EGFR and TNFR are set to 0.5 and 0.8, respectively, (2) EGFR inhibitor is added at the 10th iteration of simulation and TNFR is activated with input level 0.8 at the 20th iteration. The number of simulation iterations is set to 100. For the proposed model, the full inhibition is denoted as -1 and the perturbations can be executed at any iteration during the simulation. The degradation rate d is set to 0.2. For SimBoolNet and GINsim, the blocking effect is represented by setting the activity level of EGFR to 0 from the very first step of simulation which, to our understand, is unlikely to be a precise representation. There should be a process for the inhibitor to reduce the activity level of its target, especially when the inhibitor is not added at the beginning. The edge weights of activation and blockage are set to 0.7 and 0.8, respectively. GINsim simulation, on the other hand, does not accept parameters for edge weights and the number of iterations, and executes synchronously until the system reaches the stable state. GINsim also supports the asynchronous mode, but it is a time-consuming task due to a much larger search space than with the synchronous mode. We did not get any result from running GINsim in asynchronous mode on our network (Fig. 2) within endurable time using a desktop PC (Dell Precision T3600 workstation with Intel Xeon CPU E5–1620, 8 GB RAM and Windows 7 Professional 64–bit operating system). We have also tried other different settings of input level, edge weight and degradation rate, and the results are shown in section "Model comparison and validation with real data".

Figure 3a, b and c show the simulation results of three proteins (i.e., EGFR, TNFR and ERK) using the proposed model under the two different inputs. It can be seen that the trends of the activities of the input nodes follow an approximately sigmoidal function (the blue curves in Fig. 3a and b). When the EGFR inhibitor is added at the 10th step, the EGFR activity drops sharply within a few steps (the red curve in Fig. 3a). Consequently, the activity

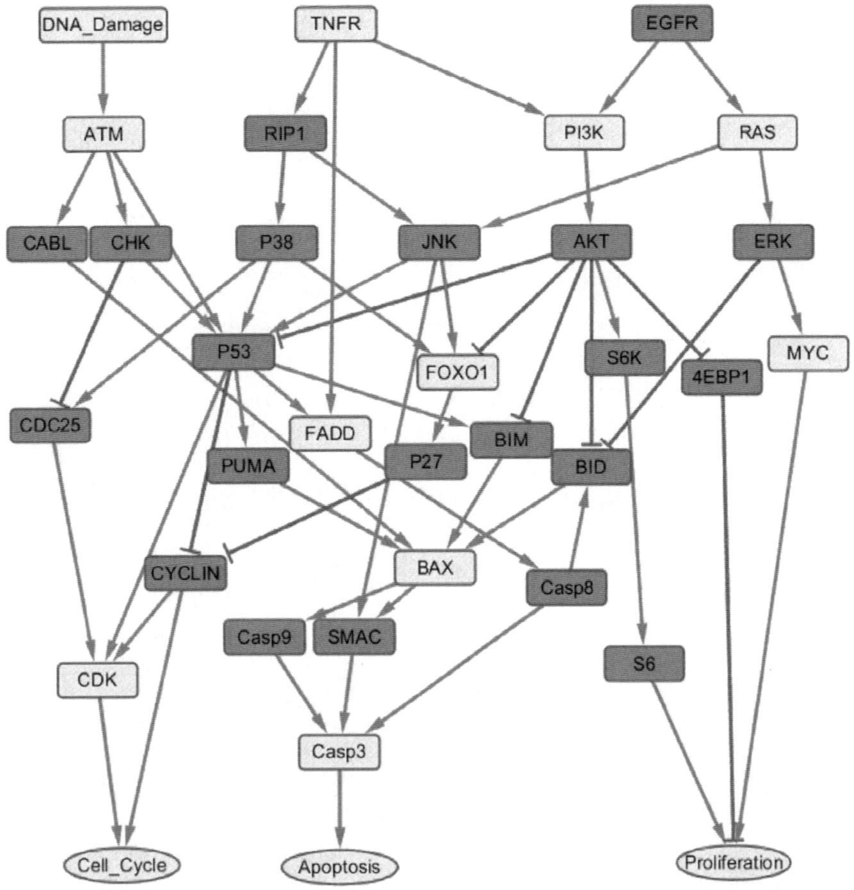

Fig. 2 Signaling pathways constructed based on the dataset in [18]. Round rectangles and ellipses represent signaling proteins (or stimuli) and cell fates, respectively. The signals that have measurements in the dataset [18] are represented by dark blue nodes. Each activation interaction is denoted as a green edge with an arrow head and each inhibition interaction is represented by a red edge with a flat-head

of its downstream node ERK (the red curve in Fig. 3c) decreases with some time delay because it takes some time for signals to be transmitted from EGFR to ERK. Under input 2, the activity of TNFR first decreases from a random initial value with a degradation rate (here is 0.2) and increases to the maximum (the input level 0.8) almost immediately at the 20th iteration (the red curve in Fig. 3b). In contrast, SimBoolNet has a limited capability of dealing with blocking effect, degradation or scheduled perturbations. We can see from the blue curves in Fig. 3d, e and f (which are outputs of SimBoolNet) that the activity levels increase monotonically from 0 to a maximum, considered as the stable state, which is unlikely to be precise in the biochemical reactions. Moreover, the inhibiting effect in SimBoolNet is represented by keeping the activity level of the target node to 0, which is unlikely to be realistic especially when the inhibiting effect should be produced in the middle of the simulation (red curves in Fig. 3d and f). Given the initial states, GINsim is able to identify the stable states [2]. However, it has similar limitations for dealing with scheduled drug additions and degradations (Fig. 3g to i).

We then compared the computational time of SimBoolNet, GINsim (in asynchronous mode) and the proposed model. In addition to the network shown in Fig. 2, a small network with 5 nodes and a large network with 500 nodes were constructed. The edges of these two networks were randomly generated. Totally, the numbers of the nodes/edges of the three networks are 5/13, 35/57 and 500/6124, respectively. The hardware being employed was Dell Precision T3600 workstation with Intel Xeon CPU E5–1620, 8 GB RAM and Windows 7 Professional 64–bit operating system. Table 1 shows the computational time of SimBoolNet, GINsim and our model on the three networks. It can be seen that the simulation with our model is faster than SimBoolNet, whereas slightly slower than GINsim when the network is small. However, GINsim has its limitation for large networks. The simulation using

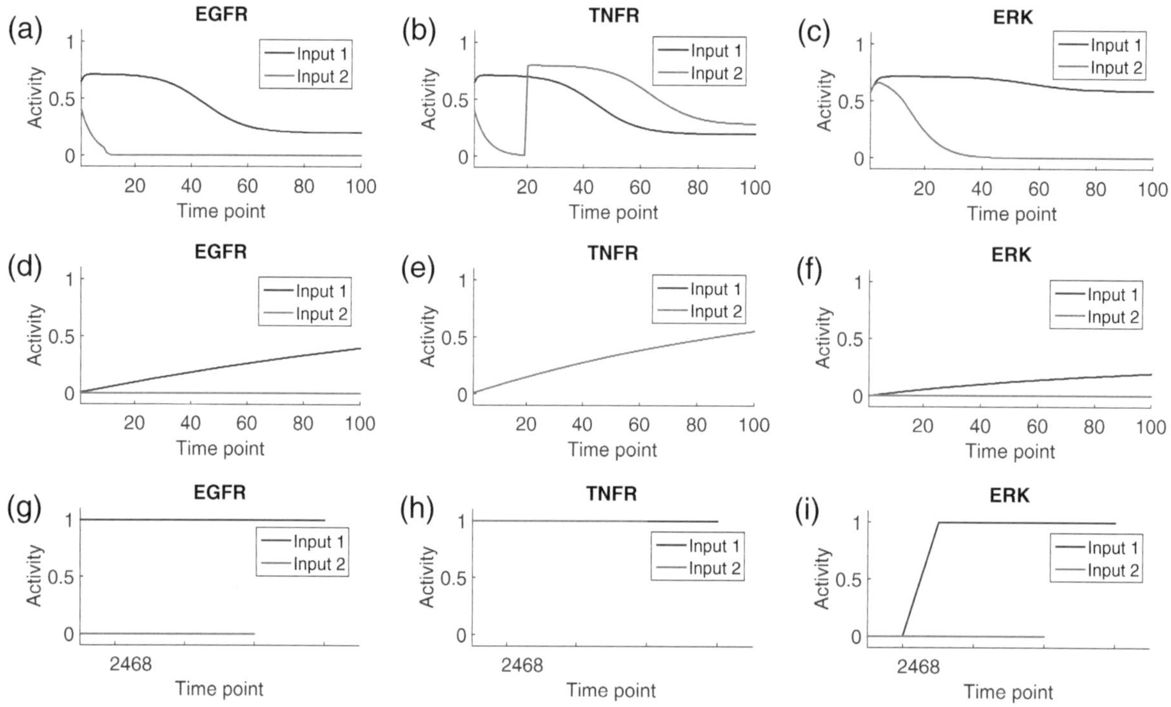

Fig. 3 Comparison of simulation-based predictions made by using SimBoolNet, GINsim and the proposed simulator. Two different inputs are used: (1) EGFR and TNFR are perturbed at the beginning of the simulation with input levels 0.5 and 0.8, respectively, and (2) EGFR is inhibited at the 10th iteration with input level -1 and TNFR is activated at the 20th iteration with input level 0.8. The edge weights of activation and blockage are set to 0.7 and 0.8, respectively, for both inputs. The simulation is executed for 100 iterations. **a**–**c** The plots of simulation results using the proposed model for EGFR, TNFR and ERK, under two different inputs. **d**–**f** The plots of SimBoolNet results under input 1 and 2, respectively. Please note that the two curves are totally overlap for TNFR in (**e**). **g**–**i** GINsim simulation results for EGFR, TNFR and ERK. Each run of GINsim simulation executes 8 iterations before reaching the stable state

GINsim (in synchronous mode) on the third network with 500 nodes did not produce any result within endurable time.

The influence of different initial values on the simulation results was explored. Figure 4a shows the simulation curves of EGFR under the aforementioned input 1. Three different initial values of EGFR, namely 0.1, 0.5 and 0.9, were selected. Conclusions can be drawn from the plot that the initial value affects the time spent in reaching the stable state, but the overall trends of the state transitions and the levels of the stable states remain stable.

Table 1 Comparison of the computational time required for SimBoolNet, GINsim and the proposed model

Number of nodes/edges	SimBoolNet	GINsim	Our model
5/13	9.51s	< 0.1s	0.09s
35/57	10.73s	< 0.1s	0.43s
500/6124	72.88s	Not applicable	10.68s

Three networks are employed for simulation, where the numbers of the nodes/edges are 5/13, 35/57 and 500/6124, respectively. The simulation using GINsim on the third network could not finish within endurable time (hence marked "Not applicable")

We went on to explore the robustness of the proposed model to the variations of edge weights. In principle, we randomly generated the edge weights to run the simulations, and then checked if the activity trend of each protein remained unchanged. For a specific input (i.e., the input levels of EGFR and TNFR are both 0.5), we first randomized the weights of all the 57 edges and ran the simulation for 100 times as the background group. For each protein, the mean activity at each time point was regarded as the background trend over time. Next, we further generated 50 groups of simulations, each group consisting of 100 simulations of randomly generated edge weights. For each group, the mean activity trend of each protein was used to calculate the correlation with the background trend. Figure 4b gives the distribution of the 50 correlations between the simulated and the background trends for the 32 non-receptor nodes (ignoring the receptors EGFR, TNFR and DNA Damage because they have no incoming edges), and the proteins are ranked based on the median of the correlations. It can be seen that 21 out of 32 proteins (i.e., from PI3K to AKT) have the medians of the correlations larger than 0.8; 10 out of 32 have

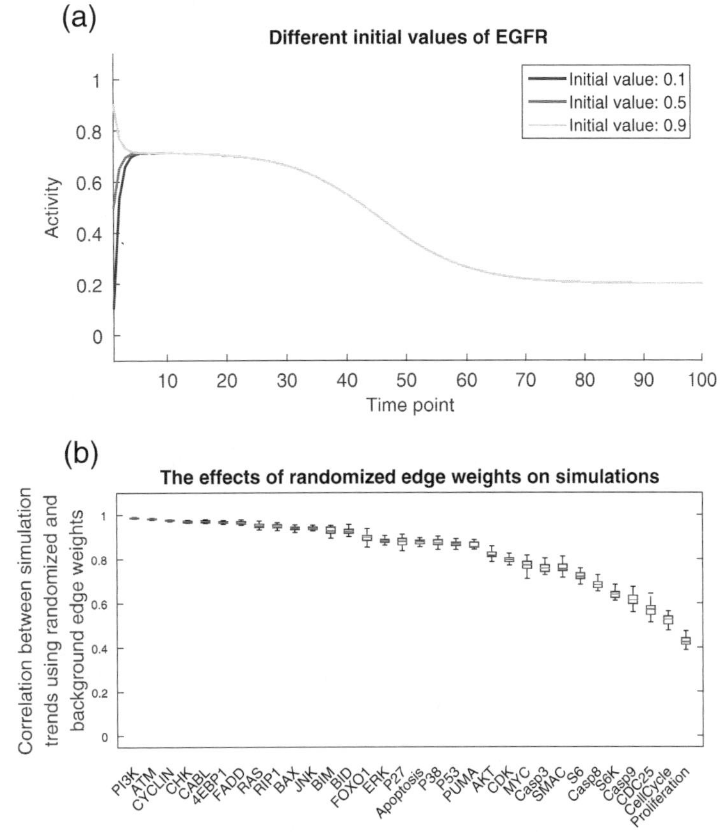

Fig. 4 a The influence of different initial values on simulations. The initial value does not change the main trend of the state transition and the level of the stable state. **b** The influence of different edge weights to simulations. Each of the 32 signals shows a small range of the correlations between the simulated and the background trends, indicating that the proposed model is robust to different edge weights

the medians falling into the interval 0.5 to 0.8; and only one (i.e., Proliferation) has the median which is lower than 0.5. Moreover, all the 32 signaling proteins show small ranges of the correlations between the simulated and the background trends, indicating that the proposed model is robust for capturing the dynamical trends of the signal transduction process under different settings of edge weights.

Model comparison and validation with real data

To estimate the performance of the proposed model, the simulated results using the network in Fig. 2 are compared with a real signaling dataset [18] containing the time-series phosphoproteomics data. In the dataset, perturbations (i.e., inhibitor of EGFR or stimuli of DNA damage or both) were applied to cells of the breast cancer cell line BT20. For each perturbation, activity levels of 32 signaling proteins (21 out of 35 are included in the network in Fig. 2) were measured at 8 time points. To simulate the perturbations, we use (*i*) half activation input signals (0.5) to represent the control situation where no stimuli or inhibitor is added; (*ii*) activation input signals (+1) to represent the addition of stimuli, i.e., the targets are fully activated; and (*iii*) inhibition input signals (−1) to represent the effect of inhibitors, i.e., the activity of the targets are suppressed.

We first simulated the dynamics of signaling without any drug addition, as a control dataset. The receptors EGFR and TNFR were selected as the input nodes with input levels both equal to 0.5. The edge weights of both activation and inhibition were set to 0.8. The simulation was executed for a number of iterations that is a multiple of 8 since there are 8 time points in the real dataset (here we chose 32 because most of the nodes reach the stable states after 30 steps). We then used Spearman correlation coefficient to measure the goodness of fit between the simulated and real data to evaluate the performance of our model. Since calculating Spearman correlation requires the two vectors to have the same length, 8 time points of the simulated data (which is the same as the real data) were extracted from the simulated time-series with equal interval, e.g., the 4th, the 8th, ... and the 32nd time points.

Table 2 Spearman correlations between simulated and real data

	SMAC	4EBP1	p53	ERK	S6	S6K	CABL
Correlation	0.95	0.88	0.69	0.71	0.31	0.26	−0.90
	Casp9	CDC25	CHK	p27	PUMA	AKT	JNK
Correlation	−0.60	0.62	0.21	−0.90	0.92	0.79	0.43
	p38	BIM	BID	RIP1	CYCLIN	Casp8	EGFR
Correlation	0.67	0.93	0.43	0.76	0.86	0.90	0.62

Table 2 gives the Spearman correlation coefficients for the 21 measured signaling proteins. We can see that for 16 out of 21 signaling proteins the Spearman correlation coefficients are relatively high (larger than 0.6) indicating that the simulation fits well to the real data. For the proteins with relatively poor correlation coefficients, it might because of over-simplification such as missing some cell line specific interactions in the network. For example, both S6 and S6K have only one incoming edge which is unlikely in the real cellular signaling network [24].

On the other hand, since the simulation is an approximately continuous process while the wet–lab experiment measured only at a few selected time points, we can scale the discrete dots in the real data to align with the 32 simulated steps by multiplying the index of each time point in the real data by 4 to match the index of each iteration in the simulation. For comparison, SimBoolNet was also employed to do the simulation under the same inputs, i.e., the input levels of EGFR and TNFR are both 0.5 and the edge weights are all 0.8. Figure 5 shows the plots of the simulated (scatter plots each with a trend line) and real data (scatter plots) for EGFR, Caspase 8, p53 and JNK. The blue and green curves represent the simulated data using the proposed model and SimBoolNet, respectively, while the red dots represent the real data (normalized to the same scale as the simulated data). In Fig. 5a, the simulation curve of EGFR captures the trend of a quick drop from a plateau to the flat bottom although there is a time delay. The slow and small decrease of the activity of Caspase 8 is also captured by our simulation as shown in Fig. 5b. For p53, the simulation shows a high decreasing rate at the beginning and the activity of p53 reaches the stable state quickly, which agrees with the trend of the real data. The gradient of the JNK simulation curve also fits the real data although the starting point of the dropping lags behind, which suggests that it is important to scale the simulated timing to accurately match the real time points. To address the issue of time scale would be an important future work. By contrast, the simulated trends of SimBoolNet are mainly monotonically increasing which cannot fit the real data well.

We then simulated the state transitions of signaling pathways under the perturbations in Table 3. For example, we gave half activating (+0.5) and full blocking (−1) signals to TNFR and EGFR, respectively, to simulate the addition of EGFR inhibitor. The edge weights of activation and inhibition were set to 0.7 and 0.8, respectively. The number of simulation iterations was set to 32, and the degradation rate 0.2.

It has been known that the in vivo drug effect on the signaling pathways is through the change of the activities of the proteins downstream of the drug targets [20, 25].

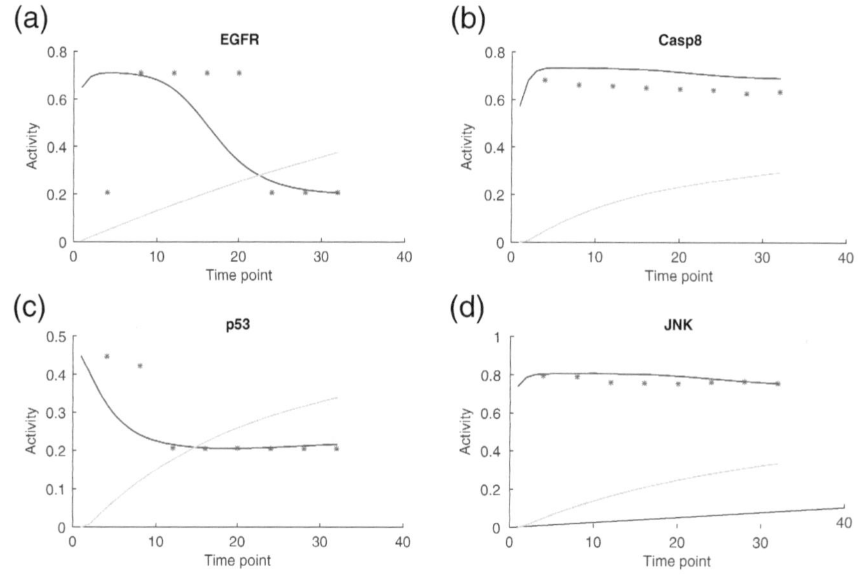

Fig. 5 The plots of the simulated and real data in the control group. The blue and green curves are the simulated data using the proposed model and SimBoolNet, respectively. The red dots are the real data. The four panels (**a**), (**b**), (**c**) and (**d**) correspond to the plots of four proteins EGFR, Casp8, p53 and JNK, respectively

Table 3 Inputs of the simulation in section "Model comparison and validation with real data"

	EGFR	TNFR	DNA damage
Control	+0.5	+0.5	random value
EGFR inhibitor	−1	+0.5	random value
DNA damage stimuli	random value	+0.5	+1
Both drugs	−1	+0.5	+1

The symbols '+' and '−' represent the perturbation types, i.e., activation and inhibition. The columns are the input nodes of simulation and the rows are various conditions including the treatments of biological experiments with drugs corresponding to wet-lab experiments in [18]

Therefore, for these downstream proteins, if the simulation data based on the perturbed inputs (rows 2 to 4 in Table 3) fit the experimental data better than the simulations using the control input (rows 1 in Table 3), we believe our simulator captures the main influences of the drugs on the networks. For example, the second and the third columns of Table 4 show the correlation coefficients between the real data and the simulations using the control input and EGFR inhibitor input (rows 1 and 2 in Table 3), respectively. It can be seen that, for the nodes downstream of EGFR (such as AKT, S6, S6K and BIM), the fitness of simulations to the drugged data (biological measurements treated with drug which targets at EGFR) is significantly improved when the control input is replaced by the EGFR inhibitor input. A similar conclusion can be drawn for the proteins downstream of DNA damage, including p53, CABL and Caspase 8, when the simulations using the control input and DNA damage stimuli as input are employed to fit the real data (the forth and the fifth columns of Table 4). Moveover, it is believed that the treatment with both drugs have a higher efficiency on killing cancer cells than that with a single drug [18] which may be explained by the characteristics of SMAC and Caspase 8 (both upstream of Caspase 3), in that the improvement of fitness is mainly achieved under the treatment of both drugs (the last two columns of Table 4). There are proteins, such as ERK and CHK (downstream of EGFR and DNA damage, respectively), that do not follow the above patterns, probably due to our insufficient knowledge about signaling pathways.

Table 4 Goodness of fit of the simulations to the real experimental measurements under drug treatments

	Control input	EGFR inhibitor	Control input	DNA damage stimuli	Control input	Both drugs
AKT	0.64	0.81	−0.40	−0.54	0.75	0.93
ERK	0.32	0.40	−0.30	−0.39	0.19	0.23
S6	0.53	0.86	0.42	0.57	0.52	0.89
S6K	0.49	0.84	0.35	0.66	0.58	0.96
4EBP1	0.67	0.67	−0.33	-0.33	0.92	0.92
BIM	0.32	0.72	0.35	0.65	−0.15	−0.38
BID	0.30	0.37	0.03	0.03	0.16	0.16
JNK	0.02	−0.07	−0.47	−0.58	−0.06	0.01
p53	−0.04	−0.45	0.21	0.83	0.01	−0.09
CABL	−0.45	−0.52	0.04	0.55	0.21	0.78
CHK	−0.12	−0.18	−0.25	−0.28	−0.30	−0.69
CDC25	0.29	0.23	0.15	0.33	0.18	0.64
Casp8	0.06	−0.06	0.14	0.72	0.26	0.68
SMAC	−0.05	0.08	−0.71	−0.87	0.48	0.99

For example, the second column is the correlation coefficients between the simulations using control input (row 1 in Table 3) and the biological measurements treated with the drug that targets EGFR

Discussion and conclusion

In this paper, we present a model to dynamically simulate the process of intracellular signal transduction. According to a phosphoproteomics dataset [18], we constructed a network, which comprises 35 nodes (21 nodes have experimental measurements) and 57 edges, to do the simulation. The state of each node is calculated based on its own previous state with a degradation rate, and the activation and inhibition effects produced by the signals transmitted from its parent nodes. Different combinations of perturbations were applied to the network. The simulation results have been evaluated with the real data, demonstrating that our simulator has the ability to grasp the main dynamical trends of signal transduction. Compared with SimBoolNet [16] and GINsim [1], the proposed model shows promising performance in revealing the graded responses, the sigmoidal biological relationships and the effects of scheduled perturbations to a signaling network. Moreover, by testing the proposed model with different values of parameters (e.g., the initial activities of the proteins and the edge weights), we have shown that our method performs robustly in revealing the dynamics of the signaling pathways when the prior knowledge of the network topology is reliable.

Studying the cell responses to extracellular perturbations is a major endeavor for biomedical research and pharmaceutical industry. With the development of high-throughput experiments, large-scale data are available to help uncover important biological mechanisms at systems level. However, most existing data-driven methods [25] have limitations in revealing underlying molecular mechanisms. Therefore, computational simulation based on the integration of prior knowledge with data shows a great potential for revealing insights into the dynamical system of signal transduction, and thus would be a valuable complement to the data-driven methods. Although the proposed model is still limited in mapping the simulation steps to the experimental time points, we believe that the integration of both knowledge and data, such as learning the edge weights from experimental data, would be a powerful approach to understanding the signal transduction networks. In addition, generalization of the present model, which uses the synchronous updating scheme, such that it is able to deal with asynchronous dynamics, e.g., updating a randomly selected node at each time point, would also be a valuable future direction.

Competing interests
The authors declare that they have no competing interests.

Authors' contributions
FZ designed study; acquired, analysed and interpreted data; implemented main experiments; drafted manuscript. HC and LNZ implemented part of experiments. HL and TMP participated in discussion, interpreted data, and revised the manuscript. JZ conceived the study, participated in conceptualization and discussion, critically reviewed and revised the manuscript and gave final approval for submission. All authors read and approved the final manuscript.

Acknowledgements
This research has been supported by MOE AcRF Tier 2 grant (ARC39/13, MOE2013-T2-1-079), Ministry of Education, Singapore. It is also supported in part by the Intramural Research Program of the National Library of Medicine, National Institutes of Health, U.S.A. The content of this publication does not necessarily reflect the views or policies of the Department of Health and Human Services, nor does mention of trade names, commercial products, or organizations imply endorsement by the U.S. Government. The publication cost for this article was funded by MOE AcRF Tier 2 grant (ARC39/13, MOE2013-T2-1-079), Ministry of Education, Singapore.

Declarations
This article has been published as part of *BMC Systems Biology* Volume 10 Supplement 1, 2016: Selected articles from the Fourteenth Asia Pacific Bioinformatics Conference (APBC 2016): Systems Biology. The full contents of the supplement are available online at http://www.biomedcentral.com/bmcsystbiol/supplements/10/S1.

Author details
[1]Biomedical Informatics Graduate Lab, School of Computer Engineering, Nanyang Technological University, Singapore 639798, Singapore. [2]Department of Industrial Engineering and Operations Research, Columbia University, New York, NY 10027, USA. [3]Bioinformatics Institute, Agency for Science, Technology and Research, Singapore 138671, Singapore. [4]Lab of Information Management, Changzhou University, Changzhou, Jiangsu 213164, China. [5]National Center for Biotechnology Information, NLM/NIH, Bethesda, MD 20894, USA. [6]Complexity Institute, Nanyang Technological University, Singapore 637723, Singapore. [7]Genome Institute of Singapore, Agency for Science, Technology and Research, Singapore 138672, Singapore.

Published: 11 January 2016

References
1. Chaouiya C, Naldi A, Thieffry D. Logical modelling of gene regulatory networks with ginsim. Methods Mol Biol. 2012;804:463–79.
2. Calzone L, Tournier L, Fourquet S, Thieffry D, Zhivotovsky B, Barillot E, et al. Mathematical modelling of cell-fate decision in response to death receptor engagement. PLoS Comput Biol. 2010;6(3):e1000702.
3. Gupta S, Bisht SS, Kukreti R, Jain S, Brahmachari SK. Boolean network analysis of a neurotransmitter signaling pathway. J Theor Biol. 2007;244(3):463–9.
4. Fumiã HF, Martins ML. Boolean network model for cancer pathways: predicting carcinogenesis and targeted therapy outcomes. PLoS One. 2013;8(7):69008.
5. Mori T, Flöttmann M, Krantz M, Akutsu T, Klipp E. Stochastic simulation of boolean rxncon models: towards quantitative analysis of large signaling networks. BMC Syst Biol. 2015;9(45).
6. Fauré A, Naldi A, Chaouiya C, Thieffry D. Dynamical analysis of a generic boolean model for the control of the mammalian cell cycle. Bioinformatics. 2006;22(14):124–31.
7. Mai Z, Liu H. Boolean network-based analysis of the apoptosis network: irreversible apoptosis and stable surviving. J Theor Biol. 2009;259(4):760–9.
8. Schlatter R, Schmich K, Vizcarra IA, Scheurich P, Sauter T, Borner C, Ederer M, et al. On/off and beyond–a boolean model of apoptosis. PLoS Comput Biol. 2009;5(12):e1000595.
9. Aldridge BB, Saez-Rodriguez J, Muhlich JL, Sorger PK, Lauffenburger DA. Fuzzy logic analysis of kinase pathway crosstalk in TNF/EGF/insulin-induced signaling. PLoS Comput Biol. 2009;5(4):e1000340.
10. Morris MK, Saez-Rodriguez J, Clarke DC, Sorger PK, Lauffenburger DA. Training signaling pathway maps to biochemical data with constrained fuzzy logic: Quantitative analysis of liver cell responses to inflammatory stimuli. PLoS Comput Biol. 2011;7(3):e1001099.
11. Novák B, Tyson JJ. A model for restriction point control of the mammalian cell cycle. J Theor Biol. 2004;230(4):567–79.
12. Tian T, Smith-Miles K. Mathematical modeling of gata-switching for regulating the differentiation of hematopoietic stem cell. BMC Syst Biol. 2014;8(1):S8.
13. Mishra SK, Bhowmick SS, Chua HE, Zhang F, Zheng J. Computational cell fate modelling for discovery of rewiring in apoptotic network for enhanced cancer drug sensitivity. BMC Syst Biol. 2015;9(Suppl 1):S4.
14. Albeck JG, Burke JM, Aldridge BB, Zhang M, Lauffenburger DA, Sorger PK. Quantitative analysis of pathways controlling extrinsic apoptosis in single cells. Mol Cell. 2008;30(1):11–25.
15. Neumann L, Pforr C, Beaudouin J, Pappa A, Fricker N, Krammer PH, et al. Dynamics within the cd95 death-inducing signaling complex decide life and death of cells. Mol Syst Biol. 2010;6:352.
16. Zheng J, Zhang D, Przytycki PF, Zielinski R, Capala J, Przytycka TM. Simboolnet–a cytoscape plugin for dynamic simulation of signaling networks. Bioinformatics. 2010;26(1):141–2.
17. Zielinski R, Przytycki PF, Zheng J, Zhang D, Przytycka TM, Capala J. The crosstalk between egf, igf, and insulin cell signaling pathways–computational and experimental analysis. BMC Syst Biol. 2009;4(3).
18. Lee MJ, Ye AS, Gardino AK, Heijink AM, Sorger PK, MacBeath G, et al. Sequential application of anticancer drugs enhances cell death by rewiring apoptotic signaling networks. Cell. 2012;149(4):780–94.
19. Sharan R, Karp RM. Reconstructing boolean models of signaling. J Comput Biol. 2013;20(3):249–57.
20. Mitsos A, Melas IN, Siminelakis P, Chairakaki AD, Saez-Rodriguez J, Alexopoulos LG. Identifying drug effects via pathway alterations using an integer linear programming optimization formulation on phosphoproteomic data. PLoS Comput Biol. 2009;5(12).
21. Zhang F, Wu M, Li XJ, Li XL, Kwoh CK, Zheng J. Predicting essential genes and synthetic lethality via influence propagation in signaling pathways of cancer cell fates. J Bioinformatics Comput Biol. 2015;11(1541002).
22. Ekins S, Nikolsky Y, Bugrim A, Kirillov E, Nikolskaya T. Pathway mapping tools for analysis of high content data. Methods Mol Biol. 2007;356: 319–350.
23. Kanehisa M, Goto S. Kegg: kyoto encyclopedia of genes and genomes. Nature. 2000;1(28):27–30.
24. Chung J, Grammer TC, Lemon KP, Kazlauskas A, Blenis J. Pdgf- and insulin-dependent pp70s6k activation mediated by phosphatidylinositol-3-oh kinase. Nature. 1994;370(6484):71–5.
25. Zhang F, Kwoh CK, Wu M, Zheng J. Data-driven prediction of cancer cell fates with a nonlinear model of signaling pathways. The Proceedings of the 5th ACM Conference on Bioinformatics, Computational Biology, and Health Informatics. New York, NY, USA: ACM; 2014, pp. 436–444. http://dl.acm.org/citation.cfm?id=2649425.

Qin et al. BMC Systems Biology 2016, 10(Suppl 1):9
DOI 10.1186/s12918-015-0250-3

BMC Systems Biology

PROCEEDINGS Open Access

Meta-analysis of sex differences in gene expression in schizophrenia

Wenyi Qin[1], Cong Liu[1], Monsheel Sodhi[2*] and Hui Lu[1,3*]

From The Fourteenth Asia Pacific Bioinformatics Conference (APBC 2016)
San Francisco, CA, USA. 11 - 13 January 2016

Abstract

Schizophrenia is a severe psychiatric disorder which influences around 1 % of the worldwide population. Differences between male and female patients with schizophrenia have been noted. There is an earlier age of onset in males compared with females with this diagnosis, and in addition, there are differences in symptom profiles between the sexes. The underlying molecular mechanism of sex difference remains unclear. Here we present a comprehensive analysis to reveal the sex differences in gene expression in schizophrenia with stringent statistics criteria. We compiled a data set consisting of 89 male controls, 90 male schizophrenia patients, 35 female controls and 32 female schizophrenia patients from six independent studies of the prefrontal cortex (PFC) in postmortem brain. When we tested for a sex by diagnosis interaction on gene expression, 23 genes were up-regulated and 23 genes were down-regulated in the male group (q-value < 0.05), several genes are related to energy metabolism, while 4 genes are located on sex chromosome. No genes were statistically significant in the female group when multiple testing correction were conducted (q-value <0.05), most likely due to the small sample size. Our protocol and results from the male group provide a starting point for identifying the underlying different mechanism between male and female schizophrenia patients.

Keywords: Dorsolateral prefrontal cortex, Microarray analysis, Postmortem brain

Introduction

Schizophrenia is a severe psychiatric disorder with a population frequency of approximately 1 % [30]. Schizophrenia is a syndrome characterized by positive symptoms such as delusions, hallucinations, disorganized speech and grossly disorganized or catatonic behavior; negative symptoms such as affective flatterning, alogia, or avolition [26, 30]. The etiology and pathophysiological mechanisms of the disorder are not well understood. Research to date indicates that schizophrenia is a multifactorial neurodevelopmental impairment of the brain that could be attributed to both genetic and environmental factors [2, 7, 21, 23, 28]. Gene expression is readout of both the genetic and the environmental factors that contribute to the pathophysiology of schizophrenia. Analysis of human postmortem brain is a powerful approach for the identification of risk factors for schizophrenia, because unlike studies of living patients, detailed molecular investigations can be performed directly in the critical brain regions of interest.

The pathophysiology of schizophrenia is likely to be different between males and females. Sex differences have been noted in several epidemiological analyses. For example, several studies indicate that men have a slightly higher incidence of schizophrenia compared with women. In addition, males have an earlier age of onset of schizophrenia, between 18–25 years of age, compared with the female age of onset which is 25–35 years [20]. The symptoms exhibited by male and female patients with schizophrenia also differ. Males tend to have a greater vulnerability to negative symptoms and traits of disorganization, while females more frequently exhibit depressive symptoms [20]. These findings suggest that

* Correspondence: mssodhi@uic.edu; huilu@uic.edu
[2]Department of Pharmacy Practice and Center for Pharmaceutical Biotechnology, University of Illinois at Chicago, 900 S Ashland Ave mc870, Chicago, IL 60607, USA
[1]Department of Bioengineering, University of Illinois at Chicago, 851 S. Morgan, Rm 218, Chicago, IL 60607, USA
Full list of author information is available at the end of the article

different underlying mechanisms of schizophrenia occur in males and females. Therefore, we have investigated sex differences in schizophrenia to gain a better understanding of the pathophysiological mechanisms underpinning this disorder.

To identify the biological factors involved in the pathogenesis of schizophrenia and how they are differentially influenced in the sexes, we have investigated microarray expression data from the prefrontal cortex (PFC) in postmortem brain. The PFC region has been strongly associated with deficits of executive function and other cognitive symptoms that occur in patients with schizophrenia. Gene expression within the PFC has been studied extensively using the microarray approach [4-6, 8, 9, 11, 13, 14, 16, 19, 24]. However, statistical analyses of gene expression data from individual small cohorts have lacked sufficient statistical power to avoid conflicting data in these different studies [17, 22]. Meta-analysis is a strategy by which these problems could be addressed, because the data from multiple studies can be combined, thus increasing the statistical power available. In this study, we have collected gene expression data generated from post-mortem cohorts of schizophrenia cases and psychiatrically healthy comparison groups that are included in publicly available databases. We have tested for sex differences in PFC gene expression in schizophrenia, using the meta-analysis paradigm.

Methods and materials

Public microarray datasets of postmortem gene expression in schizophrenia

We searched the public database and literature on the study conducted on PFC region of postmortem brains and decided to use Mistry's merged expression dataset [17] for further analysis because this combined cohort contains the largest number of samples that could be accessed from available resources. In Mistry's study, the raw image data of 306 postmortem brain samples from seven different datasets were first pooled together, Robust Multi-array Average (RMA) normalization procedure was then applied on these pooled samples to obtain normalized expression value of each probe set. Out of 306 samples, 246 are available to the public. The RMA normalized expression data (http://www.chibi.ubc.ca/wp-content/uploads/2013/02/combined.data.txt) and corresponding clinical data (http://www.chibi.ubc.ca/wp-content/uploads/2013/02/combined.design.txt) of these 246 samples are available for download on their website and will serve as a starting point in this study. The source of six studies available to the public in Mistry's dataset is summarized in Table 1. ComBat [12] batch effect adjustment was carried out in R environment. We used ComBat() function included in the "sva" package downloaded from BioConductor website (https://www.bioconductor.org/packages/release/bioc/html/sva.html) to perform batch effect correction on the original dataset. Each study is treated as a batch and default parameter setup is used in running the ComBat function.

Table 1 Public postmortem microarray datasets used in this study

Data set	Brain region	Control: Schizophrenia
Stanley Bahn [35]	Frontal BA46	31(24M7F): 34(25M9F)
Stanley AltarC [35]	Frontal BA46/10	11(7M4F): 9(8M1F)
Mclean Harvard Brain Bank [36]	Prefrontal Cortex (BA9)	26(18M8F): 19(13M6F)
Mirnics [37]	Prefrontal Cortex (BA46)	6(4M2F): 9(5M4F)
Maycox GSE17612 [16]	Anterior prefrontal cortex (BA10)	21(12M9F): 26(18M8F)
Narayan GSE21138 [19]	Frontal (BA46)	29(24M5F): 25(21M4F)

Differential expression analysis of each probe set

Expression values of each probe set were modeled using a fixed effect linear model approach, where disease status and imbalanced covariates between two groups are treated as fixed effects to be estimated from data. A model selection procedure was also employed for each probe set to address the confounding effect of imbalanced covariates. The details of the procedure will be described in next section. For each probe set, the t-statistic for the disease effect was then extracted from the selected model. P-values were computed using two-sided t-test. The resulting P-values were further converted to q-values using the *qvalue* package in R [25] which were defined as the minimum positive False Discovery Rate (q-value(t): $Pr(H = 0|T >= t)$ in Bayesian interpretation over a nested rejection region containing observed statistics t) for multiple testing correction.

Covariate adjustment

The observed covariate imbalance between two groups was analyzed to avoid generating misleading result. However modelling covariates for every gene could unnecessarily diminish statistical power if the covariate does not influence the expression of the gene. In our study, we consider only modelling imbalanced covariates for each probe set. We first obtained a probe set (gene) list where the covariate influence on the gene expression is determined with confidence. This refined gene list was generated using a previous method in a postmortem brain gene expression study, in which a correlation analysis was used to evaluate covariates such as age, post-mortem interval, brain PH etc. that influenced the expression of specific genes across multiple post-mortem normal brain datasets [18]. We extracted genes with meta-$Q \leq 0.01$ to indicate that the gene was significantly influenced by a particular covariate, and separate it into two lists: positively-correlated and negatively-correlated.

Notice that one probe set could be mapped to multiple genes, we excluded those probe sets appearing in the two lists. For each predefined covariate influenced probe set, we first modelled the expression value with a linear model including that covariate. If the direction of the fitted covariate estimates was inconsistent with the pre-refined list, we exclude this covariate and re-estimate a reduced model for this probe set. For the rest of the probe sets, no covariate adjustment was performed.

Selection of sex-specific differentially expressed genes

To identify genes that had a sex by diagnosis interaction with schizophrenia, we used a strategy similar to [3]. Each individual in our dataset is assigned to one of four subgroups: Schizophrenia Male, Control Male, Schizophrenia Female and Control Female. Individuals could also be combined into two groups based on their diagnosis: Schizophrenia and Control group, or based on their sex: Male and Female group. Differential expression analysis was first performed using the procedure described in section 2.2 within each sex. After the initial probe set list was obtained from each sex, we further eliminated those probe sets that are associated with schizophrenia regardless of sex when all of the following criteria were met: (a) the difference between Schizophrenia and Control groups was statistically significant after multiple test correction (q-value < 0.05); (b) the fold change of Schizophrenia Female vs. Control Females, and Schizophrenia Male vs Control Males should be in the same direction i.e. both higher or both lower; (c) the expression difference was not significant between Male and Female groups (defined as $p > 0.05$ between Female and Male group). After removal of these probe sets from the initial probe set list, we sorted the remaining probe sets within each sex by q-value and report the top ranked probe sets (q-value < 0.05) as sex-specific differentially expressed genes.

Function enrichment of differentially expressed genes

All differentially expressed genes, along with their Affymetrix ID numbers were imported into EASE (Expression Analysis Systematic Explorer) in DAVID (Database for Annotation, Visualization and Integrated Discovery), and were used to identify functionally significant gene classes (https://david.ncifcrf.gov/) [10]. This webserver uses statistical methods to map and identify functional gene categories (for example, Gene Ontology (GO), Kyoto Encyclopedia of Genes and Genomes (KEGG) or BioCarta), which are enriched in the significant gene list compared with their presence on the array.

Results

Batch and covariate adjustment

A total of 22277 probe sets is analyzed in this study. Before further analysis, by using hierarchical clustering analysis and Principal Component Analysis (PCA), we found it was necessary to correct for "batch effect" as samples from the same study were clustered together. ComBat [12] was used to correct this technical bias by treating each study as a batch. After ComBat adjustment, the hierarchical clustering and PCA results show that no significant clustering remained in the dataset and the adjusted expression data was suitable for further analysis (Additional file 1).

Clinical variables associated with each patient sample were examined to determine possible confounding variables. The age and postmortem interval (PMI) is well matched between schizophrenia and control groups while brain pH shows a significant difference between two groups (Table 2). Within each sex group, we observe that the brain PH is significantly different between schizophrenia males and control males, on the other hand, all covariates are well balanced in female group (Table 2). Age at death was significantly different between male and female group (Additional file 2). In the predefined brain PH list, 2413 probe sets' expressions were positively correlated with the covariate and 893 were negatively correlated; in the predefined age related list, the number is 1907 and 3028 respectively. The imbalanced covariates and the proportion of probe sets subject to covariate adjustment in each differential analysis are summarized in Table 3.

Genes with altered PFC expression in schizophrenia

To validate our analytical approach used in this study, we performed differential analysis between Schizophrenia and Control group and compared the derived gene list with two published results in which similar meta-analysis were performed [17, 22]. We identified 466 probe sets (representing 427 unique genes) that were significantly down-regulated in the schizophrenia cases relative to the controls and 312 probe sets (representing 261 unique genes) significantly up-regulated in schizophrenia with q-value < 0.05. Our results show that a large number of overlapped probe sets were observed between our gene list and the other two studies (Fig. 1). All overlapped probe sets showed the same direction of fold difference between the schizophrenia cases and controls. In comparison with result of Mistry et al. [27], 86 out of 125 probe sets (68.8 %) were also identified by our approach. In comparison with result of Santiego et al., 98 out of 160 probe sets (61.3 %) overlapped with our result [22]. Our method identified a similar proportion of the probe sets from both studies.

We also examined whether the identified 778 probe sets are associated with schizophrenia in case control studies of genetic polymorphisms. We compared our gene list with those deposited in SZGene (www.SZgene.org) database which contains the most comprehensive review

Table 2 Demographic data of postmortem subjects

	Total sample			Male group			Female group		
	Control	SCZ	p-val	Control	SCZ	p-val	Control	SCZ	p-val
Size	124	122		89	90		35	32	
Age	50.29 ± 17.2	51.09 ± 18.90	0.73	47.85 ± 16.44	48.45 ± 17.71	0.81	56.49 ± 17.89	58.53 ± 20.38	0.67
Gender	89 M:35 F	90 M:32 F	>0.05	NA	NA	NA	NA	NA	NA
Brain PH	6.48 ± 0.29	6.36 ± 0.29	0.002	6.50 ± 0.29	6.35 ± 0.26	<0.001	6.42 ± 0.29	6.40 ± 0.38	0.78
PMI[3]	25.49 ± 14.8	25.09 ± 15.9	0.84	26.12 ± 14.37	25.79 ± 16.45	0.89	23.91 ± 15.98	23.14 ± 14.34	0.84

Abbreviations: *SCZ* Schizophrenia, *p-val* p-value, *M* Male, *F* Female, *PMI* post-mortem interval
The summary demographics (mean ± s.d.) and t-test P-values for group difference are shown. For sex difference, we report the P-value generated from a chi-square test for equality of proportions

of schizophrenia association studies [31]. 80 probe sets representing 68 unique genes are found to be genetically associated with schizophrenia. Genes previously showing strong genetic evidence implicated in schizophrenia identified in this study include: regulator of G-protein signaling 4 (RGS4) [32]; discoidin domain receptor family, member 1(DDR1) [33]; and the selenium binding protein 1(SELENBP1) [34]. The full result is included in Additional file 3.

Sex differences in PFC gene expression in schizophrenia

In the male group, we first identified 138 differentially expressed probe sets with a q-value <0.05. We then removed 80 probe sets which shows differential expression regardless of sex based on the filter defined in the Method part. 50 probe sets representing 46 unique genes were identified as specifically different in male schizophrenia patients relative to male controls: 23 probe sets had lower expression (Table 4) and 27 probe sets have higher expression (Table 5) in males with schizophrenia. In the female group, we were not able to identify any differentially expressed probe sets with q-value <0.05 after multiple testing correction.

Function annotation of male-specific differentially expressed genes

The functions of 46 genes associated with schizophrenia in male patients were manually inspected. Of the genes with significantly lower expression in males with schizophrenia, several were related to energy metabolism (ATP5B, ATP5A1, MRPL23, AFG3L2, ABCG2). 4 genes (BEX1, UBL4A, CD99 and MID1) located on the sex chromosome are identified. We also detected differential expression of 4 genes in the males with schizophrenia that were previously identified by Mistry et al. [17] in which sex was not considered. These were Rho-Related BTB Domain-Containing Protein 3 (RHOBTB3), Bobby Sox homolog (BBX), H3 Histone, Family 3B (H3F3B) and pleckstrin homology domain containing, family B (evectins) member 2 (PLEKHB2). Finally, using DAVID webserver, we performed an enrichment analysis to systematically identify over-representation of biological processes or pathways that are altered in the PFC in male schizophrenia patients. After correction for multiple comparisons, we were unable to identify any significant biological process (False Discovery Rate <0.05) in the GO term Biological Process database. We achieved similarly negative results using the KEGG pathway database.

Table 3 Covariate adjustment summary

Two group comparison	Schizophrenia vs control	Schizophrenia-male vs control-male	Male vs Female
Imbalanced covariate between two groups	Brain PH	Brain PH	Age
Proportion of probe sets adjusted[a] (%)	11.87 %	11.47 %	18.00 %

[a]A total of 22777 probe sets is analyzed in this study

Discussion

We have reported the gene expression differences that show a sex by diagnosis interaction in the PFC in schizophrenia. To our knowledge, this is the first study using a meta-analytical approach to identify sex differences in this brain region in patients with schizophrenia. There are limited data on sex differences in schizophrenia at a molecular level [15, 29], although evidence from epidemiological and animal studies indicates that sex differences exist in this disorder [20, 27, 30]. Individual post-mortem gene expression studies have low statistical power to identify gene expression differences in schizophrenia. This is, most often due to the small sample sizes and moderate gene expression differences between the diagnostic groups. Meta-analysis, on the other hand, addresses this problem and increases statistical power by combining samples from different subject cohorts. The results obtained from our meta-analysis are robust at the statistical level [17]. Our findings open a new window to understand the different pathophysiological mechanisms that lead to schizophrenia in males and females.

In the differential analyses between schizophrenia and control group, we identified the most number of genes most of which overlapped with the results of two

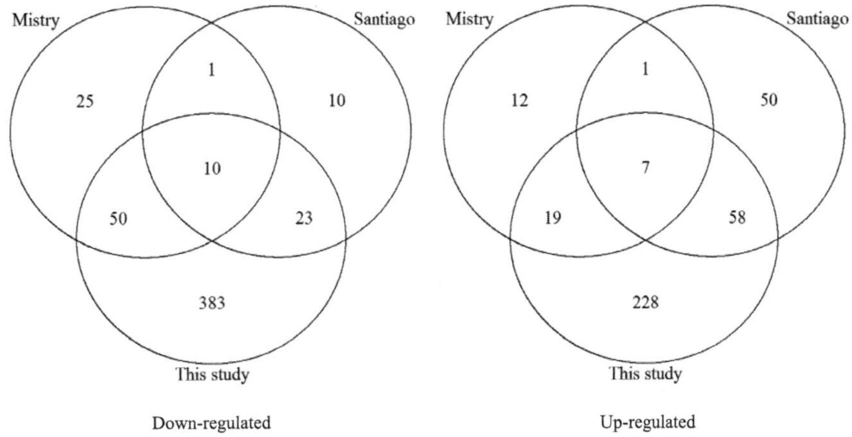

Fig. 1 Comparison of meta-analysis results among three studies

previously published studies, while the other two published studies had little overlap between each other (Fig. 1). To find the reason for the difference, we run the same procedure as in Mistry's for our dataset. We find that the main difference comes from different treatment of "batch effect". In Mistry's study, they treated each experiment date and study as batch variable. A total of 50 batch date and 6 study was modelled in their linear model framework. Introduction of too many predictor variables will decrease the degrees

Table 4 Genes with altered expression in males with schizophrenia: *downregulated probe sets*

Probe set	Gene symbol	Locus	Description	Fold difference	q-value
209735_at	ABCG2	4q22	ATP-binding cassette, sub-family G (WHITE), member 2	−1.29	0.031
208868_s_at	GABARAPL1	12p13.2	GABA(A) receptor-associated protein like 1	−1.17	0.043
208813_at	GOT1	19q24.1-q25.1	glutamic-oxaloacetic transaminase 1, soluble (aspartate aminotransferase 1)	−1.17	0.043
212878_s_at	KLC1	14q32.3	kinesin light chain 1	−1.14	0.044
208002_s_at	ACOT7	1p36	acyl-CoA thioesterase 7	−1.14	0.033
213897_s_at	MRPL23	11p15.5	mitochondrial ribosomal protein L23	−1.10	0.043
211382_s_at	TACC2	10q26	transforming, acidic coiled-coil containing protein 2	−1.10	0.033
214365_at	TPM3	1q21.2	tropomyosin 3	−1.10	0.037
213738_s_at	ATP5A1	18q21	ATP synthase, H+ transporting, mitochondrial F1 complex, alpha subunit 1, cardiac muscle	−1.10	0.031
221909_at	RNFT2	12q24.22	ring finger protein, transmembrane 2	−1.10	0.033
201322_at	ATP5B	12q13.13	ATP synthase, H+ transporting, mitochondrial F1 complex, beta polypeptide	−1.10	0.049
203272_s_at	TUSC2	3p21.3	tumor suppressor candidate 2	−1.09	0.049
218332_at	BEX1	Xq22.1	brain expressed, X-linked 1	−1.09	0.033
201077_s_at	NHP2L1	NA	NHP2 non-histone chromosome protein 2-like 1 (S. cerevisiae)	−1.09	0.041
201410_at	PLEKHB2	2q21.1	pleckstrin homology domain containing, family B (evectins) member 2	−1.08	0.033
219760_at	LIN7B	19q13.3	lin-7 homolog B (C. elegans)	−1.08	0.049
221315_s_at	FGF22	19p13.3	fibroblast growth factor 22	−1.08	0.042
221706_s_at	USE1	19p13.11	unconventional SNARE in the ER 1 homolog (S. cerevisiae)	−1.07	0.042
202967_at	GSTA4	6p12.1	glutathione S-transferase alpha 4	−1.07	0.049
207839_s_at	TMEM8B	9p13.3	transmembrane protein 8B	−1.07	0.042
221746_at	UBL4A	Xq28	ubiquitin-like 4A	−1.07	0.033
208971_at	UROD	1p34	uroporphyrinogen decarboxylase	−1.07	0.043
202486_at	AFG3L2	18p11	AFG3 ATPase family gene 3-like 2 (S. cerevisiae)	−1.06	0.044

Table 5 Genes with altered expression in males with schizophrenia: *upregulated probe sets*

Probe set	Gene symbol	Locus	Description	Fold difference	q-value
212226_s_at	PPAP2B	1p32.2	phosphatidic acid phosphatase type 2B	1.32	0.043
202975_s_at	RHOBTB3	5q15	Rho-related BTB domain containing 3	1.28	0.031
202935_s_at	SOX9	17q24.3	SRY (sex determining region Y)-box 9	1.28	0.043
202887_s_at	DDIT4	10q22.1	DNA-damage-inducible transcript 4	1.26	0.041
212859_x_at	MT1E	16q13	metallothionein 1E	1.26	0.031
200897_s_at	PALLD	4q32.3	palladin, cytoskeletal associated protein	1.24	0.031
200907_s_at	PALLD	4q32.3	palladin, cytoskeletal associated protein	1.23	0.033
209210_s_at	FERMT2	14q22.1	fermitin family member 2	1.22	0.033
213016_at	BBX	3q13.1	bobby sox homolog (Drosophila)	1.21	0.031
213158_at	unknown	3q13.2	unknown	1.21	0.042
201029_s_at	CD99	Xp22.32/Yp11.3	CD99 molecule	1.18	0.043
218350_s_at	GMNN	6p22.3	geminin, DNA replication inhibitor	1.17	0.033
214212_x_at	FERMT2	14q22.1	fermitin family member 2	1.16	0.033
209069_s_at	H3F3B	NA	H3 histone, family 3B (H3.3B)	1.16	0.037
205475_at	SCRG1	4q34.1	stimulator of chondrogenesis 1	1.15	0.041
208022_s_at	CDC14B	9q22.3	CDC14 cell division cycle 14 homolog B (S. cerevisiae)	1.14	0.039
211997_x_at	H3F3B	NA	H3 histone, family 3B (H3.3B)	1.14	0.038
41644_at	SASH1	6q24.3	SAM and SH3 domain containing 1	1.13	0.044
215811_at	Unknown gene	NA	-	1.13	0.047
209600_s_at	ACOX1	17q25.1	acyl-CoA oxidase 1, palmitoyl	1.13	0.033
202771_at	FAM38A	16q24.3	family with sequence similarity 38, member A	1.11	0.042
203636_at	MID1	Xp22	midline 1 (Opitz/BBB syndrome)	1.11	0.049
200906_s_at	PALLD	4q32.3	palladin, cytoskeletal associated protein	1.10	0.042
213342_at	YAP1	11q13	Yes-associated protein 1	1.09	0.031
210105_s_at	FYN	6q21	FYN oncogene related to SRC, FGR, YES	1.09	0.033
215823_x_at	PABPC1 /// RLIM	NA	poly(A) binding protein, cytoplasmic 1 /// ring finger protein, LIM domain interacting	1.08	0.044
210094_s_at	PARD3	10p11.21	par-3 partitioning defective 3 homolog (C. elegans)	1.08	0.041

of freedom of t-test on the estimated disease effect coefficient, thus decreasing the likelihood of rejecting the null hypothesis, leading to higher Type-II error. Based on the observation of hierarchical clustering and PCA result, we consider that modelling each study alone would be sufficient to correct the batch effect. The second reason explaining the difference is covariate adjustment. We only include a covariate in the model when its influence on the gene is confirmed with confidence; while in Mistry's study, they assumed that the covariate influenced every gene resulting in unnecessary inclusion of unrelated variables in the model. Therefore the approach taken in this study has more detection power and leads to discovery of more genes.

Our analyses identified 46 genes that were differentially expressed specifically in male patients with schizophrenia. This finding of 50 probe sets is much larger than the expected number of false positives according to our selection procedure. The expected number of false positives was calculated to be 6.9 ($138*0.05 = 6.9$). Five genes were related to energy metabolism (ATP5B, ATP5A1, MRPL23, AFG3L2, ABCG2). Genes from this function category are consistently implicated in studies of schizophrenia. Another gene that had altered expression in the male schizophrenia group encodes γ-aminobutyric acid receptor-associated protein-like 1 (GABARAPL1) is an early estrogen-induced gene that when overexpressed, interacts with GABA-A or κ-opioid receptors, and plays a role in cell proliferation and cellular metabolic processes [1]. Function enrichment analysis generates negative results for these genes, as indicated in other microarray studies of the PFC in schizophrenia [17]. These results suggest that a diverse number of molecular functions are disrupted in males with schizophrenia.

No genes could be identified in female group after multiple testing correction (q-value < 0.05). To determine if

Table 6 Genes reported by Vawter et al. [29] in this study

		Total sample		Male group		Female group		Sex difference[a]
Probe set	Gene symbol	Fold change	Q-value	P-value	Fold change	P-value	Fold change	P-value
200978_at	MDH1	−1.11	0.007	0.000491	−1.10	0.009	−1.14	0.061
207721_x_at	HINT1	−1.08	0.014	0.00051	−1.08	0.065	−1.08	0.062
205352_at	SERPINI1	−1.10	0.027	0.010158	−1.08	0.091	−1.16	0.073

[a]Two group t-test comparison between male and female group

this is due to a much smaller sample size than in male group, we randomly picked the same number of control and schizophrenia subjects from the male groups ten times and run differential analysis on these samples. The procedure was repeated 100 times. No significant difference could be detected in the expression level of any gene in the phenotype groups (data not shown). We then gradually increase the number of samples until genes could be detected with q-value < 0.05 in each run. We started to identify differentially expressed genes when there are 60 controls and 60 schizophrenia in the sample pool. This analysis showed that increasing the number of samples would improve the likelihood of identifying schizophrenia-associated genes in the females.

Very few studies have been conducted of sex differences in gene expression in the PFC in schizophrenia. Vawter et.al. [29] reported that three genes (MDH1, HINT1 and SERPINI1) had decreased expression in PFC region of 13 male schizophrenia patients compared with 11 male controls and no expression difference was observed in comparing 9 female schizophrenia patients with 10 female controls using quantitative PCR. We then extracted the corresponding probe sets from our dataset and summarized the result in Table 6. We observed significantly lower expression levels of all three genes in the schizophrenia group compared with controls (Table 6, Column 3 and 4). The expression difference of these 3 genes in was also tested in males and females separately (Table 6, Column 5 to 8). Our analysis suggests that all three genes might be altered by schizophrenia and are not related to sex difference. For MDH1, the expression is significantly decreased in the schizophrenia group of both sexes suggesting that this gene might be down regulated in schizophrenia regardless of sex. HINT1 and SERPINI1 do not show differential expression between the male and female groups (Table 6, Column 9). Our study has a larger sample size than that of Vawter et al. thus we have greater statistical power to detect small effects, and we would argue that these genes are associated with schizophrenia but are not differentially expressed between the sexes.

Conclusion

In summary, this is the most comprehensive and up-to-date analysis of sex differences in prefrontal cortex gene expression in schizophrenia. Some of our data consolidate the reports of previously published papers. Our results indicate that further investigation of sex differences in schizophrenia is required [15]. These data bring us closer to understanding the different molecular mechanisms underpinning schizophrenia in males and females, so that novel targets for antipsychotic drug development can be identified. Genes with altered expression in schizophrenia can also serve as biological markers for the disorder, so that biochemical diagnostic tools can facilitate the practice of clinical psychiatry.

Additional files

Additional file 1: Figure S1. Comparisons of hierarchical clustering and PCA results before and after ComBat adjustment. (DOCX 665 kb)

Additional file 2: Table S1. Demographic comparison between male and female group. (XLSX 9 kb)

Additional file 3: Table S2. 778 schizophrenia-affected probe sets identified in this study. (XLSX 66 kb)

Competing interests
The authors declare that they have no competing interests.

Authors' contributions
WQ, MS and HL conceived and designed the study. WQ analysed the data. WQ, CL, MS and HL interpreted the results and wrote the manuscript. All authors read and approved the final manuscript.

Acknowledgements
This work was partially supported by the Hans W. Vahlteich Fund (MS), and UIC Department of Bioengineering (WQ, CL, HL). Publication fee are paid by Department of Bioengineering at UIC and by Shanghai Jiaotong University.

Author details
[1]Department of Bioengineering, University of Illinois at Chicago, 851 S. Morgan, Rm 218, Chicago, IL 60607, USA. [2]Department of Pharmacy Practice and Center for Pharmaceutical Biotechnology, University of Illinois at Chicago, 900 S Ashland Ave mc870, Chicago, IL 60607, USA. [3]SJTU-Yale Joint Center for Biostatistics, Shanghai Jiaotong University, Shanghai, China.

Published: 11 January 2016

References
1. Boyer-Guittaut M, Poillet L, Liang Q, Bôle-Richard E, Ouyang X, Benavides GA, et al. The Role of Gabarapl1/Gec1 in Autophagic Flux and Mitochondrial Quality Control in Mda-Mb-436 Breast Cancer Cells. Autophagy. 2014;10(6).
2. Brown AS. The Environment and Susceptibility to Schizophrenia. Prog Neurobiol. 2011;93(1):23–58.
3. Cantuti-Castelvetri I, Keller-McGandy C, Bouzou B, Asteris G, Clark TW, Frosch MP, et al. Effects of Gender on Nigral Gene Expression and Parkinson Disease. Neurobiol Dis. 2007;26(3):606–14.

4. Chu TT, Liu Y, Kemether E. Thalamic Transcriptome Screening in Three Psychiatric States. J Hum Genet. 2009;54(11):665–75.
5. Dean B, Keriakous D, Scarr E, Thomas EA. Gene Expression Profiling in Brodmann's Area 46 from Subjects with Schizophrenia. Aust N Z J Psychiatry. 2007;41(4):308–20.
6. Garbett KR, Gal-Chis RO, Gaszner GA, Lewis DA, Mirnics KÁ. Transcriptome Alterations in the Prefrontal Cortex of Subjects with Schizophrenia Who Committed Suicide. Neuropsychopharmacol Hung. 2008;10(1):9–14.
7. Gejman PV, Sanders AR, Kendler KS. Genetics of Schizophrenia: New Findings and Challenges. Annu Rev Genomics Hum Genet. 2011;12:121–44.
8. Mirnics K, Middleton FA, Stanwood GD, Lewis DA, Levitt P. Disease-specific changes in regulator of G-protein signaling 4 (RGS4) expression in schizophrenia. Mol Psychiatry 2001; 6: 293–301.
9. Higgs BW, Elashoff M, Richman S, Barci B. An Online Database for Brain Disease Research. BMC Genomics. 2006;7:70.
10. Huang DW, Sherman BT, Lempicki RA. Systematic and integrative analysis of large gene lists using DAVID Bioinformatics Resources. Nature Protoc. 2009;4(1):44–57.
11. Horvath S, Janka Z, Mirnics K. Analyzing Schizophrenia by DNA Microarrays. Biol Psychiatry. 2011;69(2):157–62.
12. Johnson WE, Li C, Rabinovic A. Adjusting Batch Effects in Microarray Expression Data Using Empirical Bayes Methods. Biostatistics. 2007;8(1):118–27.
13. Katsel P, Davis KL, Gorman JM, Haroutunian V. Variations in Differential Gene Expression Patterns across Multiple Brain Regions in Schizophrenia. Schizophr Res. 2005;77(2–3):241–52.
14. Kumarasinghe N, Tooney PA, Schall U. Finding the Needle in the Haystack: A Review of Microarray Gene Expression Research into Schizophrenia. Aust N Z J Psychiatry. 2012;46(7):598–610.
15. Martins-de-Souza D, Schmitt A, Röder R, Lebar M, Schneider-Axmann T, Falkai P, et al. Sex-Specific Proteome Differences in the Anterior Cingulate Cortex of Schizophrenia. J Psychiatr Res. 2010;44(14):989–91.
16. Maycox PR, Kelly F, Taylor A, Bates S, Reid J, Logendra R, et al. Analysis of Gene Expression in Two Large Schizophrenia Cohorts Identifies Multiple Changes Associated with Nerve Terminal Function. Mol Psychiatry. 2009; 14(12):1083–94.
17. Mistry M, Gillis J, Pavlidis P. Genome-Wide Expression Profiling of Schizophrenia Using a Large Combined Cohort. Mol Psychiatry. 2013;18(2):215–25.
18. Mistry M, Pavlidis P. A Cross-Laboratory Comparison of Expression Profiling Data from Normal Human Postmortem Brain. Neuroscience. 2010;167(2):384–95.
19. Narayan S, Tang B, Head SR, Gilmartin TJ, Sutcliffe JG, Dean B, et al. Molecular Profiles of Schizophrenia in the Cns at Different Stages of Illness. Brain Res. 2008;1239:235–48.
20. Ochoa S, Usall J, Cobo J, Labad X, Kulkarni J. Gender Differences in Schizophrenia and First-Episode Psychosis: A Comprehensive Literature Review. Schizophr Res Treatment. 2012;2012:916198.
21. Owen MJ, Craddock N, O'Donovan MC. Suggestion of Roles for Both Common and Rare Risk Variants in Genome-Wide Studies of Schizophrenia. Arch Gen Psychiatry. 2010;67(7):667–73.
22. Pérez-Santiago J, Diez-Alarcia R, Callado LF, Zhang JX, Chana G, White CH, et al. A Combined Analysis of Microarray Gene Expression Studies of the Human Prefrontal Cortex Identifies Genes Implicated in Schizophrenia. J Psychiatr Res. 2012;46(11):1464–74.
23. Roth TL, Lubin FD, Sodhi M, Kleinman JE. Epigenetic Mechanisms in Schizophrenia. Biochimica et Biophysica Acta (BBA)-General Subjects. 2009; 1790(9):869–77.
24. Sequeira PA, Martin MV, Vawter MP. The First Decade and Beyond of Transcriptional Profiling in Schizophrenia. Neurobiol Dis. 2012;45(1):23–36.
25. Storey JD. The Positive False Discovery Rate: A Bayesian Interpretation and the Q-Value.The Annals of Statistics. 2003; 31(6): 2013-2035.
26. Tandon R, Gaebel W, Barch DM, Bustillo J, Gur RE, Heckers S, et al. Definition and Description of Schizophrenia in the Dsm-5. Schizophr Res. 2013;150(1):3–10.
27. Teaching Network, EURON and De Crespigny Park London. Jim Van Os, Shitij Kapur. Lancet. 2009;374:635–45.
28. Uher R. Gene–Environment Interactions in Severe Mental Illness. Frontiers in Psychiatry. 2014;5(48).
29. Vawter MP, Weickert CS, et al. Gene Expression of Metabolic Enzymes and a Protease Inhibitor in the Prefrontal Cortex Are Decreased in Schizophrenia. Neurochem Res. 2004;29(6):1245–55.
30. Wu YC, Hill RA, Gogos A, Van Den Buuse M. Sex Differences and the Role of Estrogen in Animal Models of Schizophrenia: Interaction with Bdnf. Neuroscience. 2013;239:67–83.
31. Allen NC, Bagade S, McQueen MB, Ioannidis JPA, Kavvoura FK, Khoury MJ, et al. Systematic Meta-Analyses and Field Synopsis of Genetic Association Studies in Schizophrenia: The SzGene Database. Nat Genet. 2014;40(7):827–34.
32. Prasad KMR, Chowdari KV, Nimgaonkar VL, Talkowski ME, Lewis DA, Keshavan MS. Genetic polymorphisms of the RGS4 and dorsolateral prefrontal cortex morphometry among first episode schizophrenia patients. Molec Psychiat. 2005;10:213–9.
33. Roig B, Virgos C, Franco N, Martorell L, Valero J, Costas J, et al. The discoidin domain receptor 1 as a novel susceptibility gene for schizophrenia. Mol Psychiatry. 2007;12(9):833–41.
34. Kanazawa T, Glatt SJ, Faraone SV, Hwu HG, Yoneda H, Tsuang MT. Family-based association study of SELENBP1 in schizophrenia. Schizophr Res. 2009;113(2–3):268–72.
35. Stanley Medical Research Institute. https://www.stanleygenomics.org/. Accessed 1 Feb 2013.
36. Harvard Brain Tissure Resource Center (Mclean 66 collection). http://www.brainbank.mclean.org/. Accessed 1 Feb 2013.
37. Garbett K, Gal-Chis R, Gaszner G, Lewis DA, Mirnics K. Transcriptome alterations in the prefrontal cortex of subjects withschizophrenia who committed suicide. Neuropsychopharmacol Hung. 2008;10:9–14.

Platt *et al. BMC Systems Biology* 2016, **10**(Suppl 1):10
DOI 10.1186/s12918-015-0251-2

BMC Systems Biology

PROCEEDINGS Open Access

Characterizing redescriptions using persistent homology to isolate genetic pathways contributing to pathogenesis

Daniel E. Platt[1*], Saugata Basu[2], Pierre A. Zalloua[3,4] and Laxmi Parida[1]

From The Fourteenth Asia Pacific Bioinformatics Conference (APBC 2016)
San Francisco, CA, USA. 11 - 13 January 2016

Abstract

Background: Complex diseases may have multiple pathways leading to disease. E.g. coronary artery disease evolves from arterial damage to their epithelial layers, but has multiple causal pathways. More challenging, those pathways are highly correlated within metabolic syndrome. The challenge is to identify specific clusters of phenotype characteristics (composite phenotypes) that may reflect these different etiologies. Further, GWAS seeking to identify SNPs satisfying multiple composite phenotype descriptions allows for lower false positive rates at lower α thresholds, allowing for the possibility of reducing false negatives. This may provide a window into the missing heritability problem.

Methods: We identify significant phenotype patterns, and identify fuzzy redescriptions among those patterns using Jaccard distances. Further, we construct Vietoris-Rips complexes from the Jaccard distances and compute the persistent homology associated with those. The patterns comprising these topological features are identified as composite phenotpyes, whose genetic associations are explored with logistic regression applied to pathways and to GWAS.

Results: We identified several phenotypes that tended to be dominated by metabolic syndrome descriptions, and which were distinct among the combinations of metabolic syndrome conditions. Among SNPs marking the RAAS complex, various SNPs associated specifically with different groups of composite phenotypes, as well as distinguishing between the composite phenotypes and simple phenotypes. Each of these showed different genetic associations, namely rs6693954, rs762551, rs1378942, and rs1133323. GWAS identified SNPs that associated with composite phenotypes included rs12365545, rs6847235, and rs701319. Eighteen GWAS identified SNPs appeared in combinations supported in composite combinations with greater power than for any individual phenotype.

Conclusions: We do find systematic associations among metabolic syndrome variates that show distinctive genetic association profiles. Further, the systematic characterization involves composite phenotype descriptions that allow for combined power of individual phenotype GWAS tests, yielding more significance for lower individual thresholds, permitting the exploration of SNPs that would otherwise show as false negatives.

Keywords: Pattern discovery, Redescription, Persistent homology, Genetic association, GWAS, Missing heritability

*Correspondence: watplatt@us.ibm.com
[1]Computational Biology Center, IBM T. J. Watson Research Center, 1101 Kitchawan Rd., 10598 Yorktown Hgts, NY, USA
Full list of author information is available at the end of the article

Background

Coronary artery disease (CAD) is a multifactorial disease with inherited and behavioral components. Disease progression starts with any of a diverse set of injuries to arterial epithelial layers, which may take years to develop. These pathways lead to distinctive patterns of damage, e.g. with diabetes damage distinct in character from focal plaques associated with cholesterol. CAD risk factors cluster in highly correlated conditions called metabolic syndrome with distinct etiologies and pathways. Yet, the odds ratio associated with a pathway producing CAD is identical to the odds ratio that CAD was caused by that specific pathway, leading to diluted signals. Genome Wide Association Studies (GWASs) promised to reveal which Single Nucleotide Polymorphisms (SNPs) are clear causes of CAD and other diseases. Yet, identified SNPs only account for around 10 % of CAD, leaving 90 % of the heritability component unexplained [1–3]. One possible gap is the rather large possibility of false negatives given the high threshold excluding false positives in genome-wide surveys. Also, to isolate SNPs specific to a condition, genome-wide logistic regressions explicitly subtract the effects of other risk factors by including "adjustments". Therefore, isolating genetic impact specific to CAD excludes the genetics of pathway-induced pathogenic etiology.

In this study, we sought associations connecting groups of phenotypic descriptions together against which we sought genome wide associations [4]. We identified significant phenotypic associations using pattern discovery to select combinations of factors appearing more or less often than expected by chance [5, 6]. Logistic regressions applied to such patterns then yield significant associations.

Pathway mechanisms create associations such that some phenotype condition A implies the presence of condition B, or $A \Rightarrow B$, and therefore also that $A = A \wedge B$. Therefore, the list of subjects $S(A)$ associated with condition A will be the same as the list $S(A \wedge B) = S(A) \cap S(B)$. Therefore, the list of subjects $S(A)$ associated with condition A will be the same as the list $S(A \wedge B) = S(A) \cap S(B)$. This implies two important points. First, it is important to identify patterns. Second, given patterns, we want to identify patterns that apply to the same sets of subjects, e.g. that satisfy $S(A \wedge B) = S(A)$. Further, patterns with meaningful content are those for which $P(A|B) \neq P(A)$, which may be distinguished from chance by application of a statistical test, such as a binomial test of $P(A \wedge B) \neq P(A)P(B)$. If the sample space does not resolve these features with sufficient power to distinguish from random sampling variation, associations between $A \wedge B$ and other patterns that might be revealed by clustering are not likely to be meaningful. It would be expected that any pattern that is significant will also reveal significant contributions in more detailed analyses such as logistic regression, or information-based tests to identify associations where Simpson's Pardox may be present [7], or other tests to clarify the content of these composite patterns. Secondly, those associations are also characterized by uncertainty in sampling, misclassification errors, and variability in the physiological progression of disease. This implies that the equality must be interpreted in a sense that allows for statistical variation. Therefore, we sought clusters of patterns identified by the same sets of subjects measured by Jaccard distances within a threshold typical of the variability expected for odds ratios describing the patterns. Such clusters represent the same groups of subjects that can be identified by any of a number of different patterns, or "fuzzy redescriptions", [8] reflecting underlying pathway-specific etiology. As such, the patterns each become multicomponent phenotypes suitable for a genome-wide p-value threshold $= 6.338 \times 10^{-7}$.

In the search for SNPs that may be relevant to specific pathways, these compound statements of phenotypes provide two benefits. First, they offer greater specificity identifying subgroups of subjects that are distinct from other subgroups. It is important to emphesize that while these subgroups are distinct, they may share some subjects in common with other subgroups characterized by other groups of subjects. If these patterns identify pathways, then we should be able to identify greater power for some SNPs being identified with these compound phenotypes. Second, since these compound phenotypes, and each pattern in the cluster, provides multiple tests for each SNP, the chances that a GWAS SNP would emerge by random sampling variation for these multiple tests is greatly reduced. Essentially, the threshold is equivalent to the product of the thresholds α for each of the individual tests. This reduces the level of threshold required to exclude false positivies for each individual phenotype test, which implies that those candidate SNPs excluded as false negatives for a single phenotype GWAS test with Boneferroni correction have a chance to be reconsidered multiple times at lower thresholds.

The structure of these clusters was further explored using computational topological analysis [9], seeking to identify Vietoris-Rips complexes, where the patterns are vertices, the Jaccard distances provide the filtration, and the lifetimes of the generators of interest are within the Jaccard threshold, employing JavaPlex [10] to compute annotated persistent homologies. The generator complexes are also candidate phenotypes.

We applied logistic regression to SNPs drawn from the RAAS complex, as well as GWAS predicting these composite phenotypes. We also identified SNPs individually predicting all conditions comprising composite phenotypes, requiring the joint significance to be genome wide. This greatly increases sensitivity, reducing the threshold excluding false negatives, but at the cost of requiring significance of for multiple factors.

Results and discussion

Figure 1 shows a two-way biclustering of the binary coded thresholded clinical variables, and a rainbow coded plot (red smallest, blue largest) of Euclidean distances between enrollees based on the binary coded clinical values without scaling. Given the relatively strong association between clinical risk factors among metabolic syndrome patients, and that enrollees were all in a group for which an invasive coronary catheterization was deemed appropriate by physicians and therefore likely shared metabolic syndrome risk factors, it is notable that so many localized blocks of patients share features (larger red squares on the diagonal), but only moderate to low similarity connect blocks off-diagonal. The grey two-way cluster plot gives more of a sense of how the enrollees' variables are organized within and around those blocks. This strongly suggests significant sub-organization within metabolic syndrome. Interestingly, there is not real evidence in this data set of an overall metabolic syndrome. Perhaps this is not surprising since all the subjects had been enrolled from a population of catheterized subjects, catheterization is an invasive procedure requiring adequate justification, and most subjects will have a history of metabolic syndrome as a part of their history justifying the catheterization.

Logistic regression is a very common tool employed to assess the odds ratios of various risk factors in predicting disease, and especially for sorting out dependencies and interactions among risk variables in defining these associations. These interactions are graphically visible in the two-way heirarchical clustering map. This strongly suggests that a pattern discovery algorithm as described above may identify variable configurations likely to yield significant logistic regressions. Further, the relationships among patterns based on their enrollee sets may be used to recognize relationships among clinical variables describing the disease processes, and possibly pathways and stages. Figure 2 shows an example patterns, along with how the patterns cluster according to the Jaccard distances between the lists of enrollees matching the patterns. A total of 397 patterns were generated. Figure 3 shows the Jaccard distances displayed by heatmap organized according to single linkage hierarchical clustering. The sidebar colors mark the seven leading redescription clusters (nerves), at a Jaccard threshold of 0.30.

Figure 4 shows a barcode plot of generators, and an exerpt from the report of barcode correspondences indexed by the IDs of the patterns shown in Fig. 2. It is clear that the identified generators contained within the filtration of the nerve from Fig. 2 are subsets of the nerves. At larger filtration values, the barcodes tend to combine nerves. At those ranges, redescription clusters start to merge, with lower significance relating patterns to each other. The Jaccard distance measures the fraction of members of the two enrollee lists that are not shared between the patterns. So a distance exceeding 0.5 represents a situation where any given enrollee has less than fifty percent chance of being in both clusters, which is weak for inferring clinical relationships among patterns.

Table 1 shows prediction of complex phenotypes by RAAS complex and cytochrome gene SNPs. These include *NPPA* (rs5065), *REN* (rs6693954), *AGT* (rs699), *ADRB2* (rs1042713), *CYP1A2* (rs762551, rs1378942, rs1133323), and *ACE* (rs4343). The RAAS complex is a target of a number of common drugs targeting hypertension, such as angiotension-converting-enzime (ACE) inhibitors, angiotensin-receptor-blockers (ARBs), direct renin inhibitors target this complex. A number of SNPs show activity in a number of complex phenotypes including rs6693954, rs762551, rs1378942, and rs1133323. Of these, rs6693954 shows a slightly stronger odds ratio for the DxHT cluster than for DxHT, rs762551 and rs1378942 show stronger odds ratios for DxT2D and DxHT clusters than for their simple counterparts. While rs1133323 is significant for the DxT2D cluster, it is pretty clear that the cluster dilutes the simple DxT2D association. This is even more clear since the rest of the DxT2D cluster is identical to the DxHT cluster, and rs1133323 is not significant for DxHT or its cluster. This is somewhat striking in comparison to rs762551. The subcluster of patterns associated with the DxT2D cluster complex phenotype both identifies a number of different ways to identify the same set of enrollees, which suggests there is an underlying mechanism tying them together, and there are some SNPs that relate more positively to this cluster than just to the simple phenotype form. The relationship of this cluster to the DxHT cluster explored is also informative: they are distinct, and some SNPs more strongly associate with the complex DxHT phenotype than the simple phenotype.

SNPs rs6693954 and rs1133323 show a highly significant protective association against the simple T2D, but the compound phenotype shows a weaker association. Interestingly, rs6693954 had shown a non-significant HT association, but became significant when considering the compound phenotype. SNPs rs762551 and rs1378942 showed stronger associations for compound T2D and hypertension phenotypes than the simple phenotypes. These features indicate the compound phenotypes are resolving differences in how SNPs impact physiology in this population.

GWAS was applied to the dataset with composite phenotypes yielding the results in Table 2. It is interesting to note that, though rs12365545 (*MAML2*) appears in both DxT2D and DxHT results, it appears this may be dominated by association with DxCAD with Age ≥ 60. That

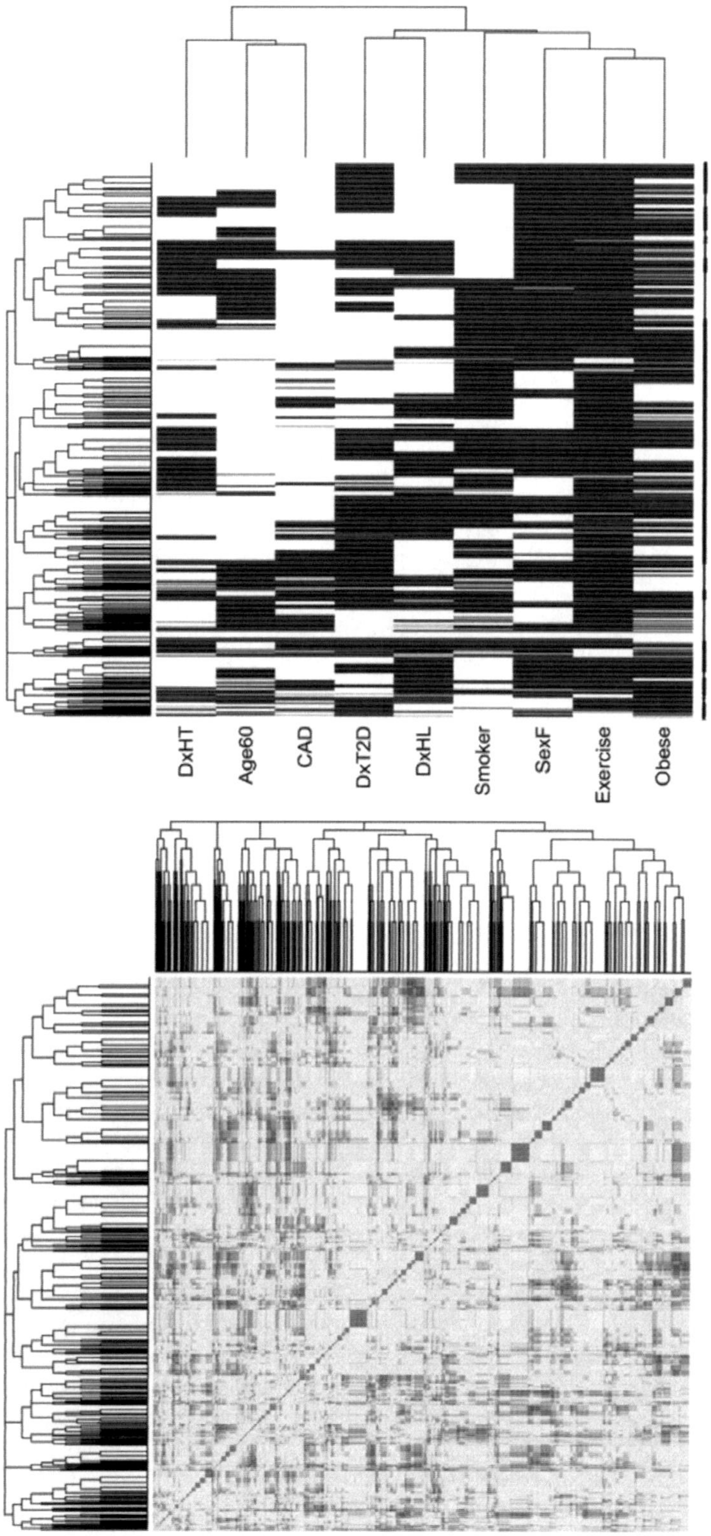

Fig. 1 Two-way hierarchical clustering of binary threshold values of clinical variables with white positive, and distances plotted in section of rainbow from red to blue

```
%Cluster 0
0; [67] 1.05673e-14, 2594.57 < 2914: Age60 = 2 & DxHT = 2
0; [145] 4.17786e-30, 2501.21 < 2973: Age60 = 2 & Smoke = 1
0; [172] 1.76366e-07, 2760.79 < 2976: DxHT = 2 & Smoke = 1
0; [174] 2.74854e-61, 1534.18 < 2136: Age60 = 2 & DxHT = 2 & Smoke = 1
0; [260] 9.4911e-17, 2414.6 < 2753: Age60 = 2 & DxHT = 2 & Exercise = 1
0; [288] 2.31702e-31, 2327.72 < 2802: Age60 = 2 & Smoke = 1 & Exercise = 1
0; [304] 2.92661e-08, 2569.29 < 2795: DxHT = 2 & Smoke = 1 & Exercise = 1
0; [367] 1.2128e-12, 2880.2 < 3179: Age60 = 2 & CAD = 2
0; [392] 3.76515e-34, 1766.64 < 2226: Age60 = 2 & DxHT = 2 & CAD = 2
0; [417] 5.91092e-37, 1703.07 < 2177: Age60 = 2 & Smoke = 1 & CAD = 2
0; [426] 9.93166e-09, 1879.82 < 2094: DxHT = 2 & Smoke = 1 & CAD = 2
0; [428] 9.67807e-69, 1044.62 < 1604: Age60 = 2 & DxHT = 2 & Smoke = 1
   & CAD = 2
0; [441] 1.8772e-13, 2680.42 < 2986: Age60 = 2 & Exercise = 1 & CAD = 2
0; [456] 1.41134e-34, 1644.1 < 2096: Age60 = 2 & DxHT = 2 & Exercise = 1
   & CAD = 2
0; [464] 5.90999e-36, 1584.94 < 2041: Age60 = 2 & Smoke = 1 & Exercise = 1
   & CAD = 2
0; [468] 4.31351e-09, 1749.43 < 1964: DxHT = 2 & Smoke = 1 & Exercise = 1
   & CAD = 2
...
%Cluster 2
0; [38] 0.0337293, 2503.32 > 2428: GenderF = 1 & DxHL = 2
0; [86] 4.76951e-14, 2282.36 < 2584: DxHL = 2 & DxHT = 2
0; [248] 0.00171293, 2329.68 > 2212: GenderF = 1 & DxHL = 2 & Exercise = 1
0; [271] 2.59223e-14, 2124.05 < 2423: DxHL = 2 & DxHT = 2 & Exercise = 1
0; [385] 1.73949e-06, 2533.62 < 2726: DxHL = 2 & CAD = 2
0; [387] 6.35428e-10, 1704.5 < 1929: GenderF = 1 & DxHL = 2 & CAD = 2
0; [398] 9.23069e-29, 1554.06 < 1955: DxHL = 2 & DxHT = 2 & CAD = 2
0; [451] 6.06477e-06, 2357.88 < 2536: DxHL = 2 & Exercise = 1 & CAD = 2
0; [452] 2.56499e-07, 1586.27 < 1767: GenderF = 1 & DxHL = 2 & Exercise = 1
   & CAD = 2
0; [458] 8.66409e-28, 1446.26 < 1830: DxHL = 2 & DxHT = 2 & Exercise = 1
   & CAD = 2
%Cluster 3
0; [20] 0.0010124, 1314.79 < 1418: Age60 = 2 & DxT2D = 2
0; [77] 1.29603e-24, 806.457 < 1093: Age60 = 2 & DxT2D = 2 & DxHT = 2
0; [157] 1.82727e-18, 777.439 < 1017: Age60 = 2 & DxT2D = 2 & Smoke = 1
0; [240] 9.55131e-06, 1223.59 < 1363: Age60 = 2 & DxT2D = 2 & Exercise = 1
0; [267] 4.73591e-29, 750.518 < 1056: Age60 = 2 & DxT2D = 2 & DxHT = 2
   & Exercise = 1
0; [298] 2.57815e-21, 723.513 < 976: Age60 = 2 & DxT2D = 2 & Smoke = 1
   & Exercise = 1
0; [377] 2.86232e-20, 895.239 < 1163: Age60 = 2 & DxT2D = 2 & CAD = 2
0; [447] 2 .36237e-23, 833.141 < 1115: Age60 = 2 & DxT2D = 2 & Exercise = 1
   & CAD = 2
...
%Cluster 5
0; [76] 1.08515e-18, 1451.24 < 1759: DxT2D = 2 & DxHT = 2
0; [266] 5.86915e-22, 1350.58 < 1679: DxT2D = 2 & DxHT = 2 & Exercise = 1
0; [376] 6.85351e-11, 1611.01 < 1844: DxT2D = 2 & CAD = 2
0; [394] 9.25943e-35, 988.149 < 1365: DxT2D = 2 & DxHT = 2 & CAD = 2
0; [446] 9.16748e-13, 1499.26 < 1749: DxT2D = 2 & Exercise = 1 & CAD = 2
0; [457] 6.64746e-38, 919.607 < 1304: DxT2D = 2 & DxHT = 2 & Exercise = 1
   & CAD = 2
%Cluster 6
0; [137] 9.26651e-07, 1379 < 1542: DxHL = 2 & Obese = 2
0; [141] 5.63167e-26, 845.842 < 1147: DxHL = 2 & DxHT = 2 & Obese = 2
0; [283] 1.73262e-07, 1283.35 < 1453: DxHL = 2 & Obese = 2 & Exercise = 1
0; [286] 3.81453e-27, 787.171 < 1087: DxHL = 2 & DxHT = 2 & Obese = 2
   & Exercise = 1
0; [413] 2.6795e-08, 938.959 < 1099: DxHL = 2 & Obese = 2 & CAD = 2
0; [462] 4.69692e-09, 873.828 < 1038: DxHL = 2 & Obese = 2 & Exercise = 1
   & CAD = 2
%Cluster 7
0; [156] 0.00827527, 1399.02 < 1481: DxT2D = 2 & Smoke = 1
0; [178] 7.10827e-22, 858.124 < 1133: DxT2D = 2 & DxHT = 2 & Smoke = 1
0; [297] 0.000659579, 1301.98 < 1409: DxT2D = 2 & Smoke = 1 & Exercise = 1
0; [306] 1.84063e-24, 798.601 < 1083: DxT2D = 2 & DxHT = 2 & Smoke = 1
   & Exercise = 1
0; [421] 5.4056e-09, 952.593 < 1122: DxT2D = 2 & Smoke = 1 & CAD = 2
0; [466] 1.5331e-10, 886.517 < 1068: DxT2D = 2 & Smoke = 1 & Exercise = 1
   & CAD = 2
...
%Cluster 9
0; [45] 3.19714e-13, 1156.58 < 1388: DxT2D = 2 & DxHL = 2
0; [90] 3.10701e-44, 709.416 < 1087: DxT2D = 2 & DxHL = 2 & DxHT = 2
0; [254] 8.6377e-16, 1076.35 < 1326: DxT2D = 2 & DxHL = 2 & Exercise = 1
0; [388] 4.70882e-28, 787.514 < 1093: DxT2D = 2 & DxHL = 2 & CAD = 2
0; [453] 1.14527e-29, 732.888 < 1039: DxT2D = 2 & DxHL = 2 & Exercise = 1
   & CAD = 2
...
```

Fig. 2 Patterns with redescription cluster identifications. First number is Fisher test for pattern list vs. cluster intersection. Second: pattern reference id. Third: binomial *p*-value. Fourth: expected count observed vs. observation marking khe tail of the binomial test evaluated. Last: list of columns and values. The Jaccard threshold was 0.25

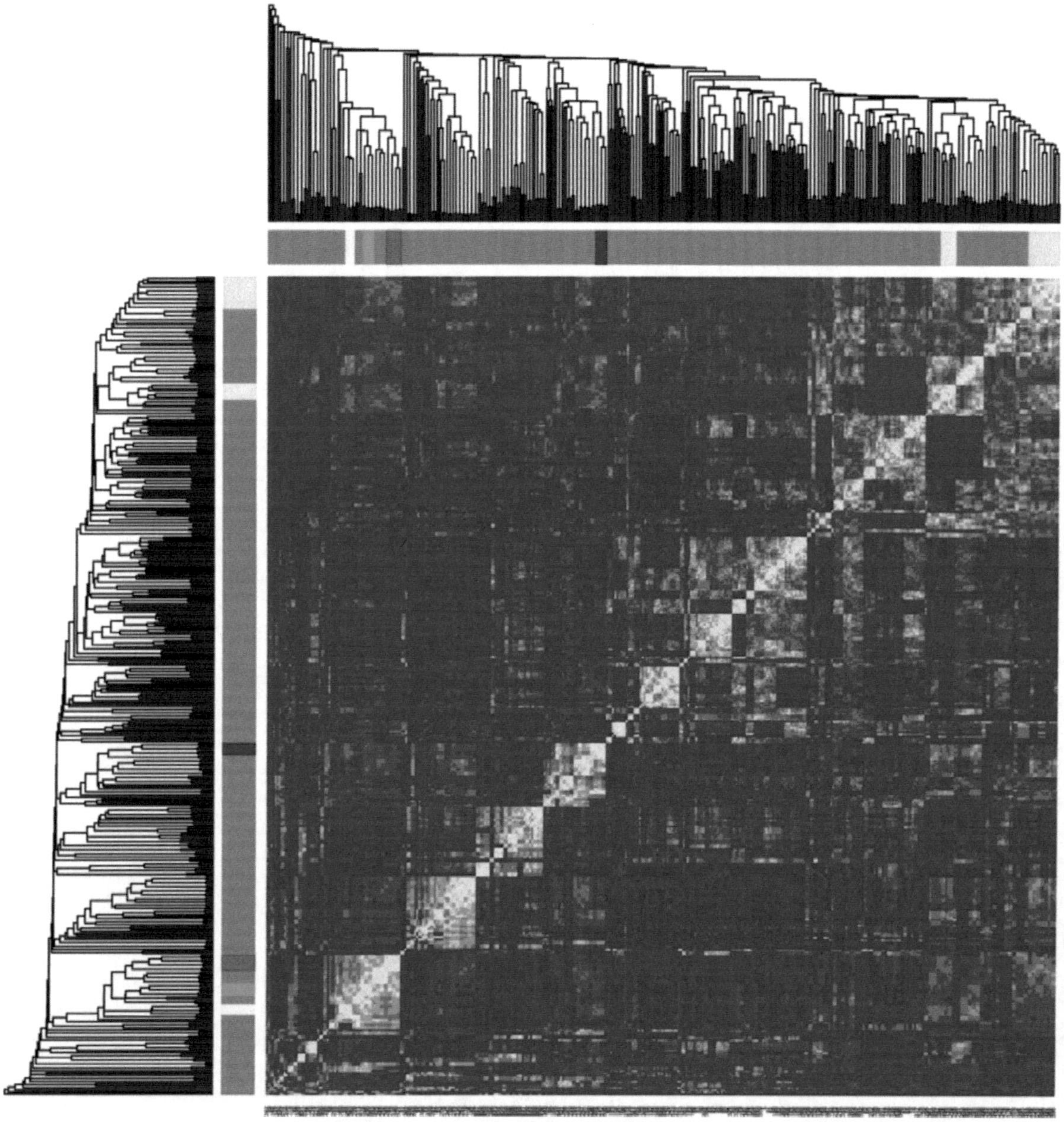

Fig. 3 Distance clustering of the redescription Jaccard distances for significant patterns. This also serves as the filtration distance in the construction of the persistent homology analysis. Sidebar colors mark six redescription clusters. Red is Cluster 3 (Age ≥ 60 ∧ DxT2D ∧ DxCAD), Orange is Cluster 5 (DxT2D ∧ DxHT ∧ DxCAD), Yellow is Cluster 7 (DxT2D *wedge* non-smoker *wedge* DxCAD), Green is Cluster 0 (Age ≥ 60 ∧ DxHT ∧ DxCAD), Cyan is Cluster 2 (male ∧ DxHL ∧ DxCAD), and Blue is Cluster 6 (Obese ∧ DxHL ∧ DxHT ∧ DxCAD)

association is a component of both, with OR = 1.42 (95 % CI 1.24 − 1.62), p-value = 3.604×10^{-7}. Yet, when predicting DxCAD by itself (no age threshold), the p-value is 0.00298. *MAML2* is mastermind-like 2 (Drosophila). It is implicated in several B cell-derived lymphomas, mucoepidermoid carcinomas, and chronic lymphocytic leukemia. It plays a role in regulation of ICN notch proteins, which have been implicated in T2D. rs6847235 (*GLRA3*) is intronic in glycine receptor alpha 3 which is a member of the ligand-gated ion channel protein family. rs701319 (intergenic) has no known clinical significance.

Table 3 shows leading composite phenotypes. The "composite p-value" is the probability that uncorrelated random samples of null associations would have produced the pattern by chance. The "pattern p-value" is the actual logistic regression p-value for that SNP for the

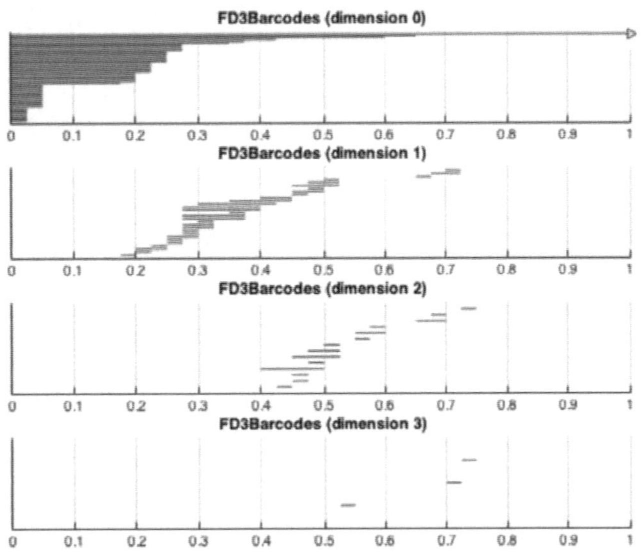


```
...
Dimension: 1
[0.175, 0.2): [377,447] + [20,377] + [20,240] + [240,447] (Cluster 3)
[0.2, 0.225): -[254,453] + [45,388] + [388,453] + [45,254] (Cluster 9)
[0.2, 0.25): [76,266] + [394,457] + [76,394] + [266,457] (Cluster 5)
[0.225, 0.25): -[260,456] + [67,260] + [392,456] + [67,392] (Cluster 0)
[0.25, 0.275): [394,457] + [446,457] + [376,446] + [376,394] (Cluster 5)
[0.25, 0.275): [141,286] + [283,286] + [137,283] + [137,141] (Cluster 6)
[0.275, 0.300): [20,157] + [20,240] + [157,298] + [240,298] (Cluster 3)
...
```

Fig. 4 Persistent homology barcode plot of generators and an exerpt of the generator simplices from dimension 1

associated pattern. The SNP with the strongest pattern association in this group is rs12101936 (intergenic). rs17107637 (*SMOC1*) is intronic. *SMOC1* is *SPARC* related modular calcium binding 1, which appears to have a role in ocular and limb development. rs3759658 (*CGRRF1*) is in the cell growth regulator with ring finger domain 1 gene. rs6926556 (intergenic), rs8113086 (intergenic), and rs878643 (intergenic) show no known clinical association. rs3019548 (*CD6*) is possibly associated with multiple sclerosis. *CD6* encodes a protein found in the outer membrane of T-lymphcytes and some other immune cells, important for continuation of T cell activiation, contains three scavenger receptor cysteine-rich domains, and has a binding site for activated leukocyte cell adhesion molecule. rs6807700 (*PLSCR5*) is in phospholipid scramblase family, member 5, with no known clinical significance. rs 118800382 is in *ZNF331* (Taneera), which has been implicated in a T2D gene expression pathway analysis. rs2088354 is in the *SLC2a13* intron, a glucose transporter. rs2992100 is in MIR4500, a non-coding region. rs3781788 is in *MMP20*, a matrix metalloproteinase. rs6080252 is in *KIF16B*, coding a kinisen-like motor protein, involved with plus end motility of early endosomes and the balance between recycling and degredation of receptors, such as EGFR, FGFR. rs6984384 is in *LOC101929576* - an uncharacterized RNA gene. rs701133 is in *GPR149* - a G protein-coupled receptor. rs818710 is in *BSPRY*, B Box and *SPRY* Domain Containing Protein. rs17077265 and rs522264 are intergenic. rs6727857 is in *SCN2A*, Sodium Channel, Voltage-Gated, Type II, Alpha Subunit. All snps were significant, though not genome wide significant, except for rs8113086.

The contrast between Tables 2 and 3 shows that standard GWAS, even treating compound phenotypes as simple phenotypes, picks out different SNPs than considering multiple tests applied by each of the compound phenotype components. This gives strong evidence that the compound approach yields SNPs that are false negatives that would otherwise be discarded as false positivies. While the *p*-values of some of the patterns ("pattern *p*-value") are not as strong as might be suggested by the composite of individual phenotypes comprising the compound phenotype ("composite *p*-value"), there are a number of highly significant SNPs worth considering that standard methods miss.

Table 1 Odds ratio logistic regression associations of RAAS complex SNPs with RAAS and cytochrome P450 1A2

Locus	DxT2D cluster	DxT2D	DxHT cluster	DxHT	DxHL cluster	DxHL
rs5065 G	0.93 (0.77–1.12) 0.443	0.94 (0.82–1.08) 0.378	1.01 (0.88–1.16) 0.863	0.99 (0.87–1.12) 0.836	0.96 (0.82–1.10) 0.542	0.96 (0.84–1.09) 0.553
rs6693954 A	0.80 (0.66–0.98) 0.029	0.74 (0.63–0.85) 6.22×10^{-5}	0.84 (0.72–0.99) 0.0321	0.91 (0.79–1.05) 0.221	0.88 (0.76–1.03) 0.120	0.91 (0.79–1.06) 0.221
rs699 A	0.84 (0.67–1.06) 0.138	0.89 (0.75–1.06) 0.182	0.94 (0.79–1.13) 0.541	0.96 (0.81–1.13) 0.632	0.92 (0.77–1.10) 0.366	0.92 (0.78–1.09) 0.333
rs1042713 A	0.93 (0.80–1.09) 0.376	0.94 (0.84–1.05) 0.277	0.99 (0.88–1.12) 0.895	1.01 (0.91–1.13) 0.798	0.98 (0.87–1.11) 0.763	1.03 (0.93–1.15) 0.564
rs762551 C	1.42 (1.19–1.69) 8.21×10^{-5}	1.23 (1.08–1.39) 0.00203	1.28 (1.12–1.47) 0.000375	1.17 (1.03–1.33) 0.0188	0.93 (0.81–1.07) 0.297	0.94 (0.83–1.07) 0.332
rs1378942 C	1.33 (1.11–1.59) 0.00154	1.19 (1.05–1.34) 0.00706	1.27 (1.11–1.46) 0.000672	1.22 (1.07–1.38) 0.00288	0.97 (0.84–1.11) 0.633	0.99 (0.87–1.12) 0.899
rs1133323 A	0.80 (0.66–0.97) 0.0245	0.78 (0.68–0.90) 0.00758	1.00 (0.86–1.16) 0.983	0.92 (0.80–1.06) 0.235	1.11 (0.96–1.30) 0.164	1.01 (0.88–1.16) 0.884
rs4343 A	1.04 (0.90–1.20) 0.618	0.94 (0.87–1.09) 0.646	0.98 (0.88–1.10) 0.785	0.93 (0.84–1.04) 0.200	1.07 (0.95–1.20) 0.273	0.98 (0.88–1.09) 0.698

Entries show locus w/ minor allele, and the odds ratio, 95 *p*-value for the SNP vs. the phenotype. DxT2D cluster = DxT2D ∧ Age ≥ 60 ∧ DxHT ∧ DxCAD, DxHT cluster = Age ≥ 60 ∧ DxHT ∧ DxCAD, DxHL cluster = DxHL ∧ Male ∧ DxCAD

Conclusions

Associations between RAAS SNPs and metabolic syndrome characters, both clustered and singly, show specificity for conditions (e.g. rs1133323 for DxT2D clustered and unclustered, rs1378942 and rs762551 for DxT2D clustered and unclustered, and DxHT clustered and unclustered), and some specific to the complex phenotype (e.g. rs6693954 for DxHT clustered, but not DxHT) but not the single form, suggesting that the complex phenotypes convey more pathway specific information than individual simple phenotypes do by themselves.

GWAS applied to combined tests from the composite phenotype components shows significant lowering of the threshold for individual tests required to identify SNPs that are significant in multiple metabolic syndrome conditions, allowing for the possibility of relieving exclusion of some of the false negatives. The SNPs identified in individual tests are distinct from composite tests, indicating that different components of the sample space are probed by applying tests for each of the phenotypes represented in a composite pattern.

Table 2 Odds ratio logistic regression associations GWAS on compound phenotypes. Entries show locus w/ minor allele, and the odds ratio, 95 % confidence interval, and *p*-value for the SNP vs. the phenotype

Locus Minor Allele	Association	OR (95 % CI) *p*-value
rs12365545 A	DxT2D	1.69 (1.38–2.07) 4.434×10^{-7}
rs12365545 A	DxHT	1.44 (1.24–1.67) 8.329×10^{-7}
rs6847235 A	DxHL	1.45 (1.25–1.68) 5.344×10^{-7}
rs701319 T	DxHL	0.57 (0.46–0.71) 6.821×10^{-7}

Genome wide *p*-value threshold = 6.338×10^{-7}

Several things may be explored in greater detail later. These include richer phenotype sets and more patterns. Another aspect is that the homology groups typically produced are sets of surfaces enclosing empty volumes, namely $H_n = \ker(\partial_n)/\mathrm{Im}(\partial_{n+1})$, including the subset $\mathrm{Im}(\partial_{n+1})/\mathrm{Im}(\partial_{n+1}) \simeq \{0\}$. These show the most topologically interesting features with conserved Betti numbers. However, other structures may be physiologically informative, such as $C_n/\ker(\partial_n)$, which would represent forms that sprawl and do not enclose volumes.

Materials and methods

Datasets

Seven thousand six hundred thirteen subjects were selected from Lebanese patients enrolled as part of a multi center cross-sectional study for the FGENT-CARD Consortium (http://www.well.ox.ac.uk/fgentcard/) [11] in a cross-sectional study. Patients were recruited from catheterized patients at the Rafic Hariri University Hospital and the Centre Hospitalier du Nord in Lebanon between May 2007 and June 2010. A questionnaire developed to measure the impact of CAD risk factors and family history was collected after informed consent was obtained from participants prior to conducting the study, as approved by the Lebanese American University institutional review board. Annotations were coded from medical charts for data such as laboratory tests, prescribed medications, and presence of other clinical conditions. Venous blood samples were drawn on EDTA. Of these, 6 were excluded due to missing information on age or sex, leaving 7607.

Nine hundred and ninety eight Lebanese Type II Diabetes (T2D) study participants were all of Lebanese origin and were recruited. A first recruitment campaign, conducted with the collaboration of the Lebanese University Medical Center, led to the recruitment of 506 subjects from the suburbs of Beirut, the capital of Lebanon. In a second campaign conducted in North Lebanon, 492 subjects were successfully recruited. Research was carried out in compliance with the Helsinki Declaration and with the approval of the LAU institutional review board and local ethics committees on human research (Reference number SMPZ08072010-4). All participants signed an informed consent and data and blood samples were obtained from each individual. By taking part in the two recruitment campaigns, participants (1) answered a detailed questionnaire, (2) gave a blood sample for DNA analysis and (3) gave a blood sample for HbA1C, fasting blood glucose (FBS), and lipid profile measures after 12 hours fasting. In this analysis, the T2D clinical data were not employed. Th e SNP data was used only to improve imputation analysis.

DNA was extracted using a standard phenol-chloroform extraction procedure. Two thousand seven hundred fifty two CAD study samples were analyzed on a number of platforms: 48 subjects' DNA samples were analyzed using Human610-Quad beadchip and Illumina (582.775); 1055 subjects were analyzed using Human610-Quad beadchip and Illumina (582.892); 928 subjects were analyzed using Human Omni EXP – 12v1 multi-use; 706 with Illumina Human 660W Quad Beadchip; 7 with HumanOmniEXP-12v1 Multi-use + Human610-Quad Beadchip and Illumina (582.892); and 8 with HumanOmniEXP-12v1 Mullti-use + Illumina Hujan660W-Quad Beadchip. The 538 T2D DNA analyses were performed using Human Omni EXP – 12v1 multi-use. Seven hundred eighty nine thousand SNPs passed QC using PLINK 2 [12], (https://www.cog-genomics.org/plink2) for data management and quality control, keeping samples with call rate *ge* 95 %, SNPs call rate $\geq$ 90 %, MAF *ge* 1 %. BEAGLE ver 4.0 [13], (http://faculty.washington.edu/browning/beagle/beagle.html) was employed to impute SNPs among inconsistent chip SNP sets. SNPs with more than 2 alleles were removed.

Descriptors were coded for analysis as follows. DxT2D indicates diagnosis of type II diabetes. DxHT indicates diagnosis of hypertension. DxHL indicates a diagnosis of hyperlipidemia. Age60 marks age 60 years or older. SexF marks female. Obese indicates BMI levels in excess of 30. Smoker implies heavy cigarette smoking or hookah. Exercise marks regular intensive exericse. CAD marks greater than 70 % occlusion.

Demographics notes

Metabolic syndrome, CAD, and T2D prevalences are rapidly increasing in the Middle East, linked to recent changes in diet and activity in the population as a whole [14]. This rapid emergence argues for the possibility of age-structured changes in dietary habits and risk factors. There may be a delay of some years between earliest epithelial damage leading to CAD, or for progression from insulin resistance to full T2D. Further, emergence of CAD, hypertension (HTN), T2D and other metabolic syndrome conditions is strongly age and sex associated, while some risk behaviors are gender and age specific associated.

Analysis

Metabolic syndrome is characterized by the strong association between obesity, hypertension, dyslipidemia, coronary artery disease, and type II diabetes mellitus. Many of these conditions emerge in older age. Among catheterized subjects, expectation would be that variables describing these conditions would cluster in well-defined patterns. To that end, hierarchical 2-way clustering was performed using R's [15] heatmap function, invoking hclust() for clustering, and unscaled Euclidean distances between enrollees across clinical variables, and between clinical variables across enrollees. Also, heatmap was used to

Table 3 SNPs present in each of the component phenotypes in compound sets, with composite *p*-value computed from individual phenotype components, pattern *p*-value computed on the compound phenotype members, odds ratio logistic regression associations GWAS for each phenotype set. Entries show locus w/ minor allele, and the odds ratio, 95 % confidence interval, and *p*-value for the SNP vs. the phenotype

Locus Minor Allele	Composite *p*-value	Pattern *p*-value	DxT2D	DxHT	DxCAD ∧ Age ≥ 60
rs12101936 A	1.079×10^{-9}	4.961×10^{-5}	0.76 (0.64–0.89) 0.000628	0.78 (0.67–0.90) 0.000615	0.80 (0.69–0.93) 0.002796
rs17107637 G	1.872×10^{-8}	0.002181	1.41 (1.14–1.75) 0.00135	1.37 (1.13–1.67) 0.00140	1.29 (1.06–1.57) 0.00992
rs3759658 A	8.914×10^{-8}	0.0304	1.34 (1.08–1.66) 0.00733	1.33 (1.09–1.61) 0.00455	1.35 (1.11–1.65) 0.00267
rs6926556 C	9.205×10^{-9}	0.0341	0.74 (0.60–0.91) 0.00438	0.72 (0.60–0.86) 0.000457	0.76 (0.63–0.92) 0.00460
rs8113086 T	2.853×10^{-7}	0.0936	0.75 (0.60–0.93) 0.00841	0.77 (0.64–0.93) 0.00760	0.75 (0.62–0.92) 0.00447
rs878643 A	4.254×10^{-8}	0.04394	0.764 (0.65–0.90) 0.00841	0.800 (0.69–0.93) 0.00760	0.819 (0.70–0.96) 0.00447
rs11880382 C	1.518×10^{-7}	0.005167	0.733 (0.64–0.84) 1.669×10^{-5}	0.901 (0.79–1.03) 0.118	0.889 (0.78–1.01) 0.077
rs2088354 T	3.133×10^{-8}	0.0144	1.679 (1.31–2.15) 3.856×10^{-5}	1.357 (1.09–1.68) 0.00557	1.175 (0.95–1.46) 0.146
rs2992100 A	4.162×10^{-8}	0.06262	0.559 (0.43–0.73) 2.35×10^{-5}	0.772 (0.60–0.99) 0.0379	1.282 (1.00–1.64) 0.0468
rs3781788 T	4.837×10^{-8}	0.001568	1.907 (1.44–2.52) 5.527×10^{-6}	1.015 (0.77–1.34) 0.913	1.436 (1.09–1.89) 0.00959
rs6080252 A	6.587×10^{-8}	0.002838	1.623 (1.28–2.06) 6.277×10^{-5}	1.303 (1.03–1.65) 0.0284	1.279 (1.02–1.61) 0.0370
rs6984384 T	3.906×10^{-8}	4.502×10^{-5}	0.742 (0.65–0.86) 4.886×10^{-5}	0.840 (0.74–0.96) 0.008405	0.895 (0.79–1.02) 0.09511
rs701133 T	3.865×10^{-7}	0.002081	1.547 (1.25–1.91) 5.57×10^{-5}	1.206 (1.00–1.46) 0.054	1.160 (0.96–1.41) 0.129
rs3019548 G	3.905×10^{-7}	0.009792	1.541 (1.20–1.98) 0.000643	1.483 (1.18–1.86) 0.000607	–
rs6807700 C	1.061×10^{-7}	0.0002289	–	1.46 (1.18–1.80) 0.000400	1.49 (1.20–1.85) 0.000265
rs17077265 T	2.80144×10^{-7}	0.0131	–	1.409 (1.23–1.61) 7.423×10^{-7}	1.062 (0.93–1.21) 0.377
rs522264 T	2.53×10^{-7}	2.238×10^{-5}	–	0.638 (0.51–0.79) 5.567×10^{-5}	0.734 (0.59–0.91) 0.00455
rs6727857 G	4.819×10^{-7}	0.000935	–	0.660 (0.56–0.78) 1.173×10^{-6}	0.932 (0.79–1.10) 0.411

Genome wide *p*-value threshold = 6.338×10^{-7}

display unscaled Euclidean distances between enrollees mapped to a rainbow segment from red through blue.

The 2-way hierarchical clustering identifies blocks of relationships among descriptors shared among groups of subjects. The structure of those blocks may be explored in greater detail. Each descriptor d in the set of descriptors D has associated with it an alphabet A_d of values that d assumes among the set of subjects S. Each subject $sinS$ has a descriptor tuple $q(s) \in Q = \times_{d \in D} A_d$. The tupple member corresponding to $d \in D$ is $q_d(s) \in A_d$ (pulldown). For each $a \in A_d$, it is possible to identify the set of subjects $Sa = q_d^{-1}(a) \subset S$ that "have" that value a of the descriptor.

The patterns $r \in R$ are comprised tuples similar to the subject tuples except that the alphabet is augmented by a wild card "*" such that $q_d^{-1}(*) = S$. Then $R \subset \times_{d \in D} (A_d \cup \{*\})$. Using the symbol r_d for the descriptor $d \in D$ on R (pulldown), it is possible to identify $r_d(r) \in A_d \cup \{*\}$. Then it is possible to identify a set of subjects $S(r)$ "matching" a pattern r as $S(r) = \bigcap_{d \in D} q_d^{-1}(r_d(r))$ The total number of possible patterns is then $\prod_{d \in D}(|A(d)| + 1) - 1$, where the pattern comprised entirely of wild cards is excluded.

Blocks that appear in multiple descriptors horizontally across the plot in the two-way hierarchical clustering may be significant, and correspond to some $S(r)$ for $r \in R$. It is expected that biological processes will tend to correlate descriptors among the sampled subjects. The difficulty is that such correlations may appear simply by chance. One way to exclude chance is to consider the fraction p of samples that would be expected given no correlation in some pattern r: $p(r) = \prod_{d \in D} \frac{|Sr_d|}{|S|}$. Then the chances of finding more than $|S_q|$ by variation in random sampling would be binomially distributed with $P = \sum_{n \geq |S(r)|} \binom{|S|}{n} p(r)^n (1 - p(r))^{|S|-n}$. Likewise, if the r are anticorrelated, then the probability of finding fewer by chance due to random sampling variation would be $P = \sum_{n \leq |S(r)|} \binom{|S|}{n} p(r)^n (1 - p(r))^{|S|-n}$.

Such patterns were identified using a pattern recognition algorithm defined in Fig. 5.

The use of a support threshold acknowledges that lower support patterns will tend to have lower statistical power. Discarding low support is an option that limits the number of weak patterns that are carried along in the computation. There are two significance levels. It is conceivable that some complex patterns may have a stronger significance level than some of the intermediate patterns. Allowing two support levels permits the retention of weaker patterns which are removed at the end with the stronger and final significance threshold.

If two patterns $A = A \cap B$, then $A \subseteq B$, or $A \Rightarrow B$. Sets of patterns that yield equal sets of enrollees are called "redescriptions" [8]. Obviously, these redescriptions imply more complex relationships among patterns and possible underlying disease mechanisms and processes. Since thresholds may not be perfect, misclassification errors, disease progression changes, etc, tends to inject variability into the analysis. Association studies may show significant association and still allow for substantial amounts of variance. For this reason, Jaccard distances measured on the list of enrollees for each pattern are computed as dissimilarities that satisfy metric conditions. Assignment of nearest neighbors by Jaccard distance below some threshold yields pairwise connectivity, from which clusters are constructed using an algorithm resembling floodfill. These clusters are "fuzzy redescriptions". Their utility emerges only if the threshold is low enough so that it would be unlikely for random processes to have constructed such clusters by chance. Fisher exact tests may be applied to assess whether random processes could have produced pairs by chance, and have been applied here between any pattern in the cluster against the intersection of all cluster members.

Rediscriptions also relate patterns to each other, some according to implications, or perhaps more generally in terms of shared physiology related to a disease process. Coronary artery disease is a "complex disease", with multiple possible pathways leading to epithelial damage in arteries, and ultimatley to formation of plaques. These clusters of descriptions may pick out phenotypic characterization of underlying pathways at specific stages of pathogenic development.

Since we have a set of patterns with a well-defined metric distance between them, it is also possible to use the Jaccard distance as a filtration index in a persistent homology computation. In this context, the fuzzy redescription clusters defined above correspond to a "nerve". One point worth noting is that the whole range of persistence is not necessarily interesting; once distances are long enough that the chances that clusters formed by chance is relatively high, there is really no information from which relationships between patterns (e.g. corresponding to $A = A \cap B$) may be derived.

Within the persistent homologies, the generators carry information about topological structures within the nerves that may reveal yet more detail about pathological development specific to disease pathways.

Given that each pair of patterns has associated with it a Jaccard distance, it is possible to think of each statement as a vertex in a simplicial complex. Each vertex is a point, in our case representing a pattern with its descriptors and list of matching subjects. Given a threshold distance, line segments connecting these vertices may be drawn if the Jaccard distance is less than the threshold distance. Those line segments may connect to close triangles, which then may be filled. If all the surfaces of adjacent triangles form a tetrahedron, then the volume may be filled.

- set $A \leftarrow \emptyset$
- for each column x
  - for each value v in column
    - construct atomic pattern p from $x = v$
    - for each pattern $a \in A$
      - construct new patern $b = a \cap p$.
      - if $|b| <$ support threshold, discard it.
      - if b's list of columns exceeds the column threshold, discard it.
      - compute the product p of the fraction of the N enrollees matching each atomic pattern in b. If all columns were independent, this would be the expected fraction of enrollees for the pattern.
      - Given an observed number of enrollees matching the pattern b, assign to pattern b the chances of seeing that many or more given $|b| > Np$, or that many or less given $|b| < Np$ will be $\sum_k \binom{N}{k} p^k (1-p)^{N-k}$ for k over the appropriate range.
      - If the probability assigned to the pattern b is below the first (tolerant) probability treshold, insert the pattern in A.
    - end for
    - insert p in A.
  - end for
- end for.
- For each pattern $a \in A$
  - if a is atomic then discard it.
  - if a's probability is above the final (strict) threshold, discard it.
- end for
- end.

Fig. 5 Algorithm used to generate patterns

This can be extended to higher dimensions. Connected segments, areas, and volumes form chains. The set of complexes of dimension n is called C_n. In this context, for a given threshold, the redescription clusters become nerves. There is a map ∂_{n+1} that extracts the boundaries of the areas in C_{n+1} and identifies those boundaries in C_n. The list C_* of all the chains C_n of dimension n is called a simplicial complex. It is possible for closed surfaces to enclose a region, yet for there to be no verticies within that region allowing for the surface to be filled. The sets of surfaces enclosing empty volumes are $H_n = \ker(\partial_n)/\mathrm{Im}(\partial_{n+1})$ form the homology of the simplicial complex, and (loosely) represents the hollow patterns.

The Jaccard distances become the basis of a "filtration" given a set of distance thresholds at which the C_n's and H_n's are evaluated. Shapes, unfilled volumes, etc that persist over some range of thresholds may reveal information about the stability of the relationships among redescription structures and their associated underlying pathological processes, yielding a finer-grained view of the structure of phenotype spaces.

Therefore, we start by computing patterns, and associate these compound patterns with phenotypes possibly marking specific processes. We compute fuzzy redescription clusters in terms of a connectivity associated with a threshold applied Jaccard distances between pattern enrollees. These redescriptions reveal relationships among patterns that may describe disease processes and pathways at specific times of development. Further, information about their topological relationships derived from persistent homology computations also reveal substructures within redescription sets (nerves). All of these pattern relationships were used to characterize phenotypes that used in logistic regressions using glm from R [15] applied to SNPs derived from RAAS complex, and to GWAS logistic regressions performed using PLINK 2.0 [12] to identify genetics related to these patterns.

We sought to test whether the compound phenotype patterns resolved evidence of pathway mechanisms more clearly than simple phenotypes. To achieve this, we applied the RAAS SNPs as described, contrasting results for compound phenotypes were more specific or less specific than the simple phenotypes for three groups of phenotypes, namely type II diabetes, hypertension, and hyperlipidemia. Second, we applied a GWAS to the complex phenotypes in the standard one-test method, and contrasted that to the list of SNPs identified through the combinatorial multi-test method.

Abbreviations

CAD: Coronary artery disease; GWAS: Genome Wide Association Study; SNP: Single Nucleotide Polymorphism; T2D: Type II Diabetes Mellitus; HTN: Hypertension; DxHT: Diagnosis of Hypertension; DxT2D: Diagnosis of Type II Diabetes Mellitus; DxHL: Diagnosis of Hyperlipidemia; DxCAD: Diagnosis of CAD (>50% occlusion in angiograpy); Age60: Age 60 or over; SexF: Sex is Female.

Competing interests

The authors declare that they have no competing interests.

Authors' contributions

LP and DEP contributed to pattern discovery methodology, implemented by DP. LP and DEP contributed to the redescription methodology, implemented by DEP. PZ and DEP contributed to the ideas of pathway explorations directed towards the missing heritability problem. DEP formulated the notion of composite phenotypes to describe pathway specificity, and the use of redescription to identify these composite phenotype clusters. PZ contributed the data. SB, LP, and DEP contributed to the use of persistent homology to identify finer grained structure in redescriptions further refining pathway phenotype structures. DEP performed the logistic regression analyses on the data. DEP conceived of the pipeline connecting redescriptions, persistent homology analysis, complex phenotypes, and improving power of GWAS in this context. All authors read and approved the final manuscript.

Acknowledgements

DEP and LP thank IBM T. J. Watson Research Center for resources and support of this study.

Declarations

All authors thank IBM for paying the publication charges for this article. This article has been published as part of BMC Systems Biology Volume 10 Supplement 1, 2016: Selected articles from the Fourteenth Asia Pacific Bioinformatics Conference (APBC 2016): Systems Biology. The full contents of the supplement are available online at http://www.biomedcentral.com/bmcsystbiol/supplements/10/S1.

Author details

[1]Computational Biology Center, IBM T. J. Watson Research Center, 1101 Kitchawan Rd., 10598 Yorktown Hgts, NY, USA. [2]Department of Mathematics, Purdue University, 150 N. University St., 47907 West Lafayette, IN, USA. [3]Graduate Studies and Research, Lebanese American University, P.O. Box 13-5053, Chouran Beirut, 1102 2801, Lebanon. [4]Department of Environmental Health, Harvard University, 401 Park Drive, Boston, MA, USA.

Published: 11 January 2016

References

1. Manolio TA, Collins FS, Cox NJ, Goldstein DB, Hindorff LA, Hunter DJ, et al. Finding the missing heritability of complex diseases. Nature. 2009;461(7265):747–53.
2. Björkegren JLM, Kovacic JC, Dudley JT, Schadt EE. Genome-wide significant loci: how important are they? J Am Coll Cardiol. 2015;65(8):830–45.
3. Platt DE, Zalloua PA. The missing heritability in coronary artery disease. In: Gene-Environment Interactions and Human Disease. Nova Science Publishers; 2015. https://www.novapublishers.com/catalog/product_info.php?products_id=55925.
4. Arnedo J, Svrakic DM, del Val C, Romero-Zaliz R, Hernández-Cuervo H, of Schizophrenia Consortium MG, et al. Uncovering the hidden risk architecture of the schizophrenias: Confirmation in three independent genome-wide association studies. Am J Psychiat. 2015;172:2:139–53.
5. Parida L. Pattern Discovery in Bioinformatics Theory and Algorithms. Mathematical and Computational Biology Series. New York: Chapman and Hall, CRC; 2008.
6. Mullins IM, Siadaty MS, Lyman J, Scully K, Garrett CT, Miller WG, et al. Data mining and clinical data repositories: insights from 667,000 patient data set. Comput Biol Med. 2006;36(12):1351–77.
7. Choy L, Ong CC, Cheong M. Modified entropy measure for detection of association rules under simpson's paradox content. arXiv:1210.1288 [stat.AP] stat/1210.1288. 2012.
8. Parida L, Ramakrishnan N. Redescription mining: Structure theory and algorithms. In: Proceedings of the 20th national conference on Artificial intelligence - Volume 2. Pittsburgh, Pennsylvania: AAAI Press; 2005. p. 837–44.
9. Edelsbrunner H, Letscher D, Zomorodian A. Topological persistence and simplification. Discrete Comput Geom. 2002;28:511–33.
10. Tausz A, Vejdemo-Johansson M, Adams H. JavaPlex: A research software package for persistent (co)homology. Software available online. 2011. http://code.google.com/javaplex.
11. Youhanna S, Platt DE, Rebeiz A, Lauridsen M, Deeb ME, Nasrallah A, et al. Parental consanguinity and family history of coronary artery disease strongly predict early stenosis. Atherosclerosis. 2010;212(2):559–63.
12. Chang CC, Chow CC, Tellier LCAM, Vattikuti S, Purcell SM, Lee JJ. Second-generation plink: rising to the challenge of larger and richer datasets. Gigascience. 2015;4:7. doi:10.1186/s13742-015-0047-8.
13. Browning SR, Browning BL. Rapid and accurate haplotype phasing and missing data inference for whole genome association studies by use of localized haplotype clustering. Am J Hum Genet. 2007;81:1084–97.
14. al-Khurdairy L, Stranges S, Kumar S, al-Daghri N, Rees K. Dietary factors and type 2 diabetes in the middle east: hat is the evidence for an association? – a systematic review. Nutrients. 2013;5(10):3871–97.
15. R Core Team. R: A Language and Environment for Statistical Computing. Vienna, Austria: R Foundation for Statistical Computing; 2013. R Foundation for Statistical Computing. ISBN 3-900051-07-0. http://www.R-project.org/.

Zheng and Sankoff *BMC Genomics* 2016, **17**(Suppl 1):1
DOI 10.1186/s12864-015-2294-6

BMC Genomics

RESEARCH Open Access

Locating rearrangement events in a phylogeny based on highly fragmented assemblies

Chunfang Zheng and David Sankoff*

From The Fourteenth Asia Pacific Bioinformatics Conference (APBC 2016)
San Francisco, CA, USA. 11 - 13 January 2016

Abstract

Background: The inference of genome rearrangement operations requires complete genome assemblies as input data, since a rearrangement can involve an arbitrarily large proportion of one or more chromosomes. Most genome sequence projects, especially those on non-model organisms for which no physical map exists, produce very fragmented assembles, so that a rearranged fragment may be impossible to identify because its two endpoints are on different scaffolds. However, breakpoints are easily identified, as long as they do not coincide with scaffold ends. For the phylogenetic context, in comparing a fragmented assembly with a number of complete assemblies, certain combinatorial constraints on breakpoints can be derived. We ask to what extent we can use breakpoint data between a fragmented genome and a number of complete genomes to recover all the arrangements in a phylogeny.

Results: We simulate genomic evolution via chromosomal inversion, fragmenting one of the genomes into a large number of scaffolds to represent the incompleteness of assembly. We identify all the breakpoints between this genome and the remainder. We devise an algorithm which takes these breakpoints into account in trying to determine on which branch of the phylogeny a rearrangement event occurred. We present an analysis of the dependence of recovery rates on scaffold size and rearrangement rate, and show that the true tree, the one on which the rearrangement simulation was performed, tends to be most parsimonious in estimating the number of true events inferred.

Conclusions: It is somewhat surprising that the breakpoints identified just between the fragmented genome and each of the others suffice to recover most of the rearrangements produced by the simulations. This holds even in parts of the phylogeny disjoint from the lineage of the fragmented genome.

Keywords: Genome rearrangements, Breakpoints, Phylogeny, Scaffold size

Background

The study of genome rearrangements generally requires a relatively complete assembled genome, consisting of a few lengthy scaffolds made up primarily of long contigs, preferably anchored to a physical map. In practice, however, *de novo* sequences of non-model organisms will consist of many short scaffolds, most containing no genes, and with no direct information on how to order the scaffolds with respect to each other [1].

Algorithmic inference of rearrangements [2] requires the possibility of identifying chromosomal fragments that have been inverted, inserted, deleted, duplicated, transposed to a new location on the chromosome or translocated from one chromosome to another. This involves identifying the whole fragment, including the two endpoints, which are in different genomic contexts before and after the rearrangement operation, and hence define one or more breakpoints. Most such fragments

*Correspondence: sankoff@uottawa.ca
Department of Mathematics and Statistics, University of Ottawa, 585 King Edward Avenue, K1N 6N5 Ottawa, Canada

cannot be detected in a short-scaffold assembly, since only the shortest of fragments generally will fit on a single scaffold.

Nevertheless, much can be learned from identifying single breakpoints, since each results from a rearrangement event, and conversely each identifiable rearrangement event gives rise to one or more breakpoints, so that the number of rearrangements is somewhat less than the number of breakpoints, depending on the evolutionary processes at play [3].

In this paper, we consider how to use breakpoint comparisons in a phylogenetic context, not by computing a distance matrix among genomes, but in locating rearrangement events on the branches of a phylogenetic tree. Prior to the eventual implementation of any practical tool for this task, we focus solely on the effect of fragmentation of just one of the genomes in the phylogeny on the integrity of rearrangement inference, with a range of degrees of fragmentation and various amounts of rearrangement. To isolate this effect, in our simulations, we will deliberately simplify the context to rearrangement by random inversion of unichromosomal genomes with the same, single-copy, gene content. And we will constrain ourselves to comparisons of the one fragmented genome in a phylogeny to all the other intact genomes.

In the course of our exposition, and especially in the conclusions, we will refer to how this simplified model would have to be amended in dealing with real genomes. Nevertheless, the findings of the present model with respect to degree of fragmentation, rate of rearrangement and phylogenetic accuracy should be pertinent to more realistic data.

Methods

From the unambiguity of gene adjacencies to the indeterminacy of block adjacency

In gene order comparisons, it is necessary to work with blocks of genes conserved in two or more genomes; trying to work with one gene at a time is not a robust procedure, especially with flowering plants, because most of these genomes have a whole genome duplication (WGD) in their history. The fractionation process ensuing from WGD deletes duplicate genes in a partially random pattern from one or the other duplicate (homeologous) chromosome, independently in two or more descendants of a duplicated genome [4]. This pattern, together with the possibility for some genes to transpose into different positions in the genome, makes it hard to identify unambiguously orthologous genes that are in the same gene order in two genomes. A set of five or ten genes in the same order, with few intervening genes, in two genomes can be confidently identified as a conserved syntenic block [5, 6].

However, the notion of block adjacency encounters a number of operational problems; the genes in a syntenic block in one genome may differ somewhat from the same block in the other genome, the minimum number of genes to establish a block is a parameter that must be determined by some empirical experimentation, as is the number of genes allowed to intervene between two pairs of orthologs within a block in the two genomes. We will avoid these practical problems in our simulations by excluding fractionation or other gene loss, duplication and small transpositions from our model.

The principle of event assignment

By assuming no evolutionary coincidences in rearrangement between large genomes, we derive the following principle.

Any breakpoint observed in two genomes must have occurred on the evolutionary path through the phylogeny leading from one of the genomes to another. Conversely, if there is an adjacency in common between the two genomes, there can have been no rearrangements affecting the corresponding scaffolds containing this adjacency in the two genomes apparent anywhere on the path between them.

In observing the distribution in a known phylogeny of which genomes conserve a given adjacency and in which it is disrupted, this principle suffices to infer the identity of the branch or a small subset of the branches where the rearrangement causing this adjacency occurred.

If a genomic comparison violates this principle, we say there is a conflict. In general a conflict will become apparent as by the appearance of a vertex reconstructed somewhere on a path between two genomes with no breakpoint.

These principles hold even if some of pairs of genomes are "can't tell" as to the whether they have a certain adjacency in common, and whether or not we are dealing with a binary or multifurcating phylogeny. In these cases, inference cannot necessarily locate a single branch where an event must have occurred, but only identify a subset of such branches.

The algorithm for the general case (with "can't tell" vertices)

We consider all rooted or unrooted trees T, binary or not, with leaves named $1, \ldots, N$. Given a set of fragmented genomes $\{g_1, \ldots, g_N\}$, we designate one at a time, say g_i as the reference genome. For each scaffold $s \in g_i$, we label each other genome g_j as $1, 2$, or 3 depending on whether the content of s is "split", according to some formal criterion as described above, among two or more scaffolds of g_j, is entirely contained in one genome of g_j, or is "can't tell". We then run the following dynamic programming algorithm.

Notation:

unrooted tree $T = (V, E)$
T_v: a subtree subtended by vertex v
possible vertex labels: {1 (split), 2 (not split), 3 ("can't tell") }
event: edge with one vertex labelled 1, the other labelled 2.
eventNumber(v): number of events in T_v:

Input:

unrooted tree T
leaf vertices labelled 1, 2, or 3.

Output:

S, a smallest possible set of event edges.

Algorithm: to label each non-leaf vertex with label 1 or 2, carry out a standard dynamic programming procedure [7] using labels 1, 2 and 3. During the traceback, replace every label 3 by a 1 or a 2, and consider the set of all possible results. This determines the set of branches on which an event could have occurred, namely the set of branches which in some solution are labeled 1 at one end and 2 at the other.

Results

All the experiments reported here were carried out on randomly generated binary unrooted trees based on seven input genomes, each having four internal branches, and all results are averages of a sample size of 100 simulations.

To produce each data sample, one of the internal nodes is chosen as a root genome. It is defined to be a single chromosome containing 1000 genes. The rearrangements we apply are inversions with two randomly chosen endpoints. (The choice of root is immaterial because of the symmetry of the inversion operation). After each simulation run, a designated genome is randomly split into a certain number of scaffolds, whose sizes are approximately geometrically distributed. Note that this is the simplest possibly case; in general, more or all of the genomes could be fragmented.

The effect of evolutionary time

In the first experiment, Fig. 1 shows that as expected, an increasing number of genome rearrangements is reflected in an increasing number of breakpoints detected. In these simulations, one rearrangement generally produces two breakpoints. With larger numbers of rearrangements, sometimes only one new breakpoint is created, the other being "re-used". When there are several branches with events in a lineage, we can usually only detect that at least one occurred.

The results of the — experiment show that, for example, with one rearrangement per branch almost 20 breakpoint events are detected on our 11-branch (7 terminal and 4 internal branches) tree but that many scaffolds do not contain a breakpoint, and some scaffolds contain several, which cannot easily be teased apart.

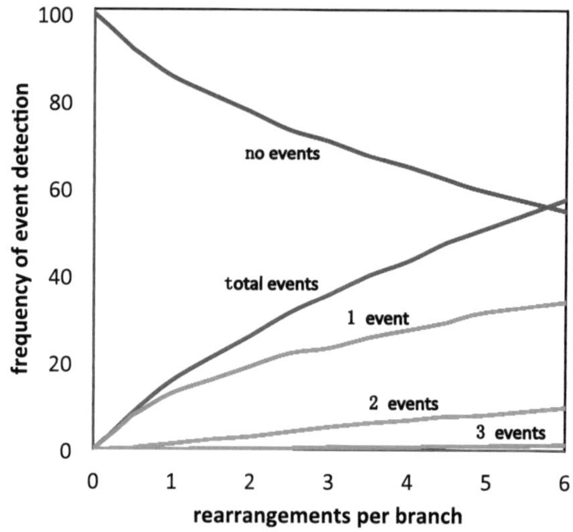

Fig. 1 Events detected as a function of amount of rearrangement. Designated genome fragmented into 100 scaffolds. "2 events" or "3 events" count rearrangements affecting a single scaffold in the designated genome. Total events = number of "1 events" + 2 × number of "2 events" + 3 × number of "3 events"

The effect of poor assembly

Figure 2 shows the results of our second experiment where increasing the number of scaffolds leads to an

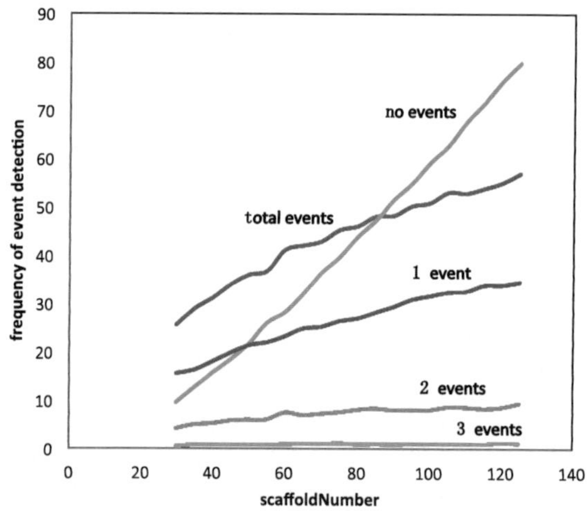

Fig. 2 Events detected as a function of amount of fragmentation. Simulations involved 5 rearrangements per branch. Increasing the number of scaffolds reduces the chance breakpoints from two rearrangements will be on a single scaffold and thus counted as a single event. It also sharply increases the number of scaffolds containing no breakpoints

increased probability that a given scaffold will contain no breakpoints. But it also decreases the probability that several rearrangements will involve the same scaffold.

Phylogeny based on the fragmented genome

The preceding experiments illustrate the use of the comparative, phylogenetic context to date most of the rearrangement events on the branch of a tree, or possibly a small subset of branches. Perhaps the main application of our method will be to actually distinguish among candidate trees, solely on the basis of rearrangements affecting the designated genome. Thus we run the algorithm for all 945 binary unrooted trees on seven genomes, not just the one T we used to generate our data, and check whether T is the most parsimonious with respect to the number of inferred events. Note that this experiment coincides with choosing the phylogeny with the smallest number of conflicts in pinpointing the rearrangement events.

In Fig. 3, we see that with even a small number of rearrangements per branch the tree T that generated the data is the best or among the 1 % best trees. In general, the trees with an equivalent or better score will be very similar in topology to T.

The effect of unidentifiable breakpoints

With real data, especially in the context of intense genomic fractionation, such as with the flowering plants, there will be a high incidence of "can't tell". By randomly re-identifying some of identified breakpoints as "can't tell", we can see the effect of degrading the data to model these real cases. Figure 4 shows that the method is stable to low levels of "can't tell", but tends to break down with increasing fragmentation when almost half the genomes exhibit "can't tell" with a designated genome scaffold.

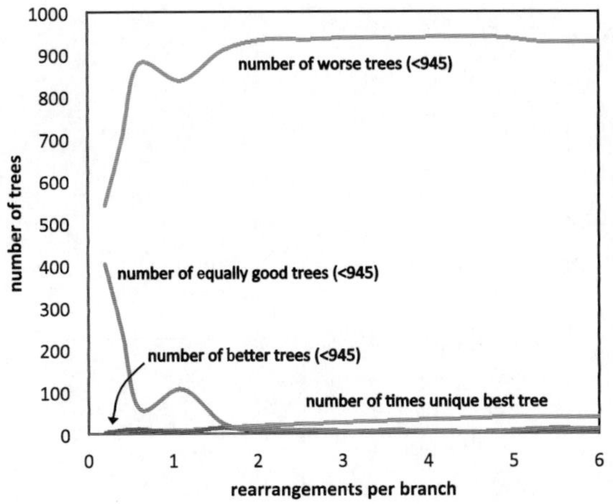

Fig. 3 Phylogenetic accuracy based on event detection. Designated genome broken into 60 scaffolds. Data-generating tree T is among the best 10 or 12, out of 945, in almost all the simulations

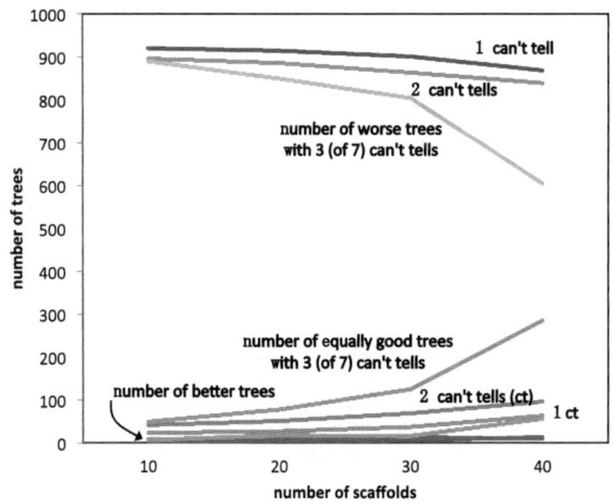

Fig. 4 Effect of "can't tells" on phylogenetic accuracy. With 3 out of 7 genomes exhibiting a "can't tell", the method becomes sensitive to the number of scaffolds

Discussion

Our simulations were set up in as simple a way as possible to explore the main potentials and the limitations of event detection based on observed breakpoints involving a single fragmented genome, the rest being completely assembled. We have studied the effects of the degree of fragmentation (number of scaffolds), rearrangement rate (number of inversions) and the rate of "can't tells". There would be many other secondary parameters to investigate in more general experiments: changing the number of genomes, allowing more or all of them to be fragmented, having larger and more heterogeneous genome sizes, and incorporating translocations, transposition, genome duplication and other rearrangement operations into a more variable and realistic model of gene order evolution, rather than the simple random inversion. Most important, confining our analysis to the phylogenetic viewpoint from a single genome is responsible for much or most of the shortfall in breakpoints and would be greatly attenuated in an analysis combining the viewpoints of all the genomes.

The most important difficulty in empirical work on this topic is determining the location and identity of breakpoints and determining if two breakpoints, on two genomes other than the designated one, are in fact the same or different. This difficulty stems from three sources, the first being fractionation, which removes gene copies more or less at random, so that single gene absences produce an apparent breakpoint; this necessitates examining neighboring syntenic blocks of genes, rather than single genes, to determine adjacencies and breakpoints, involving a whole new set of methodological problems.

The second source is the presence of numerous small, local, species-specific rearrangements, which we may or may not want to consider in a phylogenetic study spanning tens of millions of years of gene order evolution. Again, syntenic block comparisons help attenuate this problem.

Third, duplication of segments and whole genome duplications can complicate the determination.

On the theoretical level, the main interesting extensions would be the treatment of "can't tells" and of non-binary trees. Here the non-uniqueness is due not only to the distinct, conflicting, solutions occurring in special data configurations discussed above, but more generally in solutions where an event is identified as occurring somewhere on a path or other subtree, rather than on a specific branch. For phylogenetic scoring, this can be handled by splitting the unit score for the scaffold in question into fractions to redistributed among all the alternate branches.

There is a conceptual similarity between the problems we have studied here and the strategy of using comparative data to assemble contigs into scaffolds [8, 9]. In practice, however, the level of noise simulated in the current work, is greater than that in the data used for "contig fusion" or "scaffold filling", where minimizing genome rearrangement is a strategy for optimizing contig and scaffold orders.

Competing interests
The authors declare that they have no competing interests.

Authors' contributions
Both authors did the research and wrote the paper. All authors read and approved the final manuscript.

Acknowledgements
Research supported in part by grant RGPIN/8867-2008 from the Natural Sciences and Engineering Research Council of Canada. DS holds the Canada Research Chair in Mathematical Genomics.

Declerations
The publication costs for this article were partly funded by Natural Sciences and Engineering Research Council of Canada Discovery Grant number 8867-2008.

Published: 11 January 2016

References
1. Chain PSG, Grafham DV, Fulton RS, Hostetler J, Muzny D, et al. Genome project standards in a new era of sequencing. Science. 2009;326: 326:1180614.
2. Fertin G, Labarre A, Rusu I, Tannier E, Vialette S. Combinatorics of Genome Rearrangements: Cambridge: MIT Press; 2009.
3. Pevzner PA, Tesler G. Human and mouse genomic sequences reveal extensive breakpoint reuse in mammalian evolution. Proceedings of the National Academy of Sciences. 2003;100:7672–77.
4. Langham RJ, Walsh J, Dunn M, Ko C, Goff SA, Freeling M. Genomic duplication, fractionation and the origin of regulatory novelty. Genetics. 2004;166:935–45.
5. Lyons E, Freeling M. How to usefully compare homologous plant genes and chromosomes as DNA sequences. The Plant J. 2008;53:661–73. http://genomevolution.org/CoGe/.
6. Lyons E, Pedersen B, Kane J, Alam M, Ming R, Tang H, et al. Finding and comparing syntenic regions among Arabidopsis and the outgroups papaya, poplar and grape: CoGe with rosids. Plant Physiology. 2008;148: 1772–81.
7. Felsenstein J. Inferring Phylogenies, 2nd Edition. Sunderland, MA: Sinauer Associates; 2004. Chapter 2.
8. Muñoz A, Sankoff D. Rearrangement phylogeny of genomes in contig form. IEEE/ACM Transac Comput Biol Bioinforma. 2010;7:579–87.
9. Muñoz A, Zheng C, Zhu Q, Albert VA, Rounsley S, Sankoff D. Scaffold filling, contig fusion and comparative gene order inference. BMC Bioinforma. 2010;11:304.

Nariai *et al. BMC Genomics* 2016, **17**(Suppl 1):2
DOI 10.1186/s12864-015-2295-5

BMC Genomics

PROCEEDINGS Open Access

A Bayesian approach for estimating allele-specific expression from RNA-Seq data with diploid genomes

Naoki Nariai[1,2], Kaname Kojima[2], Takahiro Mimori[2], Yosuke Kawai[2] and Masao Nagasaki[2*]

From The Fourteenth Asia Pacific Bioinformatics Conference (APBC 2016)
San Francisco, CA, USA. 11 - 13 January 2016

Abstract

Background: RNA-sequencing (RNA-Seq) has become a popular tool for transcriptome profiling in mammals. However, accurate estimation of allele-specific expression (ASE) based on alignments of reads to the reference genome is challenging, because it contains only one allele on a mosaic haploid genome. Even with the information of diploid genome sequences, precise alignment of reads to the correct allele is difficult because of the high-similarity between the corresponding allele sequences.

Results: We propose a Bayesian approach to estimate ASE from RNA-Seq data with diploid genome sequences. In th e statistical framework, the haploid choice is modeled as a hidden variable and estimated simultaneously with isoform e xpression levels by variational Bayesian inference. Through the simulation data analysis, we demonstrate the effectivenes s of the proposed approach in terms of identifying ASE compared to the existing approach. We also show that our approach enables better quantification of isoform expression levels compared to the existing methods, TIGAR2, RSEM and Cufflinks. In the real data analysis of the human reference lymphoblastoid cell line GM12878, some autosomal genes were identified as ASE genes, and skewed paternal X-chromosome inactivation in GM12878 was identified.

Conclusions: The proposed method, called ASE-TIGAR, enables accurate estimation of gene expression from RNA-Seq data in an allele-specific manner. Our results show the effectiveness of utilizing personal genomic information for accurate estimation of ASE. An implementation of our method is available at http://nagasakilab.csml.org/ase-tigar.

Keywords: Allele-specific expression, RNA-Seq data, Bayesian inference

Background

Allele-specific expression (ASE) has been traditionally studied in the context of genomic imprinting, in which the expression of genes depends on whether they are paternally or maternally inherited. X-chromosome inactivation is also a form of ASE, in which one of the two alleles of the X chromosome is inactivated in female [1]. Recent studies have revealed that ASE is relatively common [2], and that many *cis*-acting sequence variants can alter gene expression in a highly context-specific manner [3]. In some cases, differences in the expression of two alleles can be predisposition to diseases, such as colorectal cancer [4]. Importantly, transcript abundances can be used as quantitative traits for identifying susceptibility loci for common diseases, such as diabetes and obesity [5, 6]. Hence, it is of our great interest to identify ASE and characterize genetic variants that are directly associated with phenotypic differences for elucidating causal mechanisms of diseases.

In order to identify allele-specific gene expression, RNA-sequencing (RNA-Seq) has now been widely used. However, there are several difficulties in measuring the amount of expressed isoforms in an allele-specific manner from RNA-Seq data given genotypes of an individual.

*Correspondence: nagasaki@megabank.tohoku.ac.jp
[2]Department of Integrative Genomics, Tohoku Medical Megabank Organization, Tohoku University, 2-1 Seiryo-machi, Aoba-ku, Sendai, Miyagi, 980-8575 Japan
Full list of author information is available at the end of the article

First, in many cases, short reads can be aligned to multiple locations of the reference genome, which poses uncertainty in quantifying gene expression levels [7]. Statistical methods that handle ambiguous alignment of reads as hidden variables have been shown to be effective in optimizing read alignments for more accurate quantification of isoforms [8–10], although the approaches do not consider isoforms in an allele-specific manner. Another difficulty is that there is a bias in alignment of reads to the reference genome if a sample has heterozygous SNPs where nucleotides are different from the reference sequence [11–13]. To avoid the bias in alignment of reads to the reference genome, one can prepare the alternative allele that includes genomic variants [14, 15], or construct diploid genomes for a specific sample [16]. Then, the best alignments of reads to the extended reference sequences are used to count the number of the paternally or maternally derived reads based on heterozygous SNP sites. However, these approaches cannot quantify isoform expression levels accurately, since only reads that align heterozygous positions are considered for ASE. To our best knowledge, there is currently no approach that can estimate ASE explicitly as well as isoform abundances in a unified statistical framework, given RNA-Seq data and diploid genomes.

In this paper, we present a novel method called ASE-TIGAR, to estimate ASE as well as gene expression levels of isoforms simultaneously from RNA-Seq data and diploid genome sequences. In the read generative model, a haploid choice is modeled as a hidden variable, and the posterior distribution for the binomial random variable is estimated by variational Bayesian inference. In order to evaluate our approach, we prepare two sets of synthetic paired-end reads (30 million reads, 100 bp × 2) with some sequencing errors, one is generated based on the null-hypothesis where there is no ASE, and the other is generated based on the alternative hypothesis where there is ASE for a certain portion of isoforms. We apply ASE-TIGAR to the simulation data and show that our method identifies more ASE isoforms than those identified with the existing approach. We also show that our method predicts isoform abundances more accurately compared to TIGAR2, RSEM and Cufflinks, which are widely used software for isoform-level quantification from RNA-Seq data. Finally, we apply our method to the RNA-Seq data obtained from the human lymphoblastoid cell line GM12878 [17] to identify autosomal genes that exhibit ASE, and investigate the balance of X-chromosome inactivation between the paternal and maternal alleles in the cell line.

Methods

ASE-TIGAR pipeline

A standard ASE-TIGAR pipeline starts from three input files, RNA-Seq data in FASTQ format, paternal and maternal cDNA sequences in FASTA format constructed from diploid genome sequences (represented as three rectangles with double lines in Fig. 1). In order to obtain cDNA sequences from the diploid genomes, "generate transcripts" function in rSeq software [18] can be used. Alternatively, it is also possible to construct diploid genome sequences from personal variants data in VCF format with vcf2diploid [16], or start from whole-genome sequencing data with the pre-processing steps described in Fig. 1 (shaded rectangles and circles). Then, RNA-Seq reads are aligned to the paternal and maternal cDNA sequences simultaneously, and all alignments are retained in BAM format. Bowtie2 [19] version 2.2.2 is used for searching all possible alignments for each read with "-k" option. Finally, ASE-TIGAR software takes the BAM file and estimate allele-specific isoform abundances after optimizing read alignments to the cDNA sequences of both alleles.

Read generative model

We use a graphical model, or Bayesian network, for representing a read generative model. For simplicity, here we describe a generative model for reads sequenced from single-end RNA-Seq libraries (Fig. 2). The model generates N independent and identically distributed reads, and each read n is associated with the three hidden variables T_n, H_n, and S_n, and the random variable R_n. The latent variable T_n represents the isoform choice of read n, and $T_n = t$ means that read n is generated from isoform t. The latent variable H_n represents the haplotype choice of read n, and $H_n = 0$ means that read n is generated from the paternal allele, whereas $H_n = 1$ means that read n is generated from the maternal allele. The latent variable S_n represents the start position of read n, and $S_n = s$ means that read n is generated from position s ($1 \leq s \leq l_{th} - L + 1$), where l_{th} is the length of isoform t of haplotype h and L is the read length. The random variable R_n is the observed data and represents the nucleotide sequences of read n. There are two model parameter vectors, $\boldsymbol{\theta}$ and $\boldsymbol{\phi}$, which represent the isoform abundances and allelic preferences for isoforms, respectively. The parameter vector $\boldsymbol{\theta} = (\theta_0, \ldots, \theta_T)'$ represents the fraction of abundance for each isoform, where $\sum_{t=0}^{t=T} \theta_t = 1$. The parameter vector $\boldsymbol{\phi} = (\phi_0, \ldots, \phi_T)'$ represents the fraction of the paternal allele for each isoform, where $0 \leq \phi_t \leq 1$.

As indicated in Fig. 2, conditional independence assumptions are used to simplify the model structure. Then, the complete likelihood of the data is decomposed as the product of conditional probabilities:

$$\begin{aligned} p(T_n, H_n, S_n, R_n \mid \boldsymbol{\theta}, \boldsymbol{\phi}) &= p(T_n \mid \boldsymbol{\theta})\, p(H_n \mid T_n, \boldsymbol{\phi}) \\ &\quad \times p(S_n \mid T_n, H_n) \\ &\quad \times p(R_n \mid T_n, H_n, S_n). \end{aligned}$$

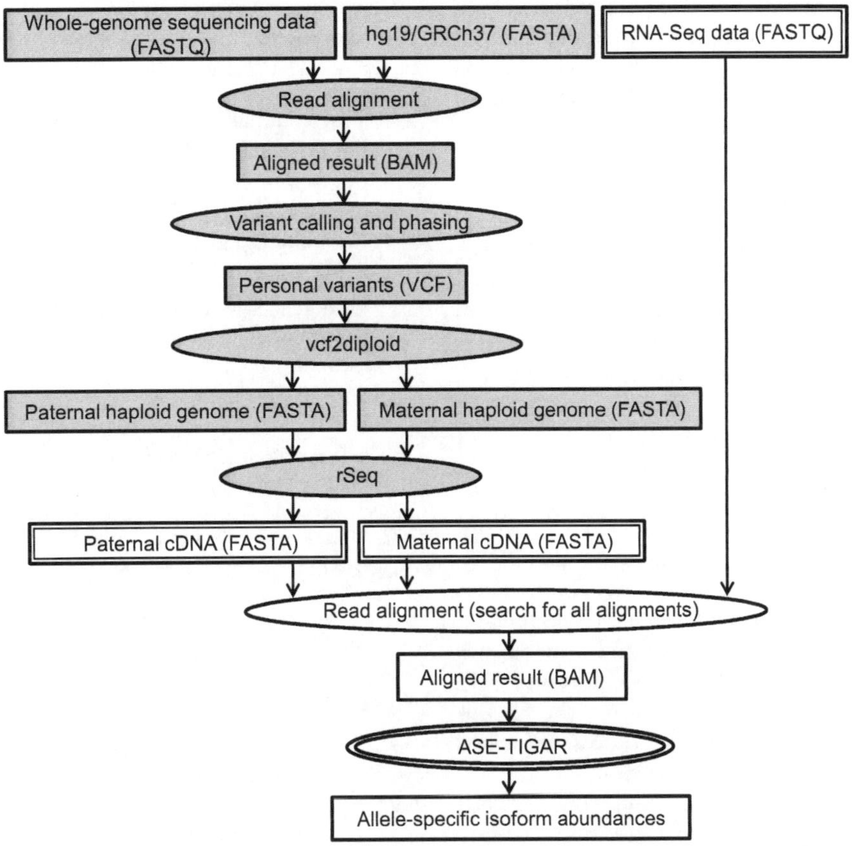

Fig. 1 ASE-TIGAR pipeline for estimating ASE. The input data for ASE-TIGAR are RNA-Seq data, paternal and maternal cDNA sequences, represented as rectangles with double lines. Alternatively, whole-genome sequencing data can be used as an input with pre-processing steps (represented as shaded rectangles and circles)

$p(T_n = t \mid \boldsymbol{\theta})$ is the probability that read n is generated from isoform t, given $\boldsymbol{\theta}$. We calculate this probability as $p(T_n = t \mid \boldsymbol{\theta}) = \theta_t$.

$p(H_n = h \mid T_n = t, \boldsymbol{\phi})$ is the probability that read n is generated from haplotype h (either paternal or maternal), given the isoform choice and $\boldsymbol{\phi}$. We calculate this probability as $p(H_n = 0 \mid T_n = t, \boldsymbol{\phi}) = \phi_t$ (if read n is generated from the paternal allele), or $p(H_n = 1 \mid T_n = t, \boldsymbol{\phi}) = 1-\phi_t$ (if read n is generated from the maternal allele).

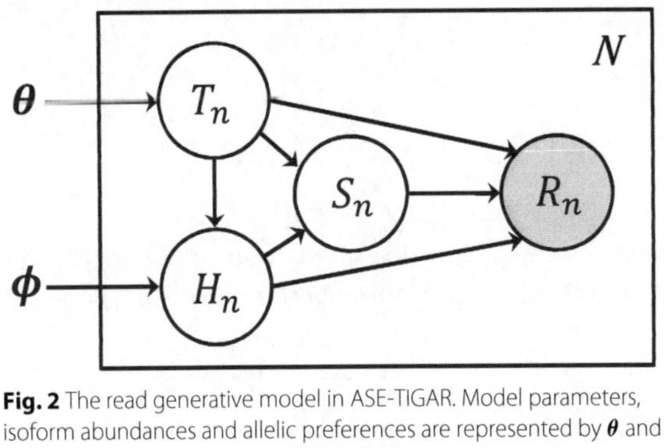

Fig. 2 The read generative model in ASE-TIGAR. Model parameters, isoform abundances and allelic preferences are represented by $\boldsymbol{\theta}$ and $\boldsymbol{\phi}$, respectively. The isoform choice, haplotype choice, and nucleotide sequence of read n are represented by T_n, H_n, and R_n, respectively

$p(S_n = s \mid T_n = t, H_n = h)$ is the probability that read n is generated from position s, given isoform t and haplotype h. We calculate this probability as $p(S_n = s \mid T_n = t, H_n = h) = 1/(l_{th} - L + 1)$.

$p(R_n \mid T_n = t, H_n = h, S_n = s)$ is the probability of observing the nucleotide sequence of read n, given the isoform choice, haplotype choice, and start position of read n. To summarize hidden variables T_n, H_n, and S_n, we introduce an indicator random variable Z_{nths}, where Z_{nths} is equal to one if $(T_n, H_n, S_n) = (t, h, s)$ and zero otherwise. Let π_n be a set of all (t, h, s) tuples for possible alignments of read n. Then, for each $(t, h, s) \in \pi_n$, we can calculate the probability of read sequence by

$$p(R_n \mid Z_{nths} = 1) = \prod_{x=1}^{L} subst\left(r_n[x], q_n[x], c_{th}[x]\right),$$

where $subst(\cdot,\cdot,\cdot)$ is the quality score dependent substitution matrix [10], $r_n[x]$ is the nucleotide character of position x of read n, $q_n[x]$ is the quality score of position x of read n, and $c_{th}[x]$ is the nucleotide character of position x of cDNA sequence of isoform t of haplotype h. The quality score dependent substitution matrix, $subst(\cdot,\cdot,\cdot)$, is either determined according to the Phred base quality score [20], or can be estimated from the best alignments of reads to the reference cDNA sequences from the RNA-Seq data.

For the cases where RNA-Seq reads are generated from paired-end libraries, and how indel errors of sequencers can be handled, the previously proposed model [10] can be naturally extended similarly to the case for the single-end data described here.

Variational Bayesian inference

We propose a Bayesian approach, in which model parameters are estimated as posterior distributions, given RNA-Seq data and prior distributions for the model parameters $\boldsymbol{\theta}$ and $\boldsymbol{\phi}$. Because full Bayesian inference involves integration over all possible hidden variable $\boldsymbol{Z}$ and is analytically intractable, we use variational Bayesian inference [21], which approximates the posterior joint distributions by assuming the factorization of latent variables and model parameters as $q(\boldsymbol{\theta},\boldsymbol{\phi},\boldsymbol{Z}) \approx q(\boldsymbol{\theta})\, q(\boldsymbol{\phi})\, q(\boldsymbol{Z})$.

For the prior distribution of $\boldsymbol{\theta}$, we use the Dirichlet distribution

$$p(\boldsymbol{\theta}) = \frac{1}{G(\boldsymbol{\alpha})} \prod_{t=0}^{T} \theta_t^{\alpha_t - 1},$$

where $\alpha_t > 0$ is a hyperparameter, $G(\boldsymbol{\alpha}) = \frac{\prod_t \Gamma(\alpha_t)}{\Gamma(\sum_t \alpha_t)}$, and $\Gamma(\cdot)$ is the gamma function. In this paper, we use a single hyperparameter α_0 for all isoforms, based on the assumption that there is no prior knowledge about relative differences in isoform abundance. The single hyperparameter α_0 controls the complexity of model parameters [22]. When $\alpha_0 \geq 1, \alpha_0 - 1$ can be interpreted as the prior count of reads that are assigned to isoforms, and when $\alpha_0 < 1$, the prior favors that some of the isoform abundances to be zero [10]. Here, we choose α_0 that maximizes the lower bound of the log marginal likelihood.

For the prior distribution of $\boldsymbol{\phi}$, we use the Beta distribution

$$p(\phi_t) = \frac{1}{B(\beta_{t,1}, \beta_{t,2})} \phi_t^{\beta_{t,1}-1} (1-\phi_t)^{\beta_{t,2}-1},$$

where $\beta_{t,1} > 0$ and $\beta_{t,2} > 0$ are hyperparameters, and $B(\cdot,\cdot)$ is the Beta function. Here, $\beta_{t,1}$ and $\beta_{t,2}$ can be interpreted as the prior counts of reads that are assigned to the paternal and maternal allele, respectively, for calculating the paternal/maternal ratio. We use $\beta_{t,1} = \beta_{t,2} = 1$ for all isoforms as a non-informative prior.

Given hyperparameters α_0, $\beta_{t,1}$, and $\beta_{t,2}$, the lower bound of the log marginal likelihood is maximized iteratively by variational Bayesian inference algorithm:

Step 1. Initialization
For each isoform t, set $\alpha_t^* = \alpha_0$, $\beta_{t,1}^* = \beta_{t,1}$, and $\beta_{t,2}^* = \beta_{t,2}$
Step 2. Update $q^*(\boldsymbol{Z})$
Compute $E_Z[Z_{nths}]$ given the current estimate of $q^*(\boldsymbol{\theta})$ and $q^*(\boldsymbol{\phi})$
Step 3. Update $q^*(\boldsymbol{\theta})$ and $q^*(\boldsymbol{\phi})$
Compute $E_\theta[\theta_t]$ and $E_\phi[\phi_t]$ given the current estimate of $q^*(\boldsymbol{Z})$
Step 4. Check for convergence
If none of the $E_\theta[\theta_t]$ has been changed more than a pre-specified threshold, exit. Otherwise, return to Step 2

In Step 2, $E_Z[Z_{nths}]$ is calculated based on the current estimate of $q^*(\boldsymbol{\theta})$ and $q^*(\boldsymbol{\phi})$ as

$$E_Z[Z_{nths}] = \begin{cases} \frac{\rho_{nths}}{\sum_{(t'h's')\in\pi_n} \rho_{nt'h's'}} & \text{if } (t,h,s) \in \pi_n, \\ 0 \text{ otherwise.} \end{cases}$$

where

$$\log\rho_{nths} = \begin{cases} E_\theta[\log\theta_t] + E_\phi[\log\phi_t] + \log p(S_n|T_n,H_n) \\ \quad + \log p(R_n|T_n,H_n,S_n) \quad \text{if } h = 0, \\ E_\theta[\log\theta_t] + E_\phi[\log(1-\phi_t)] \\ \quad + \log p(S_n|T_n,H_n) \\ \quad + \log p(R_n|T_n,H_n,S_n) \text{ otherwise.} \end{cases}$$

Note that

$$E_\theta[\log\theta_t] = \psi(\alpha_t^*) - \psi\left(\sum_t \alpha_t^*\right),$$

$$E_\phi[\log\phi_t] = \psi(\beta_{t,1}^*) - \psi(\beta_{t,1}^* + \beta_{t,2}^*),$$

$$E_\phi[\log(1-\phi_t)] = \psi(\beta_{t,2}^*) - \psi(\beta_{t,1}^* + \beta_{t,2}^*),$$

where $\psi(\cdot)$ is the digamma function.

In Step 3, $E_\theta[\theta_t]$ is calculated based on the current estimate of $q^*(\boldsymbol{Z})$ as

$$E_\theta[\theta_t] = \frac{\alpha_t^*}{\sum_{t'} \alpha_{t'}^*},$$

where

$$\alpha_t^* = \alpha_0 + \sum_{n',t'=t,h',s'} E_Z[Z_{n't'h's'}].$$

Hence, it turns out that $q^*(\boldsymbol{\theta})$ is also the Dirichlet distribution, and the prior distribution $p(\boldsymbol{\theta})$ is the conjugate prior.

Similarly, $E_\phi[\phi_t]$ is calculated based on the current estimate of $q^*(\boldsymbol{Z})$ as

$$E_\phi[\phi_t] = \frac{\beta_{t,1}^*}{\beta_{t,1}^* + \beta_{t,2}^*},$$

where

$$\beta^*_{t,1} = \beta_{t,1} + \sum_{n',t'=t,h'=0,s'} E_Z\left[Z_{n't'h's'}\right],$$

$$\beta^*_{t,2} = \beta_{t,2} + \sum_{n',t'=t,h'=1,s'} E_Z\left[Z_{n't'h's'}\right].$$

Hence, $q^*(\phi_t)$ is also the Beta distribution, and the prior distribution $p(\phi_t)$ is the conjugate prior.

In Step 4, a relative change of 10^{-3} for isoforms whose abundance parameter $E_\theta[\theta_t] > 10^{-7}$ is used as a convergence criteria.

Variational lower bound

The log marginal likelihood can be decomposed as

$$\log p(\boldsymbol{R}) = L(q) + KL(q||p),$$

where

$$L(q) = \int\int\int q(\boldsymbol{\theta},\boldsymbol{\phi},\boldsymbol{Z}) \log\frac{p(\boldsymbol{R},\boldsymbol{\theta},\boldsymbol{\phi},\boldsymbol{Z})}{q(\boldsymbol{\theta},\boldsymbol{\phi},\boldsymbol{Z})} d\boldsymbol{\theta} d\boldsymbol{\phi} d\boldsymbol{Z},$$

$$KL(q||p) = -\int\int\int q(\boldsymbol{\theta},\boldsymbol{\phi},\boldsymbol{Z}) \log\frac{p(\boldsymbol{\theta},\boldsymbol{\phi},\boldsymbol{Z}|\boldsymbol{R})}{q(\boldsymbol{\theta},\boldsymbol{\phi},\boldsymbol{Z})} d\boldsymbol{\theta} d\boldsymbol{\phi} d\boldsymbol{Z}.$$

Since $KL(q||p)$ is the Kullback-Leibler divergence between $q(\boldsymbol{\theta},\boldsymbol{\phi},\boldsymbol{Z})$ and $p(\boldsymbol{\theta},\boldsymbol{\phi},\boldsymbol{Z}|\boldsymbol{R})$, the log marginal likelihood is lower bounded by $L(q)$. With the factorization assumption $q(\boldsymbol{\theta},\boldsymbol{\phi},\boldsymbol{Z}) \approx q(\boldsymbol{\theta})q(\boldsymbol{\phi})q(\boldsymbol{Z})$, we have

$$\begin{aligned} L(q) &= E\left[\log p(\boldsymbol{R},\boldsymbol{\theta},\boldsymbol{\phi},\boldsymbol{Z})\right] - E\left[\log q(\boldsymbol{\theta},\boldsymbol{\phi},\boldsymbol{Z})\right] \\ &= E\left[\log p(\boldsymbol{R},\boldsymbol{Z}|\boldsymbol{\theta},\boldsymbol{\phi})\right] + E\left[\log p(\boldsymbol{\theta})\right] + E\left[\log p(\boldsymbol{\phi})\right] \\ &\quad - E\left[\log q(\boldsymbol{\theta})\right] - E\left[\log q(\boldsymbol{\phi})\right] - E\left[\log q(\boldsymbol{Z})\right], \end{aligned}$$

where

$$\begin{aligned} E\left[\log p(\boldsymbol{R},\boldsymbol{Z}|\boldsymbol{\theta},\boldsymbol{\phi})\right] &= \sum_{n,t,h,s} E_Z[Z_{nths}] \log\rho_{nths}, \\ E\left[\log p(\boldsymbol{\theta})\right] &= \sum_t (\alpha_0 - 1) E_\theta\left[\log\theta_t\right] - \log G(\boldsymbol{\alpha}), \\ E\left[\log p(\boldsymbol{\phi})\right] &= \sum_t \left\{(\beta_{t,1} - 1) E_\phi\left[\log\phi_t\right] + (\beta_{t,2} - 1) E_\phi\left[\log(1-\phi_t)\right]\right\} \\ &\quad - \sum_t \log B\left(\beta_{t,1},\beta_{t,2}\right), \\ E\left[\log q(\boldsymbol{\theta})\right] &= \sum_t \left(\alpha^*_t - 1\right) E_\theta\left[\log\theta_t\right] - \log G(\boldsymbol{\alpha}^*), \\ E\left[\log q(\boldsymbol{\phi})\right] &= \sum_t \left\{(\beta^*_{t,1} - 1) E_\phi\left[\log\phi_t\right] + (\beta^*_{t,2} - 1) E_\phi\left[\log(1-\phi_t)\right]\right\} \\ &\quad - \sum_t \log B\left(\beta^*_{t,1},\beta^*_{t,2}\right), \\ E\left[\log q(\boldsymbol{Z})\right] &= \sum_{n,t,h,s} E_Z\left[Z_{nths}\right] \log E_Z\left[Z_{nths}\right]. \end{aligned}$$

Results and discussion

Simulation data analysis

To evaluate the performance of the proposed method, we prepared synthetic RNA-Seq data (30 million reads, 100 bp × 2 with the mean fragment size of 400 bp and standard deviation of 40 bp) based on diploid genome sequences of NA12878, which were constructed from hg19 and publicly available from the website (http://sv.gersteinlab.org/NA12878_diploid). First, the paternal and maternal cDNA sequences were generated from the diploid genome sequences based on the UCSC gene annotations file (refFlat.txt) with rSeq (version 0.2.1) as described in Methods section. Second, 10,000 isoforms were randomly chosen and expression levels were assigned so that it follows the log-normal distribution. Then, we prepared two sets of RNA-Seq data with 0.1 % substitution, deletion, and insertion errors, one was generated based on the null hypothesis that there was no ASE, and the other was generated based on the alternative hypothesis that there were ASE for some portions of isoforms. For the null hypothesis data set, 100 % of the isoforms express the paternal and maternal alleles equally likely (50:50 chance). On the other hand, for the ASE data set, 10 % of the isoforms have the paternal-specific expression (in which the paternal allele was chosen to express with an 80 % probability, whereas the maternal allele was chosen to express with a 20 % probability), 10 % of the isoforms have the maternal-specific expression (in which the maternal allele was chosen to express with an 80 % probability, whereas the paternal allele was chosen to express with a 20 % probability), and the remaining isoforms have no ASE.

To compare with the existing approach [16], reads were aligned to the both paternal and maternal haplotypes, and the best alignments of reads were obtained. Then, for each isoform, the number of heterozygous SNPs was counted to determine the paternal/maternal ratio. On the other hand, our approach aligned reads to the both haplotypes and retained all the possible alignments with Bowtie2 specifying "-k" option. Then, ASE-TIGAR took the BAM file as input and optimized the read alignments between the paternal and maternal alleles, as well as among isoforms by variational Bayesian inference algorithm as described in Methods section. The hyperparameter α_0 was set to 0.1, which maximized the variational lower bound of the marginal log likelihood of the data.

Predicted distributions of the paternal/maternal ratio for the null and ASE hypotheses with ASE-TIGAR and the existing approach (based on the best alignments of reads to the diploid genomes) are compared with the the true distributions (Fig. 3). Note that isoforms having one or more heterozygous SNP(s) with ten or more assigned reads were considered for the comparison. Whether there is ASE or not, the predicted distributions with

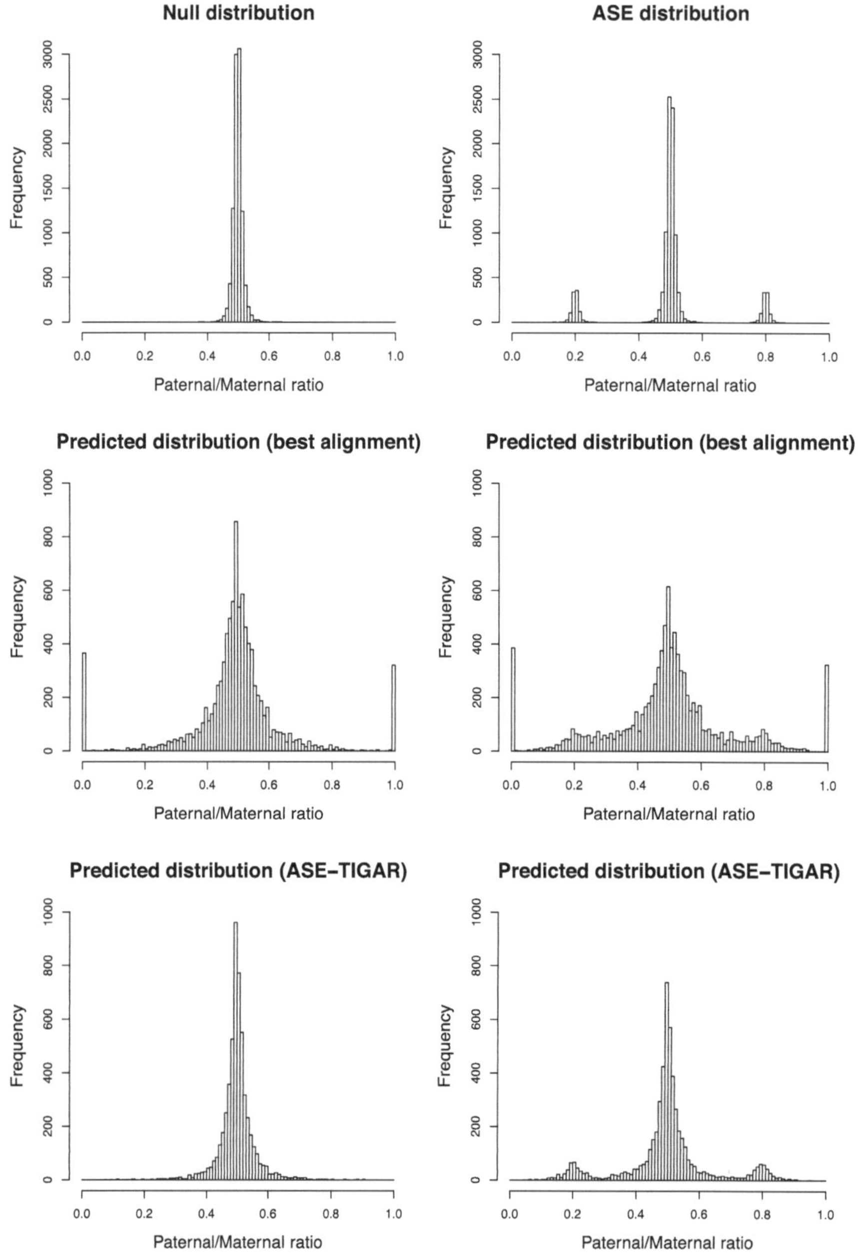

Fig. 3 Estimation of ASE from simulated data. True distributions of the paternal/maternal ratio for the null and ASE hypotheses (*top-left* and *top-right*); predicted distributions with the existing approach for the null and ASE hypotheses (*middle-left* and *middle-right*); predicted distributions with ASE-TIGAR for the null and ASE hypotheses (*bottom-left* and *bottom-right*)

ASE-TIGAR were more similar to the true distributions, particularly in the area where the paternal/maternal ratio is close to zero or one. On the contrary, the predicted distributions with the existing approach show "peaks" in those extreme area, which in fact did not exist in the true distributions. The favorable result with ASE-TIGAR came from the smoothing property of the updated beta distribution for the haplotype choice variable in the Bayesian inference, in which the prior count of one was naturally added to each allele of isoforms for calculating the paternal/maternal ratio (called as Laplace smoothing, or add-one smoothing). This feature will be especially beneficial for isoforms whose expression levels are inherently low, or when there are not many heterozygous SNPs that can be used to distinguish isoforms between paternal and maternal alleles.

Next, we evaluate the performance of quantifying isoform expression levels with ASE-TIGAR compared to existing methods using the simulation data. For comparing the performance, TIGAR2 [23], RSEM v1.2.21 [24] and Cufflinks v2.2.1 (with default options except '-u' and '-G' options) [25] are applied to the same simulation data. Note that TIGAR2, RSEM, and Cufflinks predict isoform expression levels without allelic information, and use the reference genome instead of the diploid sequences. Here, we compare the combined isoform expression levels (both paternal and maternal) predicted by ASE-TIGAR, with isoform levels predicted by TIGAR2, RSEM, and Cufflinks. The scatter-plot of the estimated isoform abundances (log of the number of reads) and the true isoform expression levels and the Pearson correlation are shown in Fig. 4. Root mean square errors were also calculated for comparison (ASE-TIGAR: 0.778, TIGAR2: 0.785, RSEM: 0.881, and Cufflinks: 1.26). The prediction accuracy with ASE-TIGAR compared to those with TIGAR2, RSEM and Cufflinks were found to be better, which proves the usefulness of ASE-TIGAR for quantifying isoform-level expression levels, in addition to identifying ASE.

Real data analysis

We applied ASE-TIGAR to the RNA-Seq data (36.5 million reads of 100 bp × 2) that was generated from the lymphoblastoid cell line GM12878 [17], which is publicly available under the NCBI SRA accession number SRX245434. This cell line was derived from the HapMap NA12878 individual, whose diploid genomes were similarly obtained and used as in the simulation data analysis.

We found that there were some autosomal genes that showed ASE from either the paternal or maternal allele (top-left in Fig. 5). In the subsequent analysis, genes were considered as ASE genes, if the paternal/maternal ratio of their isoforms were either ≥0.75 or ≤0.25. To investigate which functional categories of genes were regulated in an allele-specific manner, we used DAVID [26] to identify enriched functional categories in the autosomal 1,251 ASE genes. Enriched terms included "polymorphism", "sequence variant", and "splicing variant" (Table 1), which might be explained by genomic variations among haplotypes within the population. For example, "polymorphism" annotation means that there is at least one variant within human, that is not directly responsible for

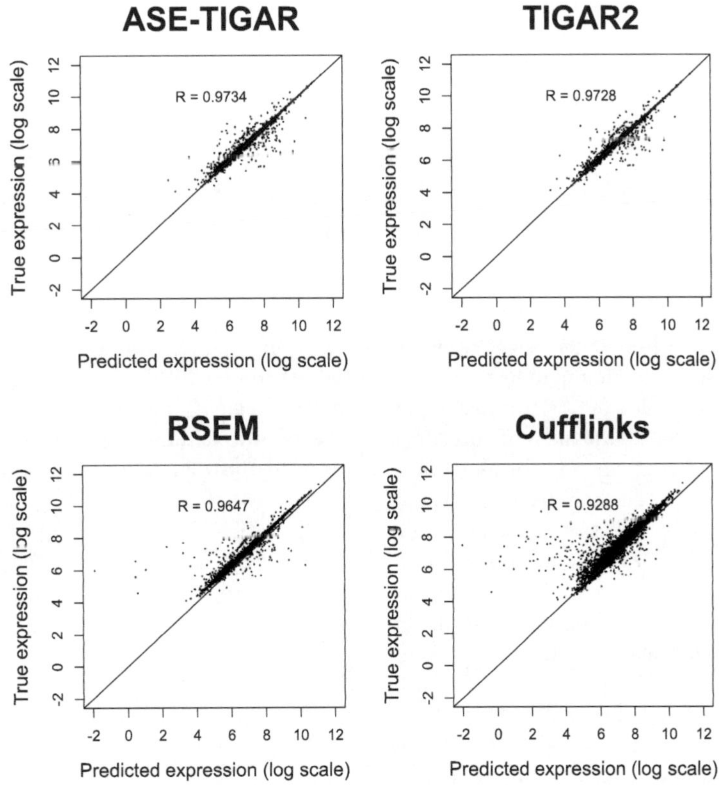

Fig. 4 Estimation of isoform expression from simulated data. Scatter plots of isoform expression levels estimated from simulated RNA-Seq data (*x*-axis) and true expression levels (*y*-axis). The Pearson correlation coefficients are calculated and shown on each plot. Predictions with ASE-TIGAR were most consistent compared to TIGAR2, RSEM, and Cufflinks

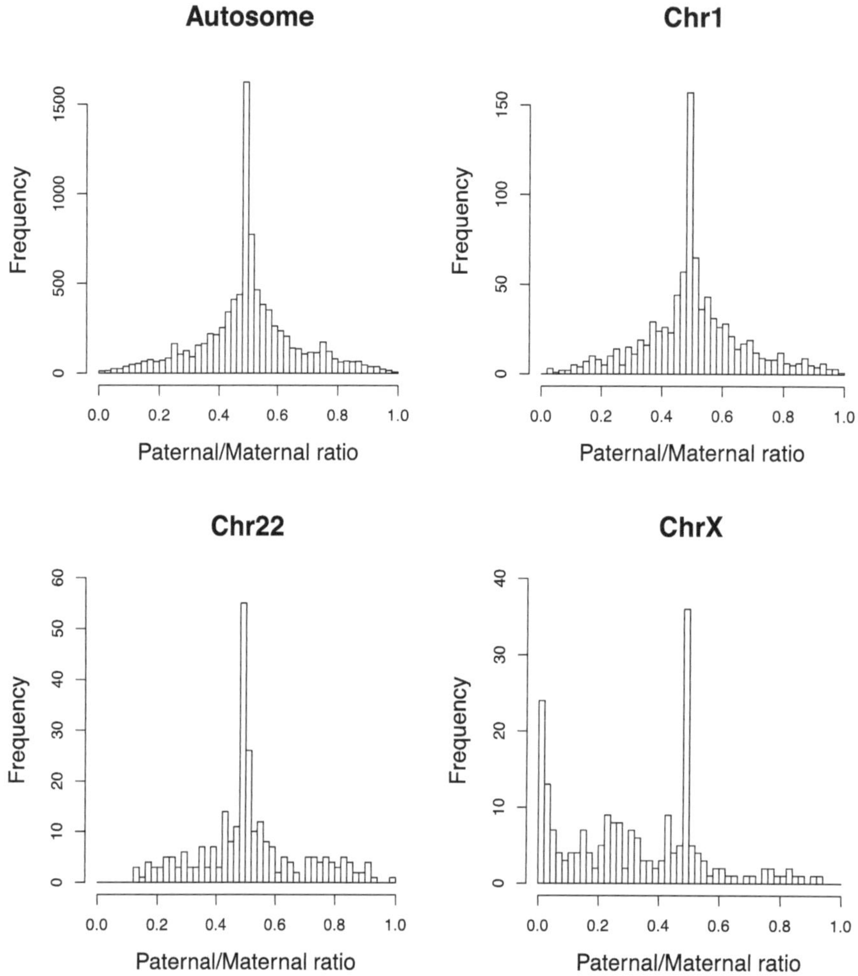

Fig. 5 Estimation of ASE from GM12878 data. Estimated distributions for autosomal genes (*top-left*), genes on chr1 (*top-right*), genes on chr22 (*bottom-left*), and genes on chrX (*bottom-right*) with ASE-TIGAR

a disease [27]. However, any functional category in the Gene Ontology Terms [28] was not found to be significant at the Bonferroni adjusted p-value of 0.001 in this analysis. When we compared overall abundances of autosomal ASE isoforms with those of autosomal isoforms without ASE, the former tend to be smaller than the latter (Fig. 6). This suggests that the lower expression from one allele due to genomic variants or other regulatory mechanisms were not compensated by the expression from the other allele in the cell line. Hence, genes showing ASE in the cell line were, in general, not likely to be house-keeping genes.

Interestingly, by looking at the paternal/maternal ratio of expressed isoforms on each chromosome, skewed X-inactivation in the paternal allele of the GM12878 cell line was observed (bottom-right in Fig. 5). This result is consistent with the findings in previous studies that showed the bias in X-chromosome inactivation by CTCF binding [29] and occupancies of RNA polymerase II [30] from ChIP-Seq data. ASE-TIGAR identified 90 maternal allele-specific isoforms on X-chromosome, whereas the existing approach based on the best alignment to the diploid genome [16] identified 76, based

Table 1 Terms enriched in the autosomal ASE genes

Category	Term	Count	*P*-value	Bonferroni
SP_PIR	Polymorphism	787	9.7E-8	5.2E-5
UP_SEQ	Sequence variant	808	1.1E-7	3.2E-4
SP_PIR	Alternative splicing	599	1.6E-6	8.6E-4
SP_PIR	Glycoprotein	231	1.8E-6	9.7E-4
UP_SEQ	Extracellular	159	9.0E-7	2.7E-3
SP_PIR	Signal	170	1.0E-5	5.6E-3
UP_SEQ	Splice variant	597	2.4E-6	7.3E-3

SP_PIR: SwissProt Protein Information Resource Keyword. UP_SEQ: UniProt Sequence Feature

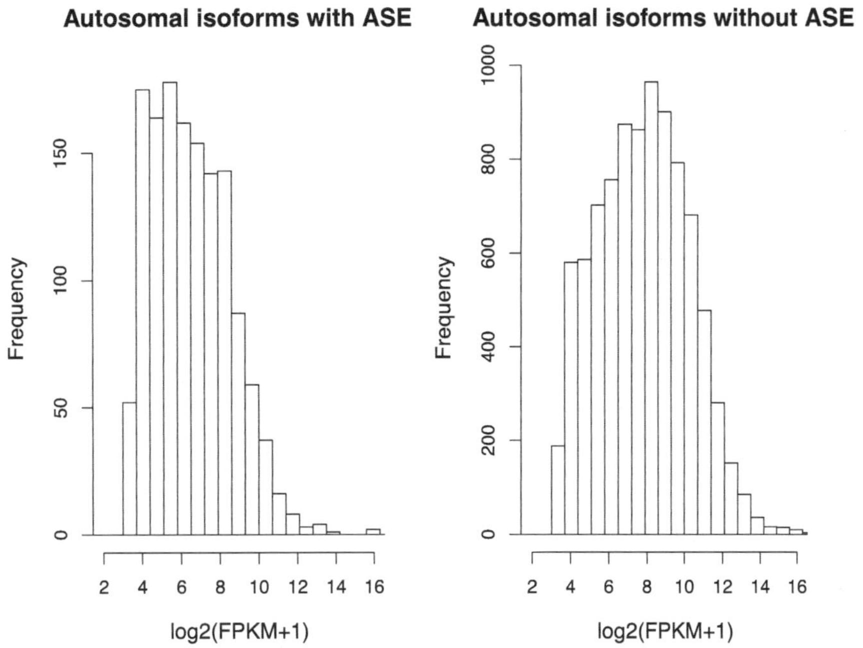

Fig. 6 Estimation of autosomal isoform expression from GM12878 data. Frequencies of abundances of autosomal isoforms with ASE (*left*), and autosomal isoforms without ASE (*right*)

on the same experimental condition in the simulation analysis.

Computational resources

Computational experiments were performed on a computer with Intel Xeon CPU E7-8837 processors (2.8 GHz) with the Red Hat Enterprise Linux Server release 6.1 operating system. ASE-TIGAR is implemented in Java and executed on 16 CPU cores with a multi-thread option. In the experiments for the simulated data sets (30 million paired-end reads), the execution time was 20 hours, and 46 GB memory was used with the Java(TM) SE Runtime Environment (build 1.8.0_45-b14).

Conclusions

In this paper, we proposed a novel method called ASE-TIGAR, a Bayesian approach to estimate ASE from RNA-Seq data with diploid genomes. Contrary to the popularly used existing methods such as TopHat-Cufflinks [25], RSEM [8], and TIGAR2 [23], personal diploid genomes are used as reference sequences in the pipeline, instead of the reference genome. Since genetic variants such as SNPs and indels are incorporated in the diploid genome sequences by construction, there will be less bias in alignment of reads compared to the conventional approaches that rely on the reference genome. In the generative model, a haplotype choice is modeled as a latent variable and estimated simultaneously with isoform abundances by variational Bayesian inference.

We showed from the simulation data analysis that ASE-TIGAR estimated ASE more consistently compared to the existing approach, in part from smoothing effect of the estimated posterior distribution of the binomial random variable that represents the fraction of the expressed paternal and maternal haplotypes. We also showed that ASE-TIGAR quantified isoform abundances more accurately compared to TIGAR2, RSEM, and Cufflinks, which is an additional benefit of ASE-TIGAR if genotypes of samples are available. In the real data analysis of human lymphoblastoid cell line GM12878, ASE was identified among relatively low-expressed genes, and that no functional GO category was found to be significantly enriched. We also observed that the paternal X-chromosome inactivation was dominant in the cell line, which was also confirmed in the previous studies [29, 30].

Although full-length transcripts can be sequenced with new sequencing technologies, such as the PacBio RS II [31], accurate estimation of ASE is challenging without enough information about isoform abundances. Currently, the Illumina platform is more suitable in quantifying isoform abundances thanks to its capacity of generating short reads in a high-throughput manner. Because the accuracy of the reference sequences is critical for our approach, it will be effective to include the obtained full-length transcript sequences as reference cDNA sequences in ASE-TIGAR pipeline combined with short reads.

As more personal whole-genome sequencing data and RNA-Seq data become available [32], ASE-TIGAR will be particularly useful to find associations between genetic variants and expression quantitative loci (eQTL). For example, links between genetic variants in transcription factor (TF) binding sites and the level of gene expression can be investigated. Incorporation of other omics data, such as ChIP-Seq data measuring CTCF binding, TF occupancies, histone modifications, or chromatin structures will be possible in the similar framework. If only a limited portion of genotypes is available for samples (such as with SNP arrays), genotype imputation with the reference panel can be considered [33]. However, there might exist imputation errors, or switching errors in phased genotypes without a complete parental genotypes, which will affect accuracies in ASE identification and isoform quantification with ASE-TIGAR. Our future work will include investigating ASE with other cell types, and the topics described above.

Competing interests
The authors declare that they have no competing interests.

Authors' contributions
NN and MN conceived the study. NN designed the computational experiments and performed the analysis. NN KK, TM, YK, and MN interpreted the results. NN, KK, TK, and YK collaborated on data collection. NN and MN wrote the manuscript. All the authors read and approved the final manuscript.

Declarations
The publication costs for this article were partly funded by MEXT Tohoku Medical Megabank Project. This article has been published as part of BMC Genomics Volume 17 Supplement 1, 2016: Selected articles from the Fourteenth Asia Pacific Bioinformatics Conference (APBC 2016): Genomics. The full contents of the supplement are available online at http://www.biomedcentral.com/bmcgenomics/supplements/17/S1.

Acknowledgements
This work was supported (in part) by MEXT Tohoku Medical Megabank Project. Super computer resources were provided by the Supercomputing services, Tohoku Medical Megabank Organization, Tohoku University, and National Institute of Genetics (NIG).

Author details
[1]Present address: Institute for Genomic Medicine, University of California, San Diego, 9500 Gilman Drive, La Jolla, California, 92093 USA. [2]Department of Integrative Genomics, Tohoku Medical Megabank Organization, Tohoku University, 2-1 Seiryo-machi, Aoba-ku, Sendai, Miyagi, 980-8575 Japan.

Published: 11 January 2011

References

1. Lyon MF. Gene action in the X-chromosome of the mouse (Mus musculus L). Nature. 1961;190:372–3.
2. Knight JC. Allele-specific gene expression uncovered. Trends Genet. 2004;20:113–6.
3. Buckland PR. Allele-specific gene expression differences in humans. Hum Mol Genet. 2004;13:R255–60. Spec No 2.
4. de la Chapelle A. Genetic predisposition to human disease: allele-specific expression and low-penetrance regulatory loci. Oncogene. 2009;28: 3345–8.
5. Schadt EE, Monks SA, Drake TA, Lusis AJ, Che N, Colinayo V, et al. Genetics of gene expression surveyed in maize, mouse and man. Nature. 2003;422:297–302.
6. Fairfax BP, Humburg P, Makino S, Naranbhai V, Wong D, Lau E, et al. Innate immune activity conditions the effect of regulatory variants upon monocyte gene expression. Science. 2014;343:6175.
7. Mortazavi A, Williams BA, McCue K, Schaeffer L, Wold B. Mapping and quantifying mammalian transcriptomes by RNA-Seq. Nat Methods. 2008;5:621–8.
8. Li B, Ruotti V, Stewart RM, Thomson JA, Dewey CN. RNA-Seq gene expression estimation with read mapping uncertainty. Bioinformatics. 2010;26:493–500.
9. Glaus P, Honkela A, Rattray M. Identifying differentially expressed transcripts from RNA-seq data with biological variation. Bioinformatics. 2012;28:1721–28.
10. Nariai N, Hirose O, Kojima K, Nagasaki M. TIGAR: transcript isoform abundance estimation method with gapped alignment of RNA-Seq data by variational Bayesian inference. Bioinformatics. 2013;29:2292–9.
11. Degner JF, Marioni JC, Pai AA, Pickrell JK, Nkadori E, Gilad Y, et al. Effect of read-mapping biases on detecting allele-specific expression from RNA-sequencing data. Bioinformatics. 2009;25:3207–12.
12. Skelly DA, Johansson M, Madeoy J, Wakefield J, Akey JM. A flexible Bayesian method for detecting allelic imbalance in RNA-seq data. Genome Res. 2011;10:1728–37.
13. León-Novelo LG, McIntyre LM, Fear JM, Graze RM. A flexible Bayesian method for detecting allelic imbalance in RNA-seq data. BMC Genomics. 2014;15:920.
14. Satya RV, Zavaljevski N, Reifman J. A new strategy to reduce allelic bias in RNA-Seq readmapping. Nucleic Acids Res. 2012;40:e127.
15. van de Geijn B, McVicker G, Gilad Y, Pritchard JK. WASP: allele-specific software for robust molecular quantitative trait locus discovery. Nat Methods. 2015;12(11):1061–3. doi: 10.1038/nmeth.3582. Epub 2015 Sep 14.
16. Rozowsky J, Abyzov A, Wang J, Alves P, Raha D, Harmanci A, et al. AlleleSeq: analysis of allele-specific expression and binding in a network framework. Mol Syst Biol. 2011;7:522.
17. Marinov GK, Williams BA, McCue K, Schroth GP, Gertz J, Myers RM, et al. From single-cell to cell-pool transcriptomes: stochasticity in gene expression and RNA splicing. Genome Res. 2014;24:496–510.
18. Jiang H, Wong WH. Statistical inferences for isoform expression in RNA-Seq. Bioinformatics. 2009;25:1026–32.
19. Langmead B, Salzberg SL. Fast gapped-read alignment with Bowtie 2. Nat Methods. 2012;9:357–9.
20. Ewing B, Hillier L, Wendl MC, Green P. Base-calling of automated sequencer traces using phred. I. Accuracy assessment. Genome Res. 1998;8:175–85.
21. Attias H. Inferring parameters and structure of latent variable models by variational bayes. In: Proceedings of the Fifteenth conference on Uncertainty in artificial intelligence. San Francisco, CA, USA: Morgan Kaufmann Publishers Inc.; 1999. p. 21–30. http://dl.acm.org/citation.cfm?id=2073799.
22. Bishop CM. Pattern Recognition and Machine Learning. New York, NY, USA: Springer Science+Business Media, LLC; 2006.
23. Nariai N, Kojima K, Mimori T, Sato Y, Kawai Y, Yamaguchi-Kabata Y, et al. TIGAR2: sensitive and accurate estimation of transcript isoform expression with longer RNA-Seq reads. BMC Genomics. 2014;15((Suppl 10)):S5.
24. Li B, Dewey CN. RSEM: accurate transcript quantification from RNA-Seq data with or without a reference genome. BMC Bioinformatics. 2011;12:323.
25. Trapnell C, Williams BA, Pertea G, Mortazavi A, Kwan G, van Baren MJ, et al. Transcript assembly and quantification by RNA-Seq reveals unannotated transcripts and isoform switching during cell differentiation. Nat Biotechnol. 2010;28:511–5.
26. Dennis G Jr, Sherman BT, Hosack DA, Yang J, Gao W, Lane HC, et al. DAVID: Database for annotation, visualization, and integrated discovery. Genome Biol. 2003;4:P3.
27. Boeckmann B, Bairoch A, Apweiler R, Blatter MC, Estreicher A, Gasteiger E, et al. The SWISS-PROT protein knowledgebase and its supplement TrEMBL in 2003. Nucleic Acids Res. 2003; 31(1):365–70.
28. Ashburner M, Ball CA, Blake JA, Botstein D, Butler H, et al. Gene ontology: tool for the unification of biology. Nat Genet. 2000;25:25–9.

29. McDaniell R, Lee BK, Song L, Liu Z, Boyle AP, Erdos MR, et al. Heritable individual-specific and allele-specific chromatin signatures in humans. Science. 2010;328:235–9.
30. Reddy TE, Gertz J, Pauli F, Kucera KS, Varley KE, Newberry KM, et al. Effects of sequence variation on differential allelic transcription factor occupancy and gene expression. Genome Res. 2012;22:860–9.
31. Tilgner H, Grubert F, Sharon D, Snyder MP. Defining a personal, allele-specific, and single-molecule long-read transcriptome. Proc Natl Acad Sci U S A. 2014;111:9869–74.
32. Chen R, Mias GI, Li-Pook-Than J, Jiang L, Lam HY, Chen R, et al. Personal omics profiling reveals dynamic molecular and medical phenotypes. Cell. 2012;148:1293–1307.
33. Howie BN, Donnelly P, Marchini J. A flexible and accurate genotype imputation method for the next generation of genome-wide association studies. PLoS Genet. 2009;5(10):e10005291728–37.

Jiang and Yu *BMC Genomics* 2016, **17**(Suppl 1):3
DOI 10.1186/s12864-015-2296-4

BMC Genomics

METHODOLOGY

Open Access

Power estimation and sample size determination for replication studies of genome-wide association studies

Wei Jiang and Weichuan Yu*

From The Fourteenth Asia Pacific Bioinformatics Conference (APBC 2016)
San Francisco, CA, USA. 11 - 13 January 2016

Abstract

Background: Replication study is a commonly used verification method to filter out false positives in genome-wide association studies (GWAS). If an association can be confirmed in a replication study, it will have a high confidence to be true positive. To design a replication study, traditional approaches calculate power by treating replication study as another independent primary study. These approaches do not use the information given by primary study. Besides, they need to specify a minimum detectable effect size, which may be subjective. One may think to replace the minimum effect size with the observed effect sizes in the power calculation. However, this approach will make the designed replication study underpowered since we are only interested in the positive associations from the primary study and the problem of the "winner's curse" will occur.

Results: An Empirical Bayes (EB) based method is proposed to estimate the power of replication study for each association. The corresponding credible interval is estimated in the proposed approach. Simulation experiments show that our method is better than other plug-in based estimators in terms of overcoming the winner's curse and providing higher estimation accuracy. The coverage probability of given credible interval is well-calibrated in the simulation experiments. Weighted average method is used to estimate the average power of all underlying true associations. This is used to determine the sample size of replication study. Sample sizes are estimated on 6 diseases from Wellcome Trust Case Control Consortium (WTCCC) using our method. They are higher than sample sizes estimated by plugging observed effect sizes in power calculation.

Conclusions: Our new method can objectively determine replication study's sample size by using information extracted from primary study. Also the winner's curse is alleviated. Thus, it is a better choice when designing replication studies of GWAS. The R-package is available at: http://bioinformatics.ust.hk/RPower.html.

Keywords: Replication study, Power, Empirical Bayes

Background

Genome-wide association studies (GWAS) are widely used to identify susceptibility variants of common diseases. Commonly, single nucleotide polymorphisms (SNPs) are genotyped across the whole genome in different individuals, and statistical methods are used to detect the associations between SNPs and disease status. According to the summary of GWAS catalog ([1], accessed [2015.05.28]), about 2000 GWAS reports related to 756 diseases/traits have been published so far, from which 14,609 associations show genome-wide significance (p-value $\leq 5 \times 10^{-8}$). More and more associations will be discovered from GWAS.

The basic statistical method used in GWAS analysis is hypothesis testing [2]. The possibilities of false positives cannot be completely removed in the analysis.

*Correspondence: eeyu@ust.hk
Department of Electronic and Computer Engineering, The Hong Kong University of Science and Technology, Clear Water Bay, Kowloon, Hong Kong, China

Hence, all findings from GWAS need to be verified. Replication study is a commonly used approach to verifying positive findings [3, 4]. If an association between one specific SNP and a certain disease has been identified in the primary study and confirmed in the replication study, we usually treat this association as true positive with a high confidence. If an association identified in the primary study cannot be confirmed in the replication study, we often suspect that it is a false positive.

The power of replication study is crucial in this validation process. If the replication study is underpowered, then the positive findings will have a low chance to be replicated. It's essential to design a replication study with enough statistical power.

How to estimate the power of a replication study in the design phase?

Traditionally, a replication study is regarded as another independent primary study. Thus, the same power calculation in the original primary study is used. For the associations identified in the primary study, a minimum effect size needs to be specified. Then, the underlying alternative distribution of test statistics is assumed to have specified effect size. The major limitation of this traditional power calculation method is that the specification of the effect size is subjective and may cause bias. Besides, no information from primary study has really been used.

One may think to plug the observed effect sizes from the primary study in the power calculation of the replication study. This power estimation approach doesn't need to specify any parameters. Since only significant associations are considered in the replication study, the observed effect sizes for those associations will tend to be overestimated [5]. This phenomenon is known as the "winner's curse" [6], which makes the estimated powers tend to have higher values.

A lot of methods have been proposed to overcome the winner's curse in effect size estimation. An incomplete list includes conditional maximum likelihood estimation (CMLE, [7–9]), bootstrap [10], full Bayesian method [11] and Empirical Bayes method (EB, [12]). Since power function is usually not a linear function of effect size, the estimators obtained by simply plugging those bias corrected effect sizes in power calculation may not achieve the best performance.

Moreover, there are two other challenges in designing replication study:

1. Due to the nonlinear nature and restricted range (limited to [0, 1]) of power function, the distribution of power is usually non-normal when effect size is normally distributed (illustrated in Fig. 1). The interval estimation of the power should consider the non-normality.
2. Since the power values of different associations in the primary study are different, a summary value is needed to determine the sample size of replication study.

This paper aims at addressing the above challenges. Our contributions are listed in the following:

1. For each association identified from the primary study, an EB based method is proposed to estimate its power in the replication study.
2. Due to the non-normality of the estimated power and the inaccuracy of the hyperparameters estimation, a novel interval estimation method combining Monte Carlo sampling and Bootstrap is proposed to estimate the corresponding credible interval of each association's power in the replication study.
3. The average power of the discovered true associations is used for determining the sample size of replication study. An weighted average method is proposed to estimate the average power. Our proposed interval estimation method can also be used to construct the credible interval of the average power.
4. Only the summary statistics of the primary study are needed when using our proposed method to design a replication study. This feature is helpful since summary statistics are more accessible than individual-level genotype data due to the privacy issue and other constraints.

The rest of this paper is organized as follows. In section 'Methods', we will introduce the Bayesian framework to estimate the power of replication studies. We will prove that Bayesian predictive power is immune to the winner's curse. Then we will present how to estimate the power with two-component mixture prior under the Bayesian framework. We will also give the details about estimation of hyperparameters, interval estimation and the estimation of average power. In section 'Results and discussion', we will first use simulation results to demonstrate that our EB based method is better than other plug-in based estimators in terms of overcoming the winner's curse and providing higher estimation accuracy. We will also demonstrate that the coverage probability of given credible interval is well-calibrated. Then we will show the sample sizes determined to replicate findings of 6 diseases from Wellcome Trust Case Control Consortium (WTCCC) [13], which are much higher than the sample sizes estimated by plugging observed effect sizes in the power calculation formula. The increased sample sizes are reasonable due to the winner's curse. In the same section, we will discuss limitations of current modeling and estimation approach. Section 'Conclusions' concludes the paper.

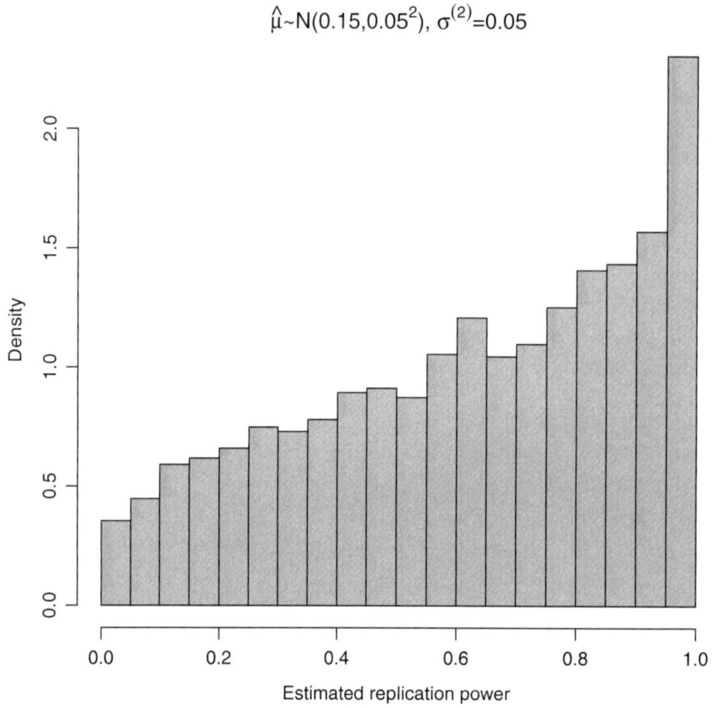

Fig. 1 When *log*-odds ratio is normally distributed, the power of replication study possesses non-normality. Assume the effect size follows normal distribution $N(0.15, 0.05^2)$. The standard error of *log*-odds ratio is $\sigma^{(2)} = 0.05$ in the replication study. The significance level of the replication study is $\alpha^{(2)} = 5 \times 10^{-3}$. The histogram of the power in the replication study possesses non-normality

Methods

We use parenthesized superscript "(j)" to denote primary study ($j = 1$) and replication study ($j = 2$). For example, we denote the sample size in the primary study as $n^{(1)}$. The sample size in the control group and case group are $n_0^{(1)}$ and $n_1^{(1)}$, respectively. The total number of SNPs genotyped in the primary study is m. Among those genotyped SNPs, the proportion of the SNPs having no association with the disease (null SNPs) is $\pi_0 (0 \le \pi_0 \le 1)$.

In both the primary study and the replication study, a contingency table can be created as in Table 1 for each genotyped SNP. With the contingency table, the logarithm of the observed odds ratio reads:

$$\widehat{\mu}^{(j)} = \log n_{00}^{(j)} - \log n_{01}^{(j)} - \log n_{10}^{(j)} + \log n_{11}^{(j)}. \quad (1)$$

Table 1 Allele based contingency table of one SNP in primary/replication study. Please see the main text for explanation of the notations

	Non-effect allele	Effect allele	Total
Control	$n_{00}^{(j)}$	$n_{01}^{(j)}$	$2n_0^{(j)}$
Case	$n_{10}^{(j)}$	$n_{11}^{(j)}$	$2n_1^{(j)}$
Total	$n_{00}^{(j)} + n_{10}^{(j)}$	$n_{01}^{(j)} + n_{11}^{(j)}$	$2n^{(j)}$

The true value of the log odds ratio μ is usually unknown. The asymptotic standard error of $\widehat{\mu}^{(j)}$ can be approximated using Woolf's method [14],

$$\sigma^{(j)} \approx \sqrt{\frac{1}{n_{00}^{(j)}} + \frac{1}{n_{01}^{(j)}} + \frac{1}{n_{10}^{(j)}} + \frac{1}{n_{11}^{(j)}}}. \quad (2)$$

To test whether there is an association between the SNP and the disease, two hypotheses are set up:

$$\mathcal{H}_0 : \mu = 0, \text{ vs. } \mathcal{H}_1 : \mu \neq 0. \quad (3)$$

Wald test can be used to examine whether the null hypothesis should be rejected. The test statistic is $z^{(j)} = \widehat{\mu}^{(j)} / \sigma^{(j)}$. The significance levels in the primary study and the replication study are fixed to α_1 and α_2, respectively.

Two-sided test is used in primary study. The rejection region is $|Z^{(1)}| > z_{\alpha_1/2}$ (We use uppercase letter to indicate a random variable), where $z_u (0 \le u \le 0.5)$ is the upper u quantile of the standard normal distribution $N(0, 1)$. For a replicated association, the test statistics in two studies should be consistent with the same sign. Hence, the test can be regarded as one-sided test in replication study. The rejection region is $sgn\left(z^{(1)}\right) Z^{(2)} > z_{\alpha_2}$, where the sign function is

$$sgn(x) = \begin{cases} 1 & \text{if } x > 0 \\ 0 & \text{if } x = 0 \\ -1 & \text{if } x < 0 \end{cases}. \quad (4)$$

Bayesian predictive power

For an association identified in primary study, the power function in replication study is defined as

$$\beta^{(2)}(\mu) = P\left(sgn\left(z^{(1)}\right)Z^{(2)} > z_{\alpha_2}|\mu, z^{(1)}, \mathcal{H}_1\right), \quad \text{where } |z^{(1)}| > z_{\alpha_1/2}. \tag{5}$$

A traditional power calculation method needs to specify a minimum detectable effect size μ_{min} first. Then, the power of replication study is $\beta^{(2)}(\mu_{min})$. Consequently, the power can be used to determine the sample size.

To incorporate information from primary study, the post-hoc method estimates the power of each association by plugging the observed effect size in Eq. (5), i.e. $\beta^{(2)}\left(\widehat{\mu}^{(1)}\right)$. This approach is widely criticized for the reason of the winner's curse. The estimated power is biased upward since only significant associations are selected in the replication study. To address this problem, a lot of methods have been proposed to overcome the winner's curse in effect size estimation [7–12]. Conditional maximum likelihood estimation (CMLE) is the most commonly used type [7–9]. In CMLE, the effect size is estimated by maximizing the likelihood conditioning on rejected region, i.e.

$$\widehat{\mu}^{(1)}_{CMLE} = arg\max_{\mu} P\left(z^{(1)}|\mu, |Z^{(1)}| > z_{\alpha_1/2}, \mathcal{H}_1\right). \tag{6}$$

Please notice that, although the selection bias can be reduced using estimator which can adjust estimated effect size, no unbiased estimator exists [11]. With estimated effect size, the power of replication study can be obtained by using $\beta^{(2)}\left(\widehat{\mu}^{(1)}_{CMLE}\right)$. The plug-in based power estimator is not optimized in terms of minimizing Bayes risk.

The Bayes risk averages the loss function over both sample space and parameter space. In terms of overcoming the winner's curse, the Bayes risk $R(\widehat{\theta})$ should be defined conditioning on rejected region and alternative hypothesis,

$$R(\widehat{\theta}) = E_{\mu, Z^{(1)}}\left(\left(\widehat{\theta} - \beta^{(2)}(\mu)\right)^2 ||Z^{(1)}| > z_{\alpha_1/2}, \mathcal{H}_1\right), \tag{7}$$

where $\widehat{\theta}$ is the power estimator of replication study. Inspired by the proof in [12], we can show that Bayesian predictive power $\eta^{(2)}$ [15] is the estimator minimizing $R(\widehat{\theta})$ (please see Appendix for detail). The Bayesian predictive power reads

$$\begin{aligned}\eta^{(2)} &= P\left(sgn\left(z^{(1)}\right)Z^{(2)} > z_{\alpha_2}|z^{(1)}, \mathcal{H}_1\right) \\ &= E_{\mu}\left(\beta^{(2)}(\mu)|z^{(1)}, \mathcal{H}_1\right),\end{aligned} \tag{8}$$

which takes the average of all power function values among all possible μ values given observed $z^{(1)}$. We will provide a detailed formula of the Bayesian predictive power under one specific prior in the following subsection.

Two-component mixture prior

In each study, the observed log odds ratio $\widehat{\mu}^{(j)}$ asymptotically follows normal distribution $N\left(\mu, \left(\sigma^{(j)}\right)^2\right)$. The underlying true value of the effect size μ is often unknown. It is widely suspected that a large proportion of SNPs with small effect sizes are associated with complex diseases [16, 17]. We use Gaussian prior to depict this pattern of the associated SNPs. For all SNPs, we use the following two-component mixture prior to describe their effect sizes:

$$\mu \sim \pi_0\delta_0 + (1 - \pi_0)N\left(0, \sigma_0^2\right), \tag{9}$$

where δ_0 is the distribution with point mass on zero and σ_0^2 is the variance of the effect sizes in associated SNPs.

With this prior, the posterior distribution of effect size μ under $\mathcal{H}_1$ is

$$\left(\mu|z^{(1)}, \mathcal{H}_1\right) \sim N\left(\lambda\widehat{\mu}^{(1)}, \lambda\left(\sigma^{(1)}\right)^2\right), \quad \text{where } \lambda = \frac{1}{1 + \left(\sigma^{(1)}/\sigma_0\right)^2}. \tag{10}$$

The Bayesian predictive power of replication study is (Detail in the Appendix):

$$\eta^{(2)} = \Phi\left(\frac{sgn\left(z^{(1)}\right)z^* - z_{\alpha_2}}{\sigma^*}\right), \tag{11}$$

where $z^* = \lambda\widehat{\mu}^{(1)}/\sigma^{(2)}$, $\sigma^* = \sqrt{1 + \lambda\left(\frac{\sigma^{(1)}}{\sigma^{(2)}}\right)^2}$ and $\Phi(x)$ is the cumulative density function (cdf) of $N(0, 1)$. By substituting observed allele frequencies from the primary study into Woolf's method, $\sigma^{(2)}$ can be approximated as

$$\sigma^{(2)} \approx \sqrt{\frac{n_0^{(1)}}{n_0^{(2)}}\left(\frac{1}{n_{00}^{(1)}} + \frac{1}{n_{01}^{(1)}}\right) + \frac{n_1^{(1)}}{n_1^{(2)}}\left(\frac{1}{n_{10}^{(1)}} + \frac{1}{n_{11}^{(1)}}\right)}. \tag{12}$$

There is an unknown hyperparameter σ_0^2 in the calculation of Bayesian predictive power. In the following subsection, we will present how to estimate σ_0^2 with Empirical Bayes approach.

Hyperparameter σ_0^2

In Empirical Bayes' thinking, we can estimate σ_0^2 by taking advantage of the shared structure of the effect size's distribution among all SNPs, which can be seen from Eq. (9). The estimator of σ_0^2 is (see Appendix for detail):

$$\widehat{\sigma}_0^2=\max\left(0,\left(\frac{\sum_{i=1}^{m}\left(z_i^{(1)}\right)^2-m\pi_0}{(1-\pi_0)}-m\right)/\sum_{i=1}^{m}\left(1/\sigma_i^{(1)}\right)^2\right). \tag{13}$$

There are two extreme cases in the above estimation:

1. If the null hypothesis is valid, then all SNPs follow a standard normal distribution with variance equal to one. When $\frac{1}{m}\sum_{i=1}^{m}\left(z_i^{(1)}\right)^2 \leq 1$, i.e., the sample variance is no bigger than one as in the null hypothesis case, we will have $\widehat{\sigma}_0^2 = 0$. In this case, the result of our EB based power estimation method will degenerate to type I error rate, which is the probability that the identified association can be replicated even when the association doesn't exist.
2. When $\frac{1}{m}\sum_{i=1}^{m}\left(z_i^{(1)}\right)^2 > 1$ but $\pi_0 = 1$, we will have $\widehat{\sigma}_0^2 = +\infty$. In this case, the above shrinkage coefficient will degenerate to $\lambda = 1$. The shrinkage effect in our EB based method will disappear.

Noticed that there is another unknown hyperparameter π_0 in calculating $\widehat{\sigma}_0^2$. The estimation of the proportion of true null hypotheses π_0 has been extensively studied [18–20]. Here we just choose Storey's method [18] for the simplicity of implementation. Let's denote the number of SNPs with p-value$>\ \gamma$ as $m_+(\gamma)$ in the primary study. Then π_0 can be estimated by using

$$\widehat{\pi}_0 = \frac{m_+(\gamma)}{m(1-\gamma)}. \tag{14}$$

There is a bias-variance tradeoff in tuning γ. An automatic procedure is proposed in [18] without tuning γ: A natural cubic spline will fit to evaluated values with different γ, then $\widehat{\pi}_0$ is the spline's value at $\gamma = 1$.

$\widehat{\sigma}_0^2$ can be calculated by plugging $\widehat{\pi}_0$ in Eq. (13). By plugging $\widehat{\sigma}_0^2$ into Eq. (11), an EB based estimator of the replication study's power can be obtained, which is denoted as $\widehat{\eta}_{EB}^{(2)}$. The corresponding credible interval can be constructed, which is presented in the following subsection.

Credible interval

From Eq. (10), the posterior distribution of log odds ratio μ under alternative hypothesis $\mathcal{H}_1$ is a normal distribution. Figure 1 shows the histogram of power values when μ is normally distributed. The shape of the histogram indicates the non-normality of the calculated power. Hence, the asymptotic approach based on normal distribution theory is not appropriate in the interval estimation of the replication study's power. The construction of the credible interval should consider the non-normality. We propose to use Monte Carlo sampling to construct the credible interval of $\beta^{(2)}(\mu)$. The credible interval is constructed with known hyperparameters σ_0^2. Since estimation error will occur in estimating σ_0^2, the constructed credible interval will have smaller coverage probability than nominal level. To incorporate the variance of the estimator $\widehat{\sigma}_0^2$, a method combining Monte Carlo sampling and Bootstrap is proposed. The test statistics from the primary study $z^{(1)}$ will be resampled N_1 times with replacement. For each run, σ_0^2 is re-estimated. Monte Carlo sampling is used to generate N_2 power values with each re-estimated σ_0^2. The credible interval is constructed among all N_1N_2 sampled power values.

Average power

Usually, multiple associations are identified in primary study. To design a replication study, a summarized value reflecting the average power of all associated SNPs is needed. A direct thinking is to average power among the identified SNPs with underlying true associations, which reads

$$\bar{\beta}^{(2)}(\mu) = \frac{1}{|S|}\sum_{i\in S}\beta^{(2)}(\mu_i), \tag{15}$$

where S is the index set of the associated SNPs identified from primary study and $|S|$ is the cardinality of S. The subscript i means that the quantity is evaluated for SNP i.

Since the index set S is unknown, we propose to use weighted average of the estimated powers $\widehat{\eta}_{EB}^{(2)}$. The local true discovery rate (*ltdr*) of each SNP is the posterior probability of being associated SNP given observed statistics, which is complementary to local false discovery rate [21]. We use *ltdr* as weight in the estimation. The estimated average power is

$$\bar{\eta}_{EB}^{(2)} = \frac{\sum_{i\in T} ltdr_i^{(1)}\widehat{\eta}_{EB,i}^{(2)}}{\sum_{i\in T} ltdr_i^{(1)}}, \tag{16}$$

where T is the index set of the SNPs identified from the primary study. The local true discovery rate of the primary study can be calculated as (See Appendix for detail):

$$ltdr^{(1)} = \frac{\pi_1\phi\left(z^{(1)}/\sqrt{1+(\widehat{\sigma}_0/\sigma^{(1)})^2}\right)}{\pi_0\phi\left(z^{(1)}\right)+\pi_1\phi\left(z^{(1)}/\sqrt{1+\left(\widehat{\sigma}_0/\sigma^{(1)}\right)^2}\right)}, \tag{17}$$

where $\phi(x)$ is the probability density function (pdf) of $N(0,1)$.

By setting the estimated average power larger than a threshold, e.g. $\bar{\eta}_{EB}^{(2)} > 80\,\%$, the sample size of replication study can be determined.

Credible interval of the average power

The proposed interval estimation method can also be used to construct the credible interval of the average power. We resample the test statistics from the primary study N_1 times. In each run, a re-estimated variance of the effect sizes in the non-null SNPs $\widehat{\sigma}_0^2$ can be obtained. For a fixed $\widehat{\sigma}_0^2$ value, we first calculate the local true discovery rate of the primary study $ltdr^{(1)}$ with Eq. (17) for each association. Then Monte Carlo sampling is used to generate N_2 sets of the power values, in each of which there are power values of the replication study for all associations identified from the primary study. In each set, an average power can be obtained by taking weighted average of those generated power values among all associations. Hence, N_2 average power values can be generated in each run. The credible interval of average power can be constructed among all N_1N_2 sampled average power values.

Results and discussion

Simulation experiments

The following questions are examined using simulation experiments:

1. Can EB based power estimator $\widehat{\eta}_{EB}^{(2)}$ perform well in terms of overcoming the winner's curse?
2. Can $\widehat{\eta}_{EB}^{(2)}$ estimate power accurately?
3. Is the corresponding credible interval well-calibrated?
4. Can weighted average estimator $\bar{\eta}_{EB}^{(2)}$ estimate average power $\bar{\beta}^{(2)}(\mu)$ accurately?

In simulation experiments, sample sizes are $n_0^{(1)} = n_1^{(1)} = 1000$ and $n_0^{(2)} = n_1^{(2)} = 500$ in primary study and replication study, respectively. The number of simulated SNPs is $m = 1 \times 10^4$. For each SNP, its minor allele frequency is uniformly distributed as $U(0.05, 0.5)$. Their effect sizes are generated from the following distribution:

$$\mu \sim 0.9\delta_0 + 0.1N(0, 0.04). \quad (18)$$

For our hypothetical disease, its prevalence is 1 %. To test the marginal association between SNPs and the disease, *log*-odds ratio test is used. The significance levels are $\alpha_1 = 5 \times 10^{-5}$ and $\alpha_2 = 5 \times 10^{-3}$ in primary study and replication study, respectively.

Figure 2 shows the histogram of the differences between observed effect sizes $\widehat{\mu}^{(1)}$ and their underlying true values μ for identified true associations. We plotted separatively for the associations with positive effect ($\mu > 0$) and negative effects ($\mu < 0$). For the associations with positive effect, the mean value of the estimated effect sizes is larger than the mean value of true effect sizes. On the contrary, the values of the estimated effect sizes tend to be smaller than their true values for the associations with negative effect. In both of these two cases, the magnitude of the observed effect sizes tend to be exaggerated, which indicates that the winner's curse generally exists in the associations identified from primary study.

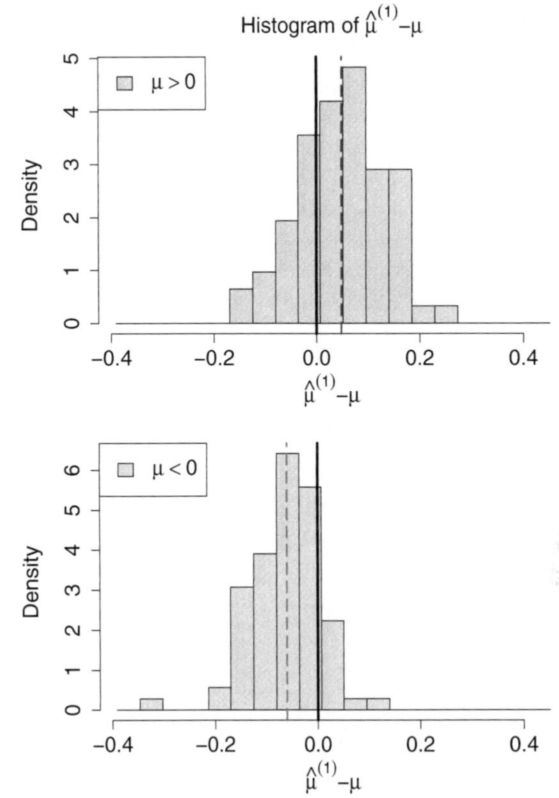

Fig. 2 The winner's curse exists in the estimation of effect size for SNPs identified from the primary study. The histograms of the estimation error $\widehat{\mu}^{(1)} - \mu$ for positive effect SNPs (SNPs with *log*-odds ratio greater than 0) and negative effect SNPs (SNPs with *log*-odds ratio smaller than 0) are plotted separately. The mean values of the estimation error is drawn with vertical dashed lines. From the figure, it can be concluded that the magnitude of the observed effect sizes $\widehat{\mu}^{(1)}$ tend to be larger than the magnitude of their true effect. The observed effect sizes tend to be exaggerated

In order to check whether our EB based power estimator $\widehat{\eta}_{EB}^{(2)}$ can overcome the winner's curse, the histogram of the differences between estimated values and true values is shown. As a comparison, we will show the corresponding histogram for power estimator by plugging in observed effect size first. A lot of methods have been proposed to overcome the winner's curse in terms of effect size estimation. CMLE and EB can be used directly in *z*-values of *log*-odds ratio test. The individual-level genotype data are also simulated so that bootstrap based bias reduction method BR2 [10] can also be used as a comparison (We modified the implementation code of BR2 so that *log*-odds ratio test can be used in the software). A direct thinking is to plug these adjusted estimators in the power calculation formula. The corresponding histograms for these three adjusted plug-in based estimators are shown as comparisons. In Fig. 3a, we use plug-in rule to estimate the replication study's power, where the observed effect size is plugged in. The estimated power is $\beta^{(2)}\left(\widehat{\mu}^{(1)}\right)$. We plot the histogram of the difference between $\beta^{(2)}\left(\widehat{\mu}^{(1)}\right)$ and the true power values $\beta^{(2)}(\mu)$ in the figure. The overestimated effect size makes the estimated replication study's power overestimated as well. Figure 3b plots the histogram of the difference between $\beta^{(2)}\left(\widehat{\mu}^{(1)}_{CMLE}\right)$ and $\beta^{(2)}(\mu)$. The winner's curse has disappeared, but there is a large downward bias in the estimated results. Equation (10) also introduces an Empirical Bayes estimator of the effect size, which reads

$$\widehat{\mu}^{(1)}_{EB} = \frac{1}{1+\left(\sigma^{(1)}/\widehat{\sigma}_0\right)^2}\widehat{\mu}^{(1)}. \tag{19}$$

Figure 3c, d plot the histogram of $\beta^{(2)}\left(\widehat{\mu}^{(1)}_{BR2}\right) - \beta^{(2)}(\mu)$ and $\beta^{(2)}\left(\widehat{\mu}^{(1)}_{EB}\right) - \beta^{(2)}(\mu)$, respectively. Large upward biases still exist in the histograms. In contrast, Fig. 3e plots the histogram of $\widehat{\eta}^{(2)}_{EB} - \beta^{(2)}(\mu)$, where $\widehat{\eta}^{(2)}_{EB}$ is our proposed EB based power estimator. The bias almost disappeared, indicating $\widehat{\eta}^{(2)}_{EB}$ is better than other estimators of the replication study's power in terms of overcoming the winner's curse. The experiment has run 5 times, and the same conclusion holds in each run. The empirical biases of these five estimators can be seen in the Table 2.

Table 3 shows the root mean square error (RMSE) of the five estimators $\beta^{(2)}\left(\widehat{\mu}^{(1)}\right)$, $\beta^{(2)}\left(\widehat{\mu}^{(1)}_{CMLE}\right)$, $\beta^{(2)}\left(\widehat{\mu}^{(1)}_{BR2}\right)$, $\beta^{(2)}\left(\widehat{\mu}^{(1)}_{EB}\right)$ and $\widehat{\eta}^{(2)}_{EB}$ in the 5 runs. We can see that $\widehat{\eta}^{(2)}_{EB}$ is better than other methods in terms of estimation accuracy.

To investigate the performance of the interval estimation, the coverage probability of 95 % credible intervals for all identified associated SNPs is shown in Table 4. The coverage probabilities by using pure Monte-carlo sampling are presented on the left side. In agreement with our analysis in the last section, the coverage probability is lower than the nominal value 95 %. The coverage probabilities by using modified method which combines Monte-carlo sampling and bootstrap are shown on the right side, which are closer to the nominal value. The credible interval given by the combined method is well-calibrated.

Figure 4 shows the average powers and their estimated results in the 5 runs. The true value of the average power is covered by 95 % credible interval in all runs. For comparison, we also show the estimated average power by using plug-in rule, i.e. $\bar{\beta}^{(2)}\left(\widehat{\mu}^{(1)}\right)$, $\bar{\beta}^{(2)}\left(\widehat{\mu}^{(1)}_{CMLE}\right)$, $\beta^{(2)}\left(\widehat{\mu}^{(1)}_{BR2}\right)$ and $\bar{\beta}^{(2)}\left(\widehat{\mu}^{(1)}_{EB}\right)$, in each run. The estimated power for each association is also weighted by its local true discovery rate *ltdr* in each average power estimator. The figure shows that the EB based average power estimator $\bar{\widehat{\eta}}^{(2)}_{EB}$ is much closer to the true value $\bar{\beta}^{(2)}(\mu)$. These results indicate that the estimated average power can be regarded as a proxy of the average power, which can be used to design replication study.

To check the performance of our method when the effect sizes of the associated SNPs are do not follow normal distribution, we also simulated data with the following distributed effect sizes:

$$\mu \sim 0.9\delta_0 + 0.1t_{5,0.2} \tag{20}$$

and

$$\mu \sim 0.9\delta_0 + 0.07N(0, 0.04) + 0.03N(0, 0.16), \tag{21}$$

where $t_{5,0.02}$ is a scaled t distribution with degree of freedom 5 and scaling factor 0.2. The distribution of the associated SNPs' effect sizes follow the Gaussian mixture model in the second case. The average empirical biases and RMSE of all estimators in these two cases are shown in Tables 5 and 6, respectively. From the tables, we can see that our method is still better in terms of overcoming winner's curse and providing higher estimation accuracy.

WTCCC datasets

To give an application example of our proposed method, we will determine the sample size of replication study used for verifying the 6 human common diseases' findings from Wellcome Trust Case Control Consortium (WTCCC). The 6 diseases include coronary artery disease, Crohn's disease, hypertension, rheumatoid arthritis, type 1 diabetes and type 2 diabetes. Each disease has 2000 cases in the dataset of the primary study. There are 3000 shared controls among all datasets. The following quality control procedure is used in the primary study's datasets:

1. Missing data control: Chiamo score is used as genotype calling accuracy in the WTCCC data. The genotypes with Chiamo score < 0.95 are regarded as missing values. The SNPs with more than 10 % missing entries are removed.
2. Minor allele frequency control: Among all samples, the SNPs with minor allele frequency < 0.05 are removed.
3. Hardy-Weinberg equilibrium control: The SNPs with p-values< 0.001 in the Hardy-Weinberg equilibrium test are removed.

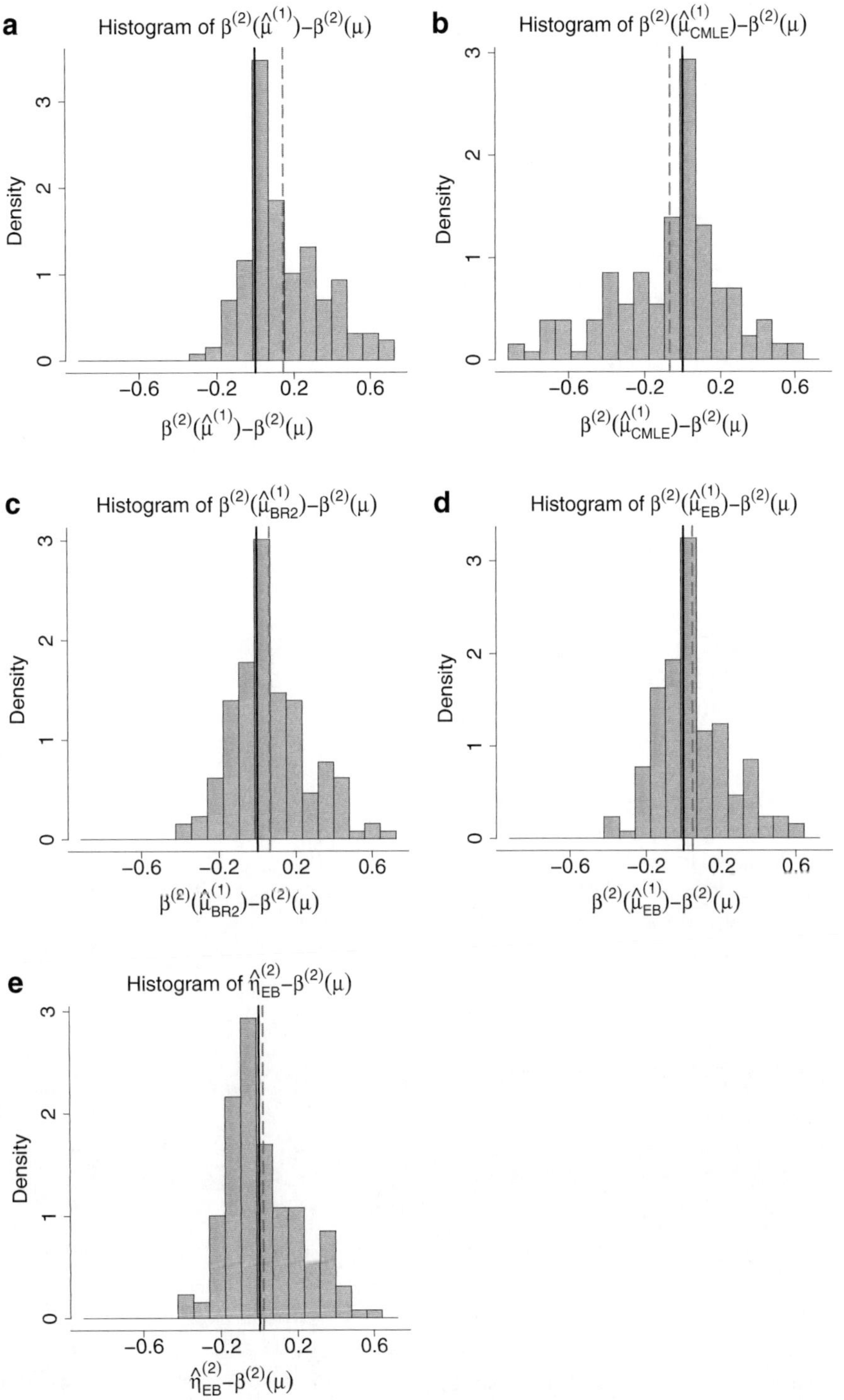

Fig. 3 The histograms of the error in power estimation for all associated SNPs identified from primary study. **a** The replication study's power is estimated by plugging the observed effect size in power calculation formula, i.e. $\beta^{(2)}\left(\widehat{\mu}^{(1)}\right)$. **b** The power is estimated by plugging the CMLE based corrected effect size $\widehat{\mu}^{(1)}_{CMLE}$ in power calculation formula, i.e. $\beta^{(2)}\left(\widehat{\mu}^{(1)}_{CMLE}\right)$. **c** The power is estimated by plugging in the BR2 estimator, i.e. $\beta^{(2)}\left(\hat{\mu}^{(1)}_{BR2}\right)$. **d** The power is estimated by plugging the EB based corrected effect size $\widehat{\mu}^{(1)}_{EB}$ in power calculation, i.e. $\beta^{(2)}\left(\widehat{\mu}^{(1)}_{EB}\right)$. **e** The power is estimated by EB based method, i.e. $\widehat{\eta}^{(2)}_{EB}$. The mean value of the estimation error is drawn with vertical dashed line. From the figure, it can be seen that $\widehat{\eta}^{(2)}_{EB}$ has the smallest bias in power estimation. The biases for these 5 estimators are 0.144, -0.068, 0.045, 0.047 and 0.021, respectively

Table 2 Empirical biases of power estimators of the replication study in the simulation experiments. The settings of the experiments can be seen in the main text

	$\beta^{(2)}\left(\widehat{\mu}^{(1)}\right)$	$\beta^{(2)}\left(\widehat{\mu}^{(1)}_{CMLE}\right)$	$\beta^{(2)}\left(\widehat{\mu}^{(1)}_{BR2}\right)$	$\beta^{(2)}\left(\widehat{\mu}^{(1)}_{EB}\right)$	$\widehat{\eta}^{(2)}_{EB}$
Run 1	0.142	−0.113	0.038	0.058	**0.032**
Run 2	0.146	−0.109	0.045	0.021	**0.001**
Run 3	0.144	−0.068	0.045	0.047	**0.021**
Run 4	0.137	−0.090	0.042	0.052	**0.026**
Run 5	0.144	−0.126	0.026	0.038	**0.016**
Average	0.142	−0.101	0.039	0.043	**0.019**

$\beta^{(2)}\left(\widehat{\mu}^{(1)}\right)$, $\beta^{(2)}\left(\widehat{\mu}^{(1)}_{CMLE}\right)$, $\beta^{(2)}\left(\widehat{\mu}^{(1)}_{BR2}\right)$ and $\beta^{(2)}\left(\widehat{\mu}^{(1)}_{EB}\right)$ are the plug-in based estimators by using observed effect size, CMLE, BR2 and EB in the effect size estimation. $\widehat{\eta}^{(2)}_{EB}$ is proposed EB-based estimator. Bold face indicates the estimator achieving the smallest bias. In the experiments, $\widehat{\eta}^{(2)}_{EB}$ behaves better than others in terms of bias reduction

The significance levels used in primary study and replication study are $\alpha_1 = 5 \times 10^{-8}$ and $\alpha_2 = 5 \times 10^{-6}$, respectively. The Control-to-Case ratio of the replication study is set to 1. The inferred hyperparameters π_0 and σ_0^2 can be seen in Table 7. With these parameters, the relationships between the estimated average power using EB based method $\bar{\eta}^{(2)}_{EB}$ and sample size of replication study $n^{(2)}$ can be seen in Fig. 5. In conclusion, to achieve 80 % average power of the replication study, we will need 6885 individuals for coronary artery disease, 8092 individuals for Crohn's disease, 10,014 individuals for hypertension, 5291 individuals for rheumatoid arthritis, 4094 individuals for type 1 diabetes and 6988 individuals for type 2 diabetes. The detail about the sample sizes needed for different values of average power can be seen in Table 8.

As a comparison, we also plot relationships between $\bar{\beta}^{(2)}\left(\widehat{\mu}^{(1)}\right)$ and $n^{(2)}$ in Fig. 5, where $\bar{\beta}^{(2)}\left(\widehat{\mu}^{(1)}\right)$ is the estimated average power by plugging in observed effect sizes. For a given sample size $n^{(2)}$, the estimated average power value using EB based method is much smaller than $\bar{\beta}^{(2)}\left(\widehat{\mu}^{(1)}\right)$. This is reasonable because $\bar{\beta}^{(2)}\left(\widehat{\mu}^{(1)}\right)$ is overestimated due to the winner's curse, which is alleviated in the EB based method. To achieve 80 % average power

Table 3 Root mean square error (RMSE) of power estimators of the replication study in the simulation experiments. The settings of the experiments can be seen in the main text

	$\beta^{(2)}\left(\widehat{\mu}^{(1)}\right)$	$\beta^{(2)}\left(\widehat{\mu}^{(1)}_{CMLE}\right)$	$\beta^{(2)}\left(\widehat{\mu}^{(1)}_{BR2}\right)$	$\beta^{(2)}\left(\widehat{\mu}^{(1)}_{EB}\right)$	$\widehat{\eta}^{(2)}_{EB}$
Run 1	0.246	0.334	0.201	0.202	**0.195**
Run 2	0.243	0.312	0.196	0.191	**0.188**
Run 3	0.247	0.303	0.203	0.198	**0.192**
Run 4	0.236	0.307	0.186	0.192	**0.186**
Run 5	0.249	0.317	0.198	0.196	**0.194**
Average	0.244	0.315	0.197	0.196	**0.191**

$\beta^{(2)}\left(\widehat{\mu}^{(1)}\right)$, $\beta^{(2)}\left(\widehat{\mu}^{(1)}_{CMLE}\right)$, $\beta^{(2)}\left(\widehat{\mu}^{(1)}_{BR2}\right)$ and $\beta^{(2)}\left(\widehat{\mu}^{(1)}_{EB}\right)$ are the plug-in based estimators by using observed effect size, CMLE, BR2 and EB in the effect size estimation. $\widehat{\eta}^{(2)}_{EB}$ is proposed EB-based estimator. Bold face indicates the estimator achieving the smallest RMSE. In the experiments, $\widehat{\eta}^{(2)}_{EB}$ behaves better than others in terms of higher estimation accuracy

Table 4 Coverage probability of the 95 % credible intervals in simulation experiments. The simulation settings can be seen in the main text

	Without Bootstrap	With Bootstrap
Run 1	0.932	0.960
Run 2	0.947	0.960
Run 3	0.918	0.943
Run 4	0.914	0.949
Run 5	0.878	0.925
Average	0.918	0.947

Column "No bootstrap" is the coverage probability of the 95 % credible intervals created by using pure Monte Carlo method. Column "Bootstrap" is the coverage probability of the 95 % credible intervals created by using the combined method of Monte Carlo sampling and bootstrap. From the experiments' results, the coverage probability of combined method is closer to the nominal value

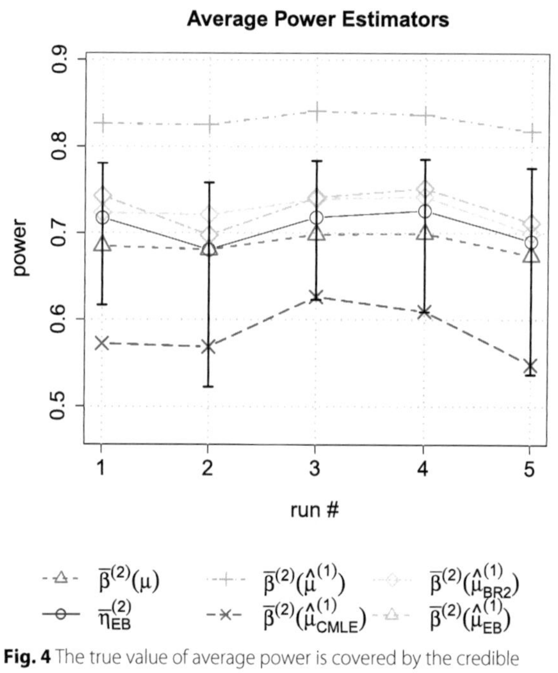

Fig. 4 The true value of average power is covered by the credible interval in simulation experiments. The EB based average power estimator $\bar{\eta}^{(2)}_{EB}$ estimates more accurately than other plug-in based average power estimators $\bar{\beta}^{(2)}\left(\widehat{\mu}^{(1)}\right)$, $\bar{\beta}^{(2)}\left(\widehat{\mu}^{(1)}_{CMLE}\right)$, $\bar{\beta}^{(2)}\left(\widehat{\mu}^{(1)}_{BR2}\right)$ and $\bar{\beta}^{(2)}\left(\widehat{\mu}^{(1)}_{EB}\right)$. Average power $\bar{\beta}^{(2)}(\mu)$ is defined in the main text. A method combining Monte Carlo method and bootstrap is proposed to create the credible intervals of the replication study's average power. In the 5 runs of the simulation experiments, the underlying true values of the average power is covered by created intervals. Compared to other plug-in based average power estimators, $\bar{\eta}^{(2)}_{EB}$ is the closest to $\bar{\beta}^{(2)}(\mu)$

Table 5 When effect sizes follow the distribution of Eq. (20), the average empirical bias and root mean square error (RMSE) of power estimators of the replication study in the simulation experiments

Average	$\bar{\beta}^{(2)}\left(\widehat{\mu}^{(1)}\right)$	$\bar{\beta}^{(2)}\left(\widehat{\mu}^{(1)}_{CMLE}\right)$	$\bar{\beta}^{(2)}\left(\widehat{\mu}^{(1)}_{BR2}\right)$	$\bar{\beta}^{(2)}\left(\widehat{\mu}^{(1)}_{EB}\right)$	$\widehat{\bar{\eta}}^{(2)}_{EB}$
Empirical Bias	0.085	−0.079	0.023	0.028	**0.003**
RMSE	0.189	0.279	0.167	0.168	**0.163**

Bold face indicates the estimator achieving the smallest value in terms of bias or RMSE. In the experiments, $\widehat{\bar{\eta}}^{(2)}_{EB}$ behaves better than others in terms of bias reduction and providing high estimation accuracy

of the replication study, the sample size needed is 3023 for coronary artery disease, 3369 for Crohn's disease, 3788 for hypertension, 2748 for rheumatoid arthritis, 2706 for type 1 diabetes and 3095 for type 2 diabetes when using $\bar{\beta}^{(2)}\left(\widehat{\mu}^{(1)}\right)$ as the estimator of average power. The sample sizes needed for other values of average power are listed in Table 9. These determined sample sizes are much smaller than the sample sizes determined by EB based method, indicating an underpowered study will be designed if we estimate power with observed effect sizes.

For coronary artery disease and type 2 diabetes, we obtained the publicly available summary statistics of the meta-analysis from two consortiums: CARDIoGRAMplusC4D Consortium [22] and DIAGRAM Consortium [23], respectively. CARDIoGRAM GWAS is a meta-analysis of 22 GWAS studies of European descent involving 22,233 cases and 64,762 controls. The odds ratio calculated from high power CARDIoGRAM GWAS will be used as underlying true odds ratio to calculate the average power of the replication study for coronary artery disease in WTCCC. The average power obtained in this manner is denoted as $\bar{\beta}^{(2)}\left(\hat{\mu}_{meta}\right)$. Figure 5a plots the relationship between $\bar{\beta}^{(2)}\left(\hat{\mu}_{meta}\right)$ and $n^{(2)}$, which is the sample size needed in the replication study. The figure shows that our EB based power estimator $\bar{\eta}^{(2)}_{EB}$ is very close to the power calculated using the results of CARDIoGRAM GWAS. Also it can be shown that $\bar{\beta}^{(2)}\left(\hat{\mu}_{meta}\right)$ is in the credible interval we estimated. DIAGRAM GWAS is a meta-analysis consisting of 12,171 type 2 diabetes cases and 56,862 controls across 12 GWAS from European descent populations. Similar to CARDIoGRAM GWAS, the allele based odds ratio calculated from DIAGRAM GWAS is used for calculating the average power of the replication study for type 2 diabetes in WTCCC. Figure 5f plots the relationship between $\bar{\beta}^{(2)}\left(\hat{\mu}_{meta}\right)$ and $n^{(2)}$. It can be shown that the result estimated by our EB based method $\bar{\eta}^{(2)}_{EB}$ is close to the power calculated using the results of DIAGRAM GWAS.

Table 6 When effect sizes follow the distribution of Eq. (21), the average empirical bias and root mean square error (RMSE) of power estimators of the replication study in the simulation experiments

Average	$\bar{\beta}^{(2)}\left(\widehat{\mu}^{(1)}\right)$	$\bar{\beta}^{(2)}\left(\widehat{\mu}^{(1)}_{CMLE}\right)$	$\bar{\beta}^{(2)}\left(\widehat{\mu}^{(1)}_{BR2}\right)$	$\bar{\beta}^{(2)}\left(\widehat{\mu}^{(1)}_{EB}\right)$	$\widehat{\bar{\eta}}^{(2)}_{EB}$
Empirical Bias	0.071	−0.081	0.015	0.033	**0.007**
RMSE	0.173	0.263	0.153	0.154	**0.150**

Bold face indicates the estimator achieving the smallest value in terms of bias or RMSE. $\widehat{\bar{\eta}}^{(2)}_{EB}$ behaves better than others in terms of bias reduction and providing high estimation accuracy in the experiments

Table 7 The estimated hyperparameters π_0 and σ_0^2 for 6 diseases of WTCCC dataset

	$\widehat{\pi}_0$	$\widehat{\sigma}_0^2$
Coronary artery disease	0.949	0.004
Crohn's disease	0.840	0.006
Hypertension	0.966	0.007
Rheumatoid arthritis	0.947	0.008
Type 1 diabetes	0.967	0.014
Type 2 diabetes	0.940	0.005

π_0 is the proportion of true null hypotheses among all SNPs. σ_0^2 is the variance of the effect sizes among all associated SNPs

If the values of the local true discovery rates $ltdr^{(1)}$ have nearly the same level for all identified associations in the primary study, the variance of the average power will be inversely proportional to the number of the associations. When the identified number is small in the primary study, the credible interval for the average power is rather wide. This can be illustrated in the study of hypertension, where there is only 1 association showing genome-wide significance. From Fig. 5c, we can see that the credible interval is rather wide. If we want to consider the credible interval for this situation, then the sample size can drastically increased.

Discussion

We propose to design replication study under the case-control setting where *log*-odds ratio test is used. The method can also be generalized to other tests within *z*-test scheme, such as regression slope test used for quantitative trait.

As described in [7], the winner's curse depends strongly on the power of primary study. For a high power primary study, most non-null SNPs will result in significant associations after random draws from the population. Hence, the bias will be small in this case. There are more and more high power studies conducted for common diseases by using pooling strategy or meta-analysis strategy, but the high power studies for rare diseases are limited. Hence, it is still helpful and necessary to propose a designing procedure for the replication study with the consideration of winner's curse.

With the development of the cost-effective sequencing technique, the targets of association studies extend from common variations to rare variants. A commonly used

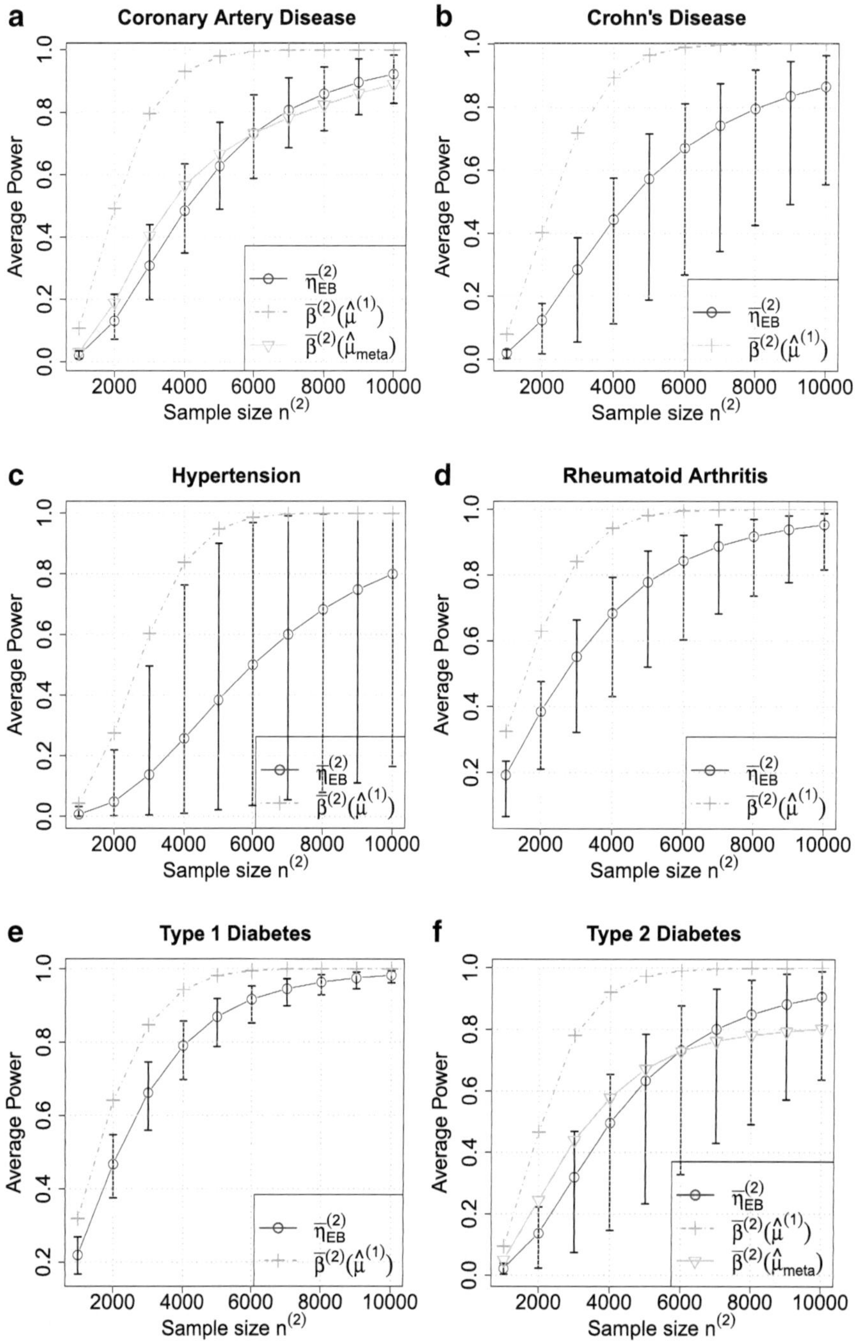

Fig. 5 The relationship between estimated average power $\bar{\eta}_{EB}^{(2)}$ and the sample size of the replication study $n^{(2)}$ for 6 diseases of the WTCCC dataset: **a** coronary artery disease, **b** Crohn's disease, **c** hypertension, **d** rheumatoid arthritis, **e** type 1 diabetes, **f** type 2 diabetes. The Control-to-Case ratio of the replication study is set to 1. The significance levels used in the primary study and the replication study are $\alpha_1 = 5 \times 10^{-8}$ and $\alpha_2 = 5 \times 10^{-6}$, respectively. As a comparison, the relationship between $\bar{\beta}^{(2)}\left(\hat{\mu}^{(1)}\right)$ and $n^{(2)}$ are also shown in the figure. For a fixed $n^{(2)}$, $\bar{\beta}^{(2)}\left(\hat{\mu}^{(1)}\right)$ is much larger than $\bar{\eta}_{EB}^{(2)}$. In (**a**) and (**f**), $\bar{\beta}^{(2)}\left(\hat{\mu}_{meta}\right)$ is the average power estimator by plugging the *log*-odds ratio obtained from high power meta-analysis study: CARDIoGRAM GWAS [22] and DIAGRAM GWAS [23], respectively. It can be shown that $\bar{\eta}_{EB}^{(2)}$ is close to $\bar{\beta}^{(2)}\left(\hat{\mu}_{meta}\right)$

strategy to discover associations with rare variants is the collapsing method [24], in which several rare variants in a certain group are pooled together to enrich the signal. For each group, a "super variant" is constructed. If *log*-odds ratio test is adopted in testing the association between "super variant" and the disease, our method can be used

Table 8 Sample size of the replication study needed for 6 diseases of WTCCC dataset when average power is estimated by EB based method. The Control-to-Case ratio of the replication study is set to 1. The significance levels used in the primary study and the replication study are $\alpha_1 = 5 \times 10^{-8}$ and $\alpha_2 = 5 \times 10^{-6}$, respectively

	50 %	60 %	70 %	80 %	90 %
Coronary artery disease	4095	4784	5652	6885	9121
Crohn's disease	4405	5252	6376	8092	11,552
Hypertension	5993	6992	8244	10,014	13,215
Rheumatoid arthritis	2666	3329	4147	5291	7357
Type 1 diabetes	2146	2640	3249	4094	5588
Type 2 diabetes	4027	4726	5633	6988	9721

directly for designing the replication study.

Some limitations of our approach need to be mentioned.

1. The assumption of our approach is that all SNPs' effect sizes are drawn independently from a two-component mixture distribution. Linkage disequilibrium widely exists in SNPs. Correlated genotype patterns can also introduce correlation between their effect sizes. The power estimation can be further improved by using correlation information in the prior set-up.
2. Our proposed method assumes the effect sizes of associated SNPs are normally distributed. This thin tail distribution may not be realistic. How to design of replication study with other heavy-tail prior needs to be discussed.

Conclusions

Replication study is commonly used to verify findings discovered from GWAS. Power analysis is essential in designing a replication study. Traditional approach will not extract information from primary study. Also it will need users to specify a parameter μ_{min}, which is subjective. Power estimation approach may address this problem, but there are several challenges in power estimation: the winner's curse, credible interval and summarization.

In this paper, we propose an EB based power estimation method to resolve these challenges. Simulation experiments show our approach is better than other plug-in based approaches in terms of overcoming the winner's curse and providing higher estimation accuracy. We also use simulation experiments to demonstrate the well calibration of the constructed credible interval. As an application example, we use our approach to determine the sample size needed in the WTCCC datasets of 6 diseases. Our approach gives an objective way to design replication study using information extracted from primary study.

Table 9 Sample size of the replication study needed for 6 diseases of WTCCC dataset when average power is estimated by plugging in observed effect sizes. The Control-to-Case ratio of the replication study is set to 1. The significance levels used in the primary study and the replication study are $\alpha_1 = 5 \times 10^{-8}$ and $\alpha_2 = 5 \times 10^{-6}$, respectively

	50 %	60 %	70 %	80 %	90 %
Coronary artery disease	2019	2290	2608	3023	3675
Crohn's disease	2270	2572	2922	3369	4058
Hypertension	2672	2988	3345	3788	4448
Rheumatoid arthritis	1553	1892	2273	2748	3465
Type 1 diabetes	1532	1856	2229	2706	3443
Type 2 diabetes	2085	2357	2676	3095	3775

Appendix

Appendix 1 — $\eta^{(2)}$ is the minimizer of $R(\widehat{\theta})$

The Bayes risk $R(\widehat{\theta})$ can be derived as follows:

$$\begin{aligned} R(\widehat{\theta}) &= E_{\mu,z^{(1)}}\left(\left(\widehat{\theta}-\beta^{(2)}(\mu)\right)^2 \big| |Z^{(1)}| > z_{\alpha_1/2}, \mathcal{H}_1\right) \\ &= \int_{-\infty}^{\infty}\left[\int_{|z^{(1)}|>z_{\alpha_1/2}} \left(\widehat{\theta}-\beta^{(2)}(\mu)\right)^2 \frac{p\left(z^{(1)}|\mu\right)}{P\left(|Z^{(1)}|>z_{\alpha_1/2}|\mu\right)} dz^{(1)}\right] \\ &\quad \times p\left(\mu\big||Z^{(1)}|>z_{\alpha_1/2},\mathcal{H}_1\right) d\mu \\ &= \frac{1}{P\left(Z^{(1)}>z_{\alpha_1/2}|\mathcal{H}_1\right)} \int_{-\infty}^{\infty}\left[\int_{|z^{(1)}|>z_{\alpha_1/2}} \left(\widehat{\theta}-\beta^{(2)}(\mu)\right)^2 \right. \\ &\quad \left. \times p\left(z^{(1)}|\mu\right) dz^{(1)}\right] p\left(\mu|\mathcal{H}_1\right) d\mu \\ &= \frac{1}{P\left(Z^{(1)}>z_{\alpha_1/2}|\mathcal{H}_1\right)} \int_{|z^{(1)}|>z_{\alpha_1/2}} \left[\int_{-\infty}^{\infty} \left(\widehat{\theta}-\beta^{(2)}(\mu)\right)^2 \right. \\ &\quad \left. \times p\left(\mu|z^{(1)},\mathcal{H}_1\right) d\mu\right] p\left(z^{(1)}\right) dz^{(1)}. \end{aligned} \tag{22}$$

The last equality is hold by Fubini's theorem.

From the last equality, it can be seen that the Bayesian predictive power $\eta^{(2)}$ is the minimizer of the expression in the brace for each value of $z^{(1)}$. Hence $\eta^{(2)}$ is also the minimizer of $R(\widehat{\theta})$.

Appendix 2 — Derivation of $\eta^{(2)}$ under two-component mixture prior

The following property of multivariate Gaussian distribution is proved in the Chapter 2 of [25], which can be used to derive $\eta^{(2)}$.

Property 1. If $\mathbf{Z}|\mu \sim N_p(\mu, \Sigma)$, and $\mu \sim N_p(\mu_0, \Sigma_0)$, then

$$\mathbf{Z} \sim N_p(\mu_0, \Sigma + \Sigma_0) \text{ and } \mu|\mathbf{z} \sim N_p(\mathbf{W}\mu_0 + (\mathbf{I} - \mathbf{W})\mathbf{z}, (\mathbf{I} - \mathbf{W})\Sigma) \quad (23)$$

where $\mathbf{W} = \Sigma(\Sigma_0 + \Sigma)^{-1}$

Because $z^{(1)} \sim N(\mu/\sigma^{(1)}, 1)$ and $(\mu|\mathcal{H}_1) \sim N(0, \sigma_0^2)$, the following can be obtained by using Property 1:

$$\left(\mu|z^{(1)}, \mathcal{H}_1\right) \sim N\left(\lambda\hat{\mu}^{(1)}, \lambda\left(\sigma^{(1)}\right)^2\right), \quad (24)$$

where $\lambda = \frac{1}{1+(\sigma^{(1)}/\sigma_0)^2}$ is a shrinkage effect factor. Under $\mathcal{H}_1$, the posterior distribution of $Z^{(2)}$ is

$$\left(Z^{(2)}|z^{(1)}, \mathcal{H}_1\right) \sim N\left(z^* = \lambda\frac{\hat{\mu}^{(1)}}{\sigma^{(2)}}, (\sigma^*)^2 = 1 + \lambda\left(\frac{\sigma^{(1)}}{\sigma^{(2)}}\right)^2\right). \quad (25)$$

Then the Bayesian predictive power of the replication study reads:

$$\eta^{(2)} = \Phi\left(\frac{sgn\left(z^{(1)}\right)z^* - z_{\alpha_2}}{\sigma^*}\right), \quad (26)$$

where $\Phi(x)$ is the cumulative density function (cdf) of $N(0, 1)$.

Appendix 3 — Derivation of the σ_0^2 estimator

By using Property 1, the marginal distribution of $Z^{(1)}$ is

$$Z^{(1)} \sim \pi_0 N(0, 1) + (1 - \pi_0)N\left(0, 1 + \left(\frac{\sigma_0}{\sigma^{(1)}}\right)^2\right), \quad (27)$$

which is a two-component Gaussian mixture model. Hence, the squared of $Z^{(1)}$ is distributed as

$$\left(Z^{(1)}\right)^2 \sim \pi_0\chi_1^2 + (1 - \pi_0)\left(1 + \left(\frac{\sigma_0}{\sigma^{(1)}}\right)^2\right)\chi_1^2, \quad (28)$$

where χ_1^2 is the 1 degree of freedom χ^2 distribution. The expectation reads

$$E\left(\left(Z^{(1)}\right)^2\right) = \pi_0 + (1 - \pi_0)\left(1 + \left(\frac{\sigma_0}{\sigma^{(1)}}\right)^2\right). \quad (29)$$

By summing over the test statistics of all SNPs, we can obtain

$$E\left(\sum_{i=1}^{m}\left(Z_i^{(1)}\right)^2\right) = m\pi_0 + (1 - \pi_0)\left(m + \sigma_0^2\sum_{i=1}^{m}\left(1/\sigma_i^{(1)}\right)^2\right), \quad (30)$$

which introduce an estimator of σ_0^2

$$\widehat{\sigma}_0^2 = \left(\frac{\sum_{i=1}^{m}\left(z_i^{(1)}\right)^2 - m\pi_0}{(1 - \pi_0)} - m\right) / \sum_{i=1}^{m}\left(1/\sigma_i^{(1)}\right)^2. \quad (31)$$

Appendix 4 — Derivation of $ltdr^{(1)}$ under a two-component mixture prior

With Eq. (27), the local true discovery rate of the primary study reads:

$$ltdr^{(1)} = \frac{\pi_1\phi\left(z^{(1)}/\sqrt{1 + \left(\widehat{\sigma}_0/\sigma^{(1)}\right)^2}\right)}{\pi_0\phi(z^{(1)}) + \pi_1\phi\left(z^{(1)}/\sqrt{1 + \left(\widehat{\sigma}_0/\sigma^{(1)}\right)^2}\right)}, \quad (32)$$

where $\phi(x)$ is the probability density function (pdf) of $N(0, 1)$.

Competing interests
The authors declare that they have no competing interests.

Authors' contributions
WJ conceived the study, developed methodology, performed data analysis and wrote the manuscript. WY conceived the study and wrote the manuscript. All authors read and approved the final manuscript.

Acknowledgements
This paper was partially supported by a theme-based research project T12-402/13N of the Hong Kong Research Grant Council (RGC). The publication costs for this article were partly funded by the Hong Kong Research Grant Council, grant number T12-402/13N.

Declarations
This article has been published as part of BMC Genomics Volume 17 Supplement 1, 2016: Selected articles from the Fourteenth Asia Pacific Bioinformatics Conference (APBC 2016): Genomics. The full contents of the supplement are available online at http://www.biomedcentral.com/bmcgenomics/supplements/17/S1.

Published: 11 January 2016

References
1. Hindorff LA, MacArthur J, Morales J, Junkins HA, Hall PN, Klemm AK, et al. A catalog of published genome-wide association studies. Available at: http://www.genome.gov/gwastudies/. Accessed [2015.05.28] (Web Resources).
2. Balding DJ. A tutorial on statistical methods for population association studies. Nat Rev Genet. 2006;7(10):781–91.
3. NCI-NHGRI Working Group on Replication in Association Studies. Replicating genotype–phenotype associations. Nature. 2007;447(7145):655–60.
4. Kraft P, Zeggini E, Ioannidis JP. Replication in genome-wide association studies. Stat Sci Rev J Inst Math Stat. 2009;24(4):561.
5. Ioannidis JP. Why most discovered true associations are inflated. Epidemiology. 2008;19(5):640–8.
6. Göring HH, Terwilliger JD, Blangero J. Large upward bias in estimation of locus-specific effects from genomewide scans. Am J Hum Genet. 2001;69(6):1357–69.
7. Zöllner S, Pritchard JK. Overcoming the winner's curse: estimating penetrance parameters from case-control data. Am J Hum Genet. 2007;80(4):605–15.

8. Zhong H, Prentice RL. Bias-reduced estimators and confidence intervals for odds ratios in genome-wide association studies. Biostatistics. 2008;9(4):621–34.
9. Ghosh A, Zou F, Wright FA. Estimating odds ratios in genome scans: an approximate conditional likelihood approach. Am J Hum Genet. 2008;82(5):1064–74.
10. Sun L, Dimitromanolakis A, Faye LL, Paterson AD, Waggott D, Bull SB. The DCCT/EDIC Research Group.: BR-squared: a practical solution to the winner's curse in genome-wide scans. Human genetics. 2011;129(5): 545–52.
11. Xu L, Craiu RV, Sun L. Bayesian methods to overcome the winner's curse in genetic studies. Ann Appl Stat. 2011;5(1):201–31.
12. Ferguson JP, Cho JH, Yang C, Zhao H. Empirical Bayes correction for the winner's curse in genetic association studies. Genet Epidemiol. 2013;37(1):60–8.
13. Wellcome Trust Case Control Consortium. Genome-wide association study of 14,000 cases of seven common diseases and 3,000 shared controls. Nature. 2007;447(7145):661–78.
14. Woolf B. On estimating the relation between blood group and disease. Ann Hum Genet. 1955;19(4):251–3.
15. Lecoutre B. Bayesian predictive procedure for designing and monitoring experiments. Bayesian Methods with Applications to Science, Policy and Official Statistics. Luxembourg: Eurostat; 2001, pp. 301–10.
16. Yang J, Benyamin B, McEvoy BP, Gordon S, Henders AK, Nyholt DR, et al. Common SNPs explain a large proportion of the heritability for human height. Nat Genet. 2010;42(7):565–9.
17. Park J-H, Wacholder S, Gail MH, Peters U, Jacobs KB, Chanock SJ, et al. Estimation of effect size distribution from genome-wide association studies and implications for future discoveries. Nat Genet. 2010;42(7): 570–5.
18. Storey JD, Tibshirani R. Statistical significance for genomewide studies. Proc Natl Acad Sci. 2003;100(16):9440–5.
19. Langaas M, Lindqvist BH, Ferkingstad E. Estimating the proportion of true null hypotheses, with application to DNA microarray data. J R Stat Soc Ser B (Stat Methodol). 2005;67(4):555–72.
20. Jin J, Cai TT. Estimating the null and the proportion of nonnull effects in large-scale multiple comparisons. J Am Stat Assoc. 2007;102(478): 495–506.
21. Efron B. Local false discovery rates. Technical Report 2005-20B. Department of Statistics, Stanford University. 2005.
22. Mehta NN. Large-scale association analysis identifies 13 new susceptibility loci for coronary artery disease. Circ Cardiovasc Genet. 2011;4(3):327–9.
23. Morris AP, Voight BF, Teslovich TM, Ferreira T, Segré AV, Steinthorsdottir V, et al. Large-scale association analysis provides insights into the genetic architecture and pathophysiology of type 2 diabetes. Nat Genet. 2012;44(9):981.
24. Bansal V, Libiger O, Torkamani A, Schork NJ. Statistical analysis strategies for association studies involving rare variants. Nat Rev Genet. 2010;11(11): 773–85.
25. Bishop CM. Pattern Recognition and Machine Learning. New York: Springer; 2006.

Nair et al. BMC Genomics 2015, 17(Suppl 1):14
DOI 10.1186/s12864-015-2297-3

BMC Genomics

PROCEEDINGS Open Access

A maximum-likelihood approach for building cell-type trees by lifting

Nishanth Ulhas Nair[1], Laura Hunter[2], Mingfu Shao[1], Paulina Grnarova[1], Yu Lin[3], Philipp Bucher[4,5†] and Bernard M. E. Moret[1,5*†]

From The Fourteenth Asia Pacific Bioinformatics Conference (APBC 2016)
San Francisco, CA, USA. 11–13 January 2016

Abstract

Background: In cell differentiation, a less specialized cell differentiates into a more specialized one, even though all cells in one organism have (almost) the same genome. Epigenetic factors such as histone modifications are known to play a significant role in cell differentiation. We previously introduce cell-type trees to represent the differentiation of cells into more specialized types, a representation that partakes of both ontogeny and phylogeny.

Results: We propose a maximum-likelihood (ML) approach to build cell-type trees and show that this ML approach outperforms our earlier distance-based and parsimony-based approaches. We then study the reconstruction of ancestral cell types; since both ancestral and derived cell types can coexist in adult organisms, we propose a lifting algorithm to infer internal nodes. We present results on our lifting algorithm obtained both through simulations and on real datasets.

Conclusions: We show that our ML-based approach outperforms previously proposed techniques such as distance-based and parsimony-based methods. We show our lifting-based approach works well on both simulated and real data.

Keywords: Cell-type trees, Histone modifications, Epigenomics, Phylogeny, Evolution, Cell-differentiation

Background

Cell differentiation is the process by which a less specialized cell becomes a more specialized one; it often proceeds in a hierarchical manner, with totipotent cells sequentially committing to fates of more restricted developmental potential [1, 2]. Epigenetic and transcription factors play a significant role in cell differentiation [3–5], therefore motivating a study of epigenetic changes across different cell types.

Arendt [6] proposed a sister-cell-type model for the hierarchical relationship between cell types. In this model [6], "novel cell types arise in pairs (sister cell types) from an ancestral cell type through sub-specialization" [2]. Under this model [6], the evolutionary relatedness of cell types is expected to be congruent with the ontogenetic hierarchy of cellular differentiation, because the "development of the sister cell types is the same up to the last stages of differentiation" [2]. The authors claim that multifunctionality has been a general feature of ancient cell types and that, with "increasing specialization during evolution, these multiple functions were then distributed in a complementary manner to sister cell types" [6].

Evolution and cell differentiation share a number of attributes. First, as mentioned before, we know that cell differentiation transforms less specialized cell types into more specialized ones. Since this transformation is unidirectional, the paths of differentiation can be represented as a tree structure, much as is done with the phylogenetic trees used to represent evolutionary histories [7]. The similarity between the two extends further: cell types

*Correspondence: bernard.moret@epfl.ch
†Equal contributors
[1] School of Computer and Communication Sciences, École Polytechnique Fédérale de Lausanne (EPFL), EPFL IC IIF LCBB, INJ 211 (Batiment INJ), Station 14, CH-1015 Lausanne, Switzerland
[5] Swiss Institute of Bioinformatics, Lausanne, Switzerland
Full list of author information is available at the end of the article

themselves have evolved into larger collections from more restricted collections in early ancestors: there are phylogenetic relationships among the various types of cells. Second, observed changes in the epigenetic state are inheritable, again much as mutations in the genome are (although, of course, through very different mechanisms and at very different scales); and finally, epigenetic traits are subject to stochastic changes. One major difference between evolution and cell differentiation is that functional changes in cell differentiation are primarily driven by programmed mutational events rather than by selection. However, the program of mutational events is itself the result of evolution, so that, as observed by [6], the cell differentiation tree often recapitulates the phylogeny of cell types.

We focus here on one important epigenetic mark—histone modification. Histones are proteins that package the DNA into nucleosomes [8]. These proteins are subjected to various types of chemical modifications, called histone modifications, such as methylation, acetylation, phosphorylation, ubiquitination, etc. These modifications alter their interactions with the DNA and thereby influence transcription and genomic function. Histone modifications have been found to vary across cell types and to play an important role in gene regulation [9]. Since histones are present in every 200 bp length of DNA, we need genome-wide high-throughput technologies to study the modifications of these proteins. ChIP-Seq is such a technology [10, 11]. The study of ChIP-Seq histone modification data can help us understand the role of histone modifications in developmental biology and cell differentiation [12].

The term "cell-type tree" was defined by our group to refer to a tree relationship between various cell types [13]. The nodes of this tree represent cell types while the edges represent directed differentiation/evolution events from one cell type to another. We know that the genome is consistent across cell types of the same individual and that it is also highly similar between individuals of the same species, but that epigenomic states of various regions of the genome differ across various cell types. These epigenomic states are believed to affect cell differentiation process through a complex interplay between histone modifications, DNA methylation, transcription factors, etc.

Kin et al. [2] recently constructed a cell-type tree using RNA-Seq data and a parsimony-based approach under assumptions very similar to ours, using the same term of "cell-type tree" to denote the "hypothetical tree-like relationship of cell types in ontogeny and phylogeny". Liang et al. [14] recently developed a statistical model for cell differentiation and applied it to ENCODE and FANTOM RNA-Seq data. As in Kin et al. work [2], they found that the RNA-Seq data contain significant tree structures. In earlier work [7], we also calculated a statistical measure to show that the distances we computed are in fact representative of a tree. Thus multiple studies on different kinds of datasets—ChIP-Seq in our case, RNA-seq in the other two papers—support the tree-like relationship of cell types and underscore the usefulness of the cell-type tree (as noted in [2]). Prior to these genome-wide computational approaches, hierarchical developmental relationships among cell types were elucidated through a series of laborious experiments involving in vitro differentiation of cell types from various stem cells [15–17].

In cell differentiation, both ancestral and derived cell types can coexist within the body. Therefore, it becomes important to be able to infer which cell types should be treated as the ancestor, or parent, of another. Our earlier work [7, 13] focused on the use of distance-based and parsimony-based phylogenetic methods to infer the tree, not the ancestors. Here we propose an ML approach to the inference of cell-type trees on histone modification data and proceed to derive a new algorithm to infer the internal nodes by a process known as lifting. (Since both ancestral and derived cell types can coexist in the body, some of the node labels should be simply "lifted"—copied—into the parent node). To our knowledge, this is the first lifting approach used in the study of cell differentiation. We also provide simulations and tests on real data, indicating that our ML approach to the inference of cell-type trees outperforms distance-based and parsimony-based building approaches and that our lifting algorithm not only works well on simulations, but also gives biologically meaningful results.

Methods

A histone-modification ChIP-Seq library contains ChIP-Seq data for one ChIP-Seq experiment. In our case, each library typically contains data for one histone modification for one replicate of a cell type.

Our approach to build cell-type trees using a ML framework is illustrated in Fig. 1. We explain the various steps below.

Model of differentiation for histone marks

We use the model of [7], in which histone marks can be independently gained or lost in regions of the genome as cells differentiate from a less specialized type to a more specialized one. This independence assumption is consistent with practice in phylogenetic inference and enormously simplifies computations.

Data preprocessing and data representation

ChIP-Seq data are converted into peak data using a peak-finder. The presence of a peak signifies the presence of a histone mark in that genomic region. We have used publicly available peak lists (from the ENCODE database) for our study.

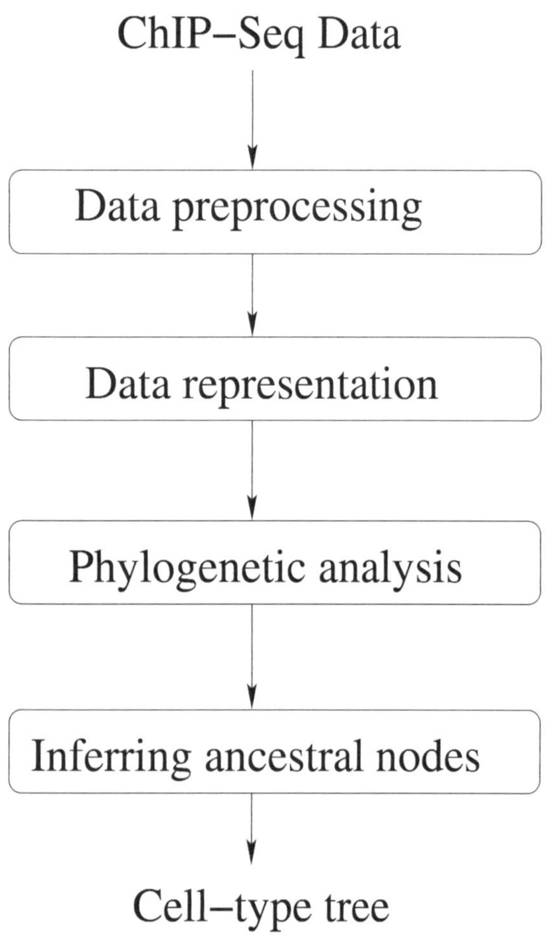

Fig. 1 Flowchart for building cell-type trees using a maximum-likelihood framework. Data preprocessing: the mapped reads in the ChIP-Seq data are used to find peaks in the data. Data representation: peak data is converted to a binary matrix using windowing/overlap representation. Phylogenetic analysis: using distance-based, parsimony-based or maximum-likelihood-based phylogenetic approach. Inferring ancestral or internal nodes: we establish a parent-child relationship between the cell types (leaf data) using a process called lifting

To represent the peak data of each cell type, we use two different data representations.

1. **Windowing representation**: The genome is divided into bins of fixed size; if the bin contains at least one peak, we code it 1, otherwise we code it 0. The coding of each library is thus independent of that of any other library (a ChIP-Seq dataset and its representation are both called libraries). We used bins of 200 bp because 147 bp of DNA wrap around the histone and two histones are connected by linker DNA of about 50 bp; thus each bin approximates the presence or absence of just one histone modification.
2. **Overlap representation**: This representation takes into account all libraries at once. The aim is to find interesting regions in the genome based on peaks. Denote the ith peak in library n as $P_i^n = [P_{iL}^n, P_{iR}^n]$, where P_{iL}^n and P_{iR}^n are the left and right endpoints (as basepair indices). Consider each peak as an interval on the genome (or on the line of real numbers) and build the interval graph defined by all peaks in all libraries. An interval graph has one vertex for each interval and an edge between two vertices whenever the two corresponding intervals overlap [13]. We simply want the connected components of the interval graph. We define an interval in the genome is an *interesting region* if and only if it corresponds to a connected component of the interval graph. More details on the overlap representation and an algorithm to identify interesting regions in linear time appear in [7].

The output of either the windowing representation or overlap representation is a string of ones and zeros to represent each data library. Both representations gave fairly similar results in earlier work [7], so in this study we chose the overlap representation, for its compactness.

ML-based phylogenetic analysis

We use an ML-based approach to build cell-type trees on the overlap data representation, carrying out the inference with the RAxML tool [18]. We run RAxML on the binary data obtained using the overlap representation and obtain a cell-type tree. For our experiments, we used the GAMMA model of rate heterogeneity (BINGAMMA) and turned on the rapid bootstrapping option (set to 100).

Inferring ancestral/internal nodes through lifting

We now describe an algorithm for inferring ancestral/internal nodes using a process called lifting, to establish a parent-child relationship between the various cell types (at the leaves). Lifting techniques have been used in the context of tree alignment problems [19]. We first infer a tree using RAxML; we then root the tree using prior biological knowledge. (One could root the tree by placing a root between two nodes of an unrooted tree connected by an edge). Denote by T the resulting rooted binary tree. We now run the lifting procedure on this tree T, so that they obey path constraints. The basic idea of the lifting procedure is to compute the likelihood of the tree after the lifting each possible leaf node (if its sibling is also a leaf) and subtract it from the likelihood of the unlifted tree, and then to actually lift the leaf with the highest probability if this difference is greater than some threshold. The old tree is now updated with the lifted tree and this procedure is continued till the lifting stops. The pseudocode for the lifting algorithm is given below.

Algorithm for lifting

1. Set $R = T$.
2. If number of leaves in R is less than or equal to 4, go to step 7.
3. Let $\mathbb{L}$ be the set of leaves whose sibling is also a leaf. For each leaf node $L \in \mathbb{L}$ we compute the likelihood/probability $P(L)$ of lifting L using the following procedure.
   (a) Divide the tree R into two smaller trees according to L, R_L^1 and R_L^2 (see example in Fig. 2). R_L^1 is built by first lifting L to its parent and then removing L and its sibling (from R). R_L^2 is a small tree of 2 nodes, the earlier leaf node L which is connected to its sibling node (parent is L).
   (b) Estimate the probabilities of R_L^1 and R_L^2 (branch lengths estimated using RAxML and probabilities calculated using built in methods in *R* software packages like *phangorn* [20]). The total probability is the product of these two probabilities.
4. Compute the best lift $L^* = \arg\max_{L \in \mathbb{L}} P(L)$.
5. Let $W = log(P(L^*)) + \alpha\ log(K) - log(P(R))$, where K is the length of the data representation sequence, α a user-defined, real-valued constant, $P(R)$ is the likelihood of tree R. W plays a role similar to a BIC criterion [21].
6. If $W > 0$, we lift, update the tree $R = R_{L^*}^1$, and mark the corresponding edge (parent of L^* and L^*) in T as lifted; we then return to step 2.
7. Output T and stop.

The output tree T is the desired cell-type tree with labelled ancestral nodes. We terminate the algorithm when the number of leaves is four or less since we use RAxML.

Results and discussion

We show the results on both real and simulated data.

Using real data

In an earlier work [7], we had shown the usefulness of using cell-type trees on different histone marks from the ENCODE project database: H3K4me3, H3K27me3, H3K4me1, H3K9me3, and H3K27ac. For testing the lifting algorithm, it is desirable to have as many samples as possible. We thus focus in this study on histone modification H3K4me3, using ChIP-Seq data for human (hg19) from the University of Washington ENCODE group [22, 23]. H3K4me3 has been assayed in the largest number of cell types and is usually associated with gene activation [24]. Table 1 gives the list of the 37 cell types for which we gathered H3K4me3 data. The cells are classified into groups based on cell type or tissue origin. (Keratinocytes (NHEK) is included in the Epithelial group). For human Embryonic Stem Cells (hESC) we have data for a differentiation time course in cell culture (day 0, 2, 5, 9, 14), so we shall use hESC T2 to mean data for hESC cells on day 2. We use only one replicate per cell type for this work.

We use the ENCODE peaks as input to our program. We convert the input data into 1s and 0s using the overlap representation. We then use RAxML for getting a maximum likelihood based tree. In this first step, we then compare our results with those obtained with a

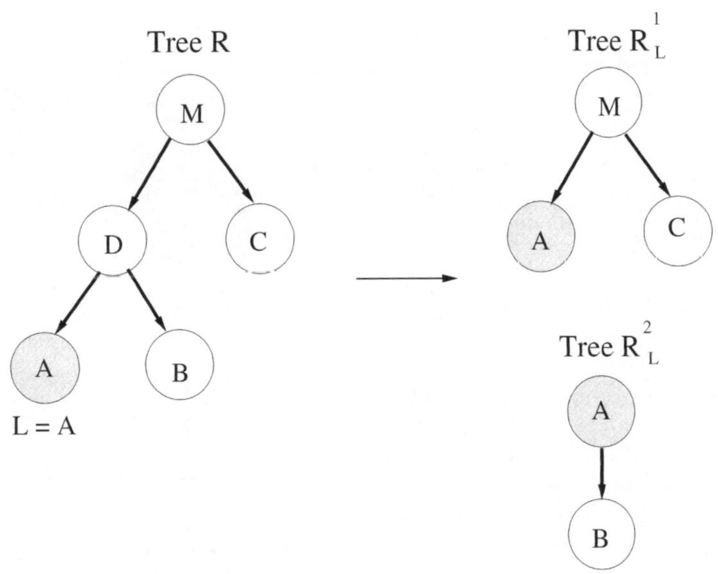

Fig. 2 Example of lifting leaf node *A* (*L* in the algorithm) in tree *R*. Tree *R* is divided into two smaller trees R_L^1 and R_L^2 as described in the algorithm

Table 1 Cell types, short description, and general group for H3K4me3 data. For details see the ENCODE website [23]

Cell name	Short description	Group
AG04449	fetal buttock/thigh fibroblast	Fibroblast
AG04450	fetal lung fibroblast	Fibroblast
AG09319	gum tissue fibroblasts	Fibroblast
AoAF	aortic adventitial fibroblast cells	Fibroblast
BJ	skin fibroblast	Fibroblast
CD14	Monocytes-CD14+ from human leukapheresis production	Blood
CD20	B cells replicate	Blood
hESC	undifferentiated embryonic stem cells	hESC
HAc	astrocytes-cerebellar	Astrocytes
HAsp	astrocytes spinal cord	Astrocytes
HBMEC	brain microvascular endothelial cells	Endothelial
HCFaa	cardiac fibroblasts- adult atrial	Fibroblast
HCF	cardiac fibroblasts	Fibroblast
HCM	cardiac myocytes	Myocytes
HCPEpiC	choroid plexus epithelial cells	Epithelial
HEEpiC	esophageal epithelial cells	Epithelial
HFF	foreskin fibroblast	Fibroblast
HFF MyC	foreskin fibroblast cells expressing canine cMyc	Fibroblast
HMEC	mammary epithelial cells	Epithelial
HPAF	pulmonary artery fibroblasts	Fibroblast
HPF	pulmonary fibroblasts isolated from lung tissue	Fibroblast
HRE	renal epithelial cells	Epithelial
HRPEpiC	retinal pigment epithelial cells	Epithelial
HUVEC	umbilical vein endothelial cells	Endothelial
HVMF	villous mesenchymal fibroblast cells	Fibroblast
NHDF Neo	neonatal dermal fibroblasts	Fibroblast
NHEK	epidermal keratinocytes	Epithelial
NHLF	lung fibroblasts	Fibroblast
RPTEC	renal proximal tubule epithelial cells	Epithelial
SAEC	small airway epithelial cells	Epithelial
SKMC	skeletal muscle cells	Skeletal muscle
WI 38	embryonic lung fibroblast cells	Fibroblast

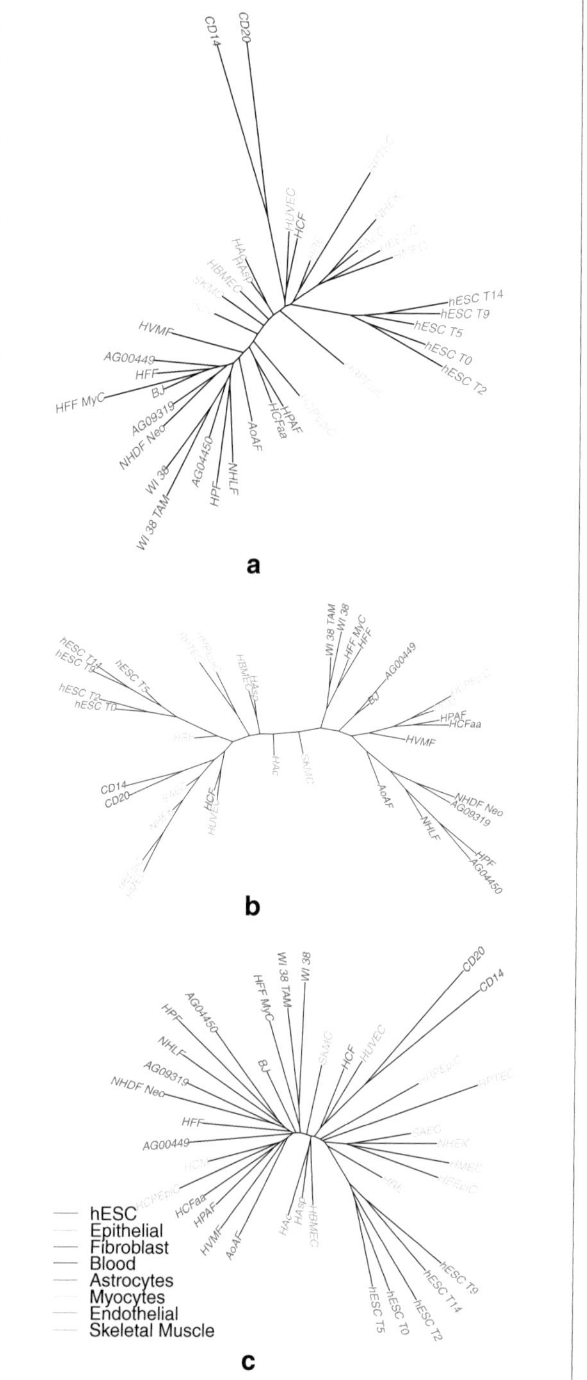

Fig. 3 Compares cell-type trees obtained on H3K4me3 data (only one replicate) using two different methods on the overlap representation: (**a**) using a maximum-likelihood based approach, (**b**) using a parsimony-based approach, (**c**) using a distance-based approach

distance-based approach (neighbor-joining [25]) and a parsimony-based approach (TNT [26]), as explained in [7]. Figure 3 summarizes the results, using color codes for the major groupings of Table 1. In order to quantify the quality of the groupings, we compute the total number of cell types in a subtree that belong to one group. Since our groups are based on cell type only, there could be many subdivisions possible within each group, therefore we choose the two largest such subtrees available for each group such that each subtree contains only the leaf nodes of that group.

Table 2 Groupings for cell-type trees on H3K4me3 data

	hESC (5)	Epithelial (8)	Fibroblast (16)	Blood (2)	Astrocytes (2)	Myocytes (1)	Endothelial (2)	Skeletal Muscle (1)
D	5,0	4,1	6,3	2,0	2,0	1,0	1,1	1,0
P	5,0	4,2	6,4	2,0	1,1	1,0	1,1	1,0
ML	5,0	6,1	15,1	2,0	2,0	1,0	1,1	1,0

Second to ninth columns show the number of cell types (of the same group) belonging to the largest and second-largest clades; the total number of cell types of that group is in the top row. Rows correspond to various methods. Overlap representation is used. ML — maximum-likelihood-based approach, P — parsimony-based approach, D — distance-based approach

Table 2 shows the results for the ML-based, parsimony-based, and distance-based methods. Cell-type trees were built without inferring ancestral/internal nodes (no lifting). The ML approach gives significantly better clades than the other two. For example, the subtrees containing Fibroblasts and Epithelial cell types are greatly improved. Of the 16 Fibroblast cell types that we considered, the ML method grouped 15 in the largest clade, a major improvement from the 6 grouped by the distance-based and parsimony-based approaches. For Epithelial cell types the improvement is from 4 to 6, out of 8 cell types in total.

We now consider a set of 19 cell types to evaluate the lifting algorithm (on H3K4me3 data) to infer ancestral nodes. The cell types include the hESC time course of 5 days (day 0, 2, 5, 9, 14), HUVEC (umbilical vein endothelial cells), HBMEC (brain microvascular endothelial cells), WI 38 (embryonic lung fibroblast cells), AG04450 (fetal lung fibroblast), HPF (pulmonary fibroblasts isolated from lung tissue). As explained in [7], we expect the following developmental pathways to occur one after the other in time during development: (1) hESC from days 0 to 14; (2) hESC to HUVEC to HBMEC; (3) hESC to WI38 to AG04550 to HPF. We now use the overlap representation and RAxML to get the cell-type tree. The tree was now rooted at the common ancestor of all the embryonic stem cells at different days (see Fig. 4a). The lifting algorithm is then used. The α parameter is set high enough such that lifting takes place as much as possible, the larger the value of α the more the number of lifts. The results are shown in Fig. 4b. We find that there is a path from hESC to HUVEC to HBMEC as expected. We also find a path from hESC to AG04550 to HPF. However we see that WI38 could not be lifted before AG04450. When we look at the data for hESC data, we see that embryonic stem cells on day 5 is an ancestor (internal node) to day 9, and day 9 is an ancestor to day 14; and we also see day 0 is an ancestor to day 2. However day 5 is incorrectly lifted above day 0. On the other hand, the fact that day 2 ends up on a side branch of the tree is not surprising in view of our previous observation [7] that day 2 shows overall increased divergence in histone modification. This most likely reflects a temporary non-specific response to the growth factor cocktail that was applied on day 0 to activate a mesoderm developmental pathway.

We repeated these experiments by picking a random number of columns from the overlap representation (sampling without replacement) and we found that our results are quite consistent, if we choose 50,000 or even only 10,000 columns (out of a total of 200,152 columns).

Using simulated data

In this paper, we use simulated studies to compare the quality of our maximum-likelihood approach and lifting algorithm. We randomly generate a set of rooted binary trees with a fixed number of leaves — containing either 12, 50, and 100 leaf nodes (random trees created using "ape" library in *R* [27]). Ten trees for each kind of tree (12, 50,

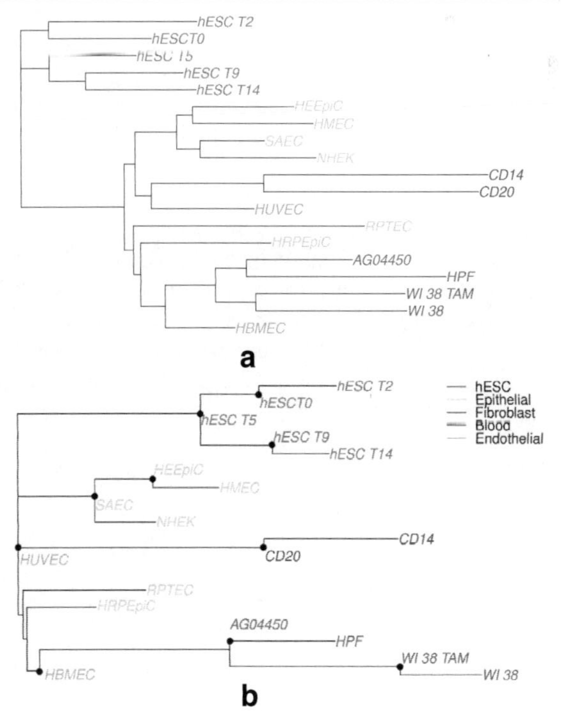

Fig. 4 Cell-type trees obtained on H3K4me3 data on a set of 19 cell types: (**a**) before lifting, (**b**) after lifting

or 100 leaf tree) are generated. Each of these trees then had their edges marked (randomly) as lifted (signifying lift from child to parent in that edge) with some probability. This is the true tree. Data representation length (K) of 1000 bits (1 or 0) for each leaf node was then simulated using each tree (used phangorn library in R).

A RAxML tree is generated on this leaf data and then this tree is given as input to the lifting program (for $\alpha = 0.1$). For the output tree, each edge which contains a lift from child to parent, is marked as lifted. This way we can compare the edges in the output tree with the true tree. Edges marked as lifted in the true tree should be shown as lifted in the output tree. Based on this, for each kind of trees (12, 50 or 100 leaf tree), we find the total number of true positives (TP), false positives (FP), true negatives (TN), false negatives (FN). We measure the following statistics [28] from this information:

(1) True positive rate $TPR = \frac{TP}{TP+FN}$.
(2) False positive rate $FPR = \frac{FP}{FP+TN}$.
(3) F-score or F_1-score $F = \frac{2TP}{2TP+FP+FN}$
(4) Accuracy $ACC = \frac{TP+TN}{TP+TN+FP+FN}$

We also compute the Robinson-Foulds metric or RF distance [29] which computes the distance between the true tree and the unlifted tree we get after applying RAxML.

The results are shown in Table 3. We see that we get high true positive rate, low false positive rate, a reasonably high F-score, and an accuracy of above 90 %. These results thus show that our lifting approach produces good quality results for trees which are both small and large. Since we expect cell-type trees to be build on only a few dozen cell types (since there are about 200 cell types in humans and we don't have data for many of these), we feel the range of tree sizes that we have considered in the simulation study is sufficient. We also get low RF distances for both 12-leaf and 50-leaf trees. The slightly higher RF distances we get for 100-leaf tree is because the data representation of 1000 is not large enough for bigger trees.

To show the robustness of our approach, we repeat the above experiments (for 12-leaf trees) by varying the length of data representation (K) per cell type. We use the following values — 500, 1000, and 5000 bits for each node. Various statistics are calculated by fixing the α threshold to 0.1. The results are shown in Table 4. We find that the different statistics are stable across different data representation lengths and the accuracy is around 90 %, and the statistics (like accuracy) improve with larger K which is expected. The RF distance is around 1 for different values of K. This shows that the threshold α is stable for a large variation of data representation lengths. We note that the choice of threshold α is dependent on the dataset used. The larger the value of α the greater the chances of lifting. For example if all the datasets for whose cell-type tree we are building are cell types which share a lineage in development, then we would like to lift as many cell types as possible; however if the dataset considered has only one or two cell types which have a lineage in cell-development then we would expect less lifting. So the value of α can be set by the biologist based on biological knowledge, since our method can be used as an exploratory tool. The value of α also depends on the length of the data representation, though it is robust over a reasonably wide range of data-representation as shown in Table 4. Now we show that for a fixed length of data representation, we can vary α parameters over a reasonably wide range. The results are shown in Table 5. In this experiment we fixed $K = 1000$, and simulated these data values for each of the 12-leaf trees and randomly chose lifting paths. We see from the

Table 3 Statistics for trees with fixed number of leaf nodes

	12-leaf	50-leaf	100-leaf
TPR	0.750	0.736	0.789
FPR	0.070	0.064	0.036
F	0.677	0.629	0.748
ACC	0.906	0.917	0.946
RF	1.3	5.900	12.20

We simulate 10 random trees (data representation length is 1000) for each of kind of tree (12, 50 or 100 leaf tree) and ran the lifting algorithm (for $\alpha = 0.1$). We then calculated the following statistics shown in the table: True positive rate (*TPR*), False positive rate (*FPR*), F-score or F_1-score (*F*), Accuracy (*ACC*), *RF* distance (*RF*)

Table 4 Statistics for trees with different length of data representations

	$K = 500$	$K = 1000$	$K = 5000$
TPR	0.783	0.750	0.880
FPR	0.088	0.070	0.067
F	0.621	0.677	0.733
ACC	0.899	0.906	0.927
RF	0.700	1.300	0.400

We simulate 10 random 12-leaf trees for varying number of data representation lengths (500, 1000 and 5000) and ran the lifting algorithm (for $\alpha = 0.1$). We then calculated the following statistics shown in the table: True positive rate (*TPR*), False positive rate (*FPR*), F-score or F_1-score (*F*), Accuracy (*ACC*), *RF* distance (*RF*)

Table 5 Statistics for trees with different values of α

	$\alpha = 0.1$	$\alpha = 1$	$\alpha = 2$
TPR	0.750	0.750	0.871
FPR	0.070	0.063	0.141
F	0.677	0.667	0.643
ACC	0.906	0.916	0.861

We simulate 10 random 12-leaf trees for data representation length of size 1000 and ran the lifting algorithm for varying values of α. We then calculated the following statistics shown in the table: True positive rate (*TPR*), False positive rate (*FPR*), F-score or F_1-score (*F*), Accuracy (*ACC*) table that even when α values vary from 0.1 to 2, the various statistics calculated have stable values. This shows the robustness of the parameters in our method.

Conclusions

We proposed a maximum-likelihood based approach to estimate cell-type trees from histone modification data. We showed that our maximum-likelihood based approach outperforms previous approaches such as distance-based or parsimony-based methods, on H3K4me3 histone modification data. We also proposed the first lifting-based approach to infer internal nodes in cell-type trees and showed the usefulness of this technique in both real and simulated data. The lifting approach is important since in cell differentiation, ancestral cell types can coexist with derived cell types in adult organisms. Our approach is easy to use and is probably the only current approach to build cell-type trees with ancestral inference. Hence we feel that our approach will be an effective way to help biologists and bioinformaticians to study the cell differentiation process. The lifting process we developed may also have many other applications, be in the study of cancer genetic data where normal cells differentiate into cancerous cells, and in other diverse fields like the evolution of languages.

Availability of supporting data

The code for this work can be downloaded from http://lcbb.epfl.ch/software.html.

Competing interests

The authors declare that they have no competing interests.

Authors' contributions

The research problem was defined by: NUN, PB, BMEM. Designed the methods and algorithms: NUN, MS, LH, PG, YL, BMEM. The experimental design was done by: NUN, LH, MS, PB, BMEM. Methods implemented by NUN, LH, PG. Carried out the experiments: NUN, LH. Analyzed the data: NUN, LH, MS, YL, PB, BMEM. NUN, BMEM wrote the paper with help from all the other authors. All authors read and approved the final manuscript.

Acknowledgements

NUN's project was funded by Swiss National Science Foundation, project subside number: 200021_121710/1. The publication costs for this article was funded by École Polytechnique Fédérale de Lausanne (EPFL).

Author details

[1]School of Computer and Communication Sciences, École Polytechnique Fédérale de Lausanne (EPFL), EPFL IC IIF LCBB, INJ 211 (Batiment INJ), Station 14, CH-1015 Lausanne, Switzerland. [2]Computer Science Department, Stanford University, Stanford, USA. [3]Department of Computer Science and Engineering, University of California, San Diego, San Diego, USA. [4]School of Life Sciences, École Polytechnique Fédérale de Lausanne (EPFL), Lausanne, Switzerland. [5]Swiss Institute of Bioinformatics, Lausanne, Switzerland.

Published: 11 January 2016

References

1. Graf T, Enver T. Forcing cells to change lineages. Nature. 2009;462(7273):587–94.
2. Kin K, Nnamani MC, Lynch VJ, Michaelides E, Wagner GP. Cell-type phylogenetics and the origin of endometrial stromal cells. Cell Rep. 2015;10(8):1398–1409.
3. Lee JH, Hart SR, Skalnik DG. Histone deacetylase activity is required for embryonic stem cell differentiation. Genesis. 2004;38(1):32–8.
4. Lister R, Pelizzola M, Kida YS, Hawkins RD, Nery JR, Hon G, et al. Hotspots of aberrant epigenomic reprogramming in human induced pluripotent stem cells. Nature. 2011;471(7336):68–73.
5. Lobe CG. Transcription factors and mammalian development. Curr Top Dev Biol. 1992;27:351–1.
6. Arendt D. The evolution of cell types in animals: emerging principles from molecular studies. Nat Rev Genet. 2008;9(11):868–82.
7. Nair NU, Lin Y, Manasovska A, Antic J, Grnarova P, Sahu AD, et al. Study of cell differentiation by phylogenetic analysis using histone modification data. BMC Bioinforma. 2014;15(1):269.
8. Nelson DL, Lehninger AL, Cox MM. Lehninger Principles of Biochemistry, 5th ed. London, England: Macmillan; 2008.
9. Berger SL. Histone modifications in transcriptional regulation. Curr Opin Genet Dev. 2002;12(2):142–8.
10. Barski A, Cuddapah S, Cui K, Roh TY, Schones DE, Wang Z, et al. High-resolution profiling of histone methylations in the human genome. Cell. 2007;129(4):823–37.
11. Mardis ER. ChIP-seq: welcome to the new frontier. Nat Methods. 2007;4(8):613–3.
12. Martin C, Zhang Y. Mechanisms of epigenetic inheritance. Curr Opin Cell Biol. 2007;19(3):266–72.
13. Nair NU, Lin Y, Bucher P, Moret BM. Phylogenetic Analysis of Cell Types using Histone Modifications. In: Proc. 13th Workshop on Algorithms in Bioinformatics (WABI), Sophia Antipolis, France, in Lecture Notes in Computer Science 8126. Berlin: Springer-Verlag; 2013. p. 326–337.
14. Liang C, Forrest AR, Wagner GP, FANTOM Consortium. The statistical geometry of transcriptome divergence in cell-type evolution and cancer. Nat Commun. 2015;6:1–6.
15. Bryder D, Rossi DJ, Weissman IL. Hematopoietic stem cells: the paradigmatic tissue-specific stem cell. Am J Pathol. 2006;169(2):338–46.
16. Pronk CJ, Rossi DJ, Månsson R, Attema JL, Norddahl GL, Chan CK, et al. Elucidation of the phenotypic, functional, and molecular topography of a myeloerythroid progenitor cell hierarchy. Cell Stem Cell. 2007;1(4):428–42.
17. Villadsen R, Fridriksdottir AJ, Rønnov-Jessen L, Gudjonsson T, Rank F, LaBarge MA, et al. Evidence for a stem cell hierarchy in the adult human breast. J Cell Biol. 2007;177(1):87–101.
18. Stamatakis A. RAxML version 8: a tool for phylogenetic analysis and post-analysis of large phylogenies. Bioinformatics. 2014;30(9):1312–1313.
19. Wang L, Jiang T, Lawler EL. Approximation algorithms for tree alignment with a given phylogeny. Algorithmica. 1996;16(3):302–15.
20. Schliep KP. phangorn: Phylogenetic analysis in R. Bioinformatics. 2011;27(4):592–3.
21. Schwarz G. Estimating the dimension of a model. Ann Stat. 1978;6(2):461–4.
22. ENCODE Project Consortium. A user's guide to the encyclopedia of DNA elements (ENCODE). PLoS Biol. 2011;9(4):1001046.
23. Rosenbloom KR, Sloan CA, Malladi VS, Dreszer TR, Learned K, Kirkup VM, et al. ENCODE Data in the UCSC Genome Browser: year 5 update. Nucleic Acids Res. 2013;41(D1):56–63.
24. Mikkelsen TS, Ku M, Jaffe DB, Issac B, Lieberman E, Giannoukos G, et al. Genome-wide maps of chromatin state in pluripotent and lineage-committed cells. Nature. 2007;448(7153):553–60.
25. Saitou N, Nei M. The neighbor-joining method: a new method for reconstructing phylogenetic trees. Mol Biol Evol. 1987;4(4):406–25.
26. Goloboff PA. Analyzing large data sets in reasonable times: solutions for composite optima. Cladistics. 1999;15(4):415–28.
27. Paradis E, Claude J, Strimmer K. APE: analyses of phylogenetics and evolution in R language. Bioinformatics. 2004;20(2):289–90.
28. Fawcett T. An introduction to ROC analysis. Pattern Recogn Lett. 2006;27(8):861–74.
29. Robinson D, Foulds LR. Comparison of phylogenetic trees. Math Biosci. 1981;53(1):131–47.

Hejase and Liu *BMC Genomics* 2016, **17**(Suppl 1):8
DOI 10.1186/s12864-015-2298-2

BMC Genomics

PROCEEDINGS Open Access

Mapping the genomic architecture of adaptive traits with interspecific introgressive origin: a coalescent-based approach

Hussein A. Hejase and Kevin J. Liu*

From The Fourteenth Asia Pacific Bioinformatics Conference (APBC 2016)
San Francisco, CA, USA. 11 - 13 January 2016

Abstract

Recent studies of eukaryotes including human and Neandertal, mice, and butterflies have highlighted the major role that interspecific introgression has played in adaptive trait evolution. A common question arises in each case: what is the genomic architecture of the introgressed traits? One common approach that can be used to address this question is association mapping, which looks for genotypic markers that have significant statistical association with a trait. It is well understood that sample relatedness can be a confounding factor in association mapping studies if not properly accounted for. Introgression and other evolutionary processes (e.g., incomplete lineage sorting) typically introduce variation among local genealogies, which can also differ from global sample structure measured across all genomic loci. In contrast, state-of-the-art association mapping methods assume fixed sample relatedness across the genome, which can lead to spurious inference. We therefore propose a new association mapping method called Coal-Map, which uses coalescent-based models to capture local genealogical variation alongside global sample structure. Using simulated and empirical data reflecting a range of evolutionary scenarios, we compare the performance of Coal-Map against EIGENSTRAT, a leading association mapping method in terms of its popularity, power, and type I error control. Our empirical data makes use of hundreds of mouse genomes for which adaptive interspecific introgression has recently been described. We found that Coal-Map's performance is comparable or better than EIGENSTRAT in terms of statistical power and false positive rate. Coal-Map's performance advantage was greatest on model conditions that most closely resembled empirically observed scenarios of adaptive introgression. These conditions had: (1) causal SNPs contained in one or a few introgressed genomic loci and (2) varying rates of gene flow – from high rates to very low rates where incomplete lineage sorting dominated as a primary cause of local genealogical variation.

Keywords: Introgression, Gene flow, Incomplete lineage sorting, Association mapping, Population structure, Phylogenomic, Coalescent, Mouse

Background

Adaptive interspecific introgression has played a key role in the evolution of novel traits in many eukaryotic organisms. Examples include hemoglobin concentration in Tibetans as an adaptation to high-altitude environments [1, 2], mimetic butterfly wing patterns [3], and pesticide resistance in house mice [4, 5]. Figure 1 illustrates the effects of adaptive introgression at the genomic sequence level. The phylogeny of three present-day species A, B, and H is shown in Fig. 2. Species A and B diverged from a most recent common ancestor at time t_1. At time t_2, a new hybrid population was formed by hybridization between species A and B. We assume that ploidy is preserved across the phylogeny. In a first generation hybrid offspring, half of the alleles in its genome are expected to come from a parent individual in species A and the other half from a parent individual in species B. A locus in a first-generation hybrid genome has ancestry either

*Correspondence: kjl@msu.edu
Department of Computer Science and Engineering, Michigan State University, 428 S. Shaw Lane, 48824 East Lansing, MI, USA

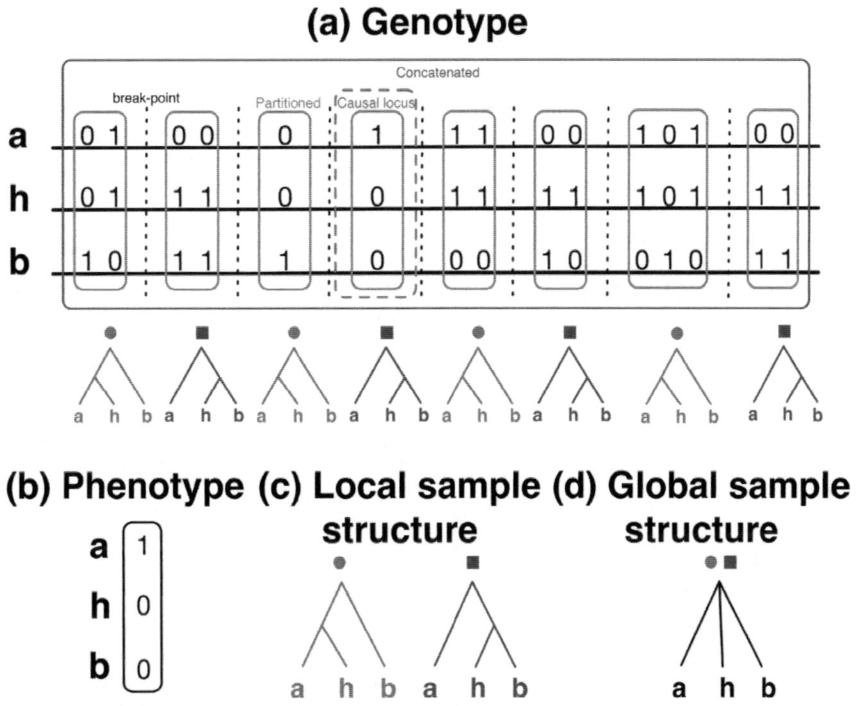

Fig. 1 Local genealogical variation: a sequence level view. The illustration shows an example outcome of evolution under the species phylogeny depicted in Fig. 2, where individual *a* is sampled from population A, individual *b* is sampled from population B, and individual *h* is sampled from population H. **a** A haploid genome sequence is shown for each of the three individuals. Different genealogies are observed for different genomic loci, depending on the specific coalescent history of each locus. Each locus is colored green or blue based on the topology of the genealogy at that locus. The ancestral and derived alleles are represented as 0 and 1, respectively. In our example, the locus marked with a dashed red box contains a causal SNP that contributes to the observed phenotype shown in (**b**). **c** Sample structure (or the evolutionary relationships between samples) in green loci differs from sample structure in blue loci. **d** In our example, global sample structure (i.e., sample structure measured across all sites) takes the form of a star tree. Notice that global sample structure differs from local sample structure in any single locus

from species A or from species B, depending upon the outcome of genetic recombination. We refer to a contiguous genomic subsequence with ancestry from a single parental species as a tract. The evolutionary history of a tract in a first-generation hybrid is represented by either the blue or the green genealogy shown in Fig. 1. Over subsequent generations, back-crossing, recombination, and incomplete lineage sorting of unlinked loci cause fragmentation of genomic tracts present in the first generation of hybrids. The net result is introgression, where genetic material from one of the parental species is incorporated into species H. Natural selection and resulting genetic hitchhiking effects further influence the outcome of introgression. Negative selection fragments and removes introgressed tracts containing maladaptive alleles; on the other hand, positive selection can maintain introgressed tracts containing adaptive alleles over time. Adaptive introgression therefore has two major effects on evolutionary relatedness: (1) local genealogies can vary substantially across different genomic loci, and (2) local genealogical variation resembles a mosaic where tracts in present-day genomes depend upon the outcome of all of these evolutionary processes acting in combination. Longer tracts reflect either more recent hybridization events or adaptive introgression. New computational methodologies have been proposed to perform phylogenomic inference under models that explicitly incorporate the complex interplay of these different evolutionary processes [6–11]. Of particular relevance is PhyloNet-HMM [11], a method that we collaboratively developed. PhyloNet-HMM is the first method capable of scalable phylogenomic inference on three or more genomes under a model that includes these evolutionary processes.

Recently in collaboration with others, we used PhyloNet-HMM to uncover genome-wide signatures of introgression between natural populations of *Mus musculus* and its sister species *Mus spretus* [5]. Dozens of introgressed genomic tracts were more than a megabase in length and inferred to be recent in origin due to the introduction of rodenticide use in and outside of the regions of sympatry between the two species. The longest of these genomic tracts – around 10 Mb in length –

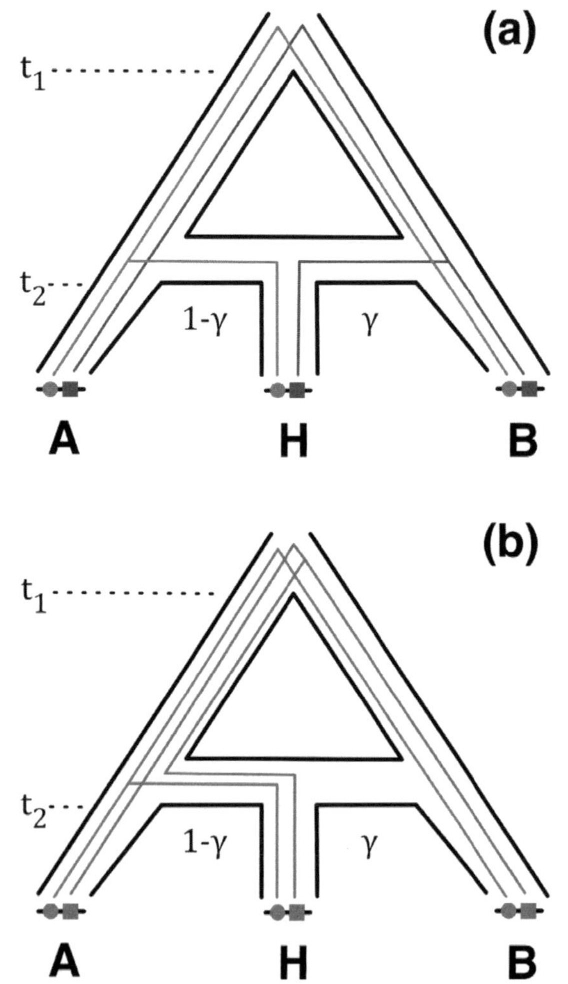

Fig. 2 Local genealogical variation: a species phylogeny view. The illustration shows two different pairs of incongruent local genealogies evolving within a species phylogeny: one pair involving incomplete lineage sorting, and the other involving hybrid origin from two different parental populations. The species phylogeny involves three populations A, B, and H. Populations A and B diverged at time t_1. At time t_2, a hybridization event between the ancestral populations of A and B occurred, giving rise to a hybrid population H. **a** The genealogies of two different loci (green and blue) are shown. A lineage in H originated from the ancestral population of B with probability γ (blue locus) or the ancestral population of A with probability $1 - \gamma$ (green locus). **b** The genealogies of two different loci (green and red) are shown. The H alleles at both loci originated from the ancestral population of A. For the green locus, the H lineage and A lineage coalesce between time t_2 and t_1. For the red locus, tracing backwards in time we find that no coalescence events occur until after time t_1, resulting in ancestral polymorphism and incomplete lineage sorting. Note that local genealogical variation can involve both topological differences (as shown here) and branch length differences

harbored mutations in the *Vkorc1* gene which are known to contribute to resistance and susceptibility to warfarin [4, 12], a widely used anticoagulant rodenticide (Fig. 3). Other introgressed tracts are suspected to harbor other adaptive alleles. The study also uncovered introgressed tracts that were putatively more ancient in origin and which had unknown functional roles. Several major open questions follow from this work: (1) What other introgressed alleles played a causative role in the evolution of the rodenticide resistance trait? (2) What other traits were involved in introgression between the mouse species, and which introgressed alleles contributed to the evolution of these traits?

Genome-wide association (GWA) mapping methods can be used to obtain important clues concerning these questions. GWA methods are widely used throughout the life sciences to investigate the genetic architecture of complex traits, particularly in natural human populations and laboratory strains of model organisms [13]. The goal of GWA mapping is to detect significant statistical associations between genomic markers and a trait of interest. An important consideration is that relatedness between sampled individuals, or sample structure, can induce spurious associations between genotypic and phenotypic characters when not properly accounted for [14]. Intuitively, genotypic and phenotypic characters evolved down a common phylogeny (or evolutionary history), introducing covariance that is distinct from covariance due to a causal relationship between a genotypic marker and trait. Depending on study design, sample structure can encompass multiple levels of relatedness. In the case of GWA studies in humans, sample structure can be due to more distant relationships from population subdivision [15], as well as less distant relationships (e.g., family relationships) [16]. In the case of GWA studies of laboratory organisms such as inbred mouse strains, sample structure can include cryptic relatedness due to their artificial origins, which typically involve a complicated breeding history [17].

The statistical methodologies used in GWA studies can also be used to investigate the genomic architecture of traits involved in adaptive introgression between different species. However, the sample structure is typically more complex than in human populations and artificial laboratory organisms such as classical and wild-derived mouse strains. Two contributing factors are: (1) greater evolutionary divergence involving heterogeneous evolutionary processes such as gene flow, recombination, lineage sorting, and natural selection, and (2) genealogical histories of different loci within genomes can vary significantly not only in terms of branch lengths but also topologies. The latter can cause sample relatedness in one locus to differ from other loci as well as the genome as a whole.

Exactly how to account for sample relatedness remains a subject of major debate. The most widely used GWA methods adopt a range of approaches, broadly categorized

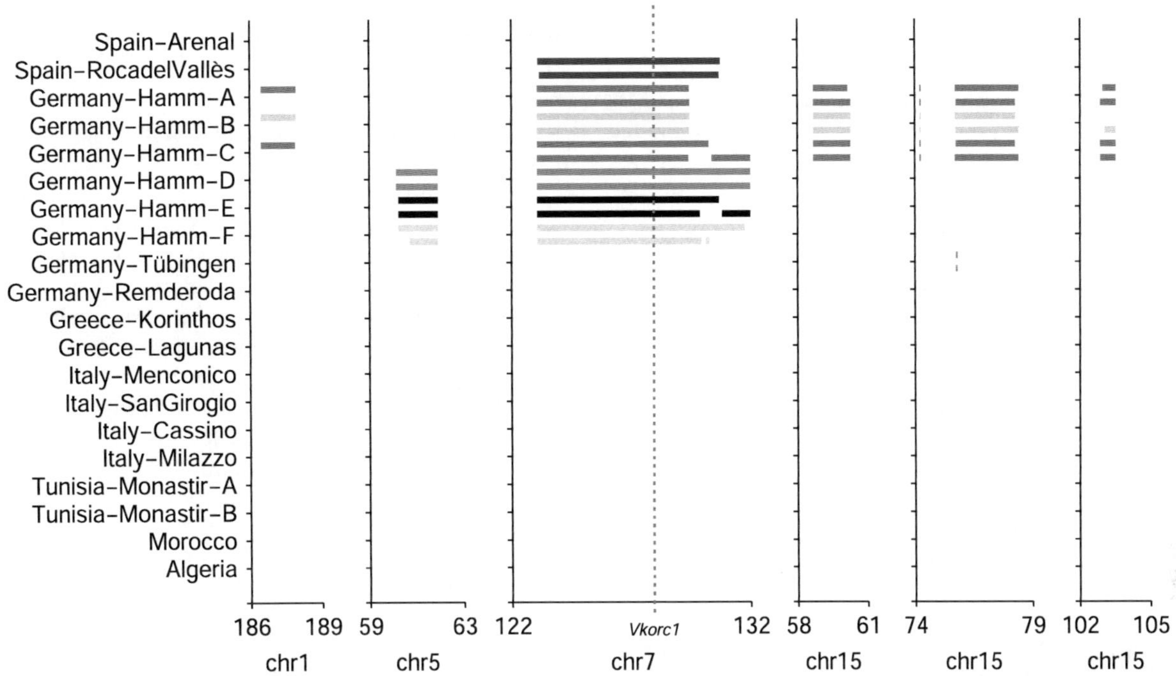

Fig. 3 Genomic tracts in *Mus musculus* samples that were inferred to originate via introgression with *M. spretus*. PhyloNet-HMM [11] was used to infer genomic signatures of interspecific introgression using the procedure described by Liu et al. [5]. Only western European and north African samples are shown. Samples are labeled by country, city/region (if multiple samples originated from the same country), and identifier (if multiple samples originated from the same city/region). The haploid chromosomes are shown for each sample in a separate track and colored a non-white color to denote introgressive origin or white otherwise; each sample is assigned a different color for visibility. The chart shows a subset of the introgressed genomic regions that were inferred to originate due to the recent selective sweep associated with anticoagulant pesticide use. The location of the *Vkorc1* gene is highlighted, which is known to contribute to anticoagulant resistance in both mice and humans [4, 12]. Panel adapted from [5]

in terms of their complexity. On one end of the spectrum, a genomic control approach involves computing an inflation factor based on the degree of sample relatedness, which is then used to correct association statistics [18]. On the other end of the spectrum are more highly parameterized fixed-effect models and mixed models where population structure is modeled using either fixed effects or random effects, respectively (reviewed in [19]). Methods based on fixed-effect models and mixed models represent the current state of the art in terms of computational efficiency and accuracy [20, 21]. Among the most accurate and efficient of these methods are EIGENSTRAT [22], EMMA [17] and its successor method EMMAX [21], and GEMMA [23]. Notably, these methods were at the center of a recent debate on exactly how to model sample structure [19, 24, 25]. Only a few existing methods explicitly adopt an evolutionary model in the form of a phylogeny. These include EMMA when used with a kinship matrix based on a Brownian motion model of phenotypic evolution on a phylogenetic tree [17] and PHYLOSTRAT [20], which uses a regression-based approach with a model that includes the bipartitions of a phylogenetic tree computed from genotypic data. Significantly, none of these methods utilize models that model variation in sample relatedness across genomes nor explicitly capture non-tree-like evolution, as in the case of species whose evolution involved adaptive introgression.

More recently, GWA studies have begun to examine admixed human populations (reviewed by [26]), introducing sample structure with greater complexity due to the non-tree-like evolutionary histories resulting from admixture. Shriner et al. [27] and Pasaniuc et al. [28] introduced BMIX and MIXSCORE, respectively, which are tests to simultaneously detect local signatures of admixture and genotypic/phenotypic association. Of the two, BMIX was shown to offer more statistical power and type I error control [27]. Shriner et al. utilized BMIX to perform association mapping on African-American populations and found several new markers that were significantly associated with fasting plasma glucose, coronary heart disease, type 2 diabetes, and breast cancer. Two aspects of the methodological design of BMIX are most relevant to our work. First, a key step in BMIX is local ancestry inference using LAMPANC [29], where each allele at a locus in an admixed genome is inferred to originate from one of two parental populations (CEU and YRI populations in the HapMap study [30]). Importantly, the inference makes use of the assumption that the parental populations are unrelated. A simulation study was performed to evaluate statistical power, which similarly used a model with unrelated parental populations. Second, an intermediate stage of the BMIX algorithm consists of stratified regression of a genotypic character corresponding to a test marker with a phenotypic character. The stratified regression technique adjusts for both local sample structure at the locus as well as global sample structure (i.e., sample structure measured across the entire genomic sequence). Our study proposes an alternative method for association mapping in the context of local variation in sample structure. We note that any such method can be used in place of their stratified regression technique (steps 3 through 5 in the algorithm shown in [27]). Other association mapping methods have been shown to be more accurate than stratified regression. These include the fixed effects model methods and mixed model methods that we consider in our study.

As suggested by [17], the true genealogical histories of the genomic loci in a GWA study could theoretically be used to construct a concise and interpretable model of complex sample structure. The genealogical histories would be especially useful where local genealogies exhibit both topological and branch length variation as is often the case in genomes with adaptively introgressed origins. In practice, local genealogical histories must be inferred. Admixture mapping methods such as LAMPANC address the specific case of inferring local genealogical variation due to genetic admixture (reviewed by [26]). In the context of interspecific introgression involving eukaryotic genomes, additional evolutionary processes have first-order effects on local genealogical variation, including incomplete lineage sorting [31]. New methods based upon the coalescent model [32] and its extensions enable efficient and accurate inference of species phylogenies and local genealogies under models that account for all of the above evolutionary processes [6, 11]. Evolutionary models provide an ideal means to understand the impact of complex sample structure on state-of-the-art association mapping methods. These evolutionary models can be utilized to account for complex sample structure in an association mapping study. We therefore introduce Coal-Map, a new association mapping method. Coal-Map is a computational pipeline which adopts an evolutionary perspective by modeling local genealogical histories in a mixed model association mapping framework. We study the performance of Coal-Map using both synthetic data and empirical data sampled from natural populations of house mouse. Coal-Map improves upon the state-of-the-art in terms of both statistical power and false positive rate.

Results

Performance study using simulated genotypic and phenotypic data

The simulation study included model conditions with adaptive gene flow. A range of genetic architectures were simulated, where one, two, or all loci contained causal markers. Figure 4 compares the performance of Coal-Map and EIGENSTRAT using receiver operating characteristic (ROC) curves on model conditions with the highest level of gene flow ($\gamma = 0.5$). Coal-Map offered better power than EIGENSTRAT across the different model conditions. The performance improvement was significant in terms of area under ROC curve (AUROC) for all model conditions (Delong et al. test [33], $\alpha = 0.05$), with P values of $2 * 10^{-14}$, $5 * 10^{-5}$, and .006 for the single-causal-locus, two-causal-loci, and all-causal-loci model conditions, respectively. As measured by AUROC, Coal-Map's performance advantage over EIGENSTRAT was largest on the single-causal-locus model condition (0.947 versus 0.889, respectively) and smaller as more loci contributed causal SNPs. This can also be seen based on each method's power at typical false positive rates. At a false positive rate (FPR) of 0.05, Coal-Map's TPR improved upon EIGENSTRAT's by 0.152 and 0.054 on the single-causal-locus and two-causal-loci model conditions, respectively; on the all-causal-loci model condition, the TPR difference between the two methods was less than 0.039.

Coal-Map's performance advantage over EIGENSTRAT was similarly observed in model conditions that involved a range of hybridization frequencies from $\gamma = 0.01$ to 0.5 and intra-locus linkage that emulated selective sweep effects (but did not directly incorporate positive selection). The synthetic traits incorporated genetic contributions from one, two, or all loci. In Fig. 5, the performance of Coal-Map and EIGENSTRAT on model conditions with the highest level of gene flow ($\gamma = 0.5$) is shown using ROC curves. Across the different trait architectures (i.e., causal SNPs drawn from one, two, or all loci in the single-causal-locus, two-causal-loci, and all-causal-loci model conditions, respectively), Coal-Map offered comparable or better power than EIGENSTRAT for a given false positive rate. The performance improvement was significant in terms of area under ROC curve (AUROC) for the single-causal-locus and two-causal-loci model conditions but not for the all-causal-loci model conditions (Delong et al. test [33], $\alpha = 0.05$). Coal-Map's performance advantage over EIGENSTRAT was largest on the single-causal-locus model condition: the methods had AUROC values of 0.938 versus 0.870, respectively, and Coal-Map's TPR improved upon EIGENSTRAT by 0.111 at an FPR of 0.05. As more loci contributed causal SNPs, Coal-Map's performance advantage was smaller. Overall, Coal-Map reported smaller p-values at causal SNPs compared to EIGENSTRAT (Fig. 6).

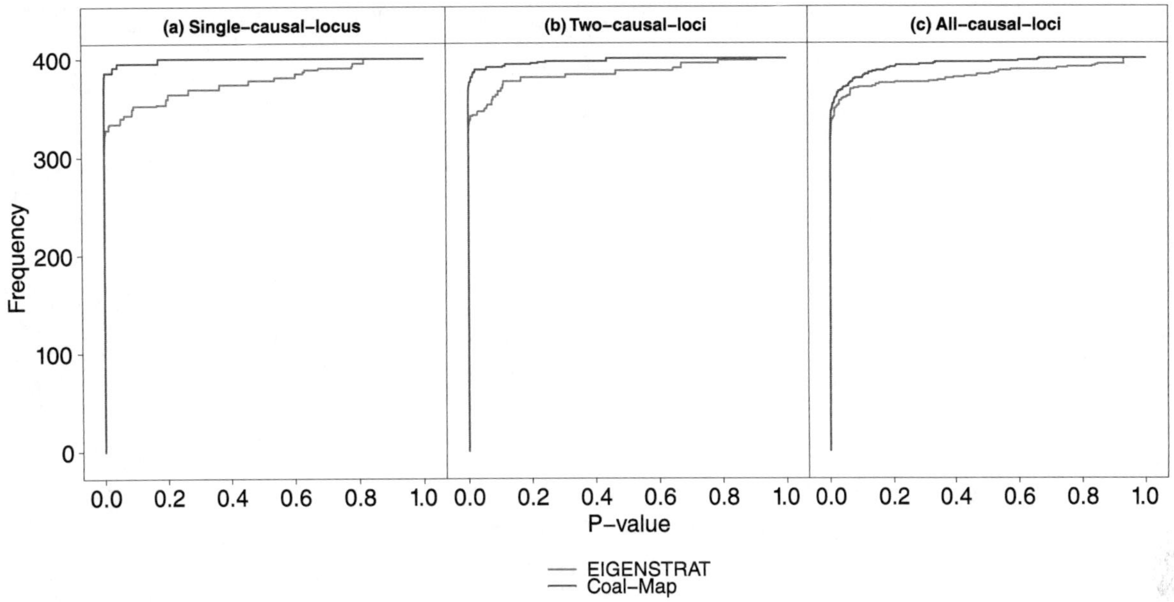

Fig. 4 Compared to EIGENSTRAT, Coal-Map has comparable or better power and false positive rate on model conditions involving adaptive gene flow (hybridization frequency $\gamma = 0.5$ and selection coefficient $s = 0.56$). True positive rate and false positive rate are shown for both methods using receiver operating characteristic (ROC) curves. **a** Results are shown for the model condition where causal SNPs are drawn from a single locus. Coal-Map and EIGENSTRAT have AUROC of 0.947 and 0.889, respectively. **b** Results are shown for the two-causal-loci model condition. Coal-Map has an AUROC of 0.897 and EIGENSTRAT has an AUROC of 0.859. **c** Results are shown for the all-causal-loci model condition, where Coal-Map and EIGENSTRAT have AUROC of 0.845 and 0.816, respectively

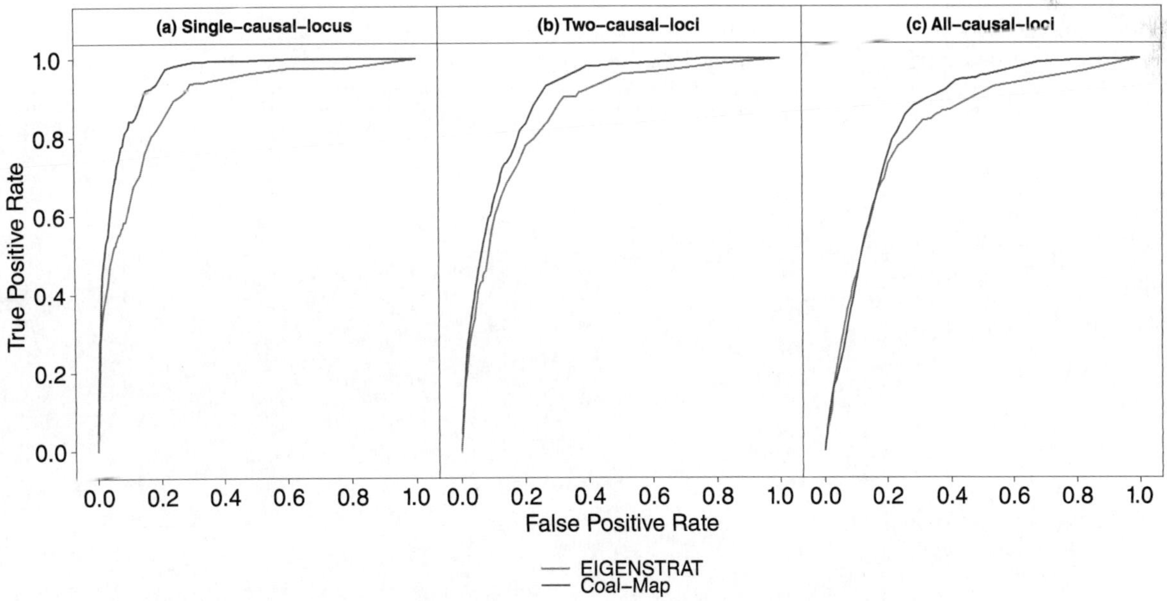

Fig. 5 Compared to EIGENSTRAT, Coal-Map has comparable or better power and false positive rate on model conditions involving neutral gene flow (hybridization frequency $\gamma = 0.5$). True positive rate and false positive rate are shown for both methods using receiver operating characteristic (ROC) curves. **a** Results are shown for the model condition where causal SNPs are drawn from a single locus. Coal-Map and EIGENSTRAT have area-under-ROC-curve (AUROC) of 0.938 and 0.870, respectively. **b** Results are shown for the two-causal-loci model condition. Coal-Map has an AUROC of 0.898 and EIGENSTRAT has an AUROC of 0.860. **c** Results are shown for the all-causal-loci model condition, where Coal-Map and EIGENSTRAT have AUROC of 0.837 and 0.827, respectively

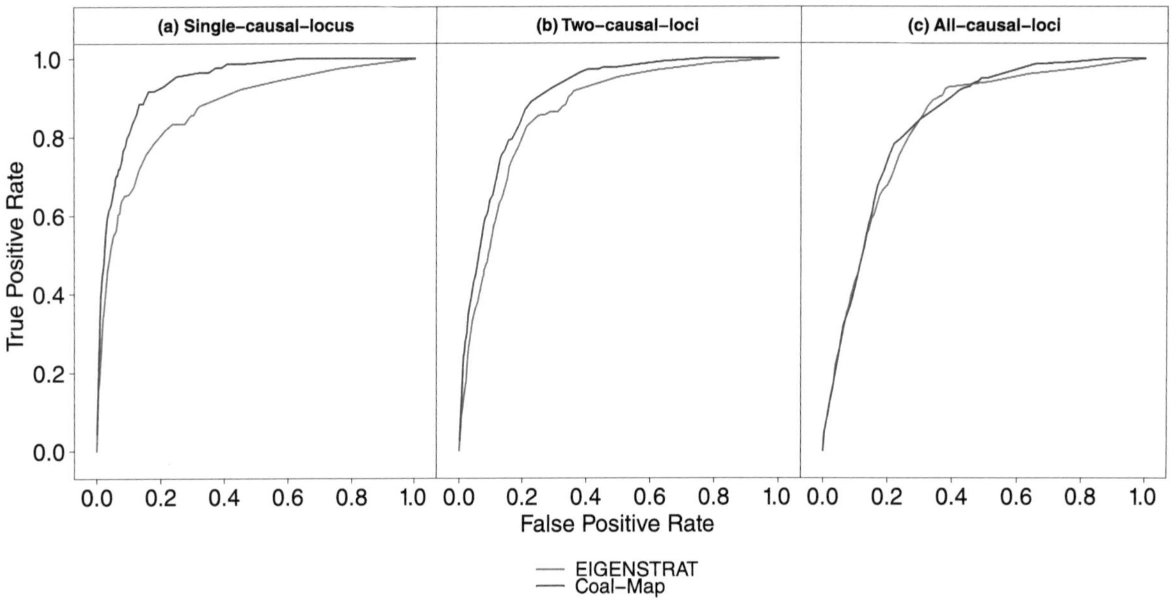

Fig. 6 The cumulative histogram of *p*-values reported by Coal-Map and EIGENSTRAT at causal SNPs is shown on model conditions involving neutral gene flow (hybridization frequency $\gamma = 0.5$). Results are shown for the **a** single-causal-locus, **b** two-causal-loci, and **c** all-causal-loci model conditions, respectively. Cumulative frequency is reported over all replicates from a model condition

As we examined single-causal-locus and two-causal-loci model conditions with smaller levels of gene flow, we observed that Coal-Map's performance advantage over EIGENSTRAT did not diminish and in fact remained roughly the same. As shown in Table 1, the performance improvement of Coal-Map over EIGENSTRAT remained significant as we decreased the gene flow parameter γ from 0.5 to 0.01 (Delong et al. test [33], $\alpha = 0.05$). Thus, on the model condition with negligible gene flow (i.e., $\gamma = 0.01$), virtually all local genealogical variation was due to

Table 1 The performance of Coal-Map and EIGENSTRAT based on area under receiver operating characteristic curve (AUROC) is compared across model conditions involving neutral evolution with incomplete lineage sorting and a wide range of gene flow

	Single-causal-locus		
Hybridization frequency γ	Coal-Map	EIGENSTRAT	Corrected q value
0.5	0.938	0.870	$< 10^{-5}$
0.25	0.935	0.882	$< 10^{-5}$
0.1	0.928	0.890	$< 10^{-5}$
0.01	0.917	0.845	$< 10^{-5}$
		Two-causal-loci	
0.5	0.898	0.860	$< 10^{-5}$
0.25	0.911	0.860	$< 10^{-5}$
0.1	0.881	0.843	$< 10^{-5}$
0.01	0.879	0.834	$< 10^{-5}$
		All-causal-loci	
0.5	0.836	0.826	0.16
0.25	0.842	0.808	0.001
0.1	0.854	0.842	0.093
0.01	0.847	0.817	0.002

On single-causal-locus and two-causal-loci model conditions, Coal-Map has AUROC that is significantly better than EIGENSTRAT (Delong et al. test [33] with Benjamini-Hochberg correction [65]; setwise $\alpha = 0.05$; $n = 20$ for each test) across different hybridization frequencies ranging from a relatively large level of gene flow ($\gamma = 0.5$) to negligible amounts of gene flow ($\gamma = 0.01$). On all-causal-loci model conditions, Coal-Map had a diminished performance advantage in terms of AUROC, and the improvement was either weakly significant or not significant (under the same test) incomplete lineage sorting. At an FPR of 0.05, Coal-Map's TPR improved upon EIGENSTRAT by 0.211, 0.171, and 0.242 on single-causal-locus model conditions with γ settings of 0.25, 0.1, and 0.01, respectively; on two-causal-loci model conditions, the corresponding TPR improvements were 0.036, 0.032, and 0.094, respectively. The performance of the two methods on the all-causal-loci model conditions was similar regardless of the amount of gene flow, as measured by either AUROC or TPR at a typical FPR level.

We explored the sensitivity of Coal-Map to the number of covariates (five and twenty) used to represent the sample structure. We found that Coal-Map's performance was robust to the number of covariates used to represent sample structure (Additional file 1: Figure S1 and S2). We also explored a trait model with only a genotypic component (i.e., lacking a random effect due to environment). We found that Coal-Map's performance advantage over EIGENSTRAT was greater than on the model conditions that included both genotypic and environmental effects (Additional file 1: Figure S4). Finally, we observed that modeling local sample structure alone resulted in a marked decrease in performance (Additional file 1: Figure S3). The resulting power and false positive rates were worse than EIGENSTRAT.

Performance study using empirical mouse genomes and simulated phenotypic data

Results from the performance study using empirical mouse genomes were consistent with the simulation study. Coal-Map's AUROC was significantly better than EIGENSTRAT on chromosomes 7 and 17 (Table 2). (Delong et al. test [33], $\alpha = 0.05$). At an FPR of 0.05, Coal-Map's TPR improvement over EIGENSTRAT was 0.076 and 0.166 for single-causal-locus traits, respectively, and 0.030 and 0.057 for two-causal-loci traits, respectively (Figs. 7 and 8). At causal SNPs in chromosomes 7 and 17, Coal-Map reported smaller p-values overall compared to EIGENSTRAT (Figs. 9 and 10, respectively). Compared to the performance study using synthetic genomes, the cumulative histograms were shifted down for both methods (i.e., fewer causal SNPs were inferred at each cutoff). We note that chromosomes 7 and 17 exhibited the greatest amount of introgression in our study (Additional file 1: Table S2). In contrast, chromosome 15 had the fewest number of introgressed sites in our study, with total length that was 68 % and 23 % smaller than those in chromosome 7 and 17, respectively; on this chromosome, the AUROC improvement of Coal-Map over EIGENSTRAT was weakly significant when mapping single-causal-locus traits and not significant for two-causal-loci traits (Table 2; Additional file 1: Figures S6 and S7).

Discussion

Our performance study utilized empirical and simulated data reflecting a wide range of evolutionary scenarios. The simulation conditions were based upon empirical studies of adaptive interspecific introgression. We consistently observed that Coal-Map had comparable or improved performance compared to EIGENSTRAT, a leading association mapping method in terms of its popularity, power, and type I error control.

One key factor that impacts Coal-Map's performance is the genetic architecture of the trait under study. Coal-Map's relative performance improvement is greatest on datasets with one or a few loci that contribute causal SNPs – which we refer to as causal loci – and is less on datasets with many causal loci. We hypothesize that the amount of local genealogical incongruence between causal loci is the main determining factor, not necessarily the number of causal loci. As incongruence becomes greater, the sample covariance contributed by any individual causal locus's local sample structure will be diminished, and global sample structure (as measured across all sites in a dataset) will predominate. We note that, in prior genomic studies of adaptive introgression [3, 5], introgressed loci were observed to have similar sample structure (i.e., similar distributions of local genealogies). Introgression of alleles that were causal for an adaptive trait and

Table 2 The performance of Coal-Map and EIGENSTRAT based on area under receiver operating characteristic curve (AUROC) is compared using empirical mouse chromosomes and simulated traits

Chromosome	Single-causal-locus			Two-causal-loci		
	Coal-Map	EIGENSTRAT	Corrected q value	Coal-Map	EIGENSTRAT	Corrected q value
7	0.964	0.928	$< 10^{-5}$	0.942	0.923	0.003
15	0.940	0.922	0.014	0.917	0.919	0.587
17	0.968	0.914	$< 10^{-5}$	0.942	0.904	$1.6 * 10^{-5}$

On the two mouse chromosomes with the greatest number of introgressed sites in our study - chromosomes 7 and 17 - Coal-Map's performance was significantly better than EIGENSTRAT for both single-causal-locus and two-causal-loci traits (Delong et al. test [33] with Benjamini-Hochberg correction [65]; setwise $\alpha = 0.05$; $n = 20$ for each test). We observed a reduced performance improvement on chromosome 15, which had relatively fewer introgressed sites: the improvement was weakly significant for single-causal-locus traits and not significant for two-causal-loci traits (using the same test)

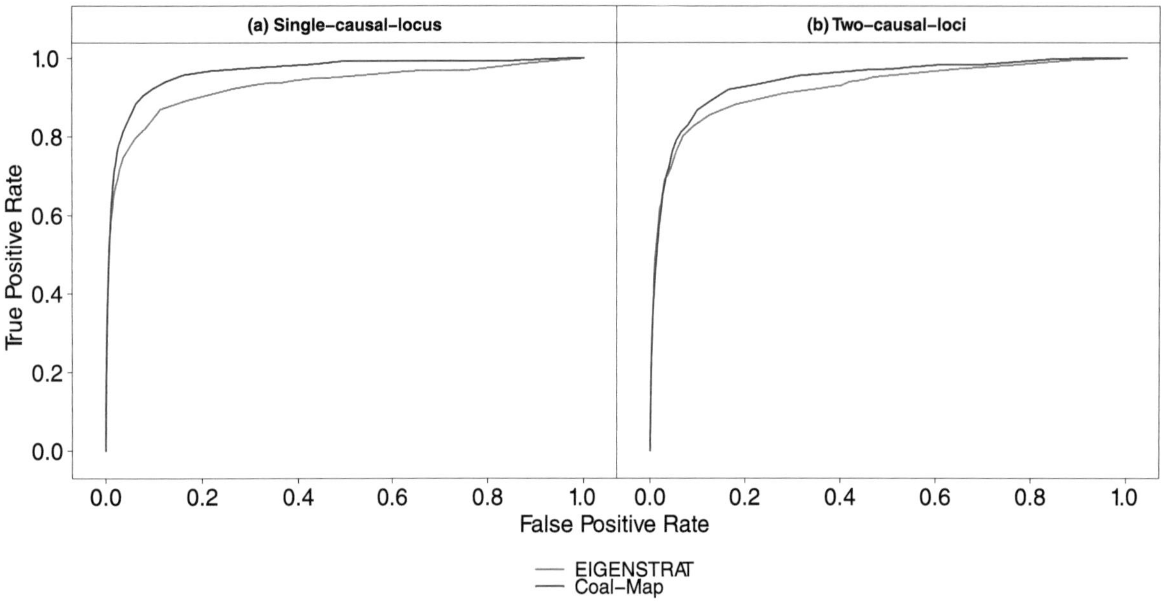

Fig. 7 In the performance study utilizing genomic data from mouse chromosome 7, Coal-Map has similar or typically better power and type I error control compared to EIGENSTRAT. Figure layout and description are otherwise identical to Fig. 4. For the single-causal-locus model condition, Coal-Map and EIGENSTRAT have AUROC of 0.965 and 0.929, respectively; for the two-causal-loci model condition, Coal-Map and EIGENSTRAT have AUROC of 0.942 and 0.923, respectively

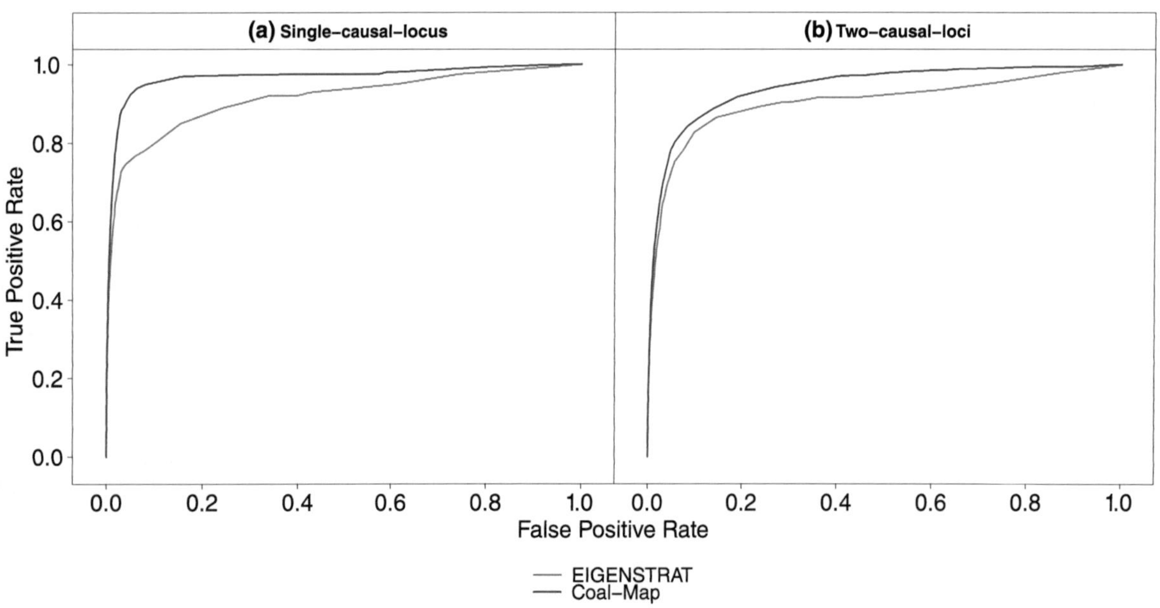

Fig. 8 In the performance study utilizing genomic data from mouse chromosome 17, Coal-Map has similar or typically better power and type I error control compared to EIGENSTRAT. Figure layout and description are otherwise identical to Fig. 7. For the single-causal-locus model condition, Coal-Map and EIGENSTRAT have AUROC of 0.968 and 0.914, respectively; for the two-causal-loci model condition, Coal-Map and EIGENSTRAT have AUROC of 0.943 and 0.904, respectively

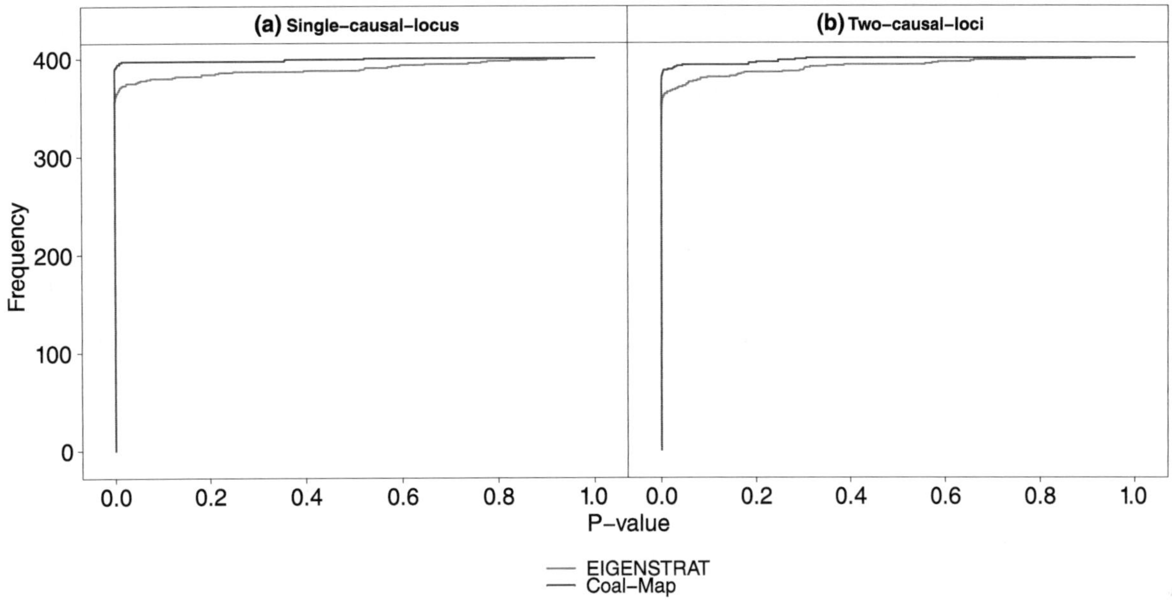

Fig. 9 The cumulative histogram of *p*-values reported by Coal-Map and EIGENSTRAT at causal SNPs is shown for the performance study utilizing genomic data from mouse chromosome 7

genetic hitchhiking of neighboring loci were hypothesized to enhance genealogical congruence between loci with common introgressive origin. An empirical example is shown in Fig. 3. We therefore anticipate that, relative to other model conditions, the single-causal-locus and two-causal-loci model conditions in our study may be most relevant to the empirical study of adaptive introgression.

In our simulation study, we found that Coal-Map's performance advantage was retained across a wide range of gene flow – even on model conditions which had virtually no gene flow. In the latter case, we attribute Coal-Map's

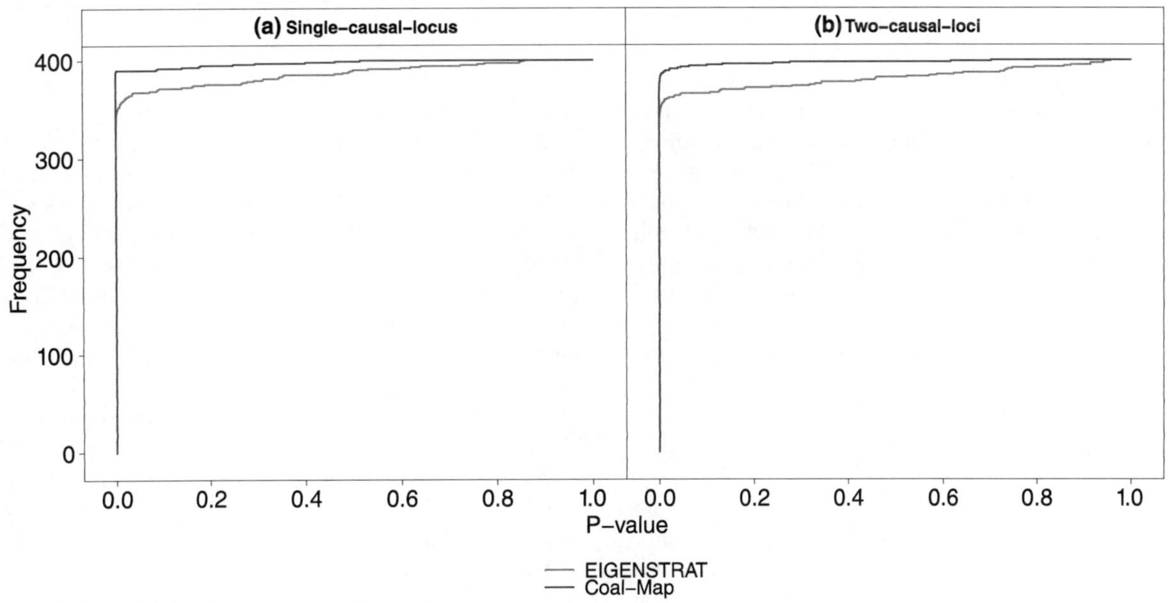

Fig. 10 The cumulative histogram of *p*-values reported by Coal-Map and EIGENSTRAT at causal SNPs is shown for the performance study utilizing genomic data from mouse chromosome 17. Figure layout and description are otherwise identical to Fig. 9

improvement to modeling the variation in local sample structure contributed by incomplete lineage sorting. The breakpoint inference stage of Coal-Map's methodological pipeline accounts for local genealogical variation due to gene flow as well as incomplete lineage sorting and other evolutionary processes. We note that the presence of incomplete lineage sorting in our performance study is unique compared to past performance studies of association mapping methods on admixed populations (e.g., the study of Shriner et al. [34], which assumed that the parental populations contributing to an admixed population were completely unrelated). Compared to the synthetic genomes, the empirical genomes contained local genealogical variation due to an even wider array of evolutionary processes, including recombination. Crucially, the breakpoint inference stage of Coal-Map's pipeline made use of PhyloNet-HMM, a probabilistic inference method that uses the coalescent model, phylogenetic networks, and hidden Markov models to account for all of these evolutionary processes acting in combination. Consistent with the simulation study, Coal-Map's performance was comparable or better than EIGENSTRAT.

The choice of the number of covariates used to represent sample structure in Coal-Map was based upon a previous algorithmic design study examining the use of fixed effects models for association mapping [22]. To further explore the ramifications of this choice, we conducted an algorithmic design experiment to explore the impact of the number of covariates used in Coal-Map's model upon its performance. We found that Coal-Map's performance was robust to this design choice.

Coal-Map's performance advantage over EIGENSTRAT was retained across different levels of environmental contribution to traits. A larger performance improvement was seen on model conditions with only a genotypic contribution to traits, which we ascribe to the lack of sample structure inherent in the additive environmental noise.

We observed that a model that only accounts for local sample structure resulted in reduced power and higher false positive rate compared to Coal-Map, which accounts for both local and global sample structure. Our finding is consistent with the findings of Shriner et al. [34]. The intuitive explanation is that local sample structure in the current partition (i.e., the partition enclosing the test SNP) should be modeled when the current partition contains causal SNPs, but otherwise not.

Conclusions

Adaptive introgression involves the complex interplay of a variety of evolutionary processes including gene flow, directional selection, recombination, lineage sorting processes that may result in incomplete lineage sorting, and sequence mutation. The need to distinguish between these differing evolutionary forces in population genomic and comparative genomic studies was emphasized by two recent reviews [31, 35], and phylogenomic inference methods are actively being developed to study the interplay of these evolutionary processes [6, 11]. One of the genomic signatures of adaptive introgression is local genealogical variation featuring introgressed genomic tracts as long as dozens of megabases. These tracts contain introgressed loci of two types: adaptive loci and nearby linked neutral loci. Sample structure at these loci can differ greatly from global sample structure (i.e., sample structure measured across all sites). Traditional approaches to association mapping account only for the latter, assuming that sample structure is mostly invariant across the genome.

We therefore introduced Coal-Map, a new association mapping method which explicitly models both local sample structure, such as arises in a genomic region containing tracts of common introgressive origin, and global sample structure. Coal-Map is a methodological pipeline that incorporates recent theoretical innovations that bridge population-level evolution under the coalescent with traditional phylogenetic models of biomolecular sequence evolution [36, 37].

We validated the performance of Coal-Map using synthetic and empirical data. The data sets in our study featured local genealogical variation due to gene flow as well as incomplete lineage sorting, sequence mutation, and (in the case of the empirical mouse genomes) recombination. We compared the performance of Coal-Map to EIGENSTRAT, a leading association mapping method. We consistently observed the same outcome across all of the datasets in our study: Coal-Map's performance in terms of power and false positive rate was comparable or better than EIGENSTRAT in all cases. Thus, Coal-Map can be generally used in place of EIGENSTRAT and related association mapping approaches that do not explicitly model local variation in evolutionary relatedness. We found that the conditions under which Coal-Map's performance was strictly better than EIGENSTRAT were those that most closely resembled empirical cases of adaptive interspecific introgression, involving: (1) traits with causal SNPs drawn from one or a few genomic loci with common introgressive origin, and (2) a range of gene flow levels. Perhaps surprisingly, Coal-Map strictly outperformed EIGENSTRAT even on model conditions where gene flow was nearly absent but incomplete lineage sorting was still a factor. We hypothesize that modeling the fine variation in sample structure due to incomplete lineage sorting may be useful to association mapping approaches outside of the multi-species evolutionary context considered in this study. These alternative contexts include traditional genome-wide association studies of human populations or inbred mouse strains.

The pipeline-based design of Coal-Map is flexible. As noted above, richer coalescent-based modeling of the evolutionary origins of local genealogical variation may permit more accurate breakpoint inference in the first stage of Coal-Map.

We explored the use of forward selection for our model selection approach, which was used to determine if local sample structure containing the test SNP was incorporated into the model or not. However, we found that forward selection is conservatively biased towards models with fewer parameters. As an alternative, we utilized an approach involving two nested models – one incorporating local sample structure and the other not – which uses a likelihood ratio test of each model against a null model (positing that the test SNP has no effect), and then selects the model with the smallest test statistic. We note that, using this approach, the resulting association score is at least as significant as a likelihood ratio test using one of the two nested models by itself. Thus, the number of positives (and therefore the number of false positives) reported using the approach can be no smaller than the number reported using one of the two nested models, and the number of false negatives can be no larger. In practice, our results suggest that the approach yields a comparable false positive rate and a substantially improved true positive rate compared to forward selection and other model selection techniques such as information criteria [38–41].

Recently, Shriner et al. and Pasaniuc et al. proposed BMIX and MIXSCORE, two methodologies for genome-wide association studies of admixed populations [27, 28]. These methodologies combine admixture mapping – which makes use of ancestry effects of admixed loci – with traditional association mapping approaches – which makes use of genotypic effects without taking admixed ancestry into account – to yield improved mapping power and type I error control. Our study is orthogonal to their studies for the following reasons. First, both approaches assume that admixed populations arise due to admixture between two completely unrelated parental populations. In general, admixture mapping approaches make use of this assumption [42, 43]. This simplifies local ancestry inference since coalescence between lineages in the "pure" parental populations is not considered and therefore incomplete sorting of these lineages is not an issue, which could otherwise confound ancestry inference [31]. In reality, any two samples share a most recent common ancestor. Not accounting for coalescence between lineages can have first-order effects upon association mapping accuracy, which we demonstrate in our performance study. Notably, our study involves evolutionary divergence greater than those seen within a single species. In a traditional association mapping study of human populations or artificial lab strains of model organisms, the population split times are smaller than the divergence times in our study and thus incomplete lineage sorting should be more likely and have an even greater effect (assuming that effective population sizes are comparable or larger). Coal-Map relaxes the simplifying assumption of pure isolation using a coalescent-based approach. A recent review has highlighted the multispecies coalescent model as an ideal means to account for evolutionary relatedness at multiple scales in functional genomic studies [44]. Second, both approaches represent local sample structure differently from Coal-Map. Both BMIX and MIXSCORE assume that population strata are discrete due to pure isolation between parental populations. In contrast, we utilize continuous phylogenetic covariates to more flexibly represent evolutionary relatedness between samples. Third, BMIX utilizes stratified regression to perform association mapping (and recall that BMIX has been shown to yield more power and comparable false positive rate compared to MIXSCORE). In contrast, mixed model approaches to association mapping have been shown to out-perform stratified regression [17]. Coal-Map makes use of the latter approaches for this reason. Fourth, a primary contribution of the two studies is the insight that combining admixture mapping with association mapping out-performs association mapping by itself. The same insight can be applied to Coal-Map, yielding a combined approach that tests for introgressed ancestry effects (without regard to specific allele effects) in combination with genotypic effects. We hypothesize that a combined approach will yield further performance improvements compared to Coal-Map, similar to BMIX and stratified regression. Fifth, the problem domains differ greatly due to the disparate evolutionary divergences involved in the studies. Compared to intraspecific admixture, adaptive interspecific introgression involves a different set of evolutionary processes with first-order effects. While the study of adaptive traits of interspecific introgressive origin and their genomic architecture is of independent interest, the two settings may be complementary for studying certain biological questions. For example, some of the introgressed loci that contribute to warfarin resistance in mice have orthologs in human populations with related cardiovascular roles. Compared to human populations, natural mouse populations have undergone much stronger recent positive selection due to the introduction of rodenticide, which can be advantageous for association mapping purposes (e.g., stronger hitchhiking effects resulting in megabases-long introgressed tract lengths).

We conclude with discussion of future research directions. To our knowledge, our study is the first to explore association mapping performance using many hundreds of genomes with adaptive interspecific introgressive origins. Compared to human genome sequence data in public databases, a relative paucity of genomic sequence data is available to explore this question. Genomic sequence data is available for less than a thousand samples from natural populations of house mice [45–47], and even less phenotypic data is available for introgressed mouse populations. The situation is similar in studies of other organisms. For example, in the studies of Counterman et al. [48] and Baxter et al. [49], approximately a hundred samples were used to examine adaptive introgression between butterfly species and its role in wing pattern mimicry. We anticipate that recent biotechnological advances will remove this limitation in the near term, especially as researchers continue to investigate adaptive interspecific introgression in eukaryotes. We also plan on exploring several aspects of Coal-Map's algorithmic design in future work. First, other model selection strategies can be used in a Coal-Map analysis (as noted above). In particular, cross-validation has been shown to perform well for model selection problems in phylogenomics [50]. This approach requires a relatively greater amount of data compared to the strategies explored in this study. Second, following MultiBLUP and Adaptive MultiBLUP's approaches to a different computational problem [44], search techniques can be used to find multiple causal loci. We anticipate two main challenges with the use of more complex models as part of a search-based approach: (1) model selection strategies will be necessary to distinguish between models that differ in the number of parameters needed to capture genetic effects from multiple causal loci, and (2) efficient techniques will be needed to learn a potentially large number of model parameters. Third, global and local sample structure could be modeled using random effects instead of fixed effects. Software libraries for generalized linear models with multiple random effects are available for this purpose [24, 44]. Fourth, a combined approach that simultaneously performs introgression breakpoint inference, local ancestry mapping, and association mapping that accounts for local sample structure may offer additional performance improvements beyond those observed in our study. Finally, Zuk et al. [51] note that one important research direction is relaxing the assumption of additive genotypic contribution to complex traits. We share their opinion that the future of association mapping will involve improved modeling of epistasis and complex traits with polygenic architectures. A promising future direction is to model multiple effects at multiple scales of systems biology [52], such as examining dependence between genotypic characters, interactomic graphs, and phenotypic characters.

Methods

Coal-Map

The input to Coal-Map consists of: (1) an n by k multiple sequence alignment $\boldsymbol{X}$ containing genotypic data for n aligned sequences and k sites, (2) local partition breakpoints $\boldsymbol{b}$ in ascending order (including trivial breakpoints corresponding to sites 1 and k), and (3) a phenotypic vector $\boldsymbol{y}$ with n observations. The ith row in alignment $\boldsymbol{X}$ and the ith entry in the phenotypic vector $\boldsymbol{y}$ correspond to the genomic sequence and phenotypic value of taxon s_i for $1 \leq i \leq n$. A local partition $\boldsymbol{X_\ell}$ is the alignment consisting of all sites in alignment $\boldsymbol{X}$ contained in the closed interval with endpoints equal to the breakpoints b_ℓ and $b_{\ell+1}$. Each local partition $\boldsymbol{X_\ell}$ represents a contiguous genomic region where introgressed tracts in the region had a common introgressive origin (i.e., originated from lineages that evolved within the same network edges in the species phylogeny $\mathcal{N}$), as inferred by either Phylonet-HMM [11] in the general case or a simpler parsimony-based alternative when certain assumptions were satisfied (see below). The output of Coal-Map consists of a score p_j that measures the statistical association between the genotypic character $\boldsymbol{x_j}$ (i.e., the jth column in X) and phenotypic vector $\boldsymbol{y}$ for $1 \leq j \leq k$.

The association score p_j for a SNP $\boldsymbol{x_j}$ is calculated under the following linear mixed model (following the notation of [23]):

$$\boldsymbol{y} = \boldsymbol{W_j}\boldsymbol{\alpha} + \boldsymbol{x_j}\beta + \boldsymbol{\epsilon}$$
$$\boldsymbol{\epsilon} \sim MVN_n(0, \tau^{-1}\boldsymbol{I_n})$$

The test SNP $\boldsymbol{x_j}$ has effect size β. The fixed effects $\boldsymbol{W_j}$ includes covariates that account for global sample structure (i.e., sample structure measured across all sites in alignment $\boldsymbol{X}$). Additional covariates that account for local sample structure (i.e., sample structure inferred within the local partition $\boldsymbol{X_\ell}$ containing the test SNP $\boldsymbol{x_j}$) may be added using a model selection approach (described below). The covariates have coefficients $\boldsymbol{\alpha}$. The error term $\boldsymbol{\epsilon}$ follows an n-dimensional multivariate normal distribution with mean 0 and variance $\tau^{-1}\boldsymbol{I_n}$. We obtained MLE estimates $\hat{\boldsymbol{\alpha}}$, $\hat{\beta}$, and $\hat{\tau}$ using the optimization procedure described in Supplementary Note 3.1.1 from [23]. The association test score p_j is computed using a likelihood-ratio test of the fitted model against a null model of no SNP effect.

We chose to model sample structure using fixed effects instead of random effects due to the popularity of fixed-effect association mapping approaches and their accuracy [19, 20, 22, 34, 53]. Global sample structure is represented using covariates $\boldsymbol{W}_{\text{global}}$, obtained as follows. Following the approach of [22], a principal components analysis is performed on the full alignment X excluding the local partition $\boldsymbol{X_\ell}$ and the top five principal components are used as the covariates $\boldsymbol{w_1}, \boldsymbol{w_2}, \ldots, \boldsymbol{w_5}$. Local sample structure in a local partition $\boldsymbol{X_\ell}$ containing the test SNP x_j is represented similarly, resulting in five additional covariates

$w_5, w_6, \ldots, w_{10}$. We chose to use at most ten covariates to represent sample structure based on the empirical findings of [22]. We also evaluated the sensitivity of Coal-Map to the number of covariates using five and twenty covariates.

The basic idea behind our use of fixed covariates to capture sample relatedness is as follows. Causal loci outside of the local partition X_ℓ (containing the test SNP) are modeled as a polygenic effect based on covariates computed from global sample structure since it aggregates over all partitions. Covariates computed from local sample structure contribute to the linear mixed model when the local partition X_ℓ contributes to the phenotypic vector y (i.e., contains causal SNPs), but otherwise not. Coal-Map selects between the two resulting models: one using the covariates $W_j^{\text{global}} = (w_1, w_2, \ldots, w_5)$ and the other using the covariates $W_j^{\text{glocal}} = (w_1, w_2, \ldots, w_{10})$. In this study we explore the use of two approaches for this task. Since the two models are nested, we use a forward selection approach (significance threshold of $P < 0.05$). We also evaluated an alternative approach which performs the likelihood-ratio test of each alternative model against a null model of no SNP effects and then chooses the model with smaller P value. Since the forward selection approach can be conservative, we focus on the latter approach.

Simulation study

Our performance evaluation of Coal-Map utilized simulated genotypic sequences and traits where local genealogical variation was due to gene flow and incomplete lineage sorting, two evolutionary processes with first-order effects in prior studies of introgressed eukaryotic genomes [1–3, 5, 31, 54]. The genotypic sequence data was simulated under the basic coalescent model [32] with instantaneous admixture [10] and a bi-allelic sequence mutation model. Simulations were run using either msms [55] to perform forward-time simulation that explicitly modeled positive selection or ms [56] to perform backward-time simulation with intra-locus linkage to emulate the genomic patterns of positive selection. Based on the the analytical and empirical findings of Neuhauser and Krone [57], both simulation approaches are expected to generate similar patterns of local genealogical variation. Furthermore, our performance study obtained consistent results for both approaches. The ms-based simulation utilized an infinite sites model of sequence mutation; the msms-based simulation utilized a sequence mutation model that allowed recurrent mutations between two alleles. Our forward-time coalescent simulation used a selection coefficient $s = 0.56$ which was based upon previously reported estimates from natural mouse populations that were involved in adaptive introgression [4]. These populations are represented by empirical samples used in our study (see below). The model phylogeny used for simulation is shown in Fig. 2. Two present-day populations A and B diverged from a most recent common ancestral population at time t_1. At time t_2, the ancestral populations of A and B hybridized to form the ancestral population of H. We based the divergence time $t_1 = 3.0$ (in coalescent units) on prior population genetic estimates in two *Mus* species (see Table 1 in [11]), corresponding to a divergence time of 1.5 Mya bp, generation time of 2 generations per year, and an effective population size of $N_e = 2.5 * 10^5$. We explored hybridization frequency values consisting of $\gamma \in \{0.5, 0.25, 0.1, 0.01\}$ in the backward-time coalescent simulations, $\gamma = 0.5$ in the forward-time coalescent simulations, mutation rate $\mu = 1.0$ in the forward-time coalescent simulations, and time $t_2 = 2.0$. Each dataset consisted of 10 loci sampled from the above model with free recombination between loci and complete linkage of sites within each locus. We used a sequence length of 250 bp/locus, resulting in 2.5 kb per simulated multiple sequence alignment. Based on the simplifying assumptions of our simulation study (infinite sites model, free recombination between loci, and complete linkage within each locus), the local partition breakpoints required as input to Coal-Map were inferred using the Four-Gamete Test [58, 59].

For each multiple sequence alignment, a quantitative trait was simulated under an extended version of the additive model used by [60] and [61], which incorporated polygenic contributions from multiple loci as follows:

$$y_i = \pi \sum_{j \in \Delta} \frac{Q_{i,j}}{|\Delta|} + (1 - \pi) N(0, 0.01),$$

where y_i is the trait value for the ith individual, π is the proportion of trait variation contributed by genotypic effects, $Q_{i,j}$ is 1 if the ith individual carries the derived allele at site j or 0 otherwise, the environmental contribution is a random effect with normal distribution $N(0, 0.01)$, and Δ is the set of causal SNPs. Twenty causal SNPs were selected uniformly at random from either one, two, or all loci, where loci were chosen uniformly at random in the case of a single or two loci contributing causal SNPs. We refer to the resulting model conditions as single-causal-locus, two-causal-loci, and all-causal-loci model conditions, respectively. Additionally, causal SNPs were selected to have minor allele frequency ranging between 0.1 and 0.3 (following the cutoffs used in [17]). In the forward-time coalescent simulations, loci containing causal markers were under positive selection and all other loci evolved in a neutral manner. For each model condition, simulation was repeated twenty times, resulting in twenty replicate datasets per model condition. Each replicate dataset was analyzed using Coal-Map and

EIGENSTRAT. EIGENSTRAT was run with default settings, where the covariates include the top ten principal components computed from the alignment $\mathbf{X}$.

Performance study using empirical mouse genomes

To better understand the performance of Coal-Map in the context of adaptive interspecific introgression, our performance study utilized genomic sequence data from a past study of genetic variation in natural *Mus musculus* and *Mus spretus* populations [5]. Detailed sample information for the 744 haploid mouse genomes is shown in Additional file 1: Table S1. The sequencing procedures were described in [5], which we briefly review here. Genomic sequence data came from two sources: (1) wild and wild-derived samples from prior studies [4, 5, 47, 62, 63] that were genotyped using the Mouse Diversity Array (following the procedure of [63]) and phased into haploid sequences, and (2) wild-derived samples with published whole-genome sequences [45]. The sequences were combined and then filtered to focus on 414,376 SNPs that were genotyped in all samples. We analyzed the genomic sequences using PhyloNet-HMM to infer genomic tracts with interspecific introgressive origin, following the procedure of [5]. Figure 3 shows the genomic tracts in the vicinity of the *Vkorc1* gene that PhyloNet-HMM inferred as introgressed in origin. Crucially, PhyloNet-HMM performs probabilistic inference to ascribe local genealogical variation to one of several evolutionary processes: interspecific introgression, incomplete lineage sorting, recombination, back mutation, and any combination thereof. Thus, local sample structure will often vary within introgressed tracts that are shared across a common subset of samples (cf. Fig. 10 in [11]).

Two main factors affecting association mapping power are effect size and minor allele frequency of causal SNPs [64] – both of which are unknown a priori. We therefore conducted a performance study using synthetic quantitative traits where these factors were specified as model parameters. We used the above trait model with genetic contribution from introgressed genomic tracts associated with the introduction of warfarin pesticide in Europe [4, 5]. Local partition breakpoints were based upon the union of the PhyloNet-HMM-inferred introgressed tracts and causal SNPs were selected from local partitions where only the genomes of rodenticide-resistant mice had introgressed tracts and contained at least 100 sites. We focused on mouse chromosomes with at least two such partitions, which consisted of chromosomes 7, 15, and 17. For each chromosome, single-causal-locus trait simulation was repeated to yield twenty replicate datasets, and similarly for two-causal-loci traits. The empirical genomic sequence data, synthetic trait data, and local partition breakpoints were provided to Coal-Map as inputs.

Availability of supporting data

Open-source software and open data are available at gitlab.msu.edu/liulab/coal-map.

Additional file

Additional file 1: Appendix. Appendix, including text, tables, and figures for performance study using empirical genomic sequence data from mouse chromosome 15, information about mouse samples used in the study, and supplementary experiments related to the algorithmic design of Coal-Map. (PDF 289 kb)

Competing interests

The authors declare that they have no competing interests.

Authors' contributions

Conceived and designed the experiments: HAH KJL. Implemented software tools: HAH. Performed the experiments: HAH. Analyzed the data: HAH KJL. Wrote the paper: HAH KJL. Both authors read and approved the final manuscript.

Acknowledgements

Faculty startup funds from Michigan State University to KJL were used to support this work and to pay for publication costs.

Declarations

This article has been published as part of BMC Genomics Volume 17 Supplement 1, 2016: Selected articles from the Fourteenth Asia Pacific Bioinformatics Conference (APBC 2016): Genomics. The full contents of the supplement are available online at http://www.biomedcentral.com/bmcgenomics/supplements/17/S1.

Published: 11 January 2016

References

1. Green RE, Krause J, Briggs AW, Maricic T, Stenzel U, Kircher M, et al. A draft sequence of the Neandertal genome. Science. 2010;328(5979): 710–22. http://www.sciencemag.org/content/328/5979/710.full.pdf.
2. Huerta-Sánchez E, Jin X, Asan, Bianba Z, Peter BM, Vinckenbosch N, et al. Altitude adaptation in Tibetans caused by introgression of Denisovan-like DNA. Nature. 2014;512(7513):194–7.
3. The Heliconious Genome Consortium. Butterfly genome reveals promiscuous exchange of mimicry adaptations among species. Nature. 2012;487(7405):94–8.
4. Song Y, Endepols S, Klemann N, Richter D, Matuschka FR, Shih CH, et al. Adaptive introgression of anticoagulant rodent poison resistance by hybridization between old world mice. Curr Biol. 2011;21(15):1296–301.
5. Liu KJ, Steinberg E, Yozzo A, Song Y, Kohn MH, Nakhleh L. Interspecific introgressive origin of genomic diversity in the house mouse. Proc Natl Acad Sci. 2015;112(1):196–201.
6. Mailund T, Halager AE, Westergaard M, Dutheil JY, Munch K, Andersen LN, et al. A new isolation with migration model along complete genomes infers very different divergence processes among closely related great ape species. PLoS Genet. 2012;8(12):1003125.
7. Mailund T, Dutheil JY, Hobolth A, Lunter G, Schierup MH. Estimating divergence time and ancestral effective population size of Bornean and Sumatran orangutan subspecies using a coalescent hidden Markov model. PLoS Genet. 2011;7(3):1001319.
8. Dutheil JY, Ganapathy G, Hobolth A, Mailund T, Uyenoyama MK, Schierup MH. Ancestral population genomics: The coalescent hidden Markov model approach. Genetics. 2009;183(1):259–74. http://www.genetics.org/content/183/1/259.full.pdf+html.
9. Hobolth A, Christensen OF, Mailund T, Schierup MH. Genomic relationships and speciation times of human, chimpanzee, and gorilla inferred from a coalescent hidden Markov model. PLoS Genet. 2007;3(2):7.
10. Durand EY, Patterson N, Reich D, Slatkin M. Testing for ancient admixture between closely related populations. Mol Biol Evol. 2011;28(8):2239–52. http://mbe.oxfordjournals.org/content/28/8/2239.full.pdf+html.

11. Liu KJ, Dai J, Truong K, Song Y, Kohn MH, Nakhleh L. An HMM-based comparative genomic framework for detecting introgression in eukaryotes. PLoS Comput Biol. 2014;10(6):1003649.
12. Rost S, Fregin A, Ivaskevicius V, Conzelmann E, Hortnagel K, Pelz HJ, et al. Mutations in *Vkorc1* cause warfarin resistance and multiple coagulation factor deficiency type 2. Nature. 2004;427(6974):537–41.
13. Marjoram P, Zubair A, Nuzhdin S. Post-GWAS: where next? more samples, more SNPs or more biology? Heredity. 2013;112(1):79–88.
14. Li C. Population subdivision with respect to multiple alleles. Ann Hum Genet. 1969;33(1):23–9.
15. Marchini J, Cardon LR, Phillips MS, Donnelly P. The effects of human population structure on large genetic association studies. Nat Genet. 2004;36(5):512–7.
16. Voight BF, Pritchard JK. Confounding from cryptic relatedness in case-control association studies. PLoS Genet. 2005;1(3):32.
17. Kang HM, Zaitlen NA, Wade CM, Kirby A, Heckerman D, Daly MJ, et al. Efficient control of population structure in model organism association mapping. Genetics. 2008;178(3):1709–23.
18. Devlin B, Roeder K. Genomic control for association studies. Biometrics. 1999;55(4):997–1004.
19. Price AL, Zaitlen NA, Reich D, Patterson N. New approaches to population stratification in genome-wide association studies. Nat Rev Genet. 2010;11(7):459–63.
20. Li M, Reilly MP, Rader DJ, Wang LS. Correcting population stratification in genetic association studies using a phylogenetic approach. Bioinformatics. 2010;26(6):798–806.
21. Kang HM, Sul JH, Service SK, Zaitlen NA, Kong S-y, Freimer NB, et al. Variance component model to account for sample structure in genome-wide association studies. Nat Genet. 2010;42(4):348–54. doi:10.1038/ng.548.
22. Price AL, Patterson NJ, Plenge RM, Weinblatt ME, Shadick NA, Reich D. Principal components analysis corrects for stratification in genome-wide association studies. Nat Genet. 2006;38(8):904–9.
23. Zhou X, Stephens M. Genome-wide efficient mixed-model analysis for association studies. Nat Genet. 2012;44(7):821–4.
24. Sul JH, Eskin E. Mixed models can correct for population structure for genomic regions under selection. Nat Rev Genet. 2013;14(4):300–0.
25. Price AL, Zaitlen NA, Reich D, Patterson N. Response to Sul and Eskin. Nat Rev Genet. 2013;14(4):300–0.
26. Seldin MF, Pasaniuc B, Price AL. New approaches to disease mapping in admixed populations. Nat Rev Genet. 2011;12(8):523–8.
27. Shriner D, Adeyemo A, Rotimi CN. Joint ancestry and association testing in admixed individuals. PLoS Comput Biol. 2011;7(12):1002325.
28. Pasaniuc B, Zaitlen N, Lettre G, Chen GK, Tandon A, Kao WHL, et al. Enhanced statistical tests for GWAS in admixed populations: Assessment using African Americans from CARe and a breast cancer consortium. PLoS Genet. 2011;7(4):1001371. doi:10.1371/journal.pgen.1001371.
29. Sankararaman S, Sridhar S, Kimmel G, Halperin E. Estimating local ancestry in admixed populations. Am J Hum Genet. 2008;82(2):290–303.
30. Gibbs RA, Belmont JW, Hardenbol P, Willis TD, Yu F, Yang H, et al. The international HapMap project. Nature. 2003;426(6968):789–96.
31. Nakhleh L. Computational approaches to species phylogeny inference and gene tree reconciliation. Trends Ecol Evol. 2013;28(12):719–28. doi:10.1016/j.tree.2013.09.004.
32. Kingman JFC. On the genealogy of large populations. J Appl Probab. 1982;19:27–43.
33. DeLong ER, DeLong DM, Clarke-Pearson DL. Comparing the areas under two or more correlated receiver operating characteristic curves: a nonparametric approach. Biometrics. 1988;44(3):837–45.
34. Shriner D. Investigating population stratification and admixture using eigenanalysis of dense genotypes. Heredity. 2011;107(5):413–20.
35. Sousa V, Hey J. Understanding the origin of species with genome-scale data: modelling gene flow. Nat Rev Genet. 2013;14(6):404–14.
36. Yu Y, Degnan JH, Nakhleh L. The probability of a gene tree topology within a phylogenetic network with applications to hybridization detection. PLoS Genet. 2012;8(4):1002660.
37. Bryant D, Bouckaert R, Felsenstein J, Rosenberg NA, RoyChoudhury A. Inferring species trees directly from biallelic genetic markers: bypassing gene trees in a full coalescent analysis. Mol Biol Evol. 2012;29(8):1917–32.
38. Akaike H. Information theory and an extension of the maximum likelihood principle. In: Proceedings of the Second International Symposium on Information Theory. Budapest, Hungary: Akademiai Kiado; 1973. p. 267–81.
39. Akaike H. A new look at the statistical model identification. IEEE Trans Autom Control. 1974;19(6):716–23.
40. Hurvich CM, Tsai CL. Regression and time series model selection in small samples. Biometrika. 1989;76(2):297–307. http://biomet.oxfordjournals.org/content/76/2/297.full.pdf+html.
41. Schwarz G. Estimating the dimension of a model. Ann Stat. 1978;6(2): 461–4.
42. Smith MW, O'Brien SJ. Mapping by admixture linkage disequilibrium: advances, limitations and guidelines. Nat Rev Genet. 2005;6(8):623–32.
43. Chakraborty R, Weiss KM. Admixture as a tool for finding linked genes and detecting that difference from allelic association between loci. Proc Natl Acad Sci. 1988;85(23):9119–23.
44. Speed D, Balding DJ. MultiBLUP: improved SNP-based prediction for complex traits. Genome Res. 2014;24(9):1550–7.
45. Keane TM, Goodstadt L, Danecek P, White MA, Wong K, Yalcin B, et al. Mouse genomic variation and its effect on phenotypes and gene regulation. Nature. 2011;477(7364):289–94. doi:10.1038/nature10413.
46. Didion J, Yang H, Sheppard K, Fu CP, McMillan L, de Villena F, et al. Discovery of novel variants in genotyping arrays improves genotype retention and reduces ascertainment bias. BMC Genomics. 2012;13(1):34. doi:10.1186/1471-2164-13-34.
47. Yang H, Wang JR, Didion JP, Buus RJ, Bell TA, Welsh CE, et al. Subspecific origin and haplotype diversity in the laboratory mouse. Nat Genet. 2011;43(7):648–55. doi:10.1038/ng.847.
48. Counterman BA, Araujo-Perez F, Hines HM, Baxter SW, Morrison CM, Lindstrom DP, et al. Genomic hotspots for adaptation: the population genetics of Müllerian mimicry in *Heliconius erato*. PLoS Genet. 2010;6(2): 1000796.
49. Baxter SW, Nadeau NJ, Maroja LS, Wilkinson P, Counterman BA, Dawson A, et al. Genomic hotspots for adaptation: the population genetics of Müllerian mimicry in the *Heliconius melpomene* clade. PLoS Genet. 2010;6(2):1000794.
50. Yu Y, Dong J, Liu KJ, Nakhleh L. Maximum likelihood inference of reticulate evolutionary histories. Proc Natl Acad Sci. 2014;111(46): 16448–53.
51. Zuk O, Hechter E, Sunyaev SR, Lander ES. The mystery of missing heritability: genetic interactions create phantom heritability. Proc Natl Acad Sci. 2012;109(4):1193–8.
52. Nuzhdin SV, Friesen ML, McIntyre LM. Genotype–phenotype mapping in a post-GWAS world. Trends Genet. 2012;28(9):421–6.
53. Alexander DH, Novembre J, Lange K. Fast model-based estimation of ancestry in unrelated individuals. Genome Res. 2009;19(9):1655–64.
54. Reich D, Green RE, Kircher M, Krause J, Patterson N, Durand EY, et al. Genetic history of an archaic hominin group from Denisova Cave in Siberia. Nature. 2010;468(7327):1053–60.
55. Ewing G, Hermisson J. MSMS: a coalescent simulation program including recombination, demographic structure and selection at a single locus. Bioinformatics. 2010;26(16):2064–5.
56. Hudson RR. Generating samples under a Wright-Fisher neutral model of genetic variation. Bioinformatics. 2002;18(2):337–8. doi:10.1093/bioinformatics/18.2.337. http://bioinformatics.oxfordjournals.org/content/18/2/337.full.pdf+html.
57. Neuhauser C, Krone SM. The genealogy of samples in models with selection. Genetics. 1997;145(2):519–34.
58. Hudson RR, Kaplan NL. Statistical properties of the number of recombination events in the history of a sample of DNA sequences. Genetics. 1985;111(1):147–64. http://www.genetics.org/content/111/1/147.full.pdf+html.
59. Wang J, Moore KJ, Zhang Q, de Villena FP-M, Wang W, McMillan L. Genome-wide compatible SNP intervals and their properties. In: Proceedings of the First ACM International Conference on Bioinformatics and Computational Biology. BCB '10. New York, NY, USA: ACM; 2010. p. 43–52. doi:10.1145/1854776.1854788. http://doi.acm.org/10.1145/1854776.1854788.
60. Besenbacher S, Mailund T, Schierup MH. Local phylogeny mapping of quantitative traits: higher accuracy and better ranking than single-marker association in genomewide scans. Genetics. 2009;181(2):747–53.
61. Long AD, Langley CH. The power of association studies to detect the contribution of candidate genetic loci to variation in complex traits. Genome Res. 1999;9(8):720–31.

62. Staubach F, Lorenc A, Messer PW, Tang K, Petrov DA, Tautz D. Genome patterns of selection and introgression of haplotypes in natural populations of the house mouse (*Mus musculus*). PLoS Genet. 2012;8(8): 1002891.
63. Yang H, Ding Y, Hutchins LN, Szatkiewicz J, Bell TA, Paigen BJ, et al. A customized and versatile high-density genotyping array for the mouse. Nat Methods. 2009;6(9):663–6. doi:10.1038/nmeth.1359.
64. Altshuler D, Daly MJ, Lander ES. Genetic mapping in human disease. Science. 2008;322(5903):881–8.
65. Benjamini Y, Hochberg Y. Controlling the false discovery rate: a practical and powerful approach to multiple testing. J R Stat Soc Series B (Methodol). 1995;57(1):289–300.

Bui et al. BMC Genomics 2016, 17(Suppl 1):9
DOI 10.1186/s12864-015-2299-1

BMC Genomics

PROCEEDINGS Open Access

SOHSite: incorporating evolutionary information and physicochemical properties to identify protein *S*-sulfenylation sites

Van-Minh Bui[1†], Shun-Long Weng[2,3,4†], Cheng-Tsung Lu[1], Tzu-Hao Chang[5], Julia Tzu-Ya Weng[1,6*] and Tzong-Yi Lee[1,6*]

From The Fourteenth Asia Pacific Bioinformatics Conference (APBC 2016)
San Francisco, CA, USA. 11 - 13 January 2016

Abstract

Background: Protein *S*-sulfenylation is a type of post-translational modification (PTM) involving the covalent binding of a hydroxyl group to the thiol of a cysteine amino acid. Recent evidence has shown the importance of *S*-sulfenylation in various biological processes, including transcriptional regulation, apoptosis and cytokine signaling. Determining the specific sites of *S*-sulfenylation is fundamental to understanding the structures and functions of *S*-sulfenylated proteins. However, the current lack of reliable tools often limits researchers to use expensive and time-consuming laboratory techniques for the identification of *S*-sulfenylation sites. Thus, we were motivated to develop a bioinformatics method for investigating *S*-sulfenylation sites based on amino acid compositions and physicochemical properties.

Results: In this work, physicochemical properties were utilized not only to identify *S*-sulfenylation sites from 1,096 experimentally verified *S*-sulfenylated proteins, but also to compare the effectiveness of prediction with other characteristics such as amino acid composition (AAC), amino acid pair composition (AAPC), solvent-accessible surface area (ASA), amino acid substitution matrix (BLOSUM62), position-specific scoring matrix (PSSM), and positional weighted matrix (PWM). Various prediction models were built using support vector machine (SVM) and evaluated by five-fold cross-validation. The model constructed from hybrid features, including PSSM and physicochemical properties, yielded the best performance with sensitivity, specificity, accuracy and MCC measurements of 0.746, 0.737, 0.738 and 0.337, respectively. The selected model also provided a promising accuracy (0.693) on an independent testing dataset. Additionally, we employed TwoSampleLogo to help discover the difference of amino acid composition among *S*-sulfenylation, *S*-glutathionylation and *S*-nitrosylation sites.

Conclusion: This work proposed a computational method to explore informative features and functions for protein *S*-sulfenylation. Evaluation by five-fold cross validation indicated that the selected features were effective in the identification of *S*-sulfenylation sites. Moreover, the independent testing results demonstrated that the proposed method could provide a feasible means for conducting preliminary analyses of protein *S*-sulfenylation. We also anticipate that the uncovered differences in amino acid composition may facilitate future studies of the extensive crosstalk among *S*-sulfenylation, *S*-glutathionylation and *S*-nitrosylation.

Keywords: *S*-sulfenylation, Sulfenic acids, Support vector machine, Physicochemical properties

* Correspondence: julweng@saturn.yzu.edu.tw; francis@saturn.yzu.edu.tw
†Equal contributors
[1]Department of Computer Science and Engineering, Yuan Ze University, Taoyuan 320, Taiwan
Full list of author information is available at the end of the article

Introduction

Post-translational modification (PTM) at the cysteine residues is essential to the dynamic functions of proteins. *S*-sulfenylation is a reversible covalent modification on the thiol group of cysteine residues by hydrogen peroxide, whereby the Cys-SH is oxidized to sulfenic (Cys-SOH), sulfinic (Cys–SO_2H) and sulfonic acids (Cys–SO_3H). More importantly, these changes contribute strongly to the regulation of protein function under both normal and oxidative stress conditions [1–4]. To date, a little more than 200 transcription factors, signaling proteins, metabolic enzymes, proteostasis regulators, and cytoskeletal components have been identified as S-sulfenylated proteins [5]. Thus, investigation of *S*-sulfenylated proteins is crucial to our understanding of the function and regulation of this oxidative post-translational modification at cysteine residues.

Several chemoproteomic approaches have been developed for identifying specific sites in proteins that undergo *S*-sulfenylation [6–10]. Nevertheless, these experimental methods are often expensive and time-consuming. Further, the lack of information about *S*-sulfenylation sites and the multiplicity of redox changes contribute to many false-positive identifications [11]. To facilitate the efficiency of experimental characterization and search for S-sulfenylated peptides, construction of a resource database is necessary, and possible with the development of powerful bioinformatics technologies. In spite of the availability of several published algorithms and public servers that aim to help analyze the reactive and oxidative sites [12, 13] of cysteine residues, there are no confirmed reports about the site-specific information of *S*-sulfenylated proteins, except for the chemical characterization of over 1000 *S*-sulfenylation sites on more than 700 proteins in cell culture experiments [11].

The present study (SOHSite) concentrated on the computational identification and characterization of *S*-sulfenylation sites. To discriminate between *S*-sulfenylation and non-*S*-sulfenylation sites, features including amino acid composition (AAC), amino acid pair composition (AAPC), position specific scoring matrix (PSSM), position weight matrix (PWM), amino acid substitution matrix (BLOSUM62), accessible surface area (ASA), and the physicochemical properties of proteins, were examined. Support vector machine [14] (SVM) was used to learn a predictive model from each feature, as well as hybrid combinations of the features. The SVM model with the best predictive performance would be selected based on the result of five-fold cross-validation. An independent testing set was applied to further evaluate the effectiveness of the chosen model. Finally, we investigated both key amino acid and hydrophobic attributes associated with *S*-sulfenylation sites.

Materials and method

Data collection and preprocessing

Figure 1 presents the analytical flowchart of SOHSite. A majority of the experimental data used in this study was obtained from the Carroll laboratory database, which stores information on experimentally verified *S*-sulfenylated cysteines in humans based on the newly discovered *S*-sulfenyl-mediated redox regulation of the transcription factor H1F1A by SIRT6 [11]. As shown in Table 1, the data were composed of 1443 positive and 10521 negative data on 987 S-sulfenylated proteins. An additional set of data was collected from RedoxDB [12], UniProtKB [15] and other published literature. From the RedoxDB database, we obtained 102 S-sulfenylated sites (positive dataset) in 92 proteins, while the rest of the cysteine residues in these proteins were considered as non-*S*-sulfenylated cysteines (negative dataset). Moreover, properties characterized by X-Ray crystallography [16] were included, resulting in 33 *S*-sulfenylated and 143 non-*S*-sulfenylated sites. Finally, a dataset comprised of 17 *S*-sulfenylated and 97 non-*S*-sulfenylated cysteines was downloaded from UniProtKB. In order to generate the training dataset, the window length of $2n + 1$ was used to extract sequence fragments centering at the *S*-sulfenylated or non-*S*-sulfenylated cysteines and containing n upstream, as well as n downstream, flanking amino acids. After extracting the sequence fragments with $2n + 1$ window length ($n = 10$), we randomly categorized the dataset into independent testing set for the evaluation of real performance and training set for the construction of *S*-sulfenylated site prediction models.

To choose the best prediction model, a five-fold cross-validation approach was utilized to evaluate the performance of discriminating *S*-sulfenylated substrate sites from non-*S*-sulfenylated cysteine residues. Then, the model yielding the best prediction performance was further examined with an independent testing dataset. To prevent overestimation on performance evaluation, homologous sequences in the training and independent testing datasets were removed using 30 % sequence similarity as the filtering threshold, with reference to MASA [17]. As shown in (Additional file 1: Figure S1), two *S*-sulfenylated proteins with more than 30 % sequence identity were defined as homologous protein sequences. Then, two homologous sequences were specified to re-align the fragment sequences using a window length of $2n + 1$, centered on the *S*-sulfenylated sites using BL2SEQ [18]. For two fragment sequences with 100 % identity, when the *S*-sulfenylated sites in the two proteins are in the same positions, only one site was kept while the other was discarded. Also, the training data had the higher priority so that if the sequence appeared in two datasets, the homologous fragment would be preserved in the training set and deducted from the testing set. The final

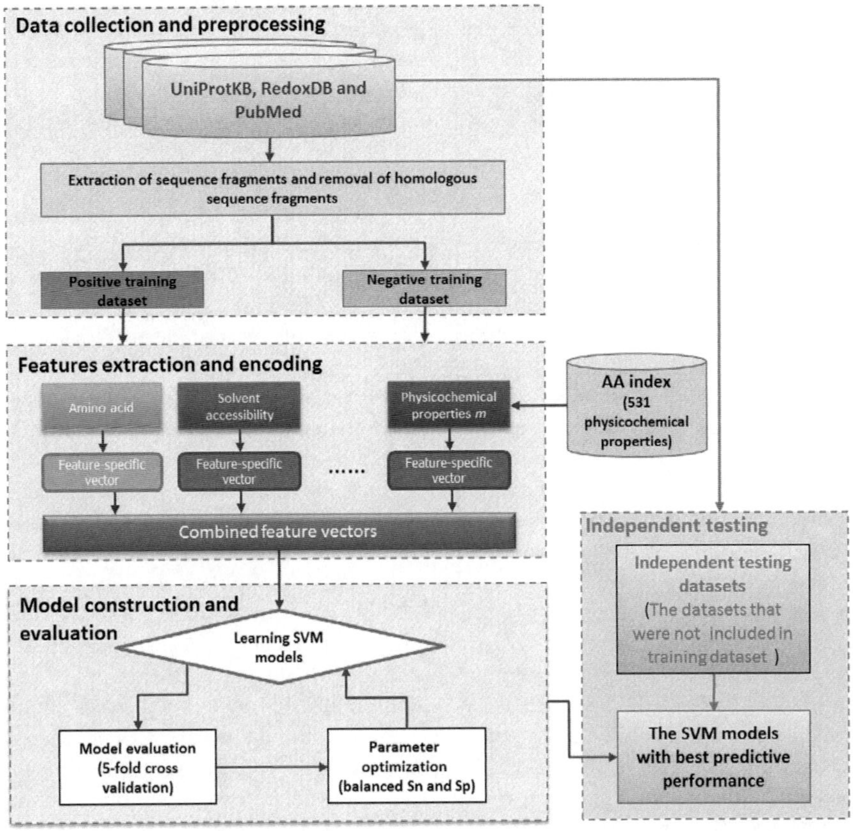

Fig. 1 Analytical flowchart of SOHSite including data collection and preprocessing, features extraction and encoding, model construction and evaluation, and independent testing

positive and negative data consisted of 1145 and 8368 for the training dataset, and 289 and 2108 for the independent testing dataset, respectively.

Features extraction and encoding

In this study, numerous sequence-based features, including amino acid composition (AAC), amino acid pair composition (AAPC), solvent-accessible surface area (ASA), amino acid substitution matrix (BLOSUM62), position-specific scoring matrix (PSSM), positional weighted matrix (PWM) and physicochemical properties (AAindex), were assessed to conduct the best prediction model. After the extraction of sequence fragments with a window size of 21-mer amino acids, each sequence was encoded based on the investigated features. The orthogonal binary coding mechanism is one of the most popular coding methods for transforming amino acids into numeric vectors, called 20-dimensional binary coding (20D) [19]. Each amino acid was represented by a vector with 20 letters. For example, alanine (A) would be encoded as "10000000000000000000" while cysteine (C) would be "01000000000000000000" and so on. For each

Table 1 Data statistics of *S*-sulfenylated and non-*S*-sulfenylated sites

Data resource	*S*-sulfenylation sites (Positive data)	Non-*S*-sulfenylation sites (Negative data)
UniProtKB 201412	17	92
RedoxDB v1 (PMID: 22833525)	102	401
Yang J et al., 2014 [11] (PMID: 25175731)	1,443	10,521
Other Literature	33	143
Combined non-redundant dataset		
Training	1,145	8,368
Independent testing	289	2,108

sequence fragment, the length of feature vectors with a window size of $2n + 1$ was set to $(2n + 1)$ x 20 to represent the flanking amino acids surrounding the *S*-sulfenylation sites. Therefore, there were a total of k vectors $\{x_i, i = 1, 2 \ldots, k\}$ corresponding to the number of *k* sequence fragments in the training and testing dataset. For a binary classification, the labels +1 and −1 were assigned to the positive and negative data, respectively.

For the representations of amino acid compositions around *S*-sulfenylation sites, the vectors x_i consisted of 21 elements for the amino acid composition (AAC) and 441 elements for the amino acid pair composition (AAPC). The 20 elements were defined as the occurrence frequencies of 20 amino acids in a sequence fragment, while the 400 elements were defined as the occurrence frequencies of 400 amino acid pairs in a sequence fragment. When the sequence fragments at the N- or C-terminus were less than 21-mer, non-existing residues were filled with "X" in the corresponding position. Therefore, a total of 21 types of amino acids and 441 types of amino acid pairs were presented in our setting. Additionally, the BLOcks Substitution Matrix (BLOSUM62) [20] was based on the alignments of amino acid sequences possessing no more than 62 % identity between two peptide sequences with 21 amino acids. With reference to the SulfoSite strategy [21], the position weighted matrix (PWM) of amino acids was determined using the non-homologous training data. The PWM described the frequency of occurrence of amino acids surrounding the *S*-sulfenylation sites, and was utilized to encode the sequence fragments. Each residue in the training dataset was represented by a matrix of $m \times w$ elements, where the window size of 21 was designated by w and the 21 elements including the 20 types of amino acids as well as the terminal signal was denoted by m. Position Specific Scoring Matrix (PSSM) profiles were generated from PSI-BLAST [22] against non-redundant sequences of *S*-sulfenylation sites. The scores were computed based on the multiple sequence alignment of proteins which may have similar structures with different amino acid compositions. In a PSSM profile, the rows in the matrix of $(2n + 1) \times 20$ elements were centered on the substrate site, with $2n + 1$ representing the window size and the number 20 denoting the position specific scores for each amino acid.

The solvent-accessible surface area (ASA), a feature representing the accessibility of an amino acid side-chain on the surface of a protein that experience post-translational modification [23], was also included as an attribute for the identification of S-sulfenylation sites. The RVP-Net [24, 25] was applied to compute the ASA value from the protein sequence due to the lack of experimentally verified tertiary structures of S-sulfenylated protein in the Protein Data Bank (PDB) [26]. Based on the information regarding the neighborhood amino acid composition, RVP-Net could predict the real ASA of a residue by using a neutral network approach. The possible mean absolute error, which was defined as the absolute difference between the predicted and experimental values of relative ASA per residue [25], was 18.0–19.5 %, for each measurement. The value of ASA represented the percentage of the solvent-accessible area of each amino acid on the protein. To compute the ASA values of all of the residues, full-length protein sequences were input into the RVP-Net. Then, the ASA values of amino acids surrounding the *S*-sulfenylation sites were extracted and normalized based on a scale from zero to one. In the investigation herein of secondary structure surrounding the SOH sites, additionally, PSIPRED [27] is employed to compute the secondary structure from the protein sequence. PSIPRED is a simple and reliable method for predicting secondary structure, which incorporates two feed-forward neural networks to analyze the output obtained from PSI-BLAST (Position Specific Iterated - BLAST) [22]. The output of PSIPRED is given in terms of "H," "E" and "C" which stand for helix, sheet and coil, respectively.

The AAindex [28] (Version 9.1) contains a total of 544 amino acid indices which specify the physicochemical properties of twenty amino acids. After the amino acid indices with the value "NA" were eliminated, the physicochemical properties of the remaining 531 were examined to determine their ability to distinguish *S*-sulfenylated sites from the non-*S*-sulfenylated sites. A set of 20 numerical values corresponding to the physicochemical property of each of the 20 amino acids was specified. The sequence fragments were transformed from AAs surrounding S-sulfenylation sites into values associated with their physicochemical properties. In order to identify the significant physicochemical properties, the F-score method [29–33] has been applied to calculate a statistical value for each position surrounding *S*-sulfenylation sites. The F-score of the *i*th physicochemical feature is defined as:

$$\text{F-score}\,(i) = \frac{\left(\bar{x}_i^{(+)} - \bar{x}_i\right)^2 + \left(\bar{x}_i^{(-)} - \bar{x}_i\right)^2}{\frac{1}{n^+-1}\sum_{k=1}^{n^+}\left(x_{k,i}^{(+)} - \bar{x}_i^{(+)}\right)^2 + \frac{1}{n^--1}\sum_{k=1}^{n^-}\left(x_{k,i}^{(-)} - \bar{x}_i^{(-)}\right)^2} \quad (1)$$

where $\bar{x}_i$, $\bar{x}_i^{(+)}$ and $\bar{x}_i^{(-)}$ denote the average value of the *i*th feature in the whole, positive, and negative data sets, respectively; n^+ denotes the number of positive data set and n^- denotes the number of negative data set; $x_{k,i}^{(+)}$ denotes the *i*th feature of the *k*th positive instance, and $x_{k,i}^{(-)}$ denotes the *i*th feature of the *k*th negative instance [34]. The prediction performances obtained from using the physicochemical properties individually were evaluated, and the properties were subsequently sorted in descending order based on the accuracy of prediction.

For the construction of predictive models, hybrid features were generated by combining two or more single features. In an attempt to obtain the highest predictive accuracy, the single features were selected based on their predictive performance. Prior to classification, the data needed to be scaled in the range of [−1, 1] to enhance the effectiveness of the results [35].

Model training and evaluation

The training data set was used for building prediction models with the support vector machine (SVM). This binary classification utilizes a kernel function to transform the input samples into a higher dimensional space and attempts to find a hyper-plane to discriminate the two classes with maximal margin and minimal error. In our study, a public SVM tool (LIBSVM) [14] was implemented to build models that could discriminate between *S*-sulfenylation and non-*S*-sulfenylation sites. The radial basis function (RBF) $K(S_i, S_j) = e^{\left(-\gamma\|s_i - s_j\|^2\right)}$ was adopted as the kernel function. Two factors were included to enhance the performance: the RBF kernel was determined by the gamma parameter, while the softness of the hyper-plane was modulated by the cost parameter.

To choose the best final model, five-fold cross-validation was carried out for each feature to evaluate the predictive performance. The training dataset was divided into five subgroups with approximately equal size. The ratio of the testing set to the training set was 1:4 and the cross-validation process was repeated five times. The five validation results were then combined to generate a single estimation. Obviously, one of the benefits of k-fold cross-validation is the improvement on the reliability of evaluation because all of the original data, including the training and testing data sets, were considered and each subset should be tested only once [36] . To estimate the predictive performance of each trained model, measures such as sensitivity (Sn), specificity (Sp), accuracy (Acc) and Matthews Correlation Coefficient (MCC) were used:

$$Sn = \frac{TP}{TP + FN} \quad (2)$$

$$Sp = \frac{TN}{TN + FP} \quad (3)$$

$$Acc = \frac{TP = TN}{TP = FP = TN = FN} \quad (4)$$

$$MCC = \frac{(TP * TN) - (FN * FP)}{\sqrt{(TP + FN) * (TN + FP) * (TP + FP) * (TN + FN)}} \quad (5)$$

where TP, TN, FP and FN represented the number of true positives, true negatives, false positives and false negatives, respectively. The MCC reflects both the sensitivity (true positive rate) and specificity (true negative rate) of a predictive model. Sometimes, accuracy is not useful when the two classes are of very different sizes [37]. Therefore, the MCC is usually regarded as a balanced measure even if the two classes are of very different sizes. Finally, after selecting the best model with the highest MCC value, an independent testing was carried out to evaluate its real predictive power.

Results and discussion

Investigation of amino acid composition of S-sulfenylation site

Through sequence-based investigation, the frequency of twenty amino acids surrounding the *S*-sulfenylation sites revealed the amino acid composition that forms the substrate environment for protein *S*-sulfenylation. Figure 2A indicates that, at *S*-sulfenylation sites, A (Alanine), R (Arginine), E (Glutamic acid), and K (Lysine) residues occur at a higher frequency, while C (Cysteine) and H (Histidine) residues have a lower frequency of occurrence. In this investigation, WebLogo [38] was utilized to compute the position-specific amino acid composition for both *S*-sulfenylation (Fig. 2B) and non-*S*-sulfenylation (Fig. 2C). However, it is difficult to compare the amino acid composition between *S*-sulfenylation and non-*S*-sulfenylation sites at a specific position. To effectively identify *S*-sulfenylation sites, this study concentrated on notable differences between *S*-sulfenylated and non-*S*-sulfenylated sites. TwoSampleLogo [39] was employed to detect statistically significant differences in position-specific amino acid composition between the positive and negative datasets. Comparing between the 1145 positive data and 8368 negative data (Fig. 2D), it is obvious that three of the four aforementioned amino acids (R, E and K) play an important role in the flanking region of *S*-sulfenylation sites. In particular, the positively charged Lysine (K) and Arginine (R) residues had the highest ratio at positions −10, −8 ~ −6, −4, −2, and +4 ~ +8 ($p < 0.01$). In contrast, at positions −1, +1 and +2 that were close to the *S*-sulfenylation sites, a lack of positively charged residues was observed, while a noticeable abundance of negatively charged resides (Glutamic acid or E) was apparent at positions −3, +1, +3, +4. Interestingly, three polar residues, serine (S), asparagine (N), and glycine (G), appeared to have a higher frequency of occurrence compared to the rest of the amino acids at position −1. For non-*S*-sulfenylated sites, however, there was an abundance of neutral amino acids, including leucine (L), cysteine (C), histidine (H), methionine (M), phenylalanine (F) and tyrosine (Y), at positions ranging from −9 to +7, while arginine (R) residue seemed to be concentrated at three positions (−1, 1 and 2) around non-*S*-sulfenylation sites. Another evident difference is the amino acid composition

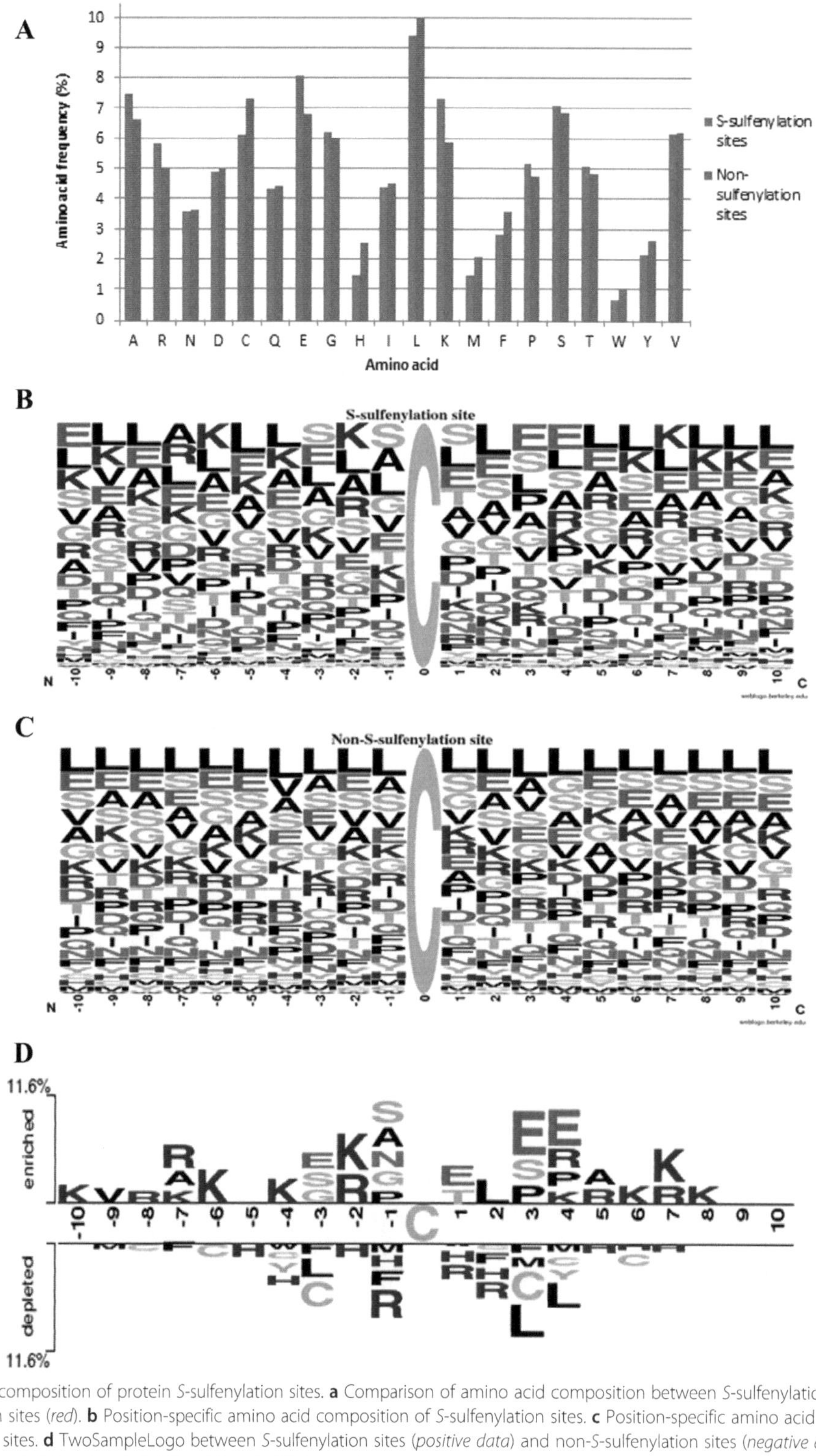

Fig. 2 Amino acid composition of protein *S*-sulfenylation sites. **a** Comparison of amino acid composition between *S*-sulfenylation sites (*blue*) and non- *S*-sulfenylation sites (*red*). **b** Position-specific amino acid composition of *S*-sulfenylation sites. **c** Position-specific amino acid composition of non-*S*-sulfenylation sites. **d** TwoSampleLogo between *S*-sulfenylation sites (*positive data*) and non-*S*-sulfenylation sites (*negative data*)

surrounding S-sulfenylation sites at positions −6, −2, −1, +3 and +4. The analytical results indicated that the positions of amino acids relative to one another in the sequence play a vital role in discriminating between *S*-sulfenylation and non-*S*-sulfenylation sites.

Computed by the RVP-Net tool, ASA was also adopted as an attribute for the identification of S-sulfenylation sites. To discover how amino acids flanking the S-sulfenylation and non-S-sulfenylation sites may differ in their interaction with solvents, a comparison was performed using the average proportion of ASA in the 21-mer window (−10 ~ +10) as illustrated in Fig. 3. Amino acids surrounding the *S*-sulfenylated sites exhibit higher ASA compared to those around non-S-sulfenylation sites. A strong evidence for hydrophilicity at the S-sulfenylated substrate sites was found because the average percentage of ASA values of the flanking residues was higher than non-S-sulfenylated cysteines. Thus, hydrophilic amino acids flanking cysteine residues may determine their modification by sulfenylation.

Performance evaluation of the trained models

Choosing suitable features is one of the crucial steps to constructing the best prediction model for the discrimination of *S*-sulfenylation sites from non-*S*-sulfenylation sites. In our study, the predictive model was trained from a variety of features, such as 20D (binary code), AAC, AAPC, BLOSUM62, ASA, PSSM, PWM and physicochemical properties. Five-fold cross-validation was performed for each model and four parameters (Sn, Sp, Acc and MCC) were included as the evaluation criteria. As shown in Table 2, the SVM model trained with PSSM feature yielded the best prediction performance: a sensitivity value of 0.71, specificity of 0.72, accuracy of 0.72, and MCC of 0.30. Slightly below the PSSM model in performance, the sensitivity, specificity, accuracy and MCC of the BLOSUM62 model were 0.68, 0.70, 0.69 and 0.26, respectively. The SVM model constructed with the ASA feature yielded the lowest predictive accuracy (0.61) and relatively lower sensitivity (0.60), specificity (0.66), and MCC (0.14).

To further investigate the physicochemical properties of S-sulfenylation sites and the adjacent amino acids, a total of 531 physicochemical properties, extracted from version 9.1 of the AAindex, were individually explored. Each physicochemical property with a significant F-score value was assessed by five-fold cross validation in order to evaluate their ability to identify *S*-sulfenylation sites. (Additional file 2: Table S1) shows the top 20 physicochemical properties ranked in a descending order according to their predictive performance. Descriptions of the 531 physicochemical properties from the AAindex indicated that four of the 20 physicochemical properties are associated with hydrophobicity. Overall, the accuracy of each of these features was over 0.6. In order to obtain a better prediction performance, the 20 highest performing physicochemical properties were combined with the best sequence-based feature (PSSM) by forward selection in (Additional file 3: Figure S2). A summary of the resulting performance is illustrated in Fig. 4, while the detailed results are presented in (Additional file 4: Table S2). With the consideration of more physicochemical properties, slight increases in specificity and accuracy were observed, while the sensitivity value fluctuated with different features. The best model yielded a sensitivity of 0.746, specificity of 0.737, accuracy of 0.738 and 0.337 MCC value. This model was built from PSSM and the top 12 most useful physicochemical properties (GUYH850101, JANJ790102, KIDA850101, FASG890101, KARP850101, EISD860102, LEVM760101, GUYH850104, GUYH850102, VINM940103, MIYS990104 and FUKS010111), and appeared to be the most effective at identifying S-sulfenylation sites among all of the models tested so far. Interestingly, three of the 12 selected indices (KIDA850101, FASG890101 and EISD860102) shared a common

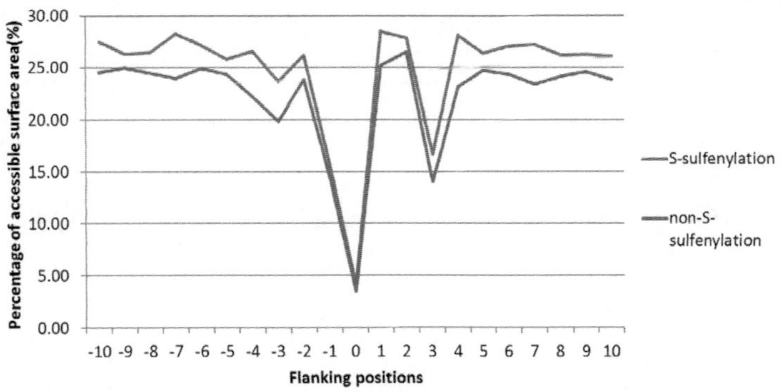

Fig. 3 Comparison of the solvent-accessible surface area between *S*-sulfenylation and non-*S*-sulfenylation sites

Table 2 Five-fold cross validation results for SVM models trained with various features individually

Training features	Sn	Sp	Acc	MCC
20D Binary code	0.66	0.68	0.68	0.23
BLOSUM62	0.68	0.70	0.69	0.26
Amino Acid Composition (AAC)	0.64	0.65	0.65	0.19
Amino Acid Pair Composition (AAPC)	0.64	0.67	0.67	0.21
Accessible Surface Area (ASA)	0.60	0.61	0.61	0.14
Secondary structure (SS)	0.56	0.56	0.56	0.08
Position Weight Matrix (PWM)	0.64	0.66	0.66	0.20
Position-specific scoring matrix (PSSM)	0.71	0.72	0.72	0.30

A total of 1145 positive data and 8368 negative data were used in the cross validation process. Sn, sensitivity; Sp, specificity; Acc, accuracy; MCC, Matthews Correlation Coefficient

physicochemical feature, which was hydrophobicity. As analyzed previously, hydrophobicity contributes significantly to the characteristics, structures and functions of these proteins.

Performance evaluation by independent testing dataset

For the classification of PTM sites, the prediction performance of the constructed models may be overestimated as a result of the overfitting to a training set. In order to evaluate the real performance of the predictive model, an independent testing data set, which was truly blind to the training data set, was utilized to further test the effectiveness of the selected model (PSSM + 12 AAindex) with the best performance in five-fold cross validation. This independent testing data set consisted of 289 positive data and 2108 negative data extracted from the original data. Figure 5 presents the comparison of the independent testing results between the model trained with the best single feature (PSSM) model and the model trained with the combination of PSSM and the top 12 physicochemical properties. The PSSM model yielded a sensitivity of 0.647, a specificity of 0.659, an accuracy of 0.657, and a MCC value of 0.205. Compared to the PSSM model, the model combining PSSM with the 12 AAindexes generated better sensitivity (0.720), specificity (0.690), accuracy (0.693), and a significantly better MCC (0.278) with p-value < 0.05. The detailed independent testing results, including true positive (TP), false negative (FN), true negative (TN) and false positive (FP), are described in (Additional file 5: Table S3). Overall, the model combining PSSM with the 12 AAindexes achieved promising predictive performance on the independent testing data set.

Functional exploitation of the S-sulfenylated proteins

In this study, the **D**atabase for **A**nnotation, **V**isualization and **I**ntegrated **D**iscovery (DAVID) [40] tool was employed to explore the functions and characteristics of *S*-sulfenylated proteins. (Additional file 6: Table S4) provides the Gene Ontology (GO) [41] annotations associated with the S-sulfenylated proteins examined in this work. According to the GO annotation of biological process, 10.8 % of the proteins were involved in RNA processing (p-value = 7.65E-32). In terms of molecular function annotated by GO, most of the proteins seemed to be involved in RNA binding (p-value = 1.06E-42) or nucleotide binding (p-value = 4.47E-35). With respect to cellular component, these *S*-sulfenylated proteins are mostly distributed in the nuclear lumen (p-value = 9.07E-44) and non-membrane-bounded organelles (p-value = 1.81E-42). Additionally, pathway analysis showed that *S*-sulfenylated proteins are involved in a variety of

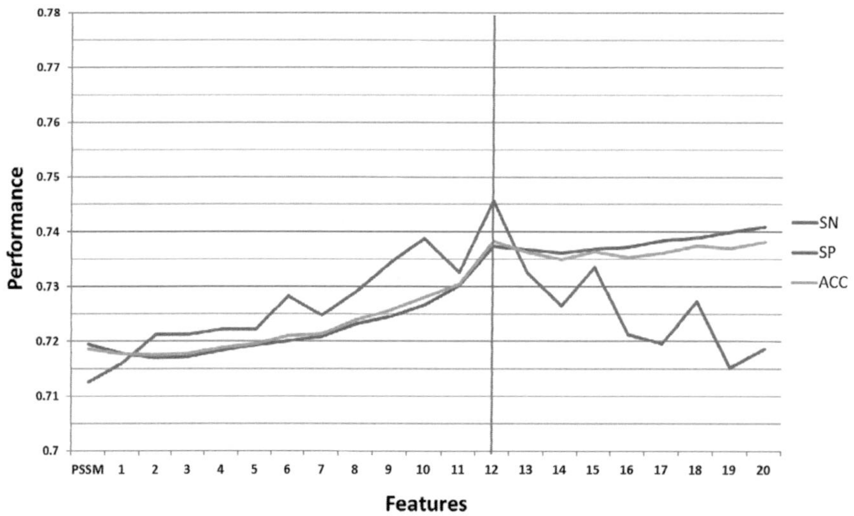

Fig. 4 The predictive performance of PSSM model combined with forward selection of the top 20 physicochemical properties

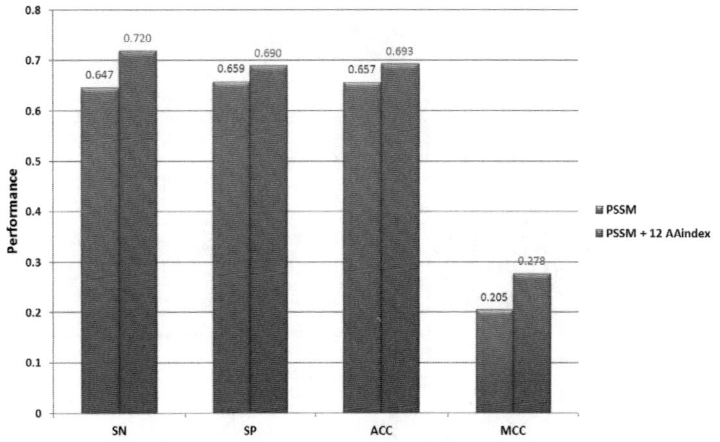

Fig. 5 Comparison of the independent testing results between PSSM model and the hybrid model combining PSSM with the top 12 physicochemical properties

pathways in (Additional file 7: Table S5). Of note, 27 of the 1096 *S*-sulfenylated proteins appeared to be strongly associated with the mechanisms underlying *Escherichia coli* infection.

It has been reported that a protein-interacting domain usually recognizes a short peptide motif on the target protein but does not bind stably until the peptide has undergone appropriate PTM [42–44]; this can create binding sites for specific protein-interaction domains that work together to carry out a specific cellular function [45]. The redox state and chemical modification of the thiol group of cysteine residues facilitate its interaction with various proteins to regulate a variety of intracellular and intercellular events [46]. Thus, information regarding the functional domains could be utilized to infer the functional roles of *S*-sulfenylation sites located in a specific protein domain. InterPro [47] is an integrated resource that provides "signatures" such as protein families, domains, and functional sites. Our investigation of the protein domains revealed that the thioredoxin-like fold domain is the most abundant functional domain in *S*-sulfenylated proteins in (Additional file 8: Table S6).

Distinguishing *S*-sulfenylation from *S*-nitrosylation or *S*-glutathionylation

The primary purpose of this study was to effectively identify *S*-sulfenylation sites from a large-scale proteome data with modified cysteine residues. However, an increasing number of cysteine-based redox modifications, such as *S*-sulfenylation, *S*-nitrosylation and *S*-glutathionylation, were reported to share the same cysteine targets [48]. Therefore, investigating the sequence characteristics among these cysteine-based PTMs is necessary for achieving reliable detection of *S*-sulfenylation sites. According to experimentally verified data obtained from dbSNO [49, 50], there exist 2212 *S*-nitrosylated proteins containing 4244 nitrosylation sites. In addition, a total of 2148 S-glutathionylated proteins harboring 3641 S-glutathionylation sites were obtained from dbGSH [51]. To avoid redundancy in data as a result of homologous sequences, sequence fragments were deleted if they were identified with 100 % similarity across the three datasets. Overall, 162 (15.74 %) of the 1029 *S*-sulfenylated proteins may undergo all three types of PTMs (Fig. 6A). In addition, percentages of the 1029 *S*-sulfenylated proteins that were also modified by *S*-nitrosylation and *S*-glutathionylation are approximately 23.23 % and 34.89 %, respectively. Investigation of the 1434 *S*-sulfenylated cysteine residues revealed that only 103 sites (7.18 %) could undergo all the three PTMs. As illustrated in Fig. 6B, the percentage of sites that are susceptible to both *S*-sulfenylation and *S*-glutathionylation (21.27 %) appeared to be higher than those that can be modified by *S*-sulfenylation and *S*-nitrosylation (14.99 %).

Taking advantage of the TwoSampleLogo, we uncovered the potential consensus motifs that may distinguish *S*-sulfenylation sites from *S*-nitrosylation or *S*-glutathionylation sites. As represented in Fig. 6C, the figure on the top panel describes notable differences in the position-specific compositions between *S*-sulfenylation and *S*-glutathionylation, while the figure on the bottom panel shows the potential amino acid composition difference between *S*-sulfenylation and *S*-nitrosylation sites. It seemed that *S*-sulfenylation sites could be recognized based on specific positions such as −10, −6, −5, −2, and +2 to +4 (p-value < 0.01). This investigation indicated a consistent composition of amino acids with Fig. 2D, particularly at positions −6, −2, +3 and +4. The frequency of positively charged groups, including R (Arginine) and K (Lysine), appeared to be significant at position −6 and −2, and that the occurrence of E

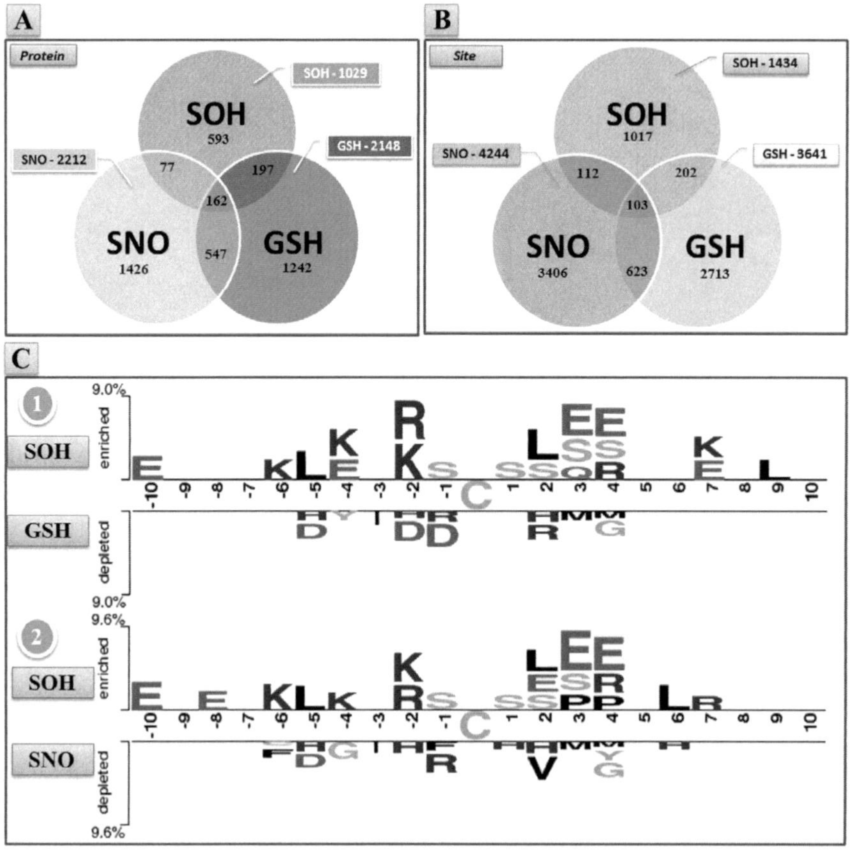

Fig. 6 Discrimination of S-sulfenylation sites from S-nitrosylation and S-glutathionylation sites. **a** Number of duplicate proteins among S-sulfenylation, S-nitrosylation and S-glutathionylation; (**b**) Number of duplicate sites among S-sulfenylation, S-nitrosylation and S-glutathionylation; (**c**) Significant differences in position-specific compositions among three PTMs as identified by TwoSampleLogo

(Glutamic acid) belonging to negatively charged residue was also apparent at positions +3 and +4.

Conclusions

This study describes a systematic investigation on the experimentally verified *S*-sulfenylation sites based on amino acid composition. The analysis of position-specific amino acids composition revealed that the most pronounced feature of *S*-sulfenylation sites is the abundance of positively charged amino acids (K and R) at surrounding positions: −10, −8 ~ −6, −4, −2, and +4 to +8. However, the depletion of positively charged residues was observed at positions −1, +1 and +2 that are close to the *S*-sulfenylation sites. Additionally, a noticeable abundance of negatively charged resides (E) was found at positions −3, +1, +3, +4. This investigation suggested that the amino acid position relative to one another in the sequence may play a vital role in discriminating between *S*-sulfenylation and non-*S*-sulfenylation sites. The comparison of position-specific amino acid composition between *S*-sulfenylation and non-*S*-sulfenylation sites also implicated that, for *S*-sulfenylation sites, distant amino acids around position −7 and +7, which may be close to *S*-sulfenylated cysteines in three-dimensional structure, harbor a notable abundance of positively charged amino acids (K and R).

In this work, the solvent accessibility and physicochemical properties were considered in the characterization of *S*-sulfenylation sites. The *S*-sulfenylation sites have a higher solvent accessibility, especially at the positions −7, +1, +2 and +4. According to the F-score measurements on 531 physicochemical properties, 20 physicochemical properties were revealed to have statistically significant differences between *S*-sulfenylation and non-*S*-sulfenylation sites. Based on the evaluation by five-fold cross-validation, the model trained with the combined features of evolutionary information (PSSM) and 12 physicochemical properties yielded the best prediction performance. Overall, our study not only determined the best prediction model for the identification of S-sulfenylation sites, but also discovered several characteristics and functions regarding S-sulfenylated proteins. Hydrophobic characteristics of

S-sulfenylated proteins seem to be crucial for the understanding of their functions. Moreover, amino acids R (Arginine), E (Glutaminc acid), and K (Lysine) may be the potential consensus motifs for specific S-sulfenylated sites.

Additional files

Additional file 1: Figure S1. Analytical flowchart of removing homologous sequences in training dataset and independent testing dataset. (DOCX 242 kb)

Additional file 2: Table S1. Top 20 best performing physicochemical properties as evaluated by five-fold cross-validation. (DOCX 17 kb)

Additional file 3: Figure S2. Methods for combining PSSM with physicochemical properties by forward selection. (DOCX 126 kb)

Additional file 4: Table S2. Five-fold cross-validation performance of combining PSSM with the top 20 physicochemical properties by forward selection. (DOCX 17 kb)

Additional file 5: Table S3. Comparison of independent testing results between the PSSM model and the hybrid model combining PSSM with the top 12 physicochemical properties. (DOCX 14 kb)

Additional file 6: Table S4. Top 10 most enriched GO categories associated with *S*-sulfenylated proteins ($p < 0.01$). (DOCX 16 kb)

Additional file 7: Table S5. Distribution of KEGG pathway annotations for *S*-sulfenylated proteins. (DOCX 15 kb)

Additional file 8: Table S6. Distribution of InterPro functional domains covering *S*-sulfenylated sites. (DOCX 17 kb)

Competing interests
The authors have declared that no competing interests exist.

Authors' contributions
TYL and JTYW conceived and supervised the project. VMB, SLW and CTL were responsible for the experimental design. VMB, CTL and THC contributed on computational analyses. VMB and TYL drafted the manuscript with revisions by JTYW. All authors read and approved the final manuscript.

Acknowledgements
The authors sincerely appreciate the Ministry of Science and Technology of Taiwan for financially supporting this research under Contract Number of MOST 103-2221-E-155-020-MY3, MOST 103-2633-E-155-002 and MOST104-2221-E-155-036-MY2.

Declarations
Publication charge for this work was funded by MOST grant under contract number of MOST 103-2221-E-155-020-MY3 and MOST 104-2221-E-155-036-MY2 to TYL.

Author details
[1]Department of Computer Science and Engineering, Yuan Ze University, Taoyuan 320, Taiwan. [2]Department of Obstetrics and Gynecology, Hsinchu Mackay Memorial Hospital, Hsin-Chu 300, Taiwan. [3]Mackay Junior College of Medicine, Nursing and Management, Taipei 112, Taiwan. [4]Department of Medicine, Mackay Medical College, New Taipei City 252, Taiwan. [5]Graduate Institute of Biomedical Informatics, Taipei Medical University, Taipei 110, Taiwan. [6]Innovation Center for Big Data and Digital Convergence, Yuan Ze University, Taoyuan 320, Taiwan.

Published: 11 January 2016

References
1. Leonard SE, Carroll KS. Chemical 'omics' approaches for understanding protein cysteine oxidation in biology. Curr Opin Chem Biol. 2011;15(1): 88–102.
2. Poole LB, Nelson KJ. Discovering mechanisms of signaling-mediated cysteine oxidation. Curr Opin Chem Biol. 2008;12(1):18–24.
3. Wani R, Qian J, Yin L, Bechtold E, King SB, Poole LB, et al. Isoform-specific regulation of Akt by PDGF-induced reactive oxygen species. Proc Natl Acad Sci U S A. 2011;108(26):10550–5.
4. Roos G, Messens J. Protein sulfenic acid formation: From cellular damage to redox regulation. Free Radic Biol Med. 2011;51(2):314–26.
5. Leonard SE, Reddie KG, Carroll KS. Mining the thiol proteome for sulfenic acid modifications reveals New targets for oxidation in cells. ACS Chem Biol. 2009;4(9):783–99.
6. Weerapana E, Wang C, Simon GM, Richter F, Khare S, Dillon MBD, et al. Quantitative reactivity profiling predicts functional cysteines in proteomes. Nature. 2010;468(7325):790–U779.
7. Wang C, Weerapana E, Blewett MM, Cravatt BF. A chemoproteomic platform to quantitatively map targets of lipid-derived electrophiles. Nat Methods. 2014;11(1):79–+.
8. Szychowski J, Mahdavi A, Hodas JJL, Bagert JD, Ngo JT, Landgraf P, et al. Cleavable biotin probes for labeling of biomolecules via azide-alkyne cycloaddition. J Am Chem Soc. 2010;132(51):18351–60.
9. Qian Y, Martell J, Pace NJ, Ballard TE, Johnson DS, Weerapana E. An isotopically tagged azobenzene-based cleavable linker for quantitative proteomics. Chembiochem. 2013;14(12):1410–4.
10. Zheng T, Jiang H, Wu P. Single-stranded DNA as a cleavable linker for bioorthogonal click chemistry-based proteomics. Bioconjug Chem. 2013; 24(6):859–64.
11. Yang J, Gupta V, Carroll KS, Liebler DC. Site-specific mapping and quantification of protein S-sulphenylation in cells. Nat Commun. 2014;5: 4776.
12. M-a S, Wang Y, Cheng H, Zhang Q, Ge W, Guo D. RedoxDB-a curated database for experimentally verified protein oxidative modification. Bioinformatics. 2012;28(19):2551–2.
13. Mucchielli-Giorgi MHM, Hazout S, Tuffery P. Predicting the disulfide bonding state of cysteines using protein descriptors. Proteins-Structure Function and Genetics. 2002;46(3):243–9.
14. Chang C-C, Lin C-J. LIBSVM: A Library for Support Vector Machines. ACM Trans Intell Syst Technol. 2011;2(3):27:1–27:27.
15. UniProt C. UniProt: a hub for protein information. Nucleic Acids Res. 2015; 43(Database issue):D204–212.
16. Furdui CM, Poole LB. Chemical approaches to detect and analyze protein sulfenic acids. Mass Spectrom Rev. 2014;33(2):126–46.
17. Shien D-M, Lee T-Y, Chang W-C, Hsu JB-K, Horng J-T, Hsu P-C, et al. Incorporating structural characteristics for identification of protein methylation sites. J Comput Chem. 2009;30(9):1532–43.
18. Tatusova TA, Madden TL. BLAST 2 Sequences, a new tool for comparing protein and nucleotide sequences. FEMS Microbiol Lett. 1999;174(2):247–50.
19. Wong YH, Lee TY, Liang HK, Huang CM, Wang TY, Yang YH, et al. KinasePhos 2.0: a web server for identifying protein kinase-specific phosphorylation sites based on sequences and coupling patterns. Nucleic Acids Res. 2007;35(Web Server issue):W588–594.
20. Henikoff S, Henikoff JG. Amino acid substitution matrices from protein blocks. Proc Natl Acad Sci U S A. 1992;89(22):10915–9.
21. Chang W-C, Lee T-Y, Shien D-M, Hsu JB-K, Horng J-T, Hsu P-C, et al. Incorporating support vector machine for identifying protein tyrosine sulfation sites. J Comput Chem. 2009;30(15):2526–37.
22. Altschul SF, Madden TL, Schaffer AA, Zhang J, Zhang Z, Miller W, et al. Gapped BLAST and PSI-BLAST: a new generation of protein database search programs. Nucleic Acids Res. 1997;25(17):3389–402.
23. Pang CNI, Hayen A, Wilkins MR. Surface accessibility of protein post-translational modifications. J Proteome Res. 2007;6(5):1833–45.
24. Ahmad S, Gromiha MM, Sarai A. RVP-net: online prediction of real valued accessible surface area of proteins from single sequences. Bioinformatics. 2003;19(14):1849–51.
25. Ahmad S, Gromiha MM, Sarai A. Real value prediction of solvent accessibility from amino acid sequence. Proteins-Structure Function and Genetics. 2003; 50(4):629–35.
26. Rose PW, Prlic A, Bi C, Bluhm WF, Christie CH, Dutta S, et al. The RCSB Protein Data Bank: views of structural biology for basic and applied research and education. Nucleic Acids Res. 2015;43(Database issue):D345–356.
27. McGuffin LJ, Bryson K, Jones DT. The PSIPRED protein structure prediction server. Bioinformatics. 2000;16(4):404–5.
28. Kawashima S, Pokarowski P, Pokarowska M, Kolinski A, Katayama T, Kanehisa M. AAindex: amino acid index database, progress report 2008. Nucleic Acids Res. 2008;36(Database issue):D202–205.

29. Nguyen VN, Huang KY, Huang CH, Chang TH, Bretana N, Lai K, et al. Characterization and identification of ubiquitin conjugation sites with E3 ligase recognition specificities. BMC bioinformatics. 2015;16 Suppl 1:S1.
30. Su MG, Huang KY, Lu CT, Kao HJ, Chang YH, Lee TY. topPTM: a new module of dbPTM for identifying functional post-translational modifications in transmembrane proteins. Nucleic Acids Res. 2014;42(Database issue): D537–545.
31. Lee TY, Chen YJ, Lu TC, Huang HD, Chen YJ. SNOSite: Exploiting Maximal Dependence Decomposition to Identify Cysteine S-Nitrosylation with Substrate Site Specificity. Plos One. 2011;6(7):e21849.
32. Lee TY, Chen SA, Hung HY, Ou YY. Incorporating distant sequence features and radial basis function networks to identify ubiquitin conjugation sites. Plos One. 2011;6(3):e17331.
33. Hsu JBK, Bretana NA, Lee TY, Huang HD. Incorporating Evolutionary Information and Functional Domains for Identifying RNA Splicing Factors in Humans. Plos One. 2011;6(11):e27567.
34. Lin C-J, Chen Y-W. Combining SVMs with various feature selection strategies. Feature Extraction. 2003;207:315–24.
35. Huang HD, Lee TY, Tzeng SW, Horng JT. KinasePhos: a web tool for identifying protein kinase-specific phosphorylation sites. Nucleic Acids Res. 2005;33(Web Server issue):W226-229.
36. Lu C-T, Chen S-A, Bretana NA, Cheng T-H, Lee T-Y. Carboxylator: incorporating solvent-accessible surface area for identifying protein carboxylation sites. J Comput Aided Mol Des. 2011;25(10):987–95.
37. Bui VM, Lu CT, Ho TT, Lee TY. MDD-SOH: Exploiting maximal dependence decomposition to identify S-sulfenylation sites with substrate motifs. Bioinformatics. 2015.
38. Crooks GE, Hon G, Chandonia JM, Brenner SE. WebLogo: A sequence logo generator. Genome Res. 2004;14(6):1188–90.
39. Vacic V, Iakoucheva LM, Radivojac P. Two Sample Logo: a graphical representation of the differences between two sets of sequence alignments. Bioinformatics. 2006;22(12):1536–7.
40. Dennis G, Sherman BT, Hosack DA, Yang J, Gao W, Lane HC, et al. DAVID: Database for annotation, visualization, and integrated discovery. Genome Biol. 2003;4(9):11.
41. Consortium TGO. The Gene Ontology: enhancements for 2011. Nucleic Acids Res. 2011;40(Database issue):D559–564.
42. Huang KY, Wu HY, Chen YJ, Lu CT, Su MG, Hsieh YC, et al. RegPhos 2.0: an updated resource to explore protein kinase-substrate phosphorylation networks in mammals. Database : the journal of biological databases and curation. 2014;2014(0):bau034.
43. Lu CT, Huang KY, Su MG, Lee TY, Bretana NA, Chang WC, et al. dbPTM 3.0: an informative resource for investigating substrate site specificity and functional association of protein post-translational modifications. Nucleic Acids Res. 2013;41(D1):D295–305.
44. Lee TY, Bo-Kai Hsu J, Chang WC, Huang HD. RegPhos: a system to explore the protein kinase-substrate phosphorylation network in humans. Nucleic Acids Res. 2011;39(Database issue):D777–787.
45. Seet BT, Dikic I, Zhou MM, Pawson T. Reading protein modifications with interaction domains. Nat Rev Mol Cell Biol. 2006;7(7):473–83.
46. Aranda E, Lopez-Pedrera C, De La Haba-Rodriguez JR, Rodriguez-Ariza A. Nitric oxide and cancer: the emerging role of S-nitrosylation. Curr Mol Med. 2012;12(1):50–67.
47. Hunter S, Jones P, Mitchell A, Apweiler R, Attwood TK, Bateman A, et al. InterPro in 2011: new developments in the family and domain prediction database. Nucleic Acids Res. 2011;40(Database issue):D306–312.
48. Couturier J, Chibani K, Jacquot JP, Rouhier N. Cysteine-based redox regulation and signaling in plants. Frontiers in plant science. 2013;4:105.
49. Chen Y-J, Lu C-T, Su M-G, Huang K-Y, Ching W-C, Yang H-H, et al. dbSNO 2. 0: a resource for exploring structural environment, functional and disease association and regulatory network of protein S-nitrosylation. Nucleic Acids Res. 2015;43(D1):D503–11.
50. Lee TY, Chen YJ, Lu CT, Ching WC, Teng YC, Huang HD, et al. dbSNO: a database of cysteine S-nitrosylation. Bioinformatics. 2012;28(17):2293–5.
51. Chen YJ, Lu CT, Lee TY, Chen YJ. dbGSH: a database of S-glutathionylation. Bioinformatics. 2014;30(16):2386–8.

Karim *et al. BMC Genomics* 2015, **17**(Suppl 1):7
DOI 10.1186/s12864-015-2300-z

BMC Genomics

RESEARCH Open Access

Identification of miRNA-mRNA regulatory modules by exploring collective group relationships

S. M. Masud Karim*, Lin Liu, Thuc Duy Le and Jiuyong Li

From The Fourteenth Asia Pacific Bioinformatics Conference (APBC 2016)
San Francisco, CA, USA. 11 - 13 January 2016

Abstract

Background: microRNAs (miRNAs) play an essential role in the post-transcriptional gene regulation in plants and animals. They regulate a wide range of biological processes by targeting messenger RNAs (mRNAs). Evidence suggests that miRNAs and mRNAs interact collectively in gene regulatory networks. The collective relationships between groups of miRNAs and groups of mRNAs may be more readily interpreted than those between individual miRNAs and mRNAs, and thus are useful for gaining insight into gene regulation and cell functions. Several computational approaches have been developed to discover miRNA-mRNA regulatory modules (MMRMs) with a common aim to elucidate miRNA-mRNA regulatory relationships. However, most existing methods do not consider the collective relationships between a group of miRNAs and the group of targeted mRNAs in the process of discovering MMRMs. Our aim is to develop a framework to discover MMRMs and reveal miRNA-mRNA regulatory relationships from the heterogeneous expression data based on the collective relationships.

Results: We propose *DIscovering COllective group REIationships* (*DICORE*), an effective computational framework for revealing miRNA-mRNA regulatory relationships. We utilize the notation of collective group relationships to build the computational framework. The method computes the collaboration scores of the miRNAs and mRNAs on the basis of their interactions with mRNAs and miRNAs, respectively. Then it determines the groups of miRNAs and groups of mRNAs separately based on their respective collaboration scores. Next, it calculates the strength of the collective relationship between each pair of miRNA group and mRNA group using canonical correlation analysis, and the group pairs with significant canonical correlations are considered as the MMRMs. We applied this method to three gene expression datasets, and validated the computational discoveries.

Conclusions: Analysis of the results demonstrates that a large portion of the regulatory relationships discovered by *DICORE* is consistent with the experimentally confirmed databases. Furthermore, it is observed that the top mRNAs that are regulated by the miRNAs in the identified MMRMs are highly relevant to the biological conditions of the given datasets. It is also shown that the MMRMs identified by *DICORE* are more biologically significant and functionally enriched.

Keywords: miRNA-mRNA regulatory modules, Collective group relationships, Group pair, Canonical correlations

*Correspondence: masud.karim@mymail.unisa.edu.au
School of Information Technology and Mathematical Sciences, University of South Australia, Mawson Lakes, SA 5095, Adelaide, Australia

Background

microRNAs (miRNAs) are a family of small (i.e. with typical length of 19–25 nucleotides) non-protein-coding RNA molecules that can play important regulatory roles in animals and plants [1, 2]. They base-pair with messenger RNAs (mRNAs) of protein-coding genes to induce mRNA degradation or translational repression [3]. The mature human miRNAs potentially target majority of the human mRNAs [4]. It has been demonstrated that miRNAs regulate a wide range of biological or cellular processes such as proliferation [5, 6], metabolism [7], differentiation [8], development [9], apoptosis [10], cellular signaling [11], and cancer development and progression [12–15].

There is a growing body of literature showing that multiple miRNAs are coordinated by forming cohesive groups to collectively regulate one or more pathways [16, 17]. The collective relationships yielded between a group of miRNAs and a group of mRNAs due to the tendency of the group formation act as a vital force in catering similar functioning miRNAs and mRNAs together. Therefore, the collective relationships between cohesive groups of miRNAs and their targeted mRNAs may provide better understandings on robust and potent regulatory relationships of miRNA-mRNA regulatory modules (MMRMs).

Several algorithms have been proposed to identify MMRMs from expression data using different approaches including Bayesian network learning [18], rule induction [19], association rule mining [20], population-based probabilistic learning [21], probabilistic graphical model [22–24], matrix factorization [25], and graph mining [17, 26]. Most of these existing methods do not consider the collective relationships between a group of miRNAs and the group of targeted mRNAs in the process of identifying MMRMs. In addition, many of them are either stochastic, or require prior knowledge such as number of modules to be identified, confirmed interactions, target site information [27].

Adapting a greedy overlapping neighborhood expansion clustering method, *ClusterONE*, which was developed to discover protein complexes from protein-protein interactions networks, Li et al. [27] proposed a clustering algorithm, *Mirsynergy* to detect synergistic miRNA regulatory modules. However, it requires and depends on the prior knowledge of confirmed gene-gene interactions. Recently Karim et al. [28] coined the notion of *collective group relationships*, and developed a method by integrating unweighted graphing mining concept and canonical correlation analysis to explore miRNA-mRNA regulatory relationships. However, it is noted that unweighted graph mining techniques are associated with limitation in representing the true interactions, and sometimes fail to capture correct regulatory relationships. Whereas weighted graph mining approaches can greatly improve the detection of the module structures [29], and hence regulatory relationships.

In this paper, we propose an effective computational framework, *DIscovering COllective group RElationships* (*DICORE*) to identify MMRMs and hence reveal miRNA-mRNA regulatory relationships from heterogeneous data. In order to extract MMRMs from the given gene expression datasets, we utilize the notion of collective group relationships, which provide MMRMs with additional quantitative strength information. The method finds a deterministic solution to the problem of discovering MMRMs from weighted bipartite graph representation of the given datasets, and rank the collective group relationships based on their strength of collective relationships. We apply *DICORE* to a dataset for Epithelial to Mesenchymal Transition, a breast cancer dataset, and a multi-class cancer dataset. Based on the knowledge from the literature, it is observed that the identified MMRMs exhibit enriched functionality with biological significance.

Methods

Problem statement

Consider two sets of variables $\mathbf{X} = \{X_1, \ldots, X_p\}$ and $\mathbf{Y} = \{Y_1, \ldots, Y_q\}$ such that $\mathbf{X} \cap \mathbf{Y} = \varnothing$, representing the attributes of two different types of objects. In this paper, $\mathbf{X}$ and $\mathbf{Y}$ refer to the expression levels of a set of miRNAs and a set of mRNAs, respectively. With their given datasets, $\mathbf{D_X}$ and $\mathbf{D_Y}$, having n matching miRNA and mRNA expression samples, our goal is to identify any $C_x \subset \mathbf{X}$ and $C_y \subset \mathbf{Y}$, such that C_x and C_y are related, as a result of miRNAs in C_x collaboratively interacting with mRNAs in C_y and vice versa. We call (C_x, C_y) a *group pair*, and the relationship between C_x and C_y a *COllective group RElationship* (in short, *CORE*). The COREs are characterized by both group pairs and the collective relationships among the two cohesive groups in group pairs. Then the group pair (C_x, C_y) is an MMRM if the strength of the CORE between C_x and C_y is significant.

In order to discover COREs, and thus to identify MMRMs, we develop a two stages method, *DIscovering CORE* (*DICORE*). Two measures, collaboration score and canonical correlations, are employed in the two stages respectively. In the following, we firstly overview the workflow of *DICORE*, and then present the details of *DICORE*, including the definition of the collaboration score and the calculation of canonical correlations.

Overview of *DICORE*

Figure 1 shows the workflow of *DICORE*. The overall workflow comprises a data pre-processing step and two main stages: (1) forming separate miRNA and mRNA groups and (2) searching for COREs.

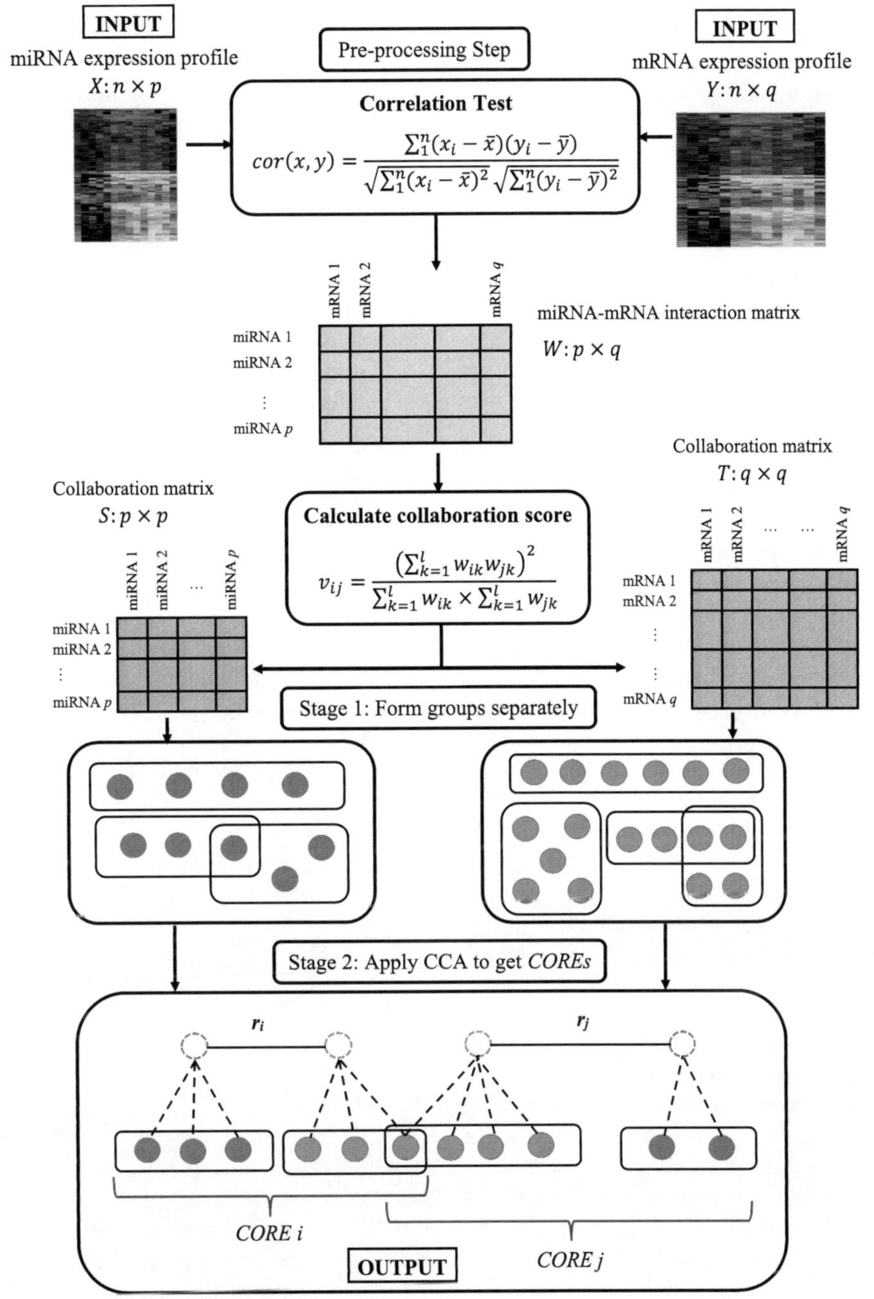

Fig. 1 *DICORE* workflow. Given the inputs of miRNA and mRNA expression profiles, we first derive an expression-based interaction weights matrix *W* using correlation test. We then compute two collaboration score matrices *S* and *T* from *W* for miRNAs and mRNAs based on their functional interaction similarities with common mRNAs (or miRNAs), respectively. Using these collaboration scores as input, we separately generate groups of miRNAs and groups of mRNAs at Stage 1 by an overlapping neighborhood expansion clustering algorithm, in which miRNAs or mRNAs are greedily added to (removed from) each cluster of miRNAs or mRNAs, respectively that maximize cohesiveness score of the cluster. Next in Stage 2, we apply canonical correlation analysis on the groups of miRNAs and groups of mRNAs to obtain significant collective group relationships, which are eventually the MMRMs with strength scores

In the data pre-processing step, *DICORE* first creates a weighted bipartite graph representation of the relationships among the individual variables of the given miRNA and mRNA expression profiles. Taking the variables as the vertices of a weighted bipartite graph *G*, a weighted edge is introduced between a miRNA variable and a mRNA variable to represent their interaction. Referring to Fig. 1, given p miRNAs and q mRNAs, let W denote the $(p \times q)$ miRNA-mRNA interaction weights matrix, where w_{ij} is the interaction weight for miRNA i targeting mRNA j. To compute miRNA-mRNA interaction weights, we calculate the *Pearson correlation coefficient* (*PCC*) [25] between each pair of miRNA and mRNA using the R built-in function, *cor*. The obtained PCCs are within the range of $[-1, 1]$, and the signed correlation coefficients provide two types of valuable information: the absolute values implying the strength of the miRNA-mRNA interactions (the higher the values, the stronger the interactions), and the signs indicating the directions of the associations. However, as the aim of the paper is to identify MMRMs (and thus to uncover miRNA-mRNA regulatory relationships), the collaboration score (explained in the next section) defined for discovering the modules considers the sum of the miRNA-mRNA correlations. In order to cater for both up and down miRNA regulations when calculating the total strength of the interactions, we use absolute values of the PCCs in the interaction weights matrix W, otherwise the signed PCCs or interaction weights will cancel out in Eq. (1).

Due to the higher possibility of dense interactions in the expression profile datasets, complete weighted graph mining may not be able to distinguish correct group structure. Accordingly we used a cutoff threshold η to trade off between the two extreme approaches namely complete unweighted graph mining and complete weighted graph mining.

At stage 1, we separately identify groups of miRNAs and groups of mRNAs. Referring to Fig. 1, based on the interaction weights matrix W, we firstly calculate the collaboration score between each pair of miRNAs and create the miRNA-miRNA collaboration matrix, S. The collaboration score between a pair of miRNAs reflects their similarity or collaboration in regulating target mRNAs (more details of collaboration scores are given in the next section). In a similar way, we compute the collaboration score between each pair of mRNAs (which implies their similarity in being regulated by miRNAs) and create the mRNA-mRNA collaboration matrix, T.

The identification of groups of miRNAs (or groups of mRNAs) is formulated as an overlapping clustering problem. Only the miRNAs (or mRNAs) that have strong collaboration between them are put in the same group, i.e. we use their collaboration scores as the similarity measure for the clustering. The clustering process is then aimed at maximizing the overall similarity of the miRNAs (or mRNAs) within the same group. We define such overall similarity within a group as the cohesiveness of a group (details of the definition is provided in the next section). The underlying clustering algorithm adapts from *ClusterONE*, which was originally developed for protein protein interaction networks [29]. Adopting the idea from [25], we discard groups with fewer than 5 mRNAs (i.e. minimum size threshold for mRNAs, $\theta_g = 5$), as they usually do not provide relevant information. Similarly, we are not interested to consider groups having more than 500 mRNAs. Additionally, in order to avoid 'star-shaped' basic network structure, we choose 3 as minimum size threshold for miRNAs, θ_m.

At stage 2, we use canonical correlation analysis to compute the strength of the collective relationships between groups of miRNAs and groups of mRNAs in terms of *canonical correlations*, and obtain COREs, which is eventually equivalent to MMRMs with additional quantitative information. We considered only the top COREs identified (i.e. the COREs with the higher canonical correlations), having minimum canonical correlation of $\rho = 0.50$.

Details of *DICORE*

In the following, we introduce the details of the collaboration score and how CCA is used to measure the strength of the collective group relationships.

The *collaboration score* expresses the degree of collaboration between two miRNAs (or between two mRNAs) considering their common interactions with mRNAs (or miRNAs). Given miRNA i, miRNA $j \, (\neq i)$ and the interaction weights matrix W, the collaboration score of the two miRNAs is calculated as follows:

$$v_{ij} = \frac{\left(\sum_{k=1}^{l} w_{ik} w_{jk} \right)^2}{\sum_{k=1}^{l} w_{ik} \times \sum_{k=1}^{l} w_{jk}}, \tag{1}$$

where l is the number of other possible components that both miRNA i and miRNA j interact with, in this case mRNAs, so $l = q$. Let S refer to the miRNA-miRNA collaboration matrix of size $p \times p$, where $s_{ij} = v_{ij}$.

Similarly, we compute the mRNA-mRNA collaboration score between mRNA i and mRNA $j \, (\neq i)$ by applying Eq. (1) on the transpose of the interaction weights matrix W, where $l = p$, the number of miRNAs. Let T refer to the mRNA-mRNA collaboration matrix of size $q \times q$, where $t_{ij} = v_{ij}$.

Notably, if W were a binary matrix, Eq. (1) became the ratio of number of target mRNAs shared between miRNA i and miRNA j over the numbers of target mRNAs possessed separately by miRNA i or miRNA j (or the ratio of number of common miRNAs regulate both mRNA i and mRNA j over the numbers of miRNAs individually regulate mRNA i or mRNA j). An miRNA (or an mRNA) i is then ranked by the total collaboration score as $\sum_{k=1}^{p} s_{ik} \left(\text{or} \sum_{k=1}^{q} t_{ik} \right)$.

Using collaboration scores as the similarity measures of pairs of miRNAs or pairs of mRNAs, miRNAs and mRNAs are clustered separately into cohesive groups by using a greedy strategy that maximize the cohesiveness score of groups. Similar to the cohesiveness defined in [29], we define *cohesiveness* score, $cs(C_i)$ for any group C_i as follows:

$$cs(C_i) = \frac{w_{int}(C_i)}{w_{int}(C_i) + w_{ext}(C_i) + \alpha * |C_i|} \quad (2)$$

where $w_{int}(C_i)$ denotes the sum of the collaboration scores of all the internal pairs of variables, i.e. each pair only contains variables within the group C_i; $w_{ext}(C_i)$ is the sum of the collaboration scores of all the external pairs, i.e. each pair contains one variable within the group C_i and one variable outside the group C_i; and $\alpha * |C_i|$ is a penalty term asserting the existence of unidentified interactions in the dataset, practically assuming that every component in C_i has α additional interactions that are undetected due to the limitations in the experimental setting.

DICORE uses *canonical correlation analysis* (*CCA*) [30] to compute the strength of the collective relationships between a group of miRNAs and a group of mRNAs in terms of the group pair's canonical correlations. CCA is commonly used for quantifying the linear association between two sets of variables. Consider $\mathcal{A} = \vec{a}'\vec{X}$, $\mathcal{B} = \vec{b}'\vec{Y}$ be the corresponding linear combinations of sets of variables $\vec{X}$ and $\vec{Y}$ respectively, where $\vec{a}$ and $\vec{b}$ are coefficient vectors. Vectors $\vec{a}$ and $\vec{b}$ are chosen such that the correlation between $\mathcal{A}$ and $\mathcal{B}$, i.e.,

$$\mathbf{r} = Corr(\mathcal{A}, \mathcal{B}) = \frac{\vec{a}'\Sigma_{XY}\vec{b}}{\sqrt{\vec{a}'\Sigma_{XX}\vec{a}}\sqrt{\vec{b}'\Sigma_{YY}\vec{b}}} \quad (3)$$

is maximized, where Σ_{XX}, Σ_{YY} and Σ_{XY} are variance of $\vec{X}$, variance of $\vec{Y}$, and covariance between $\vec{X}$ and $\vec{Y}$, respectively. The correlation $\mathbf{r}$ between the pair of linear combinations in Eq. (3) is called *canonical correlation*. Specifically, canonical correlation between a group of miRNAs and a group of mRNAs is computed using the R function *CCA* from the package PMA.

The intuition behind applying CCA is twofold. Firstly CCA captures weight scores of all interactions between all miRNAs and mRNAs in both groups of a group pair, while computing the strength of the collective interactions of the group pair. As a consequence, CCA mitigates the loss of weight scores of interactions due to the application of cutoff threshold η earlier. Secondly, it also makes it possible for a group of miRNAs (or a group of mRNAs) to be included in more than one CORE i.e. one module, if the strength of collective interactions satisfies the specified threshold.

Data collection

Three real-world gene expression datasets are used to validate *DICORE*: an NCI60 dataset for Epithelial to Mesenchymal Transition, a breast-cancer (BR) dataset, and a multi-class cancer (MCC) dataset. The pre-processed differentially expressed gene expression datasets were collected from [31].

Epithelial to Mesenchymal Transition (EMT) is a biological process that enables cells to acquire migratory mesenchymal characteristics by losing epithelial features. The EMTs are associated with embryonic development, wound healing, organ fibrosis, and in the initiation of metastasis for cancer progression. The NCI60 dataset includes 60 cancer cell lines from the National Cancer Institute (NCI). Cell lines categorized as epithelial (11 samples) and mesenchymal (36 samples) were used for this work. As a result of the differential gene expression analysis, 1154 mRNAs and 35 miRNAs were identified to be differentially expressed at significant level (adjusted p-value < 0.05, adjusted by Benjamini-Hochberg (BH) method).

The BR dataset includes expression profiles of the 50 cell lines of breast cancer. The cell lines were categorized as luminal (27 cell lines) and basal (23 cell lines). In the dataset, 89 miRNAs (adjusted p-value < 0.02) and 1500 mRNAs (adjusted p-value < 0.0001) were identified to be differentially expressed.

The MCC dataset includes samples from multiple cancers namely bladder, breast, colon, kidney, lung, pancreas, prostate and uterus. Samples of the dataset classified as normal (21 samples) and tumor (67 samples) were used in this work. In total, 62 miRNAs and 1318 mRNAs were obtained to be differentially expressed at significant level (adjusted p-value < 0.05).

The datasets are available in Additional file 1.

We used the expression data to calculate the miRNA-mRNA interaction weights matrix W. We obtained the interaction weights of W by computing the absolute values of the Pearson correlation coefficients between pairs of miRNA and mRNA.

In order to obtain the 'ground-truth' databases of experimentally confirmed miRNA-mRNA interactions, we combined the interactions from four popular interactions databases, namely DIANA-TarBase v7.0 [32], miRTarBase v4.5 [33], miRecords v2013 [34], and miRWalk v2.0 [35]. While miRWalk contains both predicted and experimentally validated miRNA-mRNA interactions, rest of the databases include high quality manually curated experimentally validated miRNA-mRNA interactions published in the literature. Recently published DIANA-TarBase v7.0 alone included more than half a million interactions utilizing cell types from 24 species. We also added a HITS-CLIP database [36], which lists the confirmed targets of two miRNAs, namely *miR-200a* and *miR-200b*. We extracted only the confirmed miRNA-mRNA interactions associated with the human miRNAs and mRNAs given in the input datasets, and removed the duplicate entries. Finally, we obtained 'ground-truth' databases of 2147, 5791, and 8733 unique miRNA-mRNA interactions for the 29 miRNAs in the NCI60 dataset (there are no confirmed interactions for the 6 miRNAs with the name prefix *hsa-miRPlus-*), 89 miRNAs in the BR dataset, and 62 miRNAs in the MCC dataset, respectively. Details of the 'ground-truth' databases are available in Additional file 2.

Results and discussions

We ran the experiment for all values for the cutoff threshold η in the range from 0 to 1 with a step size of 0.05. We only reported the summary and top results for each dataset. In each summary table, $\#C$, $\overline{mR}$, $\overline{miR}$, $\bar{r}$, and t denote the number of COREs identified, average number of mRNAs in COREs, average number of miRNAs in COREs, average strength of the COREs, and time taken for the execution in seconds, respectively. The group distributions and all COREs for all datasets are described in details on our website (visit [37]).

The NCI60 Dataset

The results obtained from the NCI60 dataset are summarized in Table 1. It is clear from the summary that potentially interesting results are obtained for the η values ranging from 0.60 to 0.85. By lowering the values of η, more miRNAs and mRNAs were added to these groups. For more in-depth analysis, we look more closely at some of the particular results.

We obtained the most informative result (in terms of the strength of COREs, and number of experimentally confirmed interactions covered) for $\eta = 0.60$, with 8 COREs involving 1 miRNA groups and 8 mRNA groups.

Table 1 Summary of results of *DICORE* on the NCI60 dataset

η	$\#C$	$\overline{mR}$	$\overline{miR}$	$\bar{r}$	t
0.45	1	6.00	35.00	0.61	1142.88
0.50	4	8.50	32.00	0.64	635.31
0.55	3	11.67	27.00	0.69	246.17
0.60	8	57.00	19.00	0.80	80.89
0.65	6	83.67	10.50	0.79	86.95
0.70	4	107.50	6.25	0.81	24.30
0.75	4	44.00	5.00	0.87	4.78
0.80	2	22.50	5.00	0.91	2.44
0.85	1	12.00	5.00	0.95	0.90

Table 2 Summary of results of *DICORE* on the BR dataset

η	$\#C$	$\overline{mR}$	$\overline{miR}$	$\bar{r}$	t
0.50	3	8.00	15.00	0.66	1938.13
0.55	44	23.57	6.18	0.63	2043.01
0.60	33	48.09	6.61	0.70	216.53
0.65	24	47.79	3.00	0.69	36.49

The only group of miRNAs 'm1N60' catered in total 19 miRNAs including the *miR-200* family. On the other hand, we got the top mRNAs group (group having highest cohesiveness) 'g1N60' having 348 mRNAs. It included *CDH1* (epithelial cadherin or in short E-cadherin, a classical member of the cadherin superfamily, which plays a vital role in EMT such that EMT is also characterized by repression of E-cadherin expression), *ZEB1* (E-cadherin transcriptional repressor, which is usually targeted by *miR-200* family), and *TWIST1* (one of the important EMT inducers).

Furthermore, another interesting result was obtained for $\eta = 0.65$. We got 6 COREs from 2 groups of miRNAs and 3 groups of mRNAs. The top miRNAs group 'm1N65' catered 14 miRNAs and is a proper subset of 'm1N60'. The second miRNAs group 'm2N65' included total 7 miRNAs including 3 miRNAs from the *miR-17* miRNA gene family, namely *miR-106b*, *miR-18a*, and *miR-18b*.

The BR dataset

From the summary given in Table 2, it is seen that higher informative results were obtained for η values from 0.55 to 0.65 from the BR dataset. The most informative result was obtained for $\eta = 0.60$. We got 33 COREs from 2 groups of miRNAs and 17 groups of mRNAs. The top group of miRNAs included *miR-221* and *miR-222*, both of them are known to play important regulation role in aggressive breast cancer [38].

The MCC dataset

Table 3 shows the results obtained by *DICORE* on the MCC dataset. The most informative result is obtained for $\eta = 0.45$. It catered all members of both *let-7* and *miR-30* miRNA gene families into the top group of miRNAs along with some other similar functioning miRNAs.

Table 3 Summary of results of *DICORE* on the MCC dataset

η	$\#C$	$\overline{mR}$	$\overline{miR}$	$\bar{r}$	t
0.40	7	15.71	35.00	0.58	354.30
0.45	10	84.70	20.40	0.58	540.36
0.50	5	30.40	29.00	0.72	12.40
0.55	1	55.00	25.00	0.80	4.18
0.60	1	19.00	9.00	0.82	1.45

Table 4 Confirmed interactions in COREs from the NCI60 for $\eta = 0.65$

ID	Confirmed interactions
C1N65	**miR-141**: BICD2, CDH1, EHF, IRF6, KLF5, PARD6B, RAB32, RAB8B, RHOD, SLC20A1, TWIST1;
	miR-148b: BIK, DDR1, ELOVL5, ERMP1, FAM84B, KLF5, MAL2, MAP1B, QKI, RAB8B, ST14, TBC1D30, TRAF4;
	miR-200a: BICD2, CDH1, EHF, ELOVL5, GRHL2, ITGB4, MAP1B, MSN, PARD6B, RAB32, TWIST1;
	miR-200b: AP1S2, ARHGAP32, CDH1, CLDN4, DSP, ELOVL5, ENSA, EPCAM, ESRP2, KIAA1949, KLF5, MAL2, MAP1B, MAPK13, MSN, OSTM1, PARD6B, QKI, SACS, SLC20A1, TINAGL1, TTL, TWIST1;
	miR-200c: AP1S2, CDH1, ENSA, ICA1, MSN, OSTM1, PARD6B, QKI, SLC20A1, ST14, TPD52L1, TWIST1, VIM;
	miR-203: ARHGAP32, CDH1, ENSA, FAM84B, OVOL1, PARD6B, TC2N, TPD52L1, VIM;
	miR-301a: AP1S2, BICD2, ERMP1, ESRP2, IRF6, MAL2, MAP1B, MAP7, PRRG4, SLC20A1, TRAF4, TTL, TWIST1;
	miR-301b: AP1S2, BICD2, MAP1B, PRRG4, TRAF4, TTL;
	miR-32: BICD2, QKI, RAB8B, RBM47, RNF43, SACS, TWIST1, VIM;
	miR-429: GRHL1, QKI, TWIST1;
	miR-590-3p: CDS1, DSP, ELOVL5, MAP1B, MRPL49, PARD6B, RAB8B, RBM47, SACS, SLC20A1;
	miR-7: ARHGAP32, DSC2, DSP, EPN3, ESRP1, F11R, FAM83H, FAM84B, GRHL1, GSR, LPAR2, MAP1B, MYO5B, PARD6B, PLS1, QKI, RAB11FIP4, S100A14, SACS, SLC29A2, TRAF4;
C2N65	**miR-141**: TMC5;
	miR-148b: EFNA1, MUC13, TBC1D30, TSKU;
	miR-200a: PLEKHF2, TMC5, TSKU,
	miR-200b: EFNA1, TACSTD2;
	miR-200c: EFNA1; **miR-301a**: PLEKHF2, TSKU;
	miR-301b: PLEKHF2; **miR-32**: PALM2-AKAP2;
	miR-429: EFNA1; **miR-7**: ANGEL1, LPAR2, TSKU;
C3N65	**miR-101**: AP1S2, BICD2, CLDN4, DLG3, MSN, RAB8B, SACS, SLC29A2, TST;
	miR-106b: BICD2, BMP4, DSP, ESRP1, F11R, GIPC2, KIAA1522, MAP7, MCF2L, MPZL2, MYO5B, OSTM1, PARD6B, PLS1, RAB8B, RBM47, S100P, SACS, SLC29A2, TBC1D30
	miR-18a: BICD2, BMP4, ELOVL5, ESRP1, INADL, MAP1B, MARK2, PARD6B;
	miR-18b: BICD2, ESRP1, INADL, MAP1B, SCNN1A;
	miR-30e: ABHD11, ANPEP, DSP, ELOVL5, FAM84B, GCNT3, GRHL2, ITGB4, MANSC1, MCF2L, OVOL1, PARD6B, PLS1, PPL, QKI, RAB32, RAB8B, SACS, SLC20A1, TC2N, VIM;
	miR-96: CDH1, CEP170, DSP, MAL2, PARD6B, RHOD, SLC20A1, TUBA1A, VIM;
C4N65	**miR-200b**: VIPR1; **miR-7**: RABGAP1L
C5N65	**miR-106b**: TBC1D30; **miR-18a**: SLC12A2;
	miR-30e: MOSC1, SLC12A2, TACSTD2;
	miR-96: EFNA1, ERBB3, PRPS1, TSKU;

miRNAs are highlighted in bold-face texts

Functional enrichment analysis of the COREs

A CORE consists of a group of miRNAs and a group of mRNAs, in which the individual interactions between miRNAs and mRNAs play a vital role. To demonstrate the effectiveness of *DICORE*, we identified the interactions in the obtained COREs and compared them with the experimentally confirmed interactions found in the 'ground-truth' databases. The confirmed interactions of the top COREs identified from the NCI60 dataset for $\eta = 0.65$ are summarized in Table 4. The confirmed interactions for the miRNAs in the *miR-200* family included in the top CORE 'C1N65' are illustrated in Fig. 2 using an example CORE, where red nodes are miRNAs and green nodes are experimentally confirmed target mRNAs. The higher number of confirmed interactions demonstrated the effectiveness of *DICORE*.

We got similar higher experimentally confirmed interactions for top COREs identified from BR and MCC datasets. The experimentally confirmed interactions for top COREs identified from the three datasets are listed in Additional file 3.

Pathway analysis of the COREs

A *biological pathway* is a group of genes that participate in a particular biological process to perform certain functionality in a cell. To find the controlling factors of a disease, it is meaningful to study the genes by considering their pathway information.

We used the GeneCodis [39] online tool at [40] to conduct pathway enrichment analysis of the COREs with the focus on significant Kyoto Encyclopedia of Genes and Genomes (KEGG) [41] pathways (adjusted

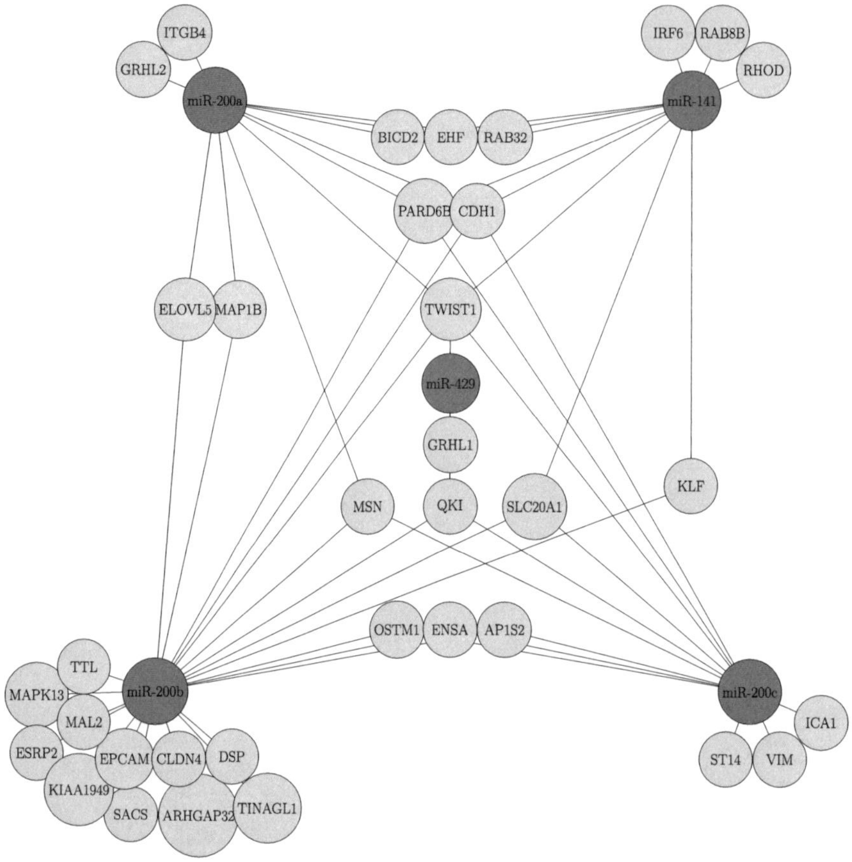

Fig. 2 Confirmed interactions for miRNAs of the *miR-200* family included in the top CORE 'C1N65' obtained from the NCI60. Red nodes are miRNAs, and green nodes are experimentally confirmed target mRNAs

p-value $<$ 0.05). We selected the top COREs 'C1N60', 'C1B60', and 'C1M45' discovered from the NCI60, BR, and MCC datasets, respectively for the analysis, and the top 7 enrichment KEGG pathways annotated with the COREs are listed respectively in Tables 5, 6 and 7 with their *p*-values, where the *p*-values are adjusted by Benjamini-Hochberg (BH) method. As shown in the tables, all the COREs are significantly associated with the KEGG pathway: *Pathways in cancer*. Since the three datasets are all cancer datasets, the results demonstrate that the identified COREs are closely related to the biological conditions of their respective datasets.

Again, we used GeneGo Metacore [42] from GeneGo Inc. to identify the pathways previously discovered in the literature that involve the mRNAs in the identified top

Table 5 Top 7 enrichment KEGG pathways for CORE 'C1N60' from the NCI60 for $\eta = 0.60$

No	KEGG Pathways	*p*-value
1	Tight junction	9.28E−08
2	Arrhythmogenic right ventricular cardiomyopathy (ARVC)	5.73E−04
3	Glutathione metabolism	3.40E−03
4	Leukocyte transendothelial migration	4.84E−03
5	Axon guidance	8.35E−03
6	Pathways in cancer	1.01E−02
7	Endocytosis	1.44E−02

Table 6 Top 7 enrichment KEGG pathways for CORE 'C1B60' from the BR for $\eta = 0.60$

No	KEGG Pathways	*p*-value
1	Inositol phosphate metabolism	2.44E−02
2	Complement and coagulation cascades	3.01E−02
3	Regulation of actin cytoskeleton	3.42E−02
4	Phosphatidylinositol signaling system	3.59E−02
5	Pathways in cancer	4.05E−02
6	ECM-receptor interaction	4.44E−02
7	Prostate cancer	4.59E−02

Table 7 Top 7 enrichment KEGG pathways for CORE 'C1M45' from the MCC for $\eta = 0.45$

No	KEGG Pathways	*p*-value
1	Vascular smooth muscle contraction	3.85E−08
2	Oocyte meiosis	6.07E−04
3	Complement and coagulation cascades	2.37E−03
4	Adherens junction	2.41E−03
5	Long-term depression	2.52E−03
6	Pathways in cancer	3.05E−03
7	Tight junction	4.46E−03

COREs. Table 8 shows the first 10 pathways as well as some other related pathways identified for another top CORE 'C1N65' from the NCI60 dataset. It confirms that 'C1N65' is highly relevant to the biological condition of the dataset. For instance, pathways number 1, 8, 11, 14 and 20 are direct pathways of the development of EMT, and others are important pathways involved in the process of EMT. Moreover, pathway number 1 includes total 12 members, of which 7 were identified in 'C1N65'.

The pathway enrichment analysis has clearly justified the use of CCA in ranking the COREs, as the top ranked COREs show higher biological significance, and represent the given datasets. The detailed information of significant pathways identified from the three datasets is summarized in Additional file 4.

Table 8 GeneGo mapped pathways for CORE 'C1N65' from the NCI60 for $\eta = 0.65$

No	Pathway maps	*p*-value
1	Development_miRNA-dependent inhibition of EMT	3.38E−12
2	Cytoskeleton remodelling_Keratin filaments	2.41E−11
3	Cell adhesion_Endothelial cell contacts by junctional mechanisms	1.03E−07
4	Cell adhesion_Tight junctions	8.05E−07
5	Cell adhesion_Gap junctions	1.54E−04
6	Development_Neural stem cell lineage commitment (schema)	3.92E−04
7	Cell cycle_Role of 14-3-3 proteins in cell cycle regulation	1.03E−03
8	Hypoxia-induced EMT in cancer and fibrosis	2.90E−03
9	LRRK2 in neurons in Parkinson's disease	3.39E−03
10	G-protein signaling_RhoA regulation pathway	3.69E−03
11	Development_TGF-β-dependent induction of EMT via SMADs	4.01E−03
14	Development_TGF-β-dependent induction of EMT via MAPK	9.18E−03
20	Development_Regulation of EMT	2.11E−02

Implication of the COREs in cancer

Since all of the input datasets included the expression profiles of miRNAs and genes associated with cancer samples, it is expected that the COREs identified from those datasets are to be related to cancer. To verify this, we used a cancer miRNA benchmark dataset of 147 miRNAs from a review article of [43]. Each of these miRNAs was reported in the literature to be dysregulated in one or more cancers.

The NCI60 dataset has 14 miRNAs from the benchmark, and except for *miR-205*, rest 13 are included in the top COREs. Both the top COREs 'C1N60' and 'C1N65' from the NCI60 dataset included 9 of the 14 miRNAs (namely *miR-141*, *miR-148b*, *miR-200a*, *miR-200b*, *miR-200c*, *miR-203*, *miR-301a*, *miR-32*, *miR-7*), which are associated with different cancers like Glioblastoma, Prostate, Lung, Bladder, Colon, Breast, Esophageal, Colorectal, Hepatocarcinoma, Ovarian, squamous cell carcinoma of tongue (SCCT), and Pancreatic.

Again, among these 147 miRNAs, 34 miRNAs are relevant to breast cancer. The BR dataset has 7 miRNAs out of these 34, of which 4 are identified in the top COREs.

On the other hand, the MCC dataset has 49 miRNAs out of the benchmark 147 miRNAs. The only CORE for $\eta = 0.60$ included 9 miRNAs, of which 8 are from the benchmark. These miRNAs are involved in verified association with breast cancer (*let-7d*, *miR-98*, *miR-101*), ovarian cancer (*let-7c*, *let-7d*, *miR-100*, *miR-126*, *miR-99a*), prostate cancer (*let-7c*), Burkitt Lymphoma cancer (*let-7c*), pancreatic cancer (*let-7d*, *miR-100*), bladder cancer (*miR-100*, *miR-195*, *miR-99a*), SCCT cancer (*miR-100*, *miR-195*, *miR-99a*), lung cancer (*miR-101*, *miR-126*), cervical cancer (*miR-126*), colon cancer (*miR-126*), and hepatocarcinoma (*miR-126*) [43]. It is interesting to note that one of the important parts of the COREs identified from MCC, i.e. the *let-7* family has special characteristics and mechanisms of tumor suppressor activity [44, 45].

Targets prediction for miRNAs of the COREs

In this section, we report a set of novel miRNA-mRNA interactions for further experiments. These miRNA-mRNA interactions identified by *DICORE* are the predicted targets of conserved miRNA families in TargetScan v6.2 [4, 46]. Fig 3 visualizes the predicted interactions in a model interaction representation of the CORE 'C1N65', where red nodes are miRNAs, yellow nodes are conserved target mRNAs, and white nodes are poorly conserved target mRNAs. Predicted (conserved) interactions for top COREs from the three databases are given in Additional file 5.

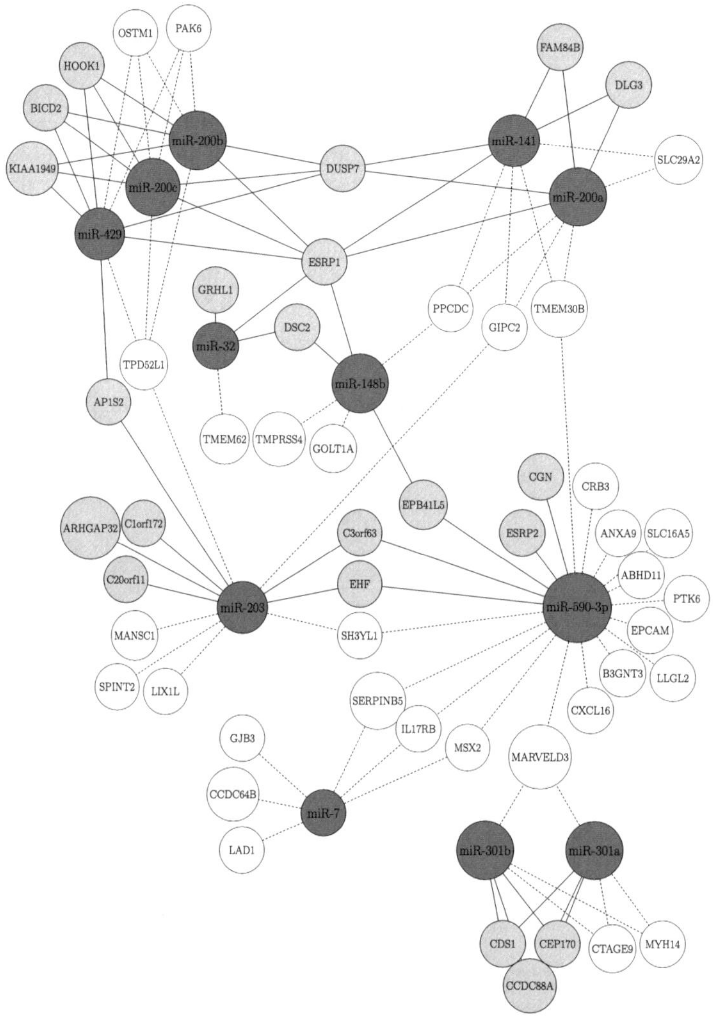

Fig. 3 Predicted interactions for miRNAs included in the top CORE 'C1N65' obtained from the NCI60. Red nodes are miRNAs, yellow and white nodes are predicted (conserved) targets and poorly conserved targets of conserved miRNA families, respectively. Solid lines and dashed lines are used to represent links between miRNAs and their conserved targets and poorly conserved targets, respectively. The interactions are predicted by both *DICORE* and TargetScan

Comparison with other methods

We summarize here the comparison study of the result of *DICORE* with the results of a few recent methods *Mirsynergy* [27], *SNMNMF* [25], and *PIMiM* [47] reported in [27]. We obtained the same ovarian cancer (OVC) dataset processed in [25]. The original miRNA and gene expression profiles for 385 ovarian cancer samples were downloaded from [48]. The expression dataset contains measurements of 559 miRNAs and 12456 mRNAs.

In case of performing a comparison study, our initial intention was to compare the result of *DICORE* with the results of two other methods, *Mirsynergy* [27] and *SNMNMF* [25] by applying them to the three cancer datasets (NCI60, BR and MCC) used for validating *DICORE*. However, both *Mirsynergy* and *SNMNMF* require as their input the gene-gene interactions (GGIs) derived from protein-protein interactions and transcription factor binding sites, which, according to [25] and [27], are to be obtained from the two datasets, BioGrid [49] and TRANSFAC [50]. Unfortunately, we could only manage to get the GGIs associated with the three cancer datasets from BioGrid. As a consequence, the results we obtained from *Mirsynergy* using these three cancer datasets were not good. Therefore to make a fair comparison with the two methods, we apply our method to the dataset (the OVC dataset) on which *Mirsynergy* and *SNMNMF* have had their results reported in the literature.

Similar to the setting used in [27], we used the absolute values of only negative interaction weights of W and same pair of values for the density thresholds, and set 2 for the penalty value in calculating cohesiveness scores. In addition, we set $\Theta = (\theta_m, \theta_g) = 2$ for both groups of miRNAs and mRNAs due to the requirement for calculating CCA, as CCA can not be applied on groups having less than 2 components.

Table 9 shows a summary of the performance of the four methods. *DICORE* identified 56 modules with an average of 8.3 miRNAs and 43.83 mRNAs per module for $\eta = 0.35$. The average strength of collective relationships is 0.61 in terms of canonical correlation among the groups. Furthermore, for $\eta = 0.30$, *DICORE* got 102 modules with 11.23 miRNAs and 73.22 mRNAs per module, and having average strength of 0.60. The average number of mRNAs identified by *Mirsynergy* is too small compared to other methods. However, average number of mRNAs identified by *DICORE* is reasonable.

We report here two interesting modules. Firstly, the module or CORE 'C9O35' consists of 4 miRNAs, namely *miR-29c**, *miR-29a*, *miR-29b*, and *miR-29c* from the same *miR-29* family. The human *miR-29* family of miRNAs is known to be associated with ovarian cancer [43, 51]. The pathway analysis of this module also shows association with cancer (see Table 10).

The module or CORE 'C17O35' included 16 miRNAs and 74 mRNAs. The module has miRNAs, namely *miR-17*, *miR-19b-1**, *miR-19b*, *miR-19a*, *miR-18b*, *miR-18a*, *miR-20a**, *miR-20a*, *miR-20b* from the polycistronic miRNA cluster *miR-17-92*, located in chromosome 13. They are considered to act as a tumor suppressor for ovarian cancer in some circumstances [52]. Furthermore, the pathway analysis of this module also illustrates association with cancer (see Table 11).

The final module structure of *Mirsynergy* is heavily depended on the initial clustering of miRNAs and the prior knowledge of gene-gene interactions. If *Mirsynergy* gets c clusters of miRNAs in the first stage, finally it will produce at most c miRNA regulatory modules. On contrary, *DICORE* separately performs clustering of miRNAs and mRNAs based on their functional interactions with mRNAs and miRNAs, respectively. This allows two distinct groups of mRNAs functioning differently to be part of different modules despite the fact that they are interacting with the same group of miRNAs. Furthermore, it also allows a group of miRNAs to interact with more than one group of miRNAs, which is common in biological sense.

Table 9 Performance of *DICORE*, *Mirsynergy*, *SNMNMF*, and *PIMiM*

Method	#C	$\overline{miR}$	$\overline{mR}$
DICORE	56	8.30	43.83
Mirsynergy	84	4.76	7.57
SNMNMF	49	4.12	81.37
PIMiM	40	4.70	67.80

Table 10 Top enrichment KEGG pathways for 'C9O35' from the OVC for $\eta = 0.35$

No	KEGG Pathways	*p*-value
1	Basal cell carcinoma	3.85E–08
2	Arginine and proline metabolism	0.0298836
3	Glutathione metabolism	0.0398112
4	Pathways in cancer	0.0426415
5	Cell cycle	0.0495754

Related works

Several computational approaches had been proposed to discover MMRMs. The concept of MMRMs was introduced by [18] to denote groups of co-expressed miRNAs and their targets mRNAs. They drew a similarity between predicting MMRMs and mining frequent itemset by mapping the set of miRNAs and the set of target mRNAs to a frequent itemset and its cover, respectively. They proposed a prediction method adopting bipartite graphs to model binding structures of the miRNAs and mRNAs at the sequence level. However, prediction based on sequence may not be sufficient to correctly predict the complex interactions.

Improved versions of this method had been proposed which also take into account coherent expression patterns between miRNAs and mRNAs, or the (anti)-correlations measured between each pair of miRNAs and mRNAs [19, 21, 26]. Joung et al. [21] integrated expression profiles of miRNAs and mRNAs with sequence information by using a biclustering approach. The approach reduced false discovery rate significantly. A rule based method was utilized by Tran et al. [19] based on the assumption that miRNAs and mRNAs of a module have similar expression patterns. However, these existing methods for discovering MMRMs suffered from several limitations. For example, Peng et al. [26] proposed a sequential integrative method based on enumerating maximal bicliques in a combined miRNA-gene network. Their method was sensitive to noise in the data, and produced too many star structures (one miRNA, many genes) which were not usable to explore miRNA combinatorial regulation.

Table 11 Top enrichment KEGG pathways for 'C17O35' from the OVC for $\eta = 0.35$

No	KEGG Pathways	*p*-value
1	Pathways in cancer	0.00679353
2	MAPK signaling pathway	0.0136845

The *functional* MMRMs (FMMRMs) are associated with MMRMs with specific biological conditions. For FMMRMs discovery, [22] and [20] proposed different methods at around the same time. Joung and Fei [22] proposed an unsupervised method which applied the author-topic model [53, 54] in bioinformatics. The method used the expression profiles of miRNAs and the putative miRNA target information, without considering the expression profiles of miRNAs. As the miRNA target information is predicted at the sequence level, it encountered similar difficulty of [18] in explaining regulation pattern of miRNAs in their target genes in the identified modules. On the other hand, Liu et al. [20] proposed a supervised method which utilized association rule mining method by associating the reverse expression patterns of miRNAs and genes with biological conditions. However, they only considered down-regulation patterns.

In order to discover FMMRMs, Liu et al. [23] applied another probabilistic graphical model, correspondence Latent Dirichlet Allocation (Corr-LDA) [55], that had been applied to automatic image annotation with caption words. By associating topics to functional modules, images to miRNAs, and words to mRNAs, respectively, the method was applied to a mouse model dataset for human breast cancer research. The method simultaneously identified FMMRMs using the expression profiles of both miRNAs and genes, with or without using target relationships between miRNAs and mRNAs. The Corr-LDA was extended and applied to identify functional regulatory module, and each module corresponds to a particular biological function. In the model, each function was represented as a latent topic, and the numeric values of expression data were converted to the counts of expression events, similar to the counts of words in a documents. Another similar semi-supervised method based on a probabilistic model which is closely associated with the Latent Dirichlet Allocation [56] was proposed in [24]. The idea of extracting topics with caption words to FMMRMs discovery by mapping topics to functional modules, documents to samples, and words to mRNAs, respectively.

The main drawback of these methods is that they did not consider the collective relationships in identifying the modules, which result in regulatory modules that may not quite correct modeling of the real biological systems. Recently Karim et al. [28] came up with the idea of collective group relationships, and proposed a method to explore miRNA-mRNA regulatory relationships. They integrated two complementary approaches associated with relationships of complex systems, namely graph mining and CCA to discover collective relationships with both quantitative and qualitative information. However, the proposed method considered unweighted graph, which are prone to make computational inaccuracy due to the approximation of many interaction weights to either 1 (interaction) or 0 (no interaction). Recently Li et al. [27] proposed a clustering algorithm, *Mirsynergy* to detect synergistic miRNA regulatory modules. They used mRNA and miRNA expression profiles, target site information and gene-gene interactions for ovarian, breast, and thyroid cancers from TCGA [57] and obtained significantly higher enrichment than existing methods. However, it partially used collective relationships in stage 1, and but in stage 2 depended on the prior knowledge of confirmed gene-gene interactions.

This paper presents a novel method that discover MMRMs by considering the collective relationships as the driving force in identifying the miRNA-mRNA regulatory relationships. Furthermore, it uses the idea of ranking the identified modules by the quantitative measure of the strength of the collective relationships between the groups in group pairs.

Conclusion

In this paper, we have used the notation of CORE, and proposed a computational framework *DICORE* to discover MMRMs. The central idea of *DICORE* is to consider the collective group relationship, and discover both the groups and collective relationships simultaneously. We have applied a greedy-based overlapping clustering approach adapted from *ClusterONE* [29] to group miRNAs and mRNAs separately based on their collective interactions with mRNAs and miRNAs respectively, and integrate CCA in order to enrich the identification of groups with both structural link information and strength of collective relationships. We have experimented on three real-world biological datasets. The experimental results have demonstrated that the proposed method *DICORE* is able to reveal correct group information with structural link information and the strength of collective relationships, and provide useful insights into the structure and functionality of the miRNA-mRNA regulatory relationships in MMRMs.

The proposed framework has also opened a few interesting research windows for further investigation. Instead of using Pearson correlation coefficient to calculate the interaction weights matrix, other approaches including statistical methods like maximal information coefficient [58], regression techniques like Lasso [59], causal inference method like IDA [31] can be applied. Considering the context of the datasets, any of the individual methods or an ensemble method [31] can be tested and reported. Furthermore, the strength of the collective interactions can be determined by applying other similar mathematical models to capture all possible association between two sets of variables. Another interesting future work will be to apply the framework to discover MMRMs from datasets obtained under different biological conditions.

Additional files

Additional file 1: Differential expression profiles of miRNAs and mRNAs for the NCI60, BR and MCC datasets. (XLSX 2785 kb)

Additional file 2: Three separate 'ground-truth' databases for the NCI60, BR and MCC datasets, respectively. (XLSX 227 kb)

Additional file 3: Experimentally confirmed interactions of the top COREs identified in the NCI60, BR and MCC datasets. (XLSX 33.5 kb)

Additional file 4: The significant pathways of top group pairs identified in the NCI60, BR and MCC datasets. (XLSX 21.9 kb)

Additional file 5: Predicted miRNA-mRNA interactions identified by *DICORE* in the NCI60, BR and MCC datasets. (XLSX 14.0 kb)

Competing interests
The authors declare that they have no competing interests.

Authors' contributions
SMMK, LL and JL conceived the research. SMMK designed and implemented the method. SMMK complied the 'ground-truth' databases, and performed the experiments and validation analysis. TDL conducted the GeneGo pathway enrichment analysis. SMMK wrote the paper. All authors read and approved the final manuscript.

Declarations
This article has been published as part of BMC Genomics Volume 17 Supplement 1, 2016: Selected articles from the Fourteenth Asia Pacific Bioinformatics Conference (APBC 2016): Genomics. The full contents of the supplement are available online at http://www.biomedcentral.com/bmcgenomics/supplements/17/S1.

Authors' information
All authors are with School of Information Technology and Mathematical Sciences, University of South Australia, Mawson Lakes, SA 5095, Adelaide, Australia.

Acknowledgements
This work was partially supported by Australian Research Council Discovery Grant DP130104090. The publication costs for this article were funded by Australian Research Council Discovery Grant DP130104090.

Published: 11 January 2016

References

1. Ambros V. microRNAs: tiny regulators with great potential. Cell. 2001;107(7):823–6. doi:10.1016/S0092-8674(01)00616-X. PMID: 11779458.
2. Ambros V. The functions of animal microRNAs. Nature. 2004;431(7006): 350–5. doi:10.1038/nature02871. PMID: 15372042.
3. Bartel DP. MicroRNAs: Target recognition and regulatory functions. Cell. 2009;136(2):215–33. doi:10.1016/j.cell.2009.01.002. PMID: 19167326.
4. Friedman RC, Farh KK-H, Burge CB, Bartel DP. Most mammalian mRNAs are conserved targets of microRNAs. Genome Res. 2009;19(1):92–105. doi:10.1101/gr.082701.108. PMID: 18955434.
5. Chen JF, Mandel EM, Thomson JM, Wu Q, Callis TE, Hammond SM, et al. The role of microRNA-1 and microRNA-133 in skeletal muscle proliferation and differentiation. Nat Genet. 2006;38(2):228–33. doi:10.1038/ng1725. PMID: 16380711.
6. Zhao Y, Samal E, Srivastava D. Serum response factor regulates a muscle-specific microRNA that targets Hand2 during cardiogenesis. Nature. 2005;436(7048):214–20. doi:10.1038/nature03817. PMID: 15951802.
7. Poy MN, Eliasson L, Krutzfeldt J, Kuwajima S, Ma X, MacDonald PE, et al. A pancreatic islet-specific microRNA regulates insulin secretion. Nature. 2004;432(7014):226–30. doi:10.1038/nature03076. PMID: 15538371.
8. Esquela-Kerscher A, Slack FJ. Oncomirs — microRNAs with a role in cancer. Nat Rev Cancer. 2006;6(4):259–69. doi:10.1038/nrc1840. PMID: 16557279.
9. Jin P, Zarnescu DC, Ceman S, Nakamoto M, Mowrey J, Jongens TA, et al. Biochemical and genetic interaction between the fragile X mental retardation protein and the microRNA pathway. Nat Neurosci. 2004;7(2): 113–7. doi:10.1038/nn1174. PMID: 14703574.
10. Xu C, Lu Y, Pan Z, Chu W, Luo X, Lin H, et al. The muscle-specific microRNAs miR-1 and miR-133 produce opposing effects on apoptosis by targeting HSP60, HSP70 and caspase-9 in cardiomyocytes. J Cell Sc. 2007;120(Pt 17):3045–52. doi:10.1242/jcs.010728. PMID: 17715156.
11. Cui Q, Yu Z, Purisima EO, Wang E. Principles of microRNA regulation of a human cellular signaling network. Mol Syst Biol. 2006;2(46):1–7. doi:10.1038/msb4100089. PMID: 16969338.
12. Bartel DP. MicroRNAs: genomics, biogenesis, mechanism, and function. Cell. 2004;116(2):281–97. doi:10.1016/S0092-8674(04)00045-5. PMID: 14744438.
13. Calin GA, Croce CM. MicroRNA signatures in human cancers. Nat Rev Cancer. 2006;6(11):857–66. doi:10.1038/nrc1997. PMID: 17060945.
14. Lu J, Getz G, Miska EA, Alvarez-Saavedra E, Lamb J, Peck D, et al. MicroRNA expression profiles classify human cancers. Nature. 2005;435(7043):834–8. doi:10.1038/nature03702. PMID: 15944708.
15. Flynt AS, Lai EC. Biological principles of microRNA-mediated regulation: shared themes amid diversity. Nat Rev Genet. 2008;9(11):831–42. doi:10.1038/nrg2455. PMID: 18852696.
16. Boross G, Orosz K, Farkas IJ. Human microRNAs co-silence in well-separated groups and have different predicted essentialities. Bioinformatics. 2009;25(8):1063–9. doi:10.1093/bioinformatics/btp018. PMID: 19131366.
17. Xu J, Li CX, Li YS, Lv JY, Ma Y, Shao TT, et al. MiRNA–miRNA synergistic network: construction via co-regulating functional modules and disease miRNA topological features. Nucleic Acids Res. 2011;39(3):825–36. doi:10.1093/nar/gkq832. PMID: 20929877.
18. Yoon S, Micheli GD. Prediction of regulatory modules comprising microRNAs and target genes. Bioinformatics. 2005;21(Suppl 2):93–100. doi:10.1093/bioinformatics/bti1116. PMID: 16204133.
19. Tran DH, Satou K, Ho TB. Finding microRNA regulatory modules in human genome using rule induction. BMC Bioinformatics. 2008;9(Suppl 12):5. doi:10.1186/1471-2105-9-S12-S5. PMID: 19091028.
20. Liu B, Li J, Tsykin A. Discovery of functional miRNA-mRNA regulatory modules with computational methods. J Biomed Inform. 2009;42(4): 685–91. doi:10.1016/j.jbi.2009.01.005. PMID: 19535005.
21. Joung JG, Hwang KB, Nam JW, Kim SJ, Zhang BT. Discovery of microRNA-mRNA modules via population-based probabilistic learning. Bioinformatics. 2007;23(9).1141–7. doi:10.1093/bioinformatics/btm045. PMID: 17350973.
22. Joung JG, Fei Z. Identification of microRNA regulatory modules in Arabidopsis via a probabilistic graphical model. Bioinformatics. 2009;25(3):387–93. doi:10.1093/bioinformatics/btn626. PMID: 19056778.
23. Liu B, Liu L, Tsykin A, Goodall GJ, Green JE, Zhu M, et al. Identifying functional miRNA-mRNA regulatory modules with correspondence latent dirichlet allocation. Bioinformatics. 2010;26(24):3105–11. doi:10.1093/bioinformatics/btq576. PMID: 20956247.
24. Zhang J, Liu B, He J, Ma L, Li J. Inferring functional miRNA-mRNA regulatory modules in epithelial-mesenchymal transition with a probabilistic topic model. Comput Biol Med. 2012;42(4):428–37. doi:10.1016/j.compbiomed.2011.12.011. PMID: 22245099.
25. Zhang S, Li Q, Liu J, Zhou XJ. A novel computational framework for simultaneous integration of multiple types of genomic data to identify microRNA-gene regulatory modules. Bioinformatics. 2011;27(13):401–9. doi:10.1093/bioinformatics/btr206. PMCID: PMC3117336.
26. Peng X, Li Y, Walters KA, Rosenzweig ER, Lederer SL, Aicher LD, et al. Computational identification of hepatitis C virus associated microRNA-mRNA regulatory modules in human livers. BMC Genomics. 2009;10(373):. doi:10.1186/1471-2164-10-373. PMID: 19671175.
27. Li Y, Liang C, Wong KC, Luo J, Zhang Z. Mirsynergy: detecting synergistic miRNA regulatory modules by overlapping neighbourhood expansion. Bioinformatics. 2014;30(18):2627–35. doi:10.1093/bioinformatics/btu373. PMID: 24894504.
28. Karim SMM, Liu L, Li J. Discovering Collective Group Relationships In: Wang H, Sharaf MA, editors. Databases Theory and Applications: Proceedings of the 25th Australasian Database Conference 2014. Switzerland: Springer; 2014. p. 110–121.
29. Nepusz T, Yu H, Paccanaro A. Detecting overlapping protein complexes in protein-protein interaction networks. Nat Methods. 2012;9(5):471–2. doi:10.1038/nmeth.1938. PMID: 22426491.

30. Hotelling H. Relations between two sets of variants. Biometrika. 1936;28(3/4):321–77. doi:10.2307/2333955.
31. Le TD, Zhang J, Liu L, Li J. Ensemble Methods for MiRNA Target Prediction from Expression Data. PLOS ONE. 2015;10(6):0131627. doi:10.1371/journal.pone.0131627. PMCID: PMC4482624.
32. Vlachos IS, Paraskevopoulou MD, Karagkouni D, Georgakilas G, Vergoulis T, Kanellos I, et al. DIANA-TarBase v7.0: indexing more than half a million experimentally supported miRNA:mRNA interactions. Nucleic Acids Res. 2015;43(Database issue):153–9. doi:10.1093/nar/gku1215. PMID: 25416803.
33. Hsu SD, Tseng YT, Shrestha S, Lin YL, Khaleel A, Chou CH, et al. miRTarBase update 2014: an information resource for experimentally validated miRNA-target interactions. Nucleic Acids Res. 2014;42(Database issue):78–85. doi:10.1093/nar/gkt1266. PMID: 24304892.
34. Xiao F, Zuo Z, Cai G, Kang S, Gao X, Li T. miRecords: an integrated resource for microRNA–target interactions. Nucleic Acids Res. 2009;37(Database issue):105–10. doi:10.1093/nar/gkn851. PMID: 18996891.
35. Dweep H, Sticht C, Pandey P, Gretz N. miRWalk–database: prediction of possible miRNA binding sites by "walking" the genes of three genomes. J Biomed Inform. 2011;44(5):839–47. doi:10.1016/j.jbi.2011.05.002. PMID: 21605702.
36. Bracken CP, Li X, Wright JA, Lawrence DM, Pillman KA, Salmanidis M, et al. Genome-wide identification of miR-200 targets reveals a regulatory network controlling cell invasion. EMBO J. 2014;33(18):2040–56. doi:10.15252/embj.201488641. PMID: 25069772.
37. DICORE - A Computational Framework. http://nugget.unisa.edu.au/~Masud/DICORE/. Access date 15 August 2015.
38. Shah MY, Calin GA. MicroRNAs miR-221 and miR-222: a new level of regulation in aggressive breast cancer. Genome Med. 2011;3(8):56. doi:10.1186/gm272. PMCID: PMC3238182.
39. Tabas-Madrid D, Nogales-Cadenas R, Pascual-Montano A. GeneCodis3: a non-redundant and modular enrichment analysis tool for functional genomics. Nucleic Acids Res. 2012;40(Web Server issue):478–83. doi:10.1093/nar/gks402. PMID: 22573175.
40. Genecodis Web Tool. http://genecodis.cnb.csic.es/analysis. Access date 15 August 2015.
41. Kanehisa M, Goto S. KEGG: Kyoto Encyclopedia of Genes and Genomes. Nucleic Acids Res. 2000;28(1):27–30. doi:10.1093/nar/28.1.27. PMID: 10592173.
42. GeneGo MetaCore (MetaCore Bioinformatics Software from Thomson Reuters). https://portal.genego.com/. Acess date 28 May 2015.
43. Koturbasha I, Zempa FJ, Pogribny I, Kovalchuk O. Small molecules with big effects: The role of the microRNAome in cancer and carcinogenesis. Mutat Res Genet Toxicol Environ Mutagen. 2011;722(2):94–105. doi:10.1016/j.mrgentox.2010.05.006. PMID: 20472093.
44. Ventura A, Jacks T. MicroRNAs and cancer: short RNAs go a long way. Cell. 2009;136(4):586–91. doi:10.1016/j.cell.2009.02.005. PMID: 19239879.
45. Lee YS, Dutta A. MicroRNAs in cancer. Annu Rev Pathol: Mech Dis. 2009;4(2010):199–227. doi:10.1146/annurev.pathol.4.110807.092222. PMID: 18817506.
46. Lewis BP, Burge CB, Bartel DP. Conserved seed pairing, often flanked by adenosines, indicates that thousands of human genes are microRNA targets. Cell. 2005;120(1):15–20. doi:10.1016/j.cell.2004.12.035. PMID: 15652477.
47. Le HS, Bar-Joseph Z. Integrating sequence, expression and interaction data to determine condition-specific miRNA regulation. Bioinformatics. 2013;29(13):89–97. doi:10.1093/bioinformatics/btt231. PMCID: PMC3694655.
48. TGCA Data Portal. http://cancergenome.nih.gov/.
49. Stark C, Breitkreutz BJ, Chatr-aryamontri A, Boucher L, Oughtred R, Livstone MS, et al. The BioGRID Interaction Database: 2011 update. Nucleic Acids Res. 2011;39(Database isse):698–704. doi:10.1093/nar/gkq1116. PMID: 21071413.
50. Wingender E, Chen X, Hehl R, Karas H, Liebich I, Matys V, et al. TRANSFAC: an integrated system for gene expression regulation. Nucleic Acids Res. 2000;28(1):316–9. doi:10.1093/nar/28.1.316. PMID: 10592259.
51. Yu PN, Yan MD, Lai HC, Huang RL, Chou YC, Lin WC, et al. Downregulation of miR-29 contributes to cisplatin resistance of ovarian cancer cells. Int J Cancer. 2013;134(3):542–51. doi:10.1002/ijc.28399. PMID: 23904094.
52. Xiang J, Wu J. Feud or Friend? The Role of the miR-17-92 Cluster in Tumorigenesis. Curr Genomics. 2010;11(2):129–35. doi:10.2174/138920210790886853. PMID: 2874222.
53. Rosen-Zvi M, Griffiths T, Steyvers M, Smyth P. The author-topic model for authors and documents. In: UAI '04 Proceedings of the 20th Conference on Uncertainty in Artificial Intelligence. Virginia, USA: AUAI Press Arlington; 2004. p. 487–94.
54. Steyvers M, Smyth P, Rosen-zvi M, Griffiths T. Probabilistic Author-Topic Models for Information Discovery. In: KDD '04 Proceedings of the 10th ACM SIGKDD Conference on Knowledge Discovery and Data Mining. New York, USA: ACM; 2004. p. 306–15.
55. Blei DM, Jordan MI. Modeling annotated data. In: SIGIR '03 Proceedings of the 26th Annual International ACM SIGIR Conference on Research and Development in Information Retrieval. New York, USA: ACM; 2003. p. 127–34.
56. Blei DM, Ng AY, Jordan MI. Latent dirichlet allocation. J Mach Learn Res. 2003;3:993–1022.
57. The Cancer Genome Atlas (TCGA) Research Network: Comprehensive genomic characterization defines human glioblastoma genes and core pathways. Nature. 2008;455(7216):1061–8.
58. Reshef DN, Reshef YA, Finucane HK, Grossman SR, McVean G, Turnbaugh PJ, et al. Detecting Novel Associations in Large Data Sets. Science. 2011;334(6062):1518–24. doi:10.1126/science.1205438. PMID: 22174245.
59. Tibshirani R. Regression Shrinkage and Selection Via the Lasso. J R Stat Soc Ser B Methodol. 1996;58(1):267–88.

Kim et al. BMC Genomics 2016, 17(Suppl 1):5
DOI 10.1186/s12864-015-2301-y

BMC Genomics

PROCEEDINGS Open Access

RDDpred: a condition-specific RNA-editing prediction model from RNA-seq data

Min-su Kim[1], Benjamin Hur[1] and Sun Kim[1,2,3*]

From The Fourteenth Asia Pacific Bioinformatics Conference (APBC 2016)
San Francisco, CA, USA. 11 - 13 January 2016

Abstract

Background: RNA-editing is an important post-transcriptional RNA sequence modification performed by two catalytic enzymes, "ADAR"(A-to-I) and "APOBEC"(C-to-U). By utilizing high-throughput sequencing technologies, the biological function of RNA-editing has been actively investigated. Currently, RNA-editing is considered to be a key regulator that controls various cellular functions, such as protein activity, alternative splicing pattern of mRNA, and substitution of miRNA targeting site. DARNED, a public RDD database, reported that there are more than 300-thousands RNA-editing sites detected in human genome(hg19). Moreover, multiple studies suggested that RNA-editing events occur in highly specific conditions. According to DARNED, 97.62 % of registered editing sites were detected in a single tissue or in a specific condition, which also supports that the RNA-editing events occur condition-specifically. Since RNA-seq can capture the whole landscape of transcriptome, RNA-seq is widely used for RDD prediction. However, significant amounts of false positives or artefacts can be generated when detecting RNA-editing from RNA-seq. Since it is difficult to perform experimental validation at the whole-transcriptome scale, there should be a powerful computational tool to distinguish true RNA-editing events from artefacts.

Result: We developed RDDpred, a Random Forest RDD classifier. RDDpred reports potentially true RNA-editing events from RNA-seq data. RDDpred was tested with two publicly available RNA-editing datasets and successfully reproduced RDDs reported in the two studies (90 %, 95 %) while rejecting false-discoveries (NPV: 75 %, 84 %).

Conclusion: RDDpred automatically compiles condition-specific training examples without experimental validations and then construct a RDD classifier. As far as we know, RDDpred is the very first machine-learning based automated pipeline for RDD prediction. We believe that RDDpred will be very useful and can contribute significantly to the study of condition-specific RNA-editing. RDDpred is available at `http://biohealth.snu.ac.kr/software/RDDpred`.

Keywords: RNA-editing, Condition-specific, Machine-learning, Random forest, RNA-seq, Systematic artefact

Background

RNA-editing: a biologically crucial regulator and highly condition-specific event

RNA-editing event is defined as a post-transcriptional RNA sequence modification [1]. Currently, there are two known RNA-editing mechanisms, performed by two different catalytic enzymes, "ADAR" (A-to-I) and "APOBEC" (C-to-U) [2, 3]. The most common type of editing in metazoans is the one catalyzed by the ADAR family of enzymes [4]. By utilizing high-throughput sequencing technologies, the biological function of RNA-editing has been actively investigated [5–7]. Currently, RNA-editing is considered to be a key regulator that controls various cellular functions including protein activity, alternative splicing pattern of mRNA and substitution of miRNA targeting site [1, 8–10].

Moreover, there are multiple studies that showed direct relation of RNA-editing to biological phenotypes. For example, Galeano's group showed that the

*Correspondence: sunkim.bioinfo@snu.ac.kr
[1] Interdisciplinary Program in Bioinformatics, Seoul National University, Seoul, Republic of Korea
[2] Department of Computer Science and Engineering, Seoul National University, Seoul, Republic of Korea
Full list of author information is available at the end of the article

editing events in glioblastoma by ADAR2 enzymes are crucial for pathogenesis and claimed that ADAR-class enzyme can be considered as a tumor-suppressor [11]. And in APOBEC3G, a type of APOBEC-class enzyme causes HIV-1 retroviral inactivation by deamination [12].

DARNED, a well-curated public RNA-editing database have more than 300-thousands editing sites detected in the human genome hg19 [13]. Interestingly, the expression patterns of editing events in different conditions varied significantly. For example, in DARNED database, 333,164 editing sites in hg19 are registered from 21 independent studies in 139 tissues. The conservation rate among tissues is strikingly low, 97.62 % of these registered sites were detected from a single tissue or condition (Fig. 1). Moreover, multiple studies suggest that RNA-editing events can be involved in condition-specific regulation of genetic functions [14, 15]. Taken together, it is reasonable to believe that RNA-editing is highly condition-specific event.

RNA-seq: an important tool for investigating condition-specific RNA-editing patterns

RNA-seq, a high-throughput sequencing of transcriptome, is a powerful method for investigating whole-transcriptome status. Since the nature of the technology is taking a snapshot of cells with massive sequencing reads, it is suitable for detecting condition-specific events in whole-transcriptome scale. Therefore, it is also suited for detecting RNA-editing events that have such condition-specific characteristics. There have been a number of studies that used RNA-seq to reveal condition-specific editing patterns in whole-transcriptome scale [5–7].

Systematic artefacts: the major huddle to detect authentic RNA-editing events from RNA-seq

Even though RNA-seq is suitable for RNA-editing detection, it is also true that the current computational pipelines of RNA-editing detection with RNA-seq have considerable false-positive risks. In 2012 Nature Biotechnology journal, an article "The difficult calls in RNA editing", reports interviews with eight prominent RNA-editing researchers. They pointed out that false-positive calling is one of the most challenging problems in RNA-editing detection with RNA-seq [16].

The false-positives caused by mis-alignment of short-reads can be termed "Systematic Artefacts" due to their inherent and reproducible characteristics. Systematic artefacts can be caused by various reasons, (a) inherent duplications/repeats within genomic sequences, (b) ambiguity caused by splicing-junctions, (c) prevalent polymorphisms between individuals and (d) shortness of sequencing reads [17, 18]. This inherent and reproducible error has been assumed to be one of the major confounding factors while detecting sequence variants [19, 20].

To assess the significance of confounding effects caused by systematic artefacts, we performed a simple simulation test in order to measure the false-detection rate of RNA-variants caused by mis-alignments. We used RNA-STAR, a state-of-art alignment tool in base-accuracy [20, 21], to evaluate inherent risks of false-positive detection in the human genome hg19 (Fig. 2). We simulated 10-millions reads from mRNA-seq [22] with 1 % simulated SNVs (513-thousand sites) and aligned them into the hg19 genome sequence. These 1 % of simulated SNVs are for representing the individual genetic

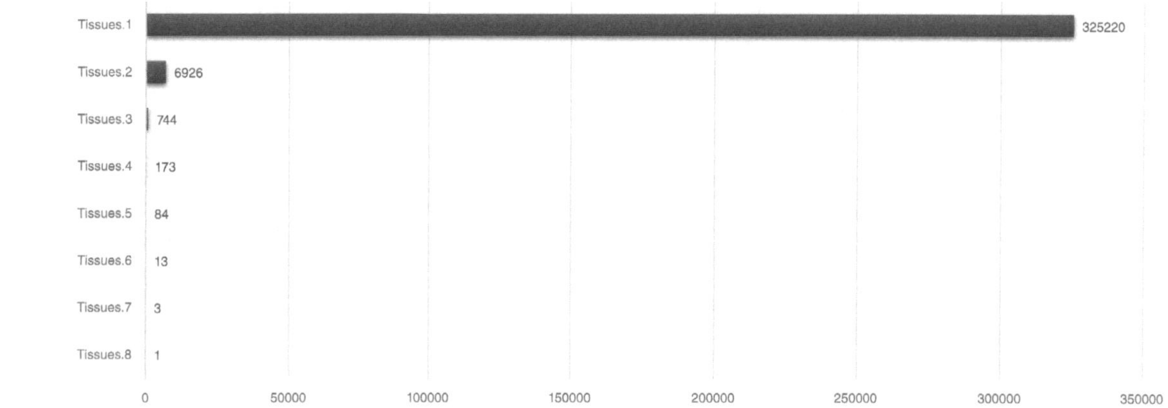

	Tissues.8	Tissues.7	Tissues.6	Tissues.5	Tissues.4	Tissues.3	Tissues.2	Tissues.1
■ Editing-sites	1	3	13	84	173	744	6926	325220

Fig. 1 Low conservation rate with various tissues: In DARNED database [13], 97.62 % of total editing-sites are detected in a single tissue. Hence, showing significantly low conservation rate

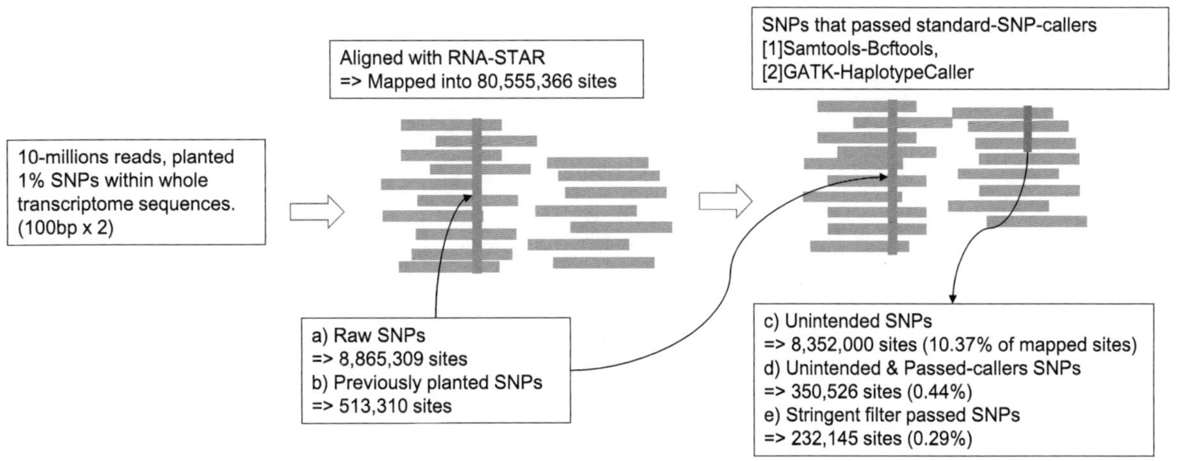

Fig. 2 Substantial amounts of systematic artefacts: As simulation-test results, significant amounts of artefacts that consist of 0.29 % of total mapped sites are detected, which turned out to be indistinguishable even by the method combined with two state-of-art variant-callers and extensive *a priori* knowledges of genomic repeats

differences including SNPs, somatic mutations, and RNA-editings.

We generated 10 simulated data and aligned them (Fig. 3). In addition to the planted sites, we discovered that average 8.3 millions SNVs were detected from each 10-million reads(while we have planted 513-thousands sites). This is clearly artefacts. Such considerable number of unintended artefacts consist of 10.37 % of total mapped sites. Moreover, 2.78 % (232-thousands sites) of these unintended artefacts cannot be excluded by standard SNP callers [23, 24] or by stringent filtering with error-inducible regions in hg19 genome [25] (Fig. 2). The result suggests that if we use 10-millions reads to detect RNA-editing, we will be confronted at least 232-thousands of artefacts that is difficult to be excluded by standard methods (Table 1).

Three distinct computational approaches that addressed the artefact-issue

To handle systematic artefacts in RNA-seq, a number of computational approaches have been developed. These can be categorized into three groups in terms of features they used: (a) *A priori* knowledge based filtering [26, 27], (b) Computational simulation of artefacts [6], (c) Machine-learning based prediction model [5, 28].

A priori knowledge based filtering used public genomic features, such as Alu repeats, genomic duplications, and pseudogenes, to assess the detected editing-sites directly. For instance, Li's group used public annotation of genomic repeats to filter out potential artefacts within the detected RDD(RNA/DNA Difference) sites [26]. On the other hand, the approach based on computational simulation of artefacts rather utilizes calculated features than public

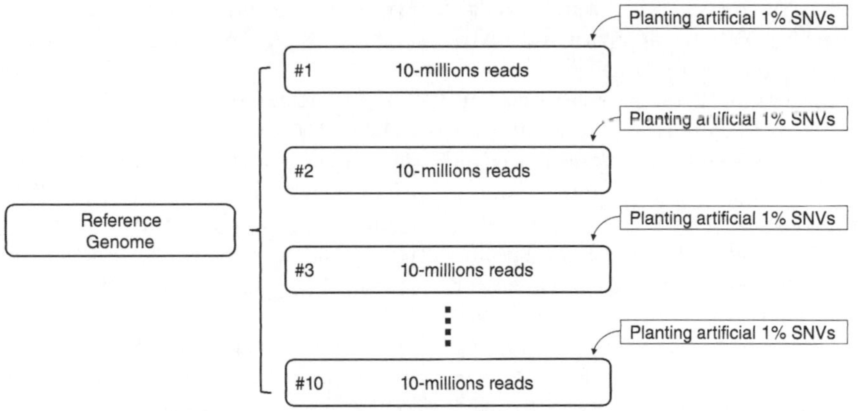

Fig. 3 Simulation dataset preparation: We generated 10 simulated sequencing reads from spliced genomes (spliced by NCBI refseq annotation). Note that each reads are generated with different sets of SNVs

Table 1 Artefact simulation results: resulted from 10-times of iterations

	Mapped reads	Mapped sites	Raw SNPs	Unintended SNPs	Caller passed	Filtered
Trial.1	9,734,787	80,552,288	8,872,433	8,358,426	350,694	231,972
Trial.2	9,735,558	80,558,479	8,878,304	8,365,007	350,670	232,496
Trial.3	9,733,473	80,568,898	8,880,681	8,366,553	350,136	231,912
Trial.4	9,733,159	80,570,416	8,879,502	8,365,311	350,442	231,539
Trial.5	9,733,939	80,545,810	8,853,408	8,339,822	350,332	232,008
Trial.6	9,733,507	80,542,007	8,838,870	8,326,074	350,917	232,649
Trial.7	9,734,222	80,555,307	8,859,741	8,346,628	350,390	232,128
Trial.8	9,735,046	80,562,701	8,874,369	8,361,655	350,807	232,063
Trial.9	9,733,971	80,555,609	8,852,720	8,339,866	350,059	232,336
Trial.10	9,734,717	80,542,143	8,863,065	8,350,655	350,809	232,347
Average	9,734,238	80,555,366	8,865,309	8,352,000	350,526	232,145

features. Peng's group used extensively simulated RNA-seq to predict inherent error-inducible regions in genome sequence and used them as a filter [6].

Unlike the filter-based methods that directly assess RDD candidate sites with pre-defined filters, machine-learning based methods generates a predictor in advance. The predictor, or machine-learning classifier, is trained to learn the differences between true and false examples. As an example, Laurent's group generated a Random Forest predictor that utilizes read-alignment patterns as attributes. With 77 attributes, Laurent's group generated a predictor and demonstrated it has 87 % of estimated accuracy by experimental validation [5]. As mentioned, since RNA-editing events are occurred highly condition-specifically, machine-learning approach might have an advantage in that they pursue more data-driven method by generating condition-specific model.

Machine-learning based RNA-editing prediction became possible

Laurent's work [5] was the first successful demonstration to show that a machine learning approach for RNA-editing prediction is both feasible and sensitive. However, to be a general-purpose model, there are several limitations. First of all, a predictor needs a training data that consists of positive and negative examples. And in Laurent's study, they collected the both training examples from additionally performed Sanger-seq [5]. However, as we emphasized, RNA-editing is a condition specific event. And, since Laurent's approach used experimentally verified training examples specific to their own conditions, the model might not be applicable in different conditions unless additional sequencing is performed. Therefore, it is more cost-efficient if we can avoid the experimental validations with utilizing the machine-learning approach.

Methods

Here we introduce RDDpred, a software package that is generally usable and do not need an experimental validation to prepare condition-specific training examples. Hence, RDDpred prepares condition-specific training data directly from input sequencing data or raw RDD candidates. In order to collect positive examples without experimental validations, we utilized two well-organized RNA-editing databases, RADAR and DARNED [13, 29]. Since we consider systematic artefacts as major cause of false-positives, we collect negative examples by utilizing the MES method that calculates the error-inducible regions within genome during alignments [6]. After collecting positive/negative examples from input data, all the remaining sites are considered as targets for prediction. RDDpred is a Random Forest predictor that utilizes 15 features that reflect the read-alignment patterns. The overall prediction scheme is illustrated in Fig. 4.

Implementation of RDDpred

We tested RDDpred in Python (2.7.3), Samtools-Bcftools (1.2.1), WEKA (3.6.12) package, in linux environment.

1) Input and output of RDDpred

RDDpred takes alignment results as input data and gives the prediction results of each SNVs, or RDD candidates as outputs. The raw RDD candidates are detected with Samtools-Bcftools pipeline [23] while the prediction model is trained by using WEKA package [30].

2) Selection of alignment tool by the user

RDDpred can take inputs from any kind of alignment methods providing BAM-format outputs. However, we recommended RNA-STAR for its high degree of overall accuracy and ultra-fast performance [20, 21].

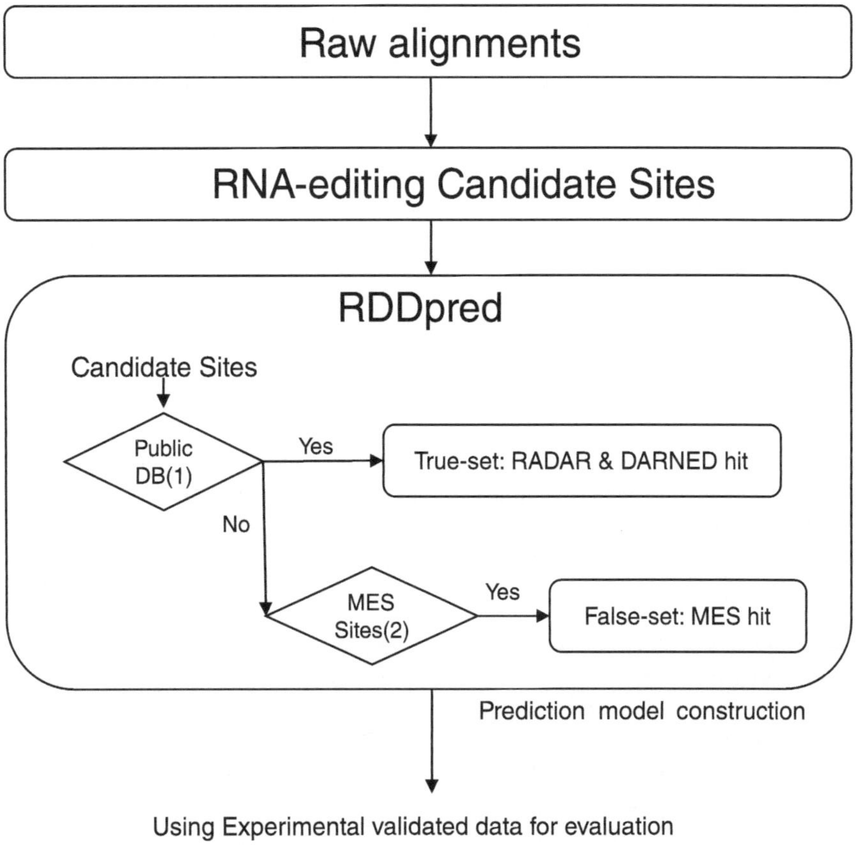

Fig. 4 RDDpred workflow: 1) RDDpred takes raw alignments or raw RDDs 2) RDDpred arranges condition-specific training-data extracting positive/negative examples by mapping raw RDDs to public database (RADAR, DARNED) and MES-predicted sites, respectively. 3) RDDpred train the condition-specific classifier with arranged training-data and predicts true-editing against artefacts

Condition-specific training data preparation

1) Positive-set of training data: utilizing public databases, RADAR and DARNED

RADAR and DARNED databases include 2.5 million, 300-thousands of curated sites respectively [13, 29]. These two databases share a considerable portion of sites, 150-thousands sites. Since the pre-known sites are already proved to have editing potential, we can use the sites matched to the consensus sites as positive examples (Fig. 4). Since RDDpred takes the positive sites as an input, users can change or supplement the sites that are considered as true events.

2) Negative-set of training data: applying MES artefact calculation method

To build a predictor, we also need negative examples. To address this issue, we utilized the MES method, a computational simulation for predicting error-inducible regions within genomes [6] (Fig. 5). With MES method, we can calculate error-inducible regions specific to the conditions, such as, SNPs combination of the samples, the experimental specification of sequencing, and the choice of alignment method. And if we calculate the error-inducible regions specific to our conditions, then we can consider them as potential error-prone regions. Therefore, just like collecting positive examples, we can reasonably assume the portion of RDD candidate sites that belong to those error-prone regions are probably systematic artefacts, thus using them as negative examples (Fig. 4).

RDDpred predictor description

1) RDDpred mainly focuses on systematic artefacts

It is known that there are various types of artefacts from RNA-seq, such as amplification errors during library construction, sequencing errors, and errors by mis-alignments. Unlike former events, errors by mis-alignments shows different characteristics that they are more reproducible. Since the errors from library construction and sequencing procedures are transient in general, they can be excluded by replicating experiments. On the other hand, since the errors by mis-alignments,

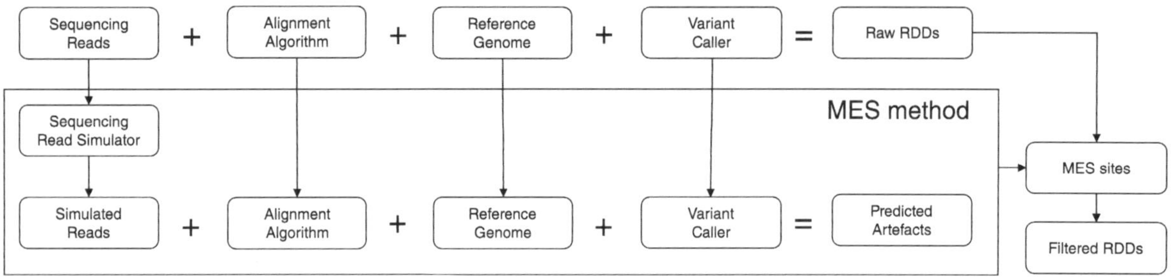

Fig. 5 MES-calculation workflow: 1) MES method simulated randomly mutated sequencing reads. 2) Then, aligns them into genome sequences with an alignment tool of interest. 3) After the alignment finishes, uses variant-caller to detect raw SNPs. 4) Picks the SNPs sites that was not planted originally, i.e., unintended artefacts

or systematic artefacts, are inherent to specific alignment method, they might not be excluded even after multiple replications. Therefore, RDDpred mainly focuses on detecting systematic artefacts with considering other artefacts as well.

2) Read-alignment pattern: a valuable source for distinguishing systematic artefacts

Read-alignment pattern is defined as local status of alignments in variant sites that has been utilized to distinguish artefacts [23]. There are at least six distinct categories of attributes calculated from read-alignments pattern. RDDpred utilizes 15 attributes of the six categories to generate a predictor, which are listed in Table 2. The six categories, such as "Read Depth", "Allele Segregation", "Mapping Quality", "Read Position", "Base Quality", and "Read strand" are known to have significant prediction power and also utilized by Samtools-Bcftools pipeline [23]. The attributes are basically measured as statistics to distinguish how the reads with variants are different with the reads without variants. RDDpred observed massive examples of positive/negative sites to learn how the statistics are represented differently between them. All of 15 attributes are calculated with Samtools-Bcftools pipeline during raw RDD detection [23]. Also we used WEKA, a data mining package, to train a prediction model [30]. Among the algorithms supported in WEKA, we chose Random-Forest algorithm, which showed the best performance in our evaluation datasets and showed significant performance in Laurent's study [5].

3) The 15 attributes for RDDpred

As mentioned, the 15 attributes are categorized into six category, (a) "Read Depth" category represents read-count in editing sites. (b) "Allele Segregation" category includes four attributes, such as VAF, SGB, FQ, and CallQual, respectively. All of these attributes are calculated from edited read-ratio against total reads. (c) "Mapping Quality" category of attributes reflects how the alignments of reads are well-performed, which utilizes alignment scores that the aligner generates. Four attributes, such as PV3, MQB, MQ0F and MQ belongs to this category. (d) "Read Position" category includes three attributes, such as VDB, RPB, and PV4, which represent how the positions of variants are biased within sequencing reads. (e) "Base Quality" category uses base-quality information generated by sequencing machine to detect whether low-quality bases are significantly biased to editing-sites. Two attributes, PV2 and BQB belongs to this category. (f) Finally, "Read

Table 2 Attributes used to train prediction model: total 15-features are calculated with samtools-bcftools(v1.2) pipeline [23]

Categories	Attributes	Description
Read depth	Read depth	Read depth
Allele segregation	VAF	Variant read ratio
Allele segregation	SGB	Segregation based metric
Allele segregation	FQ	Phred probability of all samples being the same
Allele segregation	CallQual	Variant/reference QUALity
Mapping quality	PV3	Mapping quality bias
Mapping quality	MQB	Mann-Whitney U test of Mapping Quality Bias
Mapping quality	MQ0F	Fraction of MQ0 reads
Mapping quality	MQ	Root-mean-square mapping quality of covering reads
Read position	VDB	Variant Distance Bias for filtering splice-site artefacts in RNA-seq data
Read position	RPB	Mann-Whitney U test of Read Position Bias
Read position	PV4	Tail distance bias
Base quality	PV2	Base quality bias
Base quality	BQB	Mann-Whitney U test of Base Quality Bias
Read strand	PV1	Read strand bias

Table 3 Comparison results from two different tissues which shows that RDD occurs condition specifically

	Reads	Bases	Raw RDDs	Accepted RDDs
Bahn's	115,132,348	13,815,881,760	6,856,440	105,564
Peng's	583,640,030	101,787,059,720	58,666,976	3,076,908
Fold	5.07	7.37	8.56	29.15

Strand" category includes single attribute PV1, that represents how the strands of edited reads are biased than non-edited reads (Table 2).

Results

Evaluation with two previous studies

We evaluated RDDpred with two datasets from independent studies performed by Bahn's and Peng's group, respectively [6, 7]. Both studies computationally predicted RNA-editing sites and validated them with Sanger-seq. In Bahn's study, RNA-seq produced 115,132,348 reads with 13,815,881,760 bases in human glioblastoma astrocytoma. RDDpred detected 6,856,440 raw RDDs from them and predicted 105,564 sites as true RNA-editings. In Peng's study, RNA-seq produced 583,640,030 reads with 101,787,059,720 bases in human lymphoblastoid. In this case, RDDpred detected 58,666,976 raw RDDs from them and predicted 3,076,908 sites as true RNA-editings. Note that even though both study uses human tissues, they resulted different number of RNA-editing sites, 105,564 vs. 3,076,908, which indicates that the expression patterns of RNA-editing events might be different in two experimental conditions (Table 3).

We constructed condition-specific model independently from each datasets and evaluated RDDpred separately with the corresponding validation results as test-data. And the test-data was the results of Sanger-seq by each groups. In the process of constructing each models, we carefully arranged the training-data in order not to contain any of information related to test-data. In other words, we did not use test-data for the construction of the model.

1. Training datasets
   (a) Positive examples: Predicted as positives sites by Public Databases (RADAR, DARNED)
   (b) Negative examples: Predicted as artefact sites by MES method (Peng et al. Nature biotechnology 2012)
   (c) The entries overlapped with test-data are excluded from training-data

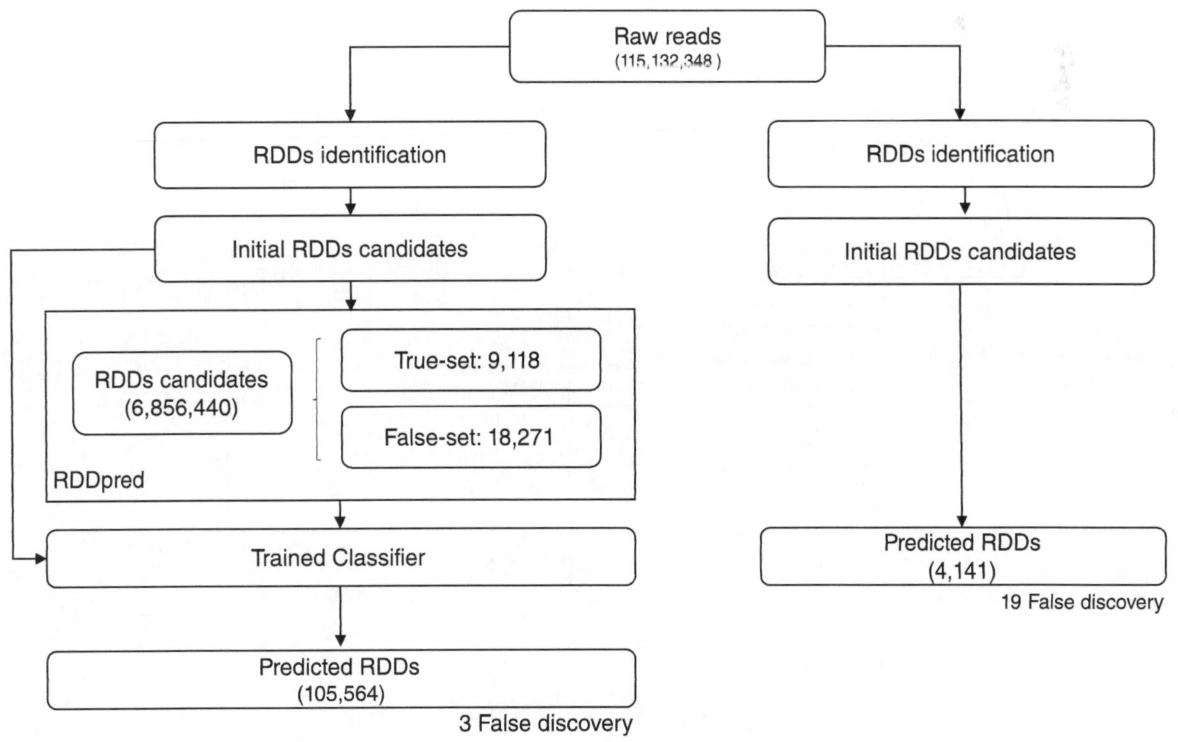

Fig. 6 Overall prediction results: RDDpred predicted 105,564 sites from 6,856,440 initial RDDs candidates, showing that the model was able to reduce potential artefacts from initial RDD candidates with less false-discovery

2. Test datasets
    (a) Positive examples: Positively detected sites by experimental validation (Sanger-seq)
    (b) Negative examples: False discovery sites proved by experimental validation (Sanger-seq)

RDDpred prediction in Bahn's dataset

RDDpred predicted 105,564 sites from 6,856,440 initial RDDs candidates, showing that not only the model was able to reproduce (95.32 %) the results but also was able to reduce potential artefacts (98.46 %) from Bahn's study. Moreover, RDDpred have successfully reject most of the false discoveries in Bahn's prediction. Resulting 84.21 % NPV(Negative Predictive Value). The following results indicate that RDDpred showed relative robustness than Bahn's by reducing potential artefacts and rejecting false-discoveries (Figs. 6 and 7).

RDDpred prediction in Peng's dataset

In Peng's study, RDDpred predicted 3,076,908 sites from 58,666,976 initial RDDs candidates, which also proves that the model was able to successfully reproduce (90.37 %) the results but also reduce potential artefacts (94.79 %) from Peng's study. Moreover, RDDpred have successfully reject most of the false discoveries in Peng's prediction. Resulting 75.86 % NPV (Negative Predictive Value). The following results indicate that RDDpred also showed relative robustness than Peng's by reducing potential artefacts and rejecting false-discoveries (Figs. 8 and 9).

Additional specification of RDDpred

To provide information about actual running-time and memory-usage of RDDpred, we monitored the resource-usage while processing the Peng's dataset. With the machine specified below, RDDpred took 18.33 hours to process 583,640,030 reads of 101,787,059,720 bases. We believe that the resources used here are not too excessive for common research groups.

1. Linux version: Linux version 2.6.32-358.el6.x86_64, CentOS release 6.4
2. Memory usage: 20 GB in maximum
3. CPU usage: 20-cores (Intel(R) Xeon(R) CPU E5645 @ 2.40 GHz)

Discussion

We developed a software package for RNA-editing prediction from RNA-seq data. RDDpred utilizes current

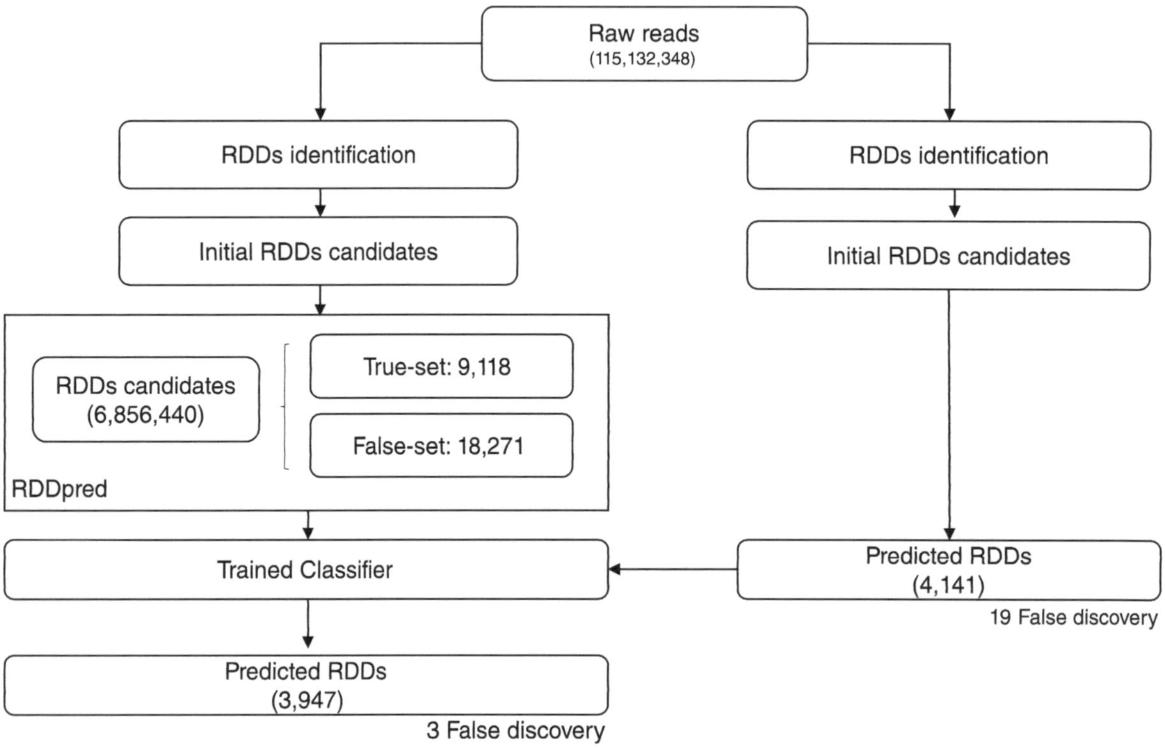

Fig. 7 Comparison with Bahn's prediction results: RDDpred predicted 3947 from 4141 sites indicating that 95.32 % of bahn's results were successfully reproduced with less false discovery, resulting 84.21 % of NPV

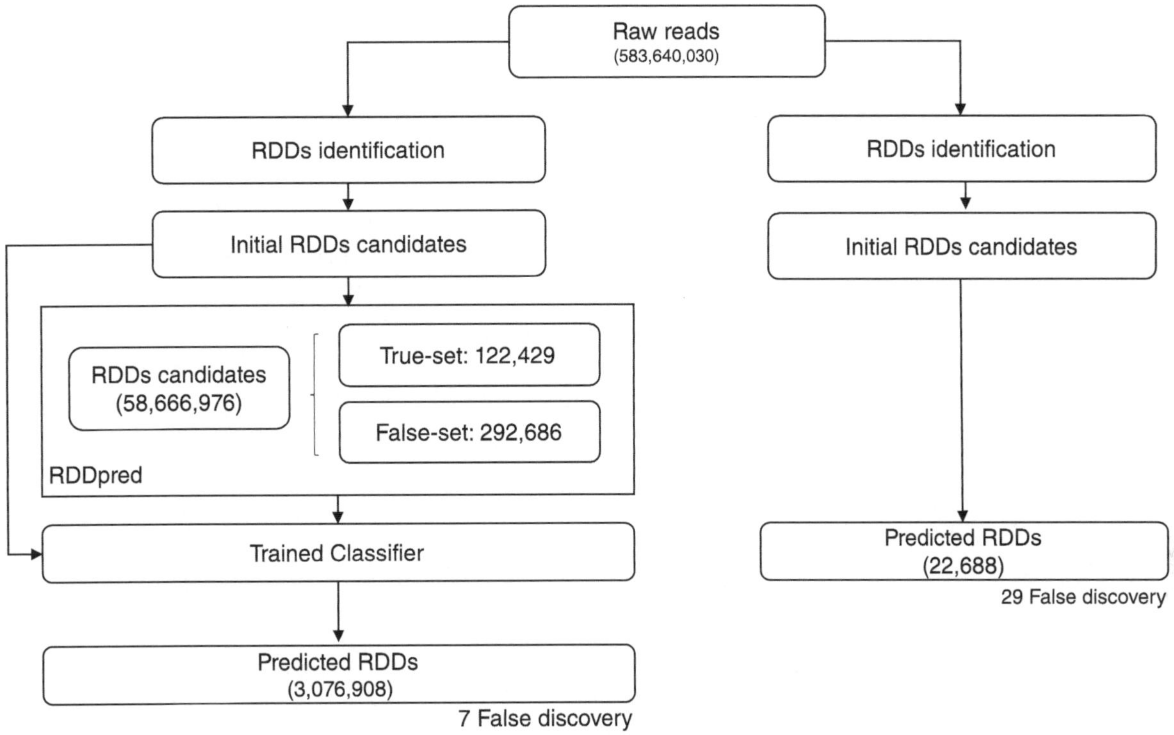

Fig. 8 Overall prediction results: RDDpred predicted 3,076,908 sites from 58,666,976 initial RDDs candidates, showing that the model was able to reduce potential artefacts from initial RDD candidates with less false-discovery

published database and methods such as RADAR, DARNED [13, 29] and MES-method [6] to build condition specific predictor. RDDpred generates a predictor that considers the experimental condition under which RNA-seq experiments are performed. As of now, there are only two studies we can compare with RDDpred. However, we successfully demonstrated that RDDpred was able to reproduce the results and reduce the false-discovery in both studies.

In order to investigate working principle of RDDpred in terms of prediction power of each categories, we calculated information-gain values with WEKA [30]. The rankings of six categories are listed in Table 4. Top 3 categories ("Read Depth", "Allele Segregation", and "Base Quality") have strong prediction power. It is well known that "Read depth" or "Allele Segregation" categories of features are generally used for assessing the authenticity of variants [5, 7]. However, we discovered that the "Base Quality" category of attributes might also have significant prediction power according to the prediction results of two datasets.

During the high-throughput sequencing, the sequencers generate bases-qualities that represent the confidence of sequencing. Therefore, unlike other five metrics, the "Base Quality" reflects the molecular status of bases that are directly recorded by sequencer. Until now, we only knew bases modified by editing enzymes are somehow recognized as guanine (or thymine for APOBEC class), but did not know how these recognitions are observed in the perspective of sequencing machines. The base-quality issue indicates that there might be some distinctions between normal and edited bases at the molecular level. Thus, it implies that more detailed recording of molecular characteristics during sequencing process might be a key to improve the accuracy of RNA-editing detection.

Conclusions

RDDpred: a useful tool for investigating condition-specific RNA-editing with RNA-seq

RNA-seq is one of the most powerful methods to investigate transcriptome and the amount of RNA-seq has recently increased nearly exponentially [31]. In spite of this rapid RNA-seq data accumulation and the recognition on important biological roles of RNA-editing, only a few studies reported RNA-editing findings due to the difficulty of getting robust profiles of RNA-editome [16]. Since it is difficult to perform the experimental validation of RNA-editing events in whole-transcriptome scale, a reliable and easily-usable prediction method is truly required.

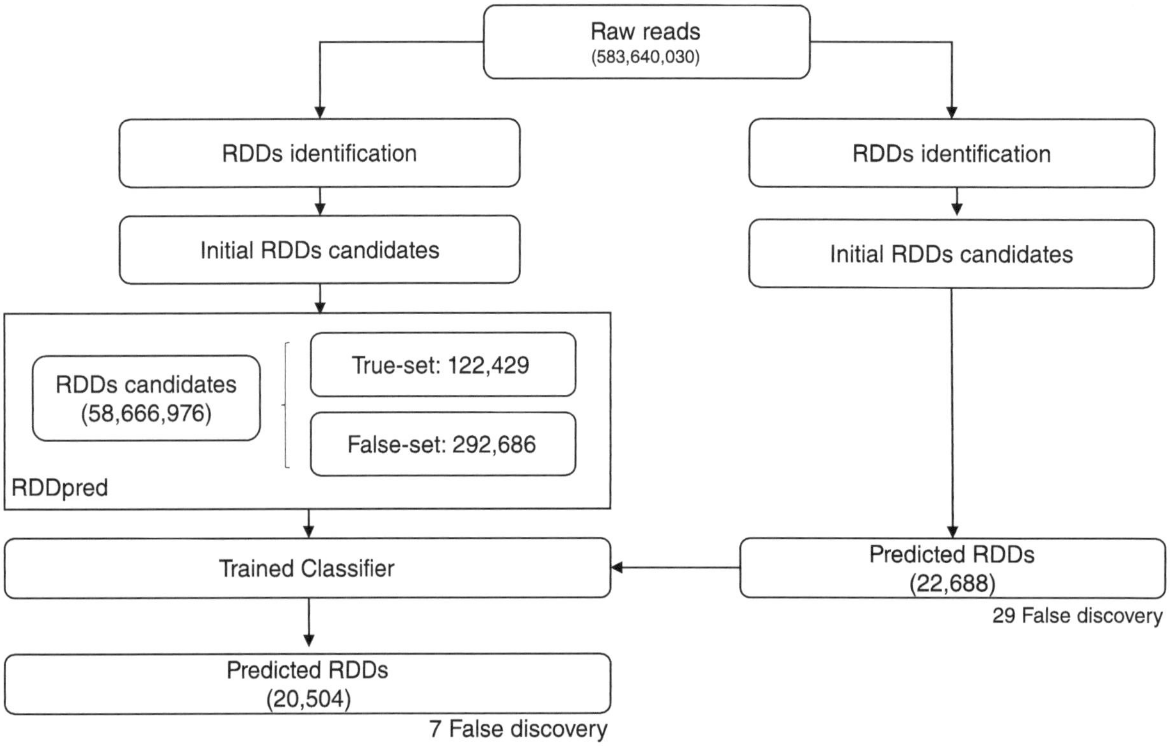

Fig. 9 Comparison with Peng's prediction results: RDDpred predicted 20,504 from 22,688 sites indicating that 90.37 % of bahn's results were successfully reproduced with less false discovery, resulting 75.86 % of NPV

RDDpred prepares training examples that are specific to the condition of input data without experimental validations. RDDpred proved good performances by reproducing the detection of two previous studies and correcting most of their false-discoveries. Moreover, as far as we know, RDDpred is the very first automated pipeline that utilizes machine-learning technique with a well-evaluated performance. Thus, we believe that RDDpred will be very useful and can contribute significantly to the study of RNA-editing. RDDpred is available at `http://biohealth.snu.ac.kr/software/RDDpred`.

Table 4 Category rankings of attributes utilized by RDDpred model: top 3 categories showed relatively strong prediction power

Category	DataA	DataB	RankA	RankB	RankMean
Read depth	0.4943	0.2515	1	2	1.5
Base quality	0.4498	0.40195	3	1	2
Allele segregation	0.457975	0.222825	2	3	2.5
Read position	0.195767	0.1019	4	4	4
Mapping quality	0.010025	0.047	6	5	5.5
Read strand	0.1584	0.0216	5	6	5.5

Competing interests

The authors declare that they have no competing interests.

Authors' contributions

SK designed and supervised the research project and edited the paper. MK developed the method, performed data analysis and wrote the manuscript. BH helped to make the website and advised and assisted the manuscript. All authors read and approved the final manuscript.

Declarations

This article has been published as part of BMC Genomics Volume 17 Supplement 1, 2016: Selected articles from the Fourteenth Asia Pacific Bioinformatics Conference (APBC 2016): Genomics. The full contents of the supplement are available online at http://www.biomedcentral.com/bmcgenomics/supplements/17/S1.

Acknowledgements

This research was supported by the Bio & Medical Technology Development Program of the NRF funded by the Korean government, MSIP(No. NRF-2014M3C9A3063541). Also, this research was supported by Next-Generation Information Computing Development Program through the National Research Foundation of Korea(NRF) funded by the Ministry of Science, ICT & Future Planning (No. NRF-2012M3C4A7033341). The publication cost will be paid by the Seoul National University Office of Research.

Author details

[1]Interdisciplinary Program in Bioinformatics, Seoul National University, Seoul, Republic of Korea. [2]Department of Computer Science and Engineering, Seoul

National University, Seoul, Republic of Korea. [3]Bioinformatics Institute, Seoul National University, Seoul, Republic of Korea.

Published: 11 January 2016

References

1. Burns CM, Chu H, Rueter SM, Hutchinson LK, Canton H, Sanders-Bush E, et al. Regulation of serotonin-2c receptor g-protein coupling by rna editing. Nature. 1997;387:303–8.
2. Keegan LP, Leroy A, Sproul D, O'Connell MA. Adenosine deaminases acting on rna (adars): Rna-editing enzymes. Genome Biol. 2004;5:209.
3. Harris RS, Petersen-Mahrt SK, Neuberger MS. Rna editing enzyme apobec1 and some of its homologs can act as dna mutators. Mol cell. 2002;10:1247–53.
4. Nishikura K. Functions and regulation of rna editing by adar deaminases. Ann Rev Biochem. 2010;79:321.
5. St Laurent G, Tackett MR, Nechkin S, Shtokalo D, Antonets D, Savva YA, et al. Genome-wide analysis of a-to-i rna editing by single-molecule sequencing in drosophila. Nat Struct Mol Biol. 2013;20:1333–39.
6. Peng Z, Cheng Y, Tan BC-M, Kang L, Tian Z, Zhu Y, et al. Comprehensive analysis of rna-seq data reveals extensive rna editing in a human transcriptome. Nat Biotechnol. 2012;30:253–60.
7. Bahn JH, Lee JH, Li G, Greer C, Peng G, Xiao X. Accurate identification of a-to-i rna editing in human by transcriptome sequencing. Genome Res. 2012;22:142–50.
8. Rueter SM, Dawson TR, Emeson RB. Regulation of alternative splicing by rna editing. Nature. 1999;399:75–80.
9. Rosenberg BR, Hamilton CE, Mwangi MM, Dewell S, Papavasiliou FN. Transcriptome-wide sequencing reveals numerous apobec1 mrna-editing targets in transcript 3′ utrs. Nat Struct Mol Biol. 2011;18:230–6.
10. Nishikura K. Editor meets silencer: crosstalk between rna editing and rna interference. Nat Rev Mol Cell Biol. 2006;7:919–31.
11. Galeano F, Rossetti C, Tomaselli S, Cifaldi L, Lezzerini M, Pezzullo M, et al. Adar2-editing activity inhibits glioblastoma growth through the modulation of the cdc14b/skp2/p21/p27 axis. Oncogene. 2013;32: 998–1009.
12. Chiu YL, Soros VB, Kreisberg JF, Stopak K, Yonemoto W, Greene WC. Cellular apobec3g restricts hiv-1 infection in resting cd4+ t cells. Nature. 2010;466:276–6.
13. Kiran A, Baranov PV. Darned: a database of rna editing in humans. Bioinformatics. 2010;26:1772–6.
14. Song W, Liu Z, Tan J, Nomura Y, Dong K. Rna editing generates tissue-specific sodium channels with distinct gating properties. J Biol Chem. 2004;279:32554–2561.
15. Miyata Y, Sugita M. Tissue-and stage-specific rna editing of rps14 transcripts in moss (physcomitrella patens) chloroplasts. J Plant Physiol. 2004;161:113–5.
16. Bass B, Hundley H, Li JB, Peng Z, Pickrell J, Xiao XG, et al. The difficult calls in rna editing. Nat Biotechnol. 2012;30:1207–9.
17. Degner JF, Marioni JC, Pai AA, Pickrell JK, Nkadori E, Gilad Y, et al. Effect of read-mapping biases on detecting allele-specific expression from rna-sequencing data. Bioinformatics. 2009;25:3207–212.
18. Heap GA, Yang JH, Downes K, Healy BC, Hunt KA, Bockett N, et al. Genome-wide analysis of allelic expression imbalance in human primary cells by high-throughput transcriptome resequencing. Hum Mol Genet. 2010;19:122–34.
19. Talwalkar A, Liptrap J, Newcomb J, Hartl C, Terhorst J, Curtis K, et al. Smash: a benchmarking toolkit for human genome variant calling. Bioinformatics. 2014;30:2787–95.
20. Engström PG, Steijger T, Sipos B, Grant GR, Kahles A, Rätsch G, et al. Systematic evaluation of spliced alignment programs for rna-seq data. Nat Methods. 2013;10:1185–91.
21. Dobin A, Davis CA, Schlesinger F, Drenkow J, Zaleski C, Jha S, et al. Star: ultrafast universal rna-seq aligner. Bioinformatics. 2013;29:15–21.
22. Pruitt KD, Brown GR, Hiatt SM, Thibaud-Nissen F, Astashyn A, Ermolaeva O, et al. Refseq: an update on mammalian reference sequences. Nucleic Acids Res. 2014;42:756–63.
23. Li H. A statistical framework for snp calling, mutation discovery, association mapping and population genetical parameter estimation from sequencing data. Bioinformatics. 2011;27:2987–993.
24. McKenna A, Hanna M, Banks E, Sivachenko A, Cibulskis K, Kernytsky A, et al. The genome analysis toolkit: a mapreduce framework for analyzing next-generation dna sequencing data. Genome Res. 2010;20:1297–303.
25. Karolchik D, Hinrichs AS, Furey TS, Roskin KM, Sugnet CW, Haussler D, et al. The ucsc table browser data retrieval tool. Nucleic Acids Res. 2004;32:493–6.
26. Li JB, Levanon EY, Yoon JK, Aach J, Xie B, LeProust E, et al. Genome-wide identification of human rna editing sites by parallel dna capturing and sequencing. Science. 2009;324:1210–13.
27. Mo F, Wyatt AW, Sun Y, Brahmbhatt S, McConeghy BJ, Wu C, et al. Systematic identification and characterization of rna editing in prostate tumors. PloS One. 2014;9(7):e101431.
28. Zhang Q, Xiao X. Genome sequence-independent identification of rna editing sites. Nat Methods. 2015;12:347–50.
29. Ramaswami G, Li JB. Radar: a rigorously annotated database of a-to-i rna editing. Nucleic Acids Res. 2014;42(D1):D109-D113.
30. Hall M, Frank E, Holmes G, Pfahringer B, Reutemann P, Witten IH. The weka data mining software: an update. ACM SIGKDD explorations newsletter. 2009;11:10–18.
31. Leinonen R, Sugawara H, Shumway M. The sequence read archive. Nucleic Acids Res. 2011;39(suppl 1):D19-D21.

Roman et al. BMC Genomics 2016, 17(Suppl 1):6
DOI 10.1186/s12864-015-2302-x

BMC Genomics

PROCEEDINGS Open Access

Medoidshift clustering applied to genomic bulk tumor data

Theodore Roman[1,2], Lu Xie[1,2] and Russell Schwartz[1,3*]

From The Fourteenth Asia Pacific Bioinformatics Conference (APBC 2016)
San Francisco, CA, USA. 11 - 13 January 2016

Abstract

Despite the enormous medical impact of cancers and intensive study of their biology, detailed characterization of tumor growth and development remains elusive. This difficulty occurs in large part because of enormous heterogeneity in the molecular mechanisms of cancer progression, both tumor-to-tumor and cell-to-cell in single tumors. Advances in genomic technologies, especially at the single-cell level, are improving the situation, but these approaches are held back by limitations of the biotechnologies for gathering genomic data from heterogeneous cell populations and the computational methods for making sense of those data. One popular way to gain the advantages of whole-genome methods without the cost of single-cell genomics has been the use of computational deconvolution (unmixing) methods to reconstruct clonal heterogeneity from bulk genomic data. These methods, too, are limited by the difficulty of inferring genomic profiles of rare or subtly varying clonal subpopulations from bulk data, a problem that can be computationally reduced to that of reconstructing the geometry of point clouds of tumor samples in a genome space. Here, we present a new method to improve that reconstruction by better identifying subspaces corresponding to tumors produced from mixtures of distinct combinations of clonal subpopulations. We develop a nonparametric clustering method based on medoidshift clustering for identifying subgroups of tumors expected to correspond to distinct trajectories of evolutionary progression. We show on synthetic and real tumor copy-number data that this new method substantially improves our ability to resolve discrete tumor subgroups, a key step in the process of accurately deconvolving tumor genomic data and inferring clonal heterogeneity from bulk data.

Keywords: Computational biology, Clustering, Tumor, Heterogeneity

Background and introduction

One of the major challenges of understanding tumor genomics is the existence of massive heterogeneity both between patients (inter-tumor heterogeneity) and within single tumors (intra-tumor heterogeneity). Clinically similar tumors may exhibit strikingly different genomic profiles [1–3] and patterns of progression [1, 2]. Even within a single tumor, there can be dramatic differences in genomic profiles at regional [4–6] and even cell-to-cell levels [7–15]. The genomes of tumors thus exhibit significant variation from individual to individual (inter-tumor heterogeneity) as well as cell-to-cell or region-to-region within single tumors (intra-tumor heterogeneity). This heterogeneity presents significant challenges to characterizing progression of individual tumors and finding the commonalities among progression processes across tumors.

New technologies have made it possible to study tumor genomics in much finer detail, but often at the expense of creating even harder computational problems in making sense of the data. Single-cell sequencing [9, 11, 13–17] has offered the promise of reconstructing tumor genomes at the cellular level, but for the moment remains noise-prone and too costly to apply to large numbers of cells in large numbers of tumors. As a result, attempts to reconstruct tumor heterogeneity in large patient populations

*Correspondence: russells@andrew.cmu.edu
[1] Computational Biology Department, School of Computer Science, Carnegie Mellon University, 5000 Forbes Ave, 15213 Pittsburgh, PA, USA
[3] Department of Biological Sciences, Mellon College of Science, Carnegie Mellon University, 4400 Fifth Avenue, 15213 Pittsburgh, PA, USA
Full list of author information is available at the end of the article

have so far largely relied on computational inferences using a class of method called "mixed membership modeling" (also known as unmixing or deconvolution) applied to genomic data from bulk tumor cell populations [18, 19]. Such methods interpret genomic data as mixtures of small numbers of major cell populations and seek to reconstruct the underlying cell populations from the mixed data. They were originally used largely as a way of "purifying" tumor data by separating contributions from tumor versus normal cells [20] but are now widely used to reconstruct clonal subpopulations and/or the progression among them from various genomic data sources (see, for example, [21–30]). Although such methods are now finding widespread use, they still fall far short of being able to reconstruct the true complexity of tumor heterogeneity within or between tumors that can be observed from true single-cell data [9, 13–15, 31, 32]. For a more complete review of computational challenges to reconstructing tumor heterogeneity, see [12].

Our work has pursued a particular agenda of reconstructing models of intra-tumor heterogeneity of individual tumors by taking advantage of inter-tumor heterogeneity across a population of tumors [19, 21, 33]. Specifically, we seek to construct a model of evolution of cell populations making up all of the tumors in a patient population, so as to explain each individual tumor as a distinct mixture of those inferred cell populations. In recent work [33], we showed that it was possible to produce a more nuanced analysis of the mixture composition of tumor cell populations by using the assumption that tumors falling into different subtypes or progression pathways will be composed of distinct mixtures of cell types. A key step in reconstructing these mixtures, and the focus of the present paper, is to partition a population of tumors into clusters, where each cluster of tumors is approximately explainable as a mixtures of a distinct subset of a common set of cell types. By finding these subsets of tumors prior to mixture reconstruction, we can infer more sophisticated mixture models accounting for this kind of clonal substructure and improve accuracy of mixture reconstruction. Given a clustering of tumors into subtypes, we can then apply methods for mixed membership modeling to reconstruct models of the cells defining these clusters. We can further organize these cell models into a tumor phylogeny, describing how evolution from a common ancestral cell type could lead to a spectrum of cells sufficient to explain all of the observed tumors across all of the clusters as distinct linear mixtures of those basic cell types.

Identifying these subgroups of tumors can be interpreted geometrically as a problem of partitioning point clouds of tumor genomic data into distinct low-dimensional subregions of a larger ambient genomic space of all tumors. That is, we can conceptualize tumors as points in a geometric space, where each gene or genomic location profiled yields a coordinate for each tumor. Distinct genomic profiles will then lie at distinct points in that high-dimensional space. Mixtures of different kinds of cells would be expected to form low-dimensional objects whose vertices correspond to the positions of the genomic profiles of the component cells. Distinct mixtures, then, would sit in distinct subregions of that high dimensional space, corresponding to mixtures of distinct sets of high-dimensional component vectors. The problem of finding tumors composed of distinct cell types can then be mathematically represented as the problem of finding point clouds that sit in these separate low-dimensional subregions. That problem itself can be posed as a special form of clustering problem. Many clustering methods have been used to study tumor genomic data, including k-means and similar methods, hierarchical methods, and Gaussian mixture models (see, for example, [34]). All of these methods are designed, however, for the problem of partitioning a high-dimensional data set into disjoint point clouds, not into discrete lower-dimensional subregions within a single contiguous point cloud.

To pose the problem more formally, we can think of each tumor sample x as a vector of genomic profiles (expression or copy number levels) at n genomic sites, $(x_1, \ldots, x_n)$. Our model depends on a hypothesis that distinct tumors will approximately evolve along similar progression pathways and thus evolve similar cell types over the course of their progression. We can then conceptualize each tumor x as a noisy mixture of a subtype of a genomic profiles of a common set of cell types $c_1 = (c_{11}, \ldots, c_{1n}), \ldots, c_k = (c_{k1}, \ldots, c_{kn})$. Any given tumor can then be modeled as a linear combination of those cell types $x \approx a_1c_1 + a_2c_2 + \ldots + a_kc_k$. This would in principle cause tumors to be approximately confined to a k-dimensional space, as is assumed in our prior unmixing approaches [19, 21]. Because tumors cluster into distinct subtypes defined by similar progression pathways, though, we would expect different tumors to sample from different subsets of that common set of cell types and thus to have different subsets of non-zero a_i coefficients. If a given tumor subtype evolves a subset of k' of those k cell types in its progression, then tumors of that subtype will approximately occupy a k'-dimensional subspace of the k-dimensional space defined by linear combinations of the full set of cell types. Our goal in the present work is to partition tumors into clusters that appear to occupy these low-dimensional subregions called subspaces so that we can more accurately infer the genomic profiles of the k' cell types corresponding to each cluster and thus build an accurate mixture model of each cluster.

In the present work, we tackle this problem of better decomposing genomic point clouds into low-dimensional subspaces for the purpose of improving deconvolution of cell subpopulations within mixed data sets. Specifically, we seek to partition a point cloud of tumors in a high-dimensional genomic space into subsets of tumors that occupy distinct low-dimensional subspaces and that can be interpreted as being composed of distinct cellular subpopulations. We use an approach based on medoidshift clustering [35], a non-parametric method for unsupervised clustering that is well suited to working with the sparse, high-dimensional point clouds characteristic of tumor genomic data. We adapt the medoidshift method to the present problem using a novel distance measure and two-stage kernel strategy designed to deal with the high noise of genomic data and the non-standard application to discovering discrete subspaces of a contiguous geometric object.

Methods

Overview of medoidshift

Medoidshift clustering is a method within the class of mode–seeking clustering algorithms [35]. It may also be considered a modification of meanshift clustering [36] that allows for cluster centers that are data points (which may be integral to data interpretation), as well as faster runtime in the case of the addition of data to a set for which medoidshift has already been computed [35]. Medoidshift clustering works by finding local centers of density as follows. Assume the data $X \in \mathbb{R}^{m \times n}$ consist of m data points in an n-dimensional ambient space. Then for a shadow kernel function $\Phi(\cdot)$, and neighborhood size $h \in \mathbb{R}$, and scaling constant $c_o \in \mathbb{R}$, we can define a measure of the contribution of a point to the center of mass of another point. For a data point $x \in X$, we can define the kernel density estimator $f(\cdot)$ at that point by

$$f(x) = c_o \sum_{i=1}^{m} \Phi\left(\left|\left|\frac{x - x_i}{h}\right|\right|^2\right) \quad (1)$$

as described in [35], where c_o is a scaling constant, $\Phi(\cdot)$ is as described in [35], and $||\cdot||$ is a distance function. Medoidshift clustering works by assigning a representative to each putative cluster using the data point nearest the center of mass of a neighborhood of points (i.e., the centroid of the neighborhood). We can therefore set the representative of x, say y, as follows, using the same notation as our previous equation:

$$y_{k+1} = \arg\min_{y \in N(x)} \sum_{i=1}^{m} ||x_i - y|| \phi\left(\left|\left|\frac{x_i - y_k}{h}\right|\right|^2\right) \quad (2)$$

where $N(x)$ is the neighborhood of x, y_{k+1} is the $(k+1)^{st}$ candidate for medoid, y_k is the kth medoid candidate, $\phi = -\Phi'$, and other variables are defined as above. y_0 can be any point within N, although in practice we choose the point with smallest distance to x in the kernel space. Sheikh et al. [35] offers a proof of convergence for the algorithm, as well as a more detailed outline of medoidshift in general. We note that strictly speaking some kernel function is always applied to the data; however, in the case that we use a linear kernel where Euclidean distances hold, for the sake of simplicity in reference, we call this approach "no kernel" or "kernel-free".

For the present applications, on both synthetic and real tumor data, we have chosen a hyperparameter neighborhood size of 1 (i.e., $h = 1$), and scaling factor of 1 (i.e., $c_0 = 1$), given that we can normalize data to the $[0, 1]$ range, because $h = 1$ encodes the belief that each cluster is approximately unit length in the space in which it is embedded. In other words, by setting $h = 1$, we allow points within unit distance, which may or may not be transformed using a kernel function, to strongly influence the choice of medoid for a point, but decrease the influence for points outside that neighborhood. For unit length clusters, this minimizes the influence on centroids for points outside that cluster. Additional preprocessing and normalization required for the real data is described in Subsection 2.4. "Validation on real tumor data".

Two-stage medoidshift clustering

Converting medoidshift into a method for decomposing a point cloud into distinct subspaces, rather than spatially separated clusters, can be accomplished with an appropriate choice of distance measure and kernel function to ensure low distance between tumors in a common subspace and high distance between those in distinct subspaces, even if they are close in the ambient genomic space by more conventional Euclidean distances. A similar problem is encountered by more general manifold learning methods. One popular method for reconstructing low–dimensional geometries embedded in higher–dimensional ambient spaces and measuring the distances in the planes and hyperplanes of the embedded manifolds is ISOMAP [37], which uses path distances in a nearest-neighbor graph to measure distance between points on low-dimensional manifolds defined by data in a high-dimensional ambient space. Tumor data would be expected to largely conform to the assumptions of the ISOMAP method [37, 38]. However, ISOMAP has been shown to have difficulty when the data occupy a low-dimensional manifold but are perturbed by ambient-dimensional noise [39], a property that is characteristic of the tumor genomic data we seek to resolve [40, 41]. We apply an approach based on a similar strategy to ISOMAP [37] for identifying low-dimensional manifolds based on path distances in a neighbor graph of the data points, but intended to deal better with the noise characteristics of genomic data.

Our approach uses a two-stage strategy to balance the desire for ISOMAP-like manifold reconstruction with tolerance to genomic noise characteristics. In the first stage we use kernel-free clustering with squared Euclidean distance intended largely to suppress noise in the data. In this stage, representatives of the data points are a reduced set of data points representing local centers of mass in the ambient space. Because medoidshift clustering is a mode–seeking algorithm [35], the representatives of points determined by medoidshift clustering after one round will be nearer to the more dense regions of the data cloud. Under our model assumptions — unbiased noise and noise variance small relative to the signal — the representatives after the first round should constitute a denoised version of the original data. Prior work [42] shows that these assumptions are reasonable for genomic copy number data, which appears to exhibit unbiased noise with variance on the order of 10 percent of the signal.

In the second stage of clustering, we use a kernel function K, where, $K = 1 - e^{D/h}$, where h is a scalar bandwidth and D is the distance between points. Distances are measured as shortest path length in the point neighbor graph, in a similar fashion to ISOMAP [37], with distances between neighboring points measured by the square of the Euclidean distance. We begin with the representatives derived from the end of the first, kernel-free stage of medoidshift clustering, and use the kernel-based medoidshift clustering until convergence to yield a small set of final representatives. The negative exponential kernel function has the effect of causing the clustering to identify extreme points of the point cloud in different subspaces as cluster representatives, rather than centroids in the more conventional sense. The combination of the specialized kernel and the square Euclidean path distance transforms the clustering into an approach for identifying points in different subspaces rather than those isolated in conventional Euclidean space. By itself, though, this specialized clustering is very sensitive to outliers in the data. Under the interpretation that the result of the first stage yields a denoised version of the data, however, the second stage will decompose the underlying geometry of the denoised point cloud in order to assign individual tumors to discrete subspaces, interpreted to represent distinct evolutionary trajectories among the possible progression pathways of the tumors.

All applications of medoidshift clustering in the present work were run on a desktop workstation with an Intel Core i7-4770K processor at 3.5 GHz per core, 32 GB of RAM, and Matlab running in 64–bit mode on Ubuntu.

Validation on synthetic data

We generated seven scenarios of synthetic data to correspond to different possible tumor progression models, each involving small numbers of possible states of progression along a tumor phylogeny capturing the progression pathways of a patient population. For each synthetic data set, we begin by defining a tumor phylogeny: an evolutionary tree describing how a set of cell types might evolve by progression from a common ancestral cell type. Each such phylogeny defines a mixture scenario, in which each path from root to leaf in the phylogeny corresponds to a mixture of cell types we would expect to see in a subtype of tumors. The length of the path will define the dimension of the corresponding point cloud (one dimension per cell type in the mixture). Sharing of cells between paths will manifest as sharing of basis vertices between subspaces. The scenarios are illustrated in Fig. 1, which provides, for each scenario, a line-graph icon describing its simplicial complex structure and a representation of the progression pathways to which that simplicial complex structure corresponds. For example, Fig. 1a, derives from an evolutionary scenario in which an initial progenitor cell type evolves to an early progression state which then evolves two late progression child states. That scenario results in two tumor subtypes sharing the progenitor and early progression state but differing in their late progression state. Geometrically, that mixture scenario would yield two three-dimensional (triangular) simplices sharing

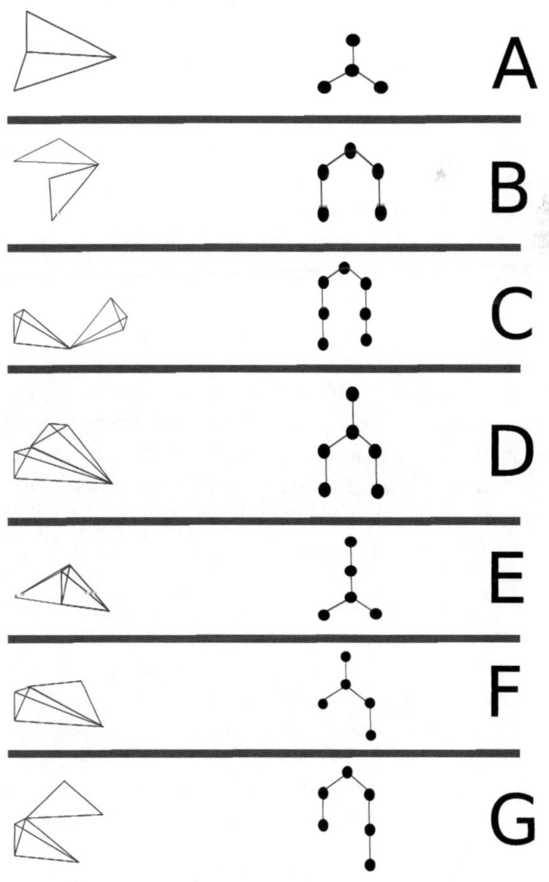

Fig. 1 Visual representation of geometric structures tested with synthetic data and corresponding evolutionary scenarios

a two-dimensional edge. The seven scenarios are as follows:

1. Two triangles sharing an edge (Fig. 1a)
2. Two triangles sharing a point (Fig. 1b)
3. Two tetrahedra sharing a point (Fig. 1c)
4. Two tetrahedra sharing an edge (Fig. 1d)
5. Two tetrahedra sharing a face (Fig. 1e)
6. A triangle and a tetrahedron joined at an edge (Fig. 1f)
7. A triangle and a tetrahedron joined at a point (Fig. 1g)

For each scenario, we generated 100 replicates with each replicate consisting of 500 points in 10 dimensions, chosen to approximate the real data after application of PCA. We varied the noise standard deviation from 0 (no noise) to 0.2 in increments of 0.05. This increasing noise was to allow us to examine the performance of the approaches in the neighborhood of the estimated 0.1 standard deviation noise of real data [42]. We then analyzed the data with three variants of clustering: medoidshift with no kernel function, medoidshift using the kernel function for every step of the medoidshift, and our proposed two-stage medoidshift. We used the ISOMAP-derived distance measure for both the no-kernel function and kernel function medoidshift clusterings. A more detailed description of the simulation technique is provided below.

Simulation procedure

To simulate a scenario in a p-dimensional principal components space, we first define a set of k basis vectors $B \subset \mathbb{R}^p$, representing genomic profiles of the cell types for that scenario. We can then describe a subsimplex in terms of a subset $b = (b_1, \ldots, b_{k'}), \subset B$ of the basis vectors, representing the subset of tumors using a particular subset of k' of the cell types. We then generate a sample tumor from such a cell population by generating a set of mixture fractions representing the amount of each cell type in each tumor. For a given tumor i we generate a mixture fraction f_{ij} for each of the k' cell types in b by defining $f_{ij} \sim unif\,[0, 1]$. We can then generate noiseless samples for each tumor i and gene j as follows:

$$y_{ij} = \frac{\sum_{r=1}^{k'} f_{ir} b_{rj}}{\sum_{r=1}^{k'} f_{ir}} \tag{3}$$

In order to simulate error in the measurements, we add then noise to the samples. For a noiseless sample y_{ik}, we construct a noisy sample x_{ij} by adding 0-biased normal noise in each dimension. That is, for $j = 1, \ldots, 10, x_{ij} = y_{ij} + \mathcal{N}(0, \sigma)$, where we enumerate over $\sigma \in \{0, 0.05, 0.1, 0.15, 0.2\}$ to represent the increasing noise scenarios visualized in Fig. 2.

We assessed accuracy of cluster assignment using the adjusted Rand index [43–45], a measure of cluster quality chosen because it deals well with situations in which the number of clusters is not known a priori and may differ between the ground truth and across inferences being assessed. The adjusted Rand index is computed as follows. Assume we have a known ground-truth partition of the data consisting of n data points; call this partition $X = X_1, X_2, \ldots, X_r$, where there are r total clusters in the ground truth. Suppose we have a partition of the data generated by the algorithm we are testing; call this partition $Y = Y_1, Y_2, \ldots, Y_s$, where there are s total clusters in the partition generated by the algorithm we are assessing. Then we can construct a matrix N to describe the amount of overlap in each of the clusters in X by rows and each of the clusters in Y by columns. More specifically, for entry $n_{ij} \in N$, where n_{ij} is the ith row, jth column, $n_{ij} = |X_i \cap Y_j|$. Finally, to account for the sizes of the clusters in the ground truth and found from the algorithm we would like to assess, we can use a variable to describe the sizes of the clusters. Let $a_i, \forall i \in 1 \ldots r$ be $\sum_{j=1 \ldots s}(n_{ij})$. Similarly, let $b_j, \forall j \in 1 \ldots s$ be $\sum_{i=1 \ldots r}(n_{ij})$. Then we can represent the adjusted Rand index of the algorithm we are testing using the following formula:

$$ARI = \frac{\sum_{i,j}\binom{n_{ij}}{2} - \left(\sum_i \binom{a_i}{2} \sum_j \binom{b_j}{2}\right)/\binom{n}{2}}{\frac{1}{2}\left(\sum_i \binom{a_i}{2} \sum_j \binom{b_j}{2}\right) - \left(\sum_i \binom{a_i}{2} \sum_j \binom{b_j}{2}\right)/\binom{n}{2}} \tag{4}$$

Validation on real tumor data

We examined two real tumor data sets, both taken from the Cancer Genome Atlas (TCGA) data repository [3], one of high-grade serous ovarian adenocarcinoma (OV) data [46] and one of lung squamous small cell carcinoma (LUSC) [47]. We applied our method to array comparative genomic hybridization (aCGH) data sets profiling copy number variation data for each tumor. We favored copy number data over expression data because copy numbers, as a measure of DNA rather than RNA, are less prone to confounding effects of cell-cell communication and from contamination by normal stromal cells [21]. aCGH data also yield relatively low noise levels, approximately 10 percent of the signal [42], which fits into the range of noise levels at which our mixture reconstruction method was previously found to provide a substantial improvement over other approaches [33]. In each case, we restricted our analysis to the autosomal chromosomes.

After downloading the copy number (CN) data, we pre-processed and compressed the data in order to load the entire dataset into memory on a workstation. Each data file from TCGA contains calls for copy number alterations, in the form of the $log2$ ratio of the alteration in the copy number of the DNA compared to normal, as well as the start and end locations for the call, and the

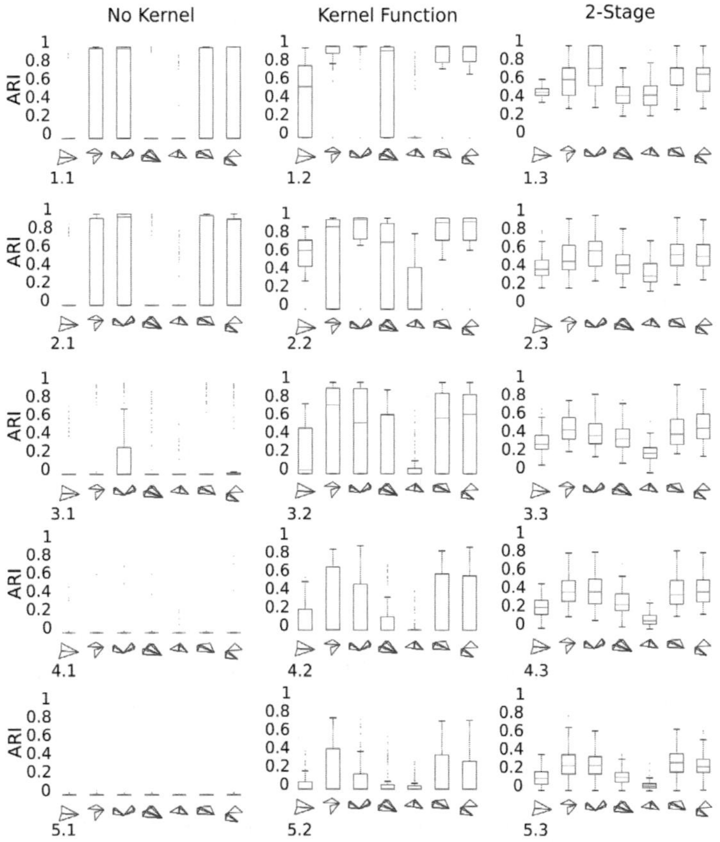

Fig. 2 Adjusted Rand indices (ARIs) for 100 replicates of synthetic data under seven mixture scenarios with varying noise. The first column (*panels x.1*) shows the performance of medoidshift without a kernel function; the second column (*panels x.2*) show the performance of using the negative exponential kernel function; and the third column (*panels x.3*) is our new 2-stage medoidshift clustering method. Each row has increasing noise; the first row (*panels 1.y*) has no noise, the second row has $\sigma = 0.05$ noise added, the third row has $\sigma = 0.1$ noise added, the fourth row has $\sigma = 0.15$ noise added, and the fifth row has $\sigma = 0.2$ noise added

chromosome on which the call is made. For each dataset, we scanned through each base pair, and whenever an amplification in copy number was observed in a sample, we designated a new unit, which we called a block. Within a block, then, each sample has a constant copy number. Then, for each chromosome in a dataset, we could represent the data in the form of an $m \times n$ matrix M, where there are m samples, and n blocks for that chromosome for that dataset. We then used principal components analysis (PCA) [48] to reduce the dimension of the data to the top 10 principal components (PCs), a way of increasing the point density, as is required by mode-seeking algorithms such as medoidshift to determine representatives [35]. Ten dimensions were chosen based on past experience that this is approximately the upper limit for the number of distinct mixture components this class of method will be able to resolve from data sets of a few hundred tumors. In order to further normalize the data into the range $[0, 1]$ for each dimension, which is built into the model as an assumption with respect to the neighborhood examined for each data point, we peformed the following computation:

$$\hat{M}_{i,j} = \frac{M_{i,j} - \min_s M_{s,j}}{\max_q M_{q,j} - \min_r M_{r,j}} \tag{5}$$

where $\hat{M}_{i,j}$ is the normalized value, and $M_{i,j}$ is the *ith* row and *jth* column of the matrix M described above. To encode all of the data of the genome into one unit, we then concatenated each chromosome's block information together, where each block in each chromosome represented additional features (columns). In terms of the data matrices M described above, if we let M^i represent the matrix M above for the *ith* chromosome, we construct a data matrix D, where D is the horizontal concatenation of M^i, $\forall i = 1 \ldots 22$, for each of the 22 non-sex chromosomes.

The ovarian cancer data set, obtained on 15 July 2015 from TCGA [3], consisted of 472 samples with both clinical and CN data. Blocking the CN data and taking the top 10 PCs of the data yielded a 472 by 10 matrix. The lung squamous small cell carcinoma dataset used was obtained from TCGA [3] on 21 July 2015, consisting of 408 samples with both clinical and CN data. Blocking the CN data and taking the top 10 PCs of the data yielded a 408 by 10 matrix.

To assess how the different clusters found in the data might become manifest in clinical data, we constructed Kaplan–Meier curves and applied the log-rank statistic to test significance of the correlation between cluster assignment and censored survival [49] for each real data set.

For the OV data, we further examined correspondence of our cluster assignments to those of the TCGA, which identified four tumor OV subtypes based on analysis of RNA expression data: differentiated, immunoreactive, mesenchymal, and proliferative. We tested significance of the overlap between our clustering and their subtyping by chi-squared test.

We further conducted a functional genomic analysis to better understand the possible biological significance of the clusters. We first converted the block coordinates back into the corresponding chromosome coordinates and then used BioMart [50] to find lists of elevated genes for each cluster. For each cluster, we used the cluster representative as determined by the two-stage clustering. For each representative, we identified each block showing at least 4-fold amplification as a block of interest at a 10kb resolution. We used Biomart [50] to extract a list of genes overlapping each block and pooled genes across blocks for each cluster. We applied DAVID [51] with default parameters to identify ontology terms overrepresented in the resulting gene lists, identifying the top cluster of DAVID functional annotations for each of the clusters found through our method.

Results

Synthetic data

We first applied our methods to synthetic data generated from the seven scenarios of Fig. 1. The adjusted Rand indices (ARIs) for each trial appear in Fig. 2. Each row of the figure shows results for all three cluster variants — no kernel, kernel, and our proposed two-stage method — across all seven scenarios at a single noise level, with noise increasing on each successive row. The kernel-free medoidshift was essentially unusuable on the synthetic data, with ARI at or near zero (no correspondence of true to inferred clusters) for all scenarios at modest noise levels. Even in the noise-free case, the kernel-free method is only occasionally able to produce non-zero ARI. The one-stage kernel-based method is reasonably effective in most scenarios at the lowest noise levels, but with very high variance even for modest noise. At higher noise levels, it, too, usually yields mean ARI at or near zero. Our proposed two-stage method yields far more consistent results. While it somewhat underperforms the one-stage kernel method in noise-free case, it is far less sensitive to increases in noise. The two-stage method gives qualitatively similar results across noise levels, with modest decreases in accuracy at each increasing noise level. Furthermore, the two-stage method yields far lower variances across replicates, suggesting it is meeting the goal of suppressing the high sensitivity to outliers produced by the negative-degree kernel function alone.

We also evaluated run time on the synthetic data. Mean run times across all scenarios and noise levels for each method are 1.2854 s, 1.2914 s, and 0.0121 s for no kernel medoidshift, kernel medoidshift, and 2-stage medoidshift respectively. The 2-stage method requires less time because the costly path-distance computations needed to derive the ISOMAP distances are only computed for cluster representatives after the first step of the two-step method, when the number of data points has been significantly reduced. The no-kernel and kernel techniques, by contrast, must compute the path costs from all points to all points.

Real tumor data

Experimental data analyzed in the manuscript were taken from The Cancer Genome Atlas [3]. Contributing centers for LUSC and OV were Harvard Medical School (LUSC and OV). Copy-number data were collected using the Aglient Human Genome CGH Microarray 244A (OV) and Aglient Human Genome CGH Custom Microarray 2x415K (LUSC). The experimental procedures used in collecting these data are as described in the primary literature from the TCGA Research Network [3].

We next analyzed the two real tumor data sets, beginning with the ovarian cancer (OV) TCGA data. We performed two–stage medoidshift clustering on the matrix of PCs of CN data, which yielded three clusters. Figure 3a shows the point cloud and cluster assignment, visualized in the space of the first three PCs. The OV data show a clear simplicial substructure, approximately described as three arms projecting from a common center. We would expect that the center would correspond approximately to normal diploid cells and the arms to mixtures of normal cells with three progression subtypes. The clustering corresponds generally, although not perfectly, to the arm structure, as intended.

To assess whether the clustering is biologically significant, we performed Kaplan-Meier survival analysis [49] to test for significant differences in survival time between pairs of clusters. We found that a comparison of cluster 1 to cluster 2 yielded a log rank test statistic chi–squared

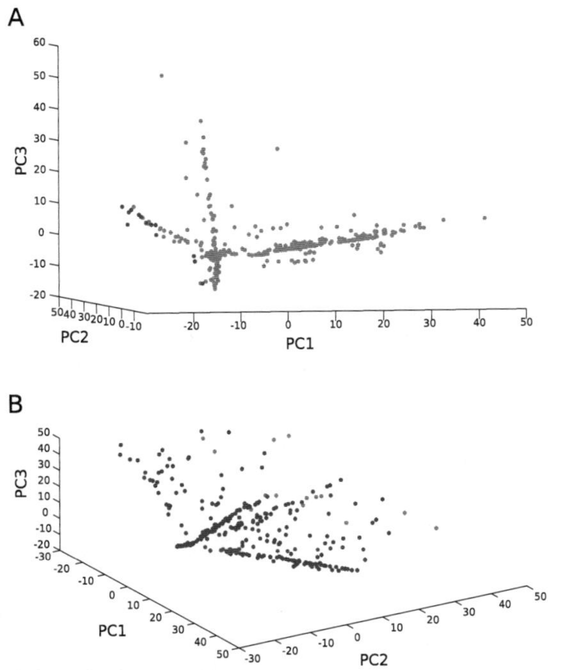

Fig. 3 Visual representation of ovarian tumor data (OV) in principal components space (panel (**a**)), and of lung squamous small cell carcinoma (LUSC) (panel (**b**)). Data are colored based on their cluster membership as determined by 2-stage medoidshift clustering

value of 12.5571, cluster 2 to cluster 3 yielded a chi-squared value of 50.5965, and cluster 1 to cluster 3 yielded a chi-squared value of 180.9646. All of the chi-squared values are strongly significant ($p < 0.01$), suggesting that the clustering into subsimplices of the simplicial complex does capture a clinically significant subgrouping of tumors.

We further compared our clusters to assignments of four molecular subtypes (differentiated, immunoreactive, mesenchymal, proliferative) derived as part of the TCGA data analysis [46] by non-negative matrix factorization clustering (NMF) [52] applied to RNAseq expression data. We obtained the TCGA subtyping information from Verhaak et al. [53]. We established a contingency table of tumors comparing our cluster assignments to the TCGA subtypes, excluding 14 samples for which no subtyping label was provided and found a χ^2 value of 15.48091566, yielding a weakly significant p-value of 0.0168. Our cluster 1 associated most strongly with the mesenchymal subtype and cluster 3 with the proliferative subtype, with cluster 2 too small for any association to be apparent.

Further ontology analysis via DAVID [51] on each of the clusters yielded enrichment for a variety of terms. In the interest of space, we focus on the most significant term grouping identified DAVID for each of our clusters. Table 1 shows these DAVID term for each of our clusters, with highly similar terms merged. We then sought to tie the enriched terms to existing analysis in the literature. The terms found appear consistent with activation of genes associated with epithelial differentiation in cluster 1, a set of pathways previously observed to be active in subsets ovarian tumors [54]. Cluster 2 shows genes that may be consistent with an immune-active subtype, although the cluster is too small for results to be significant. The top cluster 3 terms seem most associated with cytoskeletal structure and movement, which would be consistent with the proliferative subtype identification but could have many other interpretations. We also searched for correspondence between pathways dysregulated by the genes of interest from the TCGA findings [46] and pathways associated with genes found in our cluster amplicons as assessed by Entrez description [55]. Anecdotally, we found genes in several key pathways identified by the TCGA study, including NOTCH signaling and Wnt/β-catenin signaling [46], although we cannot attach significances to these assignments.

We next analyzed the lung squamous small cell carcinoma (LUSC) TCGA data set. We performed two–stage medoidshift clustering on the matrix of PCs of CN data, which yielded a result of two clusters. Figure 3b shows the point cloud and cluster assignment, again visualized in the space of the first three PCs. In this case, the clustering performs more poorly. The method captures a subset of tumors that appears to lie along the fringe of the point cloud, but misses an apparent substructure in which a large fraction of the points appear to lie along two distinct two-dimensional surfaces. We suspect the poorer performance in this case is caused by highly uneven sampling of the manifold, with points in the two-dimensonal "wings" heavily concentrated along a single one-dimensional "line" of each wing. We might expect such a geometric structure to occur if each wing represents a mixture of normal

Table 1 Summary of DAVID ontology terms most strongly associated with each ovarian (OV) and lung (LUSC) cluster

Cancer type	Cluster number	Terms
OV	1	Keratinization, small proline-rich, epidermal cell differentiation, epithelial cell differentiation
OV	2	Antigen processing, MHC class II, asthma, allograft rejection, type I diabetes mellitus, cell adhesion
OV	3	Keratin, coil 1a/b/2/12, intermediate filament, cytoskeleton, non-membrane-bound organelle
LUSC	1	Zinc finger, KRAB, C2H2, transcriptional regulation, DNA-binding, metal binding
LUSC	2	Keratin, peripherin, intermediate filament family orphan 1

cells with a single progression state found in all tumors of that wing and a second, more advanced, progression stage reached by only a minority of tumors.

We again performed Kaplan-Meier analysis on the clusters and observed a test statistic of 3.6413, which is only weakly significant ($p = 0.0339$). While the result suggests there is some meaningful separation of the data by survivability, the conclusion must be considered tentative.

The tentative results in the LUSC dataset with respect to survivability lead us to cautiously evalutate the results from ontology analysis. We again used DAVID [51] for amplified regions of the cluster representative for the LUSC dataset, and found terms in the amplified regions' gene annotations dealing with the keratin family and zinc binding, which have been noted generally as having a role in tumors [56, 57]. Some members of the keratin family have been singled out for study in lung cancers [57], although given the poorer performance of our method on the LUSC dataset, we hesitate to draw conclusions from this correspondence.

Conclusion and discussion

We have developed and implemented a novel clustering approach for partitioning mixtures of genomic data into submanifolds of interest. Our approach is meant to more accurately estimate of patterns of underlying substructure in tumor genomic data sets indicative of the shared pathways of progression between tumors that can be used to reconstruct heterogeneity within single tumors. The model is designed to exploit the geometric and statistical properties we would expect to arise in genomic point clouds produced by tumors evolving along a common set of possible progression pathways. Tests on synthetic data show the novel two-stage method proposed here is more effective than either pure kernel-based or kernel-free variants of medoidshift clustering at decomposing simplicial complex structure in the presence of ambient noise similar to what we would expect from genomic data. We believe the standard kernel-less method performs poorly because it is not designed for the task of partitioning distinct low-dimensional sub-manifolds of a contiguous higher-dimensional manifold. The negative-weight kernel method solves that problem but at the cost of a very noise-sensitive clustering unsuitable for genomic noise profiles. The proposed two-stage method overcomes these problems by using kernel-less clustering to effectively denoise data followed by the negative-weight kernel to decompose the simplicial complex structure in the denoised data.

The real tumor data shows both the power of the approach and some of its current limitations. The method is effective at decomposing simplicial complex structure evident in the ovarian cancer data, a kind of geometric structure we would predict to arise due to the nature of subtyped progression pathways found in such data. Survival analysis confirms that the derived clusters correspond to a biologically and clinically significant partitioning of tumors. The method performs more poorly on the lung squamous small cell carcinoma data, where a simplicial substructure is visually evident in the data but does not correspond well to the clustering. We attribute this problem to highly uneven sampling of tumors within the sub-manifolds, a problem that is consistent with a model of multistage progression in which tumors may be poorly sampled from late progression stages. The poorer performance in this case suggests avenues for improvement by building expectations from more sophisticated progression models into the clustering objective.

These methods might also be improved by considering other forms of genomic data. The methods are applied here to DNA copy number data and previous methods from our group have used both copy number [21] and gene expression data [19, 33]. Combining distinct data sources on a common tumor, as well as considering alternatives such as genomic methylation data, might be expected to lead to improved mixture separation relative to any one data type. Furthermore, even limited amounts of single-cell genomic data could prove valuable in improving mixture deconvolution of bulk tumor data by more precisely identifying genomic profiles of pure unmixed cell types.

We further note that this clustering is not intended to be a standalone method but rather is meant as part of a broader pipeline for deconvolving mixed genomic data and reconstructing virtual single cell profiles, mixture compositions, and progression pathways [19, 21, 33]. Additional work is still needed to better adapt the present methods for use in the full unmixing process, optimize other steps of that process, and explore ways to better solve the unified inference problem these steps collectively represent. Finally, we note that while our particular interest is analyzing tumor heterogeneity, the methods developed here may have value for analyzing other kinds of genomic data, other forms of heterogeneous cell populations, or deconvolving substructured mixtures in other problem domains.

Competing interests

The authors declare that they have no competing interests.

Authors' contributions

RS conceived the model framework and overall analysis pipeline. LX introduced the idea of using medoidshift clustering for the problem, developed the 2-stage medoidshift approach, and derived a problem-specific kernel function for the second stage. TR located the datasets, developed the synthetic data framework, executed the code, and drafted the manuscript. All authors provided feedback on the manuscript and have proofread the manuscript. All authors read and approved the final manuscript.

Acknowledgements

We are grateful to John Kang, M.D., Ph.D., for his helpful insights into tumor data sets, their associated metadata, and the interpretation of their clinical significance and to Gary Miller, Ph.D., for many helpful discussions on posing and solving the computational models and methods. This work was supported in part by U.S. National Institutes of Health awards 1R01CA140214 (RS, LX, and TR), 1R01AI076318 (RS), T32EB009403 (TR), and the Computational Biology Department at Carnegie Mellon University (LX). The publication costs for this article were funded by Carnegie Mellon University institutional resources.

Declarations

This article has been published as part of BMC Genomics Volume 17 Supplement 1, 2016: Selected articles from the Fourteenth Asia Pacific Bioinformatics Conference (APBC 2016): Genomics. The full contents of the supplement are available online at http://www.biomedcentral.com/bmcgenomics/supplements/17/S1.

Author details

[1]Computational Biology Department, School of Computer Science, Carnegie Mellon University, 5000 Forbes Ave, 15213 Pittsburgh, PA, USA. [2]Joint Carnegie Mellon/University of Pittsburgh Ph.D. Program in Computational Biology, 5000 Forbes Ave, 15213 Pittsburgh, PA, USA. [3]Department of Biological Sciences, Mellon College of Science, Carnegie Mellon University, 4400 Fifth Avenue, 15213 Pittsburgh, PA, USA.

Published: 11 January 2016

References

1. Van't Veer LJ, Dai H, Van De Vijver MJ, He YD, Hart AA, Mao M, et al. Gene expression profiling predicts clinical outcome of breast cancer. Nature. 2002;415(6871):530–6.
2. Van De Vijver MJ, He YD, van't Veer LJ, Dai H, Hart AA, Voskuil DW, et al. A gene-expression signature as a predictor of survival in breast cancer. New Eng J Med. 2002;347(25):1999–2009.
3. Weinstein JN, Collisson EA, Mills GB, Shaw KRM, Ozenberger BA, Ellrott K, et al. The cancer genome atlas pan-cancer analysis project. Nat Genet. 2013;45(10):1113–20.
4. Navin N, Krasnitz A, Rodgers L, Cook K, Meth J, Kendall J, et al. Inferring tumor progression from genomic heterogeneity. Genome Res. 2010;20(1):68–80.
5. Gerlinger M, Quezada SA, Peggs KS, Furness AJ, Fisher R, Marafioti T, et al. Ultra-deep t cell receptor sequencing reveals the complexity and intratumour heterogeneity of t cell clones in renal cell carcinomas. J Pathol. 2013;231(4):424–32.
6. Sottoriva A, Spiteri I, Piccirillo SG, Touloumis A, Collins VP, Marioni JC, et al. Intratumor heterogeneity in human glioblastoma reflects cancer evolutionary dynamics. Proc Nat Acad Sci. 2013;110(10):4009–14.
7. Pennington G, Smith C, Shackney S, Schwartz R. Expectation-maximization method for reconstructing tumor phylogenies from single-cell data. In: Computational Systems Bioinformatics Conference (CSB). Stanford, CA: Life Sciences Society; 2006. p. 371–80.
8. Pennington G, Smith CA, Shackney S, Schwartz R. Reconstructing tumor phylogenies from heterogeneous single-cell data. J Bioinform Comput Biol. 2007;5(02a):407–27.
9. Navin N, Kendall J, Troge J, Andrews P, Rodgers L, McIndoo J, et al. Tumour evolution inferred by single-cell sequencing. Nature. 2011;472(7341):90–4.
10. Tao Y, Ruan J, Yeh SH, Lu X, Wang Y, Zhai W, et al. Rapid growth of a hepatocellular carcinoma and the driving mutations revealed by cell-population genetic analysis of whole-genome data. Proc Nat Acad Sci. 2011;108(29):12042–7.
11. Hou Y, Song L, Zhu P, Zhang B, Tao Y, Xu X, et al. Single-cell exome sequencing and monoclonal evolution of a jak2-negative myeloproliferative neoplasm. Cell. 2012;148(5):873–85.
12. Nik-Zainal S, Van Loo P, Wedge DC, Alexandrov LB, Greenman CD, Lau KW, et al. The life history of 21 breast cancers. Cell. 2012;149(5): 994–1007.
13. Navin NE. Cancer genomics: one cell at a time. Genome Biol. 2014;15:452.
14. Navin NE. Delineating cancer evolution with single-cell sequencing. Sci Transl Med. 2015;7(296):296–9.
15. Wang Y, Navin NE. Advances and applications of single-cell sequencing technologies. Mol Cell. 2015;58(4):598–609.
16. Xu X, Hou Y, Yin X, Bao L, Tang A, Song L, et al. Single-cell exome sequencing reveals single-nucleotide mutation characteristics of a kidney tumor. Cell. 2012;148(5):886–95.
17. Wang D, Bodovitz S. Single cell analysis: the new frontier in 'omics'. Trends Biotechnol. 2010;28(6):281–90.
18. Quon G, Morris Q. Isolate: a computational strategy for identifying the primary origin of cancers using high-throughput sequencing. Bioinformatics. 2009;25(21):2882–9. doi:10.1093/bioinformatics/btp378.
19. Schwartz R, Shackney SE. Applying unmixing to gene expression data for tumor phylogeny inference. BMC Bioinformatics. 2010;11(1):42.
20. Etzioni R, Hawley S, Billheimer D, True LD, Knudsen B. Analyzing patterns of staining in immunohistochemical studies: application to a study of prostate cancer recurrence. Cancer Epidemiol Biomarkers Prev. 2005;14:1040–6.
21. Tolliver D, Tsourakakis C, Subramanian A, Shackney S, Schwartz R. Robust unmixing of tumor states in array comparative genomic hybridization data. Bioinformatics. 2010;26(12):106–14.
22. Roth A, Khattra J, Yap D, Wan A, Laks E, Biele J, et al. Pyclone: statistical inference of clonal population structure in cancer. Nature methods. 2014;11(4):396–398.
23. Zare H, Wang J, Hu A, Weber K, Smith J, Nickerson D, et al. Inferring clonal composition from multiple sections of a breast cancer. PLoS Comput Biol. 2014;10(7):1003703.
24. Jiao W, Vembu S, Deshwar AG, Stein L, Morris Q. Inferring clonal evolution of tumors from single nucleotide somatic mutations. BMC Bioinformatics. 2014;15(1):35.
25. Qiao Y, Quinlan AR, Jazaeri AA, Verhaak RG, Wheeler DA, Marth GT. Subcloneseeker: A computational framework for reconstructing tumor clone structure for cancer variant interpretation and prioritization. Genome Biol. 2014;15(8):443.
26. Fischer A, Vázquez-García I, Illingworth CJ, Mustonen V. High-definition reconstruction of clonal composition in cancer. Cell Reports. 2014;7(5): 1740–1752.
27. Roth A, Ding J, Morin R, Crisan A, Ha G, Giuliany R, et al. Jointsnvmix: a probabilistic model for accurate detection of somatic mutations in normal/tumour paired next-generation sequencing data. Bioinformatics. 2012;28(7):907–13.
28. Su X, Zhang L, Zhang J, Meric-Bernstam F, Weinstein JN. Purityest: estimating purity of human tumor samples using next-generation sequencing data. Bioinformatics. 2012;28(17):2265–6.
29. Goya R, Sun MG, Morin RD, Leung G, Ha G, Wiegand KC, et al. Snvmix: predicting single nucleotide variants from next-generation sequencing of tumors. Bioinformatics. 2010;26(6):730–6.
30. Li Y, Xie X. Deconvolving tumor purity and ploidy by integrating copy number alterations and loss of heterozygosity. Bioinformatics. 2014;30(15):2121–9.
31. Janocko LE, Brown KA, Smith CA, Gu LP, Pollice AA, Singh SG, et al. Distinctive patterns of her-2/neu, c-myc, and cyclin d1 gene amplification by fluorescence in situ hybridization in primary human breast cancers. Cytometry. 2001;46(3):136–49.
32. Heselmeyer-Haddad K, Berroa Garcia LY, Bradley A, Ortiz-Melendez C, Lee WJ, Christensen R, et al. Single-cell genetic analysis of ductal carcinoma *in Situ* and invasive breast cancer reveals enormous tumor heterogeneity yet conserved genomic imbalances and gain of *MYC* during progression. Am J Pathol. 2012;181(5):1807–22.
33. Roman T, Nayyeri A, Fasy B, Schwartz R. A simplicial complex-based approach to unmixing tumor progression data. BMC Bioinformatics. 2015;16:254.
34. Ding L, Wendl MC, McMichael JF, Raphael BJ. Expanding the computational toolbox for mining cancer genomes. Nat Rev Genet. 2014;15(8):556–70.
35. Sheikh YA, Khan EA, Kanade T. Mode-seeking by medoidshifts. In: Proc 11th International Conference on Computer Vision. Rio de Janeiro: IEEE; 2007. p. 1–8.
36. Comaniciu D, Meer P. Mean shift: A robust approach toward feature space analysis. Pattern Anal Mach Intell IEEE Trans. 2002;24(5):603–19.
37. Tenenbaum JB, De Silva V, Langford JC. A global geometric framework for nonlinear dimensionality reduction. Science. 2000;290(5500):2319–23.
38. Roweis ST, Saul LK. Nonlinear dimensionality reduction by locally linear embedding. Science. 2000;290(5500):2323–6.

39. Balasubramanian M, Schwartz EL. The isomap algorithm and topological stability. Science. 2002;295(5552):7.
40. Pinto D, Darvishi K, Shi X, Rajan D, Rigler D, Fitzgerald T, et al. Comprehensive assessment of array-based platforms and calling algorithms for detection of copy number variants. Nat Biotechnol. 2011;29(6):512–20.
41. Oesper L, Satas G, Raphael BJ. Quantifying tumor heterogeneity in whole-genome and whole-exome sequencing data. Bioinformatics. 2014;30(24):3532–40.
42. SEQC/MAQC-III Consortium. A comprehensive assessment of RNA-seq accuracy, reproducibility and information content by the sequencing quality control consortium. Nat Biotechnol. 2014;32(9):903–14.
43. Rand WM. Objective criteria for the evaluation of clustering methods. J Am Stat Assoc. 1971;66(336):846–50.
44. Hubert L, Arabie P. Comparing partitions. J Classif. 1985;2(1):193–218.
45. Vinh NX, Epps J, Bailey J. Information theoretic measures for clusterings comparison: is a correction for chance necessary? In: Proceedings of the 26th Annual International Conference on Machine Learning. New York: ACM; 2009. p. 1073–1080.
46. Network CGAR. Integrated genomic analyses of ovarian carcinoma. Nature. 2011;474(7353):609–15.
47. Network CGAR. Comprehensive genomic characterization of squamous cell lung cancers. Nature. 2012;489(7417):519–25.
48. Pearson K. On lines and planes of closest fit to systems of points in space. The London, Edinburgh, and Dublin Philos Mag J Sci. 1901;2:559–72.
49. Meier-Kriesche HU, Schold JD, Kaplan B. Long-term renal allograft survival: Have we made significant progress or is it time to rethink our analytic and therapeutic strategies? Am J Transpl. 2004;4(8):1289–95.
50. Smedley D, Haider S, Durinck S, Pandini L, Provero P, Allen J, et al. The BioMart community portal: an innovative alternative to large, centralized data repositories. Nucleic acids research. 2015;43(W1):W589–W598.
51. Dennis Jr G, Sherman BT, Hosack DA, Yang J, Gao W, Lane HC, et al. David: database for annotation, visualization, and integrated discovery. Genome Biol. 2003;4(5):3.
52. Lee DD, Seung HS. Algorithms for non-negative matrix factorization. In: Advances in Neural Information Processing Systems. Cambridge, MA: MIT Press; 2001. p. 556–62.
53. Verhaak RG, Tamayo P, Yang JY, Hubbard D, Zhang H, Creighton CJ, et al. Prognostically relevant gene signatures of high-grade serous ovarian carcinoma. J Clin Investig. 2013;123(1):517.
54. Godwin AK, Testa JR, Hamilton TC. The biology of ovarian cancer development. CANCER-PHILADELPHIA-. 1993;71:530.
55. Maglott D, Ostell J, Pruitt KD, Tatusova T. Entrez gene: gene-centered information at ncbi. Nucleic Acids Res. 2005;33(suppl 1):54–8.
56. Folkman J. Antiangiogenesis in cancer therapy—endostatin and its mechanisms of action. Exp Cell Res. 2006;312(5):594–607.
57. Camilo R, Capelozzi VL, Siqueira SAC, Bernardi FDC. Expression of p63, keratin 5/6, keratin 7, and surfactant-a in non-small cell lung carcinomas. Hum Pathol. 2006;37(5):542–6.

Li et al. BMC Genomics 2016, 17(Suppl 1):10
DOI 10.1186/s12864-015-2303-9

BMC Genomics

PROCEEDINGS Open Access

Epigenome overlap measure (EPOM) for comparing tissue/cell types based on chromatin states

Wei Vivian Li[1], Zahra S. Razaee[1] and Jingyi Jessica Li[1,2*]

From The Fourteenth Asia Pacific Bioinformatics Conference (APBC 2016)
San Francisco, CA, USA. 11 - 13 January 2016

Abstract

Background: The dynamics of epigenomic marks in their relevant chromatin states regulate distinct gene expression patterns, biological functions and phenotypic variations in biological processes. The availability of high-throughput epigenomic data generated by next-generation sequencing technologies allows a data-driven approach to evaluate the similarities and differences of diverse tissue and cell types in terms of epigenomic features. While ChromImpute has allowed for the imputation of large-scale epigenomic information to yield more robust data to capture meaningful relationships between biological samples, widely used methods such as hierarchical clustering and correlation analysis cannot adequately utilize epigenomic data to accurately reveal the distinction and grouping of different tissue and cell types.

Methods: We utilize a three-step testing procedure–ANOVA, t test and overlap test to identify tissue/cell-type-associated enhancers and promoters and to calculate a newly defined Epigenomic Overlap Measure (EPOM). EPOM results in a clear correspondence map of biological samples from different tissue and cell types through comparison of epigenomic marks evaluated in their relevant chromatin states.

Results: Correspondence maps by EPOM show strong capability in distinguishing and grouping different tissue and cell types and reveal biologically meaningful similarities between Heart and Muscle, Blood & T-cell and HSC & B-cell, Brain and Neurosphere, etc. The gene ontology enrichment analysis both supports and explains the discoveries made by EPOM and suggests that the associated enhancers and promoters demonstrate distinguishable functions across tissue and cell types. Moreover, the tissue/cell-type-associated enhancers and promoters show enrichment in the disease-related SNPs that are also associated with the corresponding tissue or cell types. This agreement suggests the potential of identifying causal genetic variants relevant to cell-type-specific diseases from our identified associated enhancers and promoters.

Conclusions: The proposed EPOM measure demonstrates superior capability in grouping and finding a clear correspondence map of biological samples from different tissue and cell types. The identified associated enhancers and promoters provide a comprehensive catalog to study distinct biological processes and disease variants in different tissue and cell types. Our results also find that the associated promoters exhibit more cell-type-specific
(Continued on next page)

*Correspondence: jli@stat.ucla.edu
[1]Department of Statistics, 8125 Math Sciences Bldg., University of California, Los Angeles, 90095-1554 CA, USA
[2]Department of Human Genetics, University of California, Los Angeles, 90095-7088 CA, USA

(Continued from previous page)
functions than the associated enhancers do, suggesting that the non-associated promoters have more housekeeping functions than the non-associated enhancers.

Keywords: Comparative epigenomics, Tissue and cell type characteristics, Chromatin states, Enhancers and promoters, Histone modification, Disease variants, Multiple testing

Background

While all human tissue and cell types largely preserve the biological information in the DNA sequence of the human genome, the epigenomic landscapes of different tissue and cell types vary considerably, resulting in distinct gene expression programs, biological functions and phenotypic variations. Epigenomic information, such as DNA methylation, covalent histone modifications and DNA accessibility in each tissue and cell type can be investigated using high-throughput sequencing technologies such as Bisulfite-seq, ChIP-seq and DNase-seq [1]. The genome-wide dynamics of epigenomic marks in their relevant chromatin states are considered to bridge genotypes and phenotypes, and they can promote the discovery of biologically meaningful relationships between vast cell types, tissues and lineages [2–4].

Previous research mostly relied on gene expression profiles to study the relationships of samples from different tissue and cell types [5, 6]. The 111 reference epigenomes from the NIH Roadmap Epigenomics Program [4] together with the 16 epigenomes reported by the ENCODE project [7] provided a global view of the epigenomic information covering a large variety of human tissue and cell types. ChromHMM utilized them to build a genome-wide annotation of chromatin states [8]. These large-scale datasets enabled us to study the relationships among tissue and cell types from a new perspective: the similarity of tissue and cell types in terms of histone modification marks evaluated in relevant chromatin states.

Histone modifications at enhancers and promoters in the human genome were found to both reflect and explain global cell-type-specific gene expression [7, 9]. Kundaje et al. showed that pairwise similarity matrices of diverse histone marks could be used to distinguish different subsets of the samples [2]. The similarity matrices were pairwise Pearson correlation values separately calculated for a variety of epigenomic marks. In the same work, they also performed hierarchical clustering of the 111 reference epigenomes using H3K4me1 signal in enhancers (identified by a 15-state HMM model) and showed consistent grouping of biologically similar cell and tissue types, including ES cells, iPS cells, T cells, B cells, adult brain, fetal brain, digestive systems, smooth muscle and heart. Heintzman et al. performed *k*-means clustering on chromatin modifications from both promoters and enhancers [9]. Their results suggested that the chromatin states at promoters are largely invariant across different cell types. In contrast, enhancers reveal cell-type-specificity in clustering and correlate to cell-type-specific gene expression programs on a global scale.

The recent large-scale imputation of epigenomic datasets provided a more consistent and robust resource for capturing sample relationships and dynamic epigenomic information across cell types [10]. Ernst et al. found that compared with the original data, the imputed data led to a correlation matrix of epigenomic features with a more strongly pronounced block structure, suggesting that the imputed data provided a stronger basis for clustering samples into their true tissue or cell type.

Despite the fact that hierarchical clustering and correlation analysis have been shown useful in studying the relationships of biological samples across tissue and cell types, there are many limitations in their use. In the tree representation of hierarchical clustering, it is often difficult to identify the number of groups. In correlation analysis, both Pearson and Spearman correlation coefficients usually provide a noisy correlation matrix of samples, making the detection of sample groups another challenge. Therefore, in order to find a clear correspondence map and distinct grouping of samples based on epigenomic features, we need new methods. Here we propose a new measure–Epigenome Overlap Measure (EPOM)–to distinguish different tissue and cell groups by performing a three-step testing procedure on large-scale epigenomic datasets.

Methods

We describe our method in the following three subsections. In the first subsection, we introduce how the chromatin states are defined; in the second subsection, we describe how we select the histone marks (HMs) based on their relationships with the chromatin states of interest; in the third subsection, we introduce our main three-step testing procedure. The outline of our method is illustrated in Fig. 1.

Selection of chromatin states

We study and evaluate the capacity of our method in mapping and grouping different tissue and cell types using histone marks at both enhancer and promoter regions.

In our study, we use chromatin states of genome-wide 200 bp regions learned by a 25-state multivariate Hidden

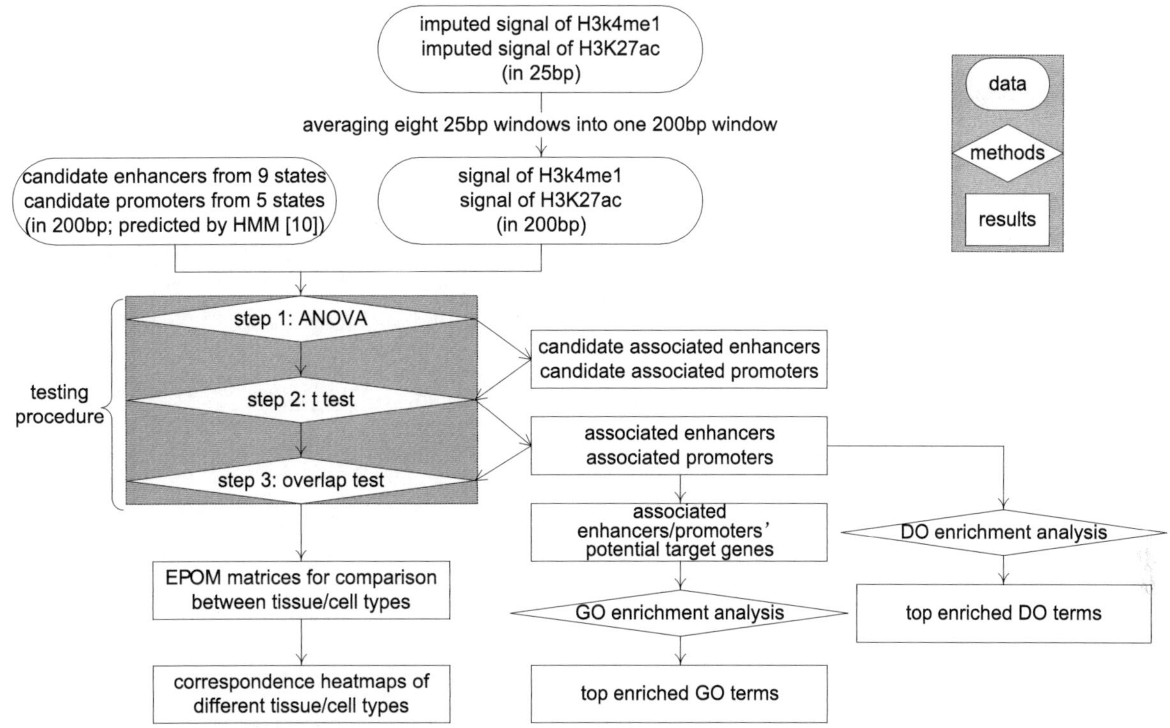

Fig. 1 Outline of EPOM. The orange box gives the legend of the outline: the rounded rectangles represent data taken as input of EPOM; the diamonds represent methods and analysis used in our work; the rectangles represent results and outcomes of our methods. More details of the testing procedure are given in Fig. 2

Markov Model (HMM) [10]. The chromatin states were learned from the imputed data of 12 marks: H3K4me1, H3K4me2, H3K4me3, H3K9ac, H3K27ac, H4K20me1, H3K79me2, H3K36me3, H3K9me3, H3K27me3, H2A.Z, and DNase, across 127 human tissue and cell types (111 from Roadmap and 16 from ENCODE) [2, 10]. Ernst and Kellis [10] interpreted biological meanings of each state through computing the overlap and neighborhood enrichments of the state with respect to various types of functional annotations such as CpG islands, exons, genes, etc. We consider their identified enhancer-related and promoter-related chromatin states (Table 1) as our candidate enhancers and candidate promoters respectively, with a length of 200 bp for each candidate enhancer (or promoter) region. Description of these chromatin states is summarized in Table 1.

Selection of histone modification marks

H3 lysine 4 monomethylation (H3K4me1) was observed to distribute in a cell-type-specific manner and associate with enhancer regions: predicted enhancers showed H3K4me1 enrichment [9, 11]. It was also verified that candidate enhancer states all shared higher frequencies of H3K4me1 than other methylation marks [12]. Another histone modification mark, H3 acetyl K27 (H3K27ac) was associated with active promoters in mammalian cells [13] and predicted enhancers [9]. Hence, we examined the signals of H3K4me1 and H3K27ac in the candidate enhancer (or promoter) regions and attempted to identify the regions where the signals can distinguish different tissue/cell types (we also extended our method to include a third mark H3K4me3 and the results are in Additional file 1). The signals of each mark are $-\log_{10}$ transformed p-values, which represent the enrichment of ChIP-seq read counts based on a Poisson distribution. A stronger signal represents a more statistically significant enrichment of histone modification [2]. The original signals are at 25 bp resolution. We compressed the signals into 200 bp resolution by taking the average of every eight 25 bp windows, so that the signals and our candidate enhancer and promoter regions can be perfectly aligned as 200 bp windows.

Testing procedure

Given the signals of H3K4me1 and H3K27ac on 124 reference epigenomes divided into 16 tissue and cell types (we excluded the three tissue and cell types that only contain one sample) and the locations of candidate enhancers and candidate promoters, we used a three-step testing procedure (please see Fig. 2) to calculate pairwise EPOM scores

Table 1 Chromatin state description

Candidate promoters		Candidate enhancers	
State number	Description	State number	Description
1	Active TSS	10	Transcription 5′ enhancer
2	Promoter upstream TSS	11	Transcription 3′ enhancer
3	Promoter downstream TSS with DNase	12	Transcription weak enhancer
4	Promoter downstream TSS	13	Active enhancer 1
22	Poised promoter	14	Active enhancer 2
23	Bivalent promoter	15	Active enhancer flank
		16	Weak enhancer 1
		17	Weak enhancer 2
		18	Enhancer acetylation only

Note: The state numbers and the description are from [10]. The state numbers were designated by the 25-state-HMM. The description was based on the enrichment of functional annotations and experimentally observed characteristics

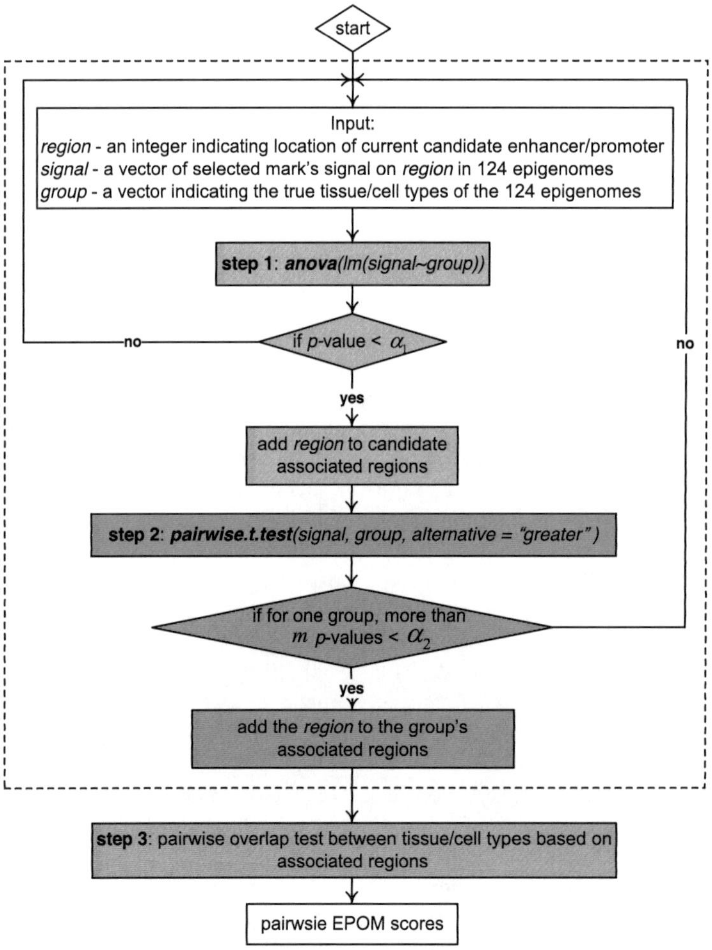

Fig. 2 Diagram of EPOM's three-step testing procedure

and study the relationships among different tissue and cell types. The 16 tissue and cell type groups are: embryonic stem cells (ESC), induced pluripotent stem cells (iPSC), ESC-derived cells (ES-deriv.), Blood & T-cells, HSC & B-cells, Mesenchymal stem cells (Mesench.), Epithelial, Neurosphere (Neurosph.), Thymus, Brain, Muscle, Heart, Smooth Muscle (Sm. Muscle), Digestive, Other, and ENCODE cell lines (ENCODE2012).

Step 1 We use ANOVA to test whether a histone mark has the same group mean signals across the 16 tissue and cell types. We denote the group mean signal of the mark at the k-th candidate enhancer (or promoter) in the i-th tissue/cell type as $\mu_{i,k}$. Then the kth null hypothesis can be expressed as

$$H_{0,k} : \mu_{1,k} = \mu_{2,k} = \cdots = \mu_{16,k} \tag{1}$$

We apply a threshold α_1 to the resulting Bonferroni-corrected p-values and refer to region k as a *candidate associated enhancer (or promoter)* if the null hypothesis $H_{0,k}$ is rejected.

Step 2 We use one-tailed t-test to perform pairwise comparison between the 16 tissue/cell types, so as to identify the associated enhancers (or promoters) of each tissue/cell type. Given two different tissue/cell types i and j and the k-th candidate associated enhancer (or promoter), the null hypothesis is

$$H_{0,ijk} : \mu_{i,k} \leq \mu_{j,k} \tag{2}$$

We apply a threshold α_2 to the resulting p-values and we consider the signal of tissue/cell type i to be significantly higher than that of tissue/cell type j on region k if the null hypothesis $H_{0,ijk}$ is rejected. For the i-th tissue/cell type, if $H_{0,ijk}$ is rejected for more than m times among all $j \neq i$, we define region k as an *associated enhancer (or promoter)* of tissue/cell type i. We separately identify the H3K4me1-based and H3K27ac-based associated enhancers and promoters of each tissue/cell type. Then we combine the information of the two histone marks by taking the union of their associated enhancers (or promoters). That is, for each tissue/cell type we take the union of the two marks' associated enhancers (or promoters) and use the union as the associated enhancers (or promoters) of that tissue/cell type.

Step 3 We perform the overlap test, described in next subsection, on the discovered associated enhancers (promoters) to calculate EPOM scores between every pair of tissue/cell types.

In this paper, we set the thresholds as $\alpha_1 = 10^{-10}$, $\alpha_2 = 0.01$, and $m = 13$ or 14. In our testing procedure, the ANOVA procedure in Step 1 aims to filter out the candidate enhancer (or promoter) regions whose HM signals do not have significant variations across all biological conditions (i.e., tissue and cell types). Step 2 consists of pairwise two-sample t-tests, which aim to find associated regions for each biological condition, such that these regions' HM signals in this condition are significantly higher than in at least m other conditions. Steps 1 and 2 are not redundant but complementary to each other. Step 1 can largely reduce the number of candidate associated regions to be tested in Step 2, so that Step 2 will find the associated regions that not only have high signals in one biological condition but also have strong signal variations across conditions. In addition, Step 1 can largely reduce computational time in Step 2, so as to increase the computational efficiency of the EPOM method. Step 2 is necessary to identify associated regions that carry cell-type-specific characteristics, because it centers on each biological condition in its search for associated regions. The two steps together ensure that the identified associated regions have strong differentiating capability of biological conditions and thus serve as good candidates for the overlap test in Step 3.

Overlap test in the three-step testing procedure

The overlap test, a procedure to check the significance of the overlap of two samples, has been demonstrated as a powerful procedure to map developmental stages from the same or different species based on transcriptomic data [14]. Here we use it to calculate pairwise EPOM scores. Given two tissue/cell types, we compare them by testing the overlap of their associated enhancers (or promoters), e.g., genomic region sets A and B. We consider the union of the associated enhancers (or promoters) of all tissue/cell types after step 2 as the population and A and B as two samples drawn from the population. The null hypothesis is that A and B are two independent samples, while the alternative hypothesis is that A and B are dependent samples. The test statistic is the number of regions shared by A and B. Given n (the population size), $|A|$ and $|B|$ (the sizes of A and B), the larger the test statistics is, the higher the likelihood that the null hypothesis will be rejected. The p-value of the test statistic is calculated as:

$$p = \sum_{i=|A\cap B|}^{\min(|A|,|B|)} \frac{\binom{n}{i}\binom{n-i}{|A|-i}\binom{n-|A|}{|B|-i}}{\binom{n}{|A|}\binom{n}{|B|}} \tag{3}$$

Then we define the EPOM score between samples A and B as

$$\text{EPOM score} = -\log_{10}(\text{Bonferroni-corrected}\ p) \tag{4}$$

The larger the EPOM score is, the more likely that A and B are dependent and the more epigenomic characteristics they share, and vice versa.

Results

From the chromatin states discovered by the 25-state HMM model [10], we identified 4,056,578 candidate enhancers and 1,401,636 candidate promoters (both in 200-bp units) from human Chromosomes 1–22 and Chromosome X. In Step 1 (ANOVA) of the testing procedure, we reduced our target regions to 1,646,842 candidate associated enhancers and 533,086 candidate associated promoters using H3K4me1, and 834,975 enhancers and 306,062 promoters using H3K27ac. Then after Step 2 (t test) of the testing procedure, the numbers of associated regions discovered for different tissue and cell types are summarized in Table 2. On one hand, our study verified that large proportion of promoters are housekeeping, which was consistent with the observation via *k*-means clustering and Pearson correlation that promoter regions

Table 2 Numbers and proportions of enhancer/promoter regions associated with various tissue/cell types

	H3K4me1		H3K27ac		Union of the two HMs	
Tissue/cell type	Enhancers	Promoters	Enhancers	Promoters	Enhancers	Promoters
			Numbers of associated regions			
ESC	43,459	8,852	13,981	4,352	51,666	11,942
iPSC	9,770	1,814	12,079	2,050	20,553	3,697
ES-deriv.	2,242	544	330	79	2,458	598
Blood & T-cell	25,8638	58,525	113,189	35,189	272,139	74,705
HSC & B-cell	29,013	19,855	12,847	9,538	37,371	26,889
Mesench.	242,345	61,975	178,065	64,313	302,647	94,113
Epithelial	4,118	1,243	481	386	4,463	1,614
Neurosph.	13,363	4,614	9,046	4,084	19,202	7,703
Thymus	10,724	2,194	11,158	2,076	18,217	3,800
Brain	152,652	22,362	149,195	30,917	209,745	40,745
Muscle	15,288	3,496	13,166	5,020	23,513	7,416
Heart	3,225	1,098	8,601	3,219	10,458	3,951
Sm. Muscle	38,548	5,386	32,615	6,257	49,460	9,107
Digestive	25,782	4,980	5,528	2,242	28,186	6,563
Other	5	0	0	1	5	1
ENCODE2012	25	13	333	18	55	30
			% of associated regions among candidate regions			
ESC	1.07	0.63	0.34	0.31	1.27	0.85
iPSC	0.24	0.13	0.30	0.15	0.51	0.26
ES-deriv.	0.06	0.04	0.01	0.01	0.06	0.04
Blood & T-cell	6.38	4.18	2.79	2.51	6.71	5.33
HSC & B-cell	0.72	1.42	0.32	0.68	0.92	1.92
Mesench.	5.97	4.42	4.39	4.59	7.46	6.71
Epithelial	0.10	0.09	0.01	0.03	0.11	0.12
Neurosph.	0.33	0.33	0.22	0.29	0.47	0.55
Thymus	0.26	0.16	0.28	0.15	0.45	0.27
Brain	3.76	1.60	3.68	2.21	5.17	2.91
Muscle	0.38	0.25	0.32	0.36	0.58	0.53
Heart	0.08	0.08	0.21	0.23	0.26	0.28
Sm. Muscle	0.95	0.38	0.80	0.45	1.22	0.65
Digestive	0.64	0.36	0.14	0.16	0.69	0.47
Other	0.00	0.00	0.00	0.00	0.00	0.00
ENCODE2012	0.00	0.00	0.01	0.00	0.00	0.00

are largely invariant across different tissue/cell types [9]. On the other hand, we did not observe the association proportion (number of associated regions divided by number of candidate regions) of promoters to be necessarily lower than the association proportion of enhancers. In ten tissue and cell types: ESC, iPSC, ES-deriv., Blood & T-cell, Mesench., Thymus, Brain, Sm. Muscle, Digestive and Other, enhancers' association proportions are about 1.1-1.9 times of promoters'; in the other six tissue and cell types: HSC & B-cell, Muscle, Epithelial, Neurosphere, Heart and ENCODE2012, enhancers' association proportions are only about 50 %–90 % of promoters'.

EPOM between different tissue/cell types

We summarize the EPOM scores of all pairwise comparisons based on the identified associated enhancers and associated promoters respectively. The two resulting matrices were plotted as heatmaps to illustrate the correspondence maps of epigenomes, as shown in Fig. 3a, b. The two heatmaps based on the associated enhancers and the associated promoters are highly consistent, showing a clear diagonal pattern corresponding to the biological groupings of tissue and cell types. The only off-diagonal element is a weak mapping between Muscle and Heart. This is not surprising since heart consists mostly of cardiac muscle cells. Figure 3e and Additional file 2 illustrate the correlation matrices based on H3K4me1 and H3K27ac signals at the candidate enhancer and promoter regions before and after our Step 1 ANOVA, respectively. These heatmaps from correlation analysis can only roughly distinguish three large groups of human tissue and cell types. Comparing the

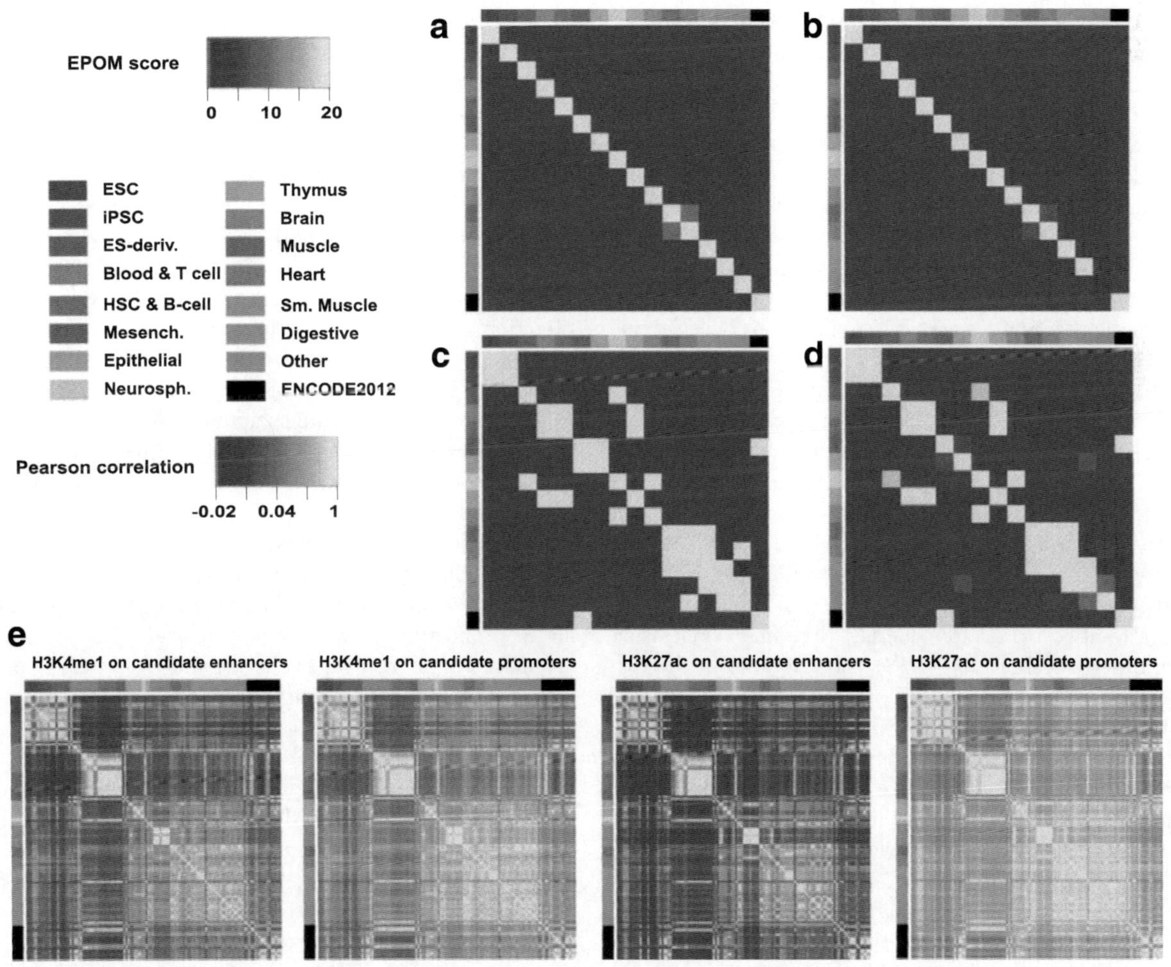

Fig. 3 a–**d** Correspondence maps of EPOM scores saturated at 20 (all the values larger than 20 are set to 20). **a** Calculated based on associated enhancers (threshold $m = 14$ in t test). **b** Calculated based on associated promoters (threshold $m = 14$ in t test). **c** Calculated based on associated enhancers (threshold $m = 13$ in t test). **d** Calculated based on associated promoters (threshold $m = 13$ in t test). **e** Correspondence maps of Pearson correlation coefficients calculated from H3K4me1 and H3K27ac signals in candidate enhancers and candidate promoters. Axis colors mark different true tissue/cell types correspondence maps established by EPOM and correlation analysis, we can see that EPOM is more efficient in capturing epigenomic characteristics of different tissue/cell types. This result also shows the necessity of including Step 2 in the testing procedure to identify cell-type-specific enhancer/promoter regions.

If in Step 2 (t test) of the testing procedure we use a lower threshold $m = 13$ instead of $m = 14$, the discovered associated enhancers and associated promoters would become less cell-type-specific. The resulting EPOM scores are consequently less distinguishable and the correspondence maps (see Fig. 3c, d) reveal subtler similarities between different tissue and cell types. The discovered off-diagonal mappings reveal biologically meaningful relationships. For example, Heart, Muscle and Smooth Muscle are grouped together; Blood & T cells and HSC & B cells are grouped together; Neurosphere is mapped to both Brain and ES-derived cells [15]; Thymus is mapped to Blood & T cells, consistent with its role in T-cell maturation and immunity: thymus is a specialized organ of the immune system and T cells mature within thymus; Thymus is also mapped to HSC & B cells, consistent with the fact that a small population of B cells develop in thymus and some HSC colonize in thymus [16]. As the associated regions become less specific from Fig. 3a, b to Fig. 3c, d the correspondence maps based on enhancers and promoters, although present slight differences, are still consistent with each other, suggesting that our identified associated promoters and enhancers have similar levels of cell/tissue specificity in terms of grouping capability.

We also calculated the EPOM matrices for each of the two histone modification marks separately to see how different the marks' abilities are to capture cell type characteristics. Instead of taking the union of two marks' associated enhancers (or promoters) in Step 2, we used H3K4me1 and H3K27ac's associated enhancers (or promoters) separately to perform the overlap test in Step 3. When using the higher threshold ($m = 14$), the results from the two marks are generally the same; when using the lower threshold ($m = 13$), the results from the two marks are still consistent, but with different scores for certain off-diagonal patterns (please see Additional file 3). To further study how different histone modification marks impact the EPOM scores, we added a third mark histone H3 lysine 4 tri-methylation (H3K4me3) to our study because H3K4me3 is acknowledged to be characteristic of actively transcribed protein-coding promoters [17]. We calculated EPOM scores based on associated enhancers or promoters identified from the three histone modification marks (see Additional file 1). The EPOM matrices still exhibit a strong diagonal pattern that is highly consistent with what we observed from H3K4me1 and H3K27ac.

Another case worth attention is how the EPOM scores change if we summarize the associated enhancers (or promoters) in Step 2 of the testing procedure by taking the intersection of associated enhancers (or promoters) identified for each mark (see Additional file 4). As expected, the diagonal pattern of EPOM matrices become stronger since less associated enhancers (or promoters) are shared among different tissue/cell types. But the significant off-diagonal mappings were still successfully identified.

Potential target genes of the associated enhancers and promoters

Gene expression programs are controlled and regulated by cell-specific changes in the activity of cis-regulatory elements, including enhancers and promoters. Although identifying and annotating these regulatory elements remains a great challenge, it is possible to infer the biological functions of these regions by analyzing the functions of their neighboring genes, which are potential target genes under their regulation [18–20]. Here we study the possible functions of the identified associated enhancers and promoters by analyzing the functions of their nearby genes, which we refer to as the potential target genes of the associated enhancers and promoters.

We related each associated enhancer or associated promoter to its nearest transcription start site (TSS) in up to 200 kb distance. Assignment of a gene to an associated enhancer was counted in both upstream and downstream directions, while assignment of a gene to an associated promoter was counted only in the promoter's downstream direction. The numbers and proportions of the potential target genes assigned to the associated enhancers and promoters are summarized in Fig. 4. The distribution of the potential target gene numbers across tissue/cell types are largely consistent: more genes are identified in Blood & T-cells, HSC & B-cells, Mesenchymal stem cells, Brain and ESC. However, the target genes of associated promoters are more cell-type-specific than those of associated enhancers: larger proportion of associated promoters' target genes are identified in unique tissue/cell types, while larger proportion of associated enhancers' target genes are shared by more than five tissue and cell types. These results suggest that although promoters are more universal to all tissue/cell types, the associated promoters, which are non-housekeeping, are more tissue/cell type specific than the associated enhancers. The associated enhancers are more largely shared by subsets of tissue/cell types. Hence, the associated promoters are better markers of tissue/cell type specificities, while the associated enhancers are better indicators to discover tissue/cell type similarities.

Noticing that real enhancers and promoters can span across regions much longer than 200 bp, we merged the adjacent associated enhancers or promoters and re-identified the potential target genes of the merged associated enhancers or promoters (see Additional file 5). With

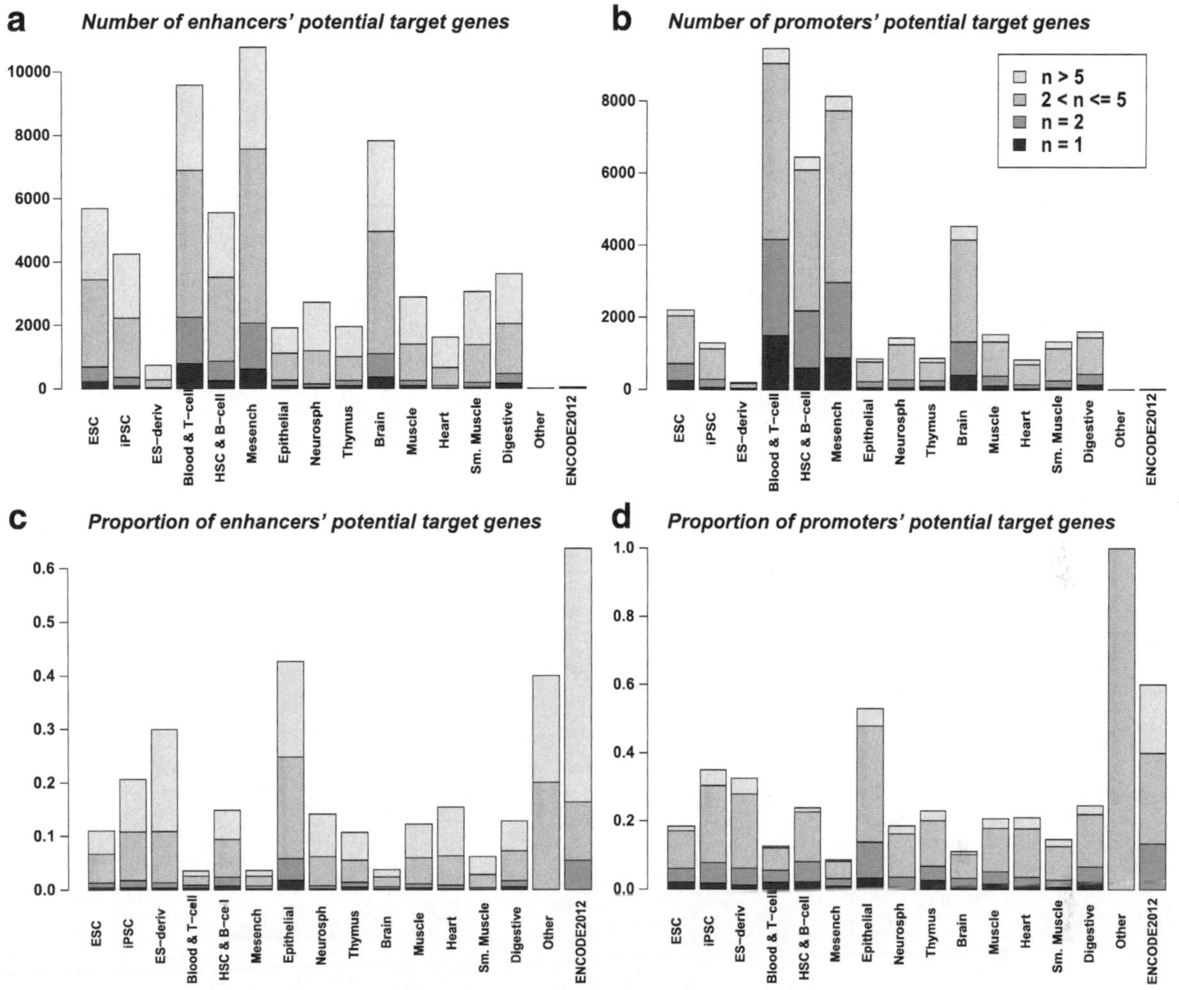

Fig. 4 a Number of associated enhancers' potential target genes in each tissue/cell type. **b** Number of associated promoters' potential target genes in each tissue/cell type. **c**–**d** Proportions are calculated as the number of regions' potential target genes divided by the number of associated regions. **c** Proportion of enhancers' potential target genes in corresponding groups as in (**a**). **d** Proportion of promoters' potential target genes in corresponding groups as in **b**. Colors of the sub-bars represent the number of tissue/cell types (*n*) in which the genes are identified, as explained in the legend

decreasing numbers of the associated enhancers and promoters, the proportions of the target genes increase (see Fig. 4 and Additional file 5); however, the distribution of the proportions across tissue/cell types remains largely the same.

Gene ontology enrichment analysis of associated enhancers and promoters

We performed gene ontology (GO) [21] enrichment analysis on the associated enhancers/promoters' potential target genes of each tissue and cell type to check which GO terms are over-represented in the associated enhancers (or promoters). We used biological process GO terms and focused on the top enriched GO terms found by an overlap test in each tissue and cell type. The heatmaps of GO enrichment scores (see Figs. 5 and 6) show that the top 10 enriched GO terms in the gene lists are distinct for each tissue/cell type, with only a small proportion shared in common. We calculated the proportion of cell-type-specific GO terms–number of specifically enriched GO terms divided by number of enriched GO terms (p-value $< 10^{-3}$)– and found that the associated promoters have around $1.5 - 6$ times cell-type-specific GO terms compared with the associated enhancers. This again implies that in terms of the 200 bp regions, the associated promoters are more cell-type-specific than associated enhancers.

The annotations of top enriched GO terms in each tissue and cell type (please see Additional files 6 and 7) verify and explain the similarity patterns discovered through

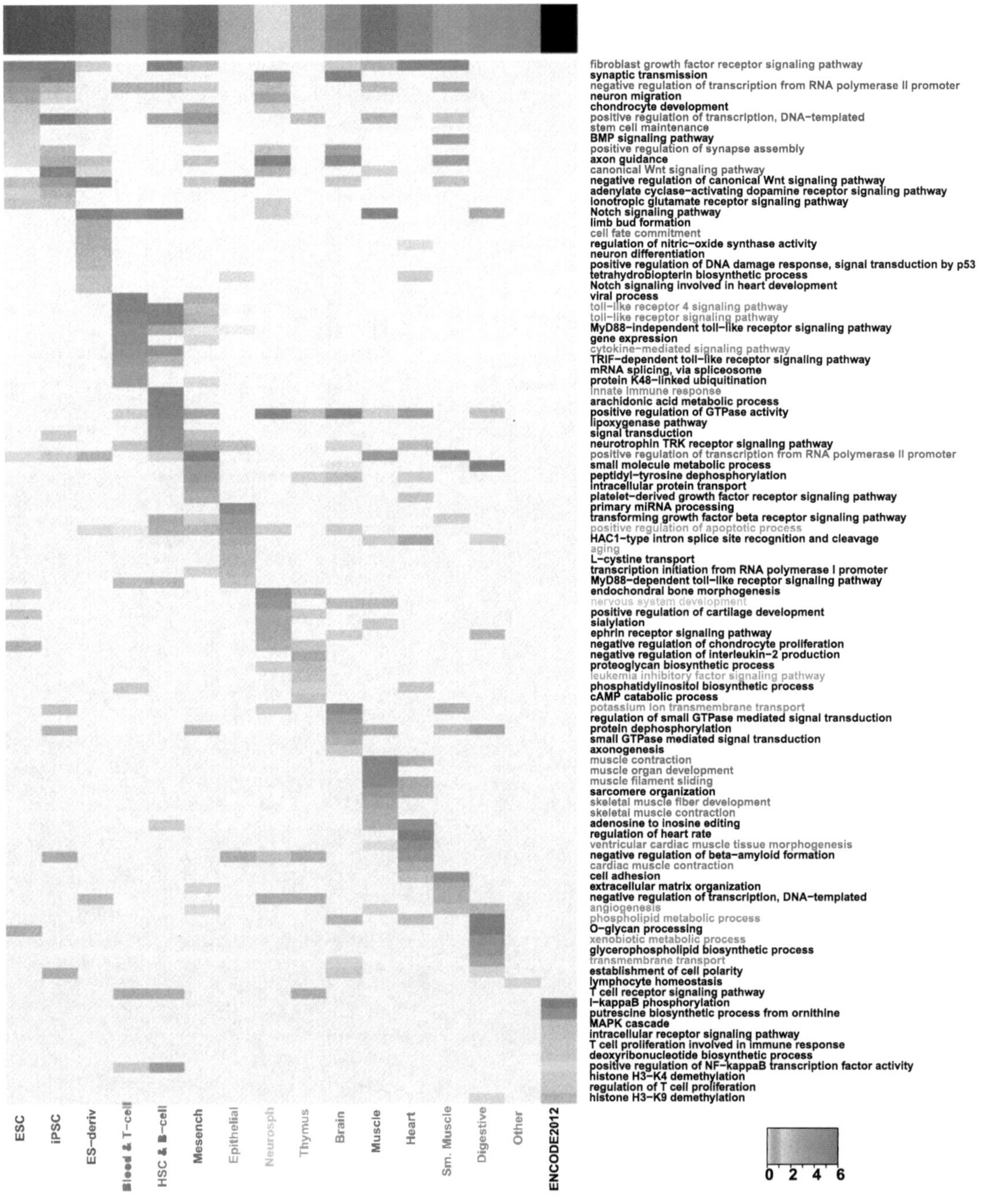

Fig. 5 GO enrichment analysis of associated enhancers' potential target genes. Shown GO terms are at least among the top 10 enriched GO terms in one tissue/cell type. Enrichment scores are calculated as $-\log_{10}$(Bonferroni corrected p − values). Darker colors represent higher scores. The enriched GO terms can be used to infer the biological functions of associated enhancers in each tissue/cell type. To illustrate this, some GO terms are marked in the same colors of the tissue/cell types (see the colorbar on top of the heatmap and the heatmap column labels) in which the terms are enriched

EPOM score matrices. For instance, we observe a mapping between Heart and Muscle through the EPOM scores (Fig. 3a). Heart and Muscle actually share six common GO terms between their top 20 enriched GO terms in associated enhancers. The common GO terms include muscle filament sliding, sarcomere organization, fibroblast growth factor receptor signaling pathway, adenosine to inosine editing, positive regulation of GTPase activity

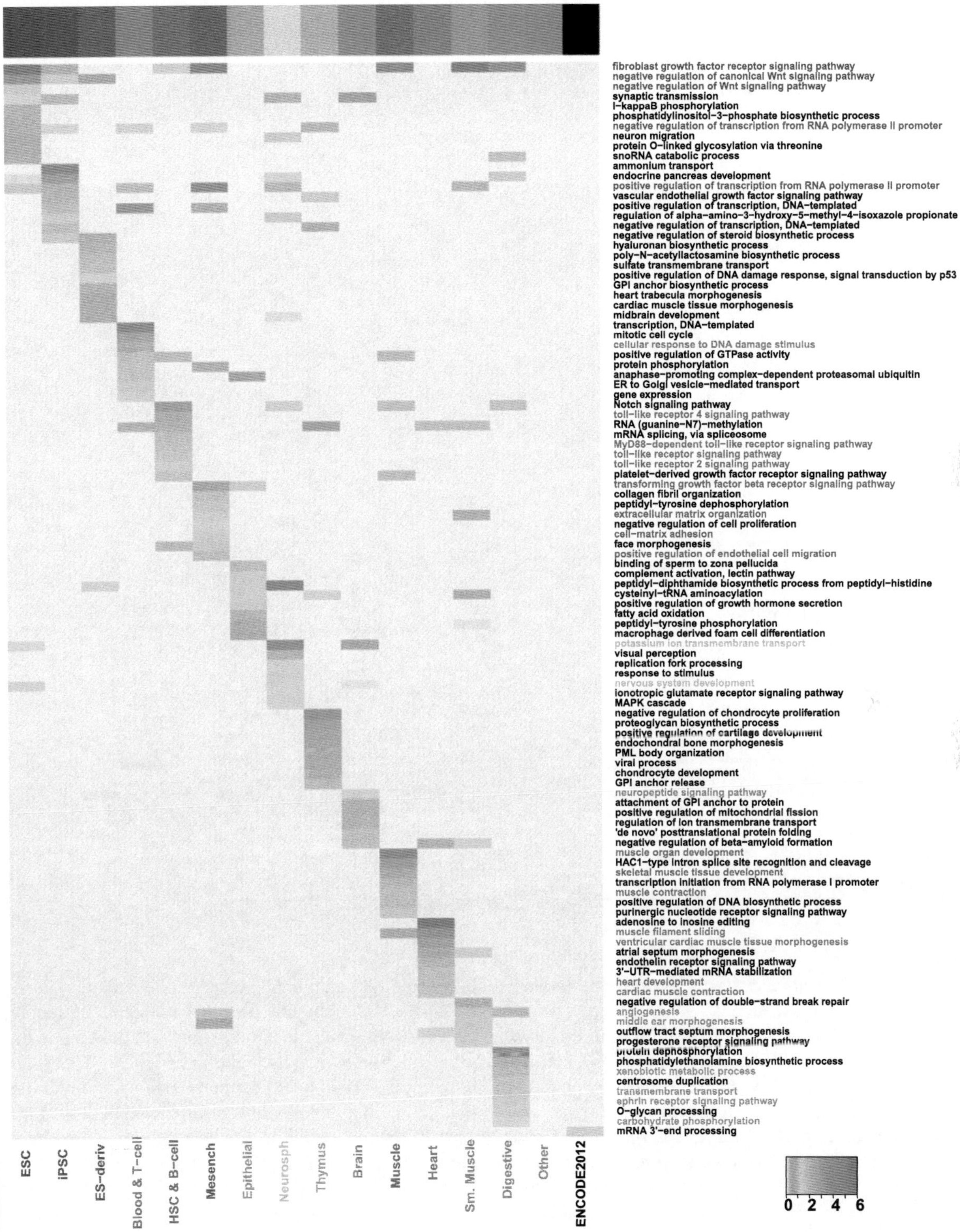

Fig. 6 GO enrichment analysis of associated promoters' potential target genes. Shown GO terms are at least among the top 10 enriched GO terms in one tissue/cell type. Enrichment scores are calculated as $-\log_{10}$(Bonferroni corrected p − values). Darker colors represent higher scores. The enriched GO terms can be used to infer the biological functions of associated promoters in each tissue/cell type. To illustrate this, some GO terms are marked in the same colors of the tissue/cell types (see the colorbar on top of the heatmap and the heatmap column labels) in which the terms are enriched

ESC	Negative regulation of transcription from RNA polymerase II promoter	Fibroblast growth factor receptor signaling pathway	Negative regulation of canonical Wnt signaling pathway, negative regulation of Wnt signaling pathway
iPSC	Positive regulation of transcription, DNA-templated, stem cell maintenance, canonical Wnt signaling pathway		Negative regulation of transcription from RNA polymerase II promoter, positive regulation of transcription from RNA polymerase II promoter
ES-deriv.	Cell fate commitment		
Blood & t-cell	Toll-like receptor 4 signaling pathway, toll-like receptor signaling pathway, cytokine-mediated signaling pathway, innate immune response		Cellular response to DNA damage stimulus
HSC & b-cell		Toll-like receptor 4 signaling pathway, toll-like receptor signaling pathway	MyD88-dependent toll-like receptor signaling pathway, toll-like receptor 2 signaling pathway
Mesench.	Positive regulation of transcription from RNA polymerase II promoter		Transforming growth factor beta receptor signaling pathway, extracellular matrix organization, cell-matrix adhesion, positive regulation of endothelial cell migration
Epithelial	Positive regulation of apoptotic process, aging		
Neurosph.	Nervous system development		Potassium ion transmembrane transport, nervous system development
Thymus	Leukemia inhibitory factor signaling pathway		
Brain	Potassium ion transmembrane transport		Neuropeptide signaling pathway
Muscle	Muscle filament sliding, skeletal muscle fiber development, skeletal muscle contraction	Muscle organ development, muscle contraction	Skeletal muscle tissue development
Heart		Ventricular cardiac muscle tissue morphogenesis, cardiac muscle contraction	Muscle filament sliding, heart development
Sm. Muscle		Angiogenesis	Middle ear morphogenesis
Digestive	Phospholipid metabolic process	Xenobiotic metabolic process, transmembrane transport	Ephrin receptor signaling pathway, carbohydrate phosphorylation

Fig. 7 Characteristic GO terms enriched in different tissue/cell types. These terms are the characteristic ones extracted from Figs. 5 and 6. The second (fourth) column lists the GO terms only enriched in associated enhancers (promoters); the third column lists the GO terms enriched in both associated enhancers and promoters

and HAC1-type intron splice site recognition and cleavage (see Fig. 5). For another example, in accordance with the mapping of Blood & T-cell and HSC & B-cell in Fig. 3c, d, they share six top enriched GO terms, including toll-like receptor signaling pathway and cytokine-mediated signaling pathway. In addition, consistent with the mapping of Neurosphere and Brain, they have six top enriched GO terms in common, including synaptic transmission, positive regulation of GTPase activity and axon guidance.

Many of the top enriched GO terms involve highly relevant functions of their corresponding tissue/cell types (see Figs. 5 and 6), proving that associated enhancers and promoters do carry important characteristics of the tissue and cell types. For example, it was observed that the DNA methylation pattern is very similar between iPSC and ESC but it is still possible to distinguish iPSC from ESC through differentially marked genomic regions [22]. In the GO enrichment analysis, we observe a great overlap between ESC's and iPSC's top enriched GO terms in associated enhancers (Fig. 5) as well as obvious distinction between ESC's and iPSC's top enriched GO terms in associated promoters (Fig. 6). Figure 8 provides a summary of the characteristic GO terms that are biologically relevant to each tissue/cell type. These terms serve as a good basis to understand the enhancer and promoter functions under different contexts [23]. In addition, the rest of top enriched terms imply potentially novel functions of enhancers and promoters in diverse tissue and cell types.

GWAS and disease ontology (DO) enrichment analysis of associated enhancers and promoters

Genome-wide association studies (GWAS) have identified millions of genetic variants associated with common traits and diseases. However, selecting informative single-nucleotide polymorphisms (SNPs) that have main effects on diverse diseases remains a great challenge [24]. It was observed that many non-coding variants associated with common diseases are concentrated in regulatory sequences on human genome [25]. As a consequence, the associated enhancers and associated promoters discovered by EPOM carry important information on cell-type-specific diseases and may serve as a potential source to promote the identification of pathogenic tissue/cell types of diverse disease disorders and the understanding of regulatory mechanisms of human disease.

To verify the associated enhancers' and promoters' ability in capturing causal genetic variants relevant to human diseases, we first globally quantify the enrichment of trait/disease associated SNPs from GWASdb [26] in each set of associated enhancers or promoters and then carry out Disease Ontology (DO) enrichment analysis to evaluate the enriched DO terms in each tissue/cell type [27]. The global test show that trait/disease associated SNPs in the GWAS catalog are strongly enriched in the associated enhancers in most tissue/cell types while the enrichment in associated promoters is less significant. Table 3 gives the enrichment score for each tissue/cell type. The results of DO enrichment analysis match the global enrichment results: among the total 426 DO terms, 122 are enriched (log(Bonferroni corrected p-values) > 0) and 85 are significantly enriched (-log(Bonferroni corrected p-values) >1) in associated enhancers of at least one tissue/cell type; 61 are enriched and 39 are significantly enriched in associated promoters of at least one tissue/cell type (please see Fig. 8 and Additional file 8).

A series of biologically meaningful relationships between diseases and tissue/cell types are identified and verified in the enrichment analysis (please see Fig. 8 and Additional file 8). In terms of associated enhancer regions, DO terms corresponding to different hypersensitivity reaction disease (celiac disease), hematopoietic system disease (lymphopenia) and immune system cancer (lymphoma and leukemia) are enriched in Blood & T-cell and HSC & B-cell; DO terms representing hepatocellular carcinoma, pancreatic cancer and a series of gastrointestinal system disease (such as ulcerative colitis and esophageal cancer) are enriched in Digestive; DO terms representing disease of mental health (such as attention deficit hyperactivity disorder , alcohol dependence and schizophrenia), major depressive disorder and neurodegenerative disease (such as Alzheimer's disease and Parkinson's disease) are enriched in Brain; Cardiovascular system disease is enriched in both Muscle and Heart; and

Table 3 GWAS enrichment scores

	$-log$(Bonferroni corrected p − values)	
Tissue/cell type	Associated enhancers	Associated promoters
ESC	0.00	0.00
iPSC	3.19	0.00
ES-deriv.	0.00	0.00
Blood & T-cell	89.01	32.25
HSC & B-cell	48.69	28.20
Mesench.	133.54	28.47
Epithelial	0.00	0.00
Neurosph.	2.90	0.00
Thymus	3.39	0.00
Brain	78.46	5.13
Muscle	11.47	0.00
Heart	3.36	0.00
Sm. Muscle	29.27	0.00
Digestive	6.96	3.49
Other	0.00	0.00
ENCODE2012	0.00	0.00

gastric adenocarcinoma (which derives from epithelial cells of glandular origin) is enriched in Epithelial. In terms of associated promoter regions, similar diseases as in associated enhancers were found to be enriched in Blood & T-cell, HSC & B-cell, Digestive and Epithelial. In addition, type 1 diabetes mellitus is also enriched in Digestive and cardiomyopathy (characterized by deterioration of the function of the heart muscle) is enriched in Muscle. Moreover, some more complicated relationships between diseases and tissue/cell types are also recovered in the DO enrichment analysis. For example, diabetes mellitus and kidney disease are found to be enriched in Heart while research have shown that both diabetes and kidney disease are high risk factors for heart disease [28, 29].

Discussion and conclusions

In this work, we propose a new measure for comparing and grouping biological samples from different tissue and cell types: Epigenomic Overlap Measure (EPOM). EPOM compares different tissue and cell types based on the similarity of histone modification marks evaluated in their relevant chromatin states. The proposed measure is calculated via a three-step testing procedure including ANOVA, t test and overlap test. Compared to traditional correlation analysis, EPOM is able to create a much clearer mapping pattern across 16 tissue and cell types. By tuning the thresholds in the testing procedure, EPOM can perform either grouping or identity mapping of biological samples based on epigenomic features. The associated enhancers and associated promoters identified

Fig. 8 DO enrichment analysis in associated enhancers in each tissue/cell type. Shown DO terms are at least significantly enriched in one tissue/cell type. Enrichment scores are calculated as $-\log_{10}$(Bonferroni corrected $p-$ values). Darker colors represent higher scores. For each tissue/cell type, its enriched DO term is marked in the same color as the tissue/cell label if there is a straightforward relationship between the DO term and the tissue/cell type

by EPOM are good indicators of tissue/cell epigenomic characteristics, and they are important genomic regions for downstream analysis such as regulatory network analysis, GO enrichment analysis and GWAS studies. Results under different settings (i.e., by taking union or intersection of the associated regions identified for different marks; by using two or three HMs together or separately using individual marks; by using 200 bp associated regions or merged longer associated regions) all demonstrate the effectiveness of our approach compared with correlation analysis in finding clear correspondence maps of biological samples. Moreover, the resulting EPOM scores reveal biologically meaningful patterns between similar tissue/cell types and confirm the belief that epigenomic landscapes are powerful resources for understanding cellular identity [30, 31]. These results imply the great potential of using EPOM to study tumor heterogeneity based on single-cell epigenomic data [32].

The EPOM method can be easily extended to study the relationships between diverse tissue/cell types based on signals of any epigenetic marks in genomic regions of interest. Here we suggest an efficient approach to systematically select epigenetic marks for EPOM if no specific marks are of prior interest. The selection will be based on the number of regions where each mark has differential signals across biological conditions. The differential regions of each mark can be found by the Step 1 (ANOVA) in our testing procedure given a specified p-value threshold, and the marks that have large numbers of differential regions will be good candidates for EPOM. The rationale behind this selection approach is that EPOM prefers the marks carrying more cell-type-specific information on the genomic regions of interest. We implement this selection approach in Additional file 9, which shows that among the eight epigenetic marks studied by the Roadmap Consortium, the three marks H3K4me1, H3K27ac and H3K4me3 we use in this work are among the top ones in terms of the numbers of differential enhancer and promoter regions.

We identified the associated enhancers/promoters' potential target genes in each tissue and cell type and used the top enriched GO terms in these genes to predict the biological functions of the associated enhancers and promoters. The results of GO enrichment analysis confirm the similarities of tissues and cell types found by EPOM and provide functional explanations for the underlying regulatory mechanisms leading to these patterns. The EPOM scores, together with the GO enrichment results, suggest that the associated enhancers and promoters have well captured the epigenomic characteristics of their corresponding tissue and cell types. An important future direction is to incorporate three-dimensional (3D) chromatin structures into the identification of the target genes of associated enhancers/promoters. The Hi-C technology makes it possible to decipher 3D chromatin structures and to thus reveal more accurate and complete interactions between genes and regulatory regions [33, 34]. However, Hi-C data are not yet available for the human tissue and cell types in our study, and without the data it is difficult to accurately infer potential target genes of associated enhancers/promoters from 3D chromatin structures [35]. In addition, better computational tools are needed for accurate 3D genome reconstruction from Hi-C data [36].

Despite the previous belief that chromatin states at promoters are largely invariant across diverse cell types [9, 37], our functional analyses on the potential target genes of the associated promoters in different tissue/cell types suggest that the non-housekeeping promoters carry cell-type-specific functions. We also found that the potential target genes of the associated enhancers are enriched with functions both specific to a single tissue/cell type or shared by a subgroup of tissue/cell types. Those associated regulatory regions identified by EPOM are key elements for understanding differential gene expression, cell differentiation and phenotypic variations.

More functional analyses based on disease ontology further confirm that the discovered associated regions carry important disease-relevant characteristics of their corresponding tissue/cell types. The identified associated enhancers and promoters can be good resources for understanding the epigenomic mechanisms of different tissue and cell types. It is a great challenge now to interpret the biological mechanisms and effects of the large amounts of identified SNPs. A common approach was to simply study the overlap between the SNPs and regulatory elements such as histone modification marks, binding sites of transcription factors and promoter regions [38]. However given that the dynamics of trait-associated variants can vary significantly in different tissue and cell types, we should carefully evaluate the enrichment of trait-associated variations in their most relevant tissues or cell types [39]. With the knowledge that our associated enhancers and promoters carry significant regulatory epigenomic features and thus represent the genomic context of their corresponding tissue and cell types better than other non-coding genomic regions, we highlight three important perspectives to make use of associated enhancers and promoters in GWAS studies. First, the identified associated enhancers and promoters provide a unique source for studying cell-type-specific disease variants and exploring disease-associated SNP functions. Although previous research showed SNP and GWAS enrichment in diverse chromatin states [12] and studied SNPs for certain selected traits [40], they did not provide a method to test the enrichment of genome-wide SNPs in cellular specific contexts. Second, the enriched DO terms can help researchers understand the dynamics of disease-related regulatory elements across diverse tissue/cell types. We can identify the potential target genes of the associated enhancers and promoters highly enriched with disease-related SNPs. Then by comparing the distinct and common target genes of each tissue/cell type and studying the regulatory networks between those genes and their associated enhancers or promoters, it is possible to shed light on the causes of cell type specific diseases as well as multi-factorial disorders. Last, the results of our study provide useful information to refine the disease ontology. Once we verify the potential target genes of the associated enhancers (or promoters) enriched with disease variants, we can update the DO terms to reflect these newly discerned genes [41].

Availability of supporting data

The epigenomic datasets supporting the results of this article are available at the web portal of the Roadmap Epigenomics Project. Both the data of the 25-state Imputation Based Chromatin State Model and the imputed signals of histone modification marks are available at http://egg2.wustl.edu/roadmap/web_portal/imputed.html#chr_imp. The data for SNP annotation is available at http://jjwanglab.org/gwasdb. The associated enhancer and promoter regions identified by EPOM are available at http://www.stat.ucla.edu/~jingyi.li/software-and-data.html or http://www.stat.ucla.edu/~jingyi.li/data/EpOM/associated_enhancers_and_promoters.tar.gz.zip.

Additional files

Additional file 1: Figure S1. Correspondence maps of EPOM scores saturated at 20. The associated enhancers (promoters) used are the unions of the associated enhancers (promoters) identified through each histone modification mark (H3K4me1, H3K27ac and H3K4me3) in step 2. (PDF 420 kb)

Additional file 2: Figure S2. Correspondence maps of Pearson correlation coefficients calculated from H3K4me1 and H3K27ac on candidate associated enhancers and candidate associated promoters after step 1 (ANOVA). (PDF 2334 kb)

Additional file 3: Figure S3. Correspondence maps of EPOM scores saturated at 20. Each heatmap plots the EPOM scores calculated from associated regions (enhancers or promoters) identified through one histone modification mark (H3K4me1 or H3K27ac) in step 2. (PDF 434 kb)

Additional file 4: Figure S4. Correspondence maps of EPOM scores saturated at 20. The associated enhancers (promoters) used to calculate EPOM scores are the intersection of the the associated enhancers (promoters) identified for the two histone modification marks. (PDF 416 kb)

Additional file 5: Figure S5. Numbers and proportions of potential target genes identified for merged associated enhancers and merged associated promoters. (PDF 493 kb)

Additional file 6: Table S1. Top 20 enriched GO terms in enhancer-associated genes in each tissue/cell type. (XLSX 15 kb)

Additional file 7: Table S2. Top 20 enriched GO terms in promoter-associated genes in each tissue/cell type. (XLSX 16 kb)

Additional file 8: Figure S6. DO enrichment analysis on associated promoter regions in each tissue/cell type. (PDF 169 kb)

Additional file 9: Table S3. Number of candidate associated enhancers (or promoters) of the eight histone modifications marks. (XLSX 10 kb)

Competing interests

The authors declare that they have no competing interests.

Authors' contributions

JJL conceived and advised the studies. WVL and ZSR prepared the data and conducted the proposed testing procedures. WVL performed GO enrichment analysis and DO enrichment analysis. WVL, ZSR and JJL drafted and edited the manuscript. All authors read and approved the final manuscript.

Declarations

This article has been published as part of *BMC Genomics* Volume 17 Supplement 1, 2016: Selected articles from the Fourteenth Asia Pacific Bioinformatics Conference (APBC 2016): Genomics. The full contents of the supplement are available online at http://www.biomedcentral.com/bmcgenomics/supplements/17/S1.

Acknowledgments

This work was supported by the start-up fund of Department of Statistics at University of California, Los Angeles, and the Hellman Fellowship from the Hellman Foundation. The publication costs for this article were funded by the Hellman Foundation. The authors would like to thank Yu-Cheng T. Yang for processing the imputed data of histone modification marks and for his ideas, indispensable advice and wise guidance. We also thank the anonymous reviewers for their valuable comments and suggestions that helped improve the manuscript.

Published: 11 January 2016

References

1. Pellegrini M, Ferrari R. Epigenetic analysis: Chip-chip and chip-seq. Methods Mol Biol. 2012;802:377–87.
2. Kundaje A, Meuleman W, Ernst J, Bilenky M, Yen A, Heravi-Moussavi A, et al. Integrative analysis of 111 reference human epigenomes. Nature. 2015;518(7539):317–30.
3. Bernstein BE, Meissner A, Lander ES. The mammalian epigenome. Cell. 2007;128(4):669–81.
4. Bernstein BE, Stamatoyannopoulos JA, Costello JF, Ren B, Milosavljevic A, Meissner A, et al. The nih roadmap epigenomics mapping consortium. Nat Biotechnol. 2010;28(10):1045–8.
5. Lee Y-s, Krishnan A, Zhu Q, Troyanskaya OG. Ontology-aware classification of tissue and cell-type signals in gene expression profiles across platforms and technologies. Bioinforma. 2013;29(23):3036–44.
6. Pettit J-B, Tomer R, Achim K, Richardson S, Azizi L, Marioni J. Identifying cell types from spatially referenced single-cell expression datasets. PLoS Comput Biol. 2014;10(9):e1003824.
7. ENCODE Project Consortium. An integrated encyclopedia of dna elements in the human genome. Nature. 2012;489(7414):57–74.
8. Ernst J, Kellis M. Chromhmm: automating chromatin-state discovery and characterization. Nat Methods. 2012;9(3):215–6.
9. Heintzman ND, Hon GC, Hawkins RD, Kheradpour P, Stark A, Harp LF, et al. Histone modifications at human enhancers reflect global cell-type-specific gene expression. Nature. 2009;459(7243):108–12.
10. Ernst J, Kellis M. Large-scale imputation of epigenomic datasets for systematic annotation of diverse human tissues. Nature Biotechnol. 2015;33(4):364–76.
11. Koch CM, Andrews RM, Flicek P, Dillon SC, Karaöz U, Clelland GK, et al. The landscape of histone modifications across 1 % of the human genome in five human cell lines. Genome Res. 2007;17(6):691–707.
12. Ernst J, Kellis M. Discovery and characterization of chromatin states for systematic annotation of the human genome. Nat Biotechnol. 2010;28(8):817–25.
13. Creyghton MP, Cheng AW, Welstead GG, Kooistra T, Carey BW, Steine EJ, et al. Histone h3k27ac separates active from poised enhancers and predicts developmental state. Proc Natl Acad Sci. 2010;107(50):21931–6.
14. Li JJ, Huang H, Bickel PJ, Brenner SE. Comparison of d. melanogaster and c. elegans developmental stages, tissues, and cells by modencode rna-seq data. Genome Res. 2014;24(7):1086–101.
15. Johansson CB, Svensson M, Wallstedt L, Janson AM, Frisén J. Neural stem cells in the adult human brain. Exp Cell Res. 1999;253(2):733–6.
16. Kissa K, Murayama E, Zapata A, Cortés A, Perret E, Machu C, et al. Live imaging of emerging hematopoietic stem cells and early thymus colonization. Blood. 2008;111(3):1147–56.

17. Hon GC, Hawkins RD, Ren B. Predictive chromatin signatures in the mammalian genome. Hum Mol Genet. 2009;18(R2):195–201.
18. Nègre N, Brown CD, Ma L, Bristow CA, Miller SW, Wagner U, et al. A cis-regulatory map of the drosophila genome. Nature. 2011;471(7339):527–31.
19. Li G, Ruan X, Auerbach RK, Sandhu KS, Zheng M, Wang P, et al. Extensive promoter-centered chromatin interactions provide a topological basis for transcription regulation. Cell. 2012;148(1):84–98.
20. Cotney J, Leng J, Oh S, DeMare LE, Reilly SK, Gerstein MB, et al. Chromatin state signatures associated with tissue-specific gene expression and enhancer activity in the embryonic limb. Genome Res. 2012;22(6):1069–80.
21. Gene Ontology Consortium. Gene ontology consortium: going forward. Nucleic Acids Res. 2015;43(D1):1049–56.
22. Bilic J, Belmonte JCI. Concise review: Induced pluripotent stem cells versus embryonic stem cells: close enough or yet too far apart? Stem Cells. 2012;30(1):33–41.
23. He B, Chen C, Teng L, Tan K. Global view of enhancer–promoter interactome in human cells. Proc Natl Acad Sci. 2014;111(21):2191–9.
24. Liang Y, Kelemen A. Statistical advances and challenges for analyzing correlated high dimensional snp data in genomic study for complex diseases. Stat Surveys. 2008;2:43–60.
25. Maurano MT, Humbert R, Rynes E, Thurman RE, Haugen E, Wang H, et al. Systematic localization of common disease-associated variation in regulatory dna. Science. 2012;337(6099):1190–5.
26. Li MJ, Wang P, Liu X, Lim EL, Wang Z, Yeager M, et al. Gwasdb: a database for human genetic variants identified by genome-wide association studies. Nucleic Acids Res. 2012;40(D1):1047–54.
27. Kibbe WA, Arze C, Felix V, Mitraka E, Bolton E, Fu G, et al. Disease ontology 2015 update: an expanded and updated database of human diseases for linking biomedical knowledge through disease data. Nucleic Acids Res. 2015;43(D1):1071–8.
28. Heart Outcomes Prevention Evaluation (HOPE) Study investigators. Effects of ramipril on cardiovascular and microvascular outcomes in people with diabetes mellitus: results of the hope study and micro-hope substudy. The Lancet. 2000;355(9200):253–9.
29. Sarnak MJ, Levey AS, Schoolwerth AC, Coresh J, Culleton B, Hamm LL, et al. Kidney disease as a risk factor for development of cardiovascular disease a statement from the american heart association councils on kidney in cardiovascular disease, high blood pressure research, clinical cardiology, and epidemiology and prevention. Circulation. 2003;108(17):2154–69.
30. Lang AH, Li H, Collins JJ, Mehta P. Epigenetic landscapes explain partially reprogrammed cells and identify key reprogramming genes. PLoS Comput Biol. 2014;10(9):e1003734.
31. Barrero MJ, Boué S, Belmonte JCI. Epigenetic mechanisms that regulate cell identity. Cell Stem Cell. 2010;7(5):565–70.
32. Shapiro E, Biezuner T, Linnarsson S. Single-cell sequencing-based technologies will revolutionize whole-organism science. Nat Rev Genet. 2013;14(9):618–30.
33. Lieberman-Aiden E, van Berkum NL, Williams L, Imakaev M, Ragoczy T, Telling A, et al. Comprehensive mapping of long-range interactions reveals folding principles of the human genome. Science. 2009;326(5950):289–93.
34. Van Berkum NL, Lieberman-Aiden E, Williams L, Imakaev M, Gnirke A, Mirny LA, et al. Hi-c: a method to study the three-dimensional architecture of genomics. J Vis Exp. 2010;39:1869.
35. Nagano T, Lubling Y, Stevens TJ, Schoenfelder S, Yaffe E, Dean W, et al. Single-cell hi-c reveals cell-to-cell variability in chromosome structure. Nature. 2013;502(7469):59–64.
36. Lesne A, Riposo J, Roger P, Cournac A, Mozziconacci J. 3d genome reconstruction from chromosomal contacts. Nat Methods. 2014;11(11):1141–3.
37. Hawkins RD, Hon GC, Yang C, Antosiewicz-Bourget JE, Lee LK, Ngo Q-M, et al. Dynamic chromatin states in human es cells reveal potential regulatory sequences and genes involved in pluripotency. Cell Res. 2011;21(10):1393–409.
38. Boyle AP, Hong EL, Hariharan M, Cheng Y, Schaub MA, Kasowski M, et al. Annotation of functional variation in personal genomes using regulomedb. Genome Res. 2012;22(9):1790–7.
39. Schmidt EM, Zhang J, Zhou W, Chen J, Mohlke KL, Chen YE, et al. Gregor: evaluating global enrichment of trait-associated variants in epigenomic features using a systematic, data-driven approach. Bioinforma. 2015;16(31):2601–6.
40. Farh KK-H, Marson A, Zhu J, Kleinewietfeld M, Housley WJ, Beik S, et al. Genetic and epigenetic fine mapping of causal autoimmune disease variants. Nature. 2015;518(7539):337–43.
41. Schriml LM, Mitraka E. The disease ontology: fostering interoperability between biological and clinical human disease-related data. Mamm Genome. 2015;26:584–9.

Ye and Li *BMC Genomics* 2016, **17**(Suppl 1):11
DOI 10.1186/s12864-015-2304-8

BMC Genomics

PROCEEDINGS Open Access

NMFP: a non-negative matrix factorization based preselection method to increase accuracy of identifying mRNA isoforms from RNA-seq data

Yuting Ye[1] and Jingyi Jessica Li[2,3*]

From The Fourteenth Asia Pacific Bioinformatics Conference (APBC 2016)
San Francisco, CA, USA. 11 - 13 January 2016

Abstract

Background: The advent of next-generation RNA sequencing (RNA-seq) has greatly advanced transcriptomic studies, including system-wide identification and quantification of mRNA isoforms under various biological conditions. A number of computational methods have been developed to systematically identify mRNA isoforms in a high-throughput manner from RNA-seq data. However, a common drawback of these methods is that their identified mRNA isoforms contain a high percentage of false positives, especially for genes with complex splicing structures, e.g., many exons and exon junctions.

Results: We have developed a preselection method called "Non-negative Matrix Factorization Preselection" (NMFP) which is designed to improve the accuracy of computational methods in identifying mRNA isoforms from RNA-seq data. We demonstrated through simulation and real data studies that NMFP can effectively shrink the search space of isoform candidates and increase the accuracy of two mainstream computational methods, Cufflinks and SLIDE, in their identification of mRNA isoforms.

Conclusion: NMFP is a useful tool to preselect mRNA isoform candidates for downstream isoform discovery methods. It can greatly reduce the number of isoform candidates while maintaining a good coverage of unknown true isoforms. Adding NMFP as an upstream step, computational methods are expected to achieve better accuracy in identifying mRNA isoforms from RNA-seq data.

Keywords: mRNA isoform discovery, Next-generation RNA sequencing (RNA-seq), Non-negative matrix factorization (NMF), Cufflinks, SLIDE

Background

In molecular biology and transcriptomics, an important question is to understand the alternative splicing process, a key step of gene transcription in diverse species, from invertebrates to mammals. Alternative splicing generates at least two mRNA isoforms (i.e., transcripts) for one gene, and these mRNA isoforms as well as their protein products can play different roles in various biological phenomena. For example, studies have found that aberrant structures or abundance of mRNA isoforms can cause various human diseases [1, 2]. Hence, systematic identification and quantification of mRNA isoforms can greatly help understand gene regulation mechanisms for both basic biology and translational medicine. For decades, mRNA isoforms of individual genes have been accurately identified and quantified by low-throughput technologies such as cDNA cloning and qPCR. However, it was

*Correspondence: jli@stat.ucla.edu
[2]Department of Statistics, 8125 Math Sciences Bldg., University of California, Los Angeles, 90095-1554 Los Angeles, CA, USA
[3]Department of Human Genetics, 695 Charles E. Young Drive South, University of California, Los Angeles, 90095-7088 Los Angeles, CA, USA
Full list of author information is available at the end of the article

impossible to conduct this task on a system-wide scale for thousands of genes simultaneously until the invention of high-throughput technologies.

More than a decade ago, microarray technologies established a high-throughput platform for identifying and quantifying mRNA isoforms of genes with known sequences. SPACE [3, 4] is a method using non-negative matrix factorization (NMF) to predict mRNA isoforms and estimate their abundance from microarray data. NMF is a popular pattern recognition method and is distinguished from other matrix factorization methods by its use of non-negativity constraints [5]. A main advantage of NMF is its interpretable factorization results, which are positive and sparse combinations of features. In the context of mRNA isoform discovery, SPACE uses NMF to find isoforms as combinations of exons and exon junctions. SPACE was demonstrated to have good power in detecting mRNA isoforms under ideal simulation settings. However, many open questions remain about how to use NMF to accurately find mRNA isoforms from real microarray data. Reasons for these questions include the following issues inherent with NMF and microarray. First, in many scenarios NMF outputs non-unique factorization results [6], making SPACE identify ambiguous mRNA isoforms. Second, NMF requires a pre-specified matrix rank for factorization results, and how to determine this rank remains a difficult question [7–9]. Third, microarray can only detect exons and exon junctions with at least partially known sequences, and thus it cannot provide information on novel exons or exon junctions. This hinders SPACE from discovering novel isoforms that contain novel exons or exon junctions.

Compared to microarray, the more recent next-generation RNA sequencing (RNA-seq) technologies yield deeper and wider insights into transcriptomes at base level resolution [10]. RNA-seq measures the expression of genome-wide exons and exon junctions, regardless of prior knowledge on their existence. Hence, unlike microarray, RNA-seq enables the identification of novel mRNA isoforms containing previously unknown exons or exon junctions. Despite its many advantages, a major drawback of Illumina RNA-seq, the most popular RNA-seq platform, is that it can only generate RNA-seq reads with short lengths ($\leq 2 \times 250$ bp for paired-end reads). Such short RNA-seq reads cannot cover a full-length mRNA isoform, making isoform identification a difficult computational problem. Since the advent of RNA-seq, various computational methods have been developed to discover mRNA isoforms from RNA-seq data. For the species with good-quality reference genomes, including human and model organisms, mainstream isoform discovery methods (e.g., Cufflinks [11], Scripture [12], IsoLasso [13] and SLIDE [14]) use RNA-seq reads first mapped to a reference genome. In this paper, our discussion focuses on these so-called "refernece-based methods", which attempt to identify full-length mRNA isoforms from mapped RNA-seq reads using various approaches. For example, Cufflinks, Scripture and IsoLasso first build graphs of exonic regions containing mapped reads, and then find mRNA isoforms as the maximal paths in the graph based on a parsimony assumption. These three methods are also often referred to as "*de novo* methods," because their isoform identification approach can completely rely on RNA-seq reads without using existing annotations. Another reference-based method SLIDE uses a different approach. It constructs a search space of possible isoforms by using known boundaries of genes and exons from existing annotations, with the possible addition of novel genes and exons inferred from RNA-seq data. SLIDE then identifies isoforms from the search space based on a regularized linear model, which accounts for the RNA-seq read generating mechanism and read distributions. Because of its use of annotations, SLIDE is also called an "annotation aided method", in contrast to the previous "*de novo* methods." Note that the recent versions of Cufflinks also added an annotation aided option, which can simulate RNA-seq reads from annotations and use the simulated reads together with RNA-seq data for isoform discovery [15]. Both *de novo* and annotation aided methods have advantages and drawbacks. The graph-based *de novo* methods are more robust to RNA-seq data noise thanks to their use of the parsimony assumption, which, however, also prevents them from finding overlapping isoforms. SLIDE, on the other hand, is capable of finding overlapping isoforms but is more sensitive to RNA-seq data noise. Despite their differences, a common obstacle for both types of methods is the large search space of possible mRNA isoforms for genes with complex splicing structures (i.e., many exons). For the graph-based *de novo* methods, a gene with many exons will lead to a graph with large numbers of nodes and edges, making it difficult to search for isoforms as maximal paths. For SLIDE, a gene with n exons will give rise to a total of $2^n - 1$ possible isoforms, placing great difficulty on the regularized linear model to find the correct isoforms. Although SLIDE is coupled with a naïve preselection procedure to filter out the isoform candidates with unsupported exon junctions by RNA-seq reads, this procedure is not ideal for lowly expressed isoforms, whose exon junctions may not be supported by reads due to the low read coverage.

Given the wide existence of RNA-seq noise and the large search space of possible isoforms, existing computational methods for isoform discovery have unsatisfactory performance in many cases. In an evaluation conducted by the RGASP consortium [16], Cufflinks only achieved precision rates of 23 %, 50 % and 47 % and recall rates of 18 %, 40 % and 39 % for discovering full-length mRNA isoforms from *H. sapiens*, *D. melanogaster*, and *C. elegans* RNA-seq data respectively. For SLIDE, the corresponding precision rates were 37 %, 80 % and 63 %, and the corresponding recall rates were 7 %, 49 % and 59 %. This evaluation showed that both Cufflinks and SLIDE, as well as other evaluated methods, have low accuracy for discovering human isoforms, mainly because many human genes have large numbers of exons and thus a huge search space of possible isoforms. In statistical literature, for high-dimensional linear models with far more features than observations, it was found that adding a screening procedure to filter out excessive features in the search space can improve the feature identification performance of Lasso [17]. Since Lasso was used in SLIDE to find isoforms as features, we are motivated to reduce the search space of possible isoforms by designing a screening (or preselection) procedure, so as to improve the performance of SLIDE and hopefully other isoform identification methods.

Also motivated by SPACE and the good properties of NMF, we propose a preselection method "Non-negative Matrix Factorization Preselection" (NMFP) to improve the accuracy of reference-based isoform discovery methods. Our goal is to use NMFP to greatly reduce the search space of isoforms, and have the selected isoform candidates cover most of the true isoforms. NMFP employs NMF to select isoform candidates for a given gene by decomposing its RNA-seq read count matrix. We demonstrate the performance of NMFP by simulation and real data studies. In the simulation, we generate RNA-seq reads from *D. melanogaster* (fly) and *M. musculus* (mouse) using Flux Simulator [18] and show that NMFP can greatly improve the accuracy of Cufflinks and SLIDE in isoform discovery. We also show the efficacy of NMFP on discovering *D. melanogaster* isoforms from real RNA-seq data. Moreover, we evaluate the robustness of NMFP to the choice of NMF ranks, the number of input samples in the read count matrix, and gene expression levels. In addition, we show that NMFP can increase the robustness of SLIDE to its regularization parameter choice.

Methods

Our proposed method NMFP shares the same root as SPACE, because they both use NMF to identify mRNA isoforms. However, unlike SPACE that aims to directly identify isoforms via NMF, NMFP has a less ambitious goal, which is to identify a set of good isoform candidates for downstream isoform discovery methods. To achieve this goal, in NMFP we use high-frequency filtering to alleviate the non-uniqueness issue of NMF results, and design a model selection approach to choose the NMF rank. Moreover, we develop a new NMF algorithm to penalize the isoforms that have conflicting exons or exon junctions.

Data processing

Suppose there are m RNA-seq data sets (samples) and a gene has n *subexons*, which were defined as the exonic regions between two adjacent splicing sites in SLIDE [14]. For this gene, we first categorize RNA-seq reads in every sample into *bins*, which are defined as two-dimensional vectors (k, l), representing the mapped reads (single-end reads or two ends of paired-end reads) with starting position in the k-th subexon and ending position in the l-th subexon, $1 \leq k \leq l \leq n$. Hence, this gene has a total number of $p = n + \binom{n}{2}$ bins. We count the numbers of reads in every bin in all samples and generate a read count matrix $\mathbf{U}$ with dimensions $p \times m$, with rows corresponding to bins and columns corresponding to samples

Since samples may have different sequencing depths and bins may have different lengths (due to different lengths of exons), to prevent the NMF results from being dominated by samples with more reads or bins with greater lengths, we normalize $\mathbf{U}$ into a matrix $\mathbf{V}$ by removing the sequencing depth and bin length effects. We conduct the normalization based on the following model

$$U_{ij} = R_j \cdot P_i \,, \tag{1}$$

where U_{ij} is the (i,j)-th entry of $\mathbf{U}$, representing the read count of the i-th bin in the j-th sample, R_j is the total number of reads mapped to this gene in the j-th sample, and P_i is the probability that a read falls into the i-th bin ($0 \leq P_i \leq 1$ and $\sum_{i=1}^{p} P_i = 1$). We assume that P_i is proportional to the length of the i-th bin. Although this assumption can be violated by the non-uniformity of read distribution due to alternative splicing and sequencing noise, it provides an approximate and simple solution to removing bin length effects from the read count matrix.

We describe the normalization process as follows.

1. Scale the read counts of each sample inverse proportionally with its total number of reads

$$U'_{ij} = U_{ij} \cdot \frac{\bar{R}}{R_j} \,,\ i = 1, 2, \ldots, p \,;\ j = 1, 2, \ldots, m \,,$$

   where $\bar{R} = \frac{1}{m} \sum_{j=1}^{m} R_j$.

2. Further scale the read counts of each bin inverse proportionally with its bin length

$$V_{ij} = U'_{ij} \cdot \frac{1}{pP_i} \,,\ i = 1, 2, \ldots, p \,;\ j = 1, 2, \ldots, m \,,$$

   and denote $\mathbf{V} = (V_{ij})$, where V_{ij} represents the normalized read count of the i-th bin in the j-th sample.

Matrix decomposition model

Li and Wong [19] proposed a model to describe the relationship between the probe intensity v and the transcript abundance t for microarray data

$$v = a \cdot t + e\,, \tag{2}$$

where a is the affinity of the probe, and e is an error term. SPACE revised the Model (2) into the following matrix model for p probes and s transcripts

$$\mathbf{V} = \mathbf{A} \cdot \mathbf{G} \cdot \mathbf{T} + \boldsymbol{\epsilon}\,, \tag{3}$$

where $\mathbf{V}$ is a $p \times m$ matrix representing the intensities of p probes in m samples; $\mathbf{A}$ is a $p \times p$ diagonal matrix representing the affinity of p probes; G is a $p \times s$ indicator matrix with binary elements indicating the existence of the p probes in s transcripts; $\mathbf{T}$ is an $s \times m$ matrix representing the abundance of s transcripts in m samples; $\boldsymbol{\epsilon}$ is a $p \times m$ error matrix.

Inspired by the Models (2) and (3), we write the following model for the relationship between normalized reads counts and unknown transcript (i.e., mRNA isoform) abundance

$$\mathbf{V} = \mathbf{W} \cdot \mathbf{H} + \boldsymbol{\epsilon}\,, \tag{4}$$

where $\mathbf{V}$ is the $p \times m$ normalized reads count matrix (from last section "Data processing"), and $\mathbf{W}$ is a $p \times s$ isoform composition matrix, whose rows and columns are bins and mRNA isoforms respectively. Each column of $\mathbf{W}$ is a p-dimensional vector with elements $\in [0, 1]$, indicating the existence of bins in an isoform, with the values subject to RNA-seq noise and biases. Theoretically, all the existing bins in an isoform should have the same read density assuming that RNA-seq reads follow a uniform distribution in that isoform. However, due to RNA-seq noise and biases, the uniformity assumption is often violated, and non-existing bins may contain erroneously mapped reads. To account for these issues, we model the existence of bins in an isoform as values between 0 and 1, where 1 indicates a bin having the maximal read intensity among all the bins, and 0 indicates a bin containing no mapped reads. $\mathbf{H}$ is a $s \times m$ isoform abundance matrix, whose rows and columns are mRNA isoforms and samples respectively. Each row of $\mathbf{H}$ is an m-dimensional vector representing the isoform's abundance levels in the m samples. Based on our definition of $\mathbf{W}$, h_{ij} (the abundance value of the i-th isoform in the j-th sample) should be equal to the maximum bin abundance of the i-th isoform in the j-th sample. $\boldsymbol{\epsilon}$ is an error matrix.

We can further decompose $\mathbf{W}$ as

$$\mathbf{W} = \mathbf{A} \cdot \mathbf{G} \tag{5}$$

where $\mathbf{A}$ is a $p \times p$ diagonal matrix with elements $\in [0, 1]$, representing the biases of read counts in different bins. Such biases are attributable to the features of the bins, such as GC contents and the bins' positions in transcripts. $\mathbf{G}$ is a $p \times s$ indicator matrix with binary values to describe the existence of bins in isoforms. Each column of $\mathbf{G}$ represents an isoform.

Modified NMF approach

With the normalized read count matrix $\mathbf{V}$, our goal is to estimate $\mathbf{G}$, whose columns will be regarded as isoform candidates. To achieve this goal, we have to first estimate $\mathbf{W}$ and $\mathbf{A}$. We adopt the SPACE approach to estimate $\mathbf{G}$. In the SPACE method [4], given an estimate $\hat{\mathbf{W}} = (\hat{W}_{ik})$, $\mathbf{A}$ was estimated as

$$\hat{\mathbf{A}} = \mathrm{diag}\left(\hat{A}_{11}, \ldots, \hat{A}_{pp}\right), \text{ where } \hat{A}_{ii} = \max_{k=1,\cdots,s} (\hat{W}_{ik})\,. \tag{6}$$

In other words, $\hat{A}_{ii}$, the bias factor of read counts in the i-th bin, is considered as $\max_k(\hat{W}_{ik})$, the largest "adjusted existence" of the i-th bin in all isoforms.

With $\hat{\mathbf{A}}$ and $\hat{\mathbf{W}}$, $\mathbf{G}$ is estimated as

$$\hat{\mathbf{G}} = (\hat{G}_{ij}), \text{ with } \hat{G}_{ij} = I\left((\hat{\mathbf{A}}^{-1} \cdot \hat{\mathbf{W}})_{ij} \geq c\right), \tag{7}$$

where $I(\cdot)$ is an indicator function, and $c \in \mathbb{R}$ is a threshold to make $\hat{\mathbf{G}}$ a binary matrix. We used $c = 0.4$ in this paper.

Prior to that, we estimate $\mathbf{W}$ by decomposing $\mathbf{V}$ using NMF. There exist many available NMF algorithms [5, 20–22], which have different preferences over the factorization results. Since our goal is to find isoform candidates by decomposing $\mathbf{V}$, one constraint we need is the compatibility of bins in every column of $\hat{\mathbf{W}}$, i.e., every isoform found by NMF. For example, Bin $(1, 3)$ and Bin $(2, 2)$ are conflicting and should not co-exist in an isoform, because the former indicates that Subexon 2 is skipped, while the latter implies the existence of Subexon 2.

To realize this constraint in NMF, we add a penalty term to the standard NMF objective function as follows

$$D(\mathbf{V}, \mathbf{W}\,\mathbf{H}) = \sum_{i=1}^{p}\sum_{j=1}^{m}\left(V_{ij}\log\frac{V_{ij}}{(\mathbf{WH})_{ij}} - V_{ij} + (\mathbf{WH})_{ij}\right) + \alpha \sum_{i \perp j}\left(\mathbf{W}\mathbf{W}^{T}\right)_{ij} \tag{8}$$

where p is the number of bins, m is the number of samples, and α is a regularization parameter controlling the weight of the penalty term (e.g., $\alpha = 0.1$). $i \perp j$ denotes that Bin i and Bin j are conflicting (See Section 1 in the Additional file 1 for more detail). We estimate $\mathbf{W}$ by minimizing the objective function $D(\mathbf{V}, \mathbf{WH})$ subject to a pre-specified $s = \text{rank}(\mathbf{W})$, which is equivalent to the number of columns of $\mathbf{W}$. How to determine the rank remains an open question. In NMFP, we choose the rank by gap statistic, which was originally designed to select the number of clusters in k-means clustering [23] (See Section 2 in the Additional file 1 for more detail). After obtaining $\hat{\mathbf{W}}$, we use Eqs. (6) and (7) to find $\hat{\mathbf{G}}$, which indicates a set of isoform candidates.

The penalty term in the objective function (8) cannot fully prevent the co-existence of conflicting bins in every isoform (i.e, column in $\hat{\mathbf{G}}$). To resolve this issue and recover potentially true isoforms, we consider both the existing and skipped statuses of the ambiguous subexons that lead to conflicting bins. For example, if both Bin (1,3) and Bin (2,2) exist, Subexon 2 is ambiguous. We will consider the cases of Subexon 2 being existing and skipped as two different isoforms. If there are n ambiguous subexons in an isoform candidate, 2^n isforms will be considered, but only those whose subexons and subexon junctions are supported by RNA-seq reads will be kept as isoform candidates.

Aggregation of multiple NMF runs

To address the issue of non-unique NMF results, we use a high-frequency filtering approach to find isoform candidates identified in a number of NMF runs. Specifically, we implement the NMF estimation procedure described in the previous two subsections for N times (e.g. $N = 100$) and aggregated the identified isoform candidates into a candidate pool. In the filtering step, we first remove the isoform candidates, which contain subexon junctions not supported by reads, from the pool. Then among the rest of candidates in the pool, we only retain the high-frequency ones, which are selected in at least r NMF runs (e.g., $r = 20$), as final isoform candidates.

The NMFP method is illustrated in Fig. 1.

Results

We conduct simulations and real-data analyses in many scenarios to demonstrate the effectiveness of NMFP in improving the performance of downstream isoform discovery methods. The following results show that NMFP can help Cufflinks and SLIDE achieve better isoform discovery accuracy. We also study the robustness of NMFP to the choice of NMF rank, the number of input samples in the read count matrix, and the gene expression levels.

Simulation results in *D. melanogaster*

We use Flux Simulator [18] to simulate RNA-seq reads of *D. melanogaster* (with reference genome dm6 and Ensembl annotation BDGP6 of release 80). We focus on 4125 genes on chr3R, in accordance with the SLIDE paper [14], which showed good performance of SLIDE on genes with 3 to 10 subexons. We simulate 50 samples, each with 10,000,000 RNA molecules and 50,000,000 paired-end reads with length 2×76 bp, from these genes' isoforms in the annotation. The isoform abundance is randomly assigned by Flux Simulator. Based on the way we define subexons (Additional file 1: Figure S1), 51.7 % (2132) genes contain 3 to 10 subexons (Table 1). Among these genes, 44.6 % (951) have more than one isoforms in the annotation. We apply NMFP to these 951 genes.

To check the effects of NMFP on improving the isoform discovery performance of SLIDE, we define a new procedure called "NMFP+SLIDE", which combines NMFP and SLIDE by inputting the isoform candidates found by NMFP into SLIDE. In the original SLIDE method, $2^n - 1$ possible isoforms are enumerated for an n-subexon gene, and SLIDE uses a simple preselection to filter out the isoforms with subexons or subexon junctions not supported by reads. NMFP+SLIDE replaces this preselection step by NMFP.

Similarly, we define "NMFP+Cufflinks", which inputs the isoform candidates found by NMFP as a GTF file into Cufflinks and uses Cufflinks with the "–GTF" option to estimate the abundance of these candidates. The candidates with non-zero estimated abundance are kept as the isoforms identified by NMFP+Cufflinks. This is different from the original Cufflinks, because it does not rely on a connectivity graph or the parsimony assumption as Cufflinks does.

To compare SLIDE with NMFP+SLIDE, and Cufflinks with NMFP+Cufflinks, we evaluate their isoform discovery performance at three different levels: the nucleotide, exon and transcript levels, similar to the RGASP evaluation [16]. In this comparison, we also evaluate the performance of NMFP alone, because it is important to check whether the isoform candidates found by NMFP have a good coverage of the true isoforms. Since SLIDE has two default values for its regularization parameter λ in the Lasso regression ($\lambda = 0.2$ for fewer isoforms and $\lambda = 0.01$ for more isoforms), we include two versions of SLIDE in this comparison: "SLIDE(fewer)" for $\lambda = 0.2$ and "SLIDE(more)" for $\lambda = 0.01$. This results in a comparison of seven methods in total:

- Cufflinks
- NMFP+Cufflinks
- SLIDE(fewer)
- NMFP+SLIDE(fewer)
- SLIDE(more)
- NMFP+SLIDE(more)
- NMFP

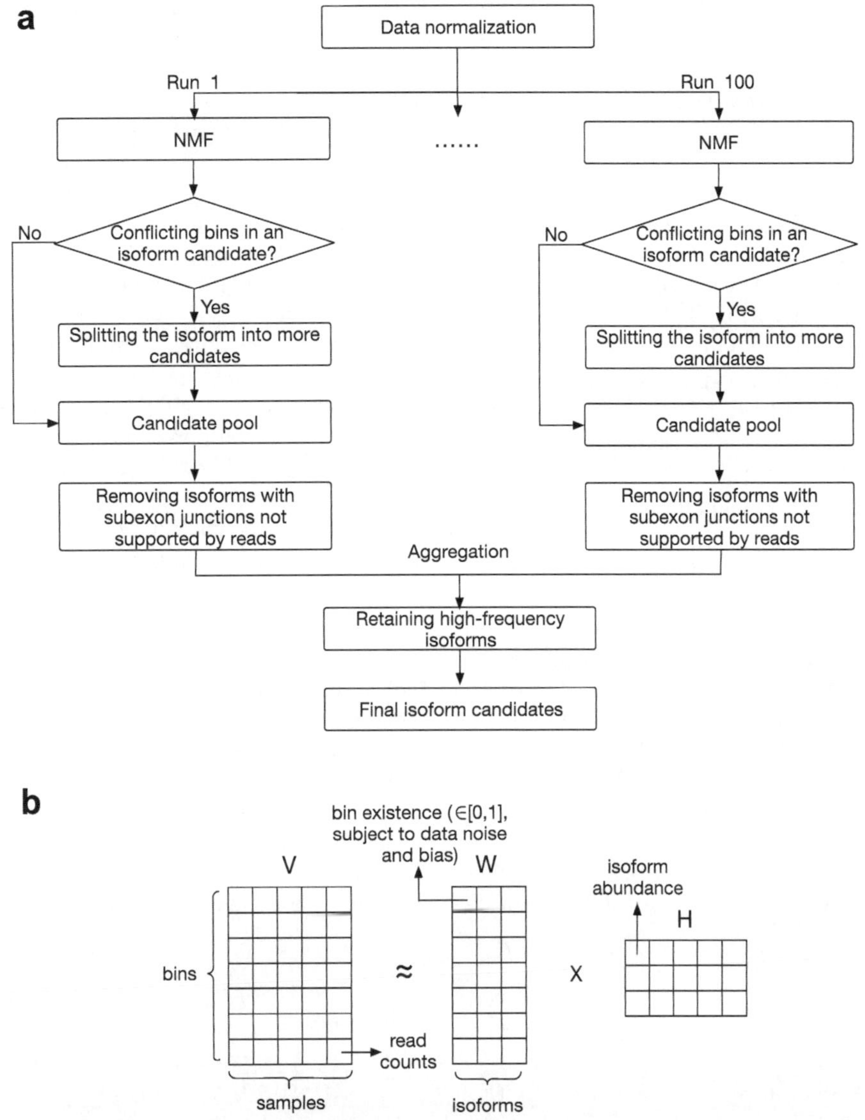

Fig. 1 a Diagram of the NMFP method. **b** Illustration of the NMF approach in NMFP

Evaluation at nucleotide level

We first compare the seven methods at the nucleotide level, that is, how many nucleotides are shared by the identified isoforms of each method and the true isoforms in the annotation. For each identified isoform, we match it to the annotated isoform that shares the most nucleotides with it. We rank the identified isoforms by their percentages of overlapping nucleotides with their matched annotated isoforms, from the highest percentage to the lowest. If there are k identified isoforms and r annotated isoforms and $k > r$, only the top r identified isoforms would be paired with an

Table 1 Profiles of 4125 genes in chr3R of *D. melanogaster*

# of subexons n	$n \leq 2$	$3 \leq n \leq 10$	$11 \leq n \leq 15$	$16 \leq n \leq 20$	$n > 20$	Sum
	1618 (39.2 %)	2132 (51.7 %)	232 (5.6 %)	79 (1.9 %)	64 (1.6 %)	4125
# of annotated isoforms q	$q = 1$	$q = 2$	$q = 3$	$q = 4$	$q > 4$	Sum
	2450 (59.4 %)	878 (21.3 %)	356 (8.6 %)	180 (4.4 %)	261 (6.3 %)	4125

annotated isoform, and the rest $k - r$ identified isoforms would be counted as false positives with zero precision rates. For each identified isoform paired with an annotated isoform, we calculate its precision rate as $\frac{\#\text{overlapping nucleotides with the paired annotated isoform}}{\#\text{nucleotides in the identified isoform}}$. The precision rate of a gene was defined as the average of the individual precision rates of its identified isoforms.

Similarly, we match each annotated isoform with an identified isoform. We calculate the recall rate of each annotated isoform with a paired identified isoform as $\frac{\#\text{ overlapping nucleotides with the paired identified isoform}}{\#\text{ nucleotides in the annotated isoform}}$. If there are more annotated isoforms than identified ones, the unpaired annotated isoforms would have zero recall rates. For a given gene, its recall rate is defined as the average of the individual recall rates of its annotated isoforms. Given the precision and recall rates of a gene, we calculate the F score as the their harmonic mean. That is, F score $= \frac{2}{1/\text{precision}+1/\text{recall}}$.

We calculate the precision and recall rates of every method in each of the 50 samples by taking the average of 951 genes. In the results, all methods except NMFP have precision rates above 0.76 in all samples. Although NMFP has low precision rates, its recall rates are as high as 0.96, implying that its identified isoform candidates has a good coverage of annotated isoforms (Fig. 2). NMFP+Cufflinks has lower precision rates but much higher recall rates than Cufflinks, because the latter identifies fewer isoforms given its parsimony assumption. SLIDE has opposite results, which show that NMFP+SLIDE have higher precision rates but lower recall rates than SLIDE, for both SLIDE(fewer) and SLIDE(more). The reason is that the original versions of SLIDE have a larger search space than NMFP+SLIDE, and thus they are likely to find more isoforms and thus cover more nucleotides given the same Lasso regularization parameter.

In terms of F scores, which combine precision and recall rates, NMFP is shown to have greatly improved the performance of Cufflinks, by increasing the F scores of approximately 0.6 for Cufflinks to more than 0.8 for NMFP+Cufflinks. Strikingly, with the addition of NMFP, Cufflinks outperforms SLIDE, which originally had better F scores than Cufflinks without NMFP [14]. NMFP+Cufflinks becomes the top performer in F scores. On the other hand, NMFP does not much improve the performance SLIDE(more) and even slightly deproves the performance SLIDE(fewer) at the nucleotide level. The reason is that the original SLIDE has reasonably good performance at the nucleotide level, though its identified isoforms are often similar to but not exactly the annotated isoforms (see the subsection "Evaluation at transcript level").

Evaluation at exon level

We next evaluate the performance of these seven methods at the exon level. If an exon in an identified isoform overlaps at least 50 % of an annotated exon, we call the two exons "overlapping". Similar to our matching scheme at the nucleotide level, we match identified isoforms with annotated isoforms at the exon level. Then for each identified isoform, its precision rate at the exon level is defined as

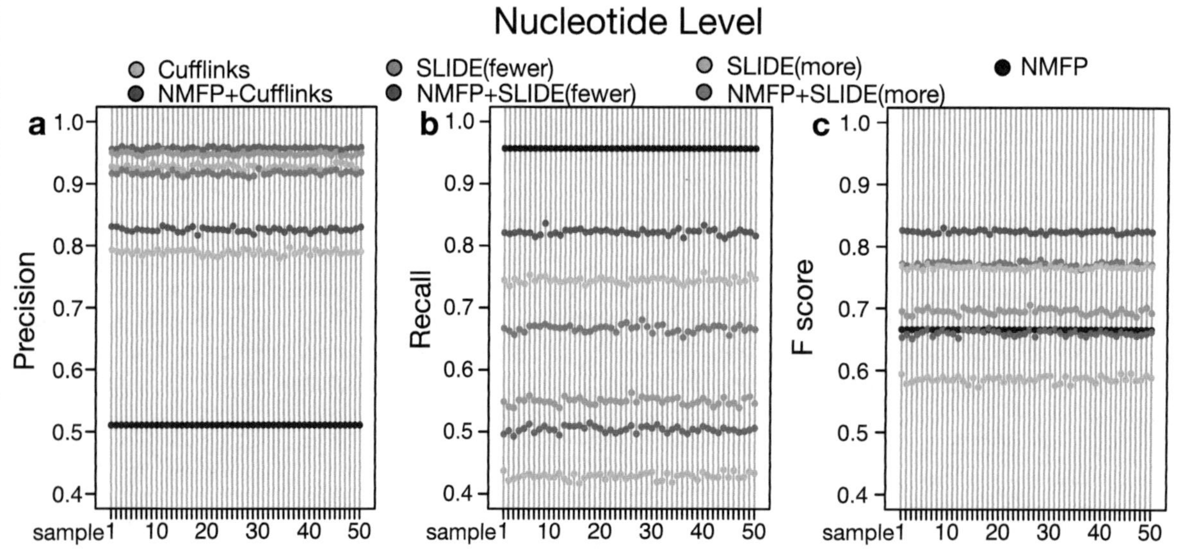

Fig. 2 Isoform discovery performance at the nucleotide level. **a** Precision rates, **b** Recall rates, and **c** F scores of the identified isoforms by seven methods (Cufflinks, NMFP+Cufflinks, SLIDE(fewer), NMFP+SLIDE(fewer), SLIDE(more), NMFP+SLIDE(more), and NMFP) on 50 simulated RNA-seq data sets of *D. melanogaster*

$\frac{\#\text{ overlapping exons with the paired annotated isoform}}{\#\text{ exons in the identified isoform}}$. Similarly for each annotated isoform, its recall rate is defined as $\frac{\#\text{ overlapping exons with the paired identified isoform}}{\#\text{ exons in the annotated isoform}}$. Then for every gene, its precision and recall rates are calculated as the average precision rate of its identified isoforms and the average recall rate of its annotated isoforms. Its F score is the harmonic mean of its precision and recall rates. For every method in every sample, the precision, recall and F score are calculated by taking the average of 951 genes.

Similar to the performance at nucleotide level, NMFP significantly improves the F scores of Cufflinks from approximately 0.6 to over 0.8, yet it does not improve the performance SLIDE (Fig. 3). The reason is that the original SLIDE has reasonably good performance at the exon level, though its identified isoforms are often similar to but not exactly the annotated isoforms (see the subsection "Evaluation at transcript level").

Evaluation at transcript level

Compared to the previous evaluation at nucleotide and exon levels, the evaluation at the transcript level is more important, as the ultimate goal of isoform discovery methods is to correctly identify full-length mRNA isoforms. At the transcript level, if an identified isoform and an annotated isoform have every exon matched by the criterion we defined for matching at the exon level, we called these two isoforms "matched". Then for every gene, its precision and recall rates at the transcript level are defined as $\frac{\#\text{ identified isoforms matched with annotated isoforms}}{\#\text{ identified isoform}}$ and $\frac{\#\text{ annotated isoforms matched with identified isoforms}}{\#\text{ annotated isoform}}$ respectively. Its F-score is the harmonic mean of its precision and recall rates. For every method in every sample, the precision, recall and F score are calculated by taking the average of 951 genes.

In the results, NMFP has high recall rates greater than 0.85, indicating that the isoform candidates found by NMFP have a good coverage of annotated transcripts (Fig. 4). Meanwhile, NMFP has precision rates above 0.4, implying that the number of isoform candidates is less than 2.5 times the number of annotated isoform. This demonstrates the effectiveness of NMFP in reducing the search space of possible isoforms. In terms of F scores at the transcript level, NMFP greatly improves the performance of both Cufflinks and SLIDE. Specifically, NMFP+Cufflinks has F scores above 0.6, while Cufflinks has F scores only below 0.5. The reason is that NMFP greatly improves the recall rates of Cufflinks. NMFP also increases the F scores of SLIDE(more) from 0.5 to approximately 0.6, because NMFP helps SLIDE(more) achieve better precision rates. These results illustrate the effectiveness of NMFP in reducing the search space for SLIDE (thus increasing the precision of SLIDE) and in removing the parsimony assumption for Cufflinks (thus increasing the recall of Cufflinks). To explain why NMFP can improve SLIDE's performance at the transcript level but not at the nucleotide or exon level, we find the reason as the unidentifiability issue of SLIDE with a large search space of possible isoforms. Due to the short RNA-seq read lengths, there may exist multiple sets of isoforms (as combinations of exons) that can well explain the observed RNA-seq reads. Hence with a large search space of isoforms, SLIDE are likely to miss the true

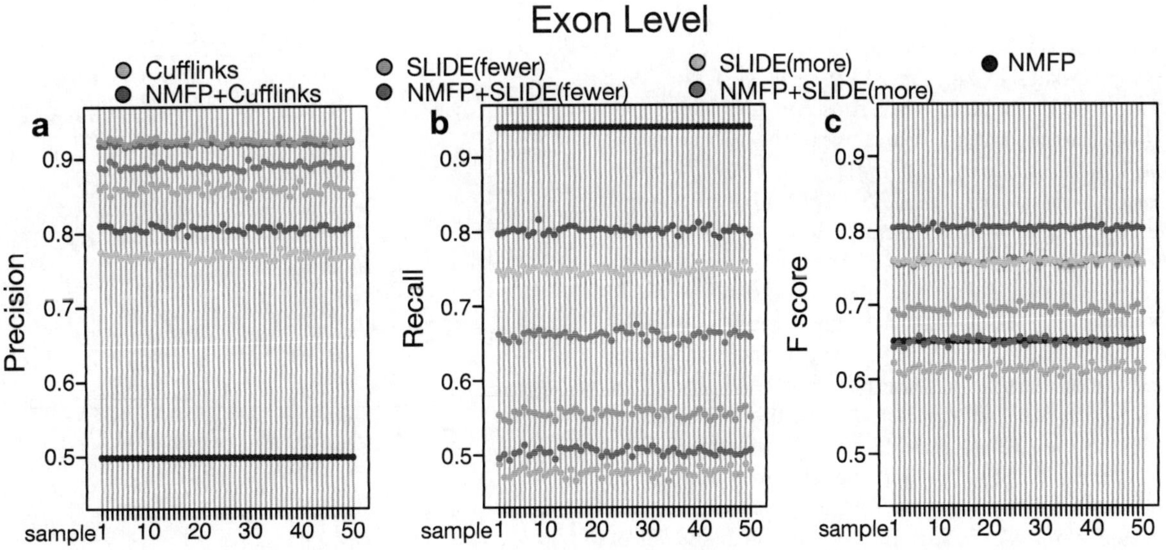

Fig. 3 Isoform discovery performance at the exon level. **a** Precision rates, **b** Recall rates, and **c** F scores of the identified isoforms by seven methods (Cufflinks, NMFP+Cufflinks, SLIDE(fewer), NMFP+SLIDE(fewer), SLIDE(more), NMFP+SLIDE(more), and NMFP) on 50 simulated RNA-seq data sets of *D. melanogaster*

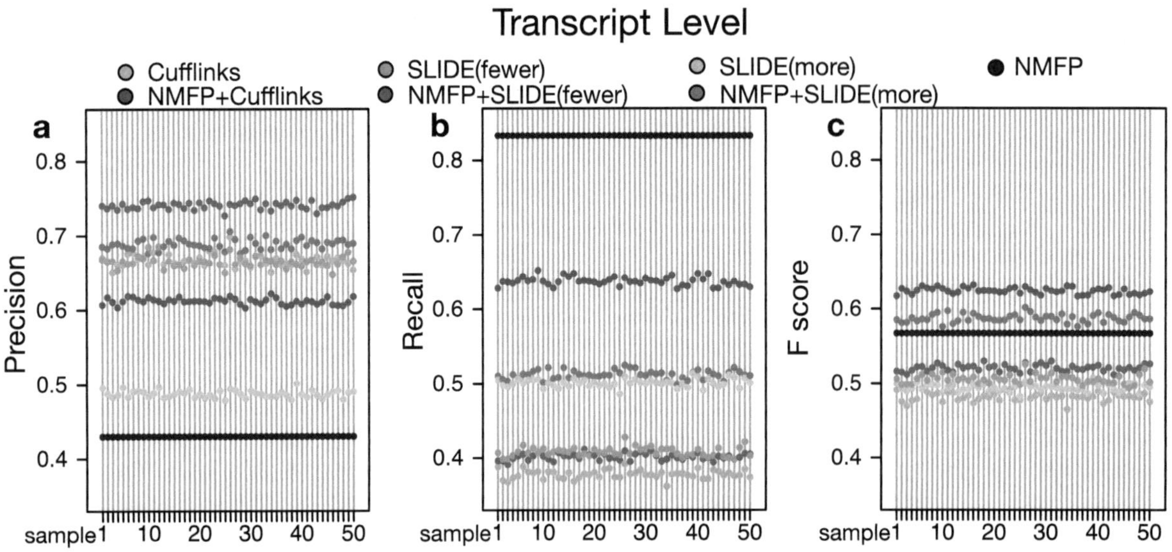

Fig. 4 Isoform discovery performance at the transcript level. **a** Precision rates, **b** Recall rates, and **c** F scores of the identified isoforms by seven methods (Cufflinks, NMFP+Cufflinks, SLIDE(fewer), NMFP+SLIDE(fewer), SLIDE(more), NMFP+SLIDE(more), and NMFP) on 50 simulated RNA-seq data sets of *D. melanogaster*

isoforms, but it still can detect nucleotides and exons reasonably well. Since NMFP can shrink the search space to have a higher concentration of true isoforms (i.e., annotated isoforms in the simulation), SLIDE would become more likely to identify the true isoforms and thus achieve better isoform discovery accuracy at the transcript level.

Robustness of NMFP to the choice of NMF rank

How to determine the matrix rank of factorization results is a difficult and open question in NMF [3, 4, 7–9]. In NMFP, the NMF rank determines the number of isoform candidates to be found after each NMF run. Our aggregation of multiple NMF runs and the subsequent high-frequency filtering can increase the robustness of NMFP to the choices of NMF rank. To illustrate this point, we use *D. melanogaster* gene *FBgn0039955* as an example and apply NMFP with different NMF ranks to the simulated RNA-seq data. This gene has 4 annotated isoforms (in Ensembl BDGP6 of release 80) and 10 subexons. In the result (Fig. 5), NMFP has a good recall rate of 0.9 for ranks ranging from 3 to 9. The precision rate and F score are also stable in this range. This result and other examples in the Additional file 1 show the robustness of NMFP to the choices of NMF rank. In practice, NMFP selects the rank based on gap statistic (See Section 2 in the Additional file 1 for more detail). Nevertheless, we note that the NMF rank cannot be too much smaller or larger than the number of true isoforms. If too small, the rank will lead to too few isoform candidates to cover the true isoforms. If too large, the identified isoform candidates will be mostly individual bins but not full-length isoforms.

Simulation results in *M. musculus*

We also evaluate the performance of NMFP in *M. musculus* (mouse) (reference genome mm10 and annotation GRCm38 of release 81), a mammalian model organism with more complex gene structures than *D. melanogaster*.

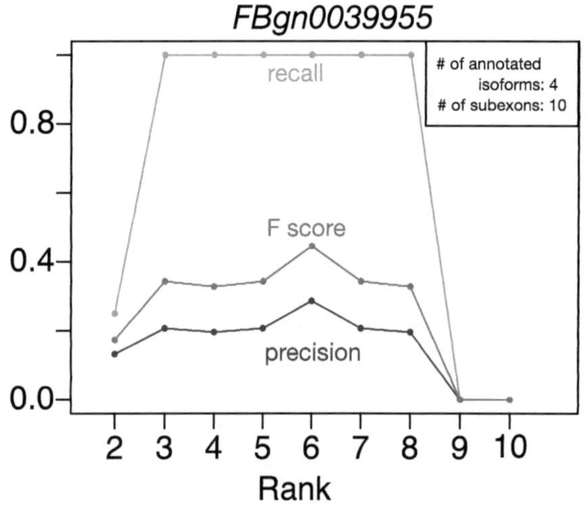

Fig. 5 Robustness of NMFP performance to NMF rank choices. Precision (*blue*), recall (*orange*) and F scores (*red*) of the identified isoform candidates by NMFP are evaluated at 9 different NMF ranks on simulated data from *D. melanogaster* gene *FBgn0039955*, which has 10 subexons and 4 annotated isoforms

We generate 100 RNA-seq samples from mouse chr1 using Flux Simulator, with paired-end reads of length 2×76 bp. The 100 samples have 10 different numbers of RNA molecules and 10 different numbers of reads. The numbers of RNA molecules range from 4,200,000 to 6,000,000, increasing in steps of 200,000. The number of reads range from 11,000,000 to 20,000,000, increasing in steps of 1,000,000. More detail on the 100 samples is summarized in Section 3.2 and Table S1 of the Additional file 1. On chr1, there are 3432 genes, among which we are interested in 852 genes with 3-10 subexons and 2-17 annotated isoforms (more detail in Additional file 1: Table S2). In this simulation, we calculate precision and recall rates and F scores at the transcript level for every method, same as what we did in the simulation in *D. melanogaster*.

Increased robustness of NMFP+SLIDE to the choices of parameter λ

By shrinking the searching space of possible isoforms, NMFP can effectively increase the robustness of SLIDE to its regularization parameter λ in the Lasso estimation. We conduct case studies on three mouse genes, for which we apply NMFP+SLIDE and SLIDE with 150 λ values, ranging from 0.001 to 0.15 by 0.001. The results show that NMFP+SLIDE has better isoform discovery accuracy than SLIDE for smaller λ values (Fig. 6 and Additional file 1: Figure S3). This finding is expected because a smaller search space does not need a large regularization parameter in the Lasso estimation. It indicates that NMFP can reduce the difficulty of choosing λ in the use of SLIDE, as users can simply choose a small λ value to obtain good isoform discovery results.

Robustness of NMFP to the number of samples

In practice, the number of RNA-seq data sets under the same biological condition can be quite small, though this issue has become less of a concern given the rapid accumulation of RNA-seq data sets in public repositories. To make NMFP a practically useful tool, it is important to evaluate its robustness to m, the number of input samples in the read count matrix $\mathbf{V}$. We apply NMFP to different numbers of simulated mouse samples, from 20 to 100 samples, following the sample order with increasing sequencing depths (for the sample order, please see Section 3.2 in the Additional file 1). As shown in Fig. 7, the NMFP results are robust to the number of samples and differences in sequencing depths.

Simulation study on the effects of expression levels

We use another simulation study to demonstrate that NMFP is capable of increasing the isoform discovery accuracy for a lowly expressed gene in a sample by leveraging other samples (Fig. 8). In this simulation, we construct a gene with 8 exons and 3 true isoforms (10111111), (10011111) and (01111111), which indicate the skipping of Exon 2, Exons 2 and 3, and Exon 1 respectively. The exon lengths are randomly sampled from $\{100, 150, 200, \ldots, 500\}$. We simulate 10 samples. In each sample, the proportions of the three isoforms are randomly simulated from $U[0, 1]$, and the three proportions are normalized to sum up to 1. Given an isoform, single-end reads with length 76 bp are assumed to follow a uniform distribution on this isoform. We fix the gene expression level in Sample 1 as 3 RPKM (Reads Per Kilobase of transcript per Million mapped reads) and vary the gene expression levels in the other nine samples from 0.5,

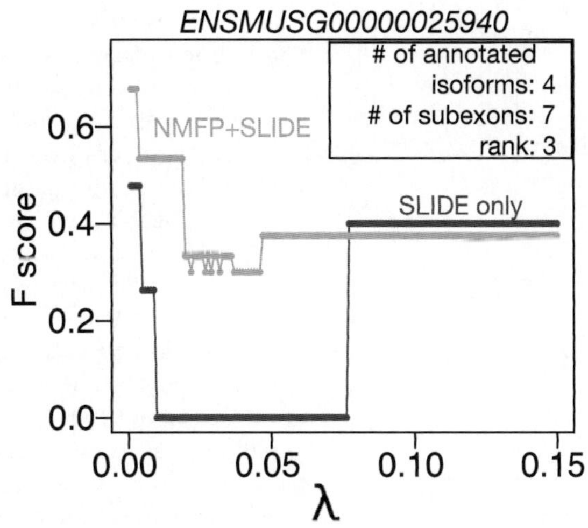

Fig. 6 NMFP increases the robustness of SLIDE to the choices of regularization parameter λ, from 0.001 to 0.15 by 0.001. F scores of NMFP+SLIDE (*orange*) and SLIDE (*blue*) were calculated at 150 values of λ used in the Lasso regression in SLIDE. NMFP used an NMF rank = 3. RNA-seq data were simulated from mouse gene *ENSMUSG00000025940*, which has 7 subexons and 4 annotated isoforms

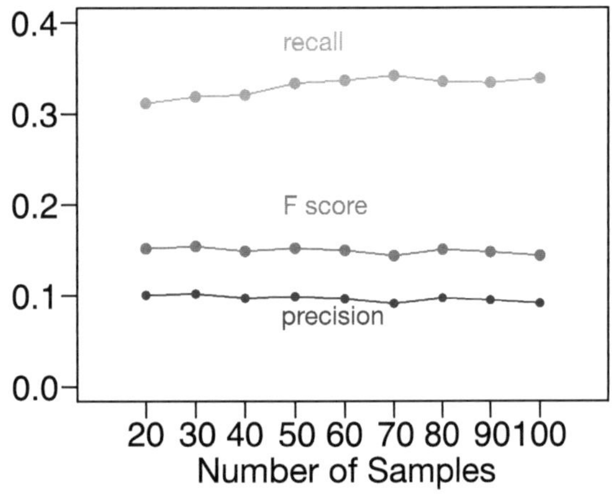

Fig. 7 Robustness of NMFP to the number of input RNA-seq samples

1, 2 to 10 RPKM in 11 different settings. In each setting, we calculate a read count matrix **U** from the gene expression levels, the isoform proportions, and the uniformity assumption. Then we add white Gaussian noise $N(0, 1)$ to **U**, and normalize the resulting matrix as **V**, for which we apply NMFP. Figure 8 depicts the relationship between the gene expression level in Samples 2–10 and the minimum number of isoform candidates found by NMFP to cover the three true isoforms. When the gene expression level is below 3 RPKM, the number of isoform candidates needed to cover the true isoforms is more than 60 with a large variance.

Once the gene expression level goes above 3 RPKM, the number of needed candidates decreases to approximately 30 and has a reduced variance. This result demonstrates that for a gene with low expression in a sample, NMFP can take advantage of other samples to help reduce the number of isoform candidates of this gene. This property makes it possible for downstream isoform discovery methods to explore isoform structures of lowly expressed genes in a sample given the availability of other samples where the genes are more highly expressed.

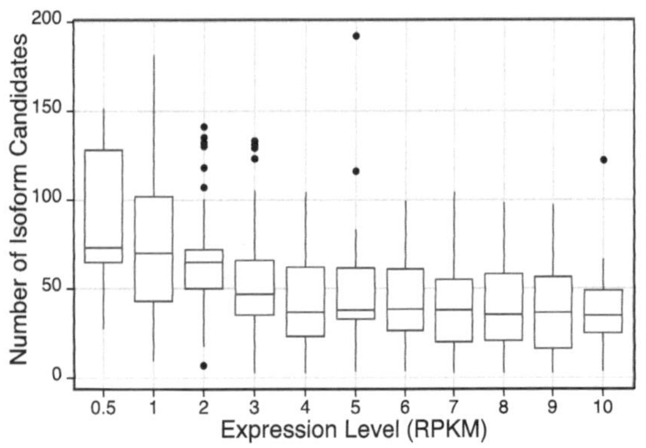

Fig. 8 Numbers of isoform candidates found by NMFP to cover true isoforms vs. gene expression levels in additional samples other than the sample of interest

Real data case study

We further demonstrate the performance of NMFP and compare the seven methods on 74 real RNA-seq data sets of *D. melanogaster* (reference genome dm5 and annotation BDGP5 of release 66). Description of the 74 data sets is in Section 5 of the Additional file 1. Since we have no knowledge on the true isoforms existing in real RNA-seq data sets, we interpret the results by studying a few genes in Integrative Genomics Viewer (IGV) [24, 25], where we plot the read distribution in an RNA-seq data set (*D. melanogaster* pupae 3 days) and the isoforms identified by the seven methods.

For *D. melanogaster* gene *FBgn0037643*, which has 11 subexons and 6 annotated isoforms, NMFP finds 20 isoform candidates, which contain 4 annotated isoforms and miss the other two by a very short subexon (of only 3 bp). Cufflinks only identifies one annotated isoform, which cannot explain the observed read distribution in the last subexon. In contrast, NMFP+Cufflinks successfully identifies four annotated isoforms, which successfully capture the last subexon. NMFP also helps SLIDE discover full-length annotated isoforms instead of isofrom fragments (Fig. 9). More real data case studies are in Section 3.4 of the Additional file 1.

Conclusion and discussions

NMF is inherently capable of providing interpretable results for the problem of mRNA isoform discovery. From an input matrix of RNA-seq read counts, NMF can find a sparse structure to indicate which bins are likely to form one isoform. However, NMF alone is not competent for recovering isoforms from RNA-seq data because of several reasons: (i) inaccurate read coverages and missing junction reads due to RNA-seq biases and mapping difficulties, (ii) non-uniqueness of NMF solutions leading to ambiguous isoform discovered, (iii) conflicting bins co-existing in discovered isoforms making the results biologically invalid, and (iv) difficulty in determining the NMF rank. Those issues can be largely alleviated or resolved when NMF is used as a method to pre-select isoform candidates rather than determining final isoforms. For (i), (ii) and (iv), since each non-unique NMF solution provides isoform candidates that may include some true isoforms among many false positives, aggregating multiple NMF solutions to find the high-frequency isoforms will increase the chance of discovering the true isoforms. We demonstrate in the simulation study of *M. musculus*, that the NMFP results are not sensitive to the

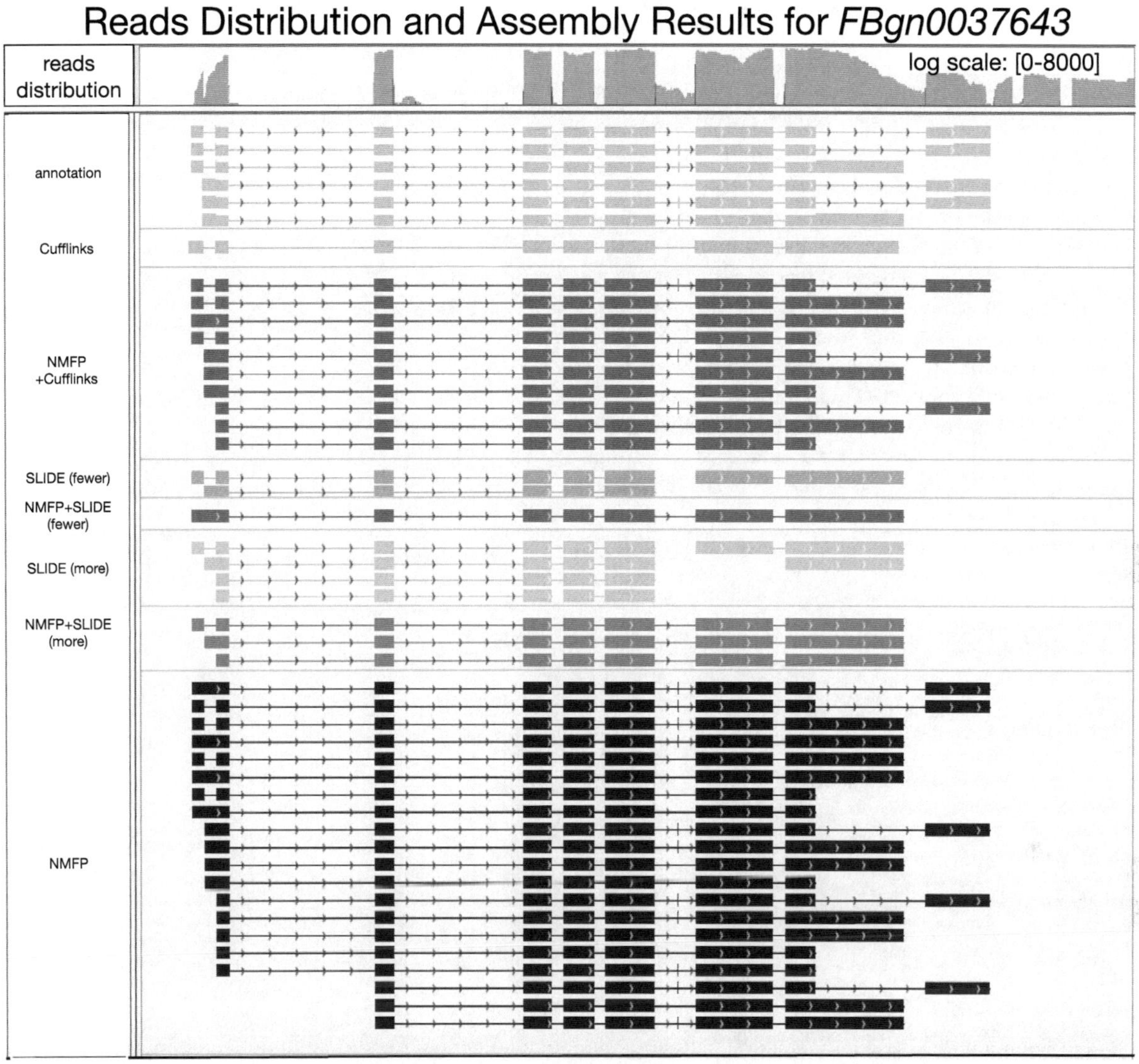

Fig. 9 RNA-seq read coverage, annotation and isoform discovery results of *FBgn0037643*

choice of NMF ranks. We address (iii) by adding a penalty term to NMF and decomposing the invalid isoform candidates with conflicting bins into biologically valid isoforms.

We propose NMFP as a preselection method for improving the accuracy of downstream isoform discovery methods. NMFP takes input from annotated genes and exon boundaries (e.g. UCSC and Ensemble annotations) and/or de novo assemblies (e.g. Cufflinks output). It outputs a small pool of isoform candidates, which are expected to well cover true isoforms and can form a reduced search space for downstream methods. From a general perspective, combination of methods with opposing advantages and drawbacks may improve over individual methods. NMFP and SLIDE are both built on the idea of selecting isoforms from all possibilities, while Cufflinks assembles transcripts directly via an overlap graph. A possible reason for NMFP+Cufflinks to become the top performer is that it combines the strengths of two different approaches: selection and assembly. This may also explain why NMFP+SLIDE, which combines two selection based methods, is outperformed by NMFP+Cufflinks, though SLIDE alone outperforms Cufflinks without NMFP. However compared to SLIDE itself, NMFP+SLIDE still has improved performance at the transcript level, and even becomes more robust to the regularization parameter choices. Overall, we demonstrate by simulation and real data studies that NMFP is a useful tool for improving isoform discovery accuracy and robustness in many aspects.

Nevertheless, NMFP still encounters several problems and there remains substantial room for improvement. The noise and bias generated during RNA-seq still has large influence on NMFP results. Some methods for denosing and debiasing may help the performance of NMFP [26, 27]. In addition, NMFP requires enough samples to implement NMF, for which normalization is important but difficult for samples generated by different platforms or from different sources. Normalization methods for RNA-seq data (e.g., [28]) may improve the performance of NMFP. Moreover, the current NMFP algorithm has high computational complexity, and optimizing the algorithm is one of our top priority. We also intend to build a database of isoform candidates found by NMFP for different species, so that researchers can conveniently use them for downstream analysis without aggregating multiple samples and going through the whole NMFP pipeline.

Additional file

Additional file 1: More detail on the methods and results; more simulation and real data results. (PDF 606 kb)

Abbreviations
NMF: non-negative matrix factorization; NMFP: non-negative matrix factorization preselection; mRNA: messenger RNA; cDNA: complementary DNA; qPCR: quantitative polymerase chain reaction.

Competing interests
The authors declare that they have no competing interests.

Authors' contributions
YY and JJL conducted simulation, processed real data,, and performed statistical analysis together. YY drafted the manuscript. JJL supervised this study. Both authors read and approved the final manuscript.

Acknowledgements
This work was supported by the start-up fund of Department of Statistics at University of California, Los Angeles, and the Hellman Fellowship from the Hellman Foundation. The publication costs for this article were funded by the Hellman Foundation. We thank Yu-Cheng T. Yang in School of Life Sciences, Tsinghua University and Zhiyi Qin and Dr. Xuegong Zhang in Department of Automation, Tsinghua University for their comments and help on this work. We also thank the three anonymous reviewers for their insightful comments that helped us improve our manuscript.

Declarations
This article has been published as part of BMC Genomics Volume 17 Supplement 1, 2016: Selected articles from the Fourteenth Asia Pacific Bioinformatics Conference (APBC 2016): Genomics. The full contents of the supplement are available online at http://www.biomedcentral.com/bmcgenomics/supplements/17/S1.

Source codes
The NMFP source codes and examples are available at http://www.stat.ucla.edu/~jingyi.li/packages/NMFP.zip.

Author details
[1]Division of Biostatistics, University of California, Berkeley, 94720 Berkeley, CA, USA. [2]Department of Statistics, 8125 Math Sciences Bldg., University of California, Los Angeles, 90095-1554 Los Angeles, CA, USA. [3]Department of Human Genetics, 695 Charles E. Young Drive South, University of California, Los Angeles, 90095-7088 Los Angeles, CA, USA.

Published: 11 January 2016

References
1. Faustino NA, Cooper TA. Pre-mRNA splicing and human disease. Genes Dev. 2003;17(4):419–37.
2. Wang G-S, Cooper TA. Splicing in disease: disruption of the splicing code and the decoding machinery. Nat Rev Genet. 2007;8(10):749–61.
3. Anton MA, Gorostiaga D, Guruceaga E, Segura V, Carmona-saez P, Pascual-montano A. SPACE : an algorithm to predict and quantify alternatively spliced isoforms using microarrays. Genome Biol. 2008;9(2):46. doi:10.1186/gb-2008-9-2-r46.
4. Anton Ma, Aramburu A, Rubio A. Improvements to previous algorithms to predict gene structure and isoform concentrations using Affymetrix Exon arrays. BMC Bioinf. 2010;11(1):578. doi:10.1186/1471-2105-11-578.
5. Lee DD, Seung HS. Learning the parts of objects by non-negative matrix factorization. Nature. 1999;401(6755):788–91. doi:10.1038/44565.
6. Donoho D, Stodden V. When Does Non-Negative Matrix Factorization Give a Correct Decomposition into Parts? In: Advances in Neural Information Processing Systems; 2003.
7. Brunet J-P, Tamayo P, Golub TR, Mesirov JP. Metagenes and molecular pattern discovery using matrix factorization. Proc Natl Acad Sci USA. 2004;101(12):4164–9. doi:10.1073/pnas.0308531101.
8. Pascual-montano A, Carmona-saez P, Chagoyen M, Tirado F, Carazo JM, Pascual-marqui RD. bioNMF: a versatile tool for non-negative matrix factorization in biology. BMC Bioinf. 2006;9:1–9. doi:10.1186/1471-2105-7-366.
9. Fogel P, Young SS, Hawkins DM, Ledirac N, Hall F, Ne CS, et al. Inferential, robust non-negative matrix factorization analysis of microarray data. Bioinformatics (Oxford, England). 2007;23(1):44–9. doi:10.1093/bioinformatics/btl550.
10. Wang Z, Gerstein M, Snyder M. RNA-seq: a revolutionary tool for transcriptomics. Nat Rev Genet. 2009;10(1):57–63.
11. Trapnell C, Williams Ba, Pertea G, Mortazavi A, Kwan G, van Baren MJ, et al. Transcript assembly and quantification by RNA-Seq reveals unannotated transcripts and isoform switching during cell differentiation. Nat Biotechnol. 2010;28(5):511–5. doi:10.1038/nbt.1621.171.
12. Guttman M, Garber M, Levin JZ, Donaghey J, Robinson J, Adiconis X, et al. Ab initio reconstruction of cell type – specific transcriptomes in mouse reveals the conserved multi-exonic structure of lincRNAs. Nat Biotechnol. 2010;28(5):503–10. doi:10.1038/nbt.1633.
13. Li W, Feng J, Jiang T. IsoLasso: a LASSO regression approach to RNA-Seq based transcriptome assembly (Extended abstract). Lect Notes Comput Sci (including subseries Lecture Notes in Artificial Intelligence and Lecture Notes in Bioinformatics). 2011;6577 LNBI:168–88. doi:10.1007/978-3-642-20036-6_18.
14. Jessica J, Jiang C-r, Brown JB, Huang H, Bickel PJ. Sparse linear modeling of next-generation mRNA sequencing (RNA-Seq) data for isoform discovery and abundance estimation. PNAS. 2011. doi:10.1073/pnas.1113972108.
15. Roberts A, Pimentel H, Trapnell C, Pachter L. Identification of novel transcripts in annotated genomes using rna-seq. Bioinformatics. 2011;27(17):2325–29.
16. Steijger T, Abril JF, Engström PG, Kokocinski F, The RGASP Consortium, Hubbard TJ, et al. Assessment of transcript reconstruction methods for RNA-seq. Nat Methods. 2013;10(12):1177–84. doi:10.1038/nmeth.2714.
17. Fan J, Lv J. Sure independence screening for ultrahigh dimensional feature space. J R Stat Soc Ser B (Stat Methodol). 2008;70(5):849–911.
18. Griebel T, Zacher B, Ribeca P, Raineri E, Lacroix V, Guigó R, et al. Modelling and simulating generic RNA-Seq experiments with the flux simulator. Nucleic Acids Res. 2012;40(20):10073–83. doi:10.1093/nar/gks666.
19. Li C, Wong WH. Model-based analysis of oligonucleotide arrays: expression index computation and outlier detection. PNAS. 2001;98:31–6. doi:10.1073/pnas.98.1.31.
20. Li SZ, Hou X, Zhang H, Cheng Q. Learning Spatially Localized , Parts-Based Representation. In: Computer Vision and Pattern Recognition, 2001. CVPR 2001. Proceedings of the 2001 IEEE Computer Society

Conference, vol. 1. IEEE; 2001. p. 207–212. http://ieeexplore.ieee.org/xpl/login.jsp?tp=&arnumber=990477&url=http%3A%2F%2Fieeexplore.ieee.org%2Fxpls%2Fabs_all.jsp%3Farnumber%3D990477.
21. Feng T, Li SZ, Shum H-y, Zhang H. Local Non-Negative Matrix Factorization as a Visual Representation. In: Proceedings of the 2nd International Conference on Development and Learning(ICDL'02); 2002. p. 178–183. http://ieeexplore.ieee.org/xpl/login.jsp?tp=&arnumber=1011835&url=http%3A%2F%2Fieeexplore.ieee.org%2Fxpls%2Fabs_all.jsp%3Farnumber%3D1011835.
22. Pascual-montano A, Carazo JM, Member S, Kochi K, Lehmann D, Pascual-marqui RD, et al. Nonsmooth Nonnegative Matrix Factorization (ns NMF). IEEE Trans Pattern Anal Mach Intell. 2006;28(3):1–13.
23. Tibshirani R, Walther G, Hastie T. Estimating the number of clusters in a data set via the gap statistics. J R Stat Soc Ser B (Stat Methodol). 2001;63(2):411–23.
24. Robinson JT, Thorvaldsdóttir H, Winckler W, Guttman M, Lander ES, Getz G, et al. Integrative genomics viewer. Nat Biotechnol. 2011;29(1): 24–6.
25. Thorvaldsdóttir H, Robinson JT, Mesirov JP. Integrative genomics viewer (igv): high-performance genomics data visualization and exploration. Briefings in bioinformatics. 2013;14:178-192. http://bib.oxfordjournals.org/content/14/2/178.long.
26. Hansen KD, Brenner SE, Dudoit S. Biases in Illumina transcriptome sequencing caused by random hexamer priming. Nucleic Acids Res. 2010;38(12):1–7. doi:10.1093/nar/gkq224.
27. Li J, Jiang H, Wong WH. Modeling non-uniformity in short-read rates in RNA-Seq data. Genome Biol. 2010;11(5):50. doi:10.1186/gb-2010-11-5-r50.
28. Risso D, Ngai J, Speed TP, Dudoit S. Normalization of RNA-seq data using factor analysis of control genes or samples. Nat Biotechnol. 2014;32(9):1–10. doi:10.1038/nbt.2931.

He *et al. BMC Genomics* 2016, **17**(Suppl 1):4
DOI 10.1186/s12864-015-2305-7

BMC Genomics

PROCEEDINGS Open Access

Identifying micro-inversions using high-throughput sequencing reads

Feifei He[1], Yang Li[2], Yu-Hang Tang[4], Jian Ma[2,3*] and Huaiqiu Zhu[1*]

From The Fourteenth Asia Pacific Bioinformatics Conference (APBC 2016)
San Francisco, CA, USA. 11 - 13 January 2016

Abstract

Background: The identification of inversions of DNA segments shorter than read length (e.g., 100 bp), defined as micro-inversions (MIs), remains challenging for next-generation sequencing reads. It is acknowledged that MIs are important genomic variation and may play roles in causing genetic disease. However, current alignment methods are generally insensitive to detect MIs. Here we develop a novel tool, MID (Micro-Inversion Detector), to identify MIs in human genomes using next-generation sequencing reads.

Results: The algorithm of MID is designed based on a dynamic programming path-finding approach. What makes MID different from other variant detection tools is that MID can handle small MIs and multiple breakpoints within an unmapped read. Moreover, MID improves reliability in low coverage data by integrating multiple samples. Our evaluation demonstrated that MID outperforms Gustaf, which can currently detect inversions from 30 bp to 500 bp.

Conclusions: To our knowledge, MID is the first method that can efficiently and reliably identify MIs from unmapped short next-generation sequencing reads. MID is reliable on low coverage data, which is suitable for large-scale projects such as the 1000 Genomes Project (1KGP). MID identified previously unknown MIs from the 1KGP that overlap with genes and regulatory elements in the human genome. We also identified MIs in cancer cell lines from Cancer Cell Line Encyclopedia (CCLE). Therefore our tool is expected to be useful to improve the study of MIs as a type of genetic variant in the human genome. The source code can be downloaded from: http://cqb.pku.edu.cn/ZhuLab/MID.

Keyword: Micro-inversion, Next-generation sequencing, Structural variation

Background

Genomic structural variations (SVs), including insertion, deletion, rearrangement, and duplication are important sources of human genetic diversity and may be responsible for human diseases [1–4]. With the rapid development of high-throughput sequencing technologies, computational tools have been developed to utilize next-generation sequencing data to identify SVs, such as BreakDancer [5], Pindel [6], PRISM [7], DELLY [8], CNVnator [9] and Gustaf [10]. It has been noted that small inversions may provide important insights into genome evolution and disease mechanisms [4, 11]. Previous studies have reported the discovery of small inversions by comparing the human genome with other mammalian genomes [12, 13]. However, despite the progress of computational tools using NGS data to identify SVs, our understanding and strategies for characterizing small inversions remain limited [14]. In this work, we focus on micro-inversions (MIs), defined as small inversions shorter than 100 bp, which are mostly ignored by existing methods.

Computational tools for SV detection mostly rely on read mapping. However, we found that reads with potential MIs, which cannot be mapped to the reference genome initially, are usually discarded as "unmapped reads" by existing alignment tools. As no mapping information is available for the initially unmapped reads, most existing tools are incapable to detect MIs shorter than the read

* Correspondence: jianma@illinois.edu; hqzhu@pku.edu.cn
[2]Department of Bioengineering, University of Illinois, Urbana, IL 61801, USA
[1]State Key Laboratory for Turbulence and Complex Systems and Department of Biomedical Engineering, and Center for Quantitative Biology, Peking University, Beijing 100871, China
Full list of author information is available at the end of the article

length. Generally speaking, existing methods use features such as read depth, read pair information, and split reads to identify SVs [15]. Yet there are still several challenges to use these features to detect MIs. The small size of MIs makes the identification difficult. The approaches used in methods based on read depth, e.g., CNVnator, would perform poorly with the small size of MIs without locating the breakpoints. The strategies in utilizing read pairs (e.g., BreakDancer, DELLY) and split reads (e.g., Pindel, PRISM, Gustaf) are also not appropriate to identify MIs less than 100 bp, because the size of SVs detected by these tools is dependent on insert size of paired-end reads and insensitive to variants as small as MIs. Additionally, to increase the applicability of the method, it is important to handle datasets with low coverage. For large-scale projects such as the 1000 Genomes Project (1KGP), there is increasing number of low coverage data [16]. For example, the sample NA12878, provided by 1KGP from two paired-end libraries with 15X coverage, has been frequently selected by variant detection tools for performance comparisons (642 deletions, 271 duplications and 30 insertions) [17]. However, most of the human individual genomes were generated with a low coverage of 2–4X in 1KGP so far. Finally, it is also challenging to identify MIs in a read with multiple variants, as other types of variants (e.g., single nucleotide variations (SNVs), small indels and other rearrangements), may also occur in nearby regions of the MI.

In this paper, we focus on MI identification from unmapped NGS short reads. Our algorithm, named MID (Micro-Inversion Detector), specifically detects MIs shorter than the read length. Moreover, MID can be applied to very low coverage samples. What makes MID different from other variant detection tools is that it is sensitive to very small size MIs, and capable of detecting MIs with multiple breakpoints in one read. Moreover, MID proves reliability in low coverage data by integrating multiple samples. The simulation results showed that MID can detect MIs efficiently and reliably using unmapped NGS short reads with low false positives. By applying MID to 1KGP data, we identified 721 MIs, 349 of which are intergenic, 342 MIs are intronic, and 30 MIs are exonic. We also applied MID to the Lung Squamous Cell Carcinoma whole exome sequencing (WXS) data from CCLE, where we identified 12 MIs.

Results

Performance of MI detection on simulated data

Until recently, few attempts have been made for MI detection using NGS data. To the best of our knowledge, only a recent tool called Gustaf [10], can detect inversions from 30 bp to 500 bp, which in principle can recognize MIs shorter than the read length. Gustaf is a tool based on a split-read approach, which records the local alignments provided by other tools and draws a split-read graph to use standard graph algorithms to evaluate relationships of the alignments. In addition, it should be noted that no suitable real datasets with confident annotation are available as a benchmark. We therefore used simulated datasets to evaluate the performance of MID. Two different types of simulated datasets were generated in the current work: one with 1,000 MIs only (Dataset 1), and the other one with 1,000 MIs surrounded by other types of variants (Dataset 2). For Dataset 1, we simulated 1,000 MIs ranging from 15 bp to 40 bp, following a normal distribution for MI size, randomly on the whole chr10 (135 Mb) from the human genome assembly hg19 [18]. For Dataset 2, we simulated 1,000 MIs surrounded by 4,000 other structural variants (including SNVs, deletions, insertions, and duplications) following a normal distribution for MI size randomly on chr10, with a size range of 15–40 bp. We also simulated Illumina paired-end short reads (in 76 bp; same as what the sample NA19213 from the 1KGP has) using Maq [19] with an error rate 0.02. Both simulated datasets had 10 different sub-datasets with coverage varying from 2X to 60X (see Table 1). We then ran the Burrows-Wheeler alignment (BWA) tool [20] in the same way as in the 1KGP to get unmapped short reads.

To have a clear view of our evaluation results, we define several metrics. If at least one simulated read generated from the reference sequence cover the MI, we call the MI as "detectable MI". If 80 % of both the detected MI and the original MI overlap, we call the MI as "correctly detected MI" [10]. We then calculate the sensitivity (SN) as the ratio of correctly detected MIs over all detectable MIs and positive predictive value (PPV) as the ratio of correctly detected MIs over all detected MIs.

As mentioned above, unmapped short reads have not been well studied by most existing tools including Gustaf. However, it is very important for the analysis of

Table 1 Sensitivity (SN), Positive Predictive Value (PPV) and Standard Deviation (SD) for MIs simulated onto chr10

	Dataset 1				Dataset 2			
Coverage	MID		Gustaf		MID		Gustaf	
	SN	PPV	SN	PPV	SN	PPV	SN	PPV
2	0.692	0.900	0.423	0.688	0.741	0.976	0.333	0.783
4	0.732	0.911	0.339	0.760	0.840	0.932	0.506	0.872
6	0.782	0.871	0.359	0.718	0.782	0.935	0.427	0.810
8	0.870	0.918	0.403	0.689	0.784	0.916	0.302	0.824
10	0.771	0.857	0.239	0.703	0.877	0.932	0.351	0.800
20	0.767	0.904	0.233	0.667	0.849	0.914	0.267	0.705
30	0.859	0.886	0.188	0.514	0.821	0.912	0.254	0.627
40	0.781	0.860	0.250	0.588	0.820	0.907	0.230	0.603
50	0.781	0.856	0.211	0.654	0.837	0.906	0.266	0.616
60	0.835	0.857	0.264	0.563	0.863	0.900	0.231	0.509
SD	0.052	0.023	0.079	0.072	0.039	0.021	0.086	0.114

personalized NGS data to study unmapped short reads so that we can exploit more SV information especially for MIs. Therefore, preprocessing of unmapped reads is needed. After running BWA to get unmapped short reads, we aligned unmapped reads to chr10 the same as what we did in MID, and then recorded the anchored alignment results for Gustaf, including the re-aligned reads and the corresponding reference sequence on chr10. Afterwards the target regions from both re-aligned read and reference were used as the input for Gustaf single-end detecting. For MID, we can directly run the whole pipeline with unmapped short reads discarded by BWA.

Table 1 shows the comparison between MID and Gustaf. Note that the sub-datasets in 2X and 4X coverage are most similar to the real data we got from the 1KGP. The high SN and high PPV of MID in both Dataset 1 and Dataset 2 demonstrates high accuracy of MID in detecting MIs, even with different variants around. Gustaf demonstrates significantly lower SN and PPV in the simulated data. Furthermore, lower standard deviation (SD) of MID compared with Gustaf in simulated data shows stable performance of MID. Overall, MID outperforms Gustaf in identifying MIs.

Identifying MIs in 1000 genomes project data

The 1KGP provides large number of individual human genomes data with NGS reads [16]. At present, 1KGP data has been widely used for SV detection by existing tools [5, 8, 16]. However, analysis on MIs is still lacking. To detect MIs, MID was applied on population-scale sequencing data from the 1KGP based on publicly released unmapped BAM files [16]. By running MID with reference human genome assembly hg19, MID reports the detailed alignment of each read containing MIs and a list of unique MIs detected. For a typical sample with a total number of 13.5 M unmapped short reads (NA19917), it takes 12 min to run MID with 16 CPU threads.

Generally speaking, low coverage raises problems for MI detection since less support from reads may lead to unreliable results. However, many datasets from large-scale projects (e.g., the 1KGP) are in low coverage. Therefore, it is very useful to develop methods for identifying MIs reliably and efficiently based on multiple low-coverage samples. It has been suggested that integrated analysis using multiple samples can be helpful in improving the reliability of SV detection [11]. We assessed the performance of MID on a total of 770 Illumina samples from the 1KGP [16], which have been categorized by different populations, and then integrated the results at the population level for more reliable and informative results. We eventually focused on 638 samples (full sample list in sample list of Additional file 1), in which MID reported at least one MI.

We calculated the number of unique MIs and the number of reads supporting each MI first. In the following analysis, if not specified, the number of MIs is equal to the number of unique MIs, either in one particular sample or in one population. Moreover, if one MI is supported by multiple reads, we calculate the number of reads containing the same MI as its "occurrence". Altogether, MID reported 2,413 occurrences of 721 MIs in 638 samples (full list of MIs and annotations refer to Additional file 1: Table S1). Of the 721 detected MIs, 349 are intergenic, 342 MIs are intronic, and 30 MIs are exonic, including five MIs overlapping with CDS regions, seven MIs overlapping with UTR regions, two MIs overlapping with both CDS and UTR regions, and the rest overlapping with miRNA, Mt_rRNA etc. (more details can be found in Additional file 1: Table S1), annotated by GENCODE [21]. Using ENCODE annotations [22], we identified 13 MIs overlapping with proximal transcription factor binding sites (TFBS), and 48 MIs overlapping with distal TFBS following ChIP peaks of transcription factors. As many regulatory elements are in intronic regions [22, 23], plus the effect of MIs extends beyond the inverted regions [24], thus 342 MIs found overlapping with introns, as well as 19 MIs found locate within the 2,000 bp upstream genomic regions of genes can also be informative for further analysis.

Furthermore, the individual samples used by MID are categorized by populations, and populations are grouped by the predominant component of ancestry as follows: East Asia (CDX, CHB, CHS, JPT, and KHV), South Asia (GIH), Europe (CEU, FIN, GBR, IBS, and TSI), America (CLM, MXL, PEL, and PUR), and Africa (YRI, LWK, ASW, and ACB) [16]. Overall MID reported 206 MIs in Americas Group, 262 MIs in East Asia Group, 22 MIs in South Asia Group, 223 MIs in Europe Group, and 284 MIs in Africa Group. Table 2 presents the overview of MIs detected in 638 individual samples, illustrating an average of 1.48 individual samples supporting one MI, ~24 % of MIs supported by at least two samples in the same population, and approximately 24 % of MIs are supported by different populations in the same ancestry category. Herein the results of MIs supported by multiple reads, individual samples, and populations suggest that our method is probably informative and reliable with an integrated view of individual samples. In addition, Additional file 1: Figure S2 shows that the length of MIs detected in 1KGP data is from 15 bp to 43 bp (mean length is 24 bp). As the majority of MIs concentrate within the size range from 18 bp to 31 bp, we assume that the incidence of MI would be higher in this size, which might be helpful for understanding the mechanism of MI. Furthermore, Additional file 1: Figure S3 shows the distribution of number of MIs across

Table 2 Overview of MIs detected in 638 individual samples from 1KGP

population	sam-num	MI-num	MI-occ	mul-sup	read-num	occ/num	mul-sup/num	read/num
MXL	48	79	153	23	175	1.94	29.11 %	2.22
PUR	48	117	167	26	185	1.43	22.22 %	1.58
CLM	26	60	83	14	91	1.38	23.33 %	1.52
PEL	20	40	58	12	63	1.45	30.00 %	1.58
Total America	142	206	296	55	514	1.44	26.70 %	2.50
CDX	40	40	65	11	71	1.63	27.50 %	1.78
CHB	26	42	52	7	57	1.24	16.67 %	1.36
CHS	41	56	72	10	79	1.29	17.86 %	1.41
JPT	58	160	325	55	356	2.03	34.38 %	2.23
KHV	31	51	81	14	96	1.59	27.45 %	1.88
Total East Asia	196	262	349	53	659	1.33	20.23 %	2.52
GIH	11	22	29	6	34	1.32	27.27 %	1.55
Total South Asia	11	22	22	0	34	1.00	0.00 %	1.55
CEU	21	67	82	10	94	1.22	14.93 %	1.40
FIN	41	72	87	6	93	1.21	8.33 %	1.29
GBR	35	80	104	16	114	1.30	20.00 %	1.43
IBS	24	56	80	14	86	1.43	25.00 %	1.54
TSI	9	28	34	5	37	1.21	17.86 %	1.32
Total Europe	130	223	303	39	424	1.36	17.49 %	1.90
YRI	42	156	239	41	255	1.53	26.28 %	1.63
LWK	54	113	180	28	193	1.59	24.78 %	1.71
ACB	14	35	46	9	48	1.31	25.71 %	1.37
ASW	49	127	262	49	286	2.06	38.58 %	2.25
Total Africa	159	284	431	90	782	1.52	31.69 %	2.75

The "sam-num" column illustrates the number of samples for each category (either population or population group); the "MI-num" column illustrates the number of different MIs detected in each population or population group; the "MI-occ" column illustrates the sum of occurrences of MIs in each population or population group; the "read-num" column illustrates the number of reads supporting MIs. For the population lines, the "mul-sup" column illustrates the number of MIs supported by at least two individual samples (named "multiple samples supported MIs") in one population, the "ooc/num" column illustrates the ratio of MI occurrence over MI number, which indicates the average number of individual samples supporting one MI in the same population, and the "mul-sup/num" column illustrates the ratio of multiple samples supported MIs over the number of all MIs. For the population group lines (which started with "Total"), the "mul-sup" column illustrates the number of MIs supported by at least two populations (named "multiple populations supported MIs") in the same population group, the "ooc/num" column illustrates the ratio of MI occurrence over MI number, which indicates the average number of populations supporting one MI in the same population group, and the "mul-sup/num" column illustrates the ratio of multiple populations supported MIs over the total number of MIs. The last "read/num" column illustrates the ratio of the number of reads containing MIs over the number of MIs, which indicates the average number of reads supporting each MI

the human chromosomes, where the number of MIs generally has positive correlation with the length of chromosomes except chr11.

To focus on the exonic MIs detected, we checked 30 MIs overlapping with exons annotated by GENCODE. For instance, in Fig. 1a, an MI overlaps with the 3' UTR of gene *PREPL* and gene *SLC3A1*, which both have strong correlation with Hypotonia-Cystinuria Syndrome [25, 26]. This MI is supported by 46 individual samples across different populations, including two samples in East Asian group, 10 samples in European group, 12 samples in Americas group, and 22 samples in African group (Fig. 1b). The chimpanzee sequence in this location is almost identical to the sequence of Neanderthal (Vi33.25 Sequence Reads) and the MI we found, which suggests that the reference human genome is inverted in this region as compared to the most recent common ancestor of the human population. However, this MI was reported as multiple nucleotide variation (GenBank:rs71416108) due to the poor understanding of MI [27]. Thus our identification of MI can be helpful for understanding genomic variants. Another MI changes 6 amino acids in the CDS region of gene *OR5I1* (Fig. 1c). These amino acids are located on the fourth transmembrane domain facing the less conserved extracellular side comparing to the intracellular side, which might cause severe influence.

Application to CCLE lung squamous cell carcinoma data

In addition to the 1KGP data, cancer genome sequencing data are widely available. A full understanding of

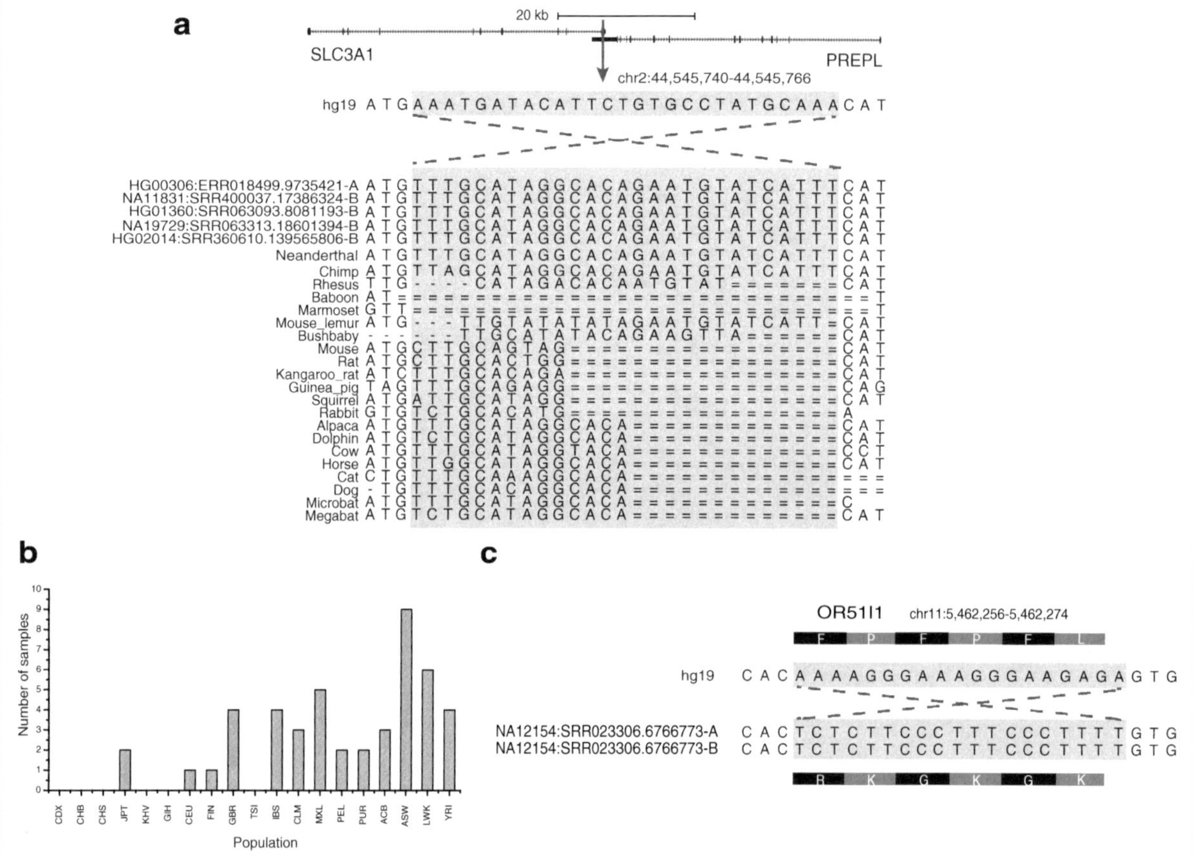

Fig. 1 MI examples in 1KGP data. **a** shows one MI supported by 46 samples which overlap with 3' UTR of gene *PREPL* and *SLC3A*; **b** shows the distribution of the MI supported by 46 samples in different populations; **c** shows one MI changing 6 amino acids of the exon in gene *OR51I1*

somatic alterations in cancer genomes is important to better understand the genetic basis for cancer development [28, 29]. Of the whole CCLE data, Lung Squamous Cell Carcinoma is a typical kind of lung cancer, which is a leading cause of death among all cancer patients [30]. In this work, we selected 14 WXS datasets of Lung Squamous Cell Carcinoma from CCLE to provide a proof-of-concept demonstration of the applicability of MID to cancer sequencing data (full sample list in Additional file 1).

We found 12 MIs overlapping with 15 genes (Additional file 1: Table S2). To be more specific, seven MIs overlap with CDS regions, one MI overlaps with one gene in CDS regions and another gene in UTR regions, and the rest locate in the CDS nearby regions, which are annotated by GENCODE. Moreover, three MIs overlap with the proximal transcription factor binding sites (TFBS) annotated by ENCODE following ChIP peaks of transcription factors (more details can be found in Additional file 1: Table S2). We also found one MI within the 2,000 bp upstream region of gene *VPS54*. In Fig. 2a, we show that an MI breaks the edge of CDS region and changes three amino acids of gene *PSRC1*, which encodes a proline-rich protein that is a target for regulation by the tumor suppressor protein p53. In addition, Fig. 2b presents an MI changing five amino acids of gene *JMJD4* and overlapping with 5'UTR of gene *SNAP47*. While *JMJD4* is a member of JmjC-domain-only family, which only contains the JmjC domain, and plays an important role in demethylation and involves in cancer diagnosis [31]. The variants and expression bias of genes in this family are related to cancer regulations, thus the change made by MI in the *JMJD4* gene may help enrich the study of *JMJD4* and related cancer diagnosis.

Discussion and conclusions

As presented above, upon test of the simulated data, MID has a steady performance of PPV for both low coverage and high coverage data, varying from 2X to 60X. Significantly, MID demonstrates high PPV in simulated dataset with low coverage 2–4X, which is the same as the coverage range used by low coverage samples from the 1KGP. In fact, the reason of our method

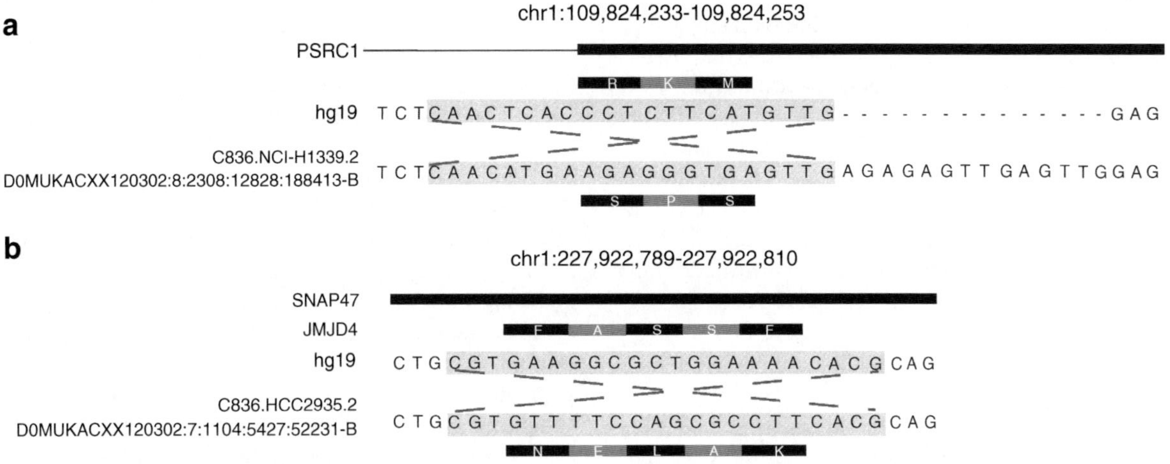

Fig. 2 MI examples in Lung Squamous Cell Carcinoma WXS data from CCLE. **a** shows one MI breaking the edge of CDS region and changing 3 amino acids of gene *PSRC1*, as well as an insertion next to the MI; **b** shows one MI changing 5 amino acids of gene *JMJD4* and overlapping with 5'UTR of gene *SNAP47*

showing stable performance on low coverage data is that the identification process is designed based on the target regions of read and reference sequence, regardless of read depth. Thus coverage bias has little influence in our method. By contrast, most existing tools require enough coverage to detect breakpoints, and only optimize good performance with high coverage data [4–10]. However, during our anchoring approach, in which the Bowtie program was called to make anchored alignment as preprocessing, higher coverage data can provide more detectable reads before identification process. Therefore we will have more reads containing possible MIs to be detected during identification, which means our strategy will benefit from higher coverage data. Therefore, although MID performs well in low coverage data, it can also benefit from data with higher coverage.

In fact, the pipeline of MID contains preprocessing and sequence mapping process. During the preprocessing of MID, we did anchored alignment by calling Bowtie. The reason we chose Bowtie [32] instead of the latest Bowtie2 [33] is that the length of paired-end reads (anchors) is shorter than 50 bp in ~100 bp NGS short reads, and Bowtie outperforms Bowtie2 in this size range, which is proved by our test and also claimed clearly in the tool clarification [33]. For sequence mapping, MID can handle small size MIs and MIs with multiple breakpoints, owing to the flexible segment mapping and scoring system during path finding. In existing tools such as BWA tool, although a number of discontinuous gaps and SNVs might be detected, it is extremely difficult to identify more complex SVs including inversions and duplications. In addition, more complicated scenario, i.e., MIs with multiple breakpoints within the read, would hardly be taken into consideration by these tools. MID uses flexible segment mapping on both strands based on *k*-mers, which is more suitable for MI detection, especially for dealing with small MIs and MIs with multiple breakpoints around. The scoring process also helps confirm the final path and distinguish incorrect matches, including palindromic sequences, and MIs as well.

In summary, we have developed a novel computational tool MID to identify MIs by mapping initially unmapped short reads back onto human genome sequence. What makes MID different from other SV detecting tools is that our approach is sensitive to very small size MIs, and capable of detecting MIs with multiple breakpoints in one read due to flexible segment mapping process, as well as scoring system in distinguishing MIs from palindromic sequence and other incorrect matches. The pipeline of MID can start from unmapped BAM files and find the optimum solution automatically rather than parameter changing. To our knowledge, MID is the first method that can efficiently and reliably identify MIs from unmapped short NGS reads. Moreover, MID is reliable on low coverage data, which is suitable for large-scale projects such as the 1KGP. We realize that the mechanism and the function of MI, as a kind of SV (as both germline and somatic alterations), are still poorly understood. Nevertheless, we expect that our tool would be useful to better understand MIs and their roles in genetic diversity and diseases. In conclusion, MID is suitable for large-scale short reads produced by present high-throughput sequencing technologies (e.g., Illumina), especially for low coverage data, and MID might have positive impact on identifying key genetic variants in human diseases with the further development of sequencing technologies.

Methods

With an input of a BAM file containing unmapped short reads, MID takes three steps to report the output of a list of MIs, as well as detailed alignment information for each MI. The three main steps for identifying MIs are: (1) Create anchored alignment, which determines the corresponding region on the reference genome for the reads that might harbor MI. (2) Perform detailed alignment between the read and the genomic region identified in Step (1). (3) Report a list of MIs with additional information on how the MI-containing reads can be aligned. The workflow of MID is shown in Fig. 3.

Anchoring and *k*-mer mapping

Anchoring

We provide an option of cutting size (0 bp as default) for both head and tail regions, which might be more error-prone, of each unmapped short read, and then select the new head and tail regions after cutting as "anchors". As for short reads of ~100 bp, we set no cutting of two ends and select the head and tail regions of each unmapped short read as "anchors" (18 bp as default), afterwards we map the two anchors onto the reference genome as two ends of one paired-end read with a length of 18 bp using Bowtie [32] to find potential MI regions in the read (Fig. 4a). Moreover, for longer short reads (e.g., ~200 bp), we recommend choosing cutting size according to sequencing quality before selecting anchors. One mismatch on the anchors generated from a short read is allowed during mapping. Take the short read data of individual sample NA19213 as an example. The length of short reads in NA19213 is 76 bp, so the original distance between two anchors in the same read is 40 bp. Owing to the possible combination of variants including unbalanced variants such as small indels, insertions, deletions and duplications, we set the range of distance between two anchors to be in a range from 15 bp to 65 bp, which is denoted as the insert size for the paired-end read in Bowtie. If the distance is shorter than 15 bp caused by a possible deletion, then there is no need to consider the read since our algorithm aims to detect MIs longer than 15 bp. If the distance is longer than 65 bp caused by a possible insertion, we suppose this match could be redundant since we focus on MI detection and the inversion length in this read can be 40 bp at most. The length of anchors can be adjusted by users. However, we suggest a minimum of 10 bp set; otherwise the pair of anchors might not have unique alignment result in this step. This process significantly reduces the potential search space for MID.

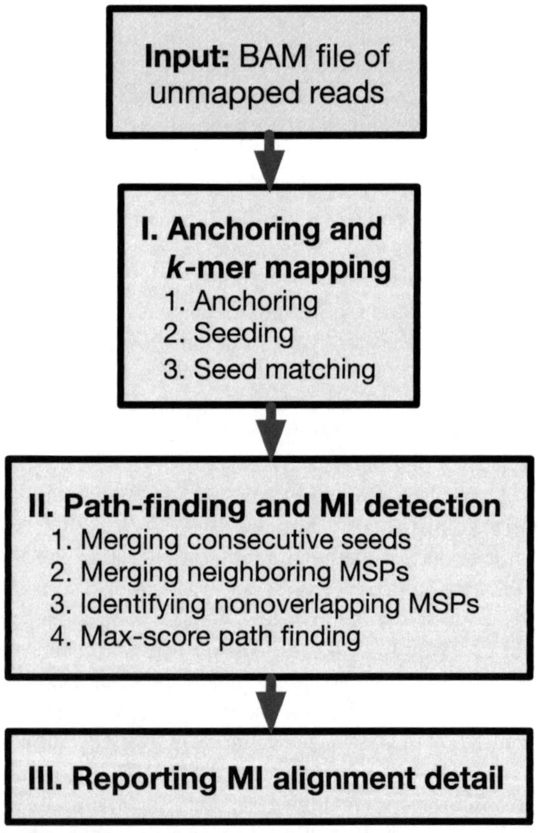

Fig. 3 The workflow of the MID program

Seeding

A *k*-mer extracted from the read is called a seed (14 bp as default). In addition to the head and tail anchors of a read, the middle part of the read is called "target region" (Fig. 4a). We then select target region of the read and extract consecutive seeds (step size 1 bp as default). We also extract consecutive seeds (step size 1 bp as default) from the reverse complement of the target region. These two groups of seeds are stored and operated separately for following steps.

Seed matching

The matching process is performed on the target regions both of the read and the reference sequence. After the anchoring process, we get the exact location of unmapped short read onto the reference sequence. We then select the target region of the initial read and its corresponding target region on the reference sequence. This seed-matching process can tolerate MIs with indels around the inversion breakpoints. We allow a maximum of i (2 as default) mismatches in a seed. In other words, if more than i mismatches occur within an interval of k, this k-mer (seed) will not be aligned. During seed matching, we compare each seed in the read target region to the reference target region and store all possible locations for each seed on the reference. To reduce the effect of a particular type of palindromic sequence tandem

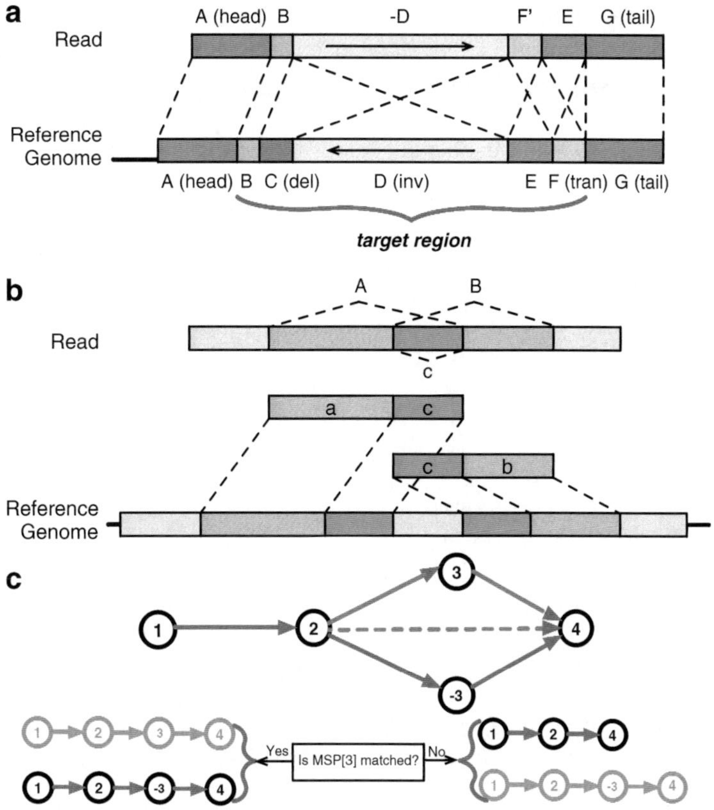

Fig. 4 Methods overview of MID. **a** shows the construction of alignment regions. Head (*A*) and tail (*G*) substrings of read serve as a pair of anchors. Segment *C* in reference sequence denotes a deletion in read, segment *D* denotes an inversion in read and segment *F'* denotes a translocation. A set of segment (*B, C, D, E, F*) in the reference sequence, as well as a set of segment (*B, −D, F', E*) in the read sequence, form the target regions. **b** shows how to use a partition and recombine strategy to transfer the pair of overlapping MSs (*A, B*) to a set of non-overlapping MSs {(*A, b*), (*a, B*), (*a, b*)}. Segment *A* and *B* have overlapping region *c*, and segment *a* and *b* are generated by cutting region *c* off the original segment *A* and *B*. **c** shows the max-score path approach. The path starting with $MSP_{[1]}$ and ending with $MSP_{[4]}$ denotes a short read to be detected, and the path in red is the max-score path. $MSP_{[1]}$, $MSP_{[2]}$, $MSP_{[3]}$, $MSP_{[4]}$ are on the forward strand, and $MSP_{[-3]}$ is on the reverse strand. MID starts with $MSP_{[1]}$ and extends the path to $MSP_{[2]}$, then if $MSP_{[3]}$ and $MSP_{[-3]}$ can both be matched, MID records both path candidates and ends with $MSP_{[4]}$, therefore we have two path candidates {1, 2, 3, 4} and {1, 2, −3, 4}, then {1, 2, 3, 4} instead of {1, 2, −3, 4} will be chosen after scoring due to the reverse penalty for $MSP_{[-3]}$ on the reverse strand. Otherwise if only $MSP_{[-3]}$ is successfully matched, we have path candidates {1, 2, 4} and {1, 2, −3, 4}, after scoring {1, 2, −3, 4} would be chosen owing to the gap penalty, since path candidate {1, 2, 4} contains much more gaps. Therefore $MSP_{[-3]}$ would be reported as detected MI

repeats (such as "AT"s or "GC"s) when matching the seed in the reverse complement to detect inversions, we discard the seeds with greater and equal to n ($n = L/4$, where L is the length of the substring sequence) "AT"s or "GC"s.

Path finding and MI identification

This step aims to obtain the best path of *k*-mer alignment. MI within a read will correspond to a reversed sub-path.

Merging consecutive seeds

One seed can be matched to multiple locations on the reference sequence due to its short length. If the number of collinear consecutive seeds on the reference is larger than a threshold (5 as default), we consider this set of seeds as a "matching segment" (MS) and the "matching segment pair" (MSP) between the read and the reference. We might get different constructs of MSPs containing the same seeds, or one MS in the read that can be matched to different locations on the reference sequence, resulting in multiple MSPs for the same MS. The information of every possible MSP is recorded and will be further evaluated in the following steps.

Merging neighboring MSPs

While we set the threshold of i (2 as default) mismatches in one seed, if the SNVs occur at the edges of MSPs and a *k*-mer covering the SNVs reach the threshold of i mismatches, then this *k*-mer fails to be aligned. To overcome the problem caused by mismatches that is possibly caused by SNVs, we merge MSPs if the neighboring MSPs are in the same orientation and within the same given distance (3 bp as default) both on the read and on the reference sequence.

Identifying non-overlapping MSPs

MSPs may overlap with each other. Since we want to find paths constructed from non-overlapping MSPs, we take a partition-and-recombine strategy to make sure all the information carried by the MSPs is well kept (Fig. 4b). Before this step, MSPs in different orientations are stored and operated separately. We then put two separate sets of MSPs together and sort them by their locations on the reference sequence. Next, we find out each pair of overlapping MSPs and partition the overlapping subsequence to get a new set of non-overlapping MSPs pairs. We then recombine the overlapping subsequence with either MSP or neither of them. In Fig. 4b, for a pair of overlapping MSs (*A*, *B*) on the read, which has overlapping subsequence *c*, we partition *A* into *a* and *c*, and partition *B* into *b* and *c*, then we recombine the MSs as {(*A*, *b*), (*a*, *B*), (*a*, *b*)}. In this way, we change the former overlapping MS (*A*, *B*) into a set of non-overlapping MSs {(*A*, *b*), (*a*, *B*), (*a*, *b*)}. For overlapping MSPs, we use the partition-and-recombination strategy to check overlapping MSs on both read and reference sequence to get non-overlapping MSP set. This procedure guarantees the transformation of MSP is lossless, because we keep all the possible combinations in the non-overlapping MSP set. The resulting non-overlapping MSPs will be used in the subsequent path finding step.

Maximum-score path finding

We generate all combinations of non-overlapping MSPs as path candidates (Fig. 4c) and calculate the alignment score of each candidate to find the path with the maximum score by dynamic programming (see definitions of variables below). In dynamic programming, we define two kinds of score. One is for the MSP itself, which includes the score for matches and mismatches. The other is for the gap penalty between two neighboring MSPs in the path. We record the maximum score $T_{[i]}$ of every path ending with $MSP_{[i]}$ based on the maximum score of any sum of $T_{[j]}$ and $GS_{[j,i]}$ $(1 < j < i)$. The complexity of our algorithm is $O(n^2)$, where n is the number of MSPs. In practice, ~85 % of n is smaller than 10. We consider match, mismatch, and gap in the similarity score [34]. Furthermore, for segments matched to the reverse strand, we use a penalty $R_{[i]}$ for the potential inversion, which aims to reduce false positives caused by palindromic sequence and other incorrect matches. In fact, $MSP_{[i]}$ of palindromic sequence can be matched to both strands and the one on the reverse strand would be added the penalty $R_{[i]}$, which results in lower $F_{[i]}$ compared with the same segment matched on the forward strand. Thus the final path finding based on maximum alignment score would select the match on the forward strand and avoid the influence of palindromic sequence.

The variables we used include:

n: The number of MSPs.
$M_{[i]}$: The number of matches in $MSP_{[i]}$.
$MS_{[i]}$: The number of mismatches in $MSP_{[i]}$.
$G_{[i,j]}$: The number of gaps between $MSP_{[i]}$ and $MSP_{[j]}$.
$R_{[i]}$: The empirical reverse penalty for MSP based on our simulation, where one segment is the reverse complement of the other.
$F_{[i]}$: The score for $MSP_{[i]}$.
$G_{[i,j]}$: The score for the gap penalty of $MSP_{[i]}$ and $MSP_{[j]}$. $GS_{[0,i]}$ refers to the penalty of gaps from the first base of the sequence to the start of $MSP_{[i]}$, and $GS_{[i,0]}$ refers to the penalty of gaps from the end of $MSP_{[i]}$ to the last base of the sequence.
$GS_{[j,i]}$: The score for the gap penalty of $MSP_{[j]}$ and $MSP_{[i]}$.
$T_{[i]}$: The maximum score of every path ending with $MSP_{[i]}$.

We define:

$$R_{[i]} = 3\left(M_{[i]} + MS_{[i]}\right)^2$$

$$F_{[i]} = \begin{cases} 100M_{[i]} - 100MS_{[i]} & (MSP_{[i]} \textit{ on the forward strand}) \\ 100M_{[i]} - 100MS_{[i]} - R_{[i]} & (MSP_{[i]} \textit{ on the reverse strand}) \end{cases}$$

$$GS_{[i,j]} = \begin{cases} 0 & (\text{no gap between } MSP_{[i]} \text{ and } MSP_{[j]}) \\ -400 - 30G_{[i,j]} & (\text{else}) \end{cases}$$

$$T_{[i]} = \begin{cases} GS_{[0,1]} + F_{[1]} & (i = 1) \\ \max\left(GS_{[0,i]}, T_{[1]} + GS_{[1,i]}, \ldots, T_{[i-1]} + GS_{[i-1,i]} + F_{[i]}\right) & (1 < i \le n) \end{cases}$$

Finally we select the maximum alignment score:

$$\max\left(T_{[i]} + GS_{[i,0]}\right), (1 \le i \le n)$$

As in Fig. 4c, the path starting with $MSP_{[1]}$ and ending with $MSP_{[4]}$ denotes a short read to be detected, and the path in red is the max-score path. $MSP_{[1]}$, $MSP_{[2]}$, $MSP_{[3]}$, $MSP_{[4]}$ are on the forward strand, and $MSP_{[-3]}$ is on the reverse strand. MID starts with $MSP_{[1]}$ and extends the path to $MSP_{[2]}$, then if $MSP_{[3]}$ and $MSP_{[-3]}$ can both be matched, MID records both path candidates and ends with $MSP_{[4]}$. Therefore we have two path candidates as follows: {1, 2, 3, 4} and {1, 2, −3, 4}. After scoring, path candidate {1, 2, 3, 4} will be chosen due to the reverse penalty for $MSP_{[-3]}$ on the reverse strand, which helps distinguish palindromic sequences. Otherwise if only $MSP_{[-3]}$ is successfully matched, we have path candidates {1, 2, 4} and {1, 2, −3, 4}, after scoring {1, 2, −3, 4} would be chosen owing to the gap penalty, since path candidate {1, 2, 4} contains much more gaps. Therefore $MSP_{[-3]}$ would be reported as detected MI.

In summary, MID is generated based on flexible segment mapping, including non-overlapping MSPs identification, which is capable of identifying very small size MIs, regardless of multiple breakpoints in one read, and scoring system which can distinguish false positives caused by palindromic sequence and other incorrect matches. In addition, although the parameter values are provided optionally, our test shows that the default option appears to generate the best overall performance for our algorithm.

Additional file

Additional file 1: Supplementary figures, sample lists used by MID and tables containing detailed information of MIs detected by MID as well as their annotations. (PDF 1727 kb)

Abbreviations

1KGP: 1000 Genomes project; CCLE: Cancer cell line encyclopedia; CDS: Coding sequence; MI: Micro-inversion; MID: Micro-inversion detector; MS: Matching segment; MSP: Matching segment pair; NGS: Next-generation sequencing; PPV: Positive predictive value; SD: Standard deviation; SN: Sensitivity; SNV: Single nucleotide variation; SV: Structural variation; TFBS: Transcription factor binding site; UTR: Untranslated region.

Competing interests

The authors declare that they have no competing interests.

Authors' contributions

JM, FFH and HQZ conceived the study. FFH developed the code and performed data analysis with help from YL and YHT. FFH, HQZ, and JM wrote the manuscript. All authors read and approved the final manuscript.

Acknowledgments

The publication costs for this article were funded by the National Key Technology Research and Design Program of China [2012BAI06B02 to HQZ]. This work was supported in part by the National Key Technology Research and Design Program of China [2012BAI06B02 to HQZ] and the National Nature Science Fund of China [91231119 and 61131003 to HQZ]; the National Institutes of Health [HG007352 and CA182360 to JM] and the National Science Foundation [1054309 and 1262575 to JM]; and the Foundation of the China Scholarship Council (CSC) to FFH.

Author details

[1]State Key Laboratory for Turbulence and Complex Systems and Department of Biomedical Engineering, and Center for Quantitative Biology, Peking University, Beijing 100871, China. [2]Department of Bioengineering, University of Illinois, Urbana, IL 61801, USA. [3]Carl R. Woese Institute for Genomic Biology, University of Illinois, Urbana, IL 61801, USA. [4]Division of Applied Mathematics, Brown University, Providence, RI 02912, USA.

Published: 11 January 2016

References

1. Baker M. Structural variation: the genome's hidden architecture. Nat Methods. 2012;9(2):133–7.
2. Botstein D, Risch N. Discovering genotypes underlying human phenotypes: past successes for mendelian disease, future approaches for complex disease. Nat Genet. 2003;33:228–37.
3. Braun EL, Kimball RT, Han K, Iuhaszvelez NR, Bonilla AJ, Chojnowski JL, et al. Homoplastic microinversions and the avian tree of life. BMC Evol Biol. 2011;11:141.
4. Medvedev P, Stanciu M, Brudno M. Computational methods for discovering structural variation with next-generation sequencing. Nat Methods. 2009;6(11 Suppl):S13–20.
5. Chen K, Wallis JW, McLellan MD, Larson DE, Kalicki JM, Pohi CS, et al. BreakDancer: an algorithm for high-resolution mapping of genomic structural variation. Nat Methods. 2009;6(9):677–81.
6. Ye K, Schulz MH, Long Q, Apweiler R, Ning Z. Pindel: a pattern growth approach to detect break points of large deletions and medium sized insertions from paired-end short reads. Bioinformatics. 2009;25(21):2865–71.
7. Jiang Y, Wang Y, Brudno M. PRISM: pair-read informed splitread mapping for base-pair level detection of insertion, deletion and structural variants. Bioinformatics. 2012;28(20):2576–83.
8. Rausch T, Zichner T, Schlattl A, Stütz AM, Benes V, Korbel JO. DELLY: structural variant discovery by integrated paired-end and split-read analysis. Bioinformatics. 2012;28(18):i333–9.
9. Abyzov A, Urban AE, Snyder M, Gerstein M. CNVnator: An approach to discover, genotype, and characterize typical and atypical CNVs from family and population genome sequencing. Genome Res. 2011;21:974–84.
10. Trappe K, Emde AK, Ehrlich HC, Reinert K. Gustaf: detecting and correctly classifying SVs in the NGS twilight zone. Bioinformatics. 2014;30(24):3484–90.
11. Lin K, Smit S, Bonnema G, Sanchez-Perez G, de Ridder D. Making the difference: integrating structural variation detection tools. Brief Bioinform. 2015;16(5):852–64.
12. Chaisson MJ, Raphael BJ, Pevzner PA. Microinversions in mammalian evolution. PNAS. 2006;103(52):19824–9.
13. Ma J, Zhang L, Suh BB, Raney BJ, Burhans RC, Kent WJ, et al. Reconstructing contiguous regions of an ancestral genome. Genome Res. 2006;16(12):1557–65.
14. Feuk L. Inversion variants in the human genome: role in disease and genome architecture. Genome Medicine. 2010;2:11.
15. Abel HJ, Duncavage EJ. Detection of structural DNA variation from next generation sequencing data: a review of informatics approaches. Cancer Genet. 2013;206:432–40.
16. The 1000 Genomes Project Consortium, Abecasis GR, Auton A, Brooks LD, DePristo MA, Durbin RM. An integrated map of genetic variation from 1,092 human genomes. Nature. 2012;491(7422):56–65.
17. Wang J, Mullighan CG, Easton J, Roberts S, Heatley SL, Ma J, et al. CREST maps somatic structural variation in cancer genomes with base-pair resolution. Nat Methods. 2011;8(8):652–4.
18. Suzuki S, Yasuda T, Shiraishi Y, Miyano S, Nagasaki M. ClipCrop: a tool for detecting structural variations with single-base resolution using soft-clipping information. BMC Bioinformatics. 2001;12 Suppl 14:S7.
19. Li H, Ruan J, Durbin R. Mapping short DNA sequencing reads and calling variants using mapping quality scores. Genome Res. 2008;18(11):1851–8.
20. Li H, Durbin R. Fast and accurate short read alignment with Burrows-Wheeler transform. Bioinformatics. 2009;25(14):1754–60.
21. Harrow J, Frankish A, Gonzalez JM, Tapanari E, Diekhans M, Kokocinski F, et al. GENCODE: The reference human genome annotation for The ENCODE Project. Genome Res. 2012;22(9):1760–74.
22. ENCODE Project Consortium. An integrated encyclopedia of DNA elements in the human genome. Nature. 2012;489:57–74.
23. Siggens L, Ekwall K. Epigenetics, chromatin and genome organization: recent advances from the ENCODE project. J Intern Med. 2014;276(3):201–14.
24. Puig M, Casillas S, Villatoro S, Caceres M. Human inversions and their functional consequences. Brief Funct Genomics. 2015;14(5):369–79.
25. Jaeken J, Martens K, Francois I, Eyskens F, Lecointre C, Derua R, et al. Deletion of PREPL, a Gene Encoding a Putative Serine Oligopeptidase, in Patients with Hypotonia-Cystinuria Syndrome. Am J Hum Genet. 2006;78(1):38–51.
26. Chabrol B, Martens K, Meulemans S, Cano A, Jaeken J, Creemers GMWM. Deletion of c2orf34, prepl and slc3a1 causes atypical hypotonia–cystinuria syndrome. J Med Genet. 2008;45(5):314–8.
27. Lew S, Sutton G, Ng PC, Feuk L, Halpern AL, Walenz BP, et al. The diploid genome sequence of an individual human. PLoS Biol. 2007;5(10), e254.
28. Mardis ER, Wilson RK. Cancer genome sequencing: a review. Hum Mol Genet. 2009;18(R2):R163–8.

29. Barretina J, Caponigro G, Stransky N, Venkatesan K, Margolin AA, Kim S, et al. The CancerCell Line Encyclopedia enables predictive modelling of anticancer drug sensitivity. Nature. 2012;483:603–7.
30. Shen C, Hui-Zhao, Wang D, Jiang G, Wang J, Zhang G. Molecular cloning, identification and analysis of lung squamous cell carcinoma-related genes. Lung Cancer. 2012;38(3):235–41.
31. Klose RJ, Kallin EM, Zhang Y. JmjC-domain-containing proteins and histone demethylation. Nat Rev Genet. 2006;7(9):715–27.
32. Langmead B, Trapnell C, Pop M, Salzberg SL. Ultrafast and memory-efficient alignment of short DNA sequences to the human genome. Genome Biol. 2009;10:R25.
33. Langmead B, Salzberg SL. Fast gapped-read alignment with Bowtie 2. Nat Methods. 2012;9:357–9.
34. Harris RS. Improved pairwise alignment of genomic DNA. Ph.D. Thesis. The Pennsylvania State University, College of Engineering; 2007.

Terai et al. BMC Genomics 2016, 17(Suppl 1):12
DOI 10.1186/s12864-015-2307-5

BMC Genomics

PROCEEDINGS Open Access

Comprehensive prediction of lncRNA–RNA interactions in human transcriptome

Goro Terai[1†], Junichi Iwakiri[2†], Tomoshi Kameda[3], Michiaki Hamada[3,4*] and Kiyoshi Asai[2,3*]

From The Fourteenth Asia Pacific Bioinformatics Conference (APBC 2016)
San Francisco, CA, USA. 11 - 13 January 2016

Abstract

Motivation: Recent studies have revealed that large numbers of non-coding RNAs are transcribed in humans, but only a few of them have been identified with their functions. Identification of the interaction target RNAs of the non-coding RNAs is an important step in predicting their functions. The current experimental methods to identify RNA–RNA interactions, however, are not fast enough to apply to a whole human transcriptome. Therefore, computational predictions of RNA–RNA interactions are desirable, but this is a challenging task due to the huge computational costs involved.

Results: Here, we report comprehensive predictions of the interaction targets of lncRNAs in a whole human transcriptome for the first time. To achieve this, we developed an integrated pipeline for predicting RNA–RNA interactions on the K computer, which is one of the fastest super-computers in the world. Comparisons with experimentally-validated lncRNA–RNA interactions support the quality of the predictions. Additionally, we have developed a database that catalogs the predicted lncRNA–RNA interactions to provide fundamental information about the targets of lncRNAs.

Keywords: RNA–RNA interactions, Interaction energy, TINCR, 1/2-sbsRNA

Introduction

Non-coding RNAs (ncRNAs), which are not translated into proteins but play essential roles in various biological processes, have been receiving increased attention [1, 2]. Among them, long non-coding RNAs (lncRNAs) have turned out to be involved in development, differentiation, epigenetic regulation and the immune system [1–3], as well as to be related to disease [4]. There are more than 20,000 lncRNAs listed in Gencode [5], but the functions of only a few lncRNAs, such as those of Xist [6] and NEAT1 [7], have been experimentally verified.

Knowing which RNAs and/or proteins are the interaction targets is essential for determining the functions of lncRNAs. The ncRNAs whose functional mechanisms have been identified are known to interact with other RNAs and/or proteins. Therefore, identifying the target RNAs or proteins with which an lncRNA interacts is the first step in characterizing the function of an lncRNA.

Several lncRNA–RNA interactions have been verified by experiment. Gong and Maquat [8] investigated the lncRNA 1/2-sbsRNA, which interacts with the 3′UTR of two mRNAs, leading to Staufen 1 (STAU1)-mediated messenger RNA decay (SMD); [9] determined the interactome of the 3.7 kilo-base lncRNA, a terminal differentiation-induced ncRNA (TINCR), by using an experimental method called RIA-seq (RNA interactome analysis with new generation sequencers). Their results suggest that TINCR interacts with many mRNAs through a sequence motif. Abdelmohsen et al. [10] suggested that the lncRNA

*Correspondence: mhamada@waseda.jp; asai@k.u-tokyo.ac.jp
†Equal contributors
[3]Biotechnology Research Institute for Drug Discovery, National Institute of Advanced Industrial Science and Technology (AIST), 2-41-6, Aomi, Koto-ku, 135-0064 Tokyo, Japan
[2]Graduate School of Frontier Sciences, University of Tokyo, 5-1-5 Kashiwanoha, Kashiwa, 277-8562 Chiba, Japan
Full list of author information is available at the end of the article

7SL interacts with TP53 mRNA, which encodes the tumor suppressor p53. In all of the above cases, the interaction partners of the lncRNAs were mRNAs. However, it is natural to investigate the possibility of lncRNA-lncRNA interactions in light of external base-pairs. In this study, we therefore focus on both lncRNA–mRNA and lncRNA–lncRNA interactions.

The experimental methods proposed to investigate RNA–RNA interactions (e.g., [9, 11]) require a specific target RNA. A similar situation occurs in CLIP-seq for RNA-protein interactions, where the target protein should be specified [12]. Hence, it is quite laborious to comprehensively determine the interactome for a large scale transcriptome across all possible pairs of RNAs under a variety of conditions (e.g., tissue, cell-type and time). These limitations emphasize the need for computational prediction of RNA–RNA interactions.

For *in-silico* predictions of non-coding RNAs, a widely accepted software tool, RNAZ, has been applied to the human genome and had an impact on non-coding RNA research [13]. However, *in-silico* predictions of RNA–RNA interactions are limited to small datasets, such as bacterial *small* RNAs (sRNAs) [14–17] because the high computational cost of predicting RNA–RNA interactions prevents us from making comprehensive predictions due to the huge combinatorial number of candidate RNA pairs in the whole human transcriptome.

For predicting interactions between two RNA molecules, an understanding of the base-pair interactions between the two RNAs is essential [15]. RNA–RNA interactions, however, are not simple processes consisting of forming base-pairs between the two RNA molecules, even if only secondary structures are considered. The two RNA molecules form a *joint* secondary structure, which involves both intra-molecular and inter-molecular base-pairs. Computational prediction of the joint secondary structures of two RNAs of length L has time complexity $O(L^4)$ to $O(L^6)$, depending on the complexity of the structures considered. It becomes infeasible to perform this calculation for all $O(N^2)$ pairs of RNAs when N (the number of RNA sequences) becomes large.

In this study, we report comprehensive predictions of lncRNA–RNA interactions in the human transcriptome (including lncRNAs and mRNAs) for the first time. To achieve this, we have developed a fast pipeline for predicting RNA–RNA interactions for a large number of RNA sequences, and have implemented the pipeline on the K computer (http://www.aics.riken.jp/en/k-computer/about/), which is one of the fastest supercomputers in the world.

In order to evaluate the proposed method, experimentally validated human lncRNA–RNA interactions were compared with the predictions of our pipeline. To avoid overfitting the pipeline to human transcriptome, the three adjustable parameters in our pipeline were determined by using the *E. coli* dataset as a training set. The results support the overall better performance of the pipeline compared with existing approaches.

As a further contribution for researchers studying ncRNAs, we have developed a database that contains all predicted lncRNA–RNA interactions; this is available for public use.

Materials and methods

For comprehensive prediction of RNA–RNA interactions, it is necessary to reduce the heavy computational costs, both for the computation for each pair of RNAs ($O(L^4)$ to $O(L^6)$), and for the number of times this computation must be performed ($O(N^2)$). In order to reduce the computational cost for each pair, it is practical to separate the calculations into those for intra-molecular base-pairs and those for inter-molecular base-pairs. In the INTARNA [18] program, which is adopted in our pipeline, *accessibility* based on secondary structural energy of each RNA sequence and the hybridization energy between the two RNA sequences are evaluated as an approximation. Another screening that can reduce the total complexity of the computation is to find the mutually reverse complementary *local* pairs of subsequences from among all the pairs of RNA sequences, followed by computing RNA–RNA interactions between those local pairs. This rough screening on huge number of combinations can itself be computationally expensive, but recent progress in computational methods of sequence analysis enables us to avoid the $O(N^2)$ combinatorial explosion of the calculations. To ensure that the above screening works effectively, the pipeline first screens each RNA sequence by its accessibility because a subsequence that forms external base-pairs should not form internal base-pairs.

Summary of the pipeline

A summary of the pipeline is shown in Fig. 1. Given two sets of RNA sequences, where one set is called the *query* RNAs (denoted by Q) and the other is called the *target* RNAs (denoted by T), our pipeline predicts RNA–RNA interactions between Q and T (the number of possible candidates of RNA–RNA interactions is $|Q| \times |T|$). In Step 1, the accessible regions of each RNA sequence are extracted using the RACCESS program [19], and tandem repeats are removed using the TANTAN program [20] in Step 2. For the subsequences screened in Steps 1 and 2, the reverse complementary 'seed matches' are detected using LAST [21, 22] in Step 3. In Step 4, the binding energies of pairs of sequences (target and query) around the seed matches are evaluated using INTARNA [18], and candidate interacting pairs are ranked by their binding energy in Step 5. Finally, in Step 6 the joint secondary structures of the interaction site with the minimum interaction energy

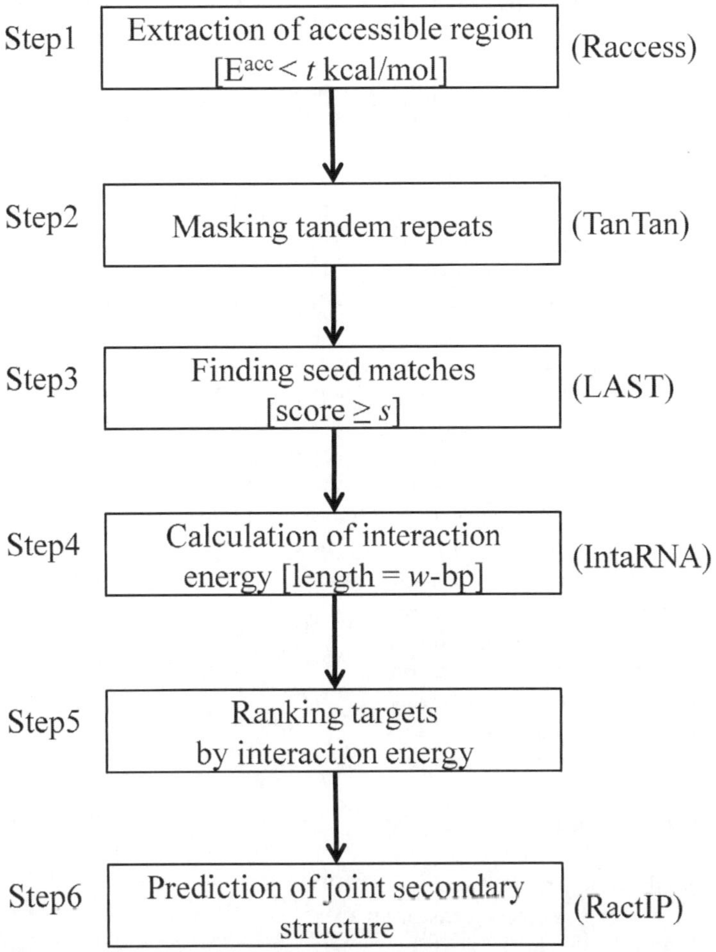

Fig. 1 Summary of our pipeline. The programs used for each step are shown in parentheses. Three key parameters (*t*, *s* and *w*) are shown in brackets; these are optimized using verified interaction data for bacteria

in each pair of RNA sequences is predicted using RACTIP [23].

Screening each RNA sequence

First, each RNA sequence is screened to find its inaccessible regions and its tandem repeats, which allows us to extract candidate subsequences to form inter-molecule interactions. This screening requires an $O(N)$ computation time, where N is the number of RNA sequences.

In Step 1, the accessible regions of each RNA sequences are extracted using the RACCESS program [19]; an *accessible* region r_a is a region of length greater than s that satisfies $\Delta E^{\mathrm{acc}}(r_a) < t$, where s and t are parameters to be specified. Accessibility is defined by

$$\Delta E^{\mathrm{acc}}(r_a) = -RT \log(P(r_a)), \tag{1}$$

where R and T are the gas constant and the temperature, respectively, and

$$P(r_a) = \sum_{y \in \mathcal{S}(x)} \exp\left(-E(y,x)/RT\right)/Z(x). \tag{2}$$

In Eq. 2, $E(y,x)$ denotes the free energy of secondary structure y of an RNA sequence x, $\mathcal{S}(x)$ is the set of possible secondary structures of x and $Z(x)$ is the partition function.

For each position in each RNA sequence (in both Q and T), we compute the accessible energy $E^{acc}(r_a)$, where r_a are the s consecutive nucleotides starting from the position, using the RACCESS program [19]. (Specifically, RACCESS [19] is executed with the `-access_len=s` option). If $E^{acc}(r_a) < t$ holds, the positions are considered to be accessible. The parameter t in our pipeline is optimized as described in a later section.

In Step 2, tandem repeat regions are masked using the TANTAN program [20] (with the default parameters). This step avoids the explosion of the number of candidates when finding the reverse complementary seed matches in Step 3.

Seed match

In our pipeline, a seed match means a pair of reverse complementary short subsequences in two RNA sequences (query and target). Specifically, a maximum non-gapped alignment is computed in which the score of `G-C` and `A-U` pairs (a.k.a. Watson-Crick base-pairs) is +1 and the score of the other pairs is −1. We consider a seed match whose integer score is more than s as a candidate RNA–RNA interaction.

In Step 3, the seed matches are detected using LAST version 250 (http://last.cbrc.jp/) [21, 22], with the options `-s0 -j1 -es -m1000000 -ls`, in which `-s0` means LAST detects only matches with reverse complementary sequences, `-j1` indicates gapless local alignment is conducted, and `-es` means that LAST reports those alignments whose score is greater than or equal to s. The option `-m1000000` is the maximum multiplicity for initial seed match, for which we use a large value because we want to obtain as many seed matches as possible to maintain sensitivity. The option `-ls` indicates that the minimum length of initial matches is at least s. The threshold s is one of the parameters of the pipeline and was adjusted by using a bacterial dataset in our computational evaluations.

Due to the scoring scheme described above, the seed matches whose lengths are more than s are found. Note that LAST employs spaced seeds as the default; it was confirmed by a homology search that this realizes a sensitive seed match [21].

Evaluation of interaction energy and prediction of the joint structure

In Step 4, the interaction energy, denoted by E^{int}, is computed using the INTARNA [18] program, with the default parameters, applied to the sequence including w base-pairs upstream and downstream of each seed match. The integer w is one of the parameters in our pipeline, and is optimized using bacterial sequences. Although calculating the optimal interaction energy between two RNA sequences entails heavy computational costs, INTARNA computes the minimum interaction energy between two *local* segments taking into account the approximated accessibility of the target sites [18]. For each pair of a query RNA sequence and a target RNA sequence, our pipeline typically predicts multiple *local* RNA–RNA interactions. In Step 5, those interactions are ranked according to E^{int}, calculated in Step 4. We introduce two ranking methods for the predicted interacting pairs with respect to a query as follows.

- MINENERGY: The minimum interaction energy among the interactions contained in each pair of RNA sequences, computed by INTARNA. This score is adequate for short RNA sequences or for cases in which the strongest local interaction is dominant.
- SUMENERGY: The sum of the local interaction energies that are lower than x kcal/mol. This score is adequate for long RNA sequences such as lncRNAs and mRNAs and for cases in which several strong interactions exist.

In Step 6, additionally, the *joint* secondary structure of the strongest interaction site in each pair of RNA sequences is calculated using RACTIP [23]. RACTIP estimates precise internal and external base-pairs by sophisticated integer programming using the principle of maximum expected accuracy [24]. This step does not affect the ranking of the predicted interactions, but information about the structural patterns of the local structures is useful for further analysis of the interactions. The predicted joint secondary structures are stored in our database, which is described later in this paper.

Computational environment

All the analyses of bacterial sequences were performed on the DELL PRECISION T7500, including 4 Intel Xeon CPU E5620 (2.40GHz, 4-cores). The analyses of Human transcriptome were performed on two computational environments as follows. Steps 3–6 in our pipeline, which require much more computational time than the other steps, were implemented on the *K* computer (http://www.aics.riken.jp/en/k-computer/about/), including 88,128 SPARC64 VIIIfx CPUs (2.0GHz, 8-cores). The other steps were calculated using the Chimera cluster system at AIST, including 176 Intel Xeon E5550 CPUs (2.53 GHz, 8-cores). In our experiments, a part of those cores is utllized (see the 'Results and discussion' section for the details).

Datasets and evaluation methods

Experimentally validated RNA–RNA interactions in E. coli

As the known interactions for training the parameters of the pipeline, 44 interacting ncRNA–mRNA pairs were taken from [15]. These pairs comprise 17 sRNAs interacting with 37 mRNAs. As non-interacting mRNA targets of those 17 query ncRNAs competing with the 37 mRNAs, we collected mRNA sequences around the start codons, because the ncRNAs are known to interact with the mRNA sequences around the start codons. About 4200 mRNA sequences of 200 bases, 150 base-pairs upstream and 50 base-pairs downstream from the start codons, were taken from the NCBI web site (http://www.ncbi.nlm.nih.gov/nuccore/NC_000913).

Human lncRNA and mRNA sequences

We used 23,898 long ncRNA (lncRNA) and 81,814 mRNA sequences obtained by the Gencode project [25] (http://

www.gencodegenes.org/releases/19.html). The average and maximum lengths of the lncRNAs were 955 and 91,677, respectively. While most human protein-coding genes have alternative transcripts, the longest mRNA transcripts were selected for each gene in our analysis. As a result, the number of mRNA sequences was reduced to 20,185, and the average and maximum lengths were 3382 and 109,224, respectively.

Evaluation methods

In this study, we use the same evaluation method as [15]. Specifically, for each query RNA, its target RNAs are sorted by rank (using either MINENERGY or SUMENERGY), and the number of true positive predictions (denoted by "nTPs" in the following) are counted for all query RNAs with a given rank.

Results and discussion

Training parameters using bacterial RNA–RNA interactions

Our pipeline includes three adjustable parameters (cf. Fig. 1 and Materials and methods): (i) the threshold for the accessibility, denoted by t; (ii) the threshold of the score of the seeds, denoted by s; and (iii) the length w of the flanking sequences around the seed region. Flanking sequences of length w upstream and downstream around a seed are used for computing the (local) interaction energy. Those three parameters affect both the prediction accuracy and the computation time. For example, smaller values of t and larger values of s reduce the computational time because the number of candidates decreases. For optimization of those parameters, the known ncRNA-mRNA interactions in *E. coli* (see Section "Datasets and evaluation methods" for the details) were used. In our pipeline, two kinds of ranking method were introduced for screening the predicted RNA–RNA interactions with respect to each query RNA sequence. The MINENERGY method was used as a ranking method for the procedures described in this section because the query RNA sequences in this dataset were short (the average length is 117.2).

We compared the running time and accuracy of our pipeline with those of a method proposed by [15], which was originally used for predicting sRNA–mRNA interactions in bacteria. Among the several pipelines proposed in Richter et al., we adopted the one based on highly accessible seeds, choosing it because its performance was better than or comparable to the other proposed pipelines (see [15] for the details).

Additional file 1: Figure S1 shows the running time for various values of w and s (Step 1, reduction of candidates by accessibility was skipped). The computation time decreased as s became larger and as w became smaller, as expected. For most of the combinations of s and w, our pipeline was faster than Richter's pipeline, whose running time for the same dataset was 9.27 h.

Additional file 1: Figure S2 shows a comparison of prediction accuracies between our pipeline and Richter's pipeline. For our pipeline, only combinations of parameters that produce a result at least five-fold faster than Richter's pipeline are plotted. Among those combinations of parameters, $s = 8$ and $w = 20$ achieved the best performance on accuracy, which was almost the same as the performance of Richter's pipeline.

Finally, we evaluated the performance of our pipeline with accessibility filtering by changing the value of t, the parameter for the threshold of accessibility, while the other parameters were held fixed at $s = 8$ and $w = 20$. Figure 2 and Additional file 1: Figure S3 show the performance and the running time, respectively, for various values of t. When $t = 4.3$ kcal/mol was used, our pipeline (1.00 h) was 9.27 times faster than Richter's pipeline (9.27 h) while the prediction performances were similar (Figure 2 and Additional file 1: Figure S3). It is interesting that the performance with $t = 4.3$ kcal/mol was better than that obtained without using accessibility filtering (Fig. 2).

In summary, when the parameters $s = 8$, $w = 20$ and $t = 4.3$ kcal/mol were used, our pipeline was about 10 times faster than Richter's pipeline while the performance was almost the same. These optimized values of the parameters, $s = 8$, $w = 20$ and $t = 4.3$ kcal/mol, were used for the comprehensive prediction of RNA–RNA interactions for the human transcriptome, described in later sections.

Comprehensive predictions for the human transcriptome

We applied our pipeline (cf. Fig. 1) to the human transcriptome and conducted comprehensive predictions of lncRNA–mRNA and lncRNA–lncRNA interactions for the 23,898 lncRNAs (the query) and 44,083 RNAs (the target; including 23,898 lncRNAs and 20,185 mRNAs) in the dataset described in the previous section. The parameters of our pipeline used in these experiments were fixed to $s = 8$, $w = 20$, $t = 4.3$. Overfitting to RNA–RNA interactions in human transcriptome was avoided, because those parameter values were optimized by using an *E. coli* dataset, as described earlier.

The step 1 (accessibility filtering) was performed on the Chimera cluster system using 128 cores, which took 1.16 h. After Step 1, 80.7 % and 79.2 % of the total lengths of lncRNA and mRNA sequences, respectively, were maintained for the next step. The step 2 (tandem repeat masking) was performed on the Chimera cluster system using a single core, which took only about one minute. After Step 2, 78.4 % and 76.7 %, respectively, were maintained. In Step 3, about 4.7×10^{10} seeds were found. In Step 4, we computed the energy for each seed with both upstream and downstream flanking sequences ($w = 20$ base-pairs). For steps 3 to 5, 40,000 cores on the K computer (http://www.

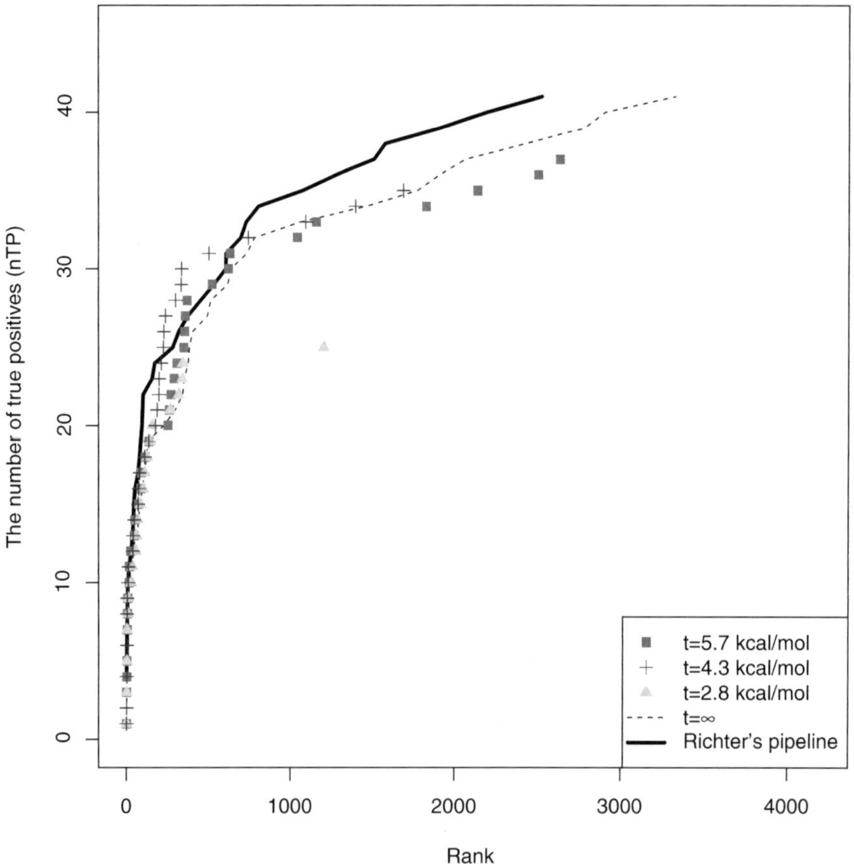

Fig. 2 The performance of our pipeline for various values of the parameter *t*, where the parameters *s* and *w* were fixed to be 8 and 20, respectively (cf. Additional file 1: Figure S2). The dashed line for "$t = \infty$" shows the performance without considering accessibility

aics.riken.jp/en/k-computer/about/), one of the fastest computers in the world, were used for the calculation, which took 45 h. For a machine with a single core, however, these processes are estimated to take 1.8×10^6 h (205 years). This indicates that it is presently infeasible to make comprehensive predictions about RNA–RNA interactions in the human transcriptome without using a supercomputer. (It is noted that Step 3–5 include several pre- and post-processes, whose running time can be neglected compared to the total computational time). Finally, in step 6, the joint secondary structure of the strongest interaction site in each pair of RNA sequences is predicted. For step 6, 16,000 cores on the K computer were used for the calculation, which took 7.3 h.

Validation of predicted interactions for the human transcriptome

We validated our predicted interactions for the human transcriptome by comparing our predictions with three reliable studies of RNA–RNA interactions [8–10], all of which include experimentally validated lncRNA–mRNA interactions.

TINCR–mRNA interactions

Recently, Kretz *et al.* [9] experimentally investigated the interactions between a specific long non-coding RNA, called TINCR, and its target mRNAs using RNA interactome analysis with high throughput sequencing (RIA-Seq). From the 11,225 target mRNAs investigated by Kretz et al. 5195 were included in the mRNA collections of our human transcriptome dataset. Among these, 1062 mRNAs were found to interact with TINCR lncRNA through at least one local RNA–RNA interaction detected as an *enriched segment* by RIA-Seq (an interaction including more enriched segments leads to more reliable TINCR–mRNA interactions; see [9] for the details). Detailed statistics of these mRNAs are shown in (Additional file 1: Table S1).

First, we compared these validated interactions with our predicted interactions, where TINCR (Ensemble ID: ENST00000448587) was included in our human transcriptome dataset. Figure 3 shows an AUC-ROC analysis of our comprehensive predictions of TINCR–mRNA interactions. Among our ranking methods (Step 5), SUMENERGY achieved better performance than

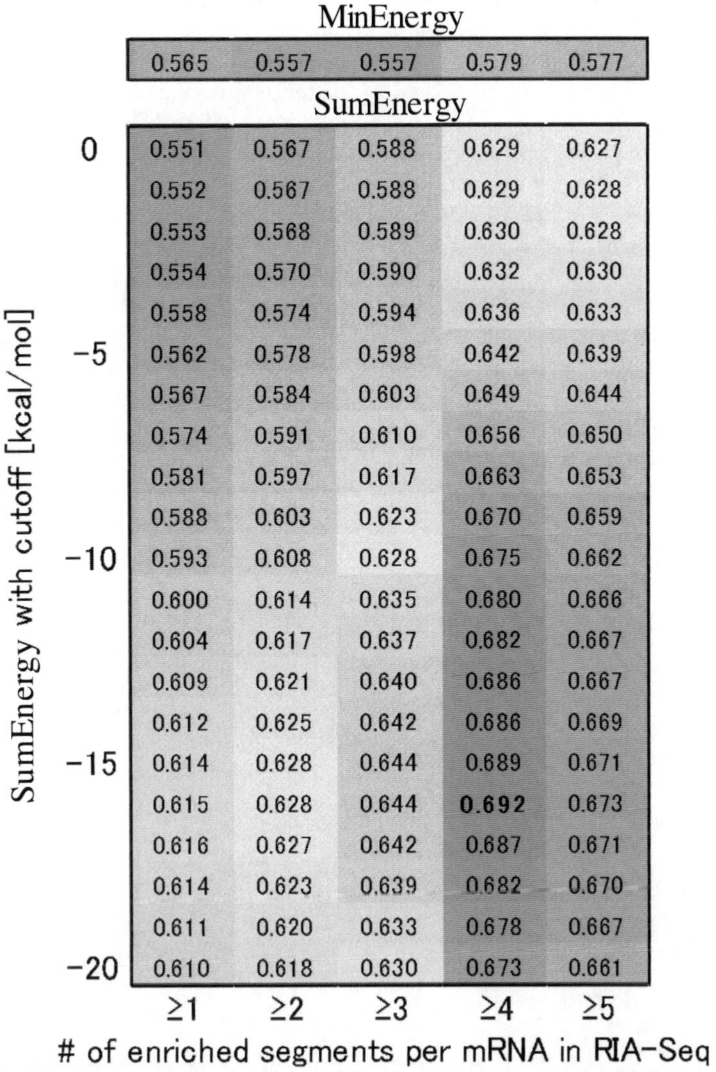

Fig. 3 AUC-ROC scores for TINCR–mRNA interactions [9]. The row shows the number of enriched segments per target mRNA in RIA-Seq experiment, which defines the positive dataset; positive datasets with a larger number include more reliable TINCR–mRNA interactions. The column indicates the energy cutoff [kcal/mol] of SUMENERGY in our comprehensive predictions. The bold value is the best score

MINENERGY. We applied various energy cutoffs to the summation process of SUMENERGY to remove the effect of weak local RNA–RNA interactions with higher energies. The results show that the performance with both a smaller energy threshold and more enriched segments in RIA-seq tends to be better (Fig. 3). These results corresponded to the fact that the validated TINCR–mRNA interactions include several local RNA–RNA interactions. In addition, the best AUC-ROC score of 0.692 (for which the energy cutoff is −16 kcal/mol and the number of enriched windows per mRNA is more than or equal to 4) shows that our prediction pipeline achieved a moderate accuracy by using only sequence information. In the predicted interactions ranked by SUMENERGY with a −16 kcal/mol cutoff (Additional file 1: Table S2), validated TINCR–mRNA interactions consisting of more than three enriched segments were frequently found with better ranking, such as rank1 (ENST00000597346), rank10 (ENST00000301067), rank25 (ENST00000269919), rank37 (ENST00000495893) and rank39 (ENST00000268489).

Finally, we analyzed the pattern of local interactions between TINCR and mRNAs in our predictions. Additional file 1: Figure S4 shows the two predicted patterns of interaction between TINCR and an mRNA: (a) TINCR–mRNA (ENST00000258704), where several local interactions between the segments including

TINCR motifs were predicted, and (b) TINCR–mRNA (ENST00000367187), where there are strong interactions between TINCR and the 3′UTR of the mRNA.

1/2-sbsRNA–mRNA interactions

Recently, Gong and Maquat [8] reported that the lncRNA 1/2-sbsRNA (ENST00000548810) directly interacts with the 3′UTRs of two mRNAs (SERPINE1 and ANKRD57). These interactions lead to Staufen 1 (STAU1)-mediated messenger RNA decay (SMD) [26].

In our predictions with the MINENERGY ranking, the ranks of SERPINE1 (ENST00000223095) and ANKRD57 (ENST00000356454) are 28 (within the top 0.06 %) and 1827 (within the top 4.14 %), respectively, among the predicted interactions with 1/2-sbsRNA (ENST00000548810) as the query. Interestingly, for this lncRNA, ranking with MINENERGY achieved better performance than ranking with SUMENERGY (where the ranks of SERPINE1 and ANKRD57 were 11,296 and 3627, respectively), while the SUMENERGY ranking achieved better performance on TINCR–mRNA interactions. This indicates that ranking by MINENERGY ranking is more appropriate than by SUMENERGY for such interactions. Figure 4 shows the interaction pattern of lncRNA–mRNA, indicating one strong interaction between a region of lncRNA and the 3′UTR in mRNA. Additionally, the joint secondary structure of the two subsequences (processed by our pipeline) of the two 1/2-sbsRNA–mRNA interactions (predicted by RactIP [23] and shown in Additional file 1: Figure S5) indicates that it includes a long anti-sense-like interaction and the binding sites are located in the 3′UTR of the mRNAs (Fig. 4). It is also noted that the interaction sites of 1/2-sbsRNA in both interactions are identical.

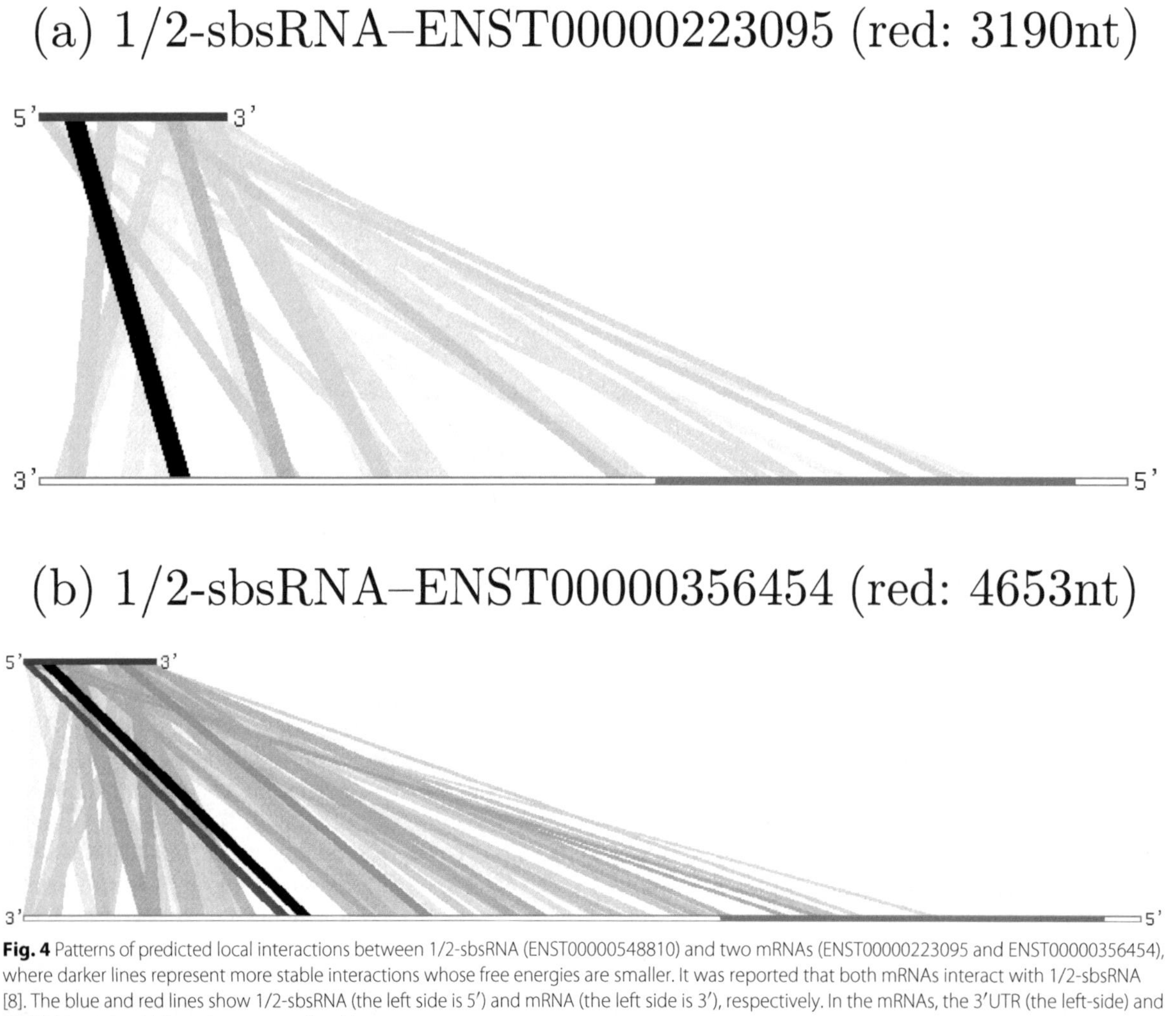

Fig. 4 Patterns of predicted local interactions between 1/2-sbsRNA (ENST00000548810) and two mRNAs (ENST00000223095 and ENST00000356454), where darker lines represent more stable interactions whose free energies are smaller. It was reported that both mRNAs interact with 1/2-sbsRNA [8]. The blue and red lines show 1/2-sbsRNA (the left side is 5′) and mRNA (the left side is 3′), respectively. In the mRNAs, the 3′UTR (the left-side) and 5′UTR (the right-side) are shown as outlined bars

7SL–TP53 interaction

As described in the introduction, Abdelmohsen *et al.* [10] found that 7SL non-coding RNA interacts with TP53 mRNA, which encodes the tumor suppressor p53. They suggested that 7SL regulates p53 translation by interacting with TP53.

The ranks of 7SL–TP53 interactions in our comprehensive predictions are 2787 for MINENERGY, 2216 for SUMENERGY and 1906 for SUMENERGY with the best energy cutoff. This shows that SUMENERGY is slightly better than MINENERGY. The result is consistent with the results of [10], who suggested multiple interaction sites for 7SL–TP53. In our pipeline, two strong interaction sites between 7SL and the 3′UTR of TP53 mRNA were predicted (Fig. 5). These two sites are consistent with two of the four interaction sites suggested by Abdelmohsen et al. Interestingly, among the top 100 candidates ranked by SUMENERGY, 71 had the strongest interaction site in the 3′UTR, suggesting that the 3′UTR of mRNAs is a common target of 7SL lncRNA.

An investigation of NEAT1–RNA interactions: a case study of lncRNA–lncRNA interactions

The biological functions of NEAT1 have been well-studied from molecular biology viewpoints (e.g., [7]), with results indicating that NEAT1 is a core molecule forming paraspeckles in the cell. There exist no reports about RNAs (mRNAs or lncRNAs) that directly interact with NEAT1, although several studies of NEAT1–protein interactions have been conducted [27]. In this section, as a case study, we investigate NEAT1–RNA interactions in our comprehensive predictions. Additional file 1: Tables S3 and S4 show lists of NEAT1–RNA interactions predicted by our pipeline, sorted by SUMENERGY and MINENERGY, respectively. (The entire lists are available from our database).

The two lists are dissimilar, indicating that each ranking method captures a different type of RNA–RNA interaction. In the list sorted by MINENERGY (Additional file 1: Table S4), most of the interactions with minimum interaction free energy are located in the 3′UTR of the mRNAs and the interaction parts in NEAT1 are similar to each other (around position 17,900). In contrast, in the list sorted by SUMENERGY (Additional file 1: Table S3), the rank of the NEAT1–NEAT1 interaction is 20. Although it requires further investigation for validation, there is a possibility that NEAT1–NEAT1 interactions occur in the formation of paraspeckles. Moreover, both lists include many NEAT1–lncRNA interactions, suggesting the possibility that NEAT1 interacts with other lncRNAs.

Database

To make our comprehensive predictions of lncRNA–RNA interactions in the human transcriptome publicly available, we have developed a database that contains all the predicted RNA–RNA interactions along with the following information: the rankings by MINENERGY and SUMENERGY, the predicted joint secondary structures, and links to the UCSC genome browser. Figure 6 shows the interface of our database, where users can search by the names (Ensemble ID or HUGO gene symbols) of the lncRNA and mRNA. Each entry includes a link to the UCSC genome browser [28] (http://genome.ucsc.edu/cgi-bin/hgTracks?db=hg19) and users can check the annotation of the target/query RNA sequences in the genomic context. The database is available at http://rtools.cbrc.jp/cgi-bin/RNARNA/index.pl.

Potential of this study

Our pipeline achieved a performance for bacterial sRNA–mRNA interactions that is comparable to the performance of existing methods (e.g., "Richter's pipeline") but with significantly faster calculation. Several promising results were found in the predicted interactions in the human transcriptome in spite of the fact that our pipeline utilizes only sequence information of RNAs. There is room, however, to improve the accuracy of the pipeline and

Fig. 5 The pattern of the predicted local interactions between 7SL RNA (ENST00000553637) and TP53 mRNA (ENST00000420246). The light blue regions show interaction sites suggested in a previous study [10]

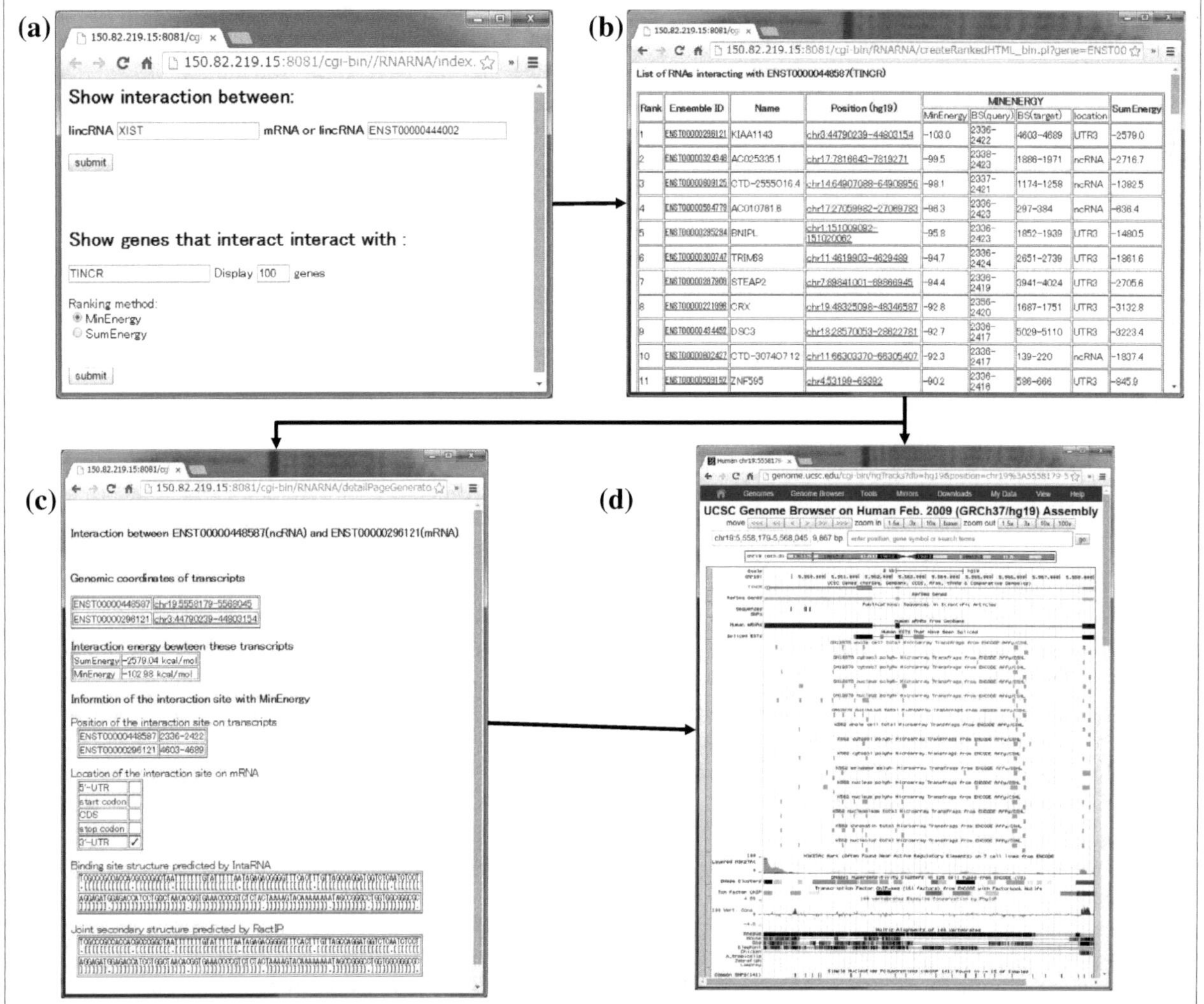

Fig. 6 User interface of our RNA–RNA interaction database of lncRNA-mRNA and lncRNA-lncRNA interactions in the human transcriptome. **a** Search window for query by gene name. **b** Search results with MINENERGY and SUMENERGY. **c** Detailed information of predicted interactions, including the joint secondary structure predicted by INTARNA [18] and RACTIP [23]. **d** Genomic context in the UCSC genome browser for target and query RNA sequences, which is opened by clicking on a link in the panel shown in (**b**)

minimize false positive predictions. In the future, we will work to combine our method with additional information, such as expressions (RNA-seq) and localizations. Extensive proteome data for humans [29] and information about protein binding sites in RNA sequences [30] can be also incorporated into those analyses. Moreover, a comparative (consensus interaction-based) approach is promising [16, 31] when homologous sequences are available. Because most lncRNAs are not phylogenetically conserved in distant species, closely related species are needed for the analysis.

In our pipeline, we used the parameters trained using sRNA-mRNAs in E.coli, which might not be optimal for predicting lncRNA-mRNA interactions in human transcriptome. It is, however, interesting that our pipeline achieved modest performance for predicting lncRNA-mRNA interactions with the parameters. This would imply a similarity between the mechanisms of mRNA recognition by sRNA and lncRNA. When the instances of lncRNA-mRNA interactions are accumulated, the similarity and difference between sRNA-mRNA and lncRNA-mRNA interaction will be elucidated. Moreover, it would be interesting to investigate the relation of two ranking methods (MINENERGY and SUMENERGY) with functions of interactions, because using SUMENERGY sometime provided better prediction results than using MINENERGY (TINCR and 7SL lncRNA), whereas the three parameters were optimized for MINENERGY.

Recently, Engreitz et al. have pointed out the importance of interactions between RNA and nascent pre-mRNAs [32]. Although in this study we focused on mature mRNA, it is desirable to make comprehensive predictions of interactions between lncRNA and pre-mRNAs, which requires much greater computational resources than used in this study.

Conclusion

In this study, we have developed a novel pipeline for predicting RNA–RNA interactions. Our pipeline was tuned and validated using an *E. coli* RNA–RNA interaction dataset and was applied to the human transcriptome using the *K* computer, one of the fastest computers in the world. We compared our predicted RNA–RNA interactions in the human transcriptome with predictions in three existing studies of RNA–RNA interactions. Moreover, we have developed a database to compile predicted interactions, which will be useful for biologists who are interested in specific lncRNAs. To the best of our knowledge, this is the first study to comprehensively predict RNA–RNA interactions in the human transcriptome.

Additional file

Additional file 1: Supplementary file (PDF). This PDF file includes supplmentary texts, figures and tables. (PDF 158 kb)

Competing interests

The authors declare that they have no competing interests.

Authors' contributions

MH and KA designed the study under the discussion with TK. GT implemented the pipeline and performed computational experiments on K computer. JI, GT and MH conducted analysis and interpretation of the results of computational experiments. MH wrote a first draft of the article. All the authors approved the final version of the article.

Acknowledgements

This research was supported in part, by a MEXT SPIRE (Supercomputational Life Science) grant to KA, by a MEXT KAKENHI Grant-in-Aid for Young Scientists (A) (No. 24680031) to MH and by a MEXT KAKENHI Grant-in-Aid for Scientific Research (A) (No. 30356357) to TK, MH and KA collectively. The online charge of this article was supported by interanal funding in Waseda University. We thank Dr. Kun Qu and Dr Paul A. Khavari for providing the TINCR dataset. We are grateful to Prof. Tetsuro Hirose, Prof. Hisanori Kiryu, Dr. Yutaka Saito and Dr. Satoshi Yamasaki for valuable discussions. A part of this work was conducted while MH was at The University of Tokyo; since 1 April 2014, MH has been at Waseda University.

Declarations

This article has been published as part of BMC Genomics Volume 17 Supplement 1, 2016: Selected articles from the Fourteenth Asia Pacific Bioinformatics Conference (APBC 2016): Genomics. The full contents of the supplement are available online at http://www.biomedcentral.com/bmcgenomics/supplements/17/S1.

Author details

[1]INTEC Inc, 1-3-3 Shinsuna Koto-ku, 136-8637 Tokyo, Japan. [2]Graduate School of Frontier Sciences, University of Tokyo, 5-1-5 Kashiwanoha, Kashiwa, 277-8562 Chiba, Japan. [3]Biotechnology Research Institute for Drug Discovery, National Institute of Advanced Industrial Science and Technology (AIST), 2-41-6, Aomi, Koto-ku, 135-0064 Tokyo, Japan. [4]Faculty of Science and Engineering, Waseda University, 55N-06-10, 3-4-1, Okubo Shinjuku-ku, 169-8555 Tokyo, Japan.

Published: 11 January 2016

References

1. Morris KV, Mattick JS. The rise of regulatory RNA. Nat Rev Genet. 2014;15(6):423–37.
2. Hirose T, Mishima Y, Tomari Y. Elements and machinery of non-coding RNAs: toward their taxonomy. EMBO Rep. 2014;15(5):489–507.
3. Turner M, Galloway A, Vigorito E. Noncoding RNA and its associated proteins as regulatory elements of the immune system. Nat Immunol. 2014;15(6):484–91.
4. Chen G, Wang Z, Wang D, Qiu C, Liu M, Chen X, et al. LncRNADisease: a database for long-non-coding RNA-associated diseases. Nucleic Acids Res. 2013;41(Database issue):983–6.
5. Derrien T, Johnson R, Bussotti G, Tanzer A, Djebali S, Tilgner H, et al. The GENCODE v7 catalog of human long noncoding RNAs: analysis of their gene structure, evolution, and expression. Genome Res. 2012;22(9): 1775–89.
6. Plath K, Mlynarczyk-Evans S, Nusinow DA, Panning B. Xist RNA and the mechanism of X chromosome inactivation. Annu Rev Genet. 2002;36:233–78.
7. Sasaki YT, Ideue T, Sano M, Mituyama T, Hirose T. MENepsilon/beta noncoding RNAs are essential for structural integrity of nuclear paraspeckles. Proc Natl Acad Sci USA. 2009;106(8):2525–30.
8. Gong C, Maquat LE. lncRNAs transactivate STAU1-mediated mRNA decay by duplexing with 3′ UTRs via Alu elements. Nature. 2011;470(7333):284–8.
9. Kretz M, Siprashvili Z, Chu C, Webster DE, Zehnder A, Qu K, et al. Control of somatic tissue differentiation by the long non-coding RNA TINCR. Nature. 2013;493(7431):231–5.
10. Abdelmohsen K, Panda AC, Kang MJ, Guo R, Kim J, Grammatikakis I, et al. 7SL RNA represses p53 translation by competing with HuR. Nucleic Acids Res. 2014;42(15):10099–111.
11. Quinn JJ, Ilik IA, Qu K, Georgiev P, Chu C, Akhtar A, et al. Revealing long noncoding RNA architecture and functions using domain-specific chromatin isolation by RNA purification. Nat Biotechnol. 2014;32(9): 933–40.
12. Cook KB, Hughes TR, Morris QD. High-throughput characterization of protein-RNA interactions. Brief Funct Genomics. 2015;14(1):74–89.
13. Washietl S, Hofacker IL, Lukasser M, Huttenhofer A, Stadler PF. Mapping of conserved RNA secondary structures predicts thousands of functional noncoding RNAs in the human genome. Nat Biotechnol. 2005;23(11): 1383–90.
14. Wright PR, Georg J, Mann M, Sorescu DA, Richter AS, Lott S, et al. CopraRNA and IntaRNA: predicting small RNA targets, networks and interaction domains. Nucleic Acids Res. 2014;42(Web Server issue): 119–23.
15. Richter AS, Backofen R. Accessibility and conservation: general features of bacterial small RNA-mRNA interactions RNA Biol. 2012;9(7):954–65.
16. Wright PR, Richter AS, Papenfort K, Mann M, Vogel J, Hess WR, et al. Comparative genomics boosts target prediction for bacterial small RNAs. Proc Natl Acad Sci USA. 2013;110(37):3487–96.
17. Backofen R, Bernhart SH, Flamm C, Fried C, Fritzsch G, Hackermüller J, et al. RNAs everywhere: genome-wide annotation of structured RNAs. J Exp Zool B Mol Dev Evol. 2007;308(1):1–25.
18. Busch A, Richter AS, Backofen R. IntaRNA: efficient prediction of bacterial sRNA targets incorporating target site accessibility and seed regions. Bioinformatics. 2008;24(24):2849–56.
19. Kiryu H, Terai G, Imamura O, Yoneyama H, Suzuki K, Asai K. A detailed investigation of accessibilities around target sites of siRNAs and miRNAs. Bioinformatics. 2011;27(13):1788–97.
20. Frith MC. A new repeat-masking method enables specific detection of homologous sequences. Nucleic Acids Res. 2011;39(4):23.
21. Kiebasa SM, Wan R, Sato K, Horton P, Frith MC. Adaptive seeds tame genomic sequence comparison. Genome Res. 2011;21(3):487–93.
22. Frith MC, Hamada M, Horton P. Parameters for accurate genome alignment. BMC Bioinformatics. 2010;11:80.
23. Kato Y, Sato K, Hamada M, Watanabe Y, Asai K, Akutsu T. RactIP: fast and accurate prediction of RNA-RNA interaction using integer programming. Bioinformatics. 2010;26(18):460–6.
24. Hamada M, Asai K. A classification of bioinformatics algorithms from the viewpoint of maximizing expected accuracy (MEA). J Comput Biol. 2012;19(5):532–49.
25. Harrow J, Frankish A, Gonzalez JM, Tapanari E, Diekhans M, Kokocinski F, et al. GENCODE: the reference human genome annotation for the ENCODE project. Genome Res. 2012;22(9):1760–74.

26. Hadjiargyrou M, Delihas N. The Intertwining of Transposable Elements and Non-Coding RNAs. Int J Mol Sci. 2013;14(7):13307–28.
27. Naganuma T, Hirose T. Paraspeckle formation during the biogenesis of long non-coding RNAs. RNA Biol. 2013;10(3):456–61.
28. Karolchik D, Barber GP, Casper J, Clawson H, Cline MS, Diekhans M, et al. The UCSC genome browser database: 2014 update. Nucleic Acids Res. 2014;42(Database issue):764–70.
29. Wilhelm M, Schlegl J, Hahne H, Moghaddas Gholami A, Lieberenz M, Savitski MM, et al. Mass-spectrometry-based draft of the human proteome. Nature. 2014;509(7502):582–7.
30. Castello A, Fischer B, Eichelbaum K, Horos R, Beckmann BM, Strein C, et al. Insights into RNA biology from an atlas of mammalian mRNA-binding proteins. Cell. 2012;149(6):1393–1406.
31. Seemann SE, Richter AS, Gesell T, Backofen R, Gorodkin J. PETcofold: predicting conserved interactions and structures of two multiple alignments of RNA sequences. Bioinformatics. 2011;27(2):211–9.
32. Engreitz JM, Sirokman K, McDonel P, Shishkin AA, Surka C, Russell P, et al. RNA-RNA interactions enable specific targeting of noncoding RNAs to nascent pre-mRNAs and chromatin sites. Cell. 2014;159(1):188–99.

Paszek and Górecki BMC Genomics 2016, 17(Suppl 1):15
DOI 10.1186/s12864-015-2308-4

BMC Genomics

PROCEEDINGS Open Access

Genomic duplication problems for unrooted gene trees

Jarosław Paszek* and Paweł Górecki

From The Fourteenth Asia Pacific Bioinformatics Conference (APBC 2016)
San Francisco, CA, USA. 11 - 13 January 2016

Abstract

Background: Discovering the location of gene duplications and multiple gene duplication episodes is a fundamental issue in evolutionary molecular biology. The problem introduced by Guigó et al. in 1996 is to map gene duplication events from a collection of rooted, binary gene family trees onto theirs corresponding rooted binary species tree in such a way that the total number of multiple gene duplication episodes is minimized. There are several models in the literature that specify how gene duplications from gene families can be interpreted as one duplication episode. However, in all duplication episode problems gene trees are rooted. This restriction limits the applicability, since unrooted gene family trees are frequently inferred by phylogenetic methods.

Results: In this article we show the first solution to the open problem of episode clustering where the input gene family trees are unrooted. In particular, by using theoretical properties of unrooted reconciliation, we show an efficient algorithm that reduces this problem into the episode clustering problems defined for rooted trees. We show theoretical properties of the reduction algorithm and evaluation of empirical datasets.

Conclusions: We provided algorithms and tools that were successfully applied to several empirical datasets. In particular, our comparative study shows that we can improve known results on genomic duplication inference from real datasets.

Keywords: Genomic duplication, Duplication episode, Reconciliation, Unrooted gene tree, Species tree

Background

Genomic duplication plays important role in evolution of life on Earth. This phenomenon have been extensively studied in the last decades for plant, bacterial and many other genomes [1–7]. Duplication events can involve individual genes, genomic segments or whole genomes. While the reconstruction of evolutionary history of individual genes is generally well established [8–13], still little is known on the inference of large genomic duplications that can span through thousands of genes families.

In this approach we propose to use the model of reconciliation in which a gene tree is reconciled with its species tree. The concept of reconciliation was introduced by Goodman [14] and formalized by Page [8] in the context of reconciling potential incongruence between a rooted gene family tree and its species tree. In this model, differences between gene and species trees are explained in terms of evolutionary events such as gene duplication, gene loss and speciation. Reconciliation can be interpreted as the embedding of a gene tree into a species tree where these evolutionary events, located in the species tree, induce a biologically consistent scenario [15]. Tree reconciliation has been extensively studied in recent decades in many theoretical and practical contexts including supertree inference, error correction and HGT detection [16–24]. In the process of reconciliation, which is relatively simple from computational point of view, each gene from a single gene family is mapped into the species tree and it is classified as a single gene duplication or related to speciation. However, the problem becomes much more complex, when a gene duplication is a part of large genomic duplications, called *multiple gene*

*Correspondence: jpaszek@mimuw.edu.pl
University of Warsaw, Institute of Informatics, Banacha 2, 02-097 Warsaw, Poland

duplication episode, in which parts of a genome are duplicated. In fact, it is known that a large duplication event is usually followed by many gene losses and gene rearrangements. In consequence, the reconstruction of large gene duplication events may be difficult.

The first approach to detect multiple gene duplication episodes from a collection of rooted gene trees was proposed by Guigó et al. [10]. In the model, for a given collection of rooted gene trees and a rooted species tree, the authors proposed heuristic to aggregate single gene duplication events into a large gene duplication. This approach was formalized and refined by Page and Cotton [25]. They formally defined the problem of *episode clustering* (EC) as the problem of locating the minimal number of locations in the species tree, where all duplications from the input gene trees can be placed. This model was applied in the context of the supertree problem by Fellows [26]. Burleigh et al. [27] and Bansal and Eulenstein [28] proposed the first polynomial time solutions for two types of the multiple gene duplication problems: the episode clustering (EC) and a more general variant of clustering called *minimum episodes* (ME). Finally, Luo et al. [29] proposed linear time and space algorithms to these problems.

While the classical reconciliation model is applicable to rooted trees only, most standard phylogenetic inference methods, like maximum likelihood, maximum parsimony or neighbour joining, infer unrooted gene family trees, and it is often difficult, to identify credible rootings. For example, outgroup rooting can result in incorrect rootings when evolutionary events cause heterogeneity in the gene trees, and rooting gene trees under the molecular clock assumption, or similarly by using midpoint rooting, also can result in error when there is a molecular rate variation throughout the tree [30, 31]. Tree reconciliation have been successfully extended to reconcile an unrooted gene tree with a rooted species tree by seeking a rooting of the unrooted gene tree that invokes the minimum number of evolutionary events such as gene duplications (D) or gene duplications and losses (DL), in the context of a given species tree [32, 33]. It is known that the rooting edges with minimal D or DL cost, induce a full subtree, called *plateau*, in the unrooted gene tree [34].

In this article we present the first solution to the open problem [27] of *unrooted episode clustering*, that is, the problem of episode clustering where the input consists of unrooted gene trees. We show that for a given set of unrooted gene trees and a species tree we can solve the unrooted episode clustering by reducing it to the rooted episode clustering problem that has a linear time complexity. Our solutions require a linear time preprocessing and a creation of at most $1 + 2^k$ collections of rooted gene trees, that is, instances of rooted EC Problem, where k is the number of input gene trees having a special topology located in the plateau of the duplication cost (formally, the condition requires two stars S2 [32]). Usually k represents a small fraction of the whole input, thus, this condition significantly reduces the complexity. In other words, we show that the problem of unrooted episode clustering is fixed parameter tractable. Finally, in a number of empirical computational experiments we show that despite the exponential worst case complexity our algorithm is able to resolve instances of the problem after the verification of at most two rooted datasets. In consequence, our solution can be efficiently applied to locate duplication clusters in collections of unrooted gene trees.

Results

Basic notation

A *species tree* is a rooted binary tree with leaves uniquely labeled by the names of species. Throughout this work, the species tree is fixed, therefore, we use S to denote it. A *rooted gene tree* is a rooted binary tree with leaves labeled by the names of species. The set of species present in T is denoted by $\mathcal{L}(T)$. The rooted tree (T_1, T_2) has two subtrees T_1 and T_2 whose roots are children of the tree root. Additionally, for nodes a and b, $a \preceq b$ means that a and b are on the same path from the root, with b being closer to the root than a. We write $a \prec b$ if $a \preceq b$ and $a \neq b$. The root of a tree T we denote by $\mathsf{root}(T)$.

Let $T = \langle V_T, E_T \rangle$ be a rooted gene tree such that $\mathcal{L}(T) \subseteq \mathcal{L}(S)$. *The least common ancestor (lca) mapping*, $\mathsf{M}_T : V_T \rightarrow V_S$, is defined as follows. If v is a leaf in T then $\mathsf{M}_T(v)$ is the leaf in S labeled by the label of v. When v is an internal node in T having two children a and b, then $\mathsf{M}_T(v)$ is the least common ancestor of $\mathsf{M}_T(a)$ and $\mathsf{M}_T(b)$ in S. An internal node $g \in V_T$ is called a *duplication* if $\mathsf{M}_T(g) = \mathsf{M}_T(a)$ for a child a of g. *The duplication cost*, denoted by $\mathsf{D}(T, S)$, is the total number of duplications in T. Each non-duplication node of T we call a *speciation*. The total number of *gene losses* required to reconcile T and S can be defined by: $\mathsf{L}(T, S) = 2\mathsf{D}(T, S) + \sum_{g \text{ is internal}, a, b \text{ children of } g}(\|\mathsf{M}_T(a), \mathsf{M}_T(b)\| - 2)$, where $\|a, b\|$ is the number of edges on the path connecting a and b in S. Finally, we can define the *duplication-loss cost* of reconciling a rooted gene tree T and a species tree S as follows: $\mathsf{DL}(T, S) = \mathsf{D}(T, S) + \mathsf{L}(T, S)$ [34]. Examples of the reconciliation are depicted in Fig. 1.

Unrooted reconciliation

The unrooted gene tree is an undirected acyclic connected graph in which each node has degree 1 (leaves) or 3 (internal nodes), and the leaves are labeled by the names of species. For an unrooted gene tree $G = \langle V_G, E_G \rangle$ and an edge $e \in E_G$, by G_e, we denote the rooting of G obtained from G by placing the root on e. Such a rooting induces the duplication cost $\mathsf{D}(G_e, S)$. We call *D-minimal*, the rooting or edges having the minimal duplication cost. It follows from the theory of unrooted reconciliation [32, 34] that

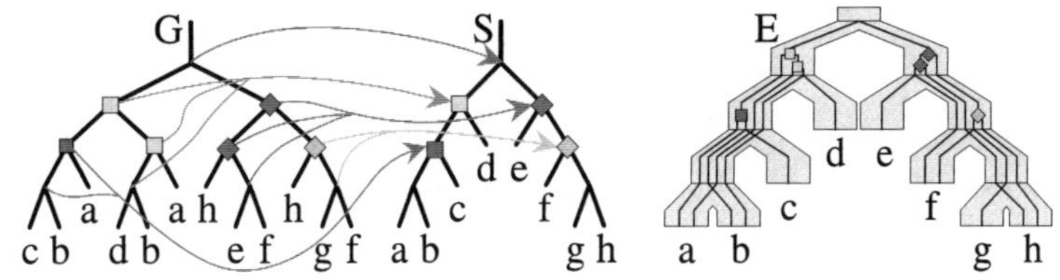

Fig. 1 Gene and species tree reconciliation. *Left*: the lca-mapping between a gene tree G and a species tree S shown for internal nodes. The decoration of nodes indicates gene duplication events. *Right*: an embedding of G into S. To reconcile these trees 6 gene duplications and 17 gene losses (not shown) are required, i.e., $\mathsf{D}(G,S) = 6$ and $\mathsf{DL}(G,S) = 23$. See also G_{4*} in Fig. 3

the set of D-minimal edges, called *D-plateau*, is a full subtree of G. The same property holds for the DL-plateau, that is, the set of edges with the minimal duplication-loss cost. We use a similar notation for DL-minimal edges, rootings and so on. The most important property of these plateaus is below.

Theorem 1 (From [34]). *DL-plateau is a subgraph of D-plateau.*

Without loss of generality we assume that every root of a gene tree is mapped into the root of S, denoted by $\top$, and both trees are non-trivial. An edge $e = \langle v, w \rangle$ of G is *empty* if the root of G_e is a speciation, i.e., $\mathsf{M}_{G_e}(v) \neq \top \neq \mathsf{M}_{G_e}(w)$. We call e *double* if $\mathsf{M}_{G_e}(v) = \top = \mathsf{M}_{G_e}(w)$. Otherwise, e is called *single*. A single edge e is called v-incoming or *w-outgoing* if $\mathsf{M}_{G_e}(v) \neq \top = \mathsf{M}_{G_e}(w)$.

Let v be an internal node of G, then a *star* with a *center* v consists of three edges, denoted by e_a, e_b and e_c, sharing v and incident to nodes a, b and c, respectively (see Fig. 2). The are several types of possible star topologies based on the above classification of edges: the *S*1 star has one v-incoming edge and two v-outgoing edges, the *S*2 star has exactly two v-outgoing edges and one empty edge, the *S*3 star has two v-outgoing edges and one double edge, the *S*4 star all 3 edges are double, and the *S*5 star has one v-outgoing edge and two double edges. The star topologies are depicted in Fig. 2.

Theorem 2 (Adopted from [32]). *For a given unrooted gene tree G, we have*

- *either G has exactly one empty edge or G has at least one double edge,*
- *if the DL-plateau of G consists of exactly one edge, then this edge is either empty or double, and all other edges are single.*
- *if the DL-plateau of G has more than one edge, then it contains all edges present in stars S4 and S5, and all other edges are single.*

Note that if a gene has an empty edge, then it has at most two stars S2 (see examples in Fig. 3).

Episode clustering problems

To model gene duplication episodes we allow to relocate a gene duplication from its lca-mapping location to one of its ancestors. In other words, we introduce mappings representing evolutionary scenarios that can differ from the scenario defined by the lca-mapping. Additionally, we require that the total number of gene duplications is minimal. To ensure biological correctness of such mappings, we introduce several conditions, e.g., time order preservation.

A mapping $\mathsf{F}_G \colon V_G \to V_S$ is called *valid* if the following conditions are satisfied:

- $\mathsf{F}_G(a) \preceq \mathsf{F}_G(b)$ if $a \preceq b$ (time consistency),
- $\mathsf{F}_G(a) = \mathsf{M}_G(a)$ for any speciation node a (fixed speciations),
- $\mathsf{F}_G(a) \succeq \mathsf{M}_G(a)$ for any duplication node a (duplication can be raised),
- $\mathsf{F}_G(a) \prec \mathsf{M}_G(b)$ for any speciation node b such that $a \prec b$ (fixed number of gene duplications).

It can be shown that every valid mapping uniquely defines an evolutionary scenario represented by a DLS-tree [15]. Additionally, every DLS-tree obtained from a

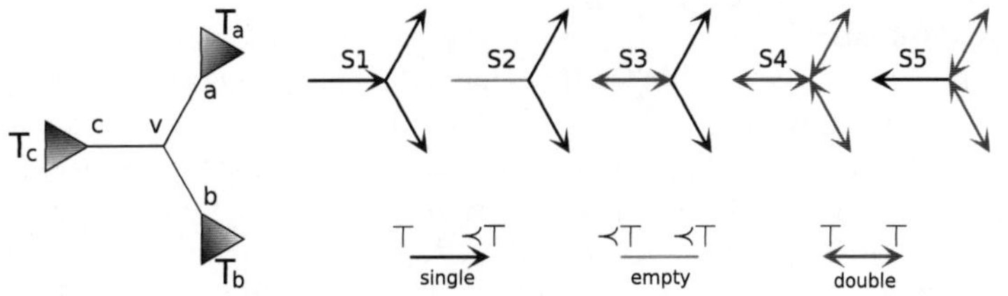

Fig. 2 Types of stars. Star topology with the center v, types of edges and stars

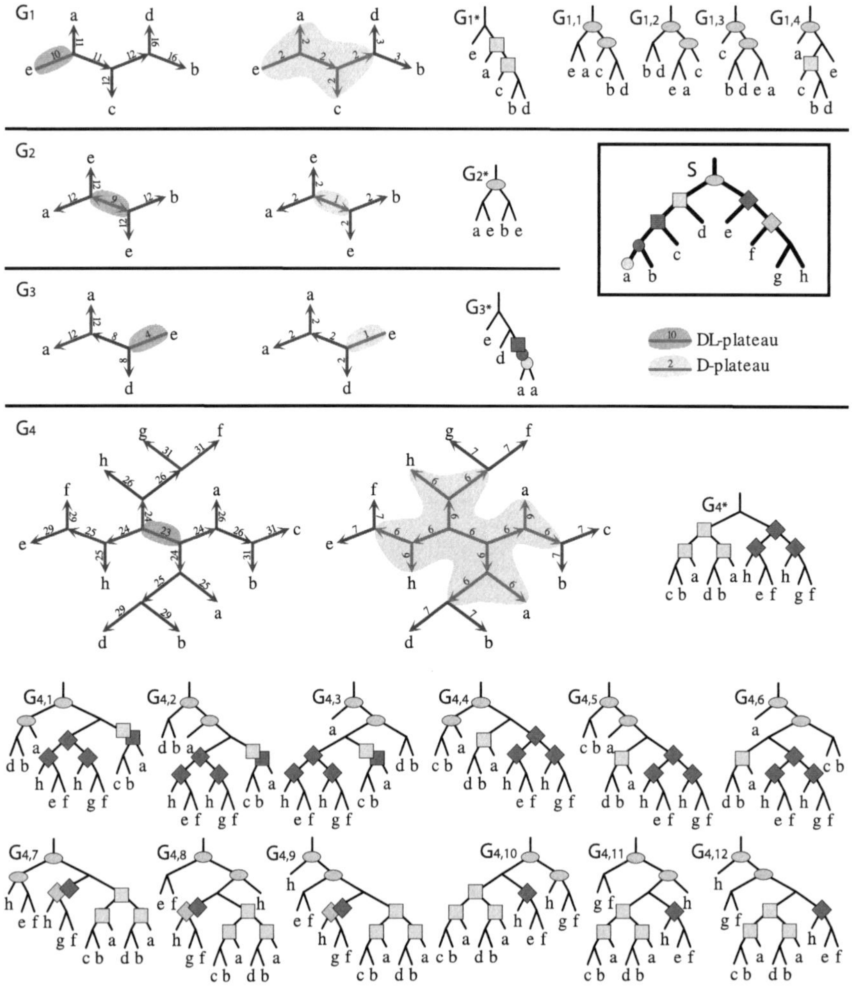

Fig. 3 An example of unrooted episode clustering. A species tree S and four unrooted gene trees G_1, G_2, G_3, G_4 with all D-minimal rootings. For every gene tree two star topologies are shown: one for the duplication-loss cost (*left*) and one for the duplications cost (*right*). Every edge of a gene tree is decorated with the corresponding cost of rooting. Every duplication node in rootings of gene trees is decorated by all possible locations (i.e., valid mappings) of its duplication cluster from optimal solutions of single-UEC. Note that the rooting G_{4*}, whose lca-mappings are shown in Fig. 1, has two duplications at $(c,(b,a))$ and $(h,(f,g))$ that are raised (here) to create two duplications clusters. Let $\{G_2, G_4\}$ be an instance of UEC Problem. Then, the $\top$-cluster, that is present in G_{2*}, contributes to the optimal solution. In such a case, the solution is induced by one of the two instances of EC problem: $\{G_{2*}, G_{4,1}\}$ or $\{G_{2*}, G_{4,7}\}$. This property is proved in Theorem 5 and in Lemma 6

valid mapping can be transformed into the optimal evolutionary scenario (i.e., lca-based scenario), by a sequence of TMOVE (i.e., lowering duplication) transformations. Please refer to [15] for more details on formal modeling of evolutionary scenarios. Observe, that the above model is more general than the model from [28].

We denote by $\mathsf{Dup}(T)$, the set of all duplication nodes in T. Let $G_1, G_2, \ldots, G_n$ be a collection of rooted gene trees. Assume that, for every $i \in \{1,2,\ldots,n\}$, F_i is a valid mapping between G_i and the species tree S. Every element $s \in \bigcup_i \mathsf{F}_i(\mathsf{Dup}_{G_i})$ denotes the location of multiple gene duplication events in S. Such locations will be called *duplication episodes.* A *duplication cluster* for s is the set of all gene duplications present in G_i's that are mapped to s. By $\top$-cluster we denote the duplication cluster whose elements are mapped to $\top$.

Problem 1 (Rooted Episode Clustering (EC)). *Given a collection of rooted gene trees* $G_1, G_2, \ldots, G_n$ *and a species tree S. Compute the minimal number of duplication episodes, denoted by* $\mathsf{EC}(G_1, G_2, \ldots, G_n, S)$*, in the set of all valid mappings* $\mathsf{F}_1, \mathsf{F}_2, \ldots, \mathsf{F}_n$ *such that* $\mathsf{F}_i\colon V_{G_i} \to V_S$.

This problem can be solved in linear-time and space [29]. In this article we solve the following problem.

Problem 2 (Unrooted Episode Clustering (UEC)). *Given a collection of unrooted gene trees* $G_1, G_2, \ldots, G_n$ *and a species tree* S. *Compute the minimal* $\mathsf{EC}(T_1, T_2, \ldots, T_n, S)$ *in the set of rooted gene trees* $\{T_1, T_2, \ldots, T_n\}$ *such that* T_i *is a rooting obtained from* G_i *by placing the root on the edge from the D-plateau.*

Observe, that we allow rootings only in the D-plateau. Otherwise, the total number of gene duplications is not minimal. By single-UEC we denote the problem UEC for a single unrooted gene tree, i.e., when $n = 1$. Every edge in an unrooted gene tree that induces the optimal solution for single-UEC will be called *optimal* (for single-UEC). For convenience, we use $\mathsf{EC}(T_1, T_2, \ldots, T_n)$ instead of $\mathsf{EC}(T_1, T_2, \ldots, T_n, S)$.

Episodes in a gene tree with an empty edge

In this Section we solve single-UEC problem for the case when the input gene tree has one empty edge.

Let v be a center of the star that contains the only DL-plateau edge in a gene tree G. This star induces three rooted subtrees T_a, T_b and T_c rooted at neighbours a, b and c, respectively, as indicated in Fig. 2. Let $\mathbb{1}$ be the indicator function, that is, $\mathbb{1}(p)$ is 1 if p is satisfied and 0 otherwise.

Lemma 1. *Let* $a_0, a_1, a_2, \ldots, a_{n+1}$ *(for* $n \geq 0$*) be the path of D-plateau nodes connecting* $v = a_0$ *and* $a_{n+1} \in T_a$ *in* G. *Let* G_n *be the D-minimal rooting induced by the edge* $\langle a_n, a_{n+1} \rangle$. *If* $e_* = \langle v, c \rangle$ *is empty then*

$$\mathsf{EC}(G_n) = \mathsf{EC}(T_1, T_2, \ldots, T_{n+1}, T_b, T_c) + \mathbb{1}(\mathsf{root}(T_i) \notin \mathsf{Dup}(G_n) \textit{ for all } i),$$

where $T_1, T_2, \ldots, T_{n+1}$ *are subtrees of* T_a *such that* $T_a = (T_1, (T_2, \ldots, (T_n, T_{n+1}) \ldots))$ *and the root of* T_{n+1} *is* a_{n+1} *(see Figs. 2 and 4).*

Proof. First we show that v is a speciation node in G_n. It follows from the fact that v is a center of S2 star and $\langle v, b \rangle$ is single. Thus, $\mathsf{M}_n(v) = \top$, $\mathsf{M}_n(c) \prec \top$ and $\mathsf{M}_n(b) \prec \top$, where M_n is the lca-mapping for G_n. From the fact that $\mathsf{M}_n(v) = \top$ we conclude that all nodes on the path connecting the parent of v with the root in G_n are mapped to $\top$, therefore, they are duplications.

Lets consider the number of duplication clusters in G_n. We have the $\top$-cluster composed of the duplication nodes $a_1, a_2, \ldots, a_n, \mathsf{root}(G_n)$ mapped to $\top$. Both T_c and T_b in G_n are under speciation node v so their clusters are disjoint with the $\top$-cluster. Finally, if the root of some T_i is a duplication then its cluster can be merged with the $\top$-cluster. Therefore, the $\top$-cluster contributes to $\mathsf{EC}(G_n)$ only if the root of T_i is a speciation for every i. Now, it is easy to conclude the final formula. □

Lemma 2. *Under the assumptions from the previous lemma, we have*

$$\mathsf{EC}(G_n) = \mathsf{EC}(G_*) + \mathbb{1}(b \in \mathsf{Dup}(G_*) \textit{ and } \mathsf{root}(T_i) \notin \mathsf{Dup}(G_*) \textit{ for all } i),$$

where G_* *is the rooting induced the empty edge* $e_* = \langle v, c \rangle$ *(see Fig. 4).*

Proof. Both rootings G_n and G_* are D-minimal. Hence, $\mathsf{D}(G_*, S) = \mathsf{D}(G_n, S)$ and, in consequence, the number of duplication nodes in $A = \{a_1, a_2, \ldots, a_n, v, \mathsf{root}(G_*)\}$ in G_* and $B = \{a_1, a_2, \ldots, a_n, v, \mathsf{root}(G_n)\}$ in G_n are equal. It follows from the properties of star S2, that in G_n node v is a speciation mapped to $\top$. Hence, all predecessors of v are duplications in G_n. Thus, we have exactly $n + 1$ duplications in B. On the other hand, by star S2, $\mathsf{root}(G_*)$ is a speciation, therefore all remaining nodes in A are duplications.

We conclude that G_n has the $\top$-cluster containing duplications from A, and G_* has a cluster (mapped below $\top$) containing duplications from B, respectively. These two

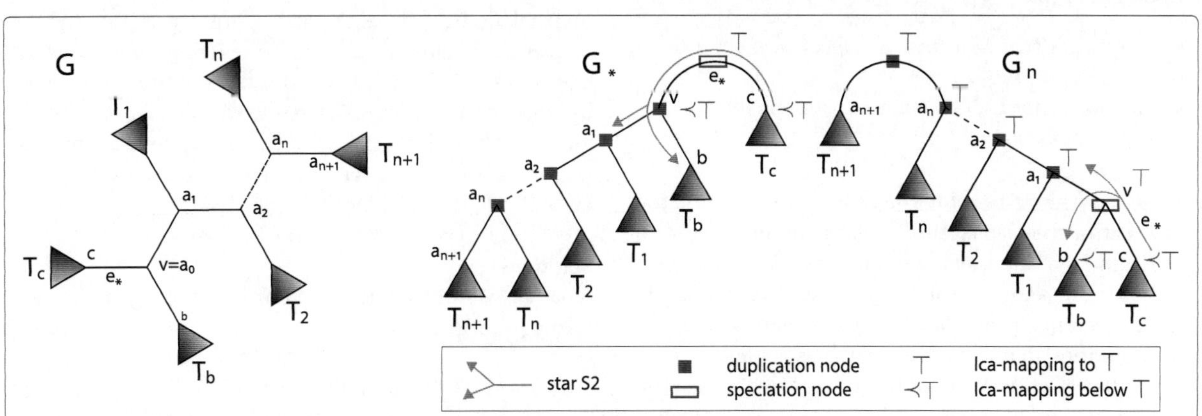

Fig. 4 Trees from Lemma 1 and 2. A gene tree G (*left*) and the rootings of G (*right*) from Lemma 1 and Lemma 2

clusters we call *high clusters*. If the root of one of T_i's is a duplication, then it can be merged with the high cluster in both rootings. Otherwise, if every root of these subtrees is a speciation then the high cluster is disjoint with clusters from $T_1, T_2, \ldots, T_{n+1}$. Moreover, if b is a duplication then the high cluster contains b in G_*. However, in G_n the cluster of b will be disjoint with the $\top$-cluster due to the speciation node v. Combining the above observations we obtain our formula. □

Lemma 1 and Lemma 2 complete the case of empty rootings. We proved that rooting on empty edge has the best EC.

Episodes in a gene tree with a double edge

We start with two technical lemmas on the properties of the plateaus.

Lemma 3. *If the DL-plateau consists of exactly one double edge then the D-plateau and the DL-plateau are equal.*

Proof. Let $\langle v, a \rangle$ be the DL-plateau edge (see Fig. 2). It follows from the property of star S3 that both v and a are mapped to $\top$ in the DL-minimal rooting and their children (if present) are mapped below $\top$. Hence, the root is a duplication, while v and a are speciation nodes. Now, it is easy to show that rooting on edge $\langle v, b \rangle$ (or $\langle v, c \rangle$) induces one additional gene duplication at v. We conclude that the only edge with the minimal duplication cost is $\langle v, a \rangle$. □

We write that a node g from unrooted gene tree G is a *super-duplication*, if g is a duplication in every rooting of G. Please recall, that the plateau is a subtree of a gene tree, thus a leaf of the D-plateau may refer to an internal node of a gene tree. For example, in Fig. 3, the D-plateau of G_1 has four leaves: one is an internal node of G_1 and others, labeled a, c, e, are leaves of G_1.

Lemma 4. *If the DL-plateau has a double edge then*

- *every leaf of the D-plateau is a speciation in every rooting from the D-plateau,*
- *and every internal node of the D-plateau is a super-duplication.*

Proof. For the first part of the proof, let us assume that v is a leaf of the D-plateau. By using the notation from Fig. 2, let v be a center of a star such that $\langle v, a \rangle$ belongs to the D-plateau. Assume that v is a duplication in every D-minimal rooting. Then, the D-minimal rooting $G_{\langle v,a \rangle}$ has one duplication in v. The edge $\langle v, b \rangle$ does not belong to D-plateau, therefore, the rooting $G_{\langle v,b \rangle}$ has at least one more duplication than $G_{\langle v,a \rangle}$. Hence, $G_{\langle v,b \rangle}$ has two duplications in v and in the root. Moreover, the root of $G_{\langle v,a \rangle}$ is not a duplication. However, this is possible only when T_a and T_v are mapped below $\top$, thus the $\langle v, a \rangle$ is an empty edge, which is a contradiction with Theorem 2. This completes the first part of the proof.

Next, if the DL-plateau consists of exactly one double edge, then, by Lemma 3 the property holds trivially. Now, we assume that the DL-plateau has more than one edge. We show that every internal node v of the DL-plateau is a super-duplication. From Theorem 2 we know that v is incident to at least two double edges. Hence, in any rooting at least one of its children is mapped to $\top$. We conclude that v is a duplication mapped to $\top$.

Let us consider a path $p = v_1, v_2, \ldots, v_n$ $(n > 1)$ connecting an internal node v_1 from the DL-plateau with a leaf v_n from the D-plateau. We show that the first $n-1$ nodes on p are duplications for every rooting placed on this path. It follows from the first part of this proof that v_1 is a super-duplication mapped to $\top$. Hence, when rooting at $\langle v_{n-1}, v_n \rangle$, we have n gene duplications: for $v_1, v_2, \ldots, v_{n-1}$ and one for the root. All edges from p are elements of the D-plateau, thus moving the root to other edges on p will preserve the total number of gene duplications.

It should be clear that the same holds when choosing other root positions. We omit the details. □

In the next lemma we show that rootings at edges of the D-plateau induce the same EC cost.

Lemma 5. *If an unrooted gene tree G has no empty edge then for any D-minimal rooting of G denoted by* G_*

$$\mathsf{EC}(G_*) = \mathsf{EC}(T_1, T_2, \ldots, T_n) + 1,$$

where $T_1, T_2, \ldots, T_n$ *are the rooted subtrees of G obtained from G by removing all internal nodes of the D-plateau.*

Proof. It follows from Lemma 4 and its proof that all internal nodes of the D-plateau are present in the $\top$-cluster in the clustering with minimal number of clusters. This cluster is separated from other duplication clusters by speciation nodes located on the border of the D-plateau. Thus, the clusters induced by optimal solution to EC for G_* are the clusters induced by optimal solution to EC of $T_1, T_2, \ldots, T_n$ plus the $\top$-cluster. □

Solutions

Now we present solutions to our unrooted episode clustering problem.

Theorem 3 (Solution to single-UEC). *For any gene tree G, an edge e is optimal for* single-UEC, *if either e is empty or e is in the D-plateau and G has a double edge.*

Proof. The first part of the proof follows immediately from Lemma 2 and the second part from Lemma 5. □

Theorem 4. *For a collection of unrooted gene trees $G_1, G_2, \ldots, G_n$, if every gene tree has a double edge then rooting every gene tree on an edge from the D-plateau yields the optimal solution for UEC.*

Proof. Assume that $n = 2$ and let G_1' and G_2' be two D-plateau rootings of G_1 and G_2, respectively. It should be clear that $\mathsf{EC}(G_1', G_2') = \mathsf{EC}(T)$, where $T = (G_1', G_2')$. Next, by Lemma 5, $\mathsf{EC}(T)$ is independent on the choice of rooting of G_1 and G_2, as long as the rootings are in the D-plateau. Therefore, we conclude that $\mathsf{EC}(T)$ is the solution to UEC Problem for G_1 and G_2. This observation can be easily generalized by induction to any n. □

Note that we cannot generalize the property stated in Theorem 4 to gene trees with empty edges. The example is shown in Fig. 3. Consider the dataset $\{G_1, G_2\}$. G_1 has five D-minimal rootings, while G_2 has exactly one. In G_{2*} we have one $\top$-cluster, therefore G_{2*} with G_{1*}, i.e., the empty edge rooting of G_1, have two duplication clusters. However, the best clusterings for $\{G_1, G_2\}$ having exactly one cluster are obtained for $G_{1,1}$, $G_{1,2}$ or $G_{1,3}$. On the other hand, the best clusterings can be also obtained for empty edge rootings, e.g. $\{G_{1,*}, G_{4,*}\}$ with cost 2 for the input $\{G_1, G_4\}$. From these examples, we see that the empty edges have different properties than double edges in the context of UEC, and we cannot generalize Theorem 4 to empty edges.

Theorem 5 (Candidate rootings for UEC). *For a collection of unrooted gene trees $\mathcal{G}$, the solution to UEC is induced by a rooting edge e of $G \in \mathcal{G}$ satisfying:*

(U1) if G has a double edge, then e is any D-minimal edge in G,

(U2) if G has an empty edge, then e is an element of star S2.

Proof. If some $G \in \mathcal{G}$ has a double edge then the property follows from Theorem 4 and Lemma 5. For gene trees with an empty edge e_* we show that any D-minimal rooting of the edge that is not adjacent to e_* can be equivalently replaced by a rooting adjacent to e_*. By using the notation from Fig. 2, let $T_a = (T_{a'}, T_{a''})$ such that a' and a'' are the roots of $T_{a'}$ and $T_{a''}$, respectively. We show that the rooting $G_{\langle v,a \rangle}$ denoted by G_a (see Fig. 5) has the same duplication episodes as the rooting $G_{a'}$ obtained for the edge $\langle a, a' \rangle$. In both rootings v is a speciation, therefore the structure of clusters present in T_b and T_c is the same in both rootings. The edge $\langle v, a \rangle$ is a-incoming, thus the roots are duplications mapped to $\top$. From the fact that $\langle a, a' \rangle$ is in the D-plateau we have that a is a duplication. Thus, every root and a induce the $\top$-cluster. Finally, if a'' is a duplication node, then in both rootings it will be a member of the $\top$-cluster. We proved these two adjacent rootings have the same structure of clusters. Therefore, it is sufficient to choose the rooting G_a instead of $G_{a'}$. This proof can be naturally extended by induction to any edge from the D-plateau. □

We conclude that for a gene tree G we have at most 5 candidates for rootings. For instance, G_4 has two stars S2 in the D-plateau, therefore we have 5 candidate rootings: the empty edge rooting $G_{4,*}$ and the rootings of adjacent edges $G_{4,1}$, $G_{4,4}$, $G_{4,7}$ and $G_{4,10}$. Note that the clusters from $G_{4,1}$ are equivalent to clusters from $G_{4,2}$ and $G_{4,3}$. Similar property holds for other candidates.

Next, we show that the condition U2 can be improved.

Lemma 6. *Under the assumptions from Theorem 5. Let the set of clusters induced by the solution to UEC contains $\top$-cluster. Then, the condition (U2) from Theorem 5 can be refined as follows:*

(U2') if e_* is the empty edge in G, then e is one among at most two non-adjacent edges such that $e = \langle x, y \rangle$ is adjacent to e_* and $\mathsf{M}_*(x) = \mathsf{M}_*(y)$, where M_* is the lca-mapping for G_*.

Proof. Let G be a gene tree with an empty edge. Let e_a be that edge from (U2'). By using the notation from Fig. 5, we compare the rooting G_* and $G_{\langle v,a \rangle}$, denoted here by G_a. We have the following clusters in G_*: the cluster C that contains c (if c is a duplication) and the cluster X that contains v (it follows from the proof of Lemma 2 that v is a duplication node). Thus, $X = \{v\} \cup A \cup B$ where A and B denote duplications from T_a and T_b, respectively. Note that C has the same contribution to EC in both rootings, which follows from the property that valid mappings of C are the same in both rootings. In G_a, A is a subset of the $\top$-cluster whose contribution to EC is already incorporated (by the assumption). The node v is a duplication in G_*. Hence, without loss of generality we assume that $M_*(a) = M_*(v)$, i.e., the rooting edge $\langle v, a \rangle$ satisfies the condition from (U2').

We have two cases depending on whether B is empty. If B is empty then G_a has "better" composition of clusters than in G_*, i.e., one cluster less then in G_* and other clusters has the same valid mappings. Otherwise, both rootings are equivalent if $M_*(b) = M_*(v)$ (B in G_a has the same valid mappings as X in G_*), or again G_a has a better structure of clusters than G_* if $M_*(b) \prec M_*(v)$ (valid mappings of X in G_* are included in valid mappings of B in G_a). Similarly, we show that G_a is also better than $G_{\langle v,b \rangle}$ (see also rootings of G_4 in Fig. 3).

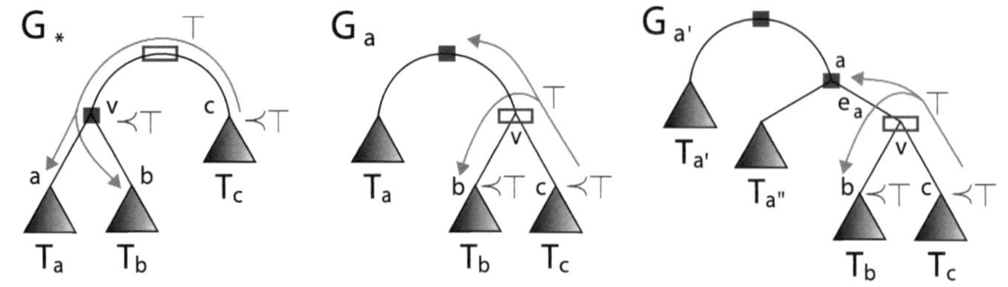

Fig. 5 Trees from Theorem 5 and Lemma 6. The rootings of *G* from Theorem 5 and Lemma 6. We use the notation G_a instead of $G_{\langle v,a\rangle}$. See Fig. 4 for a legend of the symbols used

We proved that among three rootings from the star S2 we can choose one candidate. The second edge is obtained from the second star S2 (sharing the empty edge) if it is present in the gene tree (see Theorem 2). □

From the last lemma we have at most two candidates for any gene tree from the input collection. For example, the candidate rooting $G_{4,1}$ has more flexible valid mappings than $G_{4,4}$, e.g. the duplication cluster of $((c,b),a)$ in $G_{4,1}$ has larger range of possible mappings than the duplication cluster of $((d,b),a)$ in $G_{4,4}$, while the remaining two clusters have the same locations in the species tree. Hence, for the dataset $\{G_3, G_4\}$, if the $\top$-cluster is present in solution to UEC, we have two candidates $G_{4,1}$ and $G_{4,7}$ (which is more flexible than $G_{4,10}$). Note, that the clustering costs 3 is obtained by rootings $G_{3,*}$ and $G_{4,1}$ (or $G_{4,2}$, $G_{4,3}$).

Algorithms

Algorithm 1 presents the solution to UEC problem. The correctness of this algorithm follows from Theorem 5 and Lemma 6. Algorithm 1 has two phases. In the first phase for every gene tree a set of candidate rootings is prepared with respect to the conditions (U1) and (U2'). To find optimal rootings we use a linear time algorithm (procedure FindOptEdge) based on greedy descent method that search a double or an empty edge in a gene tree [32]. Based on condition U2', we divide possible solutions into two categories depending on the presence of $\top$-cluster in an optimal clustering. If the $\top$-cluster is not present then every gene tree has an empty edge (in line 10). Otherwise, we check every possible variant of rooting candidates. Note that from Lemma 6, a gene tree has two candidates if and only if the gene tree has two stars S2 that are included in the D-plateau. Thus, the overall time complexity depends on the presence of such trees in the input. From this observation we conclude the following result.

Theorem 6. *The time complexity of Algorithm 1 is* $O(2^k(\sum_i |G_i| + |S|))$, *where k is the number of input gene trees having two stars S2 that are included in the D-plateau.*

Algorithm 1 Unrooted Episode Clustering 1

1: **Input** A binary species tree S, a collection of unrooted gene trees $G_1, G_2, \ldots, G_n$.
2: **Output** Minimal $\mathsf{EC}(T_1, T_2, \ldots, T_n, S)$ in the set of all rootings T_i of G_i such that T_i is a rooting obtained from G_i by placing the root on the edge from the D-plateau.
3: For every i compute the set of candidate rooting edges R_i:
4: e_*:=FindOptEdge(G_i)
5: **If** e_* is double: $R_i := \{e_*\}$
6: **If** e_* is empty **then for** $x \in e_*$ such that x is not a leaf.
7: **Let** c be a child of x in G_* such that $\langle x, c\rangle$ is D-minimal and not adjacent to any edge from R_i and $M_*(c) = M_*(x)$
8: $R_i := R_i \cup \{\langle x, c\rangle\}$.
9: **If** every G_i has an empty edge then $\alpha := \mathsf{EC}(T_1, T_2, \ldots, T_n)$, where T_i is the empty edge rooting of G_i **else** $\alpha := +\infty$.
10: $\beta = \min_{e_i \in R_i} \mathsf{EC}(G_{e_1}, G_{e_2}, \ldots, G_{e_n})$.
11: **Return** $\min\{\alpha, \beta\}$.
12: **Function FindOptEdge(G)**
13: **Let** $m_{x,y} = M_{G_{\langle x,y\rangle}}(x)$ *// can be computed in* $O(|G|)$ *steps [32].*
14: **Let** v be a node from V_G and let $\top$ by the lca-mapping of some rooting of G.
15: **While** there exists a node w adjacent with v such that $m_{w,v} = \top \neq m_{v,w}$
16: **do:** set $v := w$ (star S1).
17: **Return** $\langle v, w\rangle$ such that $\langle v, w\rangle$ an empty or double edge i.e., $m_{v,w} = \top = m_{w,v}$ or $m_{v,w} \neq \top \neq m_{w,v}$.

Thus, from theoretical point of view UEC is fixed parameter tractable. Later we show that k usually represents a small fraction (up to 5 %) of the whole input. For the cases when 2^k is still too large for efficient computation, we propose Algorithm 2, in which we first solve the instance of UEC for the collection of gene trees that have a unique candidate. Clearly, if there are rootings

Algorithm 2 Unrooted Episode Clustering 2

1: **Input/output** The same as in Algorithm 1.
2: **Let** $\delta = \mathsf{EC}(G'_1, G'_2, \ldots, G'_{n'})$ computed by Algorithm 1, such that $\{G'_1, G'_2, \ldots, G'_{n'}\}$ is the set of all input gene trees having a unique candidate rooting edge (i.e., $|R_i| = 1$).
3: **If** $\delta = \alpha$ **Return** α, where α is from the 9th line of Algorithm 1 computed for the whole input.
4: **For** every $e_1 \in R_1, e_2 \in R_2, \ldots, e_n \in R_n$ (i.e. candidate rootings of the whole input)
5: If $\mathsf{EC}(G_{e_1}, G_{e_2}, \ldots, G_{e_n}) = \delta$ **then Return** δ.
6: **Return** the minimal EC value computed in lines 3 and 6.

of the whole input that have the same cost, then this cost is optimal. The overall complexity of Algorithm 2 is the same as Algorithm 1, however, for large datasets this strategy appeared to be successful after checking just one additional candidate set (in lines 2–4).

Experiments

We performed several computational experiments on three empirical datasets.

Guigó dataset consists of 53 rooted gene trees from 16 Eukaryotes from [10]. This dataset was evaluated with 71 species trees from [35], known to have the total minimal duplication cost. *Génolevures* is a dataset of 4144 gene trees [33] from nine yeast genomes [36] and two species trees: one from [37] and the second one having the lowest duplication-loss cost computed by Fasturec [38]. The third dataset *TreeFam*, spanning 25 mostly animal species, consists of 1274 curated gene family trees from TreeFam v7.0 [39]. The species tree for TreeFam is based on NCBI taxonomy.

We implemented our algorithms and the algorithms for the rooted variant of EC Problem (based on [29]). In our experiments the rooting candidates were used to compare the results for UEC with the model of mappings (for rooted gene trees) proposed in [28].

We performed two series of 74 computational experiments, one for our model and one with the model described in [28]. The total running time of our program was about 7 minutes on a standard PC workstation. For every dataset we were able to find solutions to UEC by testing at most two rooted instances of input gene trees (see Algorithm 2). The summary of experiments is depicted in Table 1.

For the Guigó dataset we found four duplication clusters, while for the rooted model from [28] we located five clusters. The difference can be explained by the properties of our model that is more flexible: the input trees are unrooted and the model of valid mappings is more generic. Observe that this dataset has unique rooting candidates ($k = 0$).

Génolevures is the most complex dataset due to its size and potentially large parameter k. Despite these properties, Algorithm 2 located 17 clusters for the filtered input with all unique rooting candidates. In other words, in this filtered dataset a duplication cluster is present in every node of the species tree. Obviously, the whole input dataset has the same property. The same holds for the model from [28].

In TreeFam we located 45 clusters for the filtered dataset with unique rooting candidates. Then, Algorithm 2 found the solution having the same cost for the whole dataset (see Fig. 6). The same result was obtained for the model from [28] (see Table 1).

Conclusions

In this article we presented the first solution to the open problem of the duplication episode clustering for case when the input collection is composed of unrooted gene trees. By using theoretical properties of the unrooted reconciliation we proved that the problem has nice mathematical and computational properties. From practical point of view, we were able to provide efficient algorithms and tools that were successfully applied to locate duplication clusters in real datasets.

From the computational point of view the complexity of our algorithms depends on the parameter k, i.e., in the worst case EC Problem has to be solved 2^k times in order to find a solution to UEC. Even if k usually represents a small fraction of the whole input it can be still large, e.g. $k > 100$ for the yeast dataset, which may prohibit computation of all possible variants. Here we proposed a solution, that is based on the observation that the clustering induced from the input gene trees having unique candidates (that is, without k gene trees with non-unique variants), usually represents an optimal solution

Table 1 Experimental results

Set	# Species trees	# Leaves	# Gene trees	k	Our model		Model [28]	
					EC	% Locations	EC	% Locations
Guigó	71	16	53	0	4	12,9 %	5	16,1 %
Génolevures	1 [37]	9	4144	55	17	100 %	17	100 %
	1 [38]	9	4144	156	17	100 %	17	100 %
TreeFam	1	28	1274	67	45	81,8 %	45	81,8 %

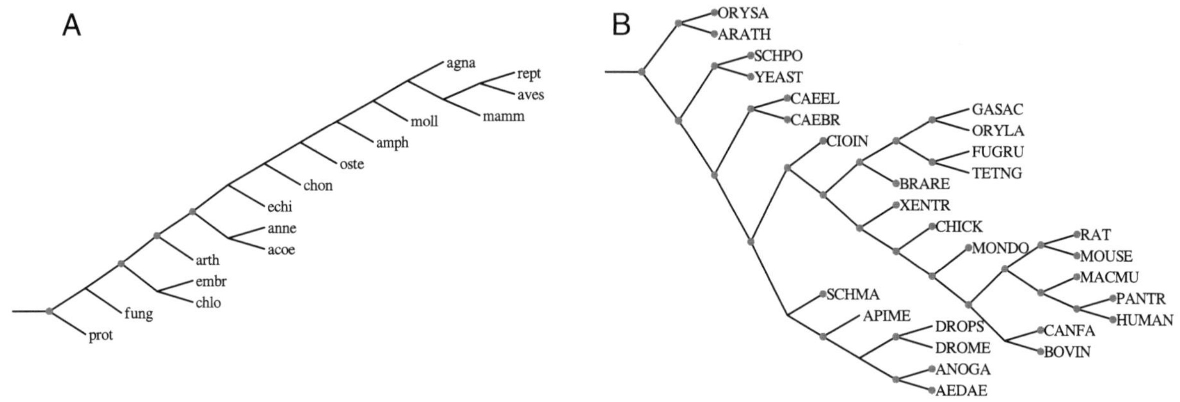

Fig. 6 Duplication clusters in empirical datasets. Duplication clusters (*marked by red circles*) inferred from experiments. **a** Guigó species tree (chosen from 71 species trees from [35] as the most biologically reasonable [40]). **b** TreeFam species tree based on NCBI taxonomy

for the whole input. Thus, the strategy that we applied in Algorithm 2, i.e., first cluster easy part and then try to incorportate the hard one by using already identified clusters, appeared to be successful even for potentially complex datasets.

Our computational experiments show that the duplication clusters are usually located in large parts of the species tree especially when the input dataset consists of thousands of gene trees. To provide more detailed information on the duplication clusters, we plan to study minimal episode problem (ME) which is a natural extension of the episode clustering problem. In the future we plan to extend the episode clustering problem by using other types of valid mappings.

Our software for solving unrooted episode clustering problem is publicly available at http://www.mimuw.edu.pl/jpaszek/uec.php.

Abbreviations
D: gene duplication; DL: Gene duplication and loss; EC: episode clustering for rooted gene trees; lca: least common ancestor; UEC: episode clustering for unrooted gene trees.

Competing interests
The authors declare that they have no competing interests.

Authors' contributions
JP and PG contributed equally to the writing of the paper. Both authors read and approved the final manuscript. JP implemented algorithms and performed all computational experiments.

Acknowledgements
We would like to thank the three reviewers for their detailed comments that allowed us to improve our paper. JP and PG were supported by the grant of NCN #2011/01/B/ST6/02777. JP was supported by the DSM funding for young researchers of the Faculty of Mathematics, Informatics and Mechanics of the University of Warsaw.

Declarations
The publication costs for this article were funded by the Polish Ministry of Science and Higher Education funding for Faculty of Mathematics, Informatics and Mechanics of the University of Warsaw.
This article has been published as part of BMC Genomics Volume 17 Supplement 1, 2016: Selected articles from the Fourteenth Asia Pacific Bioinformatics Conference (APBC 2016): Genomics. The full contents of the supplement are available online at http://www.biomedcentral.com/bmcgenomics/supplements/17/S1.

Published: 11 January 2016

References
1. Kellis M, Birren BW, Lander ES. Proof and evolutionary analysis of ancient genome duplication in the yeast Saccharomyces cerevisiae. Nature. 2004;428:617–24.
2. Guyot R, Keller B. Ancestral genome duplication in rice. Genome. 2004;47(3):610–4.
3. Vision TJ, Brown DG, Tanksley SD. The origins of genomic duplications in Arabidopsis. Science. 2000;290(5499):2114–7.
4. Costantino L, Sotiriou SK, Rantala JK, Magin S, Mladenov E, Helleday T, et al. Break-induced replication repair of damaged forks induces genomic duplications in human cells. Science. 2014;343(6166):88–91.
5. Aury JM, Jaillon O, Duret L, Noel B, Jubin C, Porcel BM, et al. Global trends of whole-genome duplications revealed by the ciliate Paramecium tetraurelia. Nature. 2006;444(7116):171–8.
6. Cui L, Wall PK, Leebens-Mack JH, Lindsay BG, Soltis DE, Doyle JJ, et al. Widespread genome duplications throughout the history of flowering plants. Genome Res. 2006;16(6):738–49.
7. Van de Peer Y, Maere S, Meyer A. The evolutionary significance of ancient genome duplications. Nat Rev Genet. 2009;10(10):725–32.
8. Page RDM. Maps between trees and cladistic analysis of historical associations among genes, organisms, and areas. Syst Biol. 1994;43(1):58–77.
9. Mirkin B, Muchnik I, Smith TF. A biologically consistent model for comparing molecular phylogenies. J Comput Biol. 1995;2(4):493–507.
10. Guigó R, Muchnik IB, Smith TF. Reconstruction of ancient molecular phylogeny. Mol Phylogenet Evol. 1996;6(2):189–213.
11. Page RDM. Extracting species trees from complex gene trees: reconciled trees and vertebrate phylogeny. Mol Phylogenet Evol. 2000;14:89–106.
12. Arvestad L, Berglund AC, Lagergren J, Sennblad B. Bayesian gene/species tree reconciliation and orthology analysis using MCMC. Bioinformatics. 2003;19(Suppl 1):i7–15.
13. Bonizzoni P, Della Vedova G, Dondi R. Reconciling a gene tree to a species tree under the duplication cost model. Theor Comput Sci. 2005;347:36–53.
14. Goodman M, Czelusniak J, Moore GW, Romero-Herrera AE, Matsuda G. Fitting the gene lineage into its species lineage, a parsimony strategy illustrated by Cladograms Constructed from Globin sequences. Syst Zool. 1979;28(2):132–63.
15. Górecki P, Tiuryn J. DLS-trees: a model of evolutionary scenarios. Theor Comput Sci. 2006;359:378–99.
16. Arvestad L, Lagergren J, Sennblad B. The gene evolution model and computing its associated probabilities. J ACM. 2009;56(2):1–44.

17. Doyon JP, Chauve C, Hamel S. Space of gene/species tree reconciliations and parsimonious models. J Comput Biol. 2009;16:1399–1418.
18. Durand D, Halldórsson BV, Vernot B. A hybrid micro-macroevolutionary approach to gene tree reconstruction. J Comput Biol. 2006;13(2):320–35.
19. Hallett MT, Lagergren J. Efficient Algorithms for Lateral Gene Transfer Problems. In: Proceedings of the Fifth Annual International Conference on Computational Biology. RECOMB '01. New York, NY, USA: ACM; 2001. p. 149–156.
20. In: Bourque G, El-Mabrouk N, editors. Comparative Genomics, RECOMB 2006 International Workshop, RCG 2006, Montreal, Canada, September 24–26, 2006, Proceedings. vol. 4205 of Lect Notes Comput Sc. Berlin, Germany: Springer; 2006.
21. Ma B, Li M, Zhang L. From gene trees to species trees. SIAM J Comput. 2000;30(3):729–52.
22. Sjostrand J, Tofigh A, Daubin V, Arvestad L, Sennblad B, Lagergren J. A Bayesian method for analyzing lateral gene transfer. Syst Biol. 2014;63(3):409–20.
23. Stolzer M, Lai H, Xu M, Sathaye D, Vernot B, Durand D. Inferring duplications, losses, transfers and incomplete lineage sorting with nonbinary species trees. Bioinformatics. 2012;28(18):i409–15.
24. Zhang L. From gene trees to species trees II: species tree inference by minimizing deep coalescence events. IEEE/ACM Trans Comput Biol Bioinform. 2011;8(6):1685–91.
25. Page RDM, Cotton JA. Vertebrate phylogenomics: reconciled trees and gene duplications. Pac Symp Biocomput. 2002;536–47.
26. Fellows M, Hallet M, Stege U. On the Multiple Gene Duplication Problem. In: 9th International Symposium on Algorithms and Computation (ISAAC'98), Lecture Notes in Computer Science 1533. Taejon, Korea: Springer Berlin Heidelberg; 1998. p. 347–356.
27. Burleigh JG, Bansal MS, Wehe A, Eulenstein O. Locating Multiple Gene Duplications through Reconciled Trees In: Vingron M, Wong L, editors. RECOMB. vol. 4955 of Lect Notes Comput Sc. Berlin, Germany: Springer; 2008. p. 273–284.
28. Bansal MS, Eulenstein O. The multiple gene duplication problem revisited. Bioinformatics. 2008;24(13):i132–8.
29. Luo CW, Chen MC, Chen YC, Yang RWL, Liu HF, Chao KM. Linear-time algorithms for the multiple gene duplication problems. IEEE/ACM Trans Comput Biol Bioinform. 2011;8(1):260–5.
30. Holland BR, Penny D, Hendy MD. Outgroup misplacement and phylogenetic inaccuracy under a molecular clock – a simulation study. Syst Biol. 2003;52:229–38.
31. Huelsenbeck JP, Bollback JP, Levine AM. Inferring the Root of a Phylogenetic Tree. Syst Biol. 2002;51(1):32–43.
32. Górecki P, Tiuryn J. Inferring phylogeny from whole genomes. Bioinformatics. 2007;23(2):e116–22.
33. Górecki P, Eulenstein O. Algorithms: simultaneous error-correction and rooting for gene tree reconciliation and the gene duplication problem. BMC Bioinformatics. 2012;13(Suppl 10):S14.
34. Górecki P, Eulenstein O, Tiuryn J. Unrooted tree reconciliation: a unified approach. IEEE/ACM Trans Comput Biol Bioinform. 2013;10(2):522–36.
35. Chang W, Górecki P, Eulenstein O. Exact solutions for species Tree Inference from discordant gene trees. J Bioinform Comput Bio. 2013;11(5): 1342005.
36. Sherman DJ, Martin T, Nikolski M, Cayla C, Souciet JL, Durrens P. Génolevures: protein families and synteny among complete hemiascomycetous yeast proteomes and genomes. Nucleic Acids Res. 2009;37(suppl 1):D550–4.
37. Dujon B. Yeasts illustrate the molecular mechanisms of eukaryotic genome evolution. Trends Genet. 2006;22(7):375–87.
38. Górecki P, Eulenstein O. GTP supertrees from unrooted gene trees: linear time algorithms for NNI based local searches. Lect Notes Comput Sc. 2012;7292:83–105.
39. Ruan J, Li H, Chen Z, Coghlan A, Coin LJ, Guo Y, et al. TreeFam 2008 Update. Nucleic Acids Res. 2008;36:D735–40.
40. Page RDM, Charleston MA. Reconciled trees and incongruent gene and species trees In: Mirkin B, McMorris FR, Roberts FS, Rzhetsky A, editors. Mathematical Hierarchies in Biology, American Mathematical Society, Providence, Rhode Island; 1997. p. 57–70.

Liu *et al. BMC Genomics* 2016, **17**(Suppl 1):13
DOI 10.1186/s12864-015-2306-6

BMC Genomics

PROCEEDINGS Open Access

Transcriptome sequencing based annotation and homologous evidence based scaffolding of *Anguilla japonica* draft genome

Yu-Chen Liu[1], Sheng-Da Hsu[1], Chih-Hung Chou[1], Wei-Yun Huang[1], Yu-Hung Chen[1], Chia-Yu Liu[1], Guan-Jay Lyu[1], Shao-Zhen Huang[1], Sergey Aganezov[2,3], Max A. Alekseyev[2], Chung-Der Hsiao[4*] and Hsien-Da Huang[1,5,6*]

From The Fourteenth Asia Pacific Bioinformatics Conference (APBC 2016)
San Francisco, CA, USA. 11 - 13 January 2016

Abstract

Background: *Anguilla japonica* (Japanese eel) is currently one of the most important research subjects in eastern Asia aquaculture. Enigmatic life cycle of the organism makes study of artificial reproduction extremely limited. Henceforth genomic and transcriptomic resources of eels are urgently needed to help solving the problems surrounding this organism across multiple fields. We hereby provide a reconstructed transcriptome from deep sequencing of juvenile (glass eels) whole body samples. The provided expressed sequence tags were used to annotate the currently available draft genome sequence. Homologous information derived from the annotation result was applied to improve the group of scaffolds into available linkage groups.

Results: With the transcriptome sequence data combined with publicly available expressed sequence tags evidences, 18,121 genes were structurally and functionally annotated on the draft genome. Among them, 3,921 genes were located in the 19 linkage groups. 137 scaffolds covering 13 million bases were grouped into the linkage groups in additional to the original partial linkage groups, increasing the linkage group coverage from 13 to 14 %.

Conclusions: This annotation provide information of the coding regions of the genes supported by transcriptome based evidence. The derived homologous evidences pave the way for phylogenetic analysis of important genetic traits and the improvement of the genome assembly.

Keywords: *Anguilla japonica*, Scaffolding, RNA-seq, Transcriptome, Genome annotation

Background

Abundance of Japanese eel, as well as other freshwater eels belongs to genus *Anguilla*, has been radically shrinking in the past decade [1]. Catadromous eels' enigmatic life cycle makes their reproduction affected deeply by anthropogenic impacts. Lack of an economical method to artificial reproduce makes this organism extremely vulnerable to overconsumption. Mature eels migrate thousands of kilometers into the open ocean to spawn eggs. Exact spawning locations of Japanese eels were hard to pinpoint and remained unknown until recently when they were found near Western Mariana Ridge. What prohibits the research from further improvement is that, eggs and larvae of *Anguilla japonica* are spread by the Kuroshio Current, making the habitats spans a wide area of Eastern Asia [2]. Cylindrical shape larva develops into transparent color leptocephalus larvae, which eventually metamorphosis into glass eels. Glass eels migrate back into the freshwater through estuaries, sometimes traveling within wet sands into the inner continent, where they spend years going through pigmentation into yellow eels, and then silver eels [3]. Such wide area of habitation potentially makes effect

* Correspondence: cdhsiao@cycu.edu.tw; bryan@mail.nctu.edu.tw
[4]Department of Bioscience Technology, Chung Yuan Christian University, Chung-Li, Taiwan, ROC
[1]Institute of Bioinformatics and Systems Biology, National Chiao Tung University, Hsin-Chu 300, Taiwan, ROC
Full list of author information is available at the end of the article

of pollution and diseases to be accumulated. Long life cycle and the spawning habit through migration make wild Anguilla eels hard to recover from the damage caused by overfishing.

Physical linkage map of Japanese eel were constructed in 2011 [4]. High throughput Sequencing technology was rarely applied upon this issue before. However, with the advances of sequencing technologies bringing down the cost and time consuming of DNA and RNA sequencing, plus the approaching extinction of fresh water eels, the field began to change. In 2010, mRNA-Seq study of deep sequencing and de novo reconstruction of European glass eel were reported as well as the hox genes of the specie [5], 2 years following that, draft genome sequence of European were also published [3]. The incorporative research of genomic and transcriptomics information from the deep sequencing should have major impacts in multiple fields. Expression profiling of both transcriptome of European eels response to environmental pollution were reported in 2012 [6]. The first draft genome of Japanese eel was assembled [7], proving that the hox genes and genomic distance of European and Japanese eels were conserved. By 2014, a ddRAD-based linkage map was published, providing 13 % coverage of the draft genome [8]. Such results left plenty of space for improvement.

Hereby, we provide a reconstructed transcriptome from whole body samples of juvenile (glass eels). The high throughput sequencing provides unprecedented amount of transcriptomic information. Instead of focusing only on certain types of tissues or organs, full transcriptome of the entire organism was sequenced. For future experimental design and guidance on ecological, physiological, artificially breeding and even toxicity resistance study of Japanese eel, such transcriptome can provide additional guidance. What's more, the massive amount of evidence provided by the transcriptome helps the complete of draft genome structural annotation. Combining transcriptome sequence data with publicly available expressed sequence

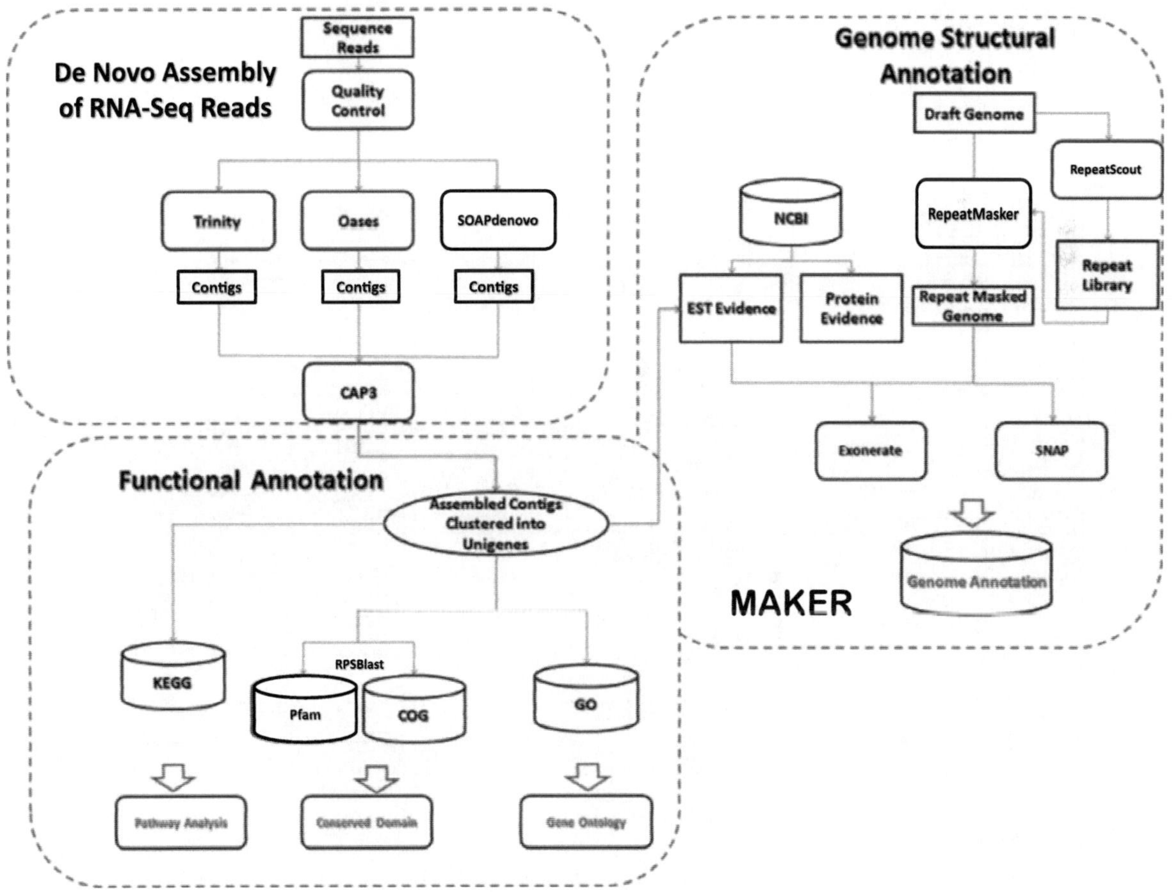

Fig. 1 Overview of the Genome Annotation Process. The overall flowchart of the genome annotation of *Anguilla Japonica* is showed in this graph. De Novo assembly of the transcriptome was treated as EST evidences, combined with other previously published information, aligned onto the draft genome. The annotation process was performed through MAKER

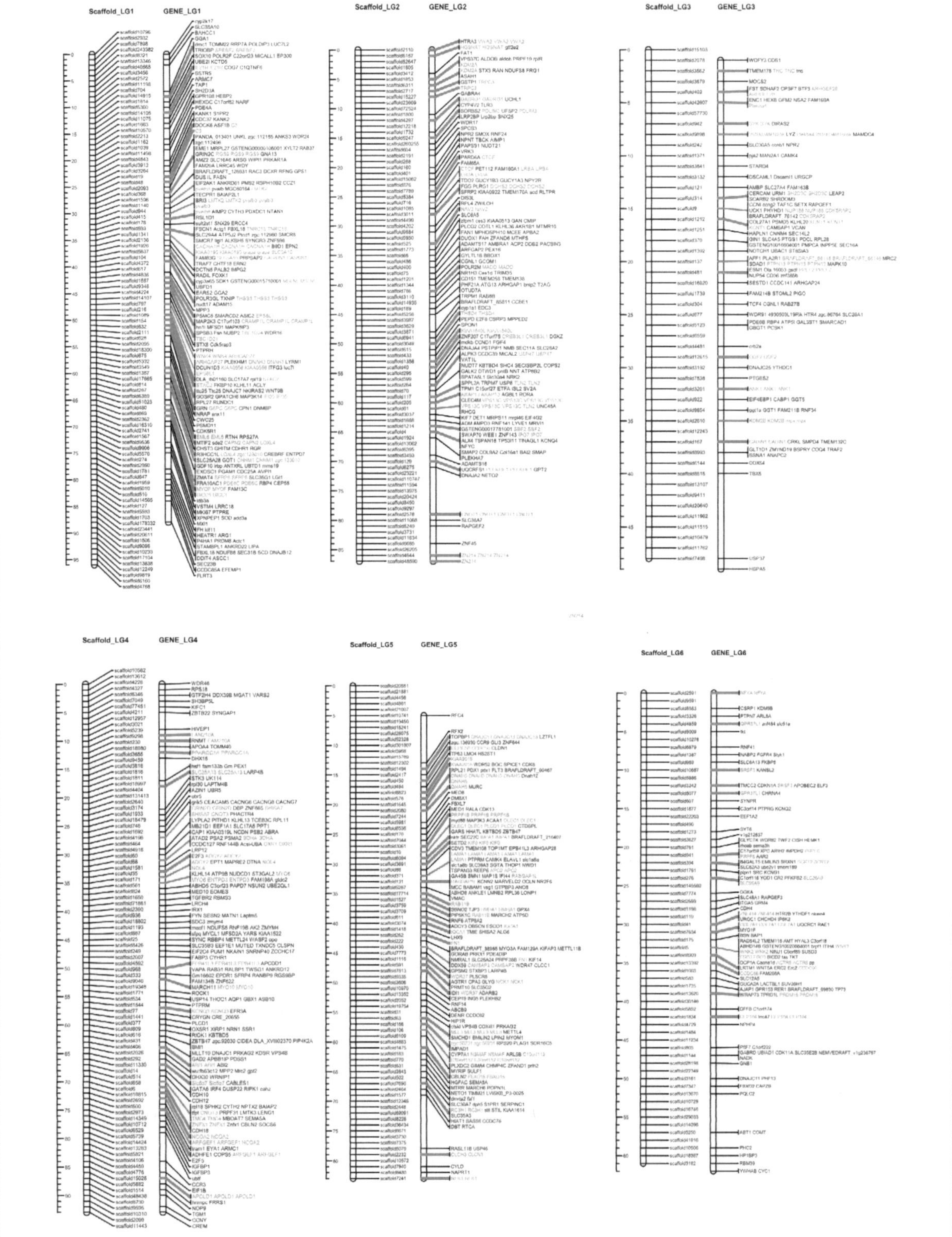

Fig. 2 (See legend on next page.)

(See figure on previous page.)
Fig. 2 Homolog Linkage Map of *Anguilla Japonica* (Linkage Groups 1 to 6). The graph shows genetic linkage maps on 6 of the 19 linkage groups. With the homologs synteny information derived from the functional annotation, we successfully allocated 137 scaffolds (13 Mb) into the established linkage group. This graph illustrates the relative position of the scaffolds. Order of the combined scaffolds was determined by an application of topological sort to combine the linkage maps of male and female linkage. Since evidence of the distance between scaffolds is not available, only the putative order was demonstrated here. The homologs gene cluster was showed in gray color

tags evidences, 18,121 genes were structurally and functionally annotated on the draft genome. The structural annotation was performed through an established pipeline, MAKER [9]. Functional annotation was based on sequence alignment. The acquired homologous evidences were further used to improve the draft genome scaffolding. Applying an improved version of scaffolding algorithm developed by Aganezov et al. [10], synteny of *Anguilla japonica* was compared to the genome of Fugu, Stickleback, Medaka, Tetraodon, Coelacanth and Zebra fish. Obtained results were integrated with previously published linkage map [8], putting 3,921 genes into the 19 linkage groups, which represent chromosomes of *Anguilla japonica*. 137 scaffolds were grouped into the linkage groups in addition to the original partial linkage groups. Phylogenetic analysis of the gene clusters correlation to thyroid hormone receptors and pigmentation were performed with MEGA 6.0 [11].

Results and discussion

Sequencing through Illumina HiSeq^{TM} 2000 generated total 85,233,812 reads, with length of 101 nucleotides. After quality control, low quality reads were trimmed and left 77,939,562 reads were left with an average length of 99.575 nucleotides (Additional file 1: Figure S1). Quality control of the sequence reads is summarized in Additional file 1: Figure S2 and Additional file 1: Table S3. Assembly were assessed through average length of unigenes, as well as quality score N50 and N90. As the result shows in Additional file 1: Table S2, average length, N50 and N90 of clustered unigenes are significantly higher than results of single De Novo assembly tools. Composition of assembled unigenes showed in Additional file 1: Table S3 demonstrates that clustered unigenes tends to have higher composition of longer nucleotides. Hence assembles generated through clustering were considered to have higher accuracy and were used for further annotation. In Additional file 1: Figure S2, we demonstrate that expression level measured with FPKM, frequency of reads per kilo base per million, distributes through all different length of assembles.

Summary of the functional annotation is listed in Additional file 1: Table S4. From the total 32,210 assembled unigenes, 16,106 were found to be aligned to known proteins in NCBI Non Redundant protein database. 10,848 of the unigenes were found to contain functional domains through RPSBLAST against NCBI Conserved Domain Database. 5641 transcripts were found to involve with system biological pathway in Kyoto Encyclopedia of Genes and Genomes (KEGG) [12]. 13,434 transcripts were annotated to certain Gene Ontology terms. Up to top 5 blast hits per query were considered in the process. Distribution of homologs belonging to other organisms were illustrated in Additional file 1: Figure S3.

Expression level of the transcripts was examined together with their frequency to be assigned to certain Gene Ontology terms. As demonstrated in Additional file 1: Figure S4, transcripts regarding enzyme regulator activity express level exceeded the total average expression level despite the fact that only four of them were found. Distribution of assembled transcripts through different KEGG pathways categories (Additional file 1: Figures S5 and S6) were also observed alongside with average expression level. As shows in Additional file 1: Figure S5, despite only few transcripts found in some pathway categories such as Circulatory system and Reaction module maps, average expression level of transcripts within these categories demonstrates potential rich activities of these pathways.

Protein functional domains found on the assembled transcripts were also viewed in their distribution alongside with expression level, demonstrated in Additional file 1: Figures S7 and S8. Protein functional domains with extreme expression level can provide guidance to the future protein- protein interaction study.

Catadromous eels' reproduction is limited by their long life cycle and migration spawning. To produce enough supply without consuming wild glass eels, development of technology that would shorten the period of time for eels to mature, and would artificially induced spawning of healthy larvae is inevitable. Hence, revealing the mechanisms of metamorphosis from leptocephalus larvae into glass eels, as well as fermentation from glass eels to mature silver eels is the key to successful artificial reproduction to supply commercial demands and keep wild eels from extinction.

In the past, transcriptomic studies of eels mainly relied on classical molecular biological experimental

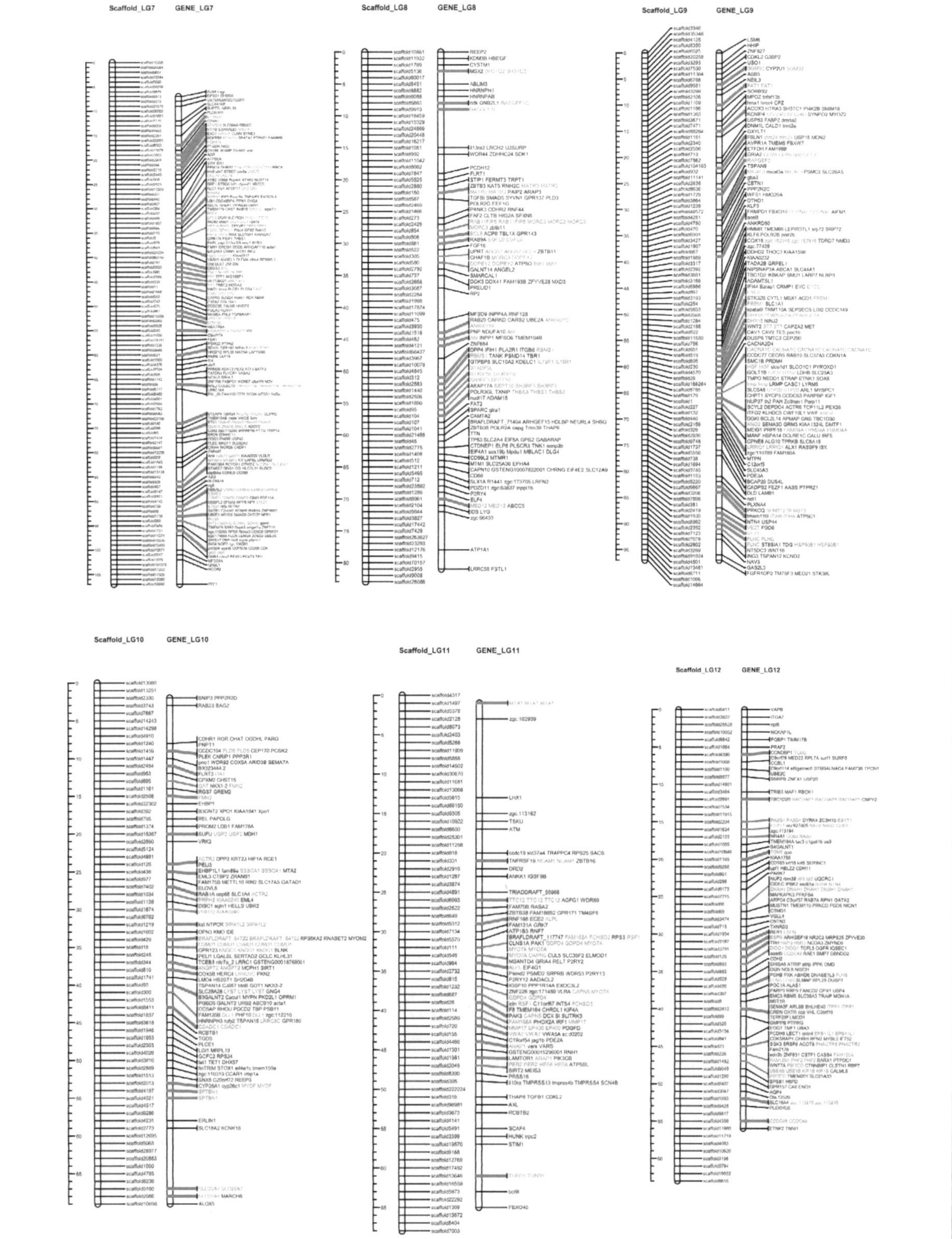

Fig. 3 (See legend on next page.)

(See figure on previous page.)
Fig. 3 Homolog Linkage Map of *Anguilla Japonica* (Linkage Groups 7 to 12). The graph shows genetic linkage maps on 6 of the 19 linkage groups. With the homologs synteny information derived from the functional annotation, we successfully allocated 137 scaffolds (13 Mb) into the established linkage group. This graph illustrates the relative position of the scaffolds. Order of the combined scaffolds was determined by an application of topological sort to combine the linkage maps of male and female linkage. Since evidence of the distance between scaffolds is not available, only the putative order was demonstrated here. The homologs gene cluster was showed in gray color

methods. Studies of various mechanisms were performed with classical molecular methods such as cloning and protein purification. Cloning and protein purification provide only partial view of the transcripts [1]. To fully capture all protein coding transcripts, a combination of next generation sequencing and new transcript assembly algorisms is necessary [13]. In 2014, a study of mRNA expression profile through RT-PCR of prolactin, growth hormone, and somatolactin of Japanese eel was reported [14]. However, researches utilizing genome information of Japanese eel, and the respective resources available for the experimental design are still limited. On the other hand, hybrids of European and American eels were found occurred naturally in Iceland [15], import of European glass eels into East Asia could trigger interspecific hybridization of *Anguilla* eel, inducing further anthropogenic impacts to this species near extinction [1]. Proven possibilities of hybrid reproduction [15], as well as the successful artificial hybrid of European and Japanese eels also bring new possibilities to the artificially reproduction [16]. However, the transcriptomic information is still limited. In 2013, the first transcriptomic study through 454 deep sequencing was performed on gill of *Anguilla japonica* [17]. Utilization of proteomic approaches and transcriptomic sequencing gave insights into the osmoregulation mechanism, providing transcriptomic view of Anguilla japonica's catadromous behavior. However the study [17] didn't correlate with the currently available draft genome.

On eel sexualize mechanism, several surveys have been conducted on ovarian steroid genesis [18]. Expression level of several genes were also found to be related to the ovarian development. Through the attempts of artificial reproduction of glass eels has been attempted since 1930s in Europe [1], only until 2003, first successful artificially induced spawning of Japanese eel was achieved through injection of salmon pituitary extracts into the female eel and human chorionic gonadotropin into male eel [19]. An unpublished successful F2 generation was declaimed in 2011 [4]. However current technology is not sufficient for large scale reproduction. Mortality of artificially cultured eels is still high.

Under current circumstance, all main stream studies of Japanese eels should inevitably focus on how to successfully improve life cycle of eels under the artificial environment to suit the existing demand. Such studies would take into consideration with all kinds of mechanisms. Metamorphosis, pigmentation and sexualize mechanisms of eels are all deeply correlated to their catadromous spawning activities, especially the metamorphosis mechanisms including reorganization of the entire body plan. Complete genome structure and transcriptome is essential for future study.

To suite such a purpose, we provide the first complete transcriptome of glass eels. Application of deep sequencing provides not only the information of homologs, but also the potential novel genes of Japanese eels. Clustering of De Novo assembled transcripts from different tools through overlapping successfully increase the assembly quality. Distribution of assembled transcripts through different species, GO terms, system biological pathway and protein functional domains of found genes were examined and demonstrated. In addition, we further provide expression level alongside the distributions. Such demonstration successfully provides the hidden information about pathways with few genes but extreme expression levels.

Without structural and functional annotation, draft genome [7] provides only limited information since *Anguilla japonica* is not an established model organism [20]. The massive amount of information provided by RNA Sequencing for our experiment makes transcriptome evidence sufficient enough to perform complete structure annotation, which identifies genes and their intron-exon structure. Before doing so, due to the fact that Eukaryotic genomes are rich with repeat sequences, a process of repeat masking needs to be carried out. Usually such a process would need a well established repeat sequence library to serve as guideline. However such a library doesn't exist for *Anguilla japonica*. In this case [20], de novo constructing a new library [21] from the draft genome is a better choice than using established model organism data base such as Repbase [22]. Structural annotation supported by RNA-seq could be done by directly assembling reads on to the genome with tool like Cufflinks. We further improved the de novo assembly into a genome based assembly following the convention guideline [20]. An overview of the annotation process can be found in Fig. 1. The automatic pipeline we used, MAKER [9], performs multiple ab initio gene

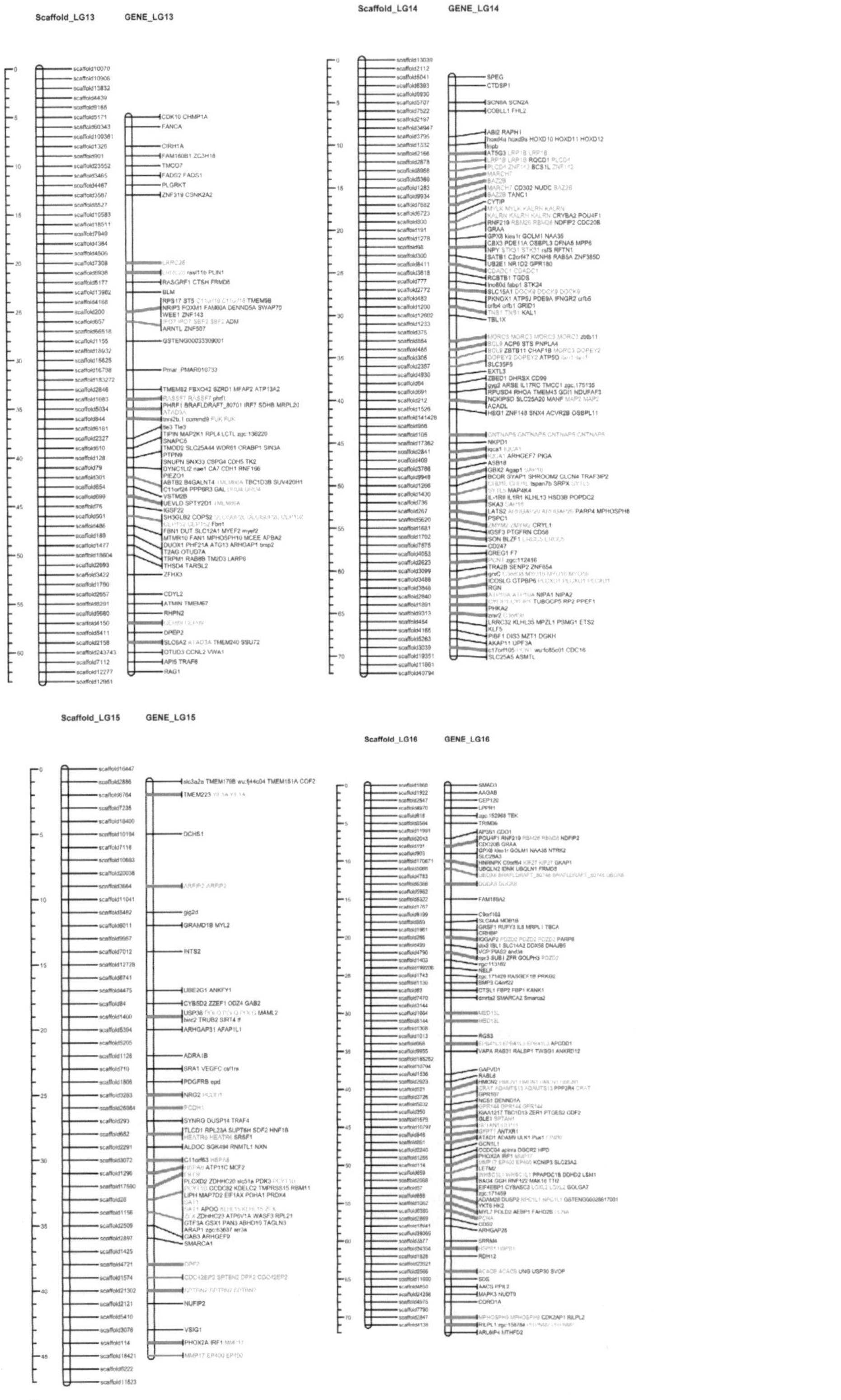

Fig. 4 (See legend on next page.)

(See figure on previous page.)

Fig. 4 Homolog Linkage Map of *Anguilla Japonica* (Linkage Groups 13 to 16). The graph shows genetic linkage maps on 4 of the 19 linkage groups. With the homologs synteny information derived from the functional annotation, we successfully allocated 137 scaffolds (13 Mb) into the established linkage group. This graph illustrates the relative position of the scaffolds. Order of the combined scaffolds was determined by an application of topological sort to combine the linkage maps of male and female linkage. Since evidence of the distance between scaffolds is not available, only the putative order was demonstrated here. The homologs gene cluster was showed in gray color

predictions, and cross verifies with the evidence driven prediction. Such prediction, backed up by the rich evidences, increase the accuracy of the structural annotation.

Many assemblies of the bone fish draft genomes were built upon SNP base linkage groups. With the homologs synteny information derived from the functional annotation, we successfully allocated 137 scaffolds (13 Mb) into the established linkage group. The relative position of the homologs and clusters on these linkage map are illustrated in Figs. 2, 3, 4 and 5. The plots were generated with Mapchart [23].

We hypothesis that metamorphosis mechanism of Anguilla japonica, like other Teleost fishes, conserve with amphibians [24], based on the fact that development and growth of fish correlates to thyroid hormone had been widely accepted [25]. Such metamorphosis of vertebrate tends to be triggered by environmental control on hypothalamo-pituitary-thyroid axis in the brain [24], and regulated by the thyroid hormone receptor on cell membrane. The environment control, in this case, would be the catadromous activity from ocean to inland. In certain developmental stages of the larval, the brain senses the environmental stress urging the metamorphosis to glass eel, and releases corticotropin release factor (CRF), which correlated with genes crh and CRHBP. CRF binds on receptors of anterior pituitary, forcing it produce thyrotrophin (TSH)

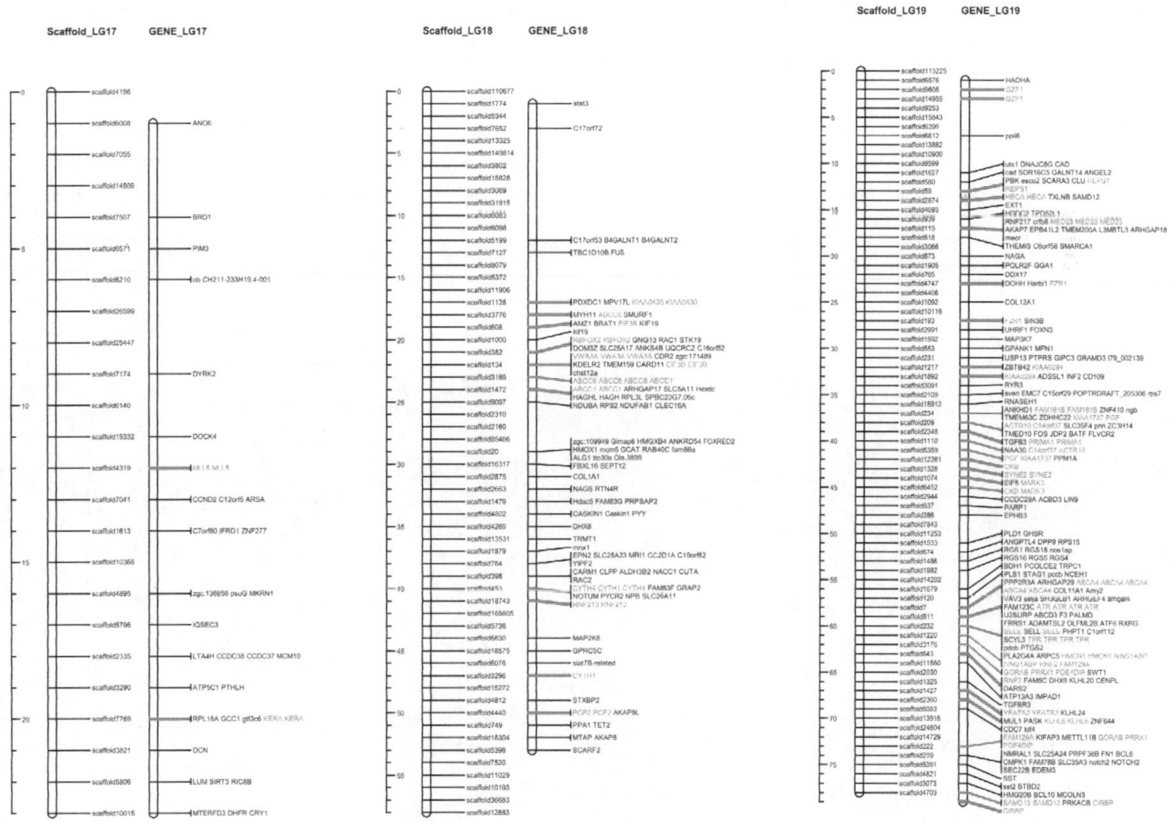

Fig. 5 Homolog Linkage Map of *Anguilla Japonica* (Linkage Groups 17 to 19). The graph shows genetic linkage maps on 3 of the 19 linkage groups. With the homologs synteny information derived from the functional annotation, we successfully allocated 137 scaffolds (13 Mb) into the established linkage group. This graph illustrates the relative position of the scaffolds. Order of the combined scaffolds was determined by an application of topological sort to combine the linkage maps of male and female linkage. Since evidence of the distance between scaffolds is not available, only the putative order was demonstrated here. The homologs gene cluster was showed in gray color

and adrenocorticotrophic hormone (ACTH). TSH triggers thyroids to secret thyroid hormone, while ACTH triggers adrenals or Interrenals to secret corticoids. From our functional annotation, we observed that the trha homolog of Japanese eel is correlated with TSH, and pomcb homolog is correlated with ACTH of Anguilla japonica. Finally, thyroid hormone receptors on the cell membrane as well as nuclear receptors within the cell regulate whether the metamorphosis would be triggered. We believe that thyroid hormone receptors of *Anguilla japonica* are regulated by the TRIP4, THRAP3, TRIP11, TRIP12, TRIP13 genes homologs, while the nuclear receptors are regulated by the genes NR4A3, NR2C2, NCOA3 and NR2F6 homologs respectfully. Phylogenetic tree of thyroid hormone receptor interactor (TRIP) family homologs is illustrated in Figs. 6 and 7. Through the analysis we found that the thyroid hormone receptor interactor family genes of Japanese eel are homologous to Asian arowana, Northern Pike, Rainbow trout, Marbled rockcod, Atlantic herring and Spotted gar. Phylogenetic tree of nuclear receptor subfamily 2 is illustrated in Figs. 8 and 9. From the graph we observed that the receptors were also closely related to homologs of same six fishes, Asian arowana, Northern Pike, Rainbow trout, Marbled rockcod, Atlantic herring and Spotted gar.

Among the homolog clusters we have found, lysosomal trafficking regulator LYST cluster in linkage group 10 (LG10) was related to the GO term pigmentation GO:0043473 [26]. Correlated scaffolds in the linkage group and phylogenetic analysis of the found homologs are illustrated in Fig. 10. From the phylogenetic tree we found that LYST homolog of Japanese eel is also close related to Asian arowana, a kind of freshwater East Asia fish with high commercial value.

While such a full transcriptome from whole body is proven to be effective on functional annotation, scope of this annotation is still limited by the sample. Since the messenger RNAs were isolated from glass eels before sexualization, certain types of hormones from gender specific tissues of silver eels such as ovarian and testicular cannot be found in such samples. On the other hand, satellite sequences and mRNAs from eggs and larval might not necessary been expressed in our samples. Despite some of a portion of this information is available on NCBI, which we used in our annotation, these factors could still limit the completeness of the annotation.

Materials and methods

Isolated whole RNA from the five glass eels were prepare for the RNA sequencing. Libraries for the RNA-Seq were sequenced through Illumina HiSeq^{TM} 2000 following the manufacturer's manual. Pair end libraries were sequenced in 101 X 2 nucleotides length reads, with 120 nucleotides adaptors. The entire fragment length was 357 nucleotides. The base calling and image

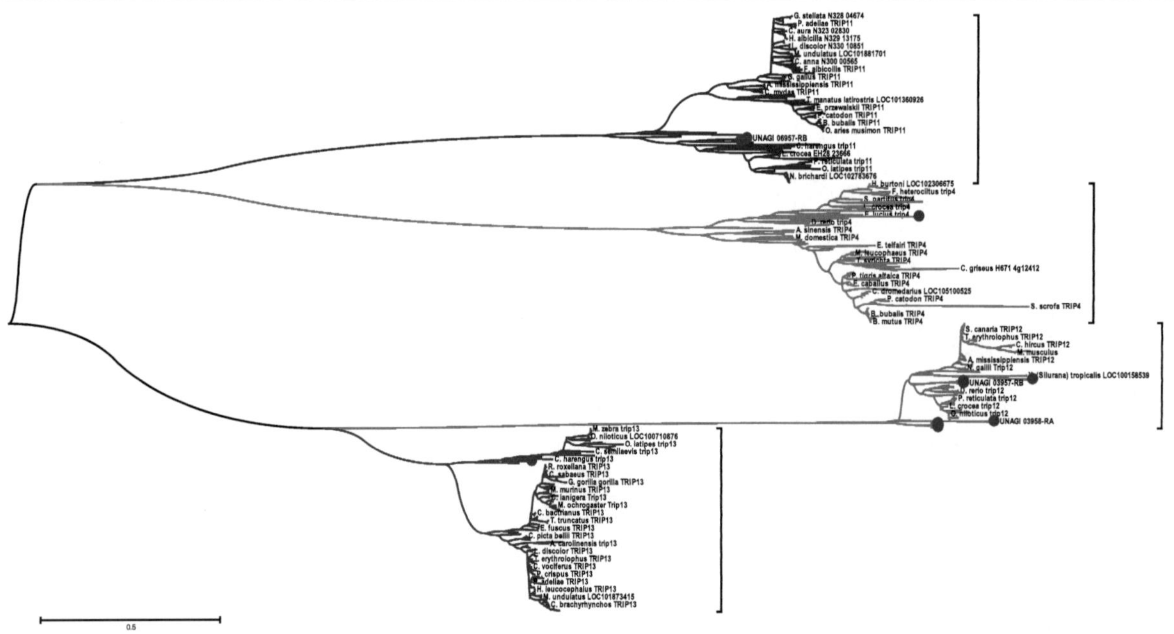

Fig. 6 Phylogenetic Tree of Thyroid Hormone Receptor Interactor (TRIP) Family Homologs. This graph illustrated the thyroid hormone receptors homologs of Anguilla Japonica. From the separate color subtree of TRIP4, TRIP11, TRIP12 and TRIP13 genes homologs, we can see that the different homologs of same gene family are distributed into different subtree

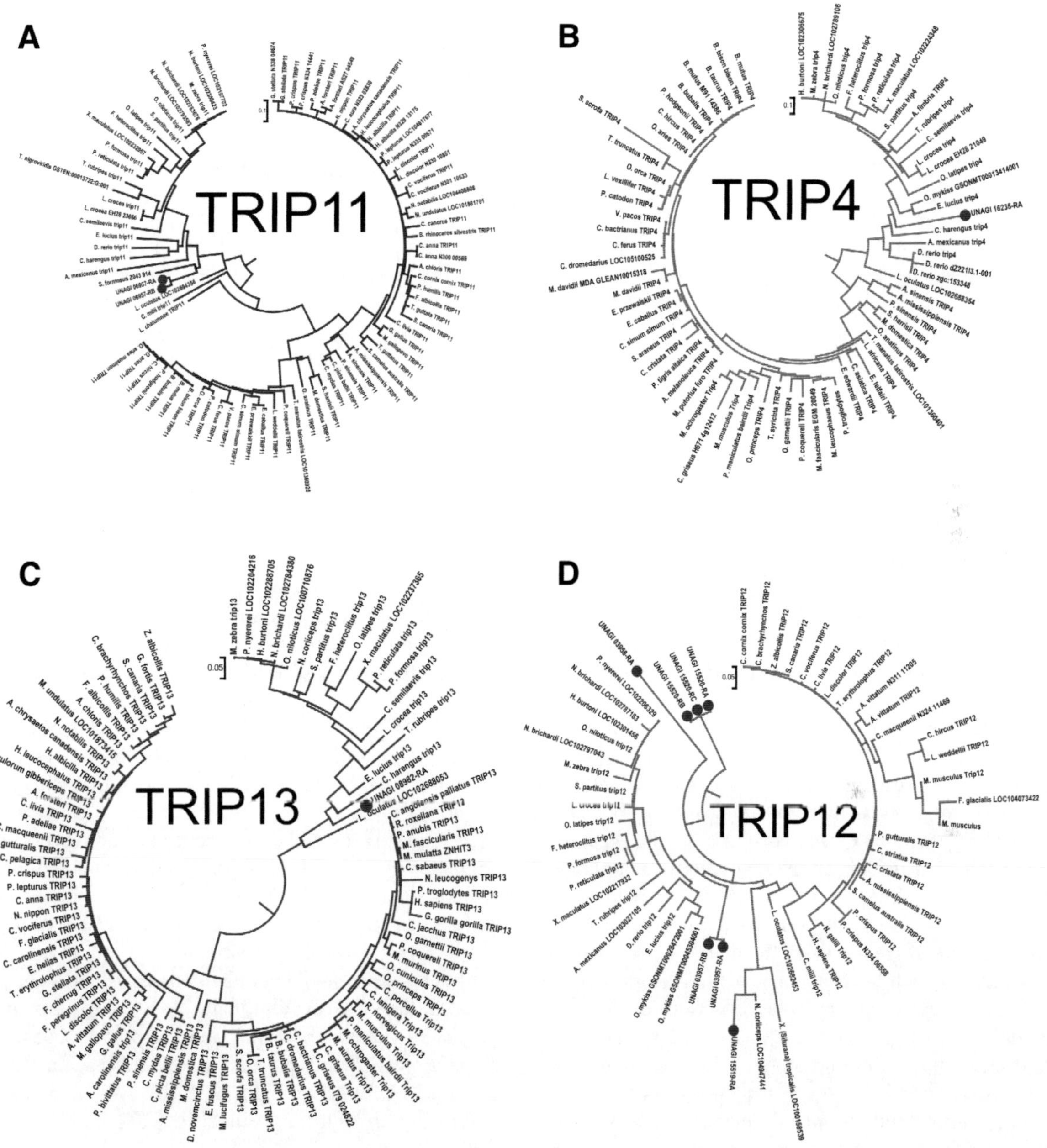

Fig. 7 Phylogenetic Trees of Thyroid Hormone Receptor Interactor (TRIP) Subfamilies. In this graph, each circular tree represents the corresponding color subtrees illustrated in Fig. 6. **a** The blue circular tree represent the TRIP11 subtree, with 2 *Anguilla Japonica* homologs marked in red dots. **b** The yellow circular tree represent the TRIP4 subtree, with 1 Anguilla Japonica homologs marked in red dot. **c** The red circular tree represent the TRIP13 subtree, with 1 Anguilla Japonica homologs marked in red dot. **d** The green circular tree represent the TRIP12 subtree, with 7 Anguilla Japonica homologs marked in red dots. Through the analysis we found that the thyroid hormone receptor interactor family genes of Japanese eel are homologous to Asian arowana, Northern Pike, Rainbow trout, Marbled rockcod, Atlantic herring and Spotted gar

analysis were done following Illumina standard pipeline. Raw reads of deep sequencing went through quality control procedure done by using FASTX-Toolkit: FASTQ/A short reads pre-processing tools, with only Quality value Phred score over 20 nucleotides remain, which means only reads with per base accuracy over 99 % were kept. Also, we trimmed the length of the reads down to 70 nucleotides for low quality reads. The transcriptome were first reconstructed through de novo assembly. To achieve maximum accuracy, we applied three different main stream de novo assembly tools : Trinity , Oases [27], and SOAPdenovo-Trans [28].

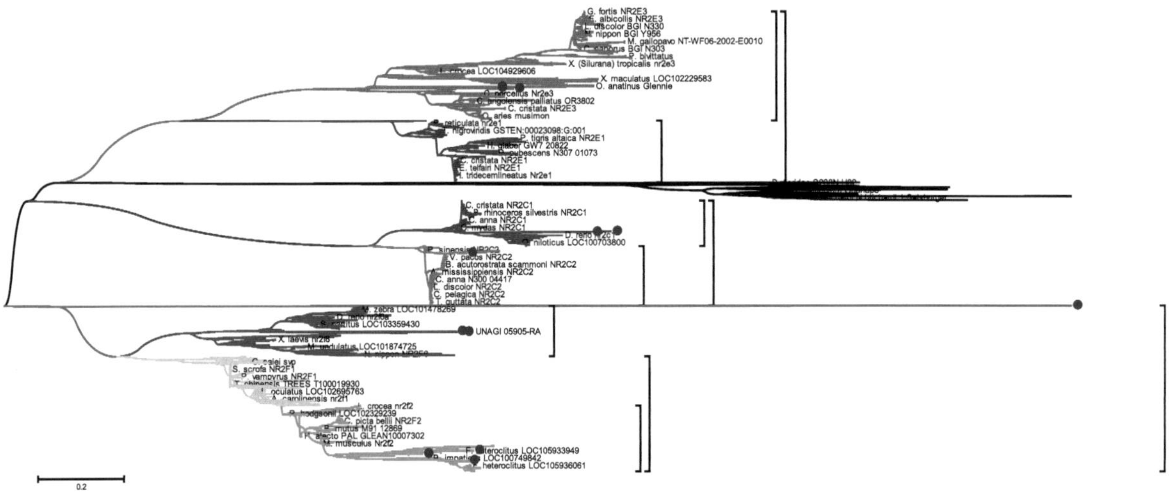

Fig. 8 Phylogenetic Tree of Nuclear Receptor Subfamily 2. This graph illustrated the nuclear receptors homologs of *Anguilla Japonica*. From the separate color circular subtree of NR2F2, NR2F6, NR2C1, NR2C2, NR2E1 and NR2E3 genes homologs, we can see that the different homologs of same gene family are distributed into different subtree

Quality controlled reads were assembled into three separate sets of contigs, with the three different tools. Trinity was applied with default settings, while Oases and SOAPdenovo-Trans were applied with multiple-kmers strategy. To further eliminate overlapping contigs, we clustered the three sets of contigs with CD-HIT-EST [29] into three sets of unigenes. Finally, we clustered unigenes with high similarity together with the tool CAP3 [30]. Quality of the assembly was estimated mainly through an average length of unigenes, as well as quality score N50 and N90. N50 represents the length of the longest unigene among the collection of unigenes equal to a half of the sum of all unigenes, while N90 means the length of the shortest unigene among the collection of unigenes equal to ninety percent of the sum of all unigenes. Maximum and minimum length of assembled unigene also serves as an index for the assessment.

Abundance of the assembled unigenes were estimated through RSEM pipeline [31]. Quantities of the transcripts were estimated through FPKM value. FPKM, frequency of reads per kilo base per million ,value was calculated through aligning reads onto assembled transcripts with Bowtie [32]. A de novo repeat library of *Anguilla japonica* was built from the draft genome [7] through RepeatScout [21]. Then, de novo assembly of the RNA-Seq data were pooled with the complete and partial CDS, EST and previously done gill RNA-seq assembly [17] of *Anguilla japonica* from NCBI as EST evidence. The EST evidence includes the public available RNA-seq data sets SRX482728, SRX247092, SRX115953 and the whole body transcriptome sequence data. Together with the known proteins from NCBI, genome structural annotation was performed through the pipeline MAKER [9]. The pipeline firstly masked the repeat sequences according to the previously build library with Repeatmasker (http://repeatmasker.org), and then perform ab initio prediction through repeat training of SNAP [33] and polished with Exonerate [34].

To find the assembled transcripts coding proteins, unigenes were blasted against NCBI non-redundant protein data base, TrEMBL and Swiss-Port [35] with BLASTX. Hits with an e value lower than 10 to negative 5, filtered by penalty estimation through the credibility of the protein, would be considered as homologs. Next, available Gene Ontology [36] terms were found listed. On the other hand, potential protein conserved domain were found through RPSBLAST against Pfam [37] and NCBI COG . To help the system biological analysis in the future, available KEGG pathways [38] were also annotated.

For scaffolding purposes, we utilized an improved version (to be described elsewhere) of the gene order based scaffolding method developed by Aganezov et al. [10]. Since this method relies on gene orders of multiple genomes, we preprocessed genomic sequences of Fugu [39], Stickleback [40], Medaka [41], Tetraodon [40], Coelacanth [42] and Zebra fish [43] in addition to *Anguilla japonica* to represent them as sequences of homologous gene (decided by respective scaffolds boundaries). Scaffolds with no homologous genes were filtered out from genomes during the preprocessing. In contrast to the original method described in Aganezov

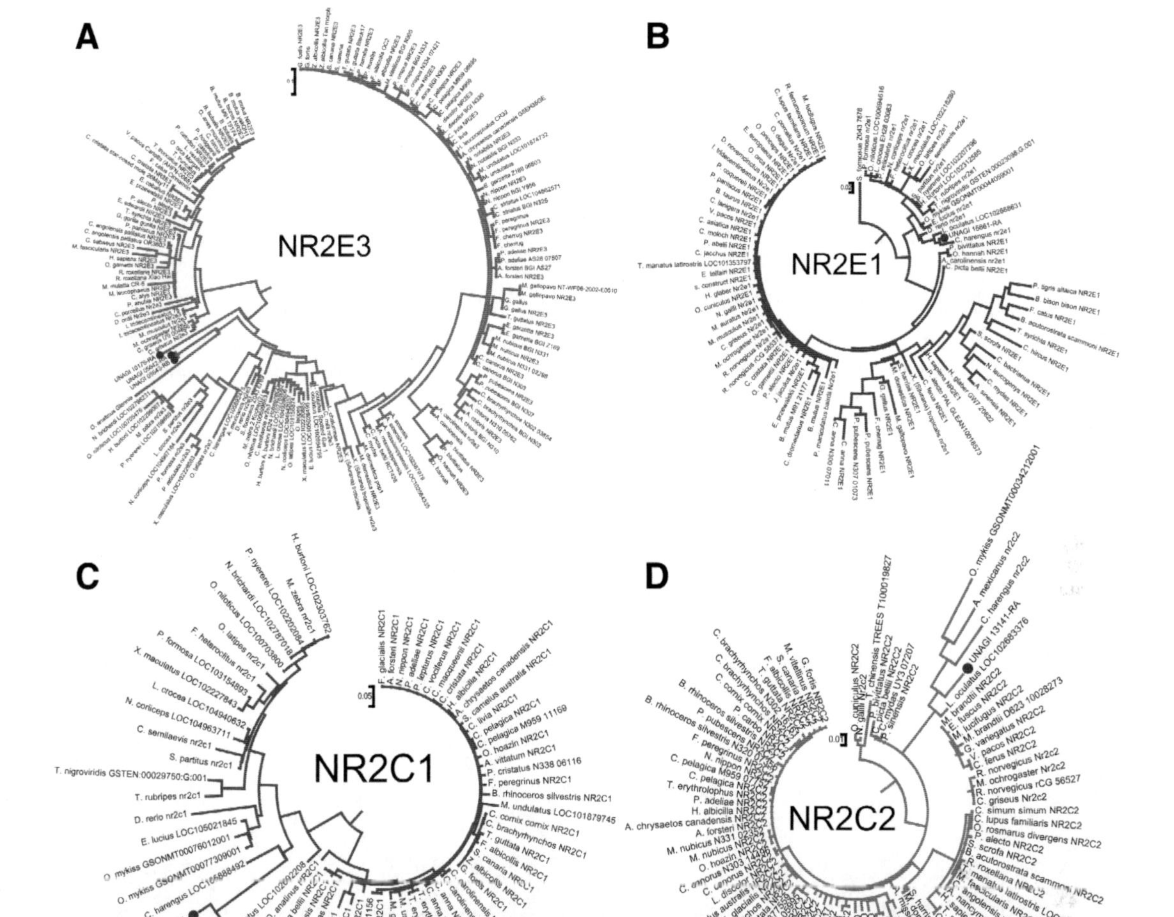

Fig. 9 (See legend on next page.)

(See figure on previous page.)

Fig. 9 Phylogenetic Trees of Nuclear Receptor Subfamily 2 Groups. In this graph, each circular tree represents the corresponding color subtrees illustrated in Fig. 8. **a** The green circular tree represent the NR2E3 subtree, with 3 *Anguilla Japonica* homologs marked in red dots. **b** The brown circular tree represent the NR2E1 subtree, with 1 *Anguilla Japonica* homolog marked in red dot. **c** The purple circular tree represent the NR2C1 subtree, with 2 *Anguilla Japonica* homologs marked in red dots. **d** The gray circular tree represent the NR2C2 subtree, with 1 *Anguilla Japonica* homolog marked in red dot. **e** The yellow circular tree represent the NR2F2 subtree, with 3 *Anguilla Japonica* homologs marked in red dots. **f** The blue circular tree represent the NR2F6 subtree, with 2 *Anguilla Japonica* homologs marked in red dots. Through the analysis we found that the thyroid hormone receptor interactor family genes of Japanese eel are homologous to Asian arowana, Northern Pike, Rainbow trout, Marbled rockcod, Atlantic herring and Spotted gar

et al. [10], the improved method accounts for gene insertions/deletions and duplications and thus no filtration for unique gene content was needed. We utilized the phylogenetic tree in Fig. 11 a. While provided with 7190 scaffolds with homologous genes on them, scaffolding method identified 525 links and assembled scaffolds respectively. These scaffolds were then mapped into male and female linkage maps provided by the Kai et al. study [8]. Order of the combined scaffolds was determined by an application of topological sort. With the scaffolds grouped into male and female linkage groups overlapped with each other, order of the scaffolds on the chromosomes can be sorted with topological sort algorithm. Phylogenetic tree of the fishes and the process of topological sort are illustrated in Fig. 11.

Conclusions and prospective works

We provide a reconstructed transcriptome from whole body samples of juvenile (glass eels). The high

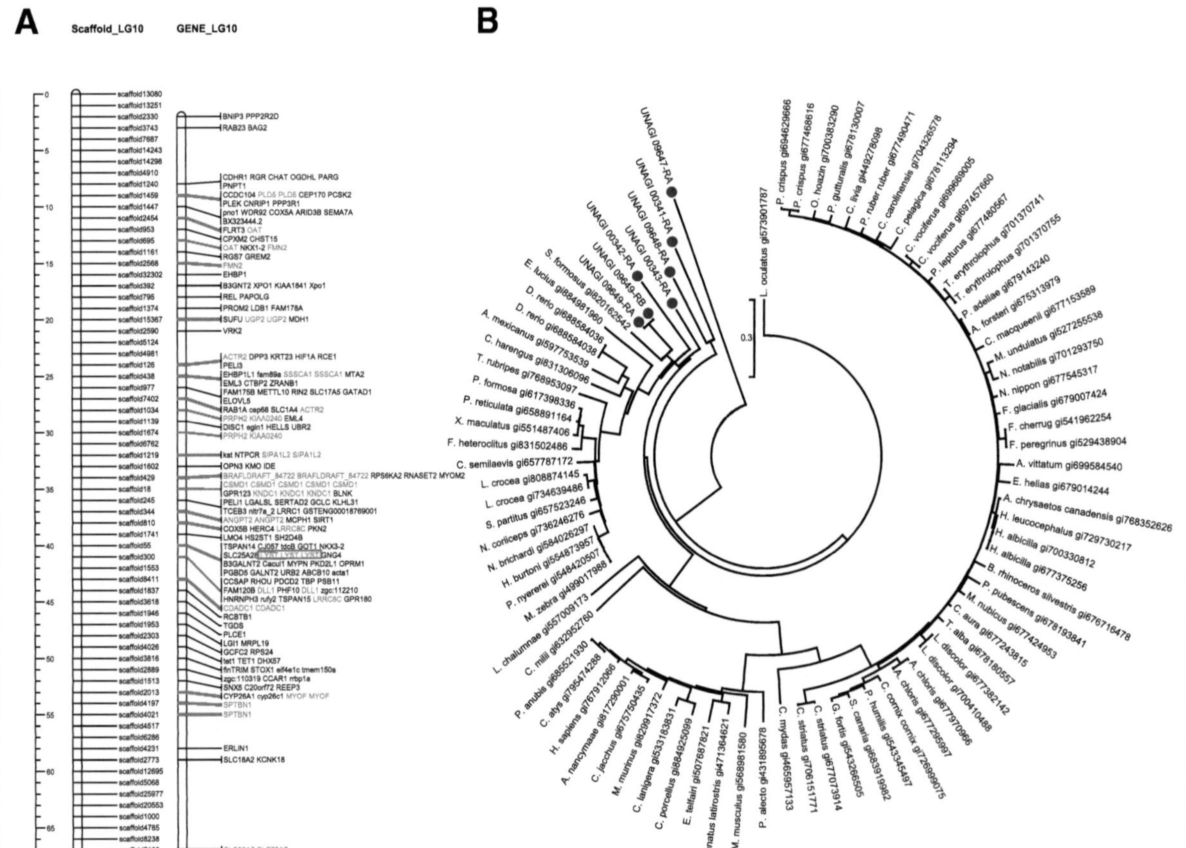

Fig. 10 Phylogenetic Tree and Linkage Location of LYST Homologs. In this graph, we found a lysosomal trafficking regulator LYST cluster (in red box) in linkage group 10 (LG10). The gene was close related to the homolog of Asian arowana. There are 4 other Anguilla Japonica homologs of LYST marked on the circular tree

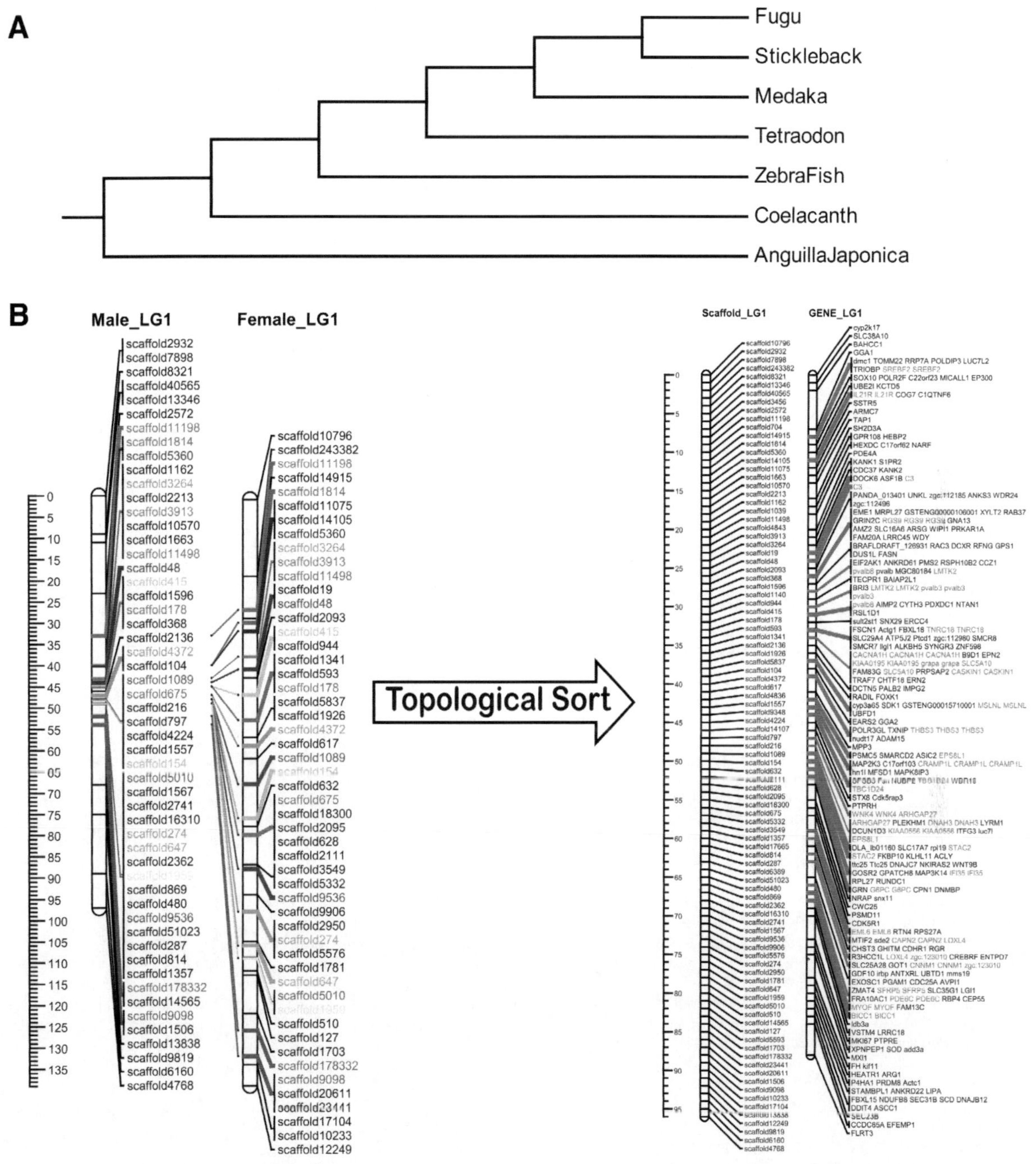

Fig. 11 Overview of the Genetic Linkage Map Building Process. **a** We utilized the following phylogenetic tree. While provided with 7190 scaffolds with homologous genes on them, scaffolding method identified 525 links and assembled scaffolds respectively. These scaffolds were then mapped into male and female linkage maps provided by the Kai et al. study. **b** The scaffolds mapped into male and female linkage groups were then sorted into single group. As illustrated in the graph, scaffolds marked with same colors provide the evidences for the general order of them on the chromosome. The order can then be sorted through topological sort algorithm

throughput sequencing provides unprecedented amount of transcriptomic information. For future experimental design and guidance on ecological, physiological, artificially breeding and even toxicity resistance study of Japanese eel, the transcriptome provide guidance. For example, expression of specific genes shows extreme patterns in glass eels and can be further compared with larvae as well as silver eels through QPCR to provide further reevaluation of the metamorphosis and pigmentation mechanism.

Availability

Link to a gtf file of the annotation, fasta files of the protein and transcripts and genetic linkage map would be available in Additional file 1.

Additional files

Additional file 1: Contains a link to the gtf file of the annotation, fasta files of the protein and transcripts and genetic linkage map. Additional file 1 also includes Figure S1 to Figure S8, Table S1 to Table S4. (DOCX 429 kb)

Competing interests

The authors declare no financial competing interests.

Authors' contributions

Y.C.L. carried out the draft genome based annotation, constructed the linkage maps, performed the phylogenetic analysis and drafted the manuscript. S.D.H., C.H.C., W.Y.H., Y.H.C., C.Y.L., G.J.L., and S.Z.H. participated in the quality control of the sequence reads, construct De Novo assembly of the transcriptome and participated in the sequence alignment based functional annotation. S.A.A. and M.A.A. conducted the comparative genomic analysis and helped the revision of the manuscript. C.D.H. prepared the samples, provided the transcriptome sequence data. C.D.H. and H.D.H. supervised the progress of the study, provided guidance in its design and coordination and helped to draft and revise the manuscript. All authors read and approved the final manuscript.

Acknowledgement

The authors would like to thank the Ministry of Science and Technology, for financially supporting this research. We also thank the UST-UCSD International Center of Excellence in Advanced Bioengineering and Veterans General Hospitals and University System of Taiwan. This work was funded by Ministry of Science and Technology, [MOST 101-2311-B-009-005-MY3, MOST 103-2628-B-009-001-MY3, MOST104-2319-B-400-002, MOST 103-2319-B-010-002 and MOST103-3111-Y-001-027], UST-UCSD International Center of Excellence in Advanced Bioengineering. sponsored by the Ministry of Science and Technology I-RiCE Program [MOST 103-2911-I-009-101], Veterans General Hospitals and University System of Taiwan (VGHUST) Joint Research Program [VGHUST103-G5-11-2], MOE ATU and the funding for open access charge: Ministry of Science and Technology of the Republic of China, Taiwan [103-2628-B-009-001-MY3] and [MOHW104-TDU-B-212-124-005]. The work of S.A. and M.A.A. was supported by the National Science Foundation under the grant No. IIS-1462107.

Author details

[1]Institute of Bioinformatics and Systems Biology, National Chiao Tung University, Hsin-Chu 300, Taiwan, ROC. [2]Computational Biology Institute & Department of Mathematics, George Washington University, Washington, DC, USA. [3]Department of Higher Mathematics, ITMO University, St. Petersburg, Russia. [4]Department of Bioscience Technology, Chung Yuan Christian University, Chung-Li, Taiwan, ROC. [5]Department of Biological Science and Technology, National Chiao Tung University, Hsin-Chu 300, Taiwan, ROC. [6]Department of Biomedical Science and Environmental Biology, Kaohsiung Medical University, Kaohsiung, Taiwan.

Published: 11 January 2016

References

1. Minegishi Y, Henkel CV, Dirks RP, van den Thillart GEEJM. Genomics in Eels - Towards Aquaculture and Biology. Marine Biotechnol. 2012;14(5):583–90.
2. Tsukamoto K, Aoyama J, Miller MJ. Migration, speciation, and the evolution of diadromy in anguillid eels. Can J Fish Aquat Sci. 2002;59(12):1989–98.
3. Henkel CV, Burgerhout E, de Wijze DL, Dirks RP, Minegishi Y, Jansen HJ, et al. Primitive Duplicate Hox Clusters in the European Eel's Genome. PloS One. 2012;7(2):e32231.
4. Nomura K, Ozaki A, Morishima K, Yoshikawa Y, Tanaka H, Unuma T, et al. A genetic linkage map of the Japanese eel (Anguilla japonica) based on AFLP and microsatellite markers. Aquaculture. 2011;310(3–4):329–42.
5. Coppe A, Pujolar JM, Maes GE, Larsen PF, Hansen MM, Bernatchez L, et al. Sequencing, de novo annotation and analysis of the first Anguilla anguilla transcriptome: EeelBase opens new perspectives for the study of the critically endangered european eel. BMC Genomics. 2010;11:635.
6. Pujolar JM, Marino IA, Milan M, Coppe A, Maes GE, Capoccioni F, et al. Surviving in a toxic world: transcriptomics and gene expression profiling in response to environmental pollution in the critically endangered European eel. BMC Genomics. 2012;13:507.
7. Henkel CV, Dirks RP, de Wijze DL, Minegishi Y, Aoyama J, Jansen HJ, et al. First draft genome sequence of the Japanese eel, Anguilla japonica. Gene. 2012;511(2):195–201.
8. Kai W, Nomura K, Fujiwara A, Nakamura Y, Yasuike M, Ojima N, et al. A ddRAD-based genetic map and its integration with the genome assembly of Japanese eel (Anguilla japonica) provides insights into genome evolution after the teleost-specific genome duplication. BMC Genomics. 2014;15(1):233.
9. Cantarel BL, Korf I, Robb SM, Parra G, Ross E, Moore B, et al. MAKER: an easy-to-use annotation pipeline designed for emerging model organism genomes. Genome Res. 2008;18(1):188–96.
10. Aganezov S, Sitdykova N, Alekseyev MA, Consortium A. Scaffold assembly based on genome rearrangement analysis. Comput Biol Chem. 2015;57:46–53.
11. Tamura K, Stecher G, Peterson D, Filipski A, Kumar S. MEGA6: molecular evolutionary genetics analysis version 6.0. Mol Biol Evol. 2013;30(12): 2725–9.
12. Kanehisa M, Goto S. KEGG: kyoto encyclopedia of genes and genomes. Nucleic Acids Res. 2000;28(1):27–30.
13. Martin JA, Wang Z. Next-generation transcriptome assembly. Nat Rev Genet. 2011;12(10):671–82.
14. Sudo R, Suetake H, Suzuki Y, Aoyama J, Tsukamoto K. Profiles of mRNA expression for prolactin, growth hormone, and somatolactin in Japanese eels, Anguilla japonica: The effect of salinity, silvering and seasonal change. Comp Biochem Physiol A Mol Integr Physiol. 2013;164(1):10–6.
15. Albert V, Jonsson B, Bernatchez L. Natural hybrids in Atlantic eels (Anguilla anguilla, A. rostrata): evidence for successful reproduction and fluctuating abundance in space and time. Mol Ecol. 2006;15(7):1903–16.
16. Okamura A, Zhang H, Utoh T, Akazawa A, Yamada Y, Horie N, et al. Artificial hybrid between Anguilla anguilla and A. japonica. J Fish Biol. 2004;64(5):1450–4.
17. Tse WK, Sun J, Zhang H, Law AY, Yeung BH, Chow SC, et al. Transcriptomic and iTRAQ proteomic approaches reveal novel short-term hyperosmotic stress responsive proteins in the gill of the Japanese eel (Anguilla japonica). J Proteomics. 2013;89:81–94.
18. Kazeto Y, Tosaka R, Matsubara H, Ijiri S, Adachi S. Ovarian steroidogenesis and the role of sex steroid hormones on ovarian growth and maturation of the Japanese eel. J Steroid Biochem Mol Biol. 2011;127(3–5):149–54.
19. Tanaka H, Kagawa H, Ohta H, Unuma T, Nomura K. The first production of glass eel in captivity: fish reproductive physiology facilitates great progress in aquaculture. Fish Physiol Biochem. 2003;28(1–4):493–7.
20. Yandell M, Ence D. A beginner's guide to eukaryotic genome annotation. Nat Rev Genet. 2012;13(5):329–42.
21. Price AL, Jones NC, Pevzner PA. De novo identification of repeat families in large genomes. Bioinformatics. 2005;21 Suppl 1:i351–358.
22. Kapitonov VV, Jurka J. A universal classification of eukaryotic transposable elements implemented in Repbase. Nat Rev Genet. 2008;9(5):411–2. author reply 414.
23. Voorrips R. MapChart: software for the graphical presentation of linkage maps and QTLs. J Hered. 2002;93(1):77–8.
24. Laudet V. The origins and evolution of vertebrate metamorphosis. Curr Biol. 2011;21(18):R726–737.
25. Power D, Llewellyn L, Faustino M, Nowell MA, Björnsson BT, Einarsdottir I, et al. Thyroid hormones in growth and development of fish. Comp Biochem Physiol C Toxicol Pharmacol. 2001;130(4):447–59.
26. Sturm RA. Molecular genetics of human pigmentation diversity. Hum Mol Genet. 2009;18(R1):R9–R17.
27. Schulz MH, Zerbino DR, Vingron M, Birney E. Oases: robust de novo RNA-seq assembly across the dynamic range of expression levels. Bioinformatics. 2012; 28(8):1086–92.

28. Luo R, Liu B, Xie Y, Li Z, Huang W, Yuan J, et al. SOAPdenovo2: an empirically improved memory-efficient short-read de novo assembler. GigaScience. 2012;1(1):18.
29. Fu L, Niu B, Zhu Z, Wu S, Li W. CD-HIT: accelerated for clustering the next-generation sequencing data. Bioinformatics. 2012;28(23):3150–2.
30. Huang X, Madan A. CAP3: A DNA sequence assembly program. Genome Res. 1999;9(9):868–77.
31. Li B, Dewey CN. RSEM: accurate transcript quantification from RNA-Seq data with or without a reference genome. BMC Bioinformatics. 2011;12:323.
32. Langmead B, Trapnell C, Pop M, Salzberg SL. Ultrafast and memory-efficient alignment of short DNA sequences to the human genome. Genome Biol. 2009;10(3):R25.
33. Korf I. Gene finding in novel genomes. BMC Bioinformatics. 2004;5(1):59.
34. Slater GS, Birney E. Automated generation of heuristics for biological sequence comparison. BMC Bioinformatics. 2005;6:31.
35. Bairoch A, Apweiler R. The SWISS-PROT protein sequence data bank and its supplement TrEMBL in 1999. Nucleic Acids Res. 1999;27(1):49–54.
36. Ashburner M, Ball CA, Blake JA, Botstein D, Butler H, Cherry JM, et al. Gene Ontology: tool for the unification of biology. Nat Genet. 2000;25(1):25–9.
37. Punta M, Coggill PC, Eberhardt RY, Mistry J, Tate J, Boursnell C, et al. The Pfam protein families database. Nucleic Acids Res. 2012;40(Database issue): D290–301.
38. Kanehisa M, Goto S, Sato Y, Furumichi M, Tanabe M. KEGG for integration and interpretation of large-scale molecular data sets. Nucleic Acids Res. 2012;40(Database issue):D109–114.
39. Aparicio S, Chapman J, Stupka E, Putnam N, Chia J-M, Dehal P, et al. Whole-genome shotgun assembly and analysis of the genome of Fugu rubripes. Science. 2002;297(5585):1301–10.
40. Jaillon O, Aury J-M, Brunet F, Petit J-L, Stange-Thomann N, Mauceli E, et al. Genome duplication in the teleost fish Tetraodon nigroviridis reveals the early vertebrate proto-karyotype. Nature. 2004;431(7011):946–57.
41. Kasahara M, Naruse K, Sasaki S, Nakatani Y, Qu W, Ahsan B, et al. The medaka draft genome and insights into vertebrate genome evolution. Nature. 2007;447(7145):714–9.
42. Amemiya CT, Alföldi J, Lee AP, Fan S, Philippe H, MacCallum I, et al. The African coelacanth genome provides insights into tetrapod evolution. Nature. 2013;496(7445):311–6.
43. Postlethwait JH, Yan Y-L, Gates MA, Horne S, Amores A, Brownlie A, et al. Vertebrate genome evolution and the zebrafish gene map. Nat Genet. 1998; 18(4):345–9.

AUTHOR INDEX

AUTHOR INDEX